Heating and Cooling
E S S E N T I A L S

by
Jerry Killinger

Illustrations by
LaDonna Killinger

Publisher
The Goodheart-Willcox Company, Inc.
Tinley Park, Illinois

About the Author

Jerry Killinger has extensive experience installing and servicing air conditioning, refrigeration, and heating systems. He has worked with industrial, commercial, and residential systems ranging from heat pumps to boilers to cooling towers to refrigerated food cases to solid-state environmental control equipment. Jerry has over 20 years' experience teaching traditional vocational classes as well as various seminars and workshops. He has conducted numerous seminars on refrigeration, air conditioning, and electricity for industrial maintenance personnel at facilities of major manufacturers across the United States.

Copyright 2003

by

THE GOODHEART-WILLCOX COMPANY, INC.

Previous Editions Copyright 1999, 1993

Library of Congress Card Catalog Number 2002025380
ISBN-13: 978-1-56637-965-6
ISBN-10: 1-56637-965-2

6 7 8 9   03 11 10

Library of Congress Cataloging-in-Publication Data

Killinger, Jerry.
    Heating and Cooling Essentials/
by Jerry Killinger; illustrations by LaDonna Killinger

        p.   cm.
    Includes index.
    ISBN 1-56637-965-2
    1. Heating.
    2. Air conditioning
I. Title.
TH7012.K55   2002
697–dc21   2002025380
                                                 CIP

Cover image courtesy of Lennox International

Introduction

Heating and Cooling Essentials provides a thorough, easily understood, and up-to-date textbook for persons entering the refrigeration, heating, and air conditioning field. The text is a practical blend of theory and on-the-job skill-building procedures.

Heating and Cooling Essentials is designed for use by first-year students in HVAC programs at vocational schools, technical schools, or community colleges. It may also be used in apprenticeship programs or adult continuing education classes, and can serve as an excellent reference for technicians already working in the field.

The content provides a solid foundation in the basics and integrates practical service procedures with technical theory. The book is a complete and accurate guide to troubleshooting and performing essential service procedures on many types of systems.

An important change in this edition of **Heating and Cooling Essentials** is the reconfiguration of the headings within each chapter to a numeric identifier system. The most important headings are identified by the chapter number, then an ascending sequence of numbers (for example, the major headings in Chapter 9 would be numbered as 9.1, 9.2, 9.3, etc.) The next level of headings, forming divisions beneath each major heading, is given an additional digit: 9.2.1, 9.2.2, 9.2.3, etc. This organization allows easy location of specific material in a chapter.

Also added in this edition is a complete chapter devoted to ductwork. It provides basic information on the planning, fabrication, and installation of the common types of distribution systems used in forced-air installations. The new chapter has been inserted as Chapter 28, immediately preceding the chapters on the various types of forced air systems.

Heating and Cooling Essentials is organized into areas of study based on a progression of concepts and systems from the simple to the complex. Learning objectives are stated at the beginning of each chapter, serving as an overview of content; a summary and review questions are provided at the end. Review questions are a convenient means of assessing whether the chapter material has been mastered, or whether additional study is needed before moving on.

The colors of the line art illustrations in the book have been selected and used consistently to represent specific aspects of a system (such as refrigerant liquids). A color key is provided to facilitate use of the book and understanding of system operations.

This textbook will provide you with a sound foundation for further study of more specialized aspects of the HVAC field.

Jerry Killinger
LaDonna Killinger

How to Use the Color Key

Colors are used throughout **Heating and Cooling Essentials** to indicate different states of gases or liquids in refrigeration systems. Various equipment components and materials are also identified with color. The following key shows what each color represents.

States of Liquids or Gases

■ High-pressure Liquid　　　　■ Low-pressure Liquid

■ High-pressure Gas　　　　■ Low-pressure Gas

Note: In some schematic representations, additional colors may be used to aid understanding. They will be identified in a key accompanying the illustration.

Components and Materials

■ Water　　　　　　　　　　　■ Iron, Steel, Other Metals

■ Water Vapor　　　　　　　　■ Evaporators and Condensers

■ Copper Tubing　　　　　　　■ Motors and Compressors

■ Wood, Building Partitions　　■ Refrigerant Oil

■ Fuses and Other　　　　　　■ Metering Devices,
Electrical Components　　　　Valves, Thermostats

→ Airflow　　　　　　　　　→ Flow, Other Fluids

Other

■ Special features, materials, or components not otherwise coded.

Contents

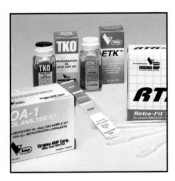

Chapter 14 ◆
Zeotropic Blends 215

Chapter 15 ◆
Troubleshooting Refrigerant Problems 229

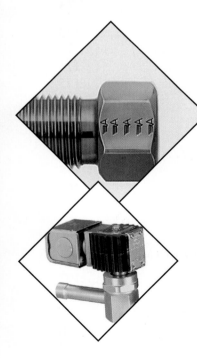

Chapter 16 ◆
Working with Refrigerant Controls 253

Chapter 17 ◆
Special-purpose Valves 269

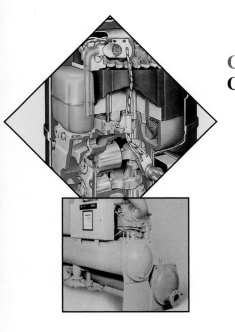

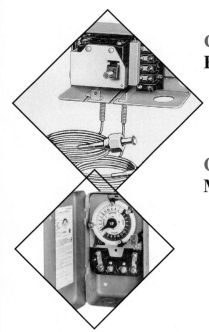

Chapter 28

Ductwork 495

Chapter 29

Gas Heat with Air Conditioning 519

Chapter 30

Oil Heat with Air Conditioning 539

Chapter 31

Electric Heat with Air Conditioning 553

Hand Tools

Objectives

After studying this chapter, you will be able to:
❑ Identify the various types of hand tools.
❑ Describe the advantages and disadvantages of each type of hand tool.
❑ Select the proper hand tool for the job.
❑ Demonstrate skills in using various hand tools.
❑ Maintain, repair, or replace damaged tools.
❑ Demonstrate good craftsmanship in the performance of daily tasks.

Important Terms

arbor	points
cheater	serrated
cheek	shank
dressed	socket
drift punch	swing space
flutes	torque
hex-head	universal joint
hexagonal	whetstone
mushroomed	

1.1 Tools for Technicians

The job of the refrigeration, heating, and air conditioning technician consists mainly of performing mechanical operations and using common tools and materials. A good technician performs each task properly the first time. Tool skills are valuable to the technician and are acquired through experience and practice.

Successful technicians select quality tools, care for them properly, and skillfully use them. Technicians who have good tool skills also have good mechanical skills. Misuse of tools indicates poor mechanical skills that can result in delays, callbacks, and injuries.

A beginning technician usually starts with a basic set of hand tools for regular use, and then gradually adds other tools as needed. Some technicians carry frequently used hand tools in a tool pouch; others prefer toolboxes. Master technicians may have several boxes of tools so they always have *exactly* the right tool for the job.

1.2 Wrenches

Wrenches are the most widely used type of hand tool. They are used to hold or turn nuts, bolts, cap screws, and other fasteners, and can be used to clamp parts together. Applying too much **torque** (turning force) can strip bolt threads or break off bolt heads. Quality wrenches are designed to keep leverage and load in safe balance. *Never* use a pipe extension or other type of *"cheater"* to increase the leverage of a wrench.

Always select a wrench with an opening that exactly fits the nut. Wrench size is determined by measuring across the jaw opening. The wrench size is stamped on the side of the tool.

1.2.1 Open-end Wrenches

An open-end wrench has an open jaw on each end, **Figure 1-1.** Each end is a different size, and the jaw opening is set at an angle. The angle permits using the wrench when there is only a small amount of room (**swing space**) to turn the nut or bolt. The wrench is turned over to obtain a new grip on the nut. Because they tend to slip, open-end wrenches work best on *loose* nuts and bolts. A slipping open-end wrench could round nut edges and cause hand injuries.

Figure 1-1. *The openings are different sizes at each end of an open-end wrench.*

1.2.2 Box-end Wrenches

A box-end wrench is closed on both ends to surround the nut or bolt head. See **Figure 1-2.** Each end has an opening of a different size. Opening sizes are stamped on the wrench. Box wrenches are available in 6-point or 12-point types. *Points* are the "teeth" that grip the edges of the nut or the bolt head. The 6-point wrench is the stronger of the two since it has more grip area. A box-end wrench will not slip easily, so it should be used on bolt heads or nuts that are tight or partially rounded.

Figure 1-2. *Box-end wrenches surround the nut or bolt head, making them much less likely to slip than open-end wrenches.*

1.2.3 Combination Wrenches

Combination wrenches combine the best features of open-end and box-end wrenches. They have one open end and one box end, **Figure 1-3.** Both ends are usually the same size. The combination wrench is designed for several tasks. The open end is useful where little swing space is available. The box end provides the grip area needed for final tightening (or loosening of tight bolts).

Box end   Open end

Figure 1-3. *The open-end and box-end of a combination wrench are usually the same size.*

1.2.4 Valve Key Wrenches

Valve key wrenches or refrigeration wrenches are designed to easily turn valve stems. The wrench in **Figure 1-4** is a ratchet-type with square openings to fit most valve stems. A quick-flip reversing lever permits instant reversal for turning service valve stems in either direction. This wrench is a must in every technician's tool kit.

Figure 1-4. *The ratchet feature of the valve key wrench makes it easy to turn a valve stem, even in tight quarters. This wrench is a necessity for every technician. (Klein Tools, Inc.)*

1.2.5 Flare-nut Wrenches

The flare-nut wrench, shown in **Figure 1-5,** is a special type of box wrench with an *opening* in the box. The opening permits the wrench to slip over tubing and onto the flare nut. Each end of the flare-nut wrench has a different-sized opening.

Figure. 1-5. *The opening in the box of the flare-nut wrench allows it to slip over tubing and fit a flare nut tightly.*

1.2.6 Adjustable Wrenches

An adjustable wrench has a movable lower jaw, **Figure 1-6.** Like an open-end wrench, this wrench tends to slip. When using an adjustable wrench, make sure it is *tightly* adjusted to the nut. Pull only in the direction that will put force on the *fixed* side of the jaw. See **Figure 1-7.**

Figure 1-6. *An adjustable wrench tends to slip, unless the movable jaw is tightened firmly on the nut or bolt head.*

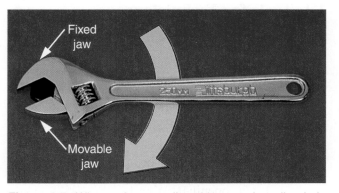

Fixed jaw

Movable jaw

Figure 1-7. *When using an adjustable wrench, pull only in the direction shown, so that force is exerted on the fixed jaw. (John Walker)*

1.2.7 Allen Wrenches

An Allen wrench, also known as a hex wrench, is a hardened steel shaft with a special *hexagonal* (6-sided) shape. It is used most often to turn setscrews. Allen wrench sets are available in "jackknife" style, as long individual wrenches, or with T-handles for applying increased torque. See **Figure 1-8.** Long wrenches allow easy access to setscrews used to anchor fan blades and pulleys to a shaft. To prevent damage to the wrench, be sure it is fully inserted in the screw before turning.

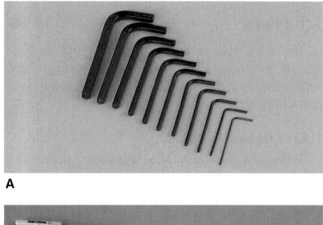

A

B

Figure 1-8. *Allen wrenches, or hex wrenches, fit hexagonal recesses in setscrews and other fasteners. A—Individual wrenches. B—T-handle wrenches.*

1.2.8 Socket Wrenches

A *socket* is an individual, cylinder-shaped box-end tool, used with a handle to perform the same turning tasks as other wrenches. One end of the socket fits over the bolt head or nut. The other end has a square opening for attaching the drive. Sockets are available in multiple sizes and are often sold in sets like the one shown in **Figure 1-9.**

As shown in **Figure 1-10,** the square drive opening of a socket may be 1/4″, 3/8″, 1/2″, or 3/4″. The drive size is a measure of the socket's capacity to withstand torque. Small drives are used for small fasteners; large drives are used for large fasteners.

Socket nut openings may be 4-point (square), 6-point, 8-point, or 12-point. The 4-point and 8-point sockets are used on nuts and bolts that have square

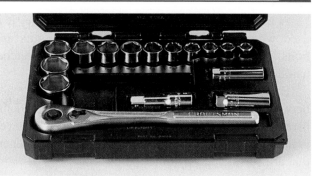

Figure 1-9. *A typical socket wrench set.*

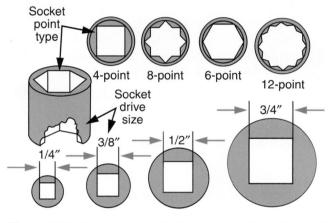

Figure 1-10. *Sockets are available in several drive sizes and point configurations.*

heads. The 6-point sockets have more grip area than 12-point sockets; both are used on fasteners with hex (6-sided) heads. However, 12-point sockets can be used with either square-head or *hex-head* bolts. Sockets of the 6-point type are most popular.

Various kinds of socket handles are made to fit into the square drive opening of the socket. A ratchet, **Figure 1-11,** is the most commonly used socket handle. Ratchets have a quick-flip reversing lever for instant reversal of socket rotation. Using excessive force can damage the ratchet mechanism.

Figure 1-11. *Ratchet handles allow tightening or loosening of fasteners when only limited swing space is available. A reversing lever allows easy change of direction.*

A breaker bar is the strongest socket handle. It is used when breaking loose extremely tight nuts and bolts. After the bolt or nut is loosened, the ratchet handle is used to finish removing the fastener.

Extension bars, **Figure 1-12,** are used to add reach between the socket and its handle. They are available in different lengths and can be combined for even greater reach. Extension bars permit reaching fasteners that are surrounded by other parts.

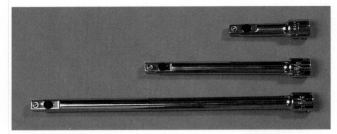

Figure 1-12. *Extension bars are used, singly or in combination, to allow tightening or removing of fasteners that otherwise could not be reached.*

A **universal joint** is a type of swivel that permits reaching around objects. It is used between the socket and drive handle and can be used with an extension bar. Do not use the universal joint at an extremely sharp angle. A sharp angle places excessive stress on the swivel.

1.2.9 Using Wrenches Safely

❑ Never use a wrench to do the job of another tool.
❑ Never use a wrench opening that is too large for the fastener.
❑ Never push a wrench beyond its capacity. Quality wrenches are designed and sized to keep leverage and intended load (torque) in safe balance. The safest wrench is a box or socket type because it is less likely to slip off the bolt head.
❑ Never expose a wrench to excessive heat.
❑ Never *push* on a wrench unless absolutely necessary. Always pull on a wrench to protect your knuckles. If you must push, use the open palm of your hand, with fingers slightly curled.
❑ Never cock or tilt an open-end wrench.
❑ Never depend on plastic-dipped handles for protection against electrical shock. These handles provide comfort and a firm grip, *not* shock protection.

1.2.10 Repair or Replace?

Repairing box, open-end, or combination wrenches is not recommended. Such wrenches should be discarded and replaced if they have bent handles, rounded or damaged box points, or jaws that are spread, nicked, or battered.

Ratchets and adjustable wrenches often can be repaired by replacing the damaged parts. However, an adjustable wrench with a bent handle or a fixed jaw that is spread or damaged should be discarded and replaced. Also discard and replace bent socket wrench handles and extensions, and cracked or battered sockets.

Worn out or broken tools are neither safe nor efficient. They can be dangerous to the user and those working nearby. Repair or replace worn tools as required. The cost of a new tool is small when compared to the cost of an injury or of time wasted by substandard performance.

1.3 Pliers ◆

Pliers are used to grip, cut, crimp, hold, or bend various materials. Each type does its particular job better than another type. Choosing the correct type of pliers improves efficiency.

1.3.1 Lineman's Pliers

Lineman's pliers, also called side-cutting pliers, are a heavy-duty tool available in various sizes, **Figure 1-13.** These pliers have a strong side-cutting feature for cutting large wires and should be in every technician's tool kit.

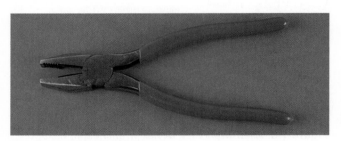

Figure 1-13. *The lineman's, or side-cutting, pliers are a basic item of the technician's toolbox. Plastic-covered handles, like those shown on these pliers, provide a better grip on the tool. They do* not *protect against electrical shock.*

1.3.2 Long-nose Pliers

Long-nose pliers, **Figure 1-14,** are available in three nose designs: needle, round, and flat (sometimes called "duck bill"). They allow the technician to reach into tight places and perform work that would be awkward or difficult with any other tool. They are available with or without side cutters. Most long-nose pliers are designed for electrical use but are useful in other ways if not abused. To avoid damage, never *pry* with these pliers or try to bend *stiff wire* with the pliers' tips. Also, avoid exposing these pliers to excessive heat; the jaw tips will bend outward and render the tool useless. The long-nose pliers recommended for refrigeration, heating, and air conditioning work are the heavy-duty 8″ (203 mm) size, with cutter, in the round pattern.

Figure 1-14. *Long-nose pliers are available in needle-, round-, and flat-nose types.*

1.3.3 Diagonal Cutting Pliers

Diagonal cutters (often called "dikes") are designed for cutting electrical wires, cotter pins, nails, and other types of fasteners. See **Figure 1-15.** When cutting wire, always cut at a right angle to avoid damage to the pliers, and do not rock them from side to side. The 8″ (203 mm), heavy-duty type is best for refrigeration, heating, and air conditioning work.

Figure 1-15. *Diagonal cutting pliers are used for heavier cutting tasks than other types of pliers.*

1.3.4 Pump Pliers

Pump pliers, also known as utility or groove-joint pliers, are used by all types of mechanics. See **Figure 1-16.** The jaws are positioned and locked into place by engaging the tongue in the proper groove. A series of grooves provides a range of jaw openings up to 4″ (102 mm). Pump pliers will not slip even under heavy pressure. They will grip round, square, flat, and hexagonal objects and can apply limited torque without damage to the work.

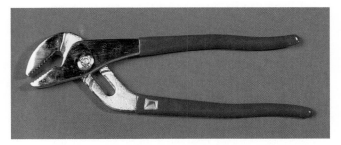

Figure 1-16. *Pump pliers provide a strong, nonslip grip on either round or square objects.*

1.3.5 Slip-joint Pliers

Slip-joint pliers, **Figure 1-17,** are an old but still widely used style. The slip joint provides two jaw positions, and the pliers may have a shear-type wire cutter. These pliers are designed for gripping, turning, and bending.

Figure 1-17. *Slip-joint pliers are an "all-purpose" tool for gripping and holding objects.*

1.3.6 Adjustable Pliers

One type of adjustable pliers, called Robo-Grip® pliers, permits one-handed adjustment of jaw size by merely squeezing the handles. See **Figure 1-18.** This type of adjustable pliers does not have a locking feature, however.

Figure 1-18. *This type of adjustable pliers offers one-handed operation. Jaw size adjustment is done automatically when the handles are squeezed.*

1.3.7 Locking Pliers

Locking pliers are available with straight or curved jaws, in a variety of sizes, **Figure 1-19.** These pliers use a compound leverage system to lock the jaws for holding various shapes and sizes of work. It is a combination tool that functions as pliers, wrench, portable vise, or clamp. Because of their clamping power, locking pliers can free both hands for doing other work (such as brazing, sawing, or drilling). Locking pliers can sometimes be used to unscrew fasteners that have stripped or rounded heads. Since they will mar surfaces, however, locking pliers should never be used on good nuts and bolts. Attempts to repair this tool are not recommended.

1.3.8 Using Pliers Safely

❑ Never use pliers to do the job of another tool.
❑ Never push pliers beyond their capacity. Bending stiff wire with light pliers or the tip of needle-nose pliers can spring or break the tool.
❑ Never expose pliers to excessive heat. Direct flame on metal can draw the temper and ruin the tool. Cutting pliers are especially vulnerable to high heat.

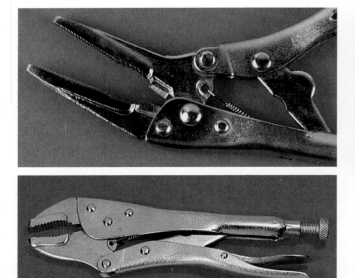

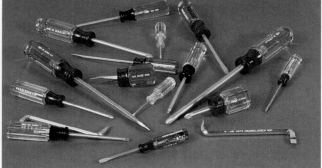

Figure 1-20. *Screwdrivers are available in a wide variety of sizes, shank lengths, and tip shapes.*

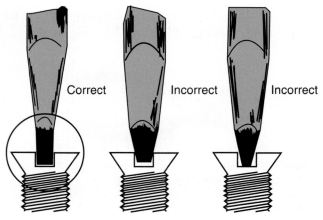

Figure 1-19. *Locking pliers are versatile and can be used to hold objects, freeing both hands for other tasks. They are available with several types of jaws.*

❑ Never cut hardened wire with ordinary pliers. Also, never rock pliers from side to side when cutting wire, nor bend the wire back and forth against the cutting knives. These practices can dull or nick cutting edges.

❑ Never cut any wire or metal unless your eyes and your fellow workers' eyes are protected. Safety goggles are an absolute must.

❑ Never depend on plastic-dipped handles to insulate you from electricity. They are intended for comfort and improved grip, not protection against electric shock.

1.3.9 Repair or Replace?

While repairing pliers is not recommended, regular maintenance is important. Dull cutting knives may be sharpened, and the hinge should receive a drop of oil occasionally. Pliers that are cracked, broken, sprung, or have nicked cutting knives should be discarded and replaced.

1.4 Screwdrivers

Screwdrivers are used to drive or withdraw such threaded fasteners as wood screws, machine screws, and self-tapping screws. Screwdrivers are available in a wide variety of shapes and sizes, **Figure 1-20.** The right tool is vital for fast, efficient driving and removal of screws. Screwdriver tips are designed to tolerate a limited amount of strain or torque. A screwdriver that is too long or too short, or does not fit the screw properly can waste time and cause damage to the fastener. See **Figure 1-21.**

Figure 1-21. *The tip of the screwdriver must fit the slot properly to avoid damaging the slot or slipping and causing injury. (State of Ohio)*

1.4.1 Special Screwdrivers

Reversible screwdrivers have a different tip at each end of the ***shank*** (shaft). The driver handle is removable and fits either end of the shank. The most common reversible screwdriver has a standard (straight-blade) tip at one end and a Phillips tip at the other.

Interchangeable bit screwdrivers, **Figure 1-22,** allow different types and sizes of bits to be mounted to suit the task at hand. Straight-blade, Phillips, Torx®, square socket, and nut-driver tips are available. Some models provide storage in the handle for a number of bits.

The screw-holding screwdriver is indispensable for working in close quarters or electrical boxes. Screw-holding drivers are used only to get the screw firmly started. After starting, a conventional driver is used for further driving. There are two types: one that holds by means of pressure against the sides of the screw slot, and one that has spring-steel fingers to grip the screw head. The relationship of the screwdriver blade and spring-steel fingers is shown in **Figure 1-23.**

Screwdrivers for Torx®, Allen, Pozidrive®, and other special screwhead patterns are available in many sizes and lengths. **Figure 1-24** shows these special patterns.

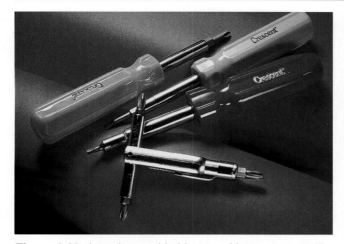

Figure 1-22. *Interchangeable bit screwdrivers are versatile, since they include allow easy selection of the tip needed for a job. The T-handle model has six tips and a pocket clip. (CooperTools)*

Figure 1-23. *In one type of screw-holding screwdriver, spring-steel fingers grip the head of the screw and hold it firmly against the blade.*

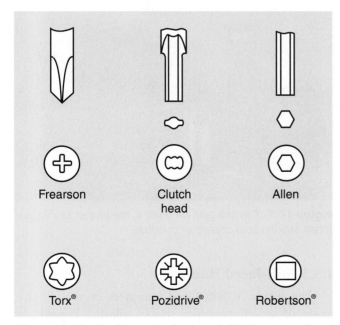

Figure 1-24. *Besides standard and Phillips screwhead patterns, there are a number of special patterns like those shown here. The Torx® pattern is widely used in the automotive industry, for example. (Vaco Products)*

1.4.2 Nut Drivers

A nut driver is a special tool with a 6-point socket at the tip of a screwdriver-type handle, **Figure 1-25.** Nut drivers are useful for removing and replacing

small hex head (6-sided) screws and nuts. Nut drivers are available in sizes ranging from 3/16″ to 5/8″. Handles are usually color-coded for quick selection of size. Hollow shaft nut drivers are considered best for general use.

1.4.3 Using Screwdrivers Safely

❏ Always fit the tool to the work. The size of the screw and the type of drive opening determine which driver to use.

❏ Never use a driver to do the job of another tool.

❏ Never push a driver beyond its capacity. For heavy work that requires a wrench to help do the turning, use a square-shank driver.

❏ Never use a driver at an angle to the screw. The driver could slip and damage the screwhead or cause injury.

❏ Never expose a driver to excessive heat.

❏ Never depend on a driver's handle or covered blade for insulation from electricity.

1.4.4 Repair or Replace?

Do not attempt to repair most types of drivers. Drivers with cracked handles, bent or twisted shafts, or worn tips should be discarded and replaced. The tip of a straight-blade driver, however, can be ***dressed*** (squared) on a bench grinder. Be careful to avoid letting the tip get hot. This will draw the temper (hardness) causing the tip to become soft and easily damaged.

1.5 Hammers

Hammers are used for various striking operations. It is important to use the correct hammer and use it properly. For example, a nail hammer should not be used to strike a chisel or other hardened tool. Likewise, a soft-faced hammer should not be used to drive a nail; nor should a sledge-type hammer be used to drive a brad.

1.5.1 Nail Hammers

Nail hammers are made in two patterns: curved claw and straight claw. See **Figure 1-26.** They are available in various head weights for different types of nailing: 16 oz. (450 g) and 20 oz. (570 g) hammers are for general use; 22 oz. (625 g) and 28 oz. (795 g) hammers are used for heavy-duty framing work. Framing hammers usually have longer handles than those for general use. Handles are made of wood, fiberglass, or steel. Never drive hardened-steel cut nails or masonry nails with a nail hammer. These nails will shatter under the force of an indirect or glancing blow. Instead, use an engineer's (sledge-type) hammer.

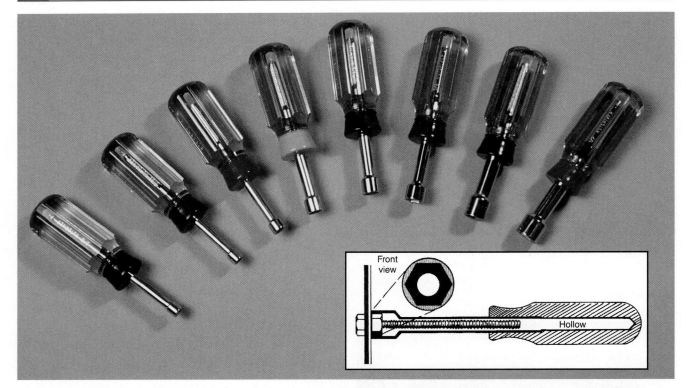

Figure 1-25. *Nut drivers, like these hollow-shaft types, usually have handles that are color-coded for easy identification.*

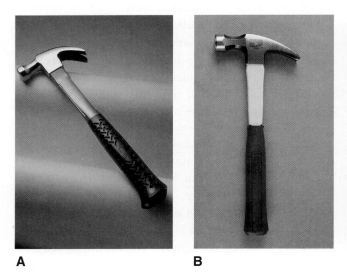

A   **B**

Figure 1-26. *Nail hammers are available in a variety of weights and in either curved- or straight-claw patterns. A—Curved claw hammer. (CooperTools) B—Straight-claw hammer.*

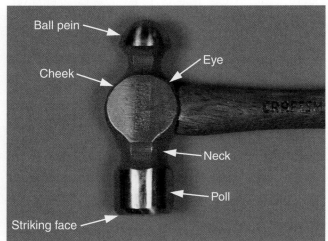

Figure 1-27. *The ball pein hammer is the proper tool to use when striking cold chisels or punches.*

1.5.2 Ball Pein Hammers

Ball pein hammers, **Figure 1-27,** are one of the most frequently used hammers. They are designed for striking chisels and punches and for riveting, shaping, and straightening metal. Ball pein hammers are available with head weights ranging from 2 oz. to 48 oz. (55 g to 1360 g), depending on the task. When striking a tool like a chisel or punch, the hammer face should be 3/8″ (9.5 mm) larger than the struck tool.

1.5.3 Soft-faced Hammers

The heads of soft-faced hammers or mallets are made of wood, rawhide, rubber, plastic, copper, or brass. See **Figure 1-28.** They are used where steel hammers would mar or damage the workpiece. Never use a soft-faced hammer to strike sharp metal objects or to drive nails or screws.

1.5.4 Engineer's Hammer

The engineer's hammer or sledge-type hammer, **Figure 1-29,** is designed for heavy-duty striking. Weights are typically 2 1/2 lb. to 4 lb. (1.1 kg to

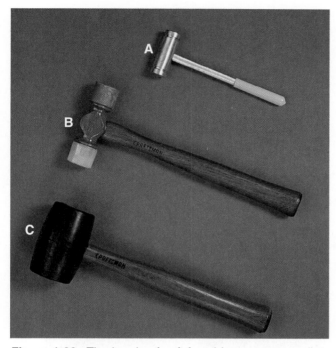

Figure 1-28. *The heads of soft-faced hammers or mallets are made from such materials as rawhide, rubber, wood, brass, or plastic. A—Brass hammer. B—Plastic-faced hammer. C—Rubber mallet.*

Figure 1-29. *The engineer's hammer, or sledge, is used for such tasks as driving hardened nails.*

1.8 kg). This type of hammer is commonly used for striking spikes, cold chisels, and hardened nails.

1.5.5 Setting Hammers

Setting hammers are designed for work on sheet metal, **Figure 1-30.** They have square corners and straight sides to form sharp corners and to close seams. The corners and sides of the setting hammer are easily chipped if incorrectly used.

1.5.6 Using Hammers Safely

❑ Always align the face of the hammer squarely with the surface being struck. Avoid glancing blows.

❑ Never use a nail hammer to strike a chisel or other hardened objects, such as masonry nails. The hammer face may chip and cause injury.

❑ Always wear safety goggles when using a hammer to drive nails or strike objects.

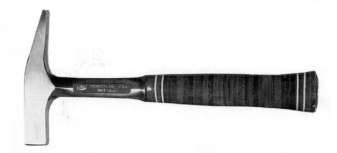

Figure 1-30. *Setting hammers have square heads that make them ideal for forming seams and corners in sheet metal. (Malco Products, Inc.)*

❑ Never strike one hammer with another hammer.

❑ Never use a hammer that has a loose or damaged handle.

❑ Never use the side or **"cheek"** of a hammer for striking any object.

❑ Never grind or redress the face of a hammer. If the hammer head is damaged, discard and replace it.

1.5.7 Repair or Replace?

Discard hammers that have damaged heads; they are dangerous to use. Damage is indicated by dents, chips, excessive wear, or a broken claw. Damaged wood or fiberglass handles can usually be replaced.

1.6 Hacksaws

Hacksaws are used to cut all types of metal objects, **Figure 1-31.** Hacksaw blades are selected according to the number of teeth per inch: 18 (coarse), 24 (medium), and 32 (fine). The blade should be fastened tightly in the frame, with the blade teeth pointed away from the handle. At least two teeth should contact the metal being cut, or the saw will catch and bind.

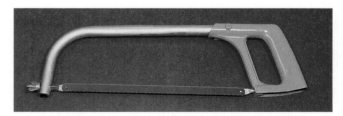

Figure 1-31. *The hacksaw is an important tool for cutting metal. Blade teeth should face away from the handle so the saw cuts on the forward stroke.*

You can use one or two hands on the saw to make cuts. Press down lightly on the forward stroke and release pressure on the backstroke. Avoid rapid strokes, or the blade will overheat and become dull. Use 50 to 60 strokes per minute.

Small (sometimes called "mini") hacksaws, **Figure 1-32,** are used for working in very confined areas or for flush-cutting. These small hacksaws are indispensable to the technician.

A

B

Figure 1-32. *For flush-cutting metal or working in close quarters, the mini-hacksaw is the ideal tool. A—A frame-type mini-hacksaw. B—A model with a contour grip.* *(Vermont American)*

1.7 Drill Bits

Various types of drill bits are used for cutting holes. Each is designed for use on a specific type of material: drill bits designed for wood cannot be used on metal or concrete; bits for metal can be used on wood, but not concrete; bits for concrete cannot be used on metal or wood.

1.7.1 Twist Drill Bits

Twist drill bits are used to cut small-diameter holes in metal or wood. Bits made of high-speed steel are most common and are sometimes carbide-tipped. A twist drill bit has three principal parts: a *shank* that is clamped into the drill chuck, a *body* with two spiral grooves called **flutes,** and a cone-shaped cutting end called the *point.* See **Figure 1-33.** The flutes act as channels for the escape of metal chips from the hole being drilled. Twist drills can be purchased separately or in sets, with drill sizes ranging from 1/16″ to 1/2″. See **Figure 1-34.**

1.7.2 Wood Bits

The *spade-type wood bit* is designed to cut straight or angled holes in wood, plastic laminate, plaster,

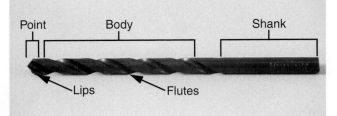

Figure 1-33. *Parts of a typical twist drill bit.*

Figure 1-34. *Twist drill bits are available individually or in sets. Most often, a technician starts with a set, then replaces bits individually as necessary.*

foam insulation, and similar materials. Like twist drills, they are available as individual bits or in sets, in sizes ranging from 1/4″ to 1 1/2″ (6 mm to 38 mm). As shown in **Figure 1-35,** sizes are stamped on each bit. Spade-type bits can be resharpened with a file or grinder. Extension bars are available for extended reach.

Auger-type wood bits are generally used for drilling large holes and are available in sizes up to 2 1/2″ (63.5 mm). The pilot screw tip draws the bit into the wood. Cutting blades can be sharpened with a file.

Figure 1-35. *Spade-type bits are used to bore medium to large holes in wood and similar nonmetal materials. Sizes are stamped on each bit.*

1.7.3 Hole Saws

High-speed hole saws are designed for cutting steel, aluminum, copper, brass, sheet metal, stainless steel, wood, or plastics. See **Figure 1-36.** The high-speed cutting edges are welded to steel bodies to produce clean, accurate holes. Shatter-resistant construction makes them safe and durable. Hole saw sets, **Figure 1-37,** are available and include sizes ranging from 9/16″ (14 mm) to as much as 5″ (127 mm). Individual hole saws are quickly attached to an **arbor** (spindle or axle) that has a twist drill for starting the pilot hole.

Figure 1-37. *Hole-saw sets include one or more arbors and a range of saw sizes for holes of different diameters. (Milwaukee Electric Tool Corp.)*

1.7.4 Masonry Bits

Masonry bits, **Figure 1-38,** are used for drilling brick, stone, concrete, and ceramic materials. Such holes are often needed to allow use of concrete anchors or similar fasteners. The special carbide tip resists dulling. Flutes remove dust from the hole so that the bit cuts at maximum efficiency without clogging. Bits may be resharpened on a bench grinder using a silicon carbide grinding wheel. Sizes range from 1/8″ to 1 1/2″ (3 mm to 38 mm).

Star drills are designed for hand-drilling holes in masonry. See **Figure 1-39.** The bit is held in place with one hand while struck with an engineer's hammer held in the other hand. The bit must be rotated slightly after each blow. The four cutting edges steadily chip into the material with each blow. Always wear safety goggles when using any form of masonry bit. *Never* use a star drill with a struck face that is chipped or **mushroomed** (flattened and spread out from being struck).

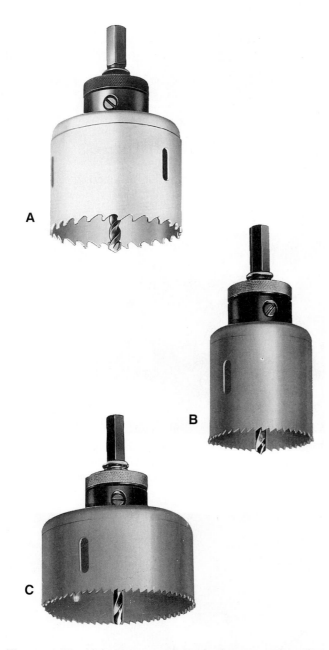

Figure 1-36. *Hole saws are attached to an arbor that includes a twist drill for boring a pilot hole. A—A carbide-tipped, deep-cutting type saw. B—A deep-cutting saw without carbide tips. C—A standard-depth hole saw. (Milwaukee Electric Tool Corp.)*

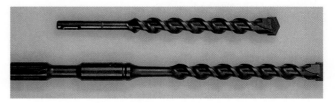

Figure 1-38. *Carbide-tipped masonry drills make fast work of holes in concrete, brick, and similar materials. Note the reduced shank on the drill at top. It allows use of the drill in a chuck with an opening smaller than the drill size. (American Tool Companies, Inc.)*

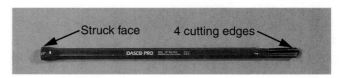

Figure 1-39. *The star drill is struck with a sledge-type hammer to make holes in concrete or masonry materials.*

1.8 Snips

Snips are used to cut sheet metal, screens, gaskets, and straps. There are four basic types in common use: straight-pattern, combination, duckbill, and aviation. Snips are made for right-handed use, although they can be used in either hand.

Other types of snips include *curved blade, bulldog, double-cut,* and *jeweler's.* These snips are used by professional metalworkers.

1.8.1 Straight Snips

Straight-pattern snips, **Figure 1-40,** are generally used for making straight-line cuts. Curved cuts can be made if the curve is not extreme. Sizes vary from 7″ to 16″ (178 mm to 406 mm) overall length. The most popular size is 10″ (254 mm).

1.8.2 Combination and Duckbill Snips

Combination snips and duckbill snips, **Figure 1-41,** can be used for straight-line cutting as well as cutting curves in either direction. When used for straight-line cutting, both types tend to bend the metal, so they require more effort than a straight-pattern snip. The duckbill snip will cut smooth curves in sheet metal in either direction, **Figure 1-42.**

1.8.3 Aviation Snips

Aviation snips are popular in all fields of metalwork. Their compound lever action requires less effort to cut metal than other types of snips, **Figure 1-43.** The blades are made of alloy steel, so they are also

Figure 1-40. *Straight-pattern snips are used primarily for straight cuts but can make gentle curves, if necessary.*

Figure 1-41. *Duckbill snips are designed for making curved cuts in sheet metal or similar materials.*

harder and tougher than regular snips. The cutting blades are **serrated** (toothed or ridged) to prevent metal from slipping out when pressure is applied. See **Figure 1-44.** Standard size for the professional aviation snip is 10″ (254 mm).

Aviation snips are available in three types of cut. The straight pattern is for cutting straight lines. The left-hand pattern is for cutting curves to the left. The right-hand pattern is for cutting curves to the right.

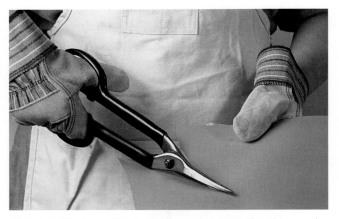

Figure 1-42. *Smoothly curved cuts in either direction can be made with duckbill snips.*

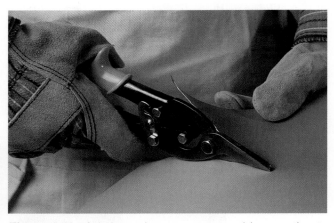

Figure 1-43. *Aviation snips use compound lever action to make cutting easier. They are available in models designed for straight cuts, left-hand curves, or right-hand curves.*

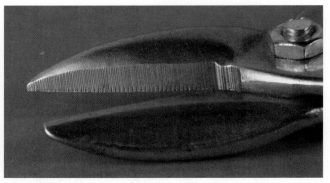

Figure 1-44. *Aviation snips are made from tough alloy steel and have serrated jaws to keep metal from slipping.*

1.8.4 Using Snips Safely

❏ Always wear safety goggles when using snips.
❏ Be careful of the sharp cutting edges on snip blades.
❏ Always wear gloves to protect against cuts from sharp metal edges.
❏ Use snips only for cutting soft metal or other soft materials.
❏ Use only hand pressure for cutting. Never hammer on the snip or use your foot to put extra pressure on the cutting edge.
❏ Perform periodic maintenance to keep snips efficient and safe. Occasionally oil the pivot bolt, and protect the cutting edges from damage. Sharpen the edges as necessary. Never try to sharpen an *aviation* snip, however. This removes the serrations that grip the metal so it does not slip.

1.9 Punches and Cold Chisels

Punches, **Figure 1-45,** and cold chisels, are called "struck tools" because they are struck with a hammer. They are made of special alloy steel that is heat-treated and drawn to provide maximum resistance to impact. They are designed to direct the force of each blow toward the center, or body, of the tool. Tools must be struck squarely. Blows that are off-center can damage the chisel or material or cause injury. Always wear eye protection when using chisels or punches.

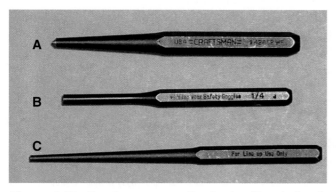

Figure 1-45. *Punches are usually struck with a ball pein hammer. A—Center punch. B—Pin punch. C—Drift punch.*

1.9.1 Punches

A center punch and the prick punch are similar, but the prick punch has a sharper point. Either one can be used to mark parts before disassembly. This ensures proper alignment when the parts are reassembled. A punch is also used to make an indentation in metal before drilling a hole. The indentation prevents the drill bit from wandering (moving).

A pin punch has a straight shank for driving pins or rivets out of a hole. A similar tool, the **drift punch,** is often used for aligning holes in different sections of material. Never use a punch with a dull, chipped, or deformed point. Such a damaged punch could slip and cause injury.

1.9.2 Cold Chisels

Cold chisels are used to cut off damaged or badly rusted nuts, bolts, rivet heads, or other fasteners, **Figure 1-46.** The angle and thickness of the cutting edge is designed to give maximum cut and durability. Never use a chisel that has a mushroomed head (striking face) since chips of metal could fly off and cause injury. Use a grinder to dress the chisel (remove the mushroomed metal and form a smooth chamfer), **Figure 1-47.** A *dull* cold chisel can slip and cause injury. Sharpen the hardened cutting edge as needed with a hand file or a **whetstone** (special sharpening stone).

Figure 1-46. *A cold chisel can be used to cut through bolts, pins, or other metal fasteners or parts.*

1.9.3 Using Punches and Cold Chisels Safely

❏ Always wear eye protection when working with chisels and punches.
❏ Always strike a punch or chisel squarely. An off-center blow could cause damage or injury.
❏ Never use a cold chisel to cut or split materials such as stone or concrete.

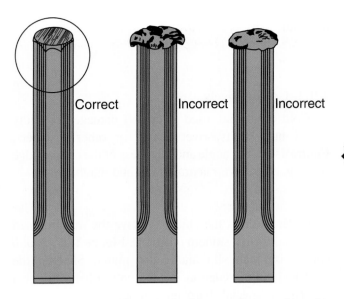

Figure 1-47. *A chisel or punch with a mushroomed head is dangerous because bits of metal can fly off when the tool is struck. A properly dressed chisel or punch is ground to a smooth chamfer as shown. (State of Ohio)*

❏ Never use a dull chisel or a punch that has a chipped or deformed point.
❏ Never use a chisel or punch with a mushroomed head. Grind off the mushroom to form a chamfer. Avoid overheating when grinding.

Summary

This chapter has described the most common hand tools used by technicians, as well as their selection and safe use. It is almost impossible to perform even the simplest repair without using some type of tool. The right tool, used properly, will give satisfactory results. The wrong tool, or an improperly used tool, will give unsatisfactory results and may cause injury.

Selecting the proper hand tool for the job depends on understanding the advantages and disadvantages of each tool. A good technician knows when, where, and why a specific tool will work better than another. For example, several different tools (pliers, socket wrench, adjustable wrench, box wrench) can be used to loosen a nut or bolt, but only one will do the job best and most safely. Take time to select or obtain the right tool. Other tools are introduced and explained in later chapters, as needed.

Test Your Knowledge

Please do not write in this text. Write your answers on a separate sheet of paper.
1. Misuse of tools indicates poor _____ skills.
2. Name six types of wrenches.
3. The types of wrenches that tend to slip are _____ and _____.
4. What is the safest type of wrench to use on a stubborn nut or bolt? Why?
5. The _____ socket is the most common.
6. *True or false?* You should always pull on a wrench, rather than push.
7. Plastic-dipped handles do *not* protect against _____.
8. List six types of pliers.
9. Safety goggles are absolutely necessary when using pliers to cut _____ or _____.
10. Name two types of special screwhead patterns.
11. Which hammer is best for striking chisels?
12. Hacksaws cut on the _____ stroke.
13. *True or false?* Twist drills can be used only on wood.
14. Which one of these basic types of snips has a serrated blade?
    a. Straight-pattern.
    b. Combination.
    c. Duckbill.
    d. Aviation.
15. Since they are hit with a hammer, punches and cold chisels are called _____ tools.

2

Fasteners

Objectives

After studying this chapter, you will be able to:
- ❏ Define thread terminology.
- ❏ Identify standard thread forms.
- ❏ Demonstrate proper use of taps and dies.
- ❏ Identify screw types and sizes.
- ❏ Identify and correctly use bolts, nuts, and washers.
- ❏ Identify and correctly use various anchors.
- ❏ Identify and correctly use setscrews, blind rivets, and threaded rods.

Important Terms

American National Acme Thread	mandrel
American National Standard Thread	masonry anchors
	minor diameter
	Molly bolts
Brown and Sharpe Worm Thread	nuts
blind hole	root
blind rivet	screw thread
bolt extractor	screws
bolt	setscrews
cap screws	shank
crest	tap
die	thread angle
flutes	threaded rod
International Thread	threads per inch
machine screws	toggle bolt
major diameter	Unified Thread
	washers

2.1 Threads

Screws, bolts, nuts, washers, and anchors are among the fasteners that should be familiar to the beginning technician. Being able to successfully use fasteners depends upon knowing and understanding:
- ❏ Thread terminology.
- ❏ Standard thread forms and thread measuring methods.
- ❏ Proper tap-drill size for a screw thread.

2.1.1 Thread Terminology

A *screw thread* is a helical ridge of uniform section formed inside a hole (such as a nut) or on the outside of a fastener (such as a screw or bolt). Threads cut on an outside surface are referred to as "external;" those cut on an inside surface are called "internal."

The following terms describe the parts of a screw thread and are shown in **Figure 2-1:**
- ❏ *Major diameter* is the widest measurement from the outside edges of the threads.
- ❏ *Minor diameter* is the smallest measurement from the inside edges of the threads.
- ❏ *Root* is the bottom area of two adjoining threads.
- ❏ *Crest* is the top edge of two adjoining threads.
- ❏ *Threads per inch* is determined by counting the number of crests (or roots) per inch of threaded section.
- ❏ *Thread angle* is the V-shaped angle of the threads. The sides of the threads normally form a 60° angle.

2.2 Standard Thread Forms

Of the standard thread forms, those most widely used in North American industry today are the American National and Unified forms. These two forms are popular because they can be mass-produced easily and economically. The thread form in widespread use in most parts of the world is International Thread, also known as the ISO Metric Thread Series.

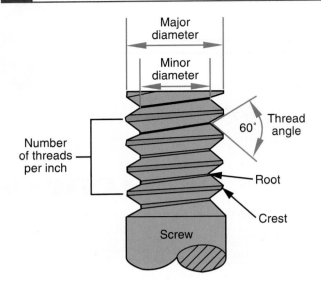

Figure 2-1. *Parts of the screw thread. The number of threads per inch determines whether a thread is coarse or fine.*

2.2.1 American National Standard

American National Standard Thread has three major series of threads: National Fine (abbreviated NF), National Coarse (NC), and National Pipe (NPT).

National Fine has a greater number of threads per inch than National Coarse, and the threads are not as deep. Both NC and NF threads maintain constant major and minor diameters.

National Pipe threads, however, are tapered (sloped) 1/16″ in diameter for every inch of threaded length. This taper causes the threads to bind together, making an increasingly tight seal as a pipe fitting is tightened. Pipe threads tend to be self-sealing due to compression of the V-threads as the fittings are connected.

All fastener components must share the same style of threads and the same number of threads per inch. For example, an NF nut cannot be screwed onto an NC bolt, **Figure 2-2.** Attempts to force mismatched threads to mate can severely damage threads and render the fastener useless. Damaged threads can sometimes be dressed by rethreading them with a tap or die of the proper size.

Figure 2-2. *Threads of different types and styles cannot be used together. The 24 threads per inch on the 1/2″ nut will not mate properly with the 27 threads per inch on the bolt.*

National Coarse threads are widely used to make bolts, nuts, and screws where quick assembly is desired in materials such as cast iron, steel, bronze, brass, aluminum, and magnesium. The National Fine thread series is also used in the manufacture of nuts, bolts, and screws. NF threads are usually specified if the thread engagement is short or the internal threads are cut into a thin-walled material. National Pipe threads are used in the assembly of steel pipe and cast iron fittings for water lines and natural gas lines. Some heavy-duty electrical conduit also uses pipe threads.

2.2.2 Unified Screw Threads

The American National screw thread form was the standard in the United States for many years. However, some industry leaders advocated a thread system that would permit interchangeability of screw threads with other countries, especially those following British standards.

British Standard Whitworth Thread was not interchangeable with American National Thread. American National had a thread angle of 60°, with roots and crests slightly flattened. The Whitworth thread had a thread angle of 55° with rounded roots and crests.

During World War II, England and Canada agreed to change the Whitworth thread angle to 60°, but to retain the rounded crests and roots. This new thread form, called **Unified Thread,** was interchangeable with American National threads. Since that time, American National Fine has been interchangeable with Unified National Fine (UNF), as have been National Coarse and Unified National Coarse (UNC).

Unified Thread is now the standard thread, but the American National coarse and fine thread series will continue to be seen on gauges, tools, and blueprints for a number of years. For this reason, it is necessary to be familiar with both systems.

2.2.3 Other Thread Forms

The standard form used in most parts of the world is **International Thread,** also known as the International Standards Organization (ISO) Metric Thread. The ISO metric thread has a 60° angle, with slightly flattened roots and crests. Use of this thread form is growing in North America. The main difference between International Thread and Unified Thread is the depth of the threads. The two thread forms are not interchangeable.

American National Acme Thread has a 29° angle and is used for feed screws, jacks, and vises.

Brown and Sharpe Worm Thread also has a 29° angle, but the depth of the root is greater than that of the Acme thread. This thread is used to mesh with worm gears and transmit motion. A self-locking feature makes it usable for winches and steering mechanisms.

2.2.4 Cutting Internal Threads

A *tap* is a tool used to cut threads inside a hole. Taps are made of special hardened steel and threaded for about half their length. The threaded grooves are called **flutes,** and the edge of each thread forms a cutting edge. See **Figure 2-3.**

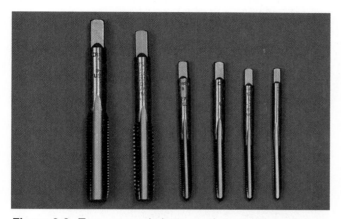

Figure 2-3. *Taps are made in many sizes and thread types.*

Taps are available in three types: taper, plug, and bottoming, **Figure 2-4.** On the *taper* tap, the first six threads are tapered for easy starting. On the *plug* tap (the most common), only three or four threads are tapered. The *bottoming* tap is not tapered. It is used to cut threads to the bottom of a hole. The size is stamped on the unthreaded portion of the tap, called the **shank.** This number indicates the outside diameter and the number of threads per inch.

Before using a tap, a pilot hole must be drilled into the workpiece. The size of the pilot hole is critical. The hole must be smaller than the tap's outside diameter to allow for cutting of full threads inside the hole. The proper pilot hole size (drill size) can be determined by consulting tables like those in **Figures 2-5** through **2-7.**

Once the pilot hole has been drilled, the tap is inserted and properly aligned with the work surface. If the tap is large, use a wrench to turn it. Small taps are usually turned with a special tap wrench, **Figure 2-8.** Use slight downward pressure to start the cut, but release the pressure once the tap threads begin to bite. Use a lightweight oil to lubricate the cutting process. Reverse the tap every two or three turns to clear metal shavings from the threads. The cutting process should continue until the tap turns smoothly with little pressure.

A **blind hole** is a pilot hole that does not penetrate completely through a component. To thread a blind hole, use a taper tap until it touches the bottom of the

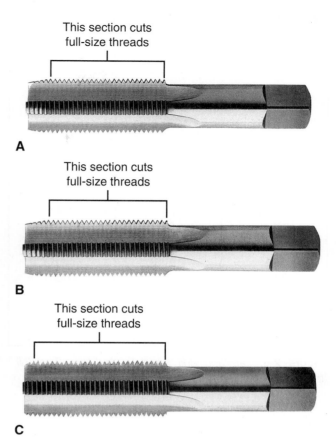

A

B

C

Figure 2-4. *Types of taps. A—A taper tap is designed for easy starting. B—A plug tap is the most common. C—A bottoming tap is used to cut threads to the bottom of a hole. (Triumph Twist Drill Co.)*

National Coarse Threads

Tap Size	Threads per Inch	Drill Size
#5	40	#38
#6	32	#36
#8	32	#29
#10	24	#25
#12	24	#16
1/4″	20	#7
5/16″	18	F
3/8″	16	5/16″
7/16″	14	U
1/2″	13	27/64″
9/16″	12	31/64″
5/8″	11	17/32″
3/4″	10	21/32″
7/8″	9	49/64″
1″	8	7/8″

Figure 2-5. *Tap and drill sizes for National Coarse (NC) threads.*

National Fine Threads

Tap Size	Threads per Inch	Drill Size
#5	44	#37
#6	40	#33
#8	36	#29
#10	32	#21
#12	28	#14
1/4″	28	#3
5/16″	24	I
3/8″	24	Q
7/16″	20	25/64″
1/2″	20	29/64″
9/16″	18	33/64″
5/8″	18	37/64″
3/4″	16	11/16″
7/8″	14	13/16″
1″	14	15/16″

Figure 2-6. *Tap and drill sizes for National Fine (NF) threads.*

National Pipe Threads

Tap Size	Threads per Inch	Drill Size
1/8″	27	11/32″
1/4″	18	7/16″
3/8″	18	19/32″
1/2″	14	23/32″
3/4″	14	15/16″
1″	11 1/2	1 5/32″
1 1/4″	11 1/2	1 1/2″
1 1/2″	11 1/2	1 23/32″
2″	11 1/2	2 3/16″

Figure 2-7. *Tap and drill sizes for National Pipe (NPT) threads.*

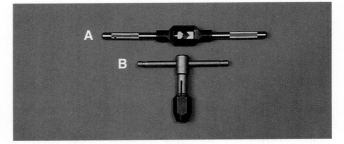

Figure 2-8. *Tap wrenches. A—A hand tap wrench provides greater leverage when using large taps. B—A T-handle tap wrench generally is used for small taps.*

Figure 2-9. *A hexagonal threading die. Round and square dies are also available.*

Figure 2-10. *A large tap and die set.*

component hole. Then change to a plug tap. Finally, complete the thread with a bottoming tap.

2.2.5 Cutting External Threads

A *die* is a tool used to cut threads around the outside of a piece of metal, such as a bolt, rod, or pipe. A threading die may be round, square, or hexagonal. See **Figure 2-9.** The thread cutters may be built-in or replaceable. The leading edges of the cutters are ground away slightly to make starting easier.

Dies are available in all sizes from very small to very large and are often sold in sets with matching taps, **Figure 2-10.** The die selected depends upon the kind of thread desired (NF, NC, or NPT) and the diameter of the material being threaded.

To use a die, first clamp the workpiece in a vise. Fasten the die into a die stock or holder, **Figure 2-11,** and place it squarely over the end of the work. Apply downward pressure while turning the die slowly to the right until the threads bite. Use lightweight oil for the cutting process, and stop applying the downward pressure once the threads are started. The die should be backed off after each half-turn to release the metal chips.

2.3 Screws

Screws fasten parts securely while, at the same time, allowing the parts to be readily removed. Screws

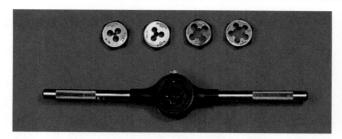

Figure 2-11. *A hexagonal-type die stock holder. A round-type stock is also made.*

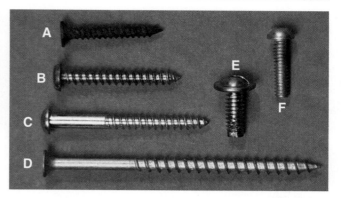

Figure 2-12. *Types of screw threads. A—Drywall. B—Sheet metal. C—Wood. D—Wood decking. E—Thread-forming. F—Machine.*

are selected according to the job and the material to be fastened. Some screws are intended for use in wood, others in thin metals, and still others in thick metals. The type of threads used will determine the holding power.

It is often necessary to drill a pilot hole before inserting a screw. The pilot hole must be smaller in diameter than the actual screw so the screw threads can grip the material.

2.3.1 Screw Sizes

Screw sizes are measured by both length and diameter. The length is designated in inches (or fractions of an inch), but the diameter is indicated by a gauge number. Screws are available in gauges ranging from 0 (about 1/16″) to 24 (about 3/8″), and are available from 1/4″ to 6″ in length. The most common screw gauge numbers are 2 through 16. For example, a 6 × 1 screw would be gauge number 6 in diameter and 1″ long. The gauge number is always listed first, and the length is second.

2.3.2 Screw Threads

Different types of threads are used for fastening different types of materials, **Figure 2-12.** For example, the threads on a wood screw are different from those on a sheet metal screw. The screw should always match the job, based on the material being fastened.

2.3.3 Screw Head Styles

Screws are available in many head styles, **Figure 2-13.** Screw heads are selected for holding ability and appearance, depending upon the application. Head styles are often available in different types of drives. The drive determines what type of screwdriver must be used with that screw. See **Figure 2-14.**

2.3.4 Self-drilling Screws

Self-drilling screws are designed to drill and thread holes in one quick and easy step. A pilot hole is not required. See **Figure 2-15.** These screws are used for anchoring sheet metal and metal duct. They have hex-type heads, slotted or unslotted, and are available in two point types, **Figure 2-16.**

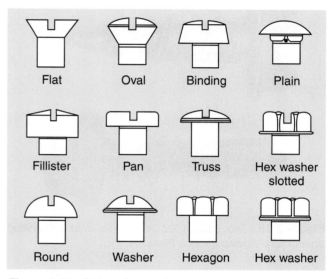

Figure 2-13. *Styles of screw heads.*

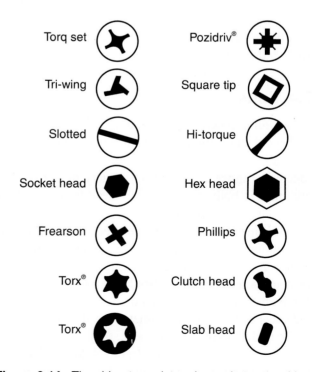

Figure 2-14. *The drive type determines what screwdriver must be used.*

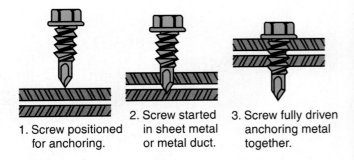

1. Screw positioned for anchoring.

2. Screw started in sheet metal or metal duct.

3. Screw fully driven anchoring metal together.

Figure 2-15. *Self-drilling sheet metal screws do not require a pilot hole.*

Figure 2-16. *Two point types are available in self-drilling sheet metal screws. (Malco Products, Inc.)*

Quick installation of self-drilling screws requires a special hex chuck driver for use with an electric drill. The driver head is magnetized to hold the screw while it is being driven. Hex chuck drivers are available in 1/4″, 5/16″, and 3/8″ sizes.

2.4 Bolts

A *bolt* is generally defined as a fastener that is used with a nut and is tightened by turning the nut. Bolts are normally used for fastening heavy metal parts together. **Figure 2-17** shows a variety of bolts. Bolts are made in both square and hex head forms and have either square or hex nuts.

Machine bolts (and threaded rod) are available in all common thread forms, with coarse threads most common. Bolt sizes are designated the same as machine screws: diameter first, then number of threads per inch, and then length. For example: 1/2-13 × 3 indicates a bolt 1/2″ in diameter, 13 threads per inch (NC), and 3″ long.

Bolt sizes ordinarily range from 1/4-20 to 1/2-13, and their lengths range from 3/8″ to 6″. Much larger sizes are available but are not stocked by most hardware stores.

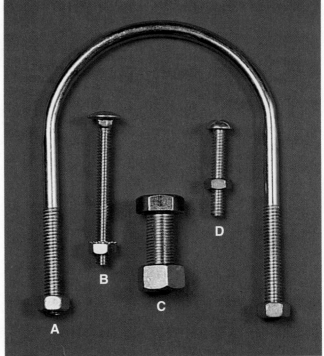

Figure 2-17. *Bolts are available in various types. A—U-bolt. B—Carriage bolt. C—Hex head bolt. D—Stove bolt.*

2.4.1 Carriage Bolts

Carriage bolts are usually used to anchor wood to metal. The head of the carriage bolt includes a countersunk square shoulder along with the round head. This square shoulder keeps the bolt from turning when the nut is tightened. Carriage bolts are also used when the head is not accessible to a wrench.

2.4.2 Stove Bolts

Stove bolts are made with slotted round, flat, or oval heads. They were originally used in stove construction but are now sold as general utility bolts. Stove bolts are very much like machine screws, except the smallest size is 1/4″ in diameter.

2.4.3 Bolt Extractors

Occasionally, a bolt or screw may be overtightened, causing the head to break off or the fastener to snap off inside a hole. In either case, the bolt or screw must be removed and replaced with a new one. Removing a broken bolt or screw is not hard, but it is time-consuming and requires the use of a ***bolt extractor.***

To remove a broken bolt or screw, you must first drill a hole in the end of the broken bolt, as shown in **Figure 2-18.** Insert a bolt extractor of the proper size in the hole, and use a wrench to turn the extractor counterclockwise. This causes the extractor to wedge (bite) into the hole and unscrew the bolt.

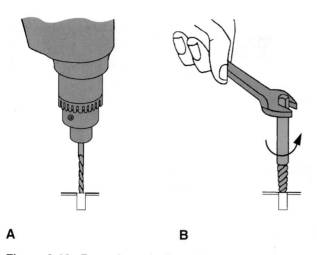

A   **B**

Figure 2-18. *Removing a broken bolt with a bolt extractor. A—Drilling a hole in a broken bolt. B—An extractor is inserted into the drilled hole and turned counterclockwise with a wrench to "back out" the broken fastener.*

2.5 Machine Screws

Machine screws fall into a unique category, filling the void between a screw and a bolt. As shown in **Figure 2-19,** they are small bolts with screw-type heads.

A machine screw is designated first by its diameter number, second by its number of threads per inch, and finally by its length. For example: 1/4-20 × 1 indicates a machine screw that is 1/4″ in diameter, has 20 threads per inch (NC), and is 1″ long. To fit this screw, a 1/4-20 machine nut would be required.

Machine screws are available in eleven diameters. Eight are in numbered sizes (like screws) and three are in fractional sizes (like bolts). As the diameter gets larger, the numbers get bigger. A table of common machine screw sizes and threads is shown in **Figure 2-20.** Machine screws are available in a variety of thread forms, head shapes, and drives, **Figure 2-21.**

2.6 Nuts

Nuts are available in a wide variety of types and sizes for use with different types of bolts and machine screws, **Figure 2-22.** Different nuts serve specialized purposes: acorn nuts for a finished appearance, wing nuts for finger tightening, and slotted or castle nuts for locking with a cotter pin or wire. Nuts are available in various finishes and degrees of hardness.

2.7 Washers

Washers are used to extend the gripping area of a fastener or to prevent a nut from loosening. See **Figure 2-23.** *Flat* washers are used to extend the surface area of a bolt head or nut. This prevents the bolt head or nut from being drawn into the surface of

Figure 2-19. *Machine screws are basically small bolts. The screw and nut must be the same diameter and have matching threads.*

Machine Screw Sizes and Threads

Machine Screw Number	Diameter (in.)	Threads per Inch	
		National Fine	National Coarse
#2	0.086	64	56
#3	0.099	56	48
#4	0.112	48	40
#5	0.125	44	40
#6	0.138	40	32
#8	0.164	36	32
#10	0.190	32	24
#12	0.216	28	24
1/4	0.250	28	20
5/16	0.3125	24	18
3/8	0.375	24	16

Figure 2-20. *Table of common machine screw sizes and threads.*

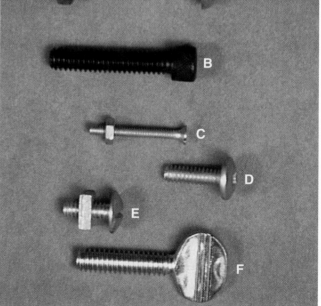

Figure 2-21. *Machine screws are available in a variety of head shapes and drives. A—Round slotted head. B—Fillister head cap screw with recessed hex drive. C—Flat slotted head. D—Pan head with Phillips drive. E—Truss slotted head. F—Thumb screw.*

Figure 2-22. *Many types of nuts are available.*

Figure 2-23. *Washers are either the flat or lock type.*

the material being fastened. *Lock* washers prevent a nut from loosening on a bolt. Lock washers are available in four styles: spring, helical, internal tooth, and external tooth.

2.8 Anchors

Various types of anchors are used to fasten materials firmly to various surfaces. The choice of anchor depends upon the surface and the load (weight of the object or material). Plastic screw anchors, expanding metal fasteners, and toggle bolts are used for solid or hollow walls; masonry anchors are used for solid surfaces, such as concrete or brick. See **Figure 2-24.**

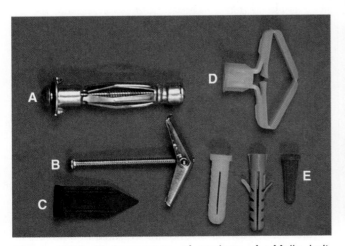

Figure 2-24. *Various types of anchors. A—Molly bolt. B—Toggle bolt. C—Drive-in-type expanding fastener. D—Plastic variation of toggle bolt. E—Several types of plastic screw anchors.*

2.8.1 Plastic Screw Anchors

Plastic screw anchors are light-duty anchors. They are used to fasten light loads to solid walls, floors, or other surfaces. As shown in **Figure 2-25,** the anchor is expanded against the sides of a hole by the screw, providing the grip needed to hold the load.

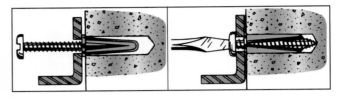

Figure 2-25. *A plastic screw anchor is inserted in a drilled hole. It is expanded and wedged against the sides of the hole as the screw is tightened. (Star Expansion Company)*

2.8.2 Hollow Wall Anchors

Special expanding metal fasteners (often called *Molly bolts*) are used to anchor relatively light objects to hollow walls. As shown in **Figure 2-26,** the fastener is inserted in a hole drilled through the wall and, as the screw is turned, the legs are pushed outward to grip the inside wall surface. A variation is a drive-in-type expanding anchor, which does not require drilling a hole. The flat metal anchor is hammered into place, then expanded by a screw. See **Figure 2-27.**

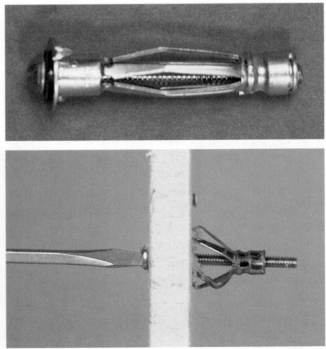

Figure 2-26. *A hollow wall anchor, top, is inserted in a drilled hole, then expanded by tightening the screw, as shown at bottom. The legs are permanently expanded inside the hollow wall, allowing the screw to be removed and replaced as necessary.*

Figure 2-27. *The screw expands the legs of a drive-in-type anchor after it is hammered into a hollow wall.*

2.8.3 Toggle Bolts

A ***toggle bolt,* Figure 2-28,** is used to fasten fairly heavy objects to a hollow wall. A hole large enough to pass the spring-loaded wings must be drilled into the wall. The wings will then spread out and bear against the inside of the wall, allowing the bolt to be screwed in, **Figure 2-29.**

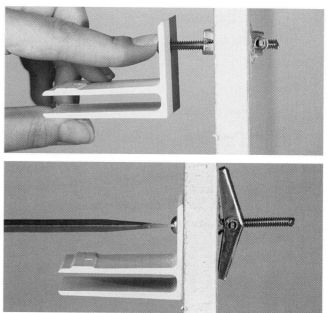

Figure 2-29. *After the toggle bolt is inserted through a hole into the hollow center of the wall, top, the wings are forced open by spring pressure. The wings are drawn against the inner wall surface by tightening the screw, bottom.*

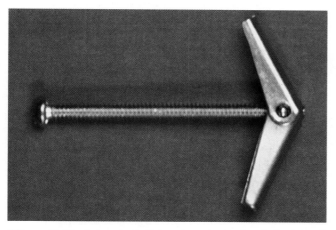

Figure 2-28. *A toggle bolt has two spring-loaded wings. It can support a heavier load than other types of wall anchors.*

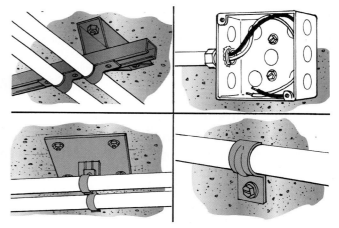

Figure 2-30. *Masonry anchors are used to fasten brackets, junction boxes, and conduit clamps to masonry walls. (ITW Ramset)*

2.8.4 Masonry Anchors

Masonry anchors are available in many forms and styles. They are often used to fasten brackets to a masonry surface, for such tasks as supporting lengths of conduit or copper tubing. See **Figure 2-30.**

Masonry anchors may be light-, medium-, or heavy-duty. Each type is available in different sizes for added strength and holding power.

In general, anchors are installed by drilling a hole the size of the anchor into the concrete or other material. When a bolt is inserted and tightened, the anchor expands and grips the inside of the hole, **Figure 2-31.** Sometimes, a special tool is required to "set" the anchor in the hole before inserting the bolt.

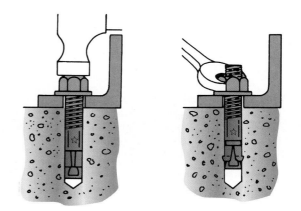

Figure 2-31. *Wedge anchors are driven into a hole drilled in masonry, then expanded by tightening the bolt. (Star Expansion Company)*

2.9 Other Fasteners

Besides screws and bolts, there are a number of other types of fasteners used in special situations or on specific jobs.

2.9.1 Setscrews

Setscrews are used to anchor pulleys and fan blades to motor shafts. As shown in **Figure 2-32,** setscrews do not have heads because they are usually recessed inside a threaded hole. Setscrews are normally driven by an Allen (hex) wrench. Setscrews that *do* have heads are properly called *cap screws.*

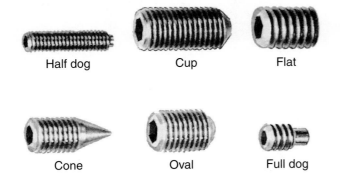

Half dog	Cup	Flat
Cone	Oval	Full dog

Figure 2-32. *Various setscrews with different point styles. (RB&W Corporation)*

2.9.2 Blind Rivets

A *blind rivet* (or pin rivet) is used to join two pieces of sheet metal for a strong connection that will not loosen under vibration. The blind rivet, **Figure 2-33,** consists of a rivet and a nail-like pin called a *mandrel.* The mandrel is designed to break off at the crimp after setting the rivet body with a hand tool, **Figure 2-34.** No special skill is required to set these rivets.

Blind rivets are available in five diameters, ranging from 3/32″ to 1/4″, with 1/8″ the most popular. They are made of steel, aluminum, stainless steel, and copper.

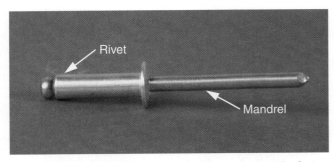

Figure 2-33. *Blind rivets are used to permanently fasten pieces of sheet metal together. The fastener consists of the rivet itself and a mandrel that is snapped off after forming a rivet head.*

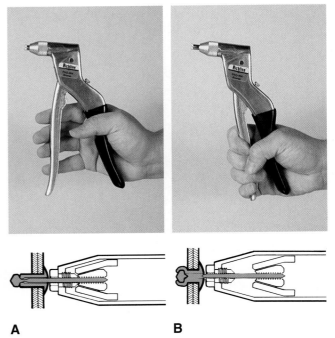

A   **B**

Figure 2-34. *Installation of a blind rivet. A—The mandrel of the blind rivet is gripped by the tool, and the rivet is inserted through the material to be fastened. B—When the tool handles are squeezed together, the mandrel is pulled forward, deforming the rivet material to form a head. The mandrel is then snapped off, leaving the rivet in place. (Malco Products, Inc.)*

2.9.3 Threaded Rod

Threaded rod (metal rod that is threaded its entire length) is used in many applications because it can be cut to any length with a hacksaw. It is used to make pipe hangers and is often combined with steel angle to suspend components from a ceiling. Threaded rod is available in many diameters, with coarse or fine threads.

◆ Summary

Manufacturers of refrigeration, heating, and air conditioning equipment use many types of fasteners. In turn, technicians routinely work with fasteners, some requiring special tools. Knowing the different types of fasteners and how they are used will mean faster repairs and installations. Since different threads are used on the same type of fasteners, failure to recognize thread differences can result in fastener damage. Damaged threads can often be repaired by using a tap or die of the proper size and type.

Service vehicles usually carry a good assortment of fasteners, but it is sometimes necessary to buy certain fasteners for a particular job. The technician must be aware of the different sizes and types of threads when ordering fasteners.

Test Your Knowledge

Please do not write in this text. Write your answers on a separate sheet of paper.

1. The widest measurement from the outside edges of a screw thread is called the _____.
2. The bottom area of two adjoining threads is the _____.
3. How are threads per inch determined?
4. Name two widely used thread forms.
5. Name the three common National series of threads.
6. Which series has tapered threads?
7. A tap is used to cut _____ threads.
8. A die is used to cut _____ threads.
9. Screw sizes are determined by _____ and _____.
10. *True or false?* Wood screws and metal screws have the same threads.
11. Machine bolts normally have _____ threads.
12. What tool is used to remove broken bolts?
13. Name three styles of lock washers.
14. Plastic screw anchors are used to fasten _____ objects to _____ walls.
15. _____ bolts and _____ bolts are used to anchor objects to hollow walls.
16. Setscrews are used to anchor _____ and _____ to motor shafts.
17. Threaded rod is cut to length with a _____.
18. Damaged threads can sometimes be repaired by using _____ and _____.

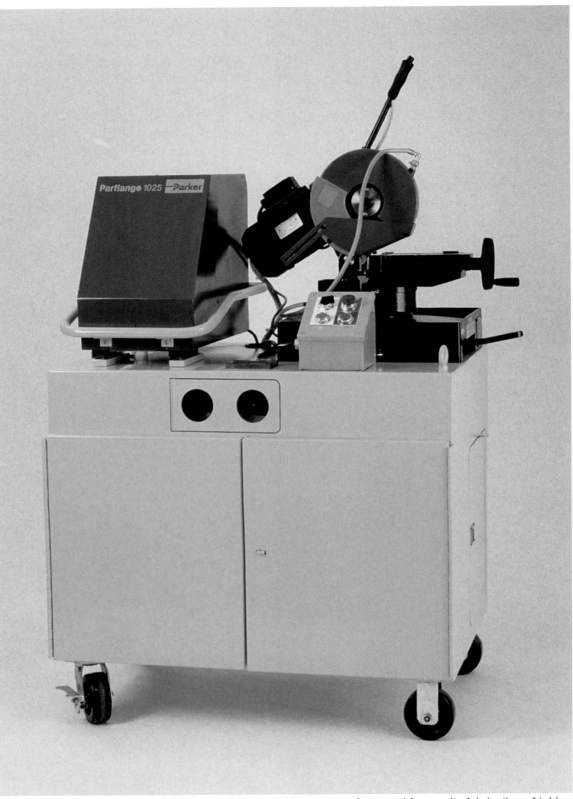

Tube-cutting, flanging, and flaring machines such as this one are often used for on-site fabricating of tubing when installing industrial systems. (Parker Hannifin)

Working with Copper Tubing

Objectives

After studying this chapter, you will be able to:
- ❏ Identify types and sizes of copper tubing.
- ❏ Select and properly use tubing tools.
- ❏ Demonstrate proper use of tube bending tools.
- ❏ Describe the steps needed to make swaged connections.
- ❏ Describe the steps needed to make flared connections.
- ❏ Identify and properly use various types of copper tubing fittings.

Important Terms

ACR tubing	offsets
annealed	outside diameter
chamfer	oxidation
compression fittings	P-trap
coupling	pressure drop
elbows	psi
flare bonnet	reamed
flare cap nut	reducing coupling
flare elbow	soft copper tubing
flare fitting	spring benders
flare nut	street elbow
flare plug	swaging
flare tee	sweat soldering
flare union	tee
flaring	telescoping
flaring block	tubing cutter
hacksaw	union
hard-drawn copper	work-hardening
tubing	wrought fittings
inside diameter	
lever-type tubing	
benders	

3.1 Copper Tubing

Copper tubing used for air conditioning and refrigeration work is called **ACR** (**A**ir **C**onditioning and **R**efrigeration) **tubing.** It differs from copper tubing used for general plumbing work. When ACR tubing is manufactured, the inside of the tubing is dehydrated to remove all moisture. The tubing is then charged (filled) with low-pressure nitrogen gas and sealed with a cap at each end to keep the tubing safe from contamination by oxygen and moisture in the air. If oxygen atoms were to combine with copper atoms (a process called **oxidation**), a layer of copper oxide would form inside the tubing. The caps also keep out dirt and other foreign matter that could contaminate a refrigeration system. Caps or plugs should be replaced after cutting a length of tubing.

Plumbing copper tubing is cheaper than ACR copper tubing, but plumbing copper is not used in refrigeration systems because it is not protected against contamination. Plumbing copper also differs from ACR copper in size. Plumbing copper is measured by its nominal **inside diameter** (ID), while ACR copper is measured by its **outside diameter** (OD). See **Figure 3-1.**

3.1.1 Types of Copper Tubing

Copper tubing is available in soft and hard-drawn types. **Soft copper tubing** has been **annealed** (softened by heating it to a bright cherry red color and permitting it to cool) so it can be bent easily. **Hard-drawn copper tubing** is rigid and cannot be bent easily unless it is first annealed. Both soft copper and hard copper are further classified according to the thickness of the tubing wall. Three thicknesses are available: K, L, and M.
- ❏ *Type K* has the thickest wall. It is used in applications where abuse or corrosion might occur. Type

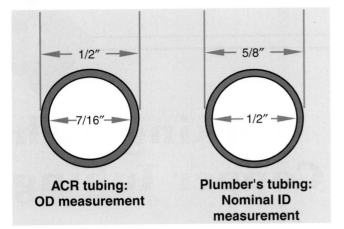

Figure 3-1. *ACR copper tubing and plumber's copper tubing are measured differently. ACR tubing is sized by actual outside diameter. Plumber's tubing is measured by nominal inside diameter — the OD usually will be 1/8″ larger than its ID.*

K is usually hard-drawn tubing and is used in commercial refrigeration systems.

❑ *Type L* is the most common tubing. It has a medium-thick wall and is used for residential and commercial applications. Most ACR tubing, both soft and hard, is L thickness. When ACR copper tubing is ordered, Type L is automatically provided unless another thickness is specified.

❑ *Type M* is a thin-walled tubing. It is not used in refrigeration systems because it does not meet safety code requirements. Some manufacturers use Type M copper tubing to construct water-carrying coils. Plumbers may use it for small drain lines and for other noncritical applications.

Soft copper tubing

Type L soft copper tubing is available from 1/8″ to 3/4″ OD and is usually sold in 50-foot coils, **Figure 3-2.** As noted earlier, these coils are dehydrated and sealed at the factory.

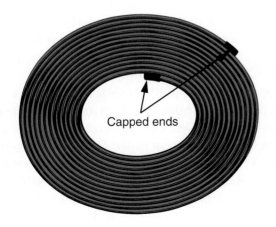

Figure 3-2. *Soft copper ACR tubing is supplied in 50-foot coils. To prevent contamination, it is dehydrated, filled with nitrogen gas, and capped at both ends. (Mueller Brass Co.)*

Because soft copper is easily bent, it must be supported with clamps or brackets every four to six feet. Soft copper tubing has a tendency to harden as a result of vibration, oxidation, and bending. This is called *work-hardening.* Work-hardened copper will crack at stress points, especially when it is flared or formed at the tubing ends. Copper that has become work-hardened can be resoftened by heating it to a bright red surface color and then allowing it to cool.

Unrolling soft copper coils. Soft copper coils should be handled with care because the tubing is easily damaged. Kinks, bends, flat spots, or dents will make the material unfit for use. Bending and kinking can be avoided by unrolling the coil properly. As shown in **Figure 3-3,** unrolling is done by supporting the coil upright with one hand and holding the free end of the tubing stationary on a flat surface with the other hand (or a foot). The coil is then rolled in a straight line to the desired length. Do not unroll an excessive amount because it is difficult to recoil the tubing without bends or kinks. After cutting off the desired length, replace the cap or plug on the end of the coil to prevent contamination.

Figure 3-3. *Uncoil soft copper tubing carefully to avoid kinking. Hold one end flat on the floor or other surface, and unroll as much as needed.*

Hard-drawn copper tubing

Hard-drawn copper tubing is hard and rigid and either Type L or Type K thickness. It comes in standard 20-foot lengths that are dehydrated, charged with nitrogen, and sealed with rubber plugs at each end, **Figure 3-4.** Hard-drawn copper tubing cannot be bent easily, so soldered or brazed fittings are used when making connections or changing directions. Because of its rigidity, hard-drawn copper tubing requires fewer supports or brackets than soft copper tubing, and assembly is quicker. It is available in sizes ranging from 3/8″ OD to over 6″ OD.

Figure 3-4. *Rubber plugs are used to seal the ends of lengths of hard-drawn copper tubing after it is dehydrated and filled with nitrogen gas. (Mueller Brass Co.)*

3.1.2 Tubing vs Pipe

In the early days of refrigeration and air conditioning, plumbers performed the installation and service work. These early systems used ammonia and other corrosive chemicals as refrigerants. Iron and steel pipe had to be used to avoid the chemical reaction that would occur if these chemicals were used with copper tubing and fittings. The development of noncorrosive refrigerants has made possible the use of copper tubing in almost all refrigeration and air conditioning systems today.

The distinction between tubing and pipe is primarily wall thickness and the resulting joining method. Tubing is considered *thin-walled* material (regardless of whether it is Type K, L, or M) compared to steel and plastic pipe. The term "tubing" is generally applied to materials such as copper and aluminum that are joined by means other than threads. Pipe is the term used to describe *thick-walled* materials, such as steel or plastic. Threads are cut into the pipe wall, allowing lengths of pipe to be joined with threaded fittings that screw into place.

Fitting different sizes of tubing inside one another is known as ***telescoping.*** It is sometimes used to perform emergency repairs when the proper fitting is not available. Telescoping can be performed with ACR tubing because of its OD sizing system and wall thickness. As shown in **Figure 3-5,** ACR 1/4″ copper tubing will fit snugly inside 3/8″ copper tubing, 3/8″ tubing will fit snugly inside 1/2″ tubing, 1/2″ will fit snugly inside 5/8″, and so on.

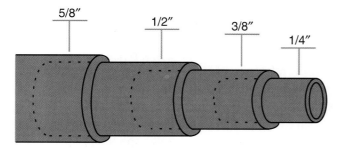

Figure 3-5. *Accurate OD sizing allows different diameters of ACR tubing to telescope inside each other.*

3.2 Working with Copper Tubing ———◆

The ability to properly perform such operations as cutting, bending, and joining copper tubing is a basic requirement for success as an HVAC technician. Careful attention to correct use of tools and the development of good work habits will result in trouble-free installations and satisfied customers.

3.2.1 Cutting Tubing

Cutting copper tubing is a simple task, but it must be performed properly. Care must be exercised not to damage the ends being cut. The most common and accurate way to cut copper tubing is to use a ***tubing cutter,*** **Figure 3-6.**

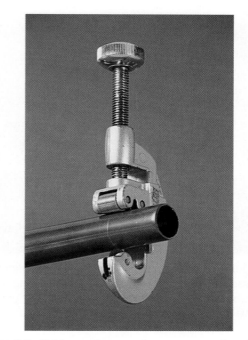

Figure 3-6. *Tubing cutters produce an accurate, clean 90° cut, resulting in strong, well-made joints.*

Tubing cutters make an accurate 90° cut on either hard or soft copper tubing. They are available in several sizes for use on different tubing size ranges or for special purposes, see **Figure 3-7.** The tubing is positioned on the cutter rollers with the cutter blade accurately located at the cut point. Rotating the cutter knob will raise or lower the cutter blade. The blade is pressed firmly against the metal, and the tool is rotated around the copper tubing. After each rotation, the knob is tightened to force the blade lower into the cut. Several rotations are needed to properly complete the cut.

Avoid excessive blade pressure, which will flatten the tubing and may cause a severe burr on the inside of the tubing ends. See **Figure 3-8.** Keep the cutter blade sharp; a dull blade will not cut a single clean groove. Instead, it will "track," making rows of shallow cuts on

Figure 3-7. *This tiny, special-purpose tubing cutter can be used to cut tubing up to 1 1/8″ OD where little clearance is available. (The Ridge Tool Company)*

Figure 3-8. *Excessive pressure from the tubing cutter blade may cause flat spots and severe burrs on the tubing. Tubing at left was cut with correct pressure; tubing at right, with excessive pressure. Note the flat spot that resulted.*

the outside of the tubing. The tubing cutter should be used only on copper or aluminum tubing. Using the cutter on steel tubing or electrical conduit will quickly dull the blade, making a replacement necessary.

A **hacksaw** is the second (and less desirable) method of cutting copper tubing. This method has two main disadvantages: it is difficult to obtain a precise 90° cut, and the sawing process creates tiny metal chips that can contaminate the inside of the tubing. To obtain the smoothest possible cut, use a saw blade with at least 32 teeth per inch. Be very careful to remove all metal chips from the tubing.

Reaming and deburring

After the copper tubing is cut, the tube ends must be **reamed,** removing burrs and scraping the ends to a flat surface. This procedure is usually performed with a pointed reamer blade built into the tubing cutter, but some technicians prefer using a pocketknife blade.

Use care to prevent copper chips or burrs from entering the tubing. Hold the tubing upside down or at an angle during the reaming process so the chips will fall to the floor. See **Figure 3-9.**

It is important to properly dress the cut tubing end. Burrs or ridges on the inside of the hole will cause problems in assembly, as will thin or uneven edges. The thickness of the tubing *at the cut* should match the thickness of the rest of the tubing, **Figure 3-10.** A little time and effort spent dressing the cut will prove worthwhile.

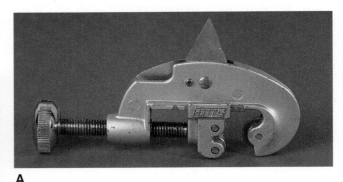

A

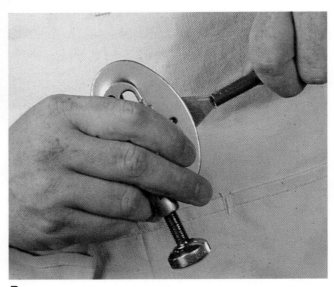

B

Figure 3-9. *Reaming. A—Reamer blade built into a tubing cutter. B—Hold the tubing with the open end down while reaming to prevent contamination of the tubing with metal chips from the burrs.*

3.2.2 Bending Copper Tubing

Several types of tubing benders are available for making accurate bends in tubing without causing flats, kinks, or dents. Flats and kinks are eliminated, not for the sake of appearance, but because they would restrict the flow of liquid or gas through the tubing.

In a refrigeration system, the tubing must carry liquid or vapor from one component to another. The

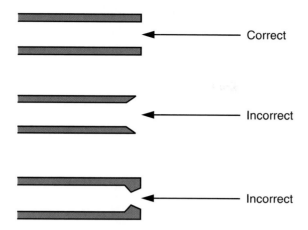

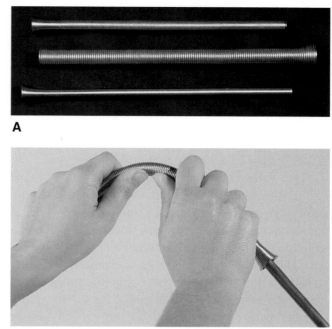

A

Figure 3-10. *Correctly and incorrectly dressed cut tubing. A—Tubing is correctly reamed, with proper wall thickness at cut end. B—Tubing is excessively reamed; wall is too thin. C—Tubing is not reamed sufficiently; wall is too thick.*

tubing is carefully sized to carry a specific flow, so the technician must be careful not to create unplanned restrictions, such as flats and kinks. Restrictions to flow reduce system efficiency, resulting in **pressure drop.** It is important to keep to a minimum any decreases in pressure in a piping system.

Hand bending

Soft copper greater than 1/4″ in diameter can sometimes be bent successfully by hand. (Smaller tubing is easily bent by hand without kinks or flats, unless the bend is very sharp.) Bending is done carefully, a little at a time, to avoid developing severe flats or kinks that must be removed. Hand bending requires skill and should be done only by experienced technicians.

Spring benders

Spring benders, Figure 3-11, provide an efficient, low-cost method to bend soft copper tubing. Spring benders are available in a variety of sizes to fit tubing from 1/4″ OD to 3/4″ OD. These springs slip over the tubing to completely cover the area of the bend. After each bend is made, the spring is slid along the tubing to the next section to be bent.

When a bend is sharp, spring benders have a tendency to bind, or stick, to the tubing. This tendency to bind can be overcome by bending the tubing a little farther than needed, then bending it back to relieve pressure on the spring. Push, *do not pull,* on the spring to remove it from the tubing. Pulling can permanently separate the spring coils, making the bender unfit for further use. If the bend is still too tight to slide the spring off the tubing, simply twist the spring to "unscrew" it.

Very little practice is needed to accomplish proper bends in smaller tubing with a spring bender. Larger

B

Figure 3-11. *Spring benders. A—Used mostly with small-diameter, soft copper tubing, spring benders are available in a number of sizes. B—Using a bending spring on soft copper tubing is a two-handed operation.*

tubing, however, requires more physical force to bend, so spring benders are used on larger tubing primarily to accomplish slight bends or curves. Although the spring bender is designed for use on the *outside* of tubing, a smaller spring is sometimes inserted into the tubing to make a bend at the tubing end.

Lever-type benders

Lever-type tubing benders, Figure 3-12, are easy to operate and are calibrated to allow accurate short-radius bends up to 180°. Some benders are designed to fit only one size of tubing, while others can be used with a range of sizes. Lever benders can be used on soft copper, aluminum, steel, stainless steel, and Types K and L hard copper tubing.

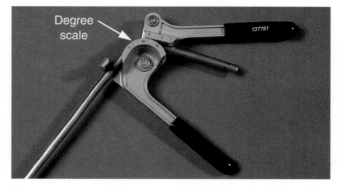

Figure 3-12. *Lever-type tubing benders can make short-radius bends of up to 180° without flattening the tubing. Note the scale used to read degrees when making bends.*

For the technician, making accurate bends in copper tubing is almost a daily task, especially on installation jobs. Bending is much faster than installing a fitting and does not present a potential leak hazard, as do fittings. Making accurate 45° and 90° bends requires some practice; lever benders are designed to make this task easier and more accurate.

3.3 Connecting Copper Tubing ───◆

The walls of copper tubing are too thin to make strong threaded connections, so other means are employed to join lengths of tubing. Two general approaches are used: permanent connections using soldering or brazing on swaged copper tubing or wrought fittings, and mechanical connections using threaded fittings that can be easily disconnected to make repairs or to replace defective parts. The ability to work with copper tubing and the various fittings used to make connections is a basic requirement for anyone entering the HVAC field.

3.3.1 Swaging Copper Tubing

Swaging (pronounced "swedging") involves enlarging the diameter of one end of a length of soft copper tubing so the end of another length can be slipped into it. See **Figure 3-13.** The connection is then soldered or brazed to make a strong, leakproof joint. Swaging is the preferred method of joining soft tubing since the process requires little time, and only one brazed joint is needed to complete the connection (compared to two joints for a fitting).

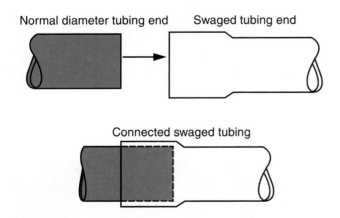

Figure 3-13. *Swaging joins tubing by enlarging the diameter of one piece so that it forms a socket for another piece. The joint is then soldered or brazed for strength and leakproofing.*

Swaging methods

Swaged connections can be made using either the punch-type joint method or the screw-type joint method. Both punch- and screw-type joints require the use of special hand tools.

Punch-type swage. The tubing is clamped into a special tool called a *flaring block* with just enough tubing protruding through the block to accomplish the enlarging process. See **Figure 3-14.** The depth of the finished swage should be equal to the original tubing diameter. For example, 1/4″ tubing is swaged 1/4″ deep, and 1/2″ tubing is swaged 1/2″ deep. Swaging punches are available in diameters ranging from 3/16″ to 7/8″.

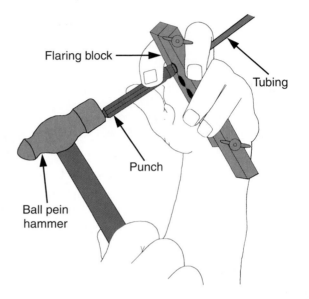

Figure 3-14. *A punch-type swaging tool uses a flaring block to hold the tubing and a ball pein hammer and a punch to stretch the tubing end.*

Swaging is done by clamping the tubing in the flaring block, then positioning the proper size swaging punch at the tubing end. A ball pein hammer is used to slowly drive the punch into the tubing end. The punch will enlarge the tubing end to the precise diameter required. The punch should be held firmly and straight during the swaging process. Striking several easy blows is the preferred technique; it permits the copper to swell slowly, resulting in a straighter and more accurate swage. (Using less force is also easier on your knuckles if you miss!) With a little practice, you will be able to determine how hard to strike the punch to obtain the desired swage.

When tubing is inserted into the swage, the fit should be tight and straight. Such a fit is best for proper soldering and brazing. Loose or crooked connections are difficult to solder or braze, resulting in a weak joint. Chapter 5 covers procedures for soldering and brazing tubing.

Screw-type swage. With this method, the tubing is clamped into the flaring block with the correct amount of tubing protruding. As shown in **Figure 3-15,** a screw-type yoke is fitted over the tubing end and

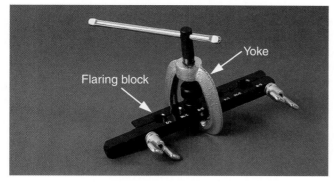

A

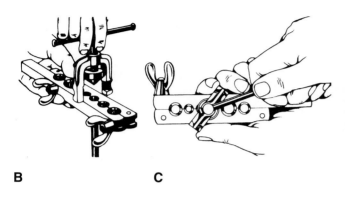

B   **C**

Figure 3-15. *Screw-type swaging tool. A—The screw-type tool consists of a yoke and a flaring block. It is easier to use in close quarters than the punch-type tool. B—The self-centering yoke is slipped over the flaring block. C—The yoke is rotated slightly to lock onto the block. This ensures centering of the adapter over the tubing. (Imperial Eastman)*

twisted to lock it onto the flaring block. By turning the handle on the yoke, the swaging adapter is slowly forced into the tubing to stretch it. Swaging adapters are interchangeable and range in OD from 1/8″ to 3/4″.

The screw-type swaging tool is often easier to use in close quarters than the punch-type tool. The screw-type tool precisely aligns the adapter with the tubing, producing nearly perfect swages every time.

Using wrought fittings

Wrought fittings are sometimes called "sweat" fittings because of the **sweat soldering** (or brazing) method used to join them to either hard-drawn or soft copper tubing. These fittings are usually made of copper, but sometimes are brass and do not have threads. They are sized according to the copper tubing used and are readily available in sizes ranging from 1/4″ to 6″.

Wrought fittings are used where the connection is permanent, such as long runs of either soft or hard copper tubing. (They are *required* when using hard-drawn copper tubing.) The brazed connection is stronger than the actual copper tubing, but the connection *must* be properly brazed and leakproof.

Wrought fittings are designed to fit tightly over copper tubing for easy soldering or brazing. Avoid dropping or roughly handling fittings since a dent in a fitting can make it unsuitable for use.

Types of fittings

Fittings are available in many different types and sizes, so almost any connection is possible. See **Figure 3-16** for examples of wrought fittings. The names and shapes of fittings are common to many trade areas (plumbing, electrical, HVAC), but the sizes, threads, and types of material often differ.

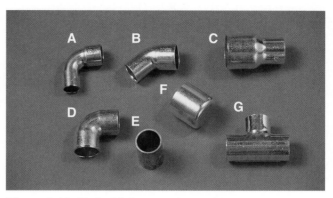

Figure 3-16. *Typical fittings used to make permanent copper tubing connections by sweat soldering or brazing. A—90° street elbow. B—45° street elbow. C—Reducing coupling. D—90° elbow. E—Coupling. F—Cap. G—Tee.*

Couplings. The fitting used to connect two lengths of hard copper tubing is the **coupling, Figure 3-17.** Couplings are available in all sizes and usually have openings of the same diameter on each end. A 5/8″ coupling, for example, is used to connect two sections of 5/8″ hard copper tubing. These fittings also can be obtained as **reducing couplings** by specifying the sizes on each end. For example: Reducing coupling, 7/8″ × 5/8″.

Figure 3-17. *Couplings are used to connect two pieces of tubing in a straight line. Most couplings have openings of the same diameter at both ends. A reducing coupling, like the one on the right, has openings of different sizes, allowing a larger-diameter piece of tubing to be connected to tubing with a smaller diameter.*

Elbows. Available in either 45° or 90° angles, *elbows* usually have female fittings on each end. See **Figure 3-18.** Elbows are used to change the direction of the tubing run. A "short-radius" elbow bends quite sharply, while a "long-radius" elbow makes a more sweeping turn. The long-radius type is preferred because it offers less restriction to the flow of refrigerant. Like couplings, elbows are sized according to the OD of the tubing on which they are used. A 1/2″ 90° elbow, for example, would have 1/2″ female openings on each end to accept 1/2″ copper tubing.

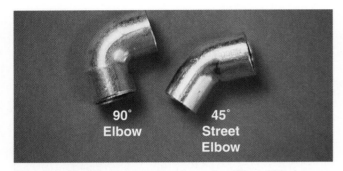

Figure 3-18. *Elbows are used to make 45° or 90° changes in direction. Standard elbows have female fittings at both ends. Street elbows have a female fitting at one end and a male fitting at the other.*

A *street elbow* contains a female end and a male end. Both openings are normally sized for the same size tubing. The male end of the street elbow is the same size as the tubing; the female end is sized to fit the outer diameter. They are used frequently in HVAC work to make *offsets* (changes in direction that extend only a short distance). See **Figure 3-19.** Street elbows reduce the number of brazed connections needed, compared to using regular elbows and short lengths of tubing.

Tees. Fittings called *tees* are used to connect a branch circuit to an existing line of copper tubing. A tee numbering system guarantees the correct size, type, and placement of an opening. As shown in **Figure 3-20,** the straight-through connections are numbered 1 and 2, and the branch is 3. With this numbering system, it is possible to order a tee with different sizes at each opening. For example: Wrought tee, 7/8″ × 5/8″ × 1/4″. (The branch would be 1/4″ OD.)

Other fittings. Wrought fittings are available to satisfy almost any need. Special fittings such as *P-traps* (named for the shape into which the tubing is bent), return bends, unions, and caps can save much time and trouble when installing HVAC equipment.

The *union* combines aspects of mechanical and permanent-type fittings. It has three parts: two shoulders and a nut for pulling the shoulders together,

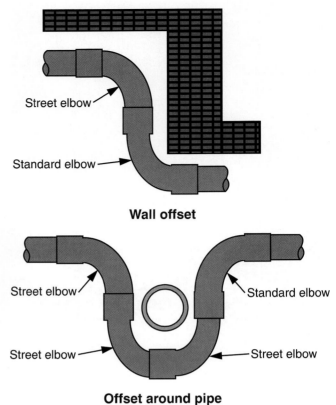

Wall offset

Offset around pipe

Figure 3-19. *Street elbows are used to make offsets (changes in direction) similar to those shown. Offsets reduce the number of brazed connections needed.*

Figure 3-20. *A standard numbering system is used for the openings in a tee. The straight-through connections are numbered 1 and 2, and the branch is assigned number 3.*

Figure 3-21. The shoulders of the union are brazed to the ends of the joined copper tubing. The nut pulls on one shoulder while screwing onto the other shoulder, providing a mechanical connection. This results in a leakproof joint that can be disconnected. A union is used in such applications as a drain line that may need to be disconnected for cleaning.

3.3.2 Using Mechanical Connections

Mechanical tubing connections can be divided into two types: those using compression fittings and those using flare fittings. Compression-type connections are rarely used on refrigeration systems since they cannot withstand the high pressure and vibration associated with such systems.

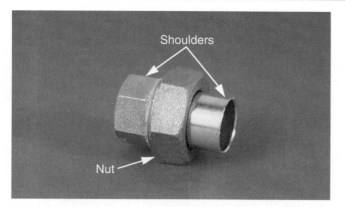

Figure 3-21. *A union combines permanent connections with a mechanical system that allows the union to be disconnected. The locking nut pulls the shoulders tightly together, making it possible to "break" the connection for cleaning or repair work.*

Flared connections, however, are very common on all types of refrigeration systems because they can withstand high pressure and some vibration. Failure of a flare connection can usually be traced to abuse or to lack of skill by the person who made it.

Compression-type connections

Compression fittings are commonly used in heating applications that involve only low pressure. They also are widely used for connecting gasoline, natural gas, propane, water, and air lines, where excessive pressure and vibration are not involved. Compression fittings are simple, efficient, and easy to assemble. See **Figure 3-22.**

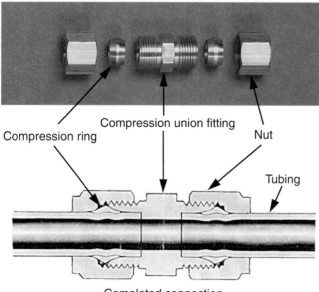

Figure 3-22. *Compression connections are made by compressing (squeezing) a sleeve or ring between a fitting and a nut. Compression fittings are used more widely in residential heating and air conditioning.*

Other than making sure tubing ends are cut square and burrs are removed, no special preparation is needed when using compression fittings. Simply slip the nut and compression ring over the tubing, then insert the tubing into the fitting until it rests against a shoulder. Slide the compression ring into position and tighten the nut.

Tightening the nut causes the ring to be compressed between mating surfaces, providing a leakproof connection. When the connection is disassembled, the ring will remain attached to the tubing and cannot be removed. The ring is often called a "sleeve" or "ferrule."

Compression fittings are *never* used on refrigeration systems because of vibration and high pressure. However, some residential air conditioning systems use special compression fittings to join tubing from the indoor unit to the outdoor unit. These field connections are simple, inexpensive, and quick.

Compression fittings. Fittings are sized according to the outside diameter of the tubing, ranging from 1/8″ to 7/16″. Nuts and compression rings needed for installation are provided with all compression fittings such as unions, elbows, and tees. Many adapters are available to connect tubing with a compression fitting on one side and a flare fitting or pipe connection on the other.

Flare-type connections

Flaring copper tubing is a process of expanding or spreading the end of the tube into a funnel shape with a 45° angle, **Figure 3-23.** All refrigeration **flare fittings** are made with a 45° angle so the tubing will fit snugly against the fitting. A flare nut is used to compress the flare against the fitting to obtain a tight, leakproof, metal-to-metal contact.

Refrigeration tubing connections must withstand at least 300 **psi** (pounds per square inch) of pressure without leaking. Because the flare connection is a mechanical, metal-to-metal contact without gaskets, it is vital that proper attention and care be given to making the flares.

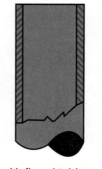
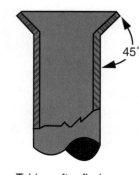

Unflared tubing   Tubing after flaring

Figure 3-23. *When copper tubing is flared, the tubing end is expanded to a funnel shape. The 45° angle matches the angle used on flare fittings.*

The tubing end should be properly reamed before attempting to make flares. Burrs or rough edges will interfere with the smooth metal-to-metal contact and permit leakage.

The tubing is clamped in a flaring block with its end protruding slightly above the **chamfer** (beveled edge) on the block's top side. A screw-type yoke with a special flaring adapter is then clamped onto the block and automatically centered above the tube. See **Figure 3-24.** Turning the screw will force the cone-shaped adapter into the tubing end, spreading it until it is formed to a 45° angle against the chamfer, **Figure 3-25.**

Flare defects. Extending the tubing too high above the chamfer will result in a flare that is too wide. This prevents the nut from sliding over the flare. If the tubing is too low in the chamfer, the result is a small flare that can pull free from the flare nut. A properly made flare will almost fill the bottom of the flare nut without binding or rubbing the threads. **Figure 3-26** shows a correctly made flare and some that are incorrectly made.

Figure 3-25. *The flaring tool's cone-shaped adapter stretches the end of the tubing and forces it against the chamfer, forming a 45° flare.*

A

B

C

Figure 3-24. *Forming a flare. A—The tubing is clamped in the flaring block. B—The yoke is slipped over the block and twisted into place. C—The adapter is screwed into the tubing to form the flare.*

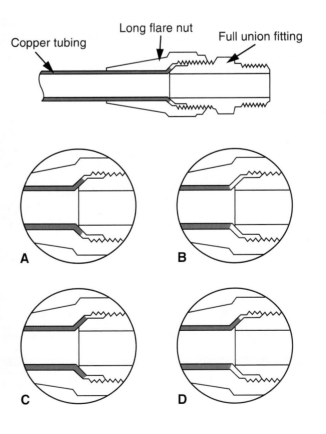

Figure 3-26. *A correctly made flare is vital to achieving a leakproof connection. A—Correctly made flare. B—Flare too small. C—Flare too large. D—Flare uneven.*

CAUTION: Overtightening the yoke screw while forming the flare will create a thin area where the adapter meets the chamfer of the block. This thin area weakens the flare and can cause it to break off when the nut is tightened.

Always remember to place the flare nut on the tubing *before* making the flare. (Every technician has, at one time or another, made a perfect flare and then discovered that the flare nut was not on the tubing.) To resolve such a problem, you must cut off the flare, install the nut on the tubing, and make a new flare. Most tubing cutters have a special groove located on the rollers, **Figure 3-27,** to permit removing only the flare.

Figure 3-27. *A groove in the roller of the tubing cutter allows just the flare to be cut off so the tubing is not shortened any more than necessary.*

Flare fittings

Flare fittings are mechanical fittings intended for use on soft copper tubing. They are usually drop-forged brass and are accurately machined to form a 45° *flare face* (area where tubing joins the fitting). See **Figure 3-28.**

Flare fittings are sized according to the copper tubing size and are commonly available from 3/16″ to 3/4″. The threads for all flare fittings are Society of Automotive Engineers (SAE) National Fine. This means they cannot be connected to other threads, such as pipe threads, National Coarse, or ISO Metric.

Flare nuts. The *flare nut* is the most frequently used fitting. Flare nuts are sized according to the hole through which the tubing is inserted. Since the hole just fits over the tubing, any flats or kinks in the tubing will interfere with the fit. Threads are located on the inside of the nut, and the bottom of the nut contains a perfect 45° flare to exactly match the flare on the tubing end. The tubing flare should almost fill the area at the bottom of the nut.

When the nut is screwed onto a fitting, the flare on the tubing is compressed between the nut and the fitting. To be leakproof to 300 psi (2000 kPa), this metal-to-metal contact requires a perfect match without burrs or ridges to interfere with the seal.

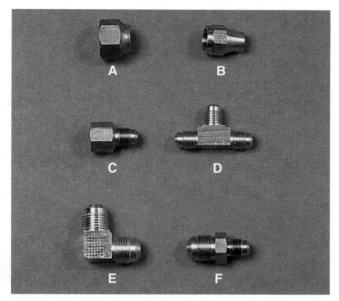

Figure 3-28. *Flare fittings are normally machined from brass. A—Short nut. B—Long nut. C—Reducing flare nut. D—Tee. E—90° elbow. F—Reducing union.*

Flare nuts can be obtained in either short or long style, Figure **3-29.** The short flare nut is the most common. The long style is used to provide more tubing support when vibration is a concern.

Figure 3-29. *Flare nuts are made in both short and long styles. Connections subjected to vibration should be made with long flare nuts for greater tubing support.*

Sometimes, a *reducing flare nut* is used to attach tubing to a fitting of another size. This eliminates the use of additional fittings and reduces the number of possible leaks.

Flare unions. A full *flare union* is a fitting used to connect two flare nuts of the same size. It is called a *full* union because each end of the fitting is the same size. A 3/8″ full union would have a 3/8″ male flare on each end. A 3/8″ full union is a quick way of saying, "3/8″ MFT × 3/8″ MFT union," (MFT = Male Flare Thread).

A reducing union performs the same function as a full union, except the sizes of the flare nuts are different. It is used to reduce from one size tubing to another, such as 1/2″ MFT × 3/8″ MFT. It is possible to obtain a union that has a male fitting on one end and a female fitting on the other. Such a union might be described as 3/8″ MFT × 3/8″ FFT (FFT = Female Flare Thread).

Flare elbows. The *flare elbow* serves the same function as the flare union while providing an accurate bend of either 45° or 90°. Flare elbows (often called "ells") can be obtained in any tubing size, conventional or reducing. For example: 90° flare elbow, 1/4″ MFT × 3/8″ MFT (or FFT).

Flare tees. The *flare tee* makes it possible to connect a branch to an existing line of copper tubing. The tee numbering system guarantees the correct size, type, and placement of the opening. With the numbering system, it is possible to order a tee with openings of different sizes. For example: flare tee, 1/2″ MFT × 3/8″ MFT × 1/4″ MFT. Position 1 is 1/2″ MFT, position 2 is reduced to 3/8″ MFT, and position 3 (the branch) is 1/4″ MFT. Of course, any of the positions on the tee can be ordered with female threads.

Flare plugs. Available for all tubing sizes, the *flare plug* is used to seal a flare nut or similar female-threaded opening. This seal can be temporary or permanent but is usually temporary until proper repairs can be made.

Flare cap nuts. A female nut used to seal off a male-threaded fitting is called a *flare cap nut.* This seal can be either permanent or temporary. The flare cap nut is used extensively on service valves and other such devices. The cap nut prevents dirt and other foreign matter from entering the system through an access fitting.

Flare bonnets. The *flare bonnet* is made of copper and is used to convert an ordinary flare nut into a cap nut. Such devices come in handy when a cap nut is not available. The bonnet is placed inside a flare nut, which can be used to cap a male-threaded fitting.

Summary

This chapter has explained the various tools and procedures for connecting copper tubing. Copper tubing is used to connect system components, which must be durable and leakproof. Skill is required to perform tubing connections that will withstand vibration and remain leakproof. Poor connections result in loss of refrigerant and failure of the system to operate properly.

Most technicians can recognize tubing sizes and fittings at a glance. They do not need to measure or double-check sizes. This familiarity comes after constantly using tubing and fittings. Likewise, consistently making swages and flares quickly and accurately the first time results from practice.

Beginning technicians are often assigned to work with copper tubing and fittings. Many never progress beyond this stage. They become professional installers of copper tubing. Their pay scale is very good because such skills are important and valuable.

Test Your Knowledge

Please do not write in this text. Write your answers on a separate sheet of paper.

1. Why is ACR tubing dehydrated, filled with gas, and capped at both ends?
2. What gas is used to charge ACR copper tubing after it has been dehydrated?
3. The two types of ACR copper tubing are _____ and _____.
4. The three thicknesses of copper tubing are Type _____, Type _____, and Type _____.
   a. J, K, L
   b. 3, 8, 12
   c. K, L, M
   d. A, P, W
5. ACR tubing is sized by its _____ diameter. Plumber's copper tubing is sized by nominal _____ diameter.
6. Name two methods for cutting copper tubing. Which is preferred? Why?
7. Why is it important to remove burrs from cut ends of copper tubing?
8. Name three methods used to bend soft copper tubing.
9. What is the advantage of swaging over joining tubing with a fitting?
10. Hard-drawn copper tubing is connected by wrought fittings, also known as _____ fittings.
11. The openings in tee fittings are numbered 1, 2, and 3. Which one is the branch opening?
12. Name the two types of mechanical connections.
13. The angle on flare fittings is _____ degrees.
   a. 30
   b. 45
   c. 60
   d. 90
14. *True or false?* Flare fittings are intended for use only on hard-drawn copper tubing.
15. The flare _____ is the most frequently used flare fitting.

Working with Pipe

Objectives

After studying this chapter, you will be able to:
- ❏ Discuss the uses for different types of steel and plastic pipe.
- ❏ Identify various sizes of pipe.
- ❏ Recognize types of fittings and use them properly.
- ❏ Select pipe tools and describe their proper use.
- ❏ Describe how to repair leaks in piping.
- ❏ Describe the process for solvent-welding rigid plastic pipe.

Important Terms

ABS	pipe die
adapters	pipe fittings
close nipple	pipe thread
corrosion	pipe vise
CPVC	PVC
galvanized	shoulder nipple
jaw-type pipe wrench	solvent cement
malleable	solvent-welding
nominal	strap-type pipe wrench
pipe compound	tapered

4.1 Uses for Pipe in HVAC

Steel or plastic pipe, depending upon the application, may be required for HVAC installation and repair jobs. Steel pipe has traditionally been used for water lines, drains, and natural gas supply lines to furnaces. However, plastic pipe is rapidly replacing steel in many applications involving water flow, drainage, and heating systems. Steel pipe is required for R-717 Ammonia systems. Plastic pipe is often used to connect certain accessories to the system. Such accessories include water cooling towers, water-cooled condensers, and condensate drains. Local and state building codes specify the acceptable uses of each type of pipe. The technician must develop and perfect the skills needed to work with both steel and plastic pipe.

4.2 Steel Pipe

Steel pipe is available in unfinished (black) or **galvanized** (zinc-coated) forms. Galvanized pipe is usually specified for water and drain lines because it is less likely to rust than unfinished pipe. Generally, black pipe is used for natural gas applications. However, local codes or installation rules will govern specific applications.

Steel pipe is available in a standard length of 21 feet. There are three grades or strengths: standard, extra strong, and double extra strong. Standard grade is acceptable for most plumbing applications.

4.2.1 Measuring Pipe

Steel and plastic pipe are measured by the inside diameter (ID), which is *not* precise and is called the **nominal** (approximate) size. As you read in Chapter 3, this method is different from the method used to measure ACR copper tubing. ACR tubing is very accurately measured by outside diameter (OD). See **Figure 4-1**.

The ID, or nominal size, is not precise because wall thicknesses of pipe vary slightly. Nominal size is, however, considered acceptable for determining sizes for pipe and pipe fittings. *Sizes given for all plumbing pipe and fittings, as well as for plumber's copper tubing, are nominal sizes.* **Figure 4-2** compares the nominal, ID, and OD sizes of steel pipe.

Nominal size copper tubing is used by plumbers for water lines and drains. If a plumber asks to borrow

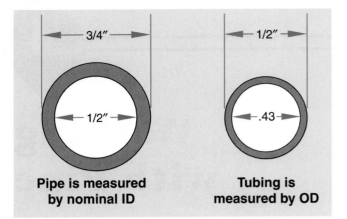

Pipe is measured by nominal ID

Tubing is measured by OD

Figure 4-1. *Pipe measurements and ACR tubing measurements differ. Pipe is measured by its nominal inside diameter (ID), while ACR tubing is measured by its more precise outside diameter (OD). Note that the wall thickness of pipe is much greater than that of tubing.*

Pipe Sizes (Inches)

Nominal Size	Inside Diameter	Outside Diameter
1/8″	0.269	0.405
1/4″	0.364	0.540
3/8″	0.493	0.675
1/2″	0.662	0.840
3/4″	0.824	1.050
1″	1.049	1.315
1 1/4″	1.380	1.660
1 1/2″	1.610	1.900

Figure 4-2. *Nominal sizes, ID, and OD of steel pipe commonly used in plumbing work.*

some 1/2″ copper tubing from you, the ACR tubing you provide would be 5/8″, since it would have the desired ID of 1/2″.

4.2.2 Pipe Threads

Pipe walls are quite thick (compared to tubing walls), making it possible to cut threads on the walls to connect pipes with fittings. **Pipe thread** is very different from other types of threads. Pipe threads are **tapered** (sloped) 1/16″ for every inch of length, **Figure 4-3.**

Pipe threads are specially formed, tapered V-threads made in a conical spiral (like the threads on a wood screw). The taper causes the threads to bind together and make a seal as a fitting is tightened. Approximately 1/2″ of threads are usable before the fitting begins to tighten on the taper; the remaining threads are mostly incomplete. A special **pipe compound** (usually referred to by plumbers as "pipe dope") is normally brushed onto the threads before

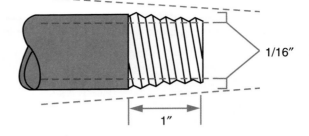

Figure 4-3. *Pipe thread tapers 1/16″ for every inch of length. Tapering helps form a tight seal between the fitting and the pipe.*

assembly to ensure a strong, leakproof seal. In some applications, Teflon™ tape is used, instead of pipe dope, to form a seal.

Pipe threads differ from National Coarse (NC) threads and National Fine (NF) threads in several ways. Both NC and NF are *nontapered*, machine-type threads and are based on the outside diameter of the fastener. Pipe thread sizes are based upon the nominal size of the pipe.

4.2.3 Tools Used with Steel Pipe

Cutting and threading steel pipe requires the use of tools specially designed for these tasks. Tools for working with steel pipe are described in the following paragraphs.

Pipe vise

A **pipe vise** is a tool used to securely hold steel pipe in position for cutting and threading. There are two basic types, the yoke vise and the chain vise. See **Figure 4-4.** Some vises mount on a bench, while others are portable and free standing on tripod-type legs. Some vises are equipped with an electric motor to rotate the pipe in either direction for easy, fast cutting and threading.

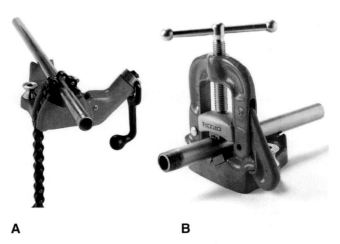

A   **B**

Figure 4-4. *Pipe vises. A—Bench-mounted chain vise. B—Bench-mounted yoke vise. (The Ridge Tool Company)*

Pipe cutter

Steel pipe is cut in the same manner as copper tubing, using a cutter rotated around the material being cut. However, the pipe cutter is larger than the tool used for tubing. It is designed for heavy-duty use, with a special blade for cutting steel pipe. See **Figure 4-5.** Pipe cutters are available in different sizes, each able to cut several different diameters of steel pipe.

When cutting steel pipe, allow about 1/2″ of extra pipe length at each end for threading. This is necessary because about 1/2″ of thread will disappear into each connecting fitting. Steel pipe can also be cut with a hacksaw, but the work must be done carefully to obtain a straight cut. Regardless of the cutting method used, the cut ends must be reamed to remove the burr on the inside of the hole. This is done with a special reaming tool, **Figure 4-6,** designed for use on steel pipe.

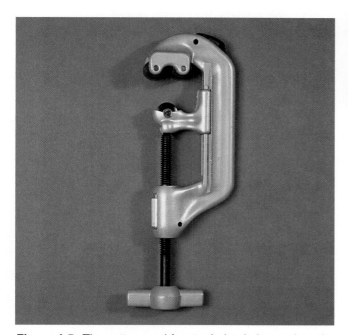

Figure 4-5. *The cutter used for steel pipe is larger than the tubing cutter. Different types are available for use with various ranges of pipe sizes.*

Figure 4-6. *This spiral pipe reamer will deburr pipe in nominal sizes ranging from 1/8″ to 2″. A ratchet handle permits use in tight quarters. (The Ridge Tool Company)*

Pipe wrench

Pipe wrenches, **Figure 4-7,** are designed for holding and turning steel pipe and fittings. They range in size from 3/8″ to 8″ and are available in several types. The regular ***jaw-type pipe wrench*** or the *chain-type pipe wrench* may be used when tooth marks from the wrench jaws are not objectionable. To avoid tooth marks, a ***strap-type pipe wrench*** is used.

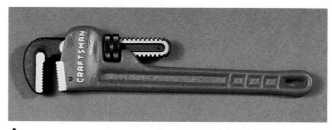

A

B

Figure 4-7. *Two basic types of pipe wrenches. A—The jaw-type pipe wrench has sharp teeth to provide a firm grip on pipe and fittings. B—A chain-type wrench is often used with larger pipe. In situations where tooth marks on the pipe or fitting would be objectionable, a similar wrench with a fabric strap, instead of a chain, may be used.*

When using the jaw-type wrench, **Figure 4-8,** it is important to maintain a gap between the back of the hook jaw and the pipe. This concentrates pressure at two points and produces maximum gripping action and rotating force. Always *pull* in the correct direction on the wrench handle. This will cause the jaw teeth to bite into the pipe. Pulling in the wrong direction will cause the wrench to lose its grip and slip around the pipe. Pipe wrenches are normally used as a pair, one for holding the pipe and the other for turning the fitting.

Pipe die

A ***pipe die*** is a special tool used to cut threads on the outside of steel pipe. As shown in **Figure 4-9,** the die is an individual tool that fits into a special handle used to turn the die on the end of the pipe. Each die is accurately sized to fit a particular size of steel pipe. When changing pipe size, the die must also be changed. The ratchet-type handle shown is reversible, allowing the die to be unscrewed from the pipe once the threads are cut. A special thread cutting oil is used to keep the die from getting too hot, which could damage the cutting edges.

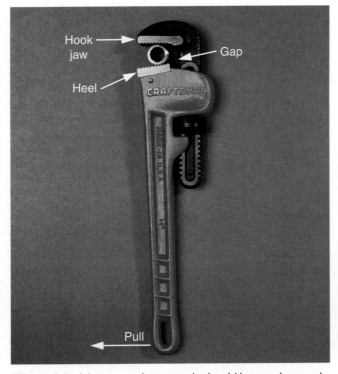

Figure 4-8. *A jaw-type pipe wrench should be used properly. By pulling in the direction shown, the teeth will grip the pipe tightly. Pulling in the opposite direction will cause the grip to loosen and the wrench to slip. Leaving a slight gap between the pipe and the back of the jaw will concentrate the force at two points for a better grip.*

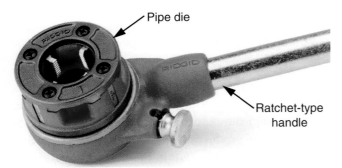

Figure 4-9. *A pipe die is used to cut threads on the outside of a pipe. A different die must be used for each pipe size. The ratchet-type handle shown can be reversed to "back off" the die after the threads have been cut. (The Ridge Tool Company)*

4.2.4 Fittings for Steel Pipe

Pipe fittings are made of annealed cast iron. Annealed cast iron fittings are called *malleable* fittings. They can withstand more bending, pounding, and internal pressure than ordinary cast iron.

Pipe fittings are very similar to those for copper tubing described in Chapter 3. The major difference is that the pipe fittings for steel are threaded rather than smooth like sweat-soldered tubing fittings. Also, pipe fittings are always sized according to the nominal size (ID) of the pipe being used. This can be confusing for someone accustomed to dealing with ACR tubing that is sized by OD, but a little practice and size comparison will quickly eliminate any confusion.

Types of fittings used with steel pipe are described in the following paragraphs. You will recall the names of fittings and their uses from Chapter 3.

Elbows

Pipe elbows, called "ells," are necessary when changing the direction of a piping run. They are readily available in all sizes and in 90° or 45° types, **Figure 4-10.** Elbows are also available as "street ells," which means one end of the elbow is female and the other end is male.

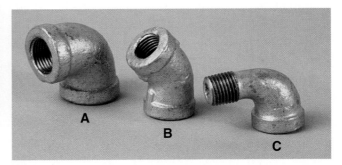

Figure 4-10. *Elbows are available in all nominal pipe sizes. A—90° elbow. B—45° elbow. C—90° street elbow.*

Couplings

A coupling is a short fitting with female threads at each end, used for connecting two lengths of pipe in a straight line. See **Figure 4-11.** Reducing couplings are used to connect two different sizes of pipe in a straight line.

Figure 4-11. *Couplings connect pipe lengths in a straight line.*

Tees

Tees are used to create a branch from an existing line, **Figure 4-12.** Pipe tees have numbered openings just like those used on copper tubing; #1 and #2 are straight through, and #3 is the branch. Pipe tees can be obtained in almost any combination of sizes, all with female threaded openings. A 1/2″ tee would have three 1/2″ openings. Each size for a reducing tee must be specified according to the numbering system. For example: 1/2″ × 3/4″ × 1/4″. (The branch would be 1/4″.)

Figure 4-12. *A tee permits the connection of a branch to a main pipe. The numbering system for tee openings is universal: Openings 1 and 2 are the main line; Opening 3 is the branch line.*

Figure 4-14. *Since all pipe threads are right-handed, a run of pipe cannot be completed unless a union is installed. The union also allows pipes to be disconnected for repairs.*

Reducing bushing

A reducing bushing has male threads on the outside and female threads on the inside. See **Figure 4-13.** It is used to reduce the size of a female opening. For example, a water valve with 3/4″ female openings can be used on 1/2″ pipe by installing reducing bushings (3/4″ × 1/2″) in each side of the valve. Reducing bushings are available in a wide variety of sizes.

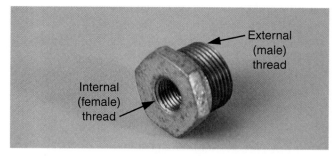

Figure 4-13. *A reducing bushing threads into a female opening (such as on an elbow) to allow a smaller pipe to be connected.*

Caps and plugs

Caps and plugs are used to seal openings. Caps have female threads and are used to close the end of a pipe. Plugs have male threads and are used to close openings in fittings. All nominal sizes are readily available.

Unions

A union, **Figure 4-14,** is a special fitting required to complete a run of pipe. Pipe threads are all right-hand threads, making it impossible to assemble or disassemble the last length of threaded pipe without a union. Pipe unions provide a means of connecting two runs of pipe while making it possible to disconnect the piping when repairs are required. A union should always be installed when the possibility exists that the pipe may need to be disconnected.

Nipples

Nipples, **Figure 4-15,** are short lengths of pipe that are threaded on each end and used to connect fittings that are close together. Nipples can be purchased in any pipe size, in lengths up to 12″. Length is measured from end to end, including threads.

A very short nipple is called a ***close nipple*** or *all-thread nipple* because it is threaded along its entire length. A ***shoulder nipple*** has a short section of unthreaded pipe. These nipples are very convenient when making short connections.

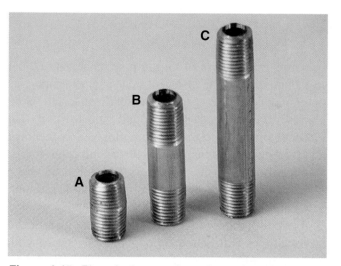

Figure 4-15. *Pipe nipples are short pieces threaded at both ends. They are useful in connecting fittings that are close together. A—Close (all-thread) nipple. B—Shoulder nipple. C—Long nipple (may be up to 12″).*

4.2.5 Repairing Steel Pipe

Leaks in steel pipe are usually caused by loose connections or by ***corrosion*** (rusting) of the pipe. Loose connections can be tightened, but rusted or otherwise damaged sections must be replaced. Rusted and leaking connections at a fitting are caused by removal of the galvanized coating during the threading procedure. The defective fitting or section usually must be removed and replaced. Pipe repair is shown in **Figure 4-16.**

Steel pipe used as a drain line sometimes can be repaired by removing the damaged section and using an adapter to replace the steel pipe with rigid plastic pipe. Check local building codes before using this repair procedure since not all codes permit the use of plastic pipe.

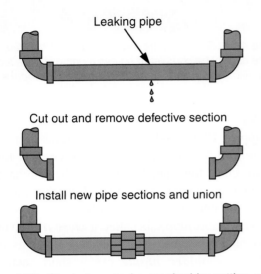

Figure 4-16. *Steel pipe can be repaired by cutting out the defective section and replacing it with two new lengths of pipe and a union.*

4.3 Rigid Plastic Pipe ◆

Growth in the use of plastic pipe has been nothing less than spectacular in recent years. Plastic pipe was once considered a "do-it-yourself" material, used chiefly for farm water systems and drains. Now, rigid plastic piping is widely used. It is employed in chemical and food processing, natural gas distribution and supply, shipboard installations, laboratory and industrial waste disposal, municipal water treatment, industrial and residential plumbing, and a host of other applications.

Despite the growing popularity of this material, it is sometimes difficult to obtain reliable, up-to-date, and comprehensive information. When in doubt, consult your local building codes before installing plastic pipe. Some communities lag behind in updating their building codes to permit the use of plastic pipe. On the other hand, building inspectors and local plumbing suppliers in many cities are well informed of the latest advances in the use of plastic piping and where its use is permitted.

4.3.1 Advantages of Plastic Pipe

Plastic pipe has become an acceptable and inexpensive means of transferring many types of liquids

and is replacing steel pipe in numerous applications. Plastic piping has outstanding resistance to nearly all acids, caustics, salt solutions, and other corrosive liquids. It does not corrode, rust, scale, or pit on inside or outside surfaces. It does not rot and resists the growth of bacteria, algae, and fungi that could cause offensive odors or create serious sanitation problems.

Since its inner wall is very smooth, plastic piping provides maximum flow rates and abrasion resistance at lowest cost. Moreover, it exhibits minimal buildup of sludge and slime.

Plastic pipe is tough and strong. It has tensile and burst strengths sufficient to handle the operating pressures that fall within the temperature range of the plastic used to make it. However, as the temperature increases, the burst point becomes lower.

Rigid plastic pipe is easier to handle, join, and install than metal piping and is less expensive. The low cost of materials and the savings in labor account for its popularity.

4.3.2 Types of Plastic Pipe

There are three types of rigid plastic pipe:

❑ *ABS (acrylonitrile-butadiene-styrene)* plastic pipe and fittings are black and used only for drain, waste, and vent piping. There are two grades of ABS, Schedule 40 and Service. Building codes generally require Schedule 40.

❑ *PVC (polyvinyl chloride)* plastic pipe is white and is used for cold water supply or drain lines. PVC pipe is rated at a temperature of 73°F (23°C) and a pressure of 100 psi (700 kPa). PVC pipe and fittings are available in diameters from 3/8" to 16", and in two grades, Schedule 40 and Schedule 80.

❑ *CPVC (chlorinated polyvinyl chloride)* plastic pipe is tan and can be used for hot and cold water lines or drains, provided local building codes permit. CPVC is available in two grades, Schedule 40 and Schedule 80. It is rated to take a water pressure of 100 psi (700 kPa) at a temperature of 180°F (82°C).

Pressure-rated plastic pipe is designed for water supply but can be used for drains. Pressure-rated pipe provides a complete system with piping of uniform strength. Plastic piping is *never* used within the refrigeration system itself but often replaces the steel pipe normally used for drains or support systems.

4.3.3 Plastic Pipe Sizes

Rigid plastic pipe and fittings are manufactured to conform to nominal pipe sizing of steel pipe, with matching outside diameters. The inside (flow) diameter of plastic pipe changes slightly with the schedule number. Both Schedule 40 and Schedule 80 plastic pipes have the same outside diameter, but Schedule 80 pipe has a thicker wall. The thicker wall of Schedule 80 plastic pipe allows it to be threaded like steel pipe. However, it can also be solvent-welded like Schedule 40 material. Schedule 80 pipe is a heavier-duty material than Schedule 40; the difference can be compared to that between Type L and Type K copper tubing.

Schedule 40 PVC and Schedule 40 CPVC are the most commonly used types of plastic pipe. Both are available in 20-foot lengths. Schedule 40 pipe and fittings are joined by using a solvent cement. *Do not* attempt to thread Schedule 40 plastic pipe — its walls are too thin.

4.3.4 Solvent-welding

The major difference between plastic pipe and steel pipe is the method used for joining. Steel pipe is joined *mechanically* using threaded fittings; plastic pipe is joined *chemically* by a process called **solvent-welding.** The solvent-welding assembly process is much faster and easier, resulting in labor savings.

Solvent-welding (commonly referred to as "cementing") is a simple operation, but each detail of the procedure is important and must be followed closely. Over 90% of plastic pipe joint failures are due to poor cementing techniques or outright carelessness. It is important to follow instructions carefully and resist shortcuts. Properly joined, PVC, ABS, and CPVC pipe and fittings produce pressure-tight joints, whether those joints are made in the shop or the field. Skill and knowledge are required to obtain joints of consistently good quality.

Using solvent cement

Solvent cement is normally purchased in one-pint metal containers with an application dauber attached to the lid. Each type of plastic pipe and fitting has a different chemical composition; a different solvent cement is required for each type and should be used *only* with that material. Solvent materials are described briefly below.

❑ ABS cement is used with all sizes of service and Schedule 40 ABS pipe and fittings.
❑ CPVC is used with all sizes of Schedule 40 and Schedule 80 CPVC pipe and fittings.
❑ PVC: *Light-duty* is used with Schedule 40 pipe and fittings 6″ or smaller in diameter. *Heavy-duty* is used with Schedule 80 pipe and fittings 6″ or smaller in diameter. *Extra heavy-duty* is used with all PVC pipe and fittings 6″ or larger in diameter.
❑ Primer is used before solvent-welding to clean and soften the bonding surfaces of rigid plastic pipe and fittings.

When not in use, solvent cement containers should always be covered to prevent evaporation. Do not use thinner; cement that is lumpy and shows signs of thickening should be discarded.

Special care must be exercised if two types of plastic pipe are to be joined. Local codes usually forbid such combinations unless a mechanical joint or special adapter is used.

WARNING: Solvent cements and primers are extremely flammable and emit dangerous vapors. They are harmful if swallowed, and fumes can cause eye irritation. Repeated or prolonged skin contact causes skin irritation. Keep away from heat, sparks, and open flame. Use only with adequate ventilation. Avoid contact with eyes, skin, and clothing. Avoid prolonged breathing of vapor. Close container after each use.

FIRST AID: In case of *skin* contact, flush with water; for *eyes*, flush with water for at least 15 minutes and seek medical attention. Wash contaminated clothing before reuse. If *swallowed*, do not induce vomiting. Call a physician immediately.

4.3.5 Pipe Preparation

Proper preparation of rigid plastic pipe for joining is important to achieving a satisfactory installation. Pipe must be cut square, burrs removed, and ends beveled to ensure joins will be properly solvent-welded and leakproof.

Cutting

Plastic pipe can be easily cut with almost any type of saw, but for best results, a fine-tooth blade should be used. For properly square end cuts, a miter box should be used, **Figure 4-17.** Pipe cutters may be used when the cutting wheel is designed for use on plastic pipe. Plastic pipe cutters, similar in appearance to pruning shears, are also available, **Figure 4-18.** These cutters are rapidly gaining popularity.

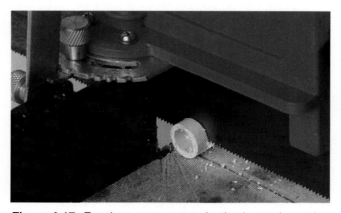

Figure 4-17. *For clean, square-cut plastic pipe ends, a miter box is recommended. A fine-tooth saw blade will give the best results.*

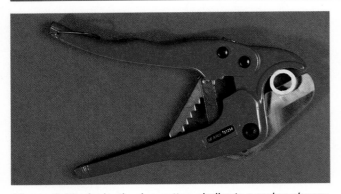

Figure 4-18. *A plastic pipe cutter, similar to pruning shears, is rapidly gaining acceptance. It has a ratchet feature that allows quick, clean, one-handed cuts for plastic pipe up to 1 1/2″ OD.*

Deburring and beveling

All burrs, chips, and other loose material must be removed from the inside and outside surfaces of the pipe before joining is attempted. Use a knife, deburring tool, or a half-round, coarse file to remove all burrs.

All pipe ends should be beveled to approximately the dimensions shown in **Figure 4-19.** Beveling makes it easier to socket the pipe in the fitting. It also minimizes the chance of wiping solvent cement from inside the fitting when the pipe is inserted.

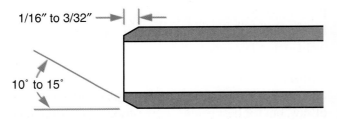

Figure 4-19. *For easier assembly of fittings to pipe, the cut end of the pipe should be beveled as shown. The bevel will also help prevent scraping of solvent from the fitting socket when the pipe is inserted. (Nibco)*

Cleaning and priming

Surfaces to be solvent-welded must be clean and dry. Use a clean cloth to wipe away all loose dirt and moisture from the inside and outside surfaces of the pipe end. Also wipe inside the socket of the fitting.

The function of the primer is to penetrate and soften the bonding surfaces of plastic pipe and fittings. Primer is a high-strength product that penetrates rapidly. It is very effective on the hard-finished, high-gloss products now being produced.

Apply primer to the pipe end with a dauber or a natural-bristle paint brush for a length approximately half the pipe diameter. Using a rag to apply primer is not recommended since repeated contact with skin may cause irritation or blistering.

Next, apply primer freely to the socket of the fitting. Keep the surface wet and the applicator in motion 5 to 15 seconds. Redip the applicator as necessary. In cold weather, allow more time for proper penetration before applying the cement to the joint.

4.3.6 Making Plastic Pipe Joints

When properly solvent-welded, plastic pipe joints will be mechanically strong and leakproof. The solvent cement used to make the joints softens the two mating surfaces, allowing them to fuse. When the solvent evaporates, the pipe and fitting will have formed one piece.

4.3.7 Applying cement

First, apply a generous coating of cement around the outside surface of the pipe to a width slightly more than the socket depth of the fitting. See **Figure 4-20.** Use the dauber supplied with the cement or a natural bristle brush (the solvent in the cement could dissolve synthetic bristles). Do not worry about excess cement on the pipe. The excess cement will be forced out as a bead when the pipe is inserted into the fitting.

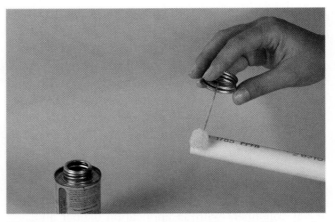

Figure 4-20. *Apply solvent cement to the end of the pipe for a distance slightly greater than the depth of the fitting's socket.*

Next, apply an even layer of cement around the inside socket of the fitting, **Figure 4-21.** Avoid applying too much cement. Excess cement inside the socket could be pushed into the fitting and form an unwanted restriction to flow through the system.

Joining pipe and fitting

Immediately after applying cement to both pieces, insert the pipe to the full socket depth of the fitting. While inserting the pipe, rotate it (or the fitting) about one-quarter turn to ensure complete and even distribution of the cement, **Figure 4-22.** Hold the joint firmly

Figure 4-21. *A full, even coat of solvent cement should be applied inside the fitting's socket. Avoid excessive application, which could restrict flow in the fitting.*

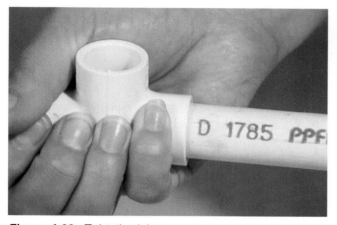

Figure 4-22. *Twist the joint one-quarter turn as it is being assembled, and hold it in position until the cement begins to set. Correctly made solvent-welded joints are leakproof and mechanically strong.*

together for a *minimum* of 20 to 30 seconds. This will prevent the pipe from moving or "backing out" of the socket until the cement has set sufficiently. Keep the newly cemented joint immobile for about two minutes after joining.

Do not attempt to solvent-weld under the following conditions:

❏ In rain (surfaces must be clean and dry).
❏ In temperatures below 40°F (4°C).
❏ In direct sunlight at temperatures above 90°F (32°C).

Do not discard empty primer or solvent cans, brushes, or daubers near plastic pipe. Concentrated fumes or drippings from such material could soften the pipe and cause it to fail.

4.3.8 Fittings for Plastic Pipe

Plastic pipe fittings are readily available in all types and sizes, and use the same names as those applied to fittings for steel pipe. See **Figure 4-23.** Do not mix different types of plastic pipe and fittings: use ABS

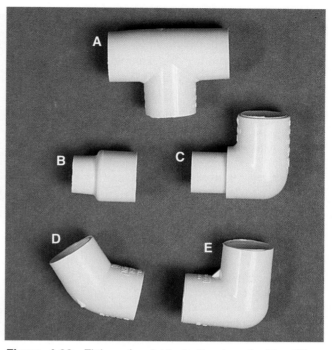

Figure 4-23. *Fittings for rigid plastic pipe. A—Tee fitting. B—Reducing coupling. C—Street elbow. D—45° elbow. E—90° elbow.*

pipe with ABS fittings, PVC pipe with PVC fittings, Schedule 40 pipe with Schedule 40 fittings, and so on. Dissimilar types of plastics will not bond together properly, allowing leaks to develop.

4.3.9 Repairing Plastic Pipe

Leaks in plastic pipe usually are repaired by cutting out the defective section of pipe and replacing it. Often, this can be done by simply inserting a coupling. An example of such a repair is shown in **Figure 4-24.** Be certain to use the same type of plastic pipe and the correct solvent cement.

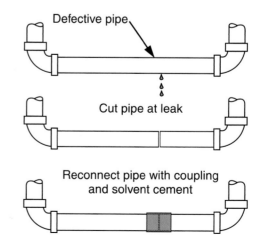

Figure 4-24. *Repairing a leak in plastic pipe involves cutting out the defect and installing a coupling or short section of new pipe.*

4.4 Adapters

Sometimes, it is necessary to connect one type of pipe or tubing to another (for example, steel pipe to PVC or copper tubing to steel pipe). Special fittings, called *adapters,* are available for making such connections. See **Figure 4-25.** These adapters can be elbows, tees, or couplings in various combinations of male or female fittings. Almost any combination is possible in adapters.

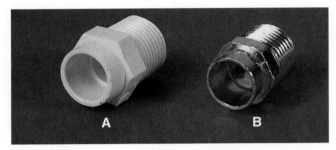

Figure 4-25. *Adapters used with plastic pipe. A—Coupling, plastic to male pipe thread. B—Coupling, tubing (solder joint) to male pipe thread.*

Summary

This chapter has explained the various tools and materials used when working with steel or plastic pipe. Such piping is often used for drains and condensate lines. Pipe (both steel and plastic) is cheaper than copper tubing and is easier to install. Piping in drainage applications is not under high pressure and is less subject to leaking.

The HVAC technician must be skilled in the use of tools and materials for working with pipe and able to perform a variety of tasks. Working with different systems and equipment eliminates boring repetition, but the technician must be prepared for this variety. Most companies start new employees on installation jobs that require the skills discussed in this chapter.

Test Your Knowledge

Please do not write in this text. Write your answers on a separate sheet of paper.

1. Name two types of steel pipe.
2. Nominal pipe sizes are measured by _____ diameter.
3. Is ACR copper tubing 1/8″ larger or 1/8″ smaller than nominal?
4. Pipe thread is tapered _____ for every inch of length.
5. What tool is used to remove burrs from steel pipe?
6. The tool used to cut threads on steel pipe is called a _____.
7. Fittings called _____ are used to change the direction of piping run.
8. What fitting is used to connect pipe in a straight line?
9. What two factors account for the popularity of rigid plastic pipe?
10. _____ plastic pipe is used for drains.
11. _____ plastic pipe is used for cold water supply and drains.
12. _____ plastic pipe is used for hot and cold water supply and drains.
13. Plastic pipe and fitting sizes are _____ sizes.
14. Plastic pipe is joined by _____ welding.
15. _____ are used to change from one type of pipe to another.

5

Soldering

Objectives

After studying this chapter, you will be able to:
❑ Select proper alloys and fluxes.
❑ Describe the use of air-fuel torches.
❑ Use fuel gas cylinders and torches properly and safely.
❑ Connect and operate pressure regulators correctly.
❑ Describe how to solder connections properly and safely.
❑ Describe air-fuel brazing.

Important Terms

acetone	liquidus
acetylene	MAPP
alloy	melting point
brazing	MPS
capillary action	plastic range
elongation	pressure regulator
epoxy	soft soldering
ferrous	solder
fillet	soldering
flow point	solidus
flux	sweat soldering
hard soldering	tare weight
heat sink	tinning

5.1 Soldering Skills

The ability to make strong, neat, leakproof solder connections is a fundamental skill of every HVAC technician. Good sweat-soldered connections must be made quickly and correctly *on the first try*. Too much heat, or not enough heat, will result in a poor connection. Trying to correct a badly made connection is more difficult than starting over. Learning to make good sweat-soldered connections is not difficult; it only takes attention to detail and practice.

5.2 What Is Soldering?

Soldering is the process of joining two pieces of base metal using a filler metal, or *alloy* (a compound of two or more metals). The alloy, normally referred to as *solder,* usually consists of tin and another low-melting-point metal, such as lead. Soldering joins the base metal pieces (which may be the same or different metals) by application of heat, without melting either piece. Soldering is an adhesion process that works much like gluing two pieces of wood together. In soldering, the alloy is the "glue;" heat melts the alloy and causes it to adhere to the surface of the base metals. See **Figure 5-1.**

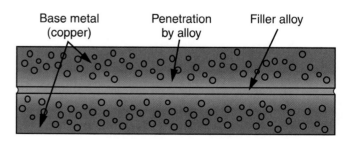

Figure 5-1. *The filler alloy melts and adheres to the surfaces of the base metal pieces, entwining (wrapping around) the molecules of the base metal to form a strong bond.*

When a soldering alloy reaches its *flow point* (the point where melted solder flows as a liquid), the molecules of the alloy penetrate and join the molecules of the base metal (without melting the base metal). For maximum joint strength, the layer of alloy between the base metal pieces should be very thin.

In *sweat soldering,* the method used most frequently in plumbing and HVAC work, the molten alloy is drawn by capillary action into the very narrow space between copper tubing and a fitting.

5.2.1 Soldering vs Brazing

Soldering joins metals at less than 840°F (448°C). Joining metals at temperatures above 840°F (448°C) is called *brazing.* The American Welding Society (AWS) established this distinction   between soldering and brazing. In the refrigeration industry, however, "soldering" is an all-inclusive term. Usually, the term *soft soldering* is applied to joining done at temperatures below 840°F (450°C), and *hard soldering* refers to the high-temperature process that AWS calls "brazing."

Two factors determine whether to soft solder or hard solder and the type of alloy to use. They are the:

❑ Physical properties of the metals being joined; and
❑ Demands that will be placed on the connection after it has been formed.

In refrigeration and air conditioning work, soft soldering is used to install copper water lines and drains. Water lines are not under high pressure nor are they subject to vibration. On the other hand, copper tubing used for refrigeration systems should be hard soldered (brazed) to withstand high pressure and vibration.

5.2.2 Soldering Alloys

Filler metals (soldering alloys) can be obtained in a variety of forms: foil, tape, sheet, wire, flux-cored or solid bar, preform, or paste. One-pound spools of solder in wire form, **Figure 5-2,** are the most popular for HVAC work. Depending upon the application, different tin-lead or tin-silver alloys are used.

The range of temperatures between the *solidus,* or melting, point (where the solder begins to melt) and the *liquidus,* or flow, point (where it flows as a liquid) is called the solder's *plastic range.* **Figure 5-3** shows the composition and plastic range of a number of common solders. A wide plastic range provides easier control of the alloy during the soldering process.

Tin-lead solders

The tin-lead solders comprise the largest group of filler metals used in soldering. They are used for joining most metals and have good corrosion resistance. Various tin-lead alloys are produced to meet the differing needs of soldering tin, copper, brass, bronze, sheet iron, and sheet steel. Soft soldering is excellent for conducting heat or electricity and for making leakproof joints. However, the soldered joint is not as strong as the base metals it brings together.

Figure 5-2. *Lead-free solders like this tin-silver type must be used for joints in any tubing that will carry drinking water. Tin-lead and tin-antimony solders are also available for various applications. (©J. W. Harris Co., Inc.)*

The tin-lead solders are *not recommended* for high-stress or vibration areas, such as those found on refrigeration systems. Stress and vibration will cause the solder to crack, permitting the joint to leak. Soft soldering is best used on condensate drain lines and similar nonstress applications.

Selecting the proper soldering alloy is very important. A numbering system indicates the percentage of each metal in the solder. In the tin-lead group, for example, 50/50 indicates 50% tin and 50% lead, and 60/40 indicates 60% tin and 40% lead. The 50/50 solder has a *melting point* of 360°F (182°C), and a *flow point* of 420°F (215°C). The 60/40 solder is slightly stronger; it melts at 360°F (182°C), and reaches its flow point at 375°F (190°C).

Tin-lead solders were formerly used by plumbers on copper water lines, but they are no longer permitted in drinking water applications. Lead-free alloys that do not present a human health hazard must be substituted. Tin-lead solders are still used in other plumbing and HVAC applications.

Tin-lead solders may contain small amounts of other metals to produce special properties. Metals commonly added to solders include antimony, bismuth, and silver. Antimony increases the strength of a solder. Where a stronger joint is desired for copper tubing, or where a lead solder is not permissible, a 95/5 tin-antimony alloy is often chosen. This solder has a melting point of 452°F (233°C) and a flow point of 464°F (240°C).

The 95/5 tin-antimony alloy should not be used on brass or galvanized (zinc-coated) metal. When antimony

Solder	Composition Percent				Temperature °F	
	Tin (Sn)	Lead (Pb)	Antimony (Sb)	Silver (Ag)	Melts (Solidus)	Flows (Liquidus)
50/50	50	50			360	420
40/60	40	60			360	460
60/40	60	40			360	375
95/5	95		5		452	464
silver solder	96			4	430	430
silver solder	94			6	430	535

Figure 5-3. *The plastic range of a solder is the span from its melting point (solidus) to its flow point (liquidus), as shown in the right-hand column of this table of common soldering alloys.*

is in a molten state, it will absorb zinc from the brass or the galvanizing, resulting in a very brittle joint. This alloy is higher in strength and **elongation** (stretch factor) than the tin-lead solders, but (like tin-lead) it has proven unsatisfactory for applications involving stress and vibration. For reasons not yet known, this alloy tends to powder over time.

Bismuth and silver are added to improve the **tinning,** or spreading, action of the solder. Bismuth is used for lower-temperature solders, while silver is used for higher-temperature solders.

Tin-silver solders

Another major group of solders are the tin-silver alloys. These silver-bearing solders are widely used because of the strong connections they form. Also, their low working temperatures eliminate the weakening of base metals caused by annealing (from high-soldering temperatures). Oxide scale formed in high temperatures is also eliminated.

The silver-bearing alloys have the ability to bond with both **ferrous** (containing iron) and nonferrous metals. These solders work well to join dissimilar metals, such as iron to copper; furthermore, they have good elongation (stretch factor) when vibration is a concern. Melting points for these alloys range from 430°F to 535°F (221°C to 279°C). For applications in the refrigeration industry involving stress or vibration, the tin-silver alloys are superior to both tin-lead and tin-antimony alloys.

5.2.3 Solder Fluxes

Flux is a multipurpose chemical. It is used to treat the clean surface of base metal, to remove oxides from filler metal, to prevent reoxidation of material, and to aid the capillary flow of filler alloy. See **Figure 5-4.** In some instances, the appearance of the flux may indicate

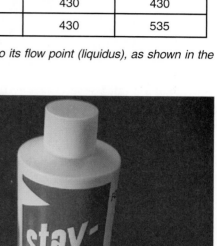

Figure 5-4. *This liquid soldering flux removes oxidation from copper tubing. It prevents new oxidation from occurring during the soldering process. (© J. W. Harris Co., Inc.)*

temperature. (Flux as a temperature guide is explained later in this chapter.) Choosing the proper flux is very important to the quality of the soldered connection. There is no universal flux suitable for all applications. Therefore, when selecting a flux for a particular alloy, it is wise to follow the manufacturer's recommendation.

Using fluxes

Thoroughly cleaning base metal joining surfaces is vital to making strong, leakproof, long-lasting joints.

The fittings and tubing must be free of oil, grease, rust, or oxides that would prevent the alloy from penetrating the base metal surfaces. Cleaning is done with a wire brush, fine sandpaper, or emery cloth, **Figure 5-5.** The joint must be cleaned well; otherwise, the alloy cannot properly flow and penetrate the base metals.

Figure 5-5. *Wire brushes, sandpaper, or emery cloth are used to clean tubing and fittings before they are soldered. The brushes are made specifically for common fitting sizes and should be used only for fittings of the specified size. (Malco)*

Flux maintains a chemically clean surface during the soldering operation. Its initial purpose is not to clean the metal but to keep the joint oxide-free. Flux dissolves the oxides that form when oxygen contacts the molten alloy or the hot base metals. It is important that the flux remain in place until soldering is completed and that it is not blown away or vaporized. Keep soldering time to a minimum; more time means more heat, and more heat means more oxide formation.

As noted, flux serves several functions, all of them necessary to providing a sound connection. Along with removing surface oxides and acting as a shield against new oxide formation, flux increases the "flow" properties of the alloy, yielding to the alloy as it melts into the joint.

The choice of flux in any soldering process is determined by the metals being joined and the filler alloy being used. Since there are many solders and fluxes, follow the manufacturer's recommendations for the best flux for the situation. Flux should be applied with care; it will prepare the way for the solder.

Special care must be exercised when using fluxes to solder copper tubing or brass fittings in a refrigeration system. Fluxes are harmful contaminants inside the tubing of a refrigeration system. Avoid excess flux; apply only the thinnest possible film.

The flux best suited to the 50/50, 95/5, and tin-silver solders is the noncorrosive type in paste form. It consists of zinc chloride (with perhaps a trace

of ammonium chloride) mixed into a petroleum jelly. When this flux is heated, it loses moisture, then melts and partly decomposes to form hydrochloric acid. The acid dissolves the oxides, floats the oxide particles ahead of the molten solder, and promotes the capillary action of the solder joint.

Flux is applied only to the portion of the tubing that is to be inserted into the fitting. Do *not* apply flux to the fitting socket. Paste-type flux should not be applied close to the cut end of the tubing. It should remain about 1/8″ (3 mm) from the end, as shown in **Figure 5-6.** The tubing is then inserted into the fitting and rotated to spread the flux evenly. This procedure prevents excess flux from entering the fitting and contaminating the refrigeration system.

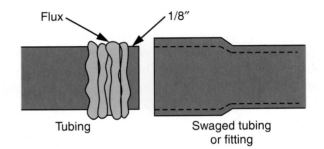

Figure 5-6. *Flux should be applied to the tubing, not the fitting. As shown, approximately 1/8″ of tubing at the cut end should not be fluxed. This will help avoid contaminating the system with soldering flux.*

Although it is commonly used in other types of soldering, an acid flux composed of 71% zinc chloride and 29% ammonium chloride should *never* be used in refrigeration work. Using such a highly corrosive flux on a refrigeration system will damage the system.

When soldering electrical connections, a noncorrosive rosin flux must be used. Rosin flux is produced from pine tar. Sometimes activators are added to rosin flux to improve its spreading action.

5.3 Fuel Gases

The melting of the alloy (filler metal) and the heating of the base metals is normally done with a fuel gas torch. In these torches, a fuel gas is mixed with atmospheric air (21% oxygen) to produce a soldering flame. Flame temperature is determined by the type of fuel gas used. **Acetylene** produces the highest temperature, followed by **MPS,** or methylacetylene-propadiene stabilized gas, (often referred to by its trade name, **MAPP**). Gases such as propane, butane, natural gas, and manufactured gas produce lower temperatures, respectively.

5.4 The Air-acetylene Torch ⬦

The air-acetylene torch, **Figure 5-7,** combines fuel gas from a cylinder with combustion air drawn from the atmosphere. Because atmospheric air contains only 21% oxygen, the air-acetylene torch produces a low-temperature flame. The low heat levels used in soldering reduce stress on the base metals and minimize the formation of oxides inside the tubing. Reduced stress and minimal oxide formation make the air-acetylene torch the ideal tool for soft soldering.

5.4.1 Acetylene

Acetylene is a colorless, gaseous hydrocarbon formed by the chemical reaction of water and calcium carbide. It is two parts carbon and two parts hydrogen, so its chemical formula is C_2H_2. It is used chiefly in soldering, brazing, and welding work. When burned with oxygen, acetylene will produce the highest flame temperatures obtainable (approximately 5660°F or 3126°C).

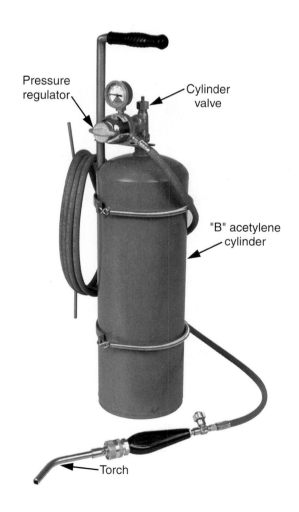

Figure 5-7. *A typical air-acetylene torch outfit for soldering. Combustion air combines with fuel gas at the tip of the air-acetylene torch. (Uniweld)*

Acetylene cylinders

Acetylene cylinders (also called "tanks") are available in various sizes, **Figure 5-8.** The smallest cylinder, size MC, contains 10 cu. ft. of acetylene. "MC" stands for Motor Car — these cylinders were originally used to fuel acetylene-burning car and motorcycle headlights. The next largest cylinder, size B, contains 40 cu. ft. of acetylene.

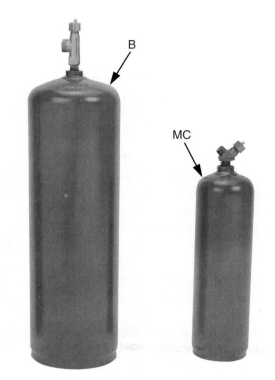

Figure 5-8. *Common acetylene cylinder sizes are B and MC. The B cylinder holds 40 pounds of acetylene; the MC cylinder holds 10 pounds. All acetylene cylinders must be tested and approved for use. (Uniweld)*

MC and B sizes are the cylinders most frequently used by service technicians, but larger cylinders are occasionally used on major installation jobs. Each acetylene cylinder is equipped with a valve that has a threaded outlet for installing a pressure regulator. The size of the pressure regulator must be compatible with the size of the cylinder.

All acetylene cylinders are manufactured to Department of Transportation (DOT) requirements. Cylinders are thoroughly tested by the DOT Bureau of Explosives before being used commercially. A fusible plug located in the bottom of each cylinder will melt at approximately 212°F (100°C) to relieve dangerous pressure buildup.

Acetylene can be unstable, unsafe, and explosive at pressures above 15 pounds per square inch (psi) or 103 kilopascals (kPa). However, the gas can be safely stored in cylinders at a higher pressure by the following method.

The cylinder is filled with a porous substance that absorbs **acetone,** a liquid solvent. Next, a precisely measured amount of acetone is introduced into the cylinder and absorbed by the porous substance. Then, the cylinder containing the porous material and acetone is weighed. The weight is stamped on the cylinder. This is the **tare weight,** or the weight of the container before it is filled with acetylene. Finally, acetylene is pumped into the cylinder, and is absorbed by the acetone. It is thought that the acetylene molecules fit in between the acetone molecules. As pressure is relieved during use, the acetylene gas is released to flow out of the torch.

Acetylene cylinders should be used in an upright position to avoid loss of acetone and poor flame quality. Each time the cylinder is refilled, the tare weight should be checked and acetone added as needed.

Safety precautions with acetylene

Safety-consciousness is a mark of good craftsmanship and is guided by common sense. Observe the following precautions for using acetylene safely:

❑ Never permit fuel gas to escape and accumulate. Acetylene concentrations ranging from 3% to 90% in air or oxygen are explosive.

❑ Never store acetylene or other fuel cylinders in a closed, unventilated area. Do not expose tanks to high temperatures or sources of ignition.

❑ When setting up an air-acetylene torch, check all fittings for leakage by spraying them with a soapy water solution. Leaks will blow bubbles. A soapy solution is made by mixing concentrated dishwashing detergent and water in a 1:1 ratio. A squirting dispenser is handy for applying the solution. Do not use the torch until all fitting leaks have been eliminated. Never use a flame to detect leaks.

❑ Never use a cylinder that is leaking acetylene. If gas leaks around the valve stem when the valve is opened, close the valve and tighten the packing nut. This will compress the packing around the valve stem and should stop the leak. If the leak does not stop, close the cylinder valve, tag the cylinder a "leaker," and return it to your supplier. Keep the "leaker" away from all sources of ignition.

❑ Never open the acetylene cylinder valve more than one-half turn.

❑ Always leave the valve key or wrench on the valve stem while a cylinder is in use, so the acetylene can be turned off quickly in case of emergency. Never use pliers or a wrench on the cylinder valves. Use only the proper valve key or valve key wrench. Always close the cylinder valve and regulator when work is completed.

❑ Handle all fuel cylinders with care. Do not drop or roll them.

❑ Never allow the torch flame or electric arc to contact the cylinder.

5.4.2 Pressure Regulators

An acetylene cylinder is pressurized to 250 psi (1725 kPa) at 70°F (21°C). A **pressure regulator,** **Figure 5-9,** reduces cylinder pressure to working pressure. The working, or flow, pressure will vary according to the type of torch tip being used, but 5 psi (34 kPa) is considered normal. Never allow the working pressure to exceed 15 psi (103 kPa).

Cylinder gauge

Flow pressure gauge

Pressure adjustment

Figure 5-9. *A two-gauge acetylene pressure regulator. The cylinder gauge shows the acetylene level, while the flow pressure gauge provides a reading of the pressure supplied to the torch. Pressure is adjusted by rotating the handle on the front of the regulator. (Uniweld)*

Pressure regulators are available with either one or two gauges. The single-gauge regulator shows working pressure (or flow pressure) traveling through the hose to the torch tip. This working pressure is adjustable by means of a knob or handle located on the front of the pressure regulator. On two-gauge models, the second gauge shows the level of acetylene in the cylinder (full, 3/4, 1/2, 1/4, or empty).

The pressure adjustment on a regulator is designed to open and close in directions opposite those of ordinary valves. Turning the adjustment knob counter-clockwise will *close* the pressure regulator, while a clockwise rotation will *open* the regulator.

Acetylene regulators and gauges are color-coded red and have special cylinder connections to eliminate confusion with other types of regulators. The outlet hose connection on the acetylene regulator is a special *left-hand threaded* male fitting. The hose fitting can be either small (size "A") or large (size "B").

5.4.3 Acetylene Hose

The female hose fitting screws onto the male fitting of the regulator outlet. All acetylene connections are left-hand threads to prevent the user from mistakenly connecting the hose to oxygen, which may be used at much higher pressure than acetylene. Left-hand threaded fittings are readily recognized because of the notches, or grooves, cut around the middle of the outside edges of the nut. See **Figure 5-10.**

Acetylene hose is a special grade designated by the letters RM. The hose is red and is measured by the inside diameter (ID). It is available in 3-, 6-, 12 1/2-, or 25-ft. lengths. The light- and medium-duty hoses are 3/16″ ID; the heavy duty hose is 1/4″ ID.

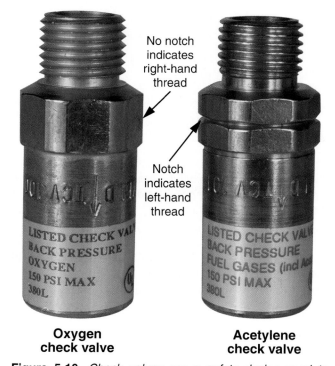

Oxygen check valve   **Acetylene check valve**

Figure 5-10. *Check valves are a safety device used to prevent the reverse flow of gases. The connectors for oxygen and acetylene are different to prevent accidentally interchanging them. The acetylene valve (right) has a left-hand thread and is identified by a groove or notch around the nut. The oxygen valve (left) has right-hand threads and no notch or groove. (Uniweld)*

The hose connector is available in two sizes. Most technicians prefer size A, a light-duty (also called "aircraft") hose. It has a 3/8-24 female nut with a left-hand thread. A size B hose is for medium- or heavy-duty use. It has a 9/16-18 female nut with a left-hand thread. Adapters are available for connecting the two fitting sizes.

5.4.4 Air-acetylene Torch Handles

One end of the torch handle has a threaded connec-tion for an acetylene hose; the other end of the torch handle has a threaded connection for a torch tip, **Figure 5-11.** The handle also contains a finger-operated valve to control and adjust gas flow. The valve opens and closes easily with a twist of two fingers. The valve should not be overtightened, or valve seat damage could occur.

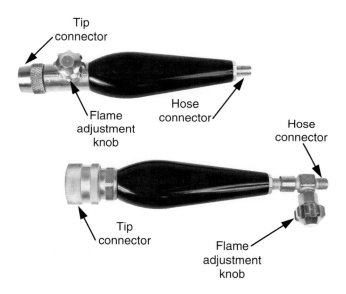

Figure 5-11. *Air-acetylene torch handles include connec-tions for the acetylene hose at one end and for the torch tip at the other. As the two samples show, a finger-operated knob for flame adjustment may be at the hose end of the handle or at the tip end. (Uniweld)*

5.4.5 Air-acetylene Torch Tips

Different sizes and styles of torch tips are available for different applications. See **Figure 5-12.** Tips can be easily interchanged by loosening a locknut on the torch handle and unscrewing the tip. No tools are needed.

5.4.6 Lighting the Torch

The procedure for lighting an air-acetylene torch is simple, but certain steps must be followed to ensure safety:

1. Inspect the equipment for damage, and make sure all valves and regulators have been turned off.

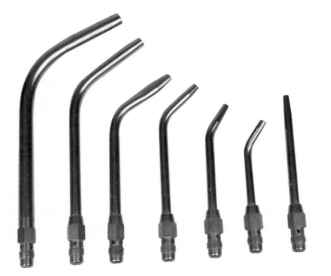

Figure 5-12. *Air-acetylene torch tips are available in a variety of sizes for different applications. (Uniweld)*

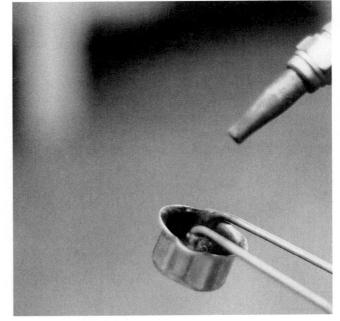

Figure 5-13. *An air-acetylene torch should never be lighted with a match. For safety, always use a flint striker (also known as a spark lighter).*

2. Open the cylinder valve slowly, using the proper valve key wrench. Open the valve one-quarter to one-half turn counterclockwise. Leave the wrench on the valve stem so the cylinder can be turned off quickly. If the pressure regulator is the two-gauge type, the contents gauge will immediately register the level of acetylene in the cylinder.

3. Open the torch handle valve one-half turn. Slowly turn the pressure-regulator adjusting screw clockwise until the delivery gauge indicates 5 psi (34 kPa) flow pressure. Vent the escaping gas safely.

4. Turn off the torch handle valve using finger-tip pressure only. The torch is now adjusted to the correct flow pressure and ready to be lit. Before lighting the torch, be sure the tip is pointed away from any object or person.

5. Slightly "crack" open the torch handle valve. Use a flint striker to ignite the acetylene gas coming out of the torch tip. Hold the striker near the end of the torch tip, but do not block the gas flow, **Figure 5-13**. A flint striker is the only safe device to use when lighting any torch. Never use a match or cigarette lighter.

6. Open the handle valve slowly until the desired flame size is obtained. The hottest part of the flame is about 1/8″ (3 mm) from the tip of the inner cone, **Figure 5-14**. Do not restrict the inner cone by holding the flame too close to the workpiece. Such a practice will interfere with proper mixing and combustion of the fuel-air mixture and result in poor flame quality. Once the workpiece is heated to a temperature that will melt the solder, back the flame away just far enough to maintain the proper flow temperature of the solder.

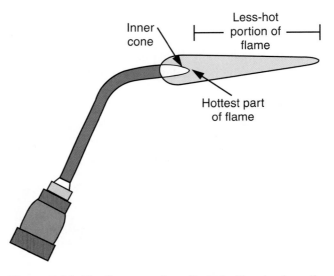

Figure 5-14. *The flame can be adjusted with a knob on the end of the handle. The hottest area of the flame is just in front of the inner cone's tip.*

The maximum temperature obtainable with this torch system is 3600°F (1982°C), regardless of the tip size. Large tips will provide more heat output, but the temperature of the flame will remain the same. The workpiece will seldom, if ever, reach this temperature, since solders will usually melt in the range of 400°F to 500°F (204°C to 260°C). Avoid overheating the base metals.

5.4.7 Shutting Off the Torch

The torch can be shut down temporarily by merely closing the handle valve to turn off the flame. It can then be laid aside until it is needed again. However, if the unit is not going to be used for a long time, it should be completely shut down as follows:

1. Close the valve on the torch handle, using fingertip pressure only. This will kill the flame.
2. Tightly close the cylinder valve.
3. Reopen the torch handle valve to bleed off (release) acetylene from the hose and regulator. Bleeding should continue until both gauges read zero.
4. Turn the adjusting screw on the regulator all the way out (counterclockwise) to close the regulator. This will help to avoid gas loss if the cylinder or torch valves leak.
5. Rewind the hose, and place the unit in safe storage.

5.5 Soldering Techniques

It is important to apply heat properly in any soldering operation. A number of heat sources are available, such as soldering irons, soldering guns, and a variety of torches, but this text will address only the air-acetylene torch and open-flame technique.

Torch selection is determined by the type and size of the soldering job. Large tips providing high heat, or even an oxyacetylene torch, may be used by skilled operators. Less-skilled operators should use the air-acetylene torch, with its soft flame and lower temperatures, to avoid overheating the base metals. A variety of tips may be used with an air-acetylene torch; more heat is obtained with large tips. Selecting the proper size tip (or flame) requires experience. In general, small tubing requires a small flame and large tubing requires a large flame. See **Figure 5-15.**

Dirty tips or faulty adjustment may produce a sooty flame that deposits carbon on the workpiece. This carbon deposit will prevent the alloy from flowing properly and result in a poor joint.

For consistently good soldering, keep in mind that the alloy will flow properly only when the joining pieces are heated equally and evenly to the solder's flow point. *Do not apply the flame directly to the alloy.* Use the flame to heat the base metals to the solder's melting point. Then, back off the flame just far enough to maintain the proper temperature until melted alloy fills the joint. Apply only enough alloy to form a bright shiny bead (a **fillet**) at the point where the pieces meet.

The solder should flow smoothly. If it forms a globule or "ball" on the work, the base metal temperature may be too low, or the joint surfaces are dirty. "Balling" can also occur if the torch flame was directed onto the solder, melting a drop from the end. It is difficult to get these globules of solder to flow properly.

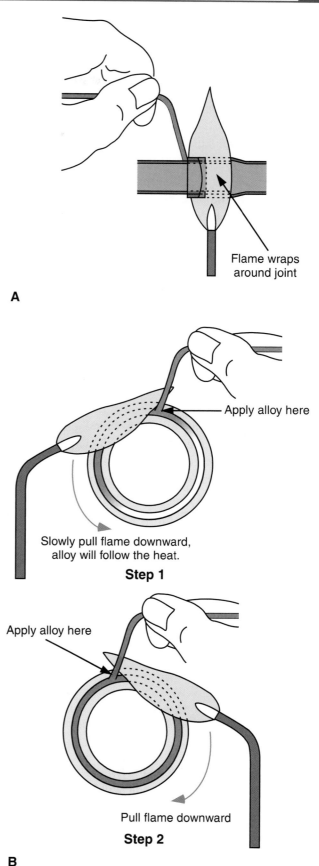

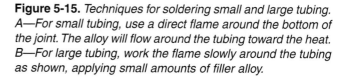

Figure 5-15. *Techniques for soldering small and large tubing. A—For small tubing, use a direct flame around the bottom of the joint. The alloy will flow around the tubing toward the heat. B—For large tubing, work the flame slowly around the tubing as shown, applying small amounts of filler alloy.*

Key points to remember when soldering are to:
- ❑ Make the joint fit.
- ❑ Clean it thoroughly.
- ❑ Apply the proper flux.
- ❑ Heat the joint evenly to the correct temperature.
- ❑ Apply the proper amount of alloy.

5.5.1 Capillary Action

All joints to be soldered should fit tightly for maximum capillary action by the alloy. **Capillary action** refers to the process where the alloy automatically fills the gap between the pieces of base metals. It draws the alloy into the full depth of the connection.

To obtain maximum capillary action (and greater strength), the entire joint should be close-fitting and well-supported to prevent movement. When the filler alloy enters the joint, the alloy molecules have a greater attraction for the base metal than for each other. The alloy is attracted to each wall of the base metal and "walks" itself into the fitting to completely fill the joint.

This capillary attraction makes it possible to solder a *vertical up* (uphill) *joint* because the alloy will be drawn up into the joint, **Figure 5-16.** When soldering an uphill joint, heat the tubing first, then apply heat to the fitting. It is important to heat both pieces evenly. If the tubing is overheated, the alloy may run downward, rather than be drawn upward into the joint. Remember, alloy flows *toward* the heat.

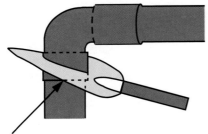

Apply alloy here

Figure 5-16. *Solder will be drawn up into a vertical joint by capillary action. The alloy will travel toward the heat.*

5.5.2 Soldering Copper Tubing

As an HVAC technician, you will frequently have to sweat solder copper tubing. Follow the procedures below to make mechanically sound and leakproof connections with copper tubing and fittings:

1. Cut tubing ends square and remove all inside and outside burrs with a reamer. (See Chapter 3.)
2. Clean the copper tubing and fitting thoroughly down to bare metal before making the soldered joint. Joining surfaces must be clean to make a sound solder joint.

3. Apply flux to cleaned tubing before soldering. Flux should be applied sparingly and kept away from the tubing end.
4. Always move the torch in short arcs; do not let it remain in one spot too long. Heat the tubing first by applying the flame next to the fitting. Heating the tubing first will conduct heat inside the fitting. Work the flame around the tubing and fitting to bring the temperature up *equally* in both pieces.
5. Use the flux as a temperature guide. Apply heat until the flux stops bubbling and becomes fluid and clear. Apply heat to the fitting, as well, to ensure the flux inside the fitting is clear. A clear flux indicates the alloy flow temperature is being reached.
6. Back the flame away slightly to reduce the heat intensity. Sweep the flame back and forth over the assembly to maintain uniform heat in both parts.
7. Apply the alloy. Heat from the base metals, not the flame, should melt the alloy. Feed the alloy into the joint between the tube and fitting. It will quickly melt and completely fill the joint area.
8. Remove the flame when the alloy begins flowing into the joint. Do not continue feeding alloy after the joint is filled.

5.5.3 Taking Apart a Soldered Joint

Clean and apply flux to the joint you want to disconnect. Heat the joint (tubing *and* fitting) evenly. When the alloy becomes fluid throughout the joint, the tubing will remove easily. Do not twist or apply force to the tubing to remove it. Twisting or force will cause severe damage to the hot fitting and tubing. To resolder the joint, clean the tubing end and the inside of the fitting. Apply flux and resolder the joint.

WARNING: Never apply heat to a line that is *under pressure.* Applying heat increases pressure; the molten alloy and fitting may be blown violently apart. Be sure to release all pressure from the tubing before heating a soldered connection.

5.5.4 Soldering Aluminum

Aluminum is now used for many applications where copper or steel were once standard. Aluminum is an excellent metal for conducting heat and is widely used in the manufacture of evaporators for domestic refrigerators. However, soldering aluminum presents special problems not normally found with other metals.

When soldering aluminum, attention must be given to such details as surface preparation, solder composition, temperature, and the application of heat. Soldering pure aluminum and the many aluminum alloys requires special techniques and materials. For example, the aluminum used in the manufacture of evaporators is very thin, so heat applied during the soldering process must be precisely controlled. Aluminum and its alloys are weak when hot (a condition called "hot-shortness"). The metal does not change color as a warning that it is approaching its melting point. Instead, it will suddenly collapse. For this reason, the metal must be properly supported during the soldering process.

The corrosion-resistance of aluminum results from oxides that form quickly on the metal's surface when it is exposed to air. To make a sound soldered connection, the oxides must be removed. Additional oxides are formed by the application of heat. Oxides have a higher melting point than aluminum. A special flux is used that will combine chemically with the oxides to form a scum, or *slag*. The slag rises to the surface of the molten solder, where it will not interfere with the soldering process.

The solder must be melted before it can flow into the joint. The melting point of the solder will determine the minimum temperature needed to make the connection. The solders used for aluminum have a higher melting point than those used for copper or steel, so aluminum must be heated to temperatures 100 to 200 degrees Fahrenheit (55 to 110 degrees Celsius) higher than copper or steel. The increased temperature is the major reason soldering aluminum and aluminum alloys is difficult.

The solders used for aluminum can be placed into four groups: zinc-based, zinc-cadmium-based, tin-zinc-based, and tin-lead-based solders. See **Figure 5-17.** These solders may contain quantities of other metals to obtain certain properties or qualities.

Soldering aluminum requires skill and proper equipment, requirements that can usually be met under carefully controlled manufacturing processes. Such control is not readily available in the field, where service and repair work must be performed. The aluminum used in the manufacture of evaporators on

domestic refrigerators presents a typical problem. The metal is very thin, which promotes good heat transfer for the cooling process. Because of the thin metal, however, these evaporators are easily punctured by sharp instruments. A puncture will result in loss of refrigerant, thus stopping the refrigerating process. In the field, soldering such a hole is almost impossible; repairing the damage would be time-consuming and expensive. The customer is usually advised to purchase a new refrigerator.

Repairs to the hole in an aluminum evaporator *can* be performed but not by soldering. **Epoxy** compounds are readily available for making such repairs in aluminum and have proven to be highly successful. See **Figure 5-18.** The epoxy kit contains all necessary materials and instructions for proper use.

Figure 5-18. *Repairs to aluminum that cannot be made in the field with soldering can often be accomplished with epoxy compounds. Epoxies harden quickly and will withstand pressure and the effects of refrigerants. (Watsco)*

The same epoxy compound can be used to repair leaks in joints between aluminum tubing and copper tubing. The original aluminum-to-copper connection is made at the factory, using welding techniques that are not available in the field. Sometimes leaks will develop at this special connection. The epoxy compounds provide a satisfactory and reliable means for repairing the joints.

5.6 Air-acetylene Brazing

Special tips have been designed for use on the regular air-acetylene soft-soldering torch to increase the working temperature of the flame, making it possible to perform some brazing operations. Some of these special tips are constructed with a gas diffuser

Aluminum

Solder	Melting Point
Zinc-based	700°F to 820°F (370°C to 440°C)
Zinc-cadmium	510°F to 750°F (265°C to 400°C)
Tin-zinc	550°F (290°C) and higher
Tin-lead	450°F (230°C) and higher

Figure 5-17. *Alloys used for soldering aluminum have melting points considerably higher than those used to solder copper or steel.*

inside. The diffuser slows the gas flow and causes the flame to start burning and expanding inside the tip. This creates a high-velocity gas stream that completes the burning process outside the tip, providing a high-heat working flame. It is important to keep a full gas flow and flame to avoid overheating the tip.

5.6.1 Torch Tips for Brazing

A major difference when using these special tips is the operating (flow) pressure of the acetylene. They are high-velocity tips that require 15 psi (103 kPa) operating pressure from the acetylene pressure regulator. See **Figure 5-19.** When using these high-velocity tips, the regulator is usually turned all the way in (fully clockwise) for maximum pressure. The tips are designed so that the flow of acetylene draws atmospheric air into the base of the tip, resulting in more oxygen mixing with the acetylene. Another design mixes the gases near the tip end, using a cupped tip to stabilize the high-velocity flame. These tips are available from various manufacturers under such trade names as "Thruster" (Uniweld), "Swirljet" (Union Carbide), "Swirl Tip" (Turbotorch), and "Ram-jet" (Airco). They are designed for use on the regular air-acetylene handle. Various sizes of tips are available to accommodate different flame sizes.

All high-velocity tips must be operated at 15 psi (103 kPa) flow pressure, which provides a high-heat flame that is usually nonadjustable. The "Thruster" tips by Uniweld are fully adjustable at the torch handle, giving the operator more flame control. The tips are adjustable from idle, to soft (medium), to high-intensity (full) operation.

Special tip advantages

Special tips offer several advantages. The flame is pencil-shaped (thin), short, and concentrated for close work. The short flame allows soldering or brazing of joints within an inch of walls or woodwork without damaging or discoloring them. It is an all-weather flame; the high velocity allows it to be used in wind, rain, and subzero temperatures.

The tips develop an efficient high-heat-transfer zone at the tip of the inner cone of the flame. This forms a concentrated heat flow that wraps around the workpiece to heat it evenly and quickly. The flame temperatures in this high-heat zone range from 1800°F (982°C) for propane, to over 2400°F (1315°C) for MAPP, to over 2600°F (1427°C) for acetylene. This gives the user a choice of tip sizes and gases to meet the heating requirements.

Special tip disadvantages

The major disadvantage of these special tips is that they cannot be used for brazing a connection located near a heat sink. A *heat sink* is any heavy metal device, such as a valve or compressor, that tends to draw heat away from the brazing area. See **Figure 5-20.**

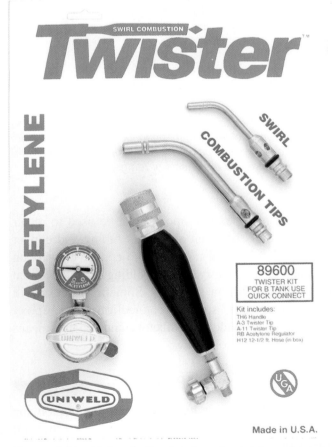

Figure 5-19. *High-velocity tips for air-acetylene torches allow the use of higher temperatures for brazing work. Various designs are available. (Uniweld)*

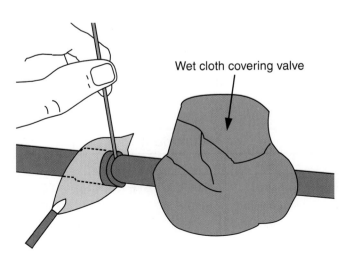

Figure 5-20. *A heavy metal component, such as a valve, will act as a heat sink and draw heat away from the area being brazed. This heat could damage the component. To prevent damage, the valve or other component is sometimes wrapped in a wet cloth. (Aeroquip)*

With these tips, even a moderate-sized heat sink will steal the heat rapidly enough to greatly prolong the brazing process. Also, the supply of acetylene is quickly exhausted because of the high-flow rates.

It is recommended that all brazing connections be performed with an oxyacetylene torch or high-heat air/gas tips, reserving soft flame air-acetylene torch tips for soft-soldering procedures. The use of air/gas tips, when possible, saves oxygen for applications requiring higher temperatures, such as cutting, welding, and brazing near heat sinks.

Tubing sizes appropriate for several high-velocity heat tips, when soldering or brazing, are listed in **Figure 5-21.** Notice the difference in the tubing sizes when changing from soldering to brazing with the same size tip. This table does not consider the presence of a possible heat sink.

Tips vs Tubing Size

Tip No.	Soft Soldering	Brazing
	Copper Tubing Size	Copper Tubing Size
#3	1/4″ to 1 1/2″	1/4″ to 7/8″
#4	1″ to 2″	5/8″ to 1 1/8″
#5	1 1/2″ to 3″	7/8″ to 1 5/8″
#6	2″ to 4″	1 1/8″ to 2 1/8″

Figure 5-21. *Tubing sizes that can be soldered or brazed with several sizes of high-velocity heat tips.*

Summary

Soldering and brazing are two processes widely used in HVAC work. This chapter described appropriate applications for each process and explored soldering in detail. Different soldering alloy compositions were described, as well as the importance and proper use of flux. Safety when working with fuel gases was emphasized.

Much of the chapter was devoted to the air-acetylene torch and related equipment. Instructions for using the torch for soldering and air-acetylene brazing were given.

Test Your Knowledge

Please do not write in this text. Write your answers on a separate sheet of paper.

1. Acetylene is used as a _____ gas for soldering and brazing procedures.
2. What is the highest flame temperature obtainable with an acetylene-oxygen mixture?
3. The fusible plug on acetylene cylinders will melt at _____°F (100°C).
4. Why should acetylene cylinders be used in an upright position?
5. The smallest acetylene cylinder is size MC. What is the next size?
6. Why should the acetylene cylinder valve key or wrench be left on the valve during use?
7. Explain the operating principle of the air-acetylene torch.
8. Acetylene cylinders are pressurized to _____ psi (1725 kPa) at 70°F (_____°C).
9. Pressure regulators are used to reduce cylinder pressure to _____ pressure.
10. What color is acetylene hose?
11. What is special about threads for acetylene connections?
12. The hottest area of the air-acetylene flame is _____″ from the _____ of the inner cone.
13. The maximum flame temperature obtainable with an air-acetylene torch is 3600°F (_____°C).
14. To distinguish it from brazing, soldering is defined as the joining of metals at temperatures less than _____°F (450°C).
15. What is the purpose of soldering flux?
16. Name three types of soldering alloys.
17. The _____ should be melted by heat from the base metal, not from the torch flame.
18. What is the flow pressure of the fuel gas for the air-acetylene torch?
19. Aside from the type of thread, how does an acetylene hose connector differ from a connector used for an oxygen hose?
20. The acetylene cylinder valve should never be opened more than _____ to _____ turn.

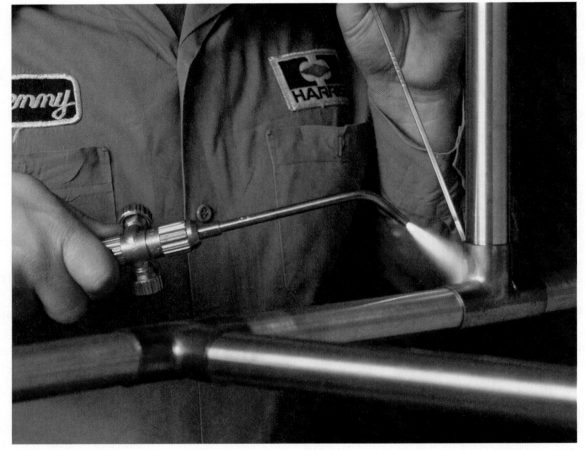

Silver brazing copper pipe joints, using an oxyfuel gas torch as the source of heat. (© J.W. Harris Co., Inc.)

Brazing and Flame-cutting

Objectives

After studying this chapter, you will be able to:
- ❏ Describe the brazing process.
- ❏ Select appropriate brazing alloys and fluxes.
- ❏ Select and use cylinders and torches.
- ❏ Connect and operate oxyacetylene equipment for brazing metal.
- ❏ Set up, adjust, and use an oxyacetylene torch for cutting metal.

Important Terms

backfire	melting temperature
brazing	multiflame tip
cadmium-bearing	neutral flame
alloys	oxidizing flame
carburizing flame	oxyacetylene torch
cutting oxygen lever	oxygen regulator
downhill joint	preheat flame
filler metals	preheat valve
flashback	uphill joint
flow temperature	vertical down joint
hard solders	vertical up joint
horizontal joint	

6.1 What Is Brazing?

Brazing and soldering are very similar metal-joining processes. The same skills are required, and proper use of the flame is critical to each. Brazing and soldering differ in the temperatures and alloys used to join metals. By definition, **brazing** joins metals above 840°F (448°C). However, in practice, it is usually accomplished at about 1200°F (648°C).

Like soldering, brazing is performed by heating the base metals to a specified temperature *above* the melting point of the filler alloy, **Figure 6-1,** but *below* the melting point of the base metals (for example, copper melts at 1981°F or 1083°C). The melted filler alloy is distributed between the close-fitting base metal pieces by capillary attraction.

Figure 6-1. *Filler alloys for use in brazing come in a variety of shapes, sizes, and compositions. (©J. W. Harris Co., Inc.)*

Brazing is the principal method of assembling heating, air conditioning, and refrigeration equipment. A typical large system may contain hundreds of brazed joints. Brazing permits joining of similar and dissimilar metals, thin and thick sections, and metals with different melting points. Moreover, a brazed joint is stronger than the base metals, tolerant of vibration without cracking, leakproof under high pressures, and noncorrosive. All these attributes make brazing the preferred method for joining tubing on refrigeration systems. In fact, many local, state, and federal codes *require* brazing on all refrigeration systems.

Some local codes also require that licensed technicians make the brazed connections. A license is usually obtained by paying a fee and submitting test samples of properly brazed connections. Leakproof connections are important, so they are never entrusted to a beginner.

6.2 Heat Sources for Brazing

Heating the base metals for brazing is normally done with a torch. Flame temperature is determined by the type of fuel gas used. Oxygen mixed with any fuel gas will give the highest attainable flame temperature.

The most versatile heat source for brazing is an oxygen-acetylene torch, often referred to as an **oxyacetylene torch.** It provides a heat range that handles all types of joining work, regardless of the materials.

Using an oxyacetylene torch properly and safely, and achieving good results, is not difficult. Unsafe conditions or poor results are usually due to an improperly adjusted flame and incorrect pressure. The technician should make every effort to acquire good habits working with this tool and maintain these habits with every use.

6.2.1 Where the Oxyacetylene Process Is Used

The oxyacetylene process is used in many fields where metal work is involved. When acetylene is burned with oxygen, it produces a flame so hot it can melt, fuse, or cut any metal.

In the hands of a skilled operator, the oxyacetylene torch makes it possible to quickly solder, braze, weld, or cut all metals. The intensely hot flame concentrates considerable heat in a small area, so the brazing process can be performed quickly.

WARNING: The presence of oxygen rapidly increases the rate of burning of almost any ignited material, especially oil and grease. Also, a concrete surface on which material is being cut will chip explosively when overheated by an oxyacetylene flame. The flying concrete chips can cause injury.

6.2.2 Dressing for Safety

An oxyacetylene torch flame can reach almost 6000°F (3316°C); the workpiece can reach almost 3000°F (1649°C). The process can produce flying sparks, molten metal, slag, fumes, and intense light rays, all of which can harm the torch operator or persons nearby.

For your safety, take a practical "head-to-toe" approach to protective clothing. Wear hair and head coverage, and welding and cutting goggles with safety-tempered dark lenses (shade 5 is standard).

Wear thick gloves and proper shoes. Do not wear any flammable articles or clothing that has been exposed to flammables (oil, grease, wax, or solvent). Sparks and molten material will find their way to unprotected areas, so protect yourself thoroughly before starting.

6.3 Brazing Alloys

Brazing alloys usually contain a certain amount of silver, which has led to the term "silver soldering." However, the term is incorrect. Silver-bearing solder is an alloy of tin and silver with a melting point of 430°F (221°C). This *soldering* alloy should not be confused with the silver-bearing *brazing* alloys that have melting temperatures ranging from 1200°F to 1500°F (593°C to 816°C). These brazing alloys are sometimes called **hard solders,** a term that is commonly used but not universally accepted.

Brazing alloys are available in a variety of forms. The preferred form is a 20″ (50.8 cm) length of 1/8″ (3 mm) flat rod. See **Figure 6-2**. The rod is packaged in 1 lb. (0.45 kg) or 5 lb. (2.27 kg) tubes. Larger containers are also available.

As noted in Chapter 5, the term **melting temperature** (*solidus*) refers to the temperature at which an alloy starts to melt, while **flow temperature** (*liquidus*) is the temperature at which the alloy becomes a liquid. The difference between the solidus and liquidus temperatures is called the *plastic range.* When compared to soldering alloys, brazing alloys have a relatively wide plastic range.

Since they are alloys composed of different metals with different melting points, the **filler metals** used in brazing may separate during the heating process. The metal with the lowest melting temperature in the alloy may flow into the joints, leaving the higher-melting-point metals behind. This causes a change in the color and strength of the alloy and makes it difficult to get the remaining alloy to flow. Heating the base metals quickly prevents separation of the alloy and minimizes oxidation.

Separation of filler metals is a feature that can be advantageous when brazing a loose connection. The higher-melting-point alloys will span the gap. However, the connection will not be as strong as one made with an alloy whose metals did not separate. Silver-brazing alloys generally melt at approximately 1200°F (593°C) and flow at about 1500°F (816°C), which gives a plastic range of about 300 degrees Fahrenheit (165 degrees Celsius).

6.3.1 Alloy Compositions

Brazing alloys are composed of varying combinations of copper (Cu), silver (Ag), phosphorus (P), and

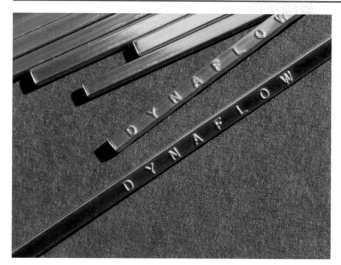

A

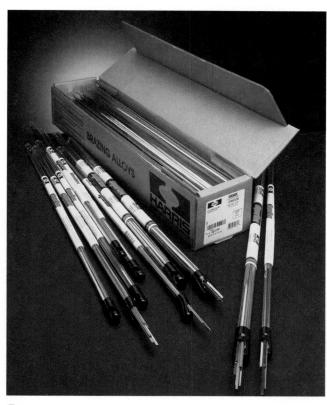

B

Figure 6-2. *Brazing rods. A—Flat 1/8″ (3 mm) brazing rod is the form preferred by many technicians. B—Brazing rod is packaged in 1 lb. (0.45 kg) tubes, as shown here. Also available are 5 lb. (2.27 kg) tubes and larger containers. (©J. W. Harris Co., Inc.)*

zinc (Zn). The composition of a given brazing alloy determines its use, strength, and cost. Manufacturers have their own numbering system for identification of any particular alloy. However, the American Welding Society (AWS) uses a special numbering system to identify brazing alloys by their composition, regardless of manufacturer. The AWS number consists of the letter B to indicate a brazing alloy, followed by the

chemical abbreviations of the metals used in the alloy and an identifying number. See **Figure 6-3.**

Copper-phosphorus alloy (AWS BCuP-2)

This is a low-cost alloy that is about 93% copper and 7% phosphorus. It is sometimes called *phos-copper* and eliminates the need for a flux when brazing copper to copper; the phosphorus acts as a deoxidizing agent and flux. When joining copper tubing to a brass or bronze fitting, however, a flux *will* be needed. Do not use phos-copper alloy for joining copper to steel since a very brittle joint will result.

When using phos-copper for connecting copper to copper, the surfaces of the connection should be cleaned to bare metal. AWS BCuP-2 alloy has a plastic range of 1185°F to 1500°F (641°C to 816°C).

Copper-phosphorus-silver alloy (AWS BCuP-5)

Many alloys of silver have been developed to provide different melting points, flow properties, strengths, and colors. Silver-brazing is one of the best methods for making connections in refrigeration components that are leakproof and able to withstand severe conditions.

An alloy of 80% copper, 5% phosphorous, and 15% silver (AWS BCuP-5) has been the industry leader for many years. This alloy allows the creation of highly ductile and very strong connections when brazing copper to copper, without using a flux. In addition, with use of a flux, this alloy can braze copper to brass, bronze, or steel. AWS BCuP-5 alloy melts at 1190°F (643°C) and flows at 1500°F (816°C).

Copper-phosphorus-silver alloy (AWS BCuP-4)

Silver-bearing alloys are expensive, and the higher the silver content, the more expensive the alloy. Alloys with silver content lower than 15% have proven economical and have nearly the same characteristics as the 15% alloys. An example is AWS BCuP-4, which contains 6% silver. This alloy was developed to perform nearly as well as the higher silver-content alloys. It is widely used in the HVAC industry because of its economy. AWS BCuP-4 alloy melts at 1190°F (643°C) and flows at 1480°F (804°C).

6.3.2 Alloy Additives

The metals zinc and cadmium are used as additives in brazing alloys that are 35% or more silver. These additives have superior capillary-attraction properties and form excellent brazes on all metals, except aluminum and magnesium. They also lower the melting and flow temperatures of the alloys. Despite these benefits, **cadmium-bearing alloys** should be avoided whenever possible.

◆─────────────────────────────── **Chemical Composition of Brazing Alloys**

AWS Number	Copper (Cu)%	Phosphorus (P)%	Silver (Ag)%	Zinc (Zn)%	Tin (Sn)%	°F	
						Melts	Flows
BCuP-2	92.88	7.12				1310	1485
BCuP-3	88.75	6.25	5			1190	1485
BCuP-4	87.00	7.00	6			1190	1480
BCuP-5	80.00	5.00	15			1190	1500
BAg-5	38.00		30	25		1250	1370
BAg-7	22.00		56	17	5	1145	1200
BAg-20	38.00		30	32		1250	1410

Figure 6-3. *The American Welding Society's identification system for brazing alloys assigns a specific letter/number combination to each alloy. This table shows the chemical compositions and the melting and flow (solidus and liquidus) temperatures for some common brazing alloys.*

WARNING: Cadmium is a toxic metal when molten. It emits highly poisonous cadmium oxide fumes that can cause illness or death. Adequate ventilation must be provided when using cadmium-bearing alloys.

Cadmium-free alloys that meet all brazing requirements are available, **Figure 6-4.** HVAC technicians should adopt cadmium-free brazing alloys wherever possible to protect their health. The danger of cadmium poisoning is real and the risk is unnecessary. The nontoxic alloys eliminate the danger of cadmium oxide fumes and can be used on both ferrous and nonferrous metals.

6.4 Fluxes for Brazing ───────────◆

Common fluxes for silver brazing are mixtures of boric acid, borates, fluorides, and fluoborates. See **Figure 6-5.** These chemical powders form a paste when water or alcohol is added. The container usually has an application brush, and if the paste becomes too dry, it may be thinned with water and stirred. The flux must be kept clean to avoid contaminating the joint.

By watching the behavior of the flux, you can determine the temperature at different stages of the brazing process. The flux will dry out when the water boils off at 212°F (100°C). When the temperature

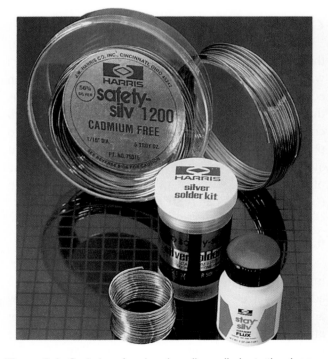

Figure 6-4. *Cadmium-free brazing alloys eliminate the danger of poisonous cadmium oxide fumes generated when using cadmium-bearing filler alloy. (© J. W. Harris Co., Inc.)*

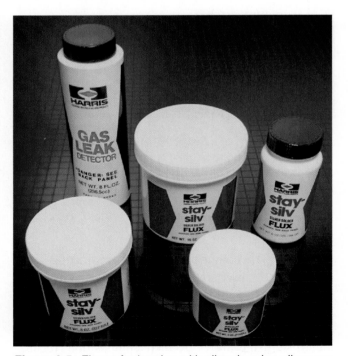

Figure 6-5. *Fluxes for brazing with silver-bearing alloys are packaged in containers of various sizes. Most packages have a built-in applicator brush. (© J. W. Harris Co., Inc.)*

reaches about 600°F (316°C), the flux will start to bubble and turn milky white. At about 1100°F (593°C), the flux will turn into a clear liquid. At this point, you should apply the filler alloy.

The heating process should be performed as evenly and quickly as possible to reduce oxide formation. Brazing that is performed quickly reduces stress on the base metals. **Figure 6-6** lists visual clues to various stages in the heating process.

6.5 Fuel Gas Cylinders

Oxygen and acetylene cylinders are sized according to the amount of gas in cubic feet the cylinder will hold. The cylinder sizes are designated by a letter/number combination. The table in **Figure 6-7** matches oxygen and acetylene cylinders by size.

HVAC service technicians seldom use the large fuel gas cylinders common in welding operations. Technicians prefer smaller cylinders that can be easily transported to rooftops, basements, and other locations where heating, ventilation, and air conditioning equipment is installed. See **Figure 6-8.** The small

oxyacetylene units perform the same duty as the large units, except the operating pressures are different, and the supply of oxygen and acetylene does not last as long in the small units. To overcome the supply problem, technicians stock one or two extra cylinders for additional capacity. Carrying stands and two-wheel carts are available to accept the small cylinders.

An oxyacetylene outfit that uses small cylinders, regulators, hoses, and torches is called a light-duty, or *aircraft-type*, unit. It is preferred by many service technicians because of its portability. All the components are matched and have 3/8″ (size "A") fittings at the regulator, hoses, and torch handle.

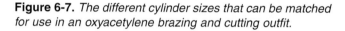

Cylinder Sizes for Matched Systems

Acetylene		Oxygen
10 cu. ft. (MC)	equals	20 cu. ft. (AA or R-Oxy)
40 cu. ft. (B)	equals	40 cu. ft. (A)
60 cu. ft. (#2)	equals	60 cu. ft. (J)
60 cu. ft. (#2)	also equals	80 cu. ft. (JJ)

Figure 6-7. *The different cylinder sizes that can be matched for use in an oxyacetylene brazing and cutting outfit.*

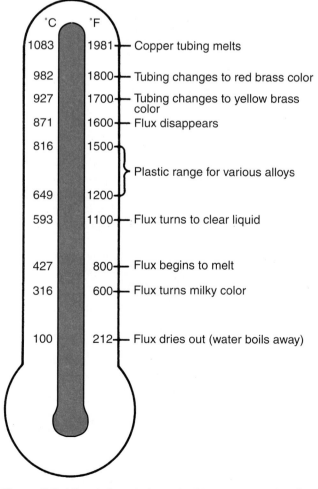

Figure 6-6. *Visual clues to important temperatures involved in the brazing process.*

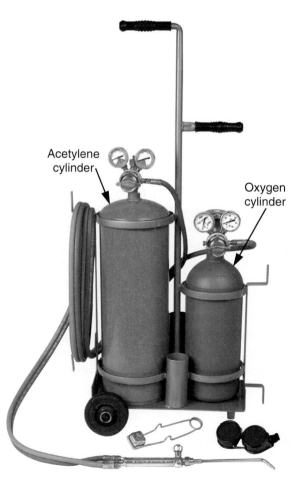

Figure 6-8. *A portable oxyacetylene outfit with a size B (40 cu. ft.) acetylene cylinder and a size R (20 cu. ft.) oxygen cylinder. (Uniweld)*

The small oxyacetylene outfits also require small brazing and cutting tips. These, in turn, require small operating (or flow) pressures. Follow the manufacturer's recommendations for the type of equipment you are operating.

6.5.1 Cylinder Pressures

Oxygen cylinders are charged to 2200 psi at 70°F (15 169 kPa at 21°C), but the actual pressure in the cylinder will vary with the temperature. All gases expand when heated and contract when cooled. If the temperature exceeds 70°F (21°C), the pressure in a full cylinder will rise above 2200 psi. A safety relief device is built into oxygen cylinder valves to protect against explosion from extremely high pressures. Such pressures might occur if the cylinders were exposed to fire. Oxygen cylinders should never be used or stored where they might become overheated.

6.5.2 Oxygen Cylinder Safety

An oxygen cylinder valve is designed to operate at high pressures, **Figure 6-9**. An iron cap screws down over the valve to protect it during shipment or handling. This cap should always be in place when the cylinder is not in use.

The cylinder valve has a *double seat* that prevents leakage around the valve stem when the valve is fully opened for operation. To open and close the valve, hand pressure is sufficient. A wrench should not be used.

Observe the following precautions when using or handling oxygen cylinders:
- ❑ Carefully secure cylinders to keep them from falling over.
- ❑ Never store cylinders and equipment in unventilated and confined spaces, in closed vehicles, or near any source of heat or ignition.
- ❑ Close cylinder valves securely when not in service or when empty.
- ❑ Avoid dangerous pressure imbalance and cylinder contamination by never allowing any cylinder (especially oxygen) to become completely empty.
- ❑ For maximum protection against contamination, install external reverse-flow check valves on the regulator or torch handle. See **Figure 6-10**.
- ❑ Never allow oil or grease to come in contact with oxygen cylinder valves. Oxygen reacts violently with oil and grease.
- ❑ Do not use a cylinder that has a leaking valve. Carefully move the cylinder outdoors, and notify your gas supplier.
- ❑ Do not expose cylinders to torch flames or electric arcs.

- ❑ Never use a cylinder, full or empty, as a roller or support. The cylinder wall could be damaged and rupture or explode.

6.5.3 Oxygen Regulators

The working (or flow) pressure at the torch tip is considerably lower than the available cylinder pressure. Obviously, a device must be provided to reduce

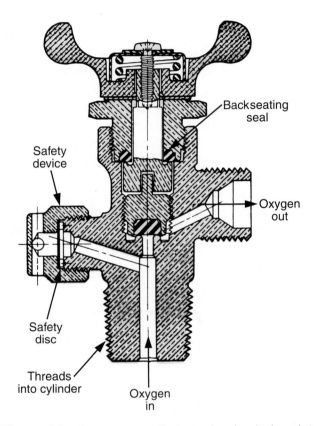

Figure 6-9. *An oxygen cylinder valve is designed to withstand high pressures. It has a double-seating feature (backseating seal) that prevents leakage around the stem when the valve is fully opened. Note the safety disc. It will rupture if oxygen pressure becomes dangerously high.*

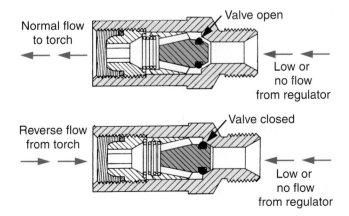

Figure 6-10. *Check valves are designed to prevent contamination of fuel gas cylinders by allowing flow in only one direction. Reverse flow causes the valve to close. (Uniweld)*

high cylinder pressure to working pressure. Also, to maintain a steady and uniform flame, the pressure of the gases reaching the torch tip must not vary (despite steadily decreasing cylinder pressure). An *oxygen regulator,* **Figure 6-11,** performs both these important functions.

The regulator is adjustable to permit any flow pressure, and it will maintain that pressure without further adjustment until the cylinder approaches empty. It is easy to recognize a nearly empty cylinder by variations in the flame or by checking the pressure gauge. Always replace a cylinder *before* it becomes empty and unsafe.

The oxygen regulator is attached directly to the oxygen cylinder valve and contains an outlet for connecting the hose leading to the torch handle. Most regulators have two gauges. One, the high-pressure gauge, indicates the pressure in the cylinder; the other, the flow-pressure gauge, indicates the pressure being supplied to the torch handle.

A pressure-adjusting screw on the front of the regulator sets flow (working) pressure. When this

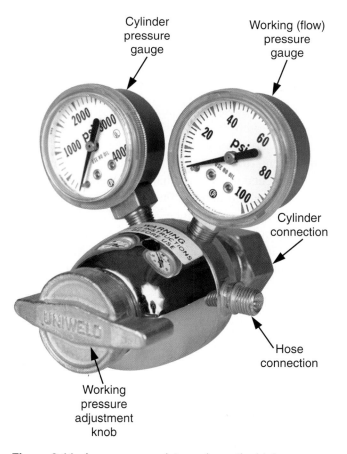

Figure 6-11. *An oxygen regulator reduces the high-pressure flow of gas from the cylinder to a much lower-pressure flow through the hose and torch. It has separate gauges for the cylinder pressure and the working (flow) pressure. Adjustment of the working pressure is done with a knob on the front of the regulator. (Uniweld)*

screw is turned completely to the left (counterclockwise), the valve is closed and no gas can pass through the regulator. As the screw is turned to the right (clockwise), the regulator opens and permits gas to flow to the torch handle. Gas flow is immediately indicated on the flow gauge. Working pressure is changed by turning the adjusting screw until the desired flow is registered on the gauge.

Adjusting flow pressure

Pressure cannot be adjusted properly unless the gas is actually flowing. The torch handle valve must be open to permit the gas to be released into the atmosphere. Adjust the flow pressure as if the torch were lit and operating. After the pressure is adjusted, turn off the torch handle valve to stop the flow.

There must be separate regulators for oxygen and acetylene. The two regulators are *not* interchangeable. The regulators and gauges are color-coded, green for oxygen and red for acetylene. They also have special cylinder connections to prevent an acetylene regulator from being connected to an oxygen cylinder, or vice versa. Always return both regulator screws to the *off* ("backed out" or counterclockwise) position when shutting down the outfit.

Regulator safety

The following procedures for regulators should be followed each time the equipment is set up and used:

❑ Use a regulator only with the gas for which it is intended. Oxygen regulators must be used only for oxygen service.

❑ Keep regulators and cylinder connections free of dirt, dust, grease, and oil. Do not use a regulator if it is damaged or has oil or grease on it.

❑ Do not *crack* (slightly open) fuel gas or oxygen cylinders near a flame or any source of ignition. Make sure you are in a well-ventilated area, and stand clear of the valve outlet.

❑ Make sure the threads engage properly when connecting the regulator to the cylinder valve. Tighten it with a wrench, but do not use excessive force.

❑ To shut off the regulator, turn the pressure-adjusting screw counterclockwise until tension is fully released. The regulator should always be shut off when not in use to avoid gas loss if the cylinder or torch valves leak.

❑ Never stand directly in front of or behind regulators when opening a cylinder valve. Open the valve slowly; open fully only after the high-pressure gauge indicator stops moving. Open acetylene valves one-quarter to one-half turn. Open oxygen valves fully to seal the double-seated valve.

6.5.4 Oxyacetylene Hoses

Oxyacetylene hoses combine specially reinforced rubber and fabric layers to make them strong and flexible. Only hoses specifically made for welding and cutting should be used. Two lengths of hose, one for oxygen and one for acetylene, connect the regulators to the torch handle. To prevent confusion, the hoses are color-coded, green for oxygen and red for acetylene. To further guard against interchanging the hoses, the oxygen fittings (both male and female) have right-hand threads, while the acetylene fittings (male and female) have left-hand threads and a notch cut into the female connector. It is not possible to screw a left-hand nut onto a right-hand fitting.

Twin hoses, **Figure 6-12,** are designed for easy handling. They are furnished in either 12.5′, 25′, or 50′ lengths. The most popular is the 12.5′ length with an inside diameter of 3/16″.

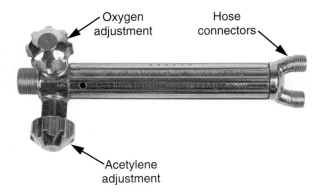

Figure 6-13. *The torch handle has separate adjustment knobs for acetylene and oxygen flow. To prevent mismatching of hoses, oxygen connectors have a right-hand thread; acetylene connectors have a left-hand thread. (Uniweld)*

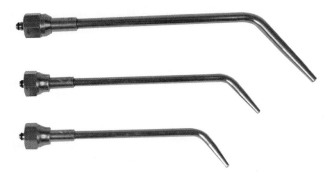

Figure 6-14. *Welding tips come in different sizes. The tip is hand-screwed onto the torch handle. (Uniweld)*

Figure 6-12. *Twin color-coded hoses (red for acetylene, green for oxygen) are convenient and less likely to tangle than individual hoses. (Uniweld)*

6.5.5 Torch Handle

Nuts on the oxyacetylene hoses screw onto matching threaded fittings on the base of the torch handle, **Figure 6-13.** These hose connections should be firmly tightened with a wrench. The torch handle also contains two finger-operated valves to adjust the flow of oxygen and of acetylene to the gas mixer in the welding tip or cutting attachment.

Oxyacetylene tips

Various welding tips are available to fit the torch handle, **Figure 6-14.** Tips are selected according to the type or size of flame desired. Tip sizes range from #000 (smallest) to #5 (largest). Sizes #0, #2, and #4 are commonly used; #4 is preferred. The tip is screwed onto the torch handle only by hand, using firm pressure. Do not use a wrench. The tip is equipped with special rubber O-rings to provide a leakproof seal when attached to the torch handle. Overtightening this connection could damage the O-rings.

Welding tips produce a small, concentrated, intensely hot flame. These tips are used to fuse steel or, by using a filler welding rod, to gas weld. Welding tips are frequently used for brazing copper to copper or copper to steel tubing, but care must be exercised to prevent overheating the metal.

Copper melts at 1981°F (1083°C), so it is very easy to melt the copper tubing. The flame must be moved, or "backed away," from the metal to prevent overheating and making a hole in the tubing. The hottest part of the flame is at the tip of the inner cone, as shown in **Figure 6-15.** The temperature of the metal can be controlled by rotating the flame or by pulling the inner cone away from the workpiece.

The proper tip size depends entirely upon the flame size desired. Many factors are involved in determining the proper tip size for a given application. Factors include the type of metal, thickness of the metal, operation (brazing or welding), and skill of operator.

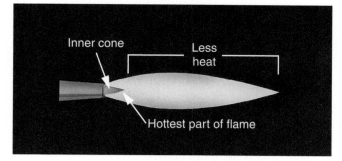

Figure 6-15. *The hottest part of a neutral oxyacetylene flame is just ahead of the inner cone. Heating of the metal can be controlled by moving the inner cone closer or farther away. (Lucas-Milhaupt)*

Trial and error, or some practice with the different tips, will reveal the tip size that produces the desired flame. In general, thick metals or large copper tubing require large tips; thin metals or small copper tubing require small tips. The very small tips are suitable for delicate work, and are used by jewelers.

Multiflame tip

Most technicians prefer to braze with a **multiflame tip, Figure 6-16,** which produces several small flames in one tip. The flames tend to wrap around the tubing. This feature is very useful because the flames heat the metal to the desired temperature quickly without overheating it.

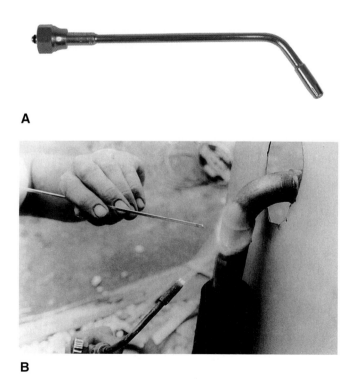

A

B

Figure 6-16. *Multiflame tip. A—A typical multiflame tip. (Uniweld) B—The multiflame tip helps complete brazing jobs quickly since the flame tends to wrap around the fitting for rapid, even heating.*

The flames produced by such a tip are not large or long but do provide intense heat to the workpiece. Such flames permit a skilled operator to rapidly braze on any tubing size. The multiflame tip is the only tip used for *all* types of brazing jobs. The small multiflame tip (#15) will successfully braze 1/8″ to 3″ (3 mm to 76 mm) copper tubing with relative ease. The same tip will braze larger tubing, but heating will take longer. The next tip size (#30) is recommended for best results in such cases.

Multiflame tips and single tips have some similarities and differences. The amount of gases required to operate the multiflame tip is slightly greater than the amount for a single tip, but the difference is not significant. The flow pressures required for the multiflame tip are the same as those for a single tip. The holes in the multiflame tip are very small. In fact, it requires several of these small flames to equal the diameter of the single tip flame. However, the multiflame tip can wrap around the tubing and distribute heat more evenly to the workpiece. This important feature speeds the brazing process and helps prevent hot spots.

6.5.6 Backfire and Flashback

Brazing tips are designed to operate at the pressure recommended by the manufacturer. Tip sizes from #0 to #5 operate at 5 psig (34 kPa) for both oxygen and acetylene, then adjust to conform to the desired type of flame. If the flame produced by this procedure is undesirable, the unit should be shut down and the tip size changed. *Do not change flow pressures at the regulator.* Improper operation of the torch tip may cause the flame to extinguish with a loud cracking sound. This is called a **backfire.** A backfire may be caused by touching the brazing tip against the workpiece, but the most common cause is flow pressures that are too low.

Flashback is a condition in which the flame burns back inside the tip; sometimes, it may extend back through the hoses and to the regulators. This condition is revealed by a shrill hissing or squealing sound. When flashback occurs, the unit should immediately be shut down and allowed to cool before relighting. Flashback is an indication that something is *very wrong.* This condition can be caused by a clogged orifice, but it is more commonly caused by incorrect oxygen and acetylene pressures.

Backfire and flashback are conditions usually brought about by the operator attempting to soften the flame by choking the flow pressures, either at the handle valves or the regulators. A skilled operator *can* produce a softer flame by adjusting the handle valves, but this procedure has its limits. Such an operator can produce a flame that is suitable for soft soldering, but a less-skilled person should resort to the *air-acetylene* torch for a softer flame.

6.6 The Oxyacetylene Flame ◆

There are three types of oxyacetylene flames, **Figure 6-17.** The different flames are obtained by controlling the amount of oxygen and acetylene supplied to the torch tip:

- ❏ A ***carburizing flame*** (also called a "reducing" flame) results from supplying excess acetylene.
- ❏ A ***neutral flame*** results from supplying equal amounts of oxygen and acetylene.
- ❏ An ***oxidizing flame*** results from supplying an excess amount of oxygen.

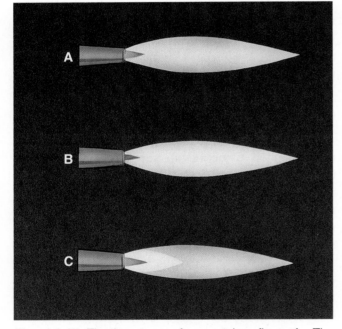

Figure 6-17. *The three types of oxyacetylene flame. A—The neutral flame has equal amounts of oxygen and acetylene. B—The oxidizing flame has excess oxygen. C—The carburizing flame has excess acetylene. (Lucas-Milhaupt)*

An oxidizing flame (excess oxygen) will burn the base metal and produce oxides that interfere with the flow of alloy. A carburizing flame (excess acetylene) results and deposits carbon on the base metals, interfering with the brazing process. The *neutral* flame, which heats the base metal and neither carburizes nor oxidizes, is the correct flame to use when brazing.

A neutral flame results from using an equal (1:1) mixture of oxygen and acetylene. In this union of two gases, the oxygen burns up the carbon and hydrogen in the acetylene so the flame releases only heat and harmless gases.

Flame Appearance

The *carburizing* flame has three distinct sections:
- ❏ A blue inner cone.
- ❏ A yellow-white "acetylene feather."
- ❏ An outer flame.

The *neutral* flame has two distinct sections:
- ❏ A slightly rounded, blue-white inner cone.
- ❏ An outer flame. (The acetylene feather has disappeared into the inner cone.)

The *oxidizing* flame has two distinct sections:
- ❏ A very sharp, pale-blue inner cone.
- ❏ A shorter, more ragged outer flame than shown by a neutral flame.

The difference between the neutral and oxidizing flames is very slight; not much difference is indicated in the inner cone. Care should be taken to remove *only* the acetylene feather when adjusting for a neutral flame. It is better to err on the acetylene side (a slightly carburizing flame) than on the oxygen side (an oxidizing flame).

Although the neutral flame is said to result from supplying equal amounts of the two gases, a *perfectly neutral flame* requires 2 1/2 parts oxygen to 1 part acetylene. The regulators are adjusted to provide the 1:1 ratio in the inner cone of the flame, and *atmospheric air* provides the other 1 1/2 parts oxygen needed to complete the combustion process.

6.7 Preparing to Use the Oxyacetylene Torch ◆

The procedure for getting started with the oxyacetylene torch consists of opening the cylinder valves and adjusting the pressure regulators for the tip size being used. The specific steps in this procedure are to:

1. Inspect the equipment for any damage. Make certain all valves and regulators are turned off.
2. Open the acetylene cylinder valve slowly, turning it counterclockwise with the proper wrench. Open the valve only one-quarter to one-half turn. Leave the wrench on the valve stem so the cylinder can be turned off quickly in case of an emergency.
3. Open the acetylene valve on the torch handle one full turn. Slowly rotate the adjusting screw on the acetylene regulator clockwise until the flow pressure gauge indicates 5 psi of pressure. Close the acetylene torch handle valve, using only finger-tight pressure.
4. Open the oxygen cylinder valve very slowly, turning it counterclockwise until the regulator's cylinder (high-pressure) gauge reaches its maximum setting. When opening the oxygen cylinder valve, always stand to one side of the regulator in case of a malfunction. The cylinder valve must be opened slowly to avoid damage to

the regulator in case of a sudden surge of about 2200 psi (15 180 kPa) of pressure. After the high-pressure gauge reaches its maximum reading, continue to turn the oxygen cylinder valve until it is fully open. This will backseat the valve and prevent leakage around the stem.

5. Open the oxygen valve on the torch handle one full turn. Next, turn the oxygen-regulator-adjusting screw clockwise until the flow pressure gauge indicates a pressure of 5 psi (34.5 kPa). Close the torch handle valve, using only finger-tight pressure (too much pressure on these sensitive valves can damage the ball seat).

6.7.1 Lighting the Torch

These are the proper procedures for lighting an oxyacetylene torch and adjusting to a neutral flame:

1. Crack (open slightly) the acetylene torch handle valve. Use a flint lighter (also called a spark lighter) to ignite the acetylene gas coming out of the torch tip. Hold the flint lighter near the end of the tip, but do not cover the end. A flint lighter is the only safe device to use when lighting a torch. Never use matches or a cigarette lighter. **CAUTION:** When lighting a torch, always be sure the torch tip is pointed away from any ignition source or object or person that could be harmed by the flame.

2. Slowly open the acetylene handle valve. When sufficient acetylene is flowing, the flame will stop releasing soot.

3. Slowly open the oxygen valve on the torch handle. As the oxygen is fed into the flame, the inner cone will develop, and the acetylene feather will appear. As the amount of oxygen is increased, the acetylene feather will draw back into the inner cone. When the acetylene feather disappears, the inner cone will lose its blurred edge and become round and smooth. At this point, you have a *neutral flame*. If more oxygen is added, a sharp pale blue inner cone will appear, indicating an *oxidizing flame*. A neutral or slight carburizing flame is preferred for brazing.

6.7.2 Shutting Down the Torch

If the torch is to be shut down for a short time, simply cut off the gas flow by closing the torch handle valves. Then, lay the torch aside (on a nonflammable surface) until it is needed again.

If the unit is not going to be used for a long time, the system should be shut down completely following these steps:

1. Close the hand valves on the torch handle in the proper order: Turn off the oxygen first to avoid a backfire into the mixer. Then, quickly shut off the acetylene valve to avoid forming soot. NOTE: If the acetylene is shut off first, the flame can burn back to the oxygen supply. The acetylene, however, cannot burn back without oxygen. This step will also prevent formation of soot inside the mixer and torch that might plug the passages.

2. Tightly close both cylinder valves.

3. Reopen each valve on the torch handle separately to bleed the gases from the hoses and regulators. Continue until regulator gauges read zero.

4. Close the adjusting screw on each regulator by turning it fully counterclockwise. This will help avoid gas loss if cylinder or torch valves leak.

5. Close both valves on the torch handle. Do not overtighten; they should be only finger-tight.

6. Rewind the hose, and place the unit in safe storage.

6.8 Making Brazed Connections ——◆

Like soldering (discussed in Chapter 5), brazing is a four-step process:

1. Thoroughly clean the base metal pieces.
2. Apply flux when appropriate.
3. Heat the base metal pieces evenly to the melting point of the alloy.
4. Apply alloy to the joint.

6.8.1 Cleaning Base Metal

Dirty metal surfaces will prevent the alloy from traveling into the joint and penetrating the metal. Brazed connections depend upon capillary action to draw the alloy into the joint. Dirt on metal surfaces prevents capillary action and keeps the alloy from flowing or adhering properly. It is much like applying paint to a dirty wall. The paint does not flow properly, and when the dirt falls off, so does the paint.

The cleaning technique is the same as for soldering: use sandpaper, emery cloth, a wire brush, or other abrasive to remove dirt and oxidation. The cleaned metal surface should shine.

6.8.2 Using Flux

In brazing, as in soldering, flux is a chemical compound applied to joint surfaces to prevent oxidation of the metal surfaces as they are heated. Flux is normally supplied in paste form and is applied with a small brush just before brazing. It melts and becomes active during the heating process, absorbing oxide and floating it away from the flowing alloy.

Flux is seldom used in refrigeration and air conditioning system brazing that joins *copper to copper*. The flux is eliminated by using an alloy that contains phosphorus, which acts as a fluxing agent. When brazing dissimilar metals, such as copper to brass or copper to steel, however, use of the phosphorus-bearing alloy will result in a brittle joint. For this reason, a different alloy is used, and a flux *is* required.

6.8.3 Brazing Procedure

Even heating of the base metal pieces is important so the alloy will flow equally well on both metal surfaces and completely fill the joint. Follow these steps for even heating during the brazing process:

1. Adjust the torch for a neutral flame. The flame size should be large enough to envelop as much of the connection as possible.
2. Begin heating the tubing about 1/2″ to 1″ (1.25 cm to 2.5 cm) away from the fitting.
3. After the tubing is heated, shift the flame to the fitting. Once the fitting is heated, move the flame steadily back and forth from the tubing to the fitting. Do not hold the flame in one spot — this causes localized overheating. Continue heating until the assembly reaches the alloy melting temperature.
4. When the assembly reaches the proper temperature, pull the flame back a little. Apply the filler alloy firmly against the tubing at the connection. If the assembly has been properly heated, the alloy will melt and completely penetrate and fill the joint. The alloy will always flow toward the hottest area.
5. After the joint has been completed, make one final pass of the flame around the connection to ensure proper flow and penetration of the alloy.

6.8.4 Types of Joints

There are three general types of brazing connections, identified by the direction the filler alloy flows: **downhill joints, uphill joints**, and **horizontal joints**. Recommended procedures for making each of these connections follow.

Downhill joint

In this type of joint (also known as a **vertical down joint**), the fitting is *below* the point where the alloy is applied, **Figure 6-18**. Bring the entire joint area to brazing temperature quickly and evenly. Heat the tubing first, then the fitting. When brazing temperature is reached, apply a little extra heat to the fitting as the alloy is melting. Since alloy always flows toward the heat, this will help it penetrate the fitting.

Uphill joint

In this type of joint (also known as a **vertical up joint**), the fitting is *above* the point where the alloy is applied, **Figure 6-19**. Start by heating the tubing, then

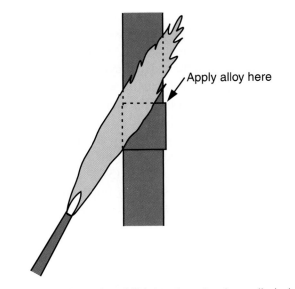

Figure 6-18. *In a downhill joint, the alloy is applied above the fitting and is drawn downward into the joint.*

transfer the heat to the fitting. Sweep the flame back and forth from the fitting to the tubing, all around the joint area. When brazing temperature is reached, keep the flame on the fitting while applying alloy to the connection. This heating pattern will draw the alloy upward into the joint. Do not overheat the tubing since this will cause alloy to run downward, rather than be drawn upward into the joint.

Horizontal joint

The tubing and the fitting are on the *same level* in this type of joint. Heat both the tubing and the fitting quickly and evenly. When brazing temperature is reached, apply the alloy to the top of joint, as shown in **Figure 6-20**. With proper heating, the combination of gravity and capillary action will draw the alloy into the

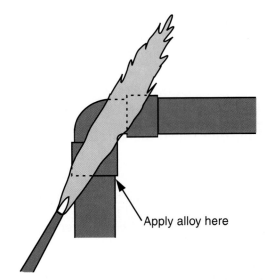

Figure 6-19. *Applying heat to the fitting after the entire joint area has been heated will help draw the alloy upward into the uphill joint.*

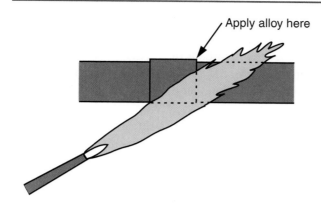
Apply alloy here

Figure 6-20. *On a horizontal joint, alloy is applied at the top so a combination of gravity and capillary action will draw the alloy into the joint.*

fitting and completely around the tubing. On small-diameter tubing, the flame can be directed at the bottom of the joint while alloy is applied at the top.

When heating assemblies for brazing, extra heat may be needed for sections that have heavier mass or greater thickness. The heavy section will heat more slowly. Also, different metals conduct heat at different rates. Copper, for example, is a good thermal conductor; it carries heat away more rapidly than steel. *Never heat base metals to the point where they begin to melt!*

6.8.5 Troubleshooting Brazed Connections

Brazing is a skill that requires practice to develop and maintain. Sometimes, the process fails to produce satisfactory results. Refer to the following to troubleshoot brazing problems.

PROBLEM: The alloy melts and forms a fillet but does not flow into the joint.

CAUSES:
❑ Tubing is hot, but the inside is not up to brazing temperature. (Review the heating procedure. Alloy always travels to the heat.)
❑ Flux was destroyed by excessive heat. (Use less heat or apply a heavier coating of flux.)

PROBLEM: Alloy balls up and does not flow into the joint.

CAUSES:
❑ Base metal is not up to brazing temperature; the alloy was melted by the flame.
❑ Joint was overheated and destroyed the flux.
❑ Base metal pieces were not cleaned.

PROBLEM: Alloy flows away from, rather than into, the joint.

CAUSES:
❑ Fitting is not heated to brazing temperature.
❑ Flame is directed away from the fitting.

Overcoming heat sink problems

The oxyacetylene process is especially useful when brazing tubing connections to a valve, compressor, or other heavy metal object. Typically during brazing, a heavy metal object will act as a "heat sink," absorbing applied heat and drawing it away from the brazing area. The heat loss prevents the connection from achieving brazing temperature quickly, as desired. Worse, the excessive heat may damage the object that acts as a heat sink.

The oxyacetylene torch makes it possible to overcome heat sink problems. The intense flame heats the brazing area rapidly, allowing the operation to be completed quickly, before the heat sink can absorb much heat.

When the brazing operation involves a heat sink that could be damaged, precautions must be taken. Common practice is to wrap the heat sink with damp rags, **Figure 6-21.** Commercial products are also available to apply to the heat sink before brazing. Both methods will adequately protect the heat sink from damage if brazing is not prolonged.

6.9 Oxyacetylene Cutting ———◆

A special cutting attachment, **Figure 6-22,** is used in place of the regular brazing tip when *cutting* with an oxyacetylene outfit. Like a brazing tip, the attachment is hand-tightened. No wrench is needed.

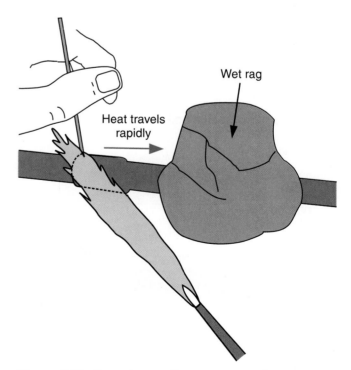
Wet rag
Heat travels rapidly

Figure 6-21. *If a valve or other component that acts as a heat sink might be damaged by the heat of a brazing operation, precautions must be taken. Wrapping the heat sink in wet rags is one method for keeping the temperature down.*

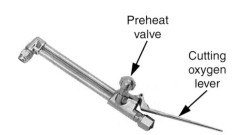

Figure 6-22. *The cutting attachment is threaded to fit on the end of the torch handle (in place of a brazing tip). The preheat valve allows adjustment of the oxygen for the preheat flames. The cutting oxygen lever, when squeezed, permits a strong flow of oxygen through the center hole of the cutting tip. This helps melt or burn away the metal being cut.*

The cutting attachment consists of an interchangeable tip, wheel valve (***preheat valve***), and lever-actuated valve (***cutting oxygen lever***). Cutting tips are available in a variety of types and sizes to serve special needs. The size #0 "general purpose" cutting tip will cut steel up to 1/2″ (12.7 mm) in thickness and is preferred by most technicians.

The tip is connected to the cutting attachment by a threaded nut; a wrench is required to make it gas-tight.

6.9.1 Cutting Torch Operation

The cutting tip, shown in **Figure 6-23,** has several small holes surrounding a single large hole in the center. The purpose of the small holes is to produce a series of small neutral flames (called the ***preheat flame***). The small flames heat the metal to be cut until it glows bright orange. This occurs at about 1625°F (885°C), the ignition point of steel.

Once the metal glows orange, the cutting lever is depressed slowly to start a stream of oxygen flowing through the center hole of the cutting tip. The stream of oxygen burns, melts, and blows away the slag and

molten metal from the line of the cut. As the torch moves, the preheat flame heats and cleans the metal while the steady stream of oxygen keeps the burning and melting process going. If the torch is not moved, the cutting action will stop.

6.9.2 Preparing the Cutting Torch for Use

Check the equipment to be certain all parts are in good operating condition and all valves and regulators are turned off. Put on welding goggles, gloves, and other required protective clothing. Before starting to cut, inspect the area to be certain it is free of combustible materials. Make sure there is no chance the sparks or slag produced by the cutting operation will start a fire. It is good practice to use a shield of some fireproof material to protect your legs and feet from the sparks and slag. It also prevents possible injury when the piece of metal you cut drops off.

Follow these steps to properly prepare the oxyacetylene torch:
1. Slowly open the acetylene cylinder valve one-quarter to one-half turn.
2. Open the acetylene valve on the torch mixing handle one-half turn. Adjust the acetylene regulator to permit 5 psig (34 kPa) of flow pressure. Close the torch handle valve finger-tight.
3. Slowly open the oxygen cylinder valve all the way to backseat the valve stem and prevent leakage.
4. Fully open the torch handle oxygen valve. This will transfer control of the oxygen to the preheat valve and the cutting lever.
5. Open the preheat valve. Adjust the oxygen pressure regulator 20 psig to 35 psig (172 kPa to 241 kPa) for tip sizes #00 or #0 and approximately 25′ (7.6m) of hose. The table in **Figure 6-24** relates oxygen pressures to cutting tip sizes.
6. Close the preheat valve. The cutting torch is now ready to light.

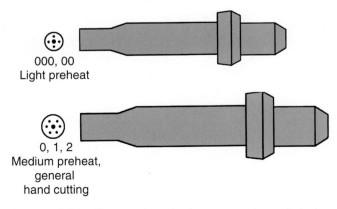

000, 00
Light preheat

0, 1, 2
Medium preheat,
general
hand cutting

Figure 6-23. *The cutting tip has several small holes surrounding a large central opening. The small holes are for the preheat flames; the center is for the cutting oxygen jet. Generally, the number of holes for preheat flames increases as the tip size increases.*

Cutting Tip Pressure Settings

Metal Thickness	Tip Size	Oxygen (psig)	Acetylene (psig)
1/8″	000	20–25	5
1/4″	00	20–25	5
3/8″	0	25–30	5
1/2″	0	30–35	5
3/4″	1	30–40	5
1″	2	35–50	6

Figure 6-24. *The larger the cutting tip size, the higher the oxygen pressure used, as shown in this table. Also note the relationship of tip size to thickness of the metal being cut.*

Lighting the cutting torch

Follow these steps to safely light the oxyacetylene cutting torch:

1. Open the torch handle acetylene valve one-half turn. Quickly light the flame at the cutting tip with a flint lighter (not matches). Open the valve until a gap appears between the flame and the tip end. Slowly close the valve until the gap is eliminated. This sets the correct acetylene gas flow for the tip size, one that will maintain a stable preheat flame.

2. Slowly open the preheat oxygen valve until the acetylene feathers disappear into the sharp blue-white inner cones of the small preheat flames. This establishes a correct neutral flame.

3. Squeeze the cutting oxygen lever while readjusting the preheat flames to remove the feathers and restore the correct flame cones. See **Figure 6-25.**

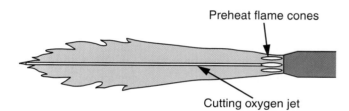

Preheat flame cones

Cutting oxygen jet

Figure 6-25. *Readjust preheat flame cones while releasing a jet of oxygen by squeezing the cutting oxygen lever. (Smith Equipment, Division of Tescom Corp.)*

Using the cutting tip

Hold the torch in your right hand so you have instant and positive control of the cutting oxygen lever. To steady the torch, close your left hand into a fist, and rest the torch handle on it.

To start the cut, hold the cutting tip straight down, facing the surface of the metal. Position the preheat flame about 1/8″ (3 mm) above the metal's surface. Starting at the edge of the metal to be cut, hold the torch steady until this spot has been heated to a bright orange. Slowly press down on the cutting oxygen lever. When cutting starts, there will be a shower of sparks from the metal. The cutting oxygen lever should be pressed down all the way. Begin moving the torch slowly and steadily along the line of cut. The motion of the torch should be slow enough for the cut to penetrate completely through the metal without excessive oxidation or melting.

If the torch is moved too slowly, the preheat flame will tend to melt the edges of the cut, producing a ragged appearance. It is also possible the metal could fuse again. If the torch is moved too fast, however, the cutting action will not penetrate all the way through

the plate, and the cutting will stop. If cutting action should stop, release the cutting oxygen lever; then begin again at the point where the cut stopped.

To avoid flame backfiring (loud popping) or flashback (squealing) inside the attachment, it is important to maintain the correct gas flow to the cutting tip. Backfiring and flashback can also be caused by a plugged, dirty, damaged, or loose cutting tip. If backfiring or flashback occur, shut down the unit, and check the tip for plugged holes or seat damage. Clear and clean holes with the proper size tip cleaner, **Figure 6-26.** *Never* use a cutting tip with plugged holes or a damaged seat.

Do not try to reduce the heat for cutting thin sections by "choking down" or "starving" the preheat flames. Instead, use full, correct preheat flames with the tip tilted so the flames hit the surface at an angle. Move the tip fast enough to prevent overheating and excess melting. Be sure to use the correct small size tip for thin metal to avoid overheating and melting. Refer to **Figure 6-24** for the relationship between metal thickness and tip size.

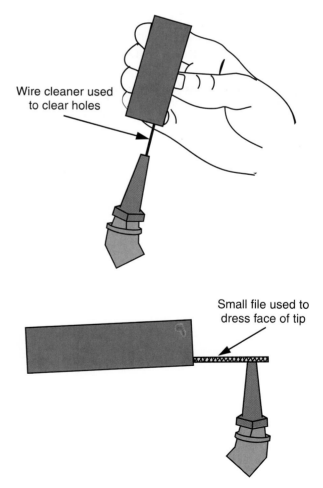

Wire cleaner used to clear holes

Small file used to dress face of tip

Figure 6-26. *A tip cleaner should be used to clear and clean holes in the cutting tip. Several sizes of tip cleaner may be needed.*

Shutting down the cutting torch

When you finish cutting, shut down the torch by following the steps described earlier in this chapter. Remember to shut off the oxygen valve first to avoid backfiring and flashbacks. Also, be sure to bleed all pressure from hoses and regulators by opening the torch handle valves after the cylinder valves have been closed.

Summary

Every HVAC technician must be able to make good brazed connections, a skill routinely used in installation and service procedures. The information presented in this chapter will lead to the development of good brazing skills and safe use of oxyacetylene torches.

Various methods and procedures for brazing connections were explained in this chapter. The different alloys and fluxes were also described. Proper use and operation of the oxyacetylene torch, the recommended torch for performing brazing operations, was emphasized.

This chapter explained how to select the proper tips and how to operate valves and regulators to achieve the desired flame. Types of flames and proper methods of applying heat to a connection were covered. Proper use of the cutting torch was also described.

Test Your Knowledge

Please do not write in this text. Write your answers on a separate sheet of paper.

1. Brazing is the process of joining metals at temperatures above _____ °F (448°C).
2. Copper will melt at _____°F (_____°C).
3. Give at least three reasons brazing is the preferred procedure for joining tubing on refrigeration systems.
4. Brazing should be performed with a(n) _____ torch, while soldering is done with a(n) _____ .
5. What is the AWS number for the brazing alloy containing 93% copper and 7% phosphorus?
6. The copper-phosphorus brazing alloy is good for connecting copper to _____ without need for a flux.
7. The silver-bearing brazing alloys can be used to join copper to steel but require a _____.
8. The oxyacetylene outfit with small cylinders that is preferred by many service technicians is called an _____ type.
9. Oxygen cylinders are charged to a pressure of 2200 psi (_____ kPa).
10. When using oxygen, why should the cylinder valve be fully opened?
11. When brazing with an oxyacetylene torch, what type of flame should be used?
    a. Carburizing.
    b. Oxidizing.
    c. Neutral.
    d. Reducing.
12. What are the flow pressure regulator settings for oxygen and acetylene when using an oxyacetylene torch for brazing?
13. The wheel valve on the cutting attachment is called the _____ valve.
14. When oxyacetylene cutting, the preheat flame is used to heat the metal to about _____°F (885°C).
15. When using the cutting torch, the base metal should be preheated to a(n) _____ color before pressing the oxygen cutting lever.
16. What happens to the preheated metal when the cutting oxygen lever on the torch is squeezed?
17. What is the flow pressure regulator setting normally used for oxygen with the oxyacetylene cutting torch?

Mathematics for Technicians

Objectives

After studying this chapter, you will be able to:
- ❏ Read the place values of numbers.
- ❏ Work with fractions and decimals.
- ❏ Make linear, two-, and three-dimensional measurements.
- ❏ Convert Fahrenheit temperature readings to Celsius and vice versa.
- ❏ Use percentages to calculate profit margin on parts.

Important Terms

angle	mixed numbers
area of a circle	mixed decimal
Celsius	numbers
circumference	number
cubic measure	numerator
decimal numbers	percent
decimal point	pitch
degree	place value
denominator	profit margin
diameter	proper fractions
digits	protractor
dimensions	quotient
equivalent	radius
exponent	remainder
factor	rough measurements
Fahrenheit	square measure
formula	square units
fractions	surface area
height	units
improper fractions	vertex
length	volume
linear measure	whole number
lowest terms	

7.1 Using Math

Math skills are essential in the refrigeration, heating, and air conditioning field. Making accurate measurements, properly sizing parts, installing equipment, and completing work orders and time sheets are daily duties requiring computational skills. Apprentices and advanced technicians alike depend upon their math abilities for successful job performance and advancement.

Can you calculate a time sheet to the quarter-hour? How many gallons of paint are required to cover 240 sq. ft.? How many bags of ready mix concrete are needed to pour a concrete pad $2' \times 4' \times 6''$? What is the difference between a 45° and a 90° elbow? These are just a few of the mathematical operations you might be required to perform on the job. This chapter will help you refresh your math skills in preparation for a position as a refrigeration, heating, and air conditioning technician.

7.2 Reading Place Values

A **number** is a figure or word that denotes quantity. The numbers 0 to 9 are called **digits**. They are used alone, or in combination, to indicate "how many." When digits are used in combination with each other, each one is given a special position called a **place value**. Although the digits change, the positions never vary. Working with numbers is much easier when you understand place value.

Place value is shown in **Figure 7-1**. The number shown reads, *seven million, six hundred fifty-four thousand, three hundred twenty-one*. The smallest number, positioned farthest to the right, is a single digit from 0 to 9 (in this case, the digit is 1). This position is called **units**. Other numbers are added to the left of the units position. The next position has a

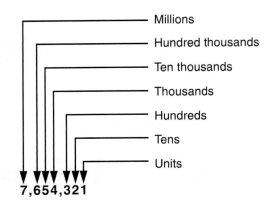

Figure 7-1. *Place value is important in reading whole numbers. The positions of the digits, from right to left, determine their place values.*

value of *ten*, and the next to the left of that has a value of *one hundred*. Thus, the number 321 is made up of 3 hundreds (300), 2 tens (20), and one unit (1), or 300 + 20 + 1 = 321. To make reading and writing large numbers easier, groups of three digits are separated by commas, starting from the units place.

7.3 Whole Numbers

Numbers can take different forms, including the forms of whole numbers, proper fractions, improper fractions, mixed numbers and decimals, all which will be discussed in this chapter. A **whole number** is any positive number that contains no fractional parts; it is a complete unit. The numbers 0, 1, 2, 3, 4, etc. are whole numbers.

7.4 Fractions

Fractions are a means of indicating parts of a whole number. For instance, a 1/4-inch mark on a ruler or tape measure designates a fraction (or part) of a *whole* inch. A payroll time sheet may be rounded off to the nearest 15 minutes or 1/4 of a *whole* hour.

Fractions are either written with one number above another number, or with the two numbers separated by a diagonal line, as shown in **Figure 7-2**. The top number on a fraction (called the **numerator**) indicates the number of parts available. The bottom number (called the **denominator**) indicates the number of parts required to make a whole.

A fraction is read by stating the numerator first, then the denominator. The fraction 5/16 is read, *five-sixteenths.* The numerator (**5**) indicates there are 5 parts available; the denominator (**16**) indicates 16 parts are required to make one whole.

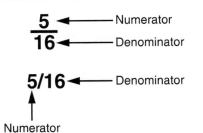

Figure 7-2. *Two ways to write a fraction: one number above another, or side-by-side separated by a diagonal line. The number on top or to the left is called the numerator; the one on the bottom or to the right is the denominator.*

7.4.1 Proper Fractions

When the numerator of a fraction is *smaller* than the bottom number, the fraction is termed "proper." For example, the fractions 1/4, 3/8, 1/2, and 7/16 are **proper fractions**.

7.4.2 Improper Fractions

In an **improper fraction**, the numerator is the same or larger than the denominator. Examples are 3/3, 9/8, 7/6, 19/16, 5/4, 8/8. These fractions are called "improper" because the top number is *equal to* or *greater* than the amount required to make a whole. When the numerator and denominator are the same, the fraction is **equivalent** (equal) to the whole number **1**. Thus, 3/3, 4/4, 5/5, 8/8, and 16/16 are all equivalent to 1. For ease of understanding, improper fractions should be converted to whole or mixed numbers.

7.4.3 Mixed Numbers

In a **mixed number**, a whole number and a fraction are used together. Examples are 3 1/2, 6 5/8, 4 2/3, 8 9/16. Mixed numbers are read by inserting the word "and" between the whole number and the proper fraction. Thus, 6 5/8 is read, *six and five-eighths.*

To convert an improper fraction (such as 7/4) to a mixed number, *divide* the numerator by the denominator. Dividing the numerator (**7**) by the denominator (**4**) yields a **quotient** (result) of 1 and a **remainder** (the number left over) of **3**. The quotient becomes the whole number, and the remainder becomes the numerator. It is placed above the original denominator. Thus, the improper fraction 7/4 can be converted to the mixed number 1 3/4.

7.4.4 Reducing Fractions

Reducing a fraction to its **lowest terms** means changing the numerator and the denominator to the smallest possible numbers without changing the *value* of the fraction. It is done by dividing the numerator and denominator by a whole number that divides evenly into both. For example, 5/10 is equal to 1/2 because both numbers can be evenly divided by **5**. The fraction 1/2 cannot be reduced any further, it is in its *lowest terms*. Reducing fractions to their lowest terms makes them easier to understand and use.

In the same way, 25/100 can be reduced to 5/20 by dividing both numbers by **5**. This fraction could be further reduced to 1/4 by again dividing both numbers by **5**. The fraction 1/4 is the fraction 25/100 reduced to lowest terms.

Figure 7-3 illustrates how fractions can be reduced. A whole pizza is cut into six pieces. This can be expressed in fractional form as 6/6. If one piece (1/6) were removed, 5/6 would remain. If two pieces (2/6) were removed, you could say **1/3** of the pizza is missing. The fraction 2/6 can be reduced to its lowest terms, **1/3**, if you divide the numerator and denominator by 2. If 1/3 of the pizza is missing, then 2/3 (4/6 reduced to lowest terms) remain. Even though we are now measuring the pizza in different units (thirds, instead of sixths) the units are equivalent: either 3/3 or 6/6 equal a whole pizza.

If three of the original six pieces of pizza are removed, 3/6 of the pizza is gone. This fraction can be reduced to lowest terms, **1/2**, by dividing the numerator and denominator by 3. Again, the units are equivalent: 2/2, 3/3, or 6/6 equal a whole pizza.

7.5 Decimal Numbers ◆

Decimal numbers are easier to work with than fractions and are simply another way of expressing fractions, or parts, of numbers. The number of parts is always located to the right of the decimal point. Our money system makes use of decimals. For example, five dollars and fifty cents is written $5.50. The whole number **5** occupies the unit position, while **50** represents 50 parts of a whole dollar. Since a dollar can be broken down to 100 parts (cents or pennies), the decimal number 0.50 stands for *half* that amount.

7.5.1 Reading Decimal Numbers

Actually, 0.50 should be read, *fifty one hundredths* because of place value. Positions of digits appearing to the right of a decimal point are given specific values, just as they are to the left for whole numbers. The place value of the digit farthest to the right indicates how many parts are required to make a whole.

The number shown in **Figure 7-4** indicates a very small quantity because the digit farthest to the right (**4**) is located in the *ten thousandths* position. This decimal number reads, *one thousand two hundred thirty-four ten thousandths*. Why? Because when reading decimal numbers, the position of the digit farthest to the right indicates how many parts are needed to equal one whole (in this case, 10,000). The actual digits indicate how many parts are available (1234).

Three digits following a decimal indicate that 1000 parts are required to equal one whole. For example, 0.187 is one hundred eighty-seven *one thousandths*. In the same way, 0.75 is seventy-five *one hundredths*, and 0.2 is two *tenths*. Zeros are used as place holders for unoccupied positions: twenty-five *ten thousandths* = 0.0025.

7.5.2 Mixed Decimal Numbers

Decimal numbers with digits before *and* after the decimal point, such as 16.75, are referred to as ***mixed decimal numbers***. The digits to the left of the decimal point give the number of wholes; the digits to

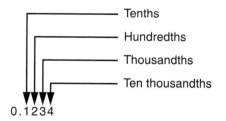
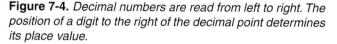

Figure 7-4. *Decimal numbers are read from left to right. The position of a digit to the right of the decimal point determines its place value.*

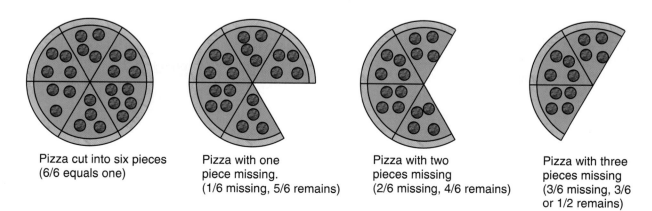

Pizza cut into six pieces (6/6 equals one)

Pizza with one piece missing. (1/6 missing, 5/6 remains)

Pizza with two pieces missing (2/6 missing, 4/6 remains)

Pizza with three pieces missing (3/6 missing, 3/6 or 1/2 remains)

Figure 7-3. *Slices of a pizza provide examples of reducing fractions to lowest terms.*

the right of the decimal point give the fractional parts. The mixed decimal number 16.75 is read, *sixteen and seventy-five one hundredths*. The word "and" is used to indicate the **decimal point**. In industry, the word "point" is often used instead. For example, the number 3.485 is read, *three point four eight five*.

7.5.3 Expressing Fractions of Time in Decimals

A practical application of both fractions and decimal numbers is a payroll time sheet for keeping track of hours or fractions of hours worked.

For simplicity, many companies "round off" time to the nearest quarter hour. For example, 10 minutes would be rounded to 1/4 hour (15 minutes). Twenty minutes also would be rounded to 1/4 hour. The fractions 1/4, 1/2, and 3/4 can also be expressed in decimal terms, as 0.25, 0.50, and 0.75. Converting fractions to decimals makes it easier to calculate a worker's pay simply because decimals are easier to multiply than fractions. If a technician's pay rate is $10 per hour, and the technician works 25 hours and 45 minutes, it is much easier to multiply $10 by 25.75 than by 25 45/60 (or, in simplest terms, 25 3/4).

If a person works more than 40 hours per week (or 8 hours per day in some states), he or she usually qualifies for overtime pay for the extra hours. Overtime pay is usually 1.5 (one and a half) times the regular hourly rate. Pay for work on Sundays or holidays may be double time, or 2 times the regular hourly rate. For example, if a technician's regular pay is $10 per hour, the overtime rate would be $15 per hour, and the double-time rate would be $20 per hour.

The following example shows time worked by a technician earning $10 per hour:

Monday:	9.5 hours
Tuesday:	10.25 hours
Wednesday:	8.0 hours
Thursday:	11.75 hours
Friday:	11.25 hours
Saturday:	9.0 hours
Total hours worked during week:	59.75

40 regular hours @ $10 = $400
19.75 overtime hours @ $15 = $296.25
Total pay amount for week = $696.25

7.6 Measuring Dimensions ———◆

The application of fractions and decimal numbers to heating and cooling work becomes apparent when the technician must measure dimensions. **Dimensions** are measurements of length, width, and depth. (Sometimes, the terms "height" or "thickness" are

used in place of "depth.") New employees are often assigned to installation projects where their skills in making accurate measurements are quickly tested.

Measurements are made in either the US Conventional or SI Metric systems. The US Conventional system uses basic units of inch, foot, and yard dimensions; SI Metric uses the millimeter, centimeter, and meter. Most nations of the world, other than the United States, use the SI Metric system. The units of the two systems are compared below.

1 inch = 2.54 centimeters
(25.4 millimeters or 0.0254 meter)

1 foot = 30.48 centimeters
(304.8 millimeters or 0.3048 meter)

1 yard = 91.44 centimeters
(914.4 millimeters or 0.9144 meter)

7.6.1 Linear (Length) Measure

A measurement in one dimension is **linear measure**, or the distance from one point to another. **Length**, width, and depth are all linear measures. So are the radius, diameter, and circumference of a circle, and measurements around the perimeter of any other flat shape. See **Figure 7-5**. A typical use of the length dimension would be measuring a piece of electrical conduit to fit a specific installation. Tools used to measure length include rulers, steel tape measures, yardsticks or metersticks, chains (for land measurement in *rods* or *meters*), and odometers (for measurements in miles or kilometers).

In refrigeration, heating, and air conditioning work, length is normally measured in compound units of feet, inches, and fractions of an inch (for example, 6 feet, 2 1/4 inches) or in meters, centimeters, or millimeters. In SI Metric, compound units are not used. Instead, dimensions are given in decimal fractions (for example, 2.3 meters).

1 foot	= 12 inches
	= 0.3048 meter
1 yard	= 3 feet
	= 36 inches
	= 0.9144 meter
1 rod	= 16 1/2 feet
	= 5 1/2 yards
	= 5.029 meters
1 statute mile	= 5280 feet
	= 1760 yards
	= 320 rods
	= 1.6093 kilometers

Figure 7-5. *Common linear measuring units in the US Conventional system and their equivalents in the SI Metric system.*

Abbreviations for the units of length are commonly used: "ft." for feet, "in." for inches, "m" for meter, "cm" for centimeter, and "mm" for millimeter. On architectural drawings using US Conventional measurements, symbols may be used: (′) indicates feet, and (″) indicates inches. For example, a drawing might show a length measurement as:

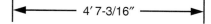

This measurement in compound units reads, *four feet, seven and three-sixteenths inches*. The inches are mixed numbers; therefore, the word "and" is used to separate whole inches from fractional inches.

The equivalent measurement in the SI Metric system would be shown as:

This measurement is read, *one hundred forty point one eight centimeters*.

Using measuring tools

Folding rules and measuring tapes are frequently used to obtain accurate measurements. Such measuring devices are illustrated in **Figure 7-6**. The ability to read measuring tools accurately is vital to successful job performance. Incorrect measurements waste time and materials, resulting in poor craftsmanship and cost overruns.

Folding rules are usually 6′ (approximately 2 m) long. Tape measures are available in a variety of lengths, ranging from 6′ to 100′. Many have both

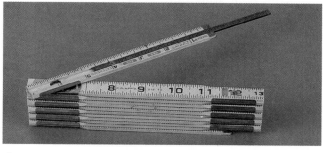

A

B

Figure 7-6. *Common measuring devices. A—Carpenter's folding wood rule. B—Steel measuring tape.*

conventional and metric divisions. The blades of these tape measures are usually steel, **Figure 7-7**. In the 50′ and 100′ (15 m and 30 m) lengths, nonmetallic blades reinforced with fiberglass are sometimes offered. Also available are tapes equipped with a battery and small lightbulb for reading measurements in dimly lighted areas, such as attics and basements.

Figure 7-7. *Measuring tapes, such as these 50′-length models, usually have steel blades for durability. Some long tapes, however, have nonmetallic blades. (CooperTools)*

Reading a US Conventional ruler

Each foot on a standard ruler or tape is calibrated to contain exactly 12 inches, numbered from 1 to 12, as shown in **Figure 7-8**. Measurements in inches can sometimes be expressed as fractions of a foot. A measurement of six inches, for example, may be read as *1/2 foot*. Since one foot contains 12 inches, or 12/12, six inches equal 6/12. As described earlier, 6/12 can be reduced to its lowest terms, 1/2. In the same way, 4 inches equal 4/12, which reduces to 1/4 foot, and 9 inches equal 9/12, which reduces to 3/4 foot.

Sixteenths. Generally, an inch is divided into sixteen equal parts, called *sixteenths*. Therefore, one inch equals 16/16, as shown in **Figure 7-9**. Because sixteenths are tiny segments, very short lines are used on the ruler to indicate these divisions.

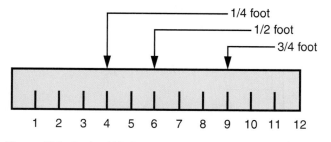

Figure 7-8. *In the US Conventional system, each foot of a ruler or tape is divided into 12 equal inches. Measurements can also be expressed as fractions of a foot, as shown.*

Eighths. Two tiny sixteenth divisions equal 2/16, which reduced to lowest terms is 1/8. As shown in **Figure 7-10**, the mark for *eighths* is slightly longer than the one for sixteenths.

Fourths. Two eighths (or four sixteenths) equal one-fourth of an inch. The longer mark used to divide the inch into *fourths* is shown in **Figure 7-11**.

Halves. The *half-inch* mark is longer than the marks for the fourths, eighths, or sixteenths divisions, as shown in **Figure 7-12**.

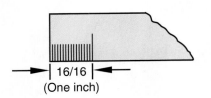

16/16
(One inch)

Figure 7-9. *Sixteenths are the smallest division of an inch shown on most rulers or measuring tapes.*

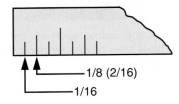

1/8 (2/16)
1/16

Figure 7-10. *One-eighth inch is equivalent to two-sixteenths. Eighth marks on measuring devices are slightly longer or darker than sixteenth marks.*

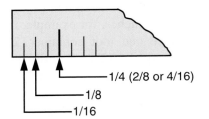

1/4 (2/8 or 4/16)
1/8
1/16

Figure 7-11. *One-fourth inch is equivalent to two-eighths or four-sixteenths. The fourth marks are longer and darker than eighth or sixteenth marks.*

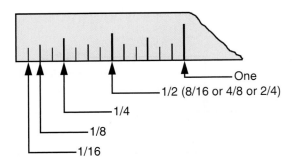

One
1/2 (8/16 or 4/8 or 2/4)
1/4
1/8
1/16

Figure 7-12. *Half-inch marks are longer and darker than fourth-inch marks, but shorter and usually less dark than marks indicating each full inch. A number is usually placed at each inch mark, as well.*

Reading an SI Metric ruler

Metric rulers usually show only two divisions, centimeters and millimeters. Measuring tapes normally indicate meter divisions, as well.

The SI Metric system is based on units of 10. The smallest unit on the ruler is the *millimeter*. As shown in **Figure 7-13**, ten millimeters make up one *centimeter*, which is set off with a longer line. (Approximately 2 1/2 centimeters equal one inch.) Ten centimeters, in turn, make up a *decimeter* (this unit is not widely used and is seldom even marked on rulers or tapes). Ten decimeters make up one *meter*. One hundred centimeters ("centi" means hundred) or 1000 millimeters ("milli" means thousand) are also equal to one meter.

Rough measurements

Rough measurements are distance estimates that are simpler, but less accurate, than measurements made with a ruler or tape. Rough measurements are often satisfactory for quickly estimating job requirements but should be as accurate as possible. With proper preparation, you can use various body parts as standards for obtaining rough measurements.

Hand measurements. Measure from the base of the palm to the end of each finger. Unless you have unusually small hands, one of these measurements should be very close to 6″ (152.4 mm). Also, measure across the palm and remember the distance. Hand measurements are useful for short lengths of copper tubing.

Body height and arms. Measure and remember your height (floor to top of head) and your height with an arm and fingers extended upward. Also, measure and remember the distance from the floor to your belt line. With both arms fully extended sideways, have someone measure the distance from fingertips to fingertips. Also, with an arm extended forward, measure from your fingertips to the middle of your chest. These measurements can prove very useful.

Stepping off distances. You can "step off" yards or meters quite accurately with some preparation and practice. To perfect this time-saving method of rough measurement, lay a yardstick or meterstick on the floor, and line up the heel of your left shoe with the end of the stick nearest you. Next, place your right shoe in front of the left shoe, touching heel-to-toe. Finally, move the left shoe in front of the right shoe. Again, touch heel-to-toe. Note the measurement on the stick next to the toe of your left shoe; it should be near the end of the yardstick or meterstick. For many adults, three shoe-lengths will be approximately equal to a yard or meter. The remaining distance between the toe of your left shoe and the end of the stick will be a guide for the amount of space to allow between

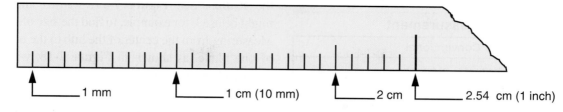

Figure 7-13. *On a meter stick or metric ruler, each centimeter is divided into 10 millimeters. Note that approximately 2.5 centimeters (more precisely, 2.54 cm) is equal to 1 inch.*

shoe-lengths when "stepping off" a measurement. Try stepping off a room or hallway, and then check the measurement with a tape. Practice until some degree of accuracy is obtained. Accurate stepping off can save much time and trouble.

7.6.2 Square (Area) Measure

A measurement of two dimensions, length and width, is known as ***square measure***. A two-dimensional measurement gives **surface area**. To find the area of a rectangular surface, multiply the first dimension (length) by the second dimension (width). See **Figure 7-14**. To find the area of a triangle, multiply the length by the width, then divide by 2. (Finding the area of a circle is covered later in this chapter.)

Surface area measurements aid in placing equipment and determining the amount of paint or tile needed to cover floors, walls, and ceilings.

The surface area of any shape is expressed in ***square units:*** square inches or square feet, square centimeters or square meters, and so on. The units

used for land area, acres and hectares, are inherently square measures, so there is no need to use the word "square" when referring to them.

Reading exponents or power numbers

An **exponent**, or *power*, is an abbreviation used to indicate a multiplication process. The exponent is a numeral placed above and to the right of a *base* number. For example, in the number 10^2, the numeral 2 is an exponent or power. It tells how many times the base number *10* (also called a **factor**) is used for multiplying. Thus, 10^2 actually means 10×10. The factor (10) was used twice. In math terms, a number that is multiplied by itself (10×10) is said to be "squared." Thus, 10^2 can be read, *10 squared or 10 to the second power.*

Exponents can express any number and are particularly useful for expressing large whole numbers. See **Figure 7-15**. For example, 20^5 means the factor 20 must be used five times, or $20 \times 20 \times 20 \times 20 \times 20$. The result is $20^5 = 3,200,000$.

Converting to a common unit

A mixture of feet and inches cannot be used when multiplying measurements to find surface area. Neither can a mixture of millimeters and centimeters, or centimeters and meters. All measurements must be converted to the same unit (a *common unit*) before multiplying. See **Figure 7-16**.

For example, to find the area of a floor measuring $10' \ 6'' \times 12' \ 4''$, the dimensions must be changed to a common unit, inches. Since there are 12 inches in a foot, multiply the number of feet by 12, then add the inches:

$$10' \ 6'' = (10' \times 12'' = 120'' + 6'' = 126'')$$
$$12' \ 4'' = (12' \times 12'' = 144'' + 4'' = 148'')$$

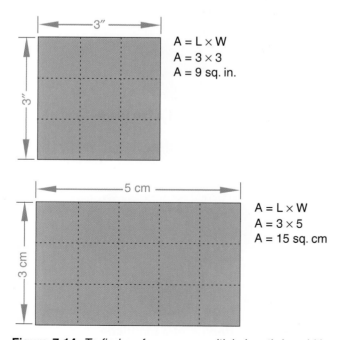

$$A = L \times W$$
$$A = 3 \times 3$$
$$A = 9 \text{ sq. in.}$$

$$A = L \times W$$
$$A = 3 \times 5$$
$$A = 15 \text{ sq. cm}$$

Figure 7-14. *To find surface area, multiply length by width. The result is expressed in square units. In this figure, one area is in square inches, the other in square centimeters.*

$20^2 = 400 \ (20 \quad 20 = 400)$

$20^3 = 8,000 \ (20 \quad 20 \quad 20 = 8,000)$

$20^4 = 160,000 \ (20 \quad 20 \quad 20 \quad 20 = 160,000)$

$20^5 = 3,200,000 \ (20 \quad 20 \quad 20 \quad 20 \quad 20 = 3,200,000)$

Figure 7-15. *An example of how exponents are used to express large numbers.*

Area Measurement

US Conventional
1 square foot = 144 square inches
1 square yard = 9 square feet
SI Metric
1 square centimeter = 100 square millimeters
1 square meter = 10,000 square centimeters

Figure 7-16. *Common measuring units for area in both US Conventional and SI Metric systems.*

Multiplying 126″ × 148″ yields a surface area of 18,648 sq. in. Since large areas are usually given in square feet, the square inches should be converted to square feet. This is done by dividing by 144. Each square foot contains 12″ by 12″, or 144 sq. in. Dividing 18,648 by 144 results in a surface area of 129.5 sq. ft.

The same problem can be solved by converting all measurements to feet and the decimal equivalents of fractions of feet: 10′ 6″ equal 10.5′, and 12′ 4″ equal 12.3333′. Thus:

10.5′ × 12.3333′ = 129.4996 sq. ft. (rounded to 129.5 for convenience)

Converting *metric* dimensions to a common unit is simpler since all metric units are multiples of 10. For example, imagine you must find the area of a countertop that is 85 cm wide and 2 m long. First, put the measurements into a common unit; either convert the 2 m dimension to centimeters, or the 85 cm dimension to a decimal fraction of a meter.

First, we will use centimeters as the common unit. Since there are 100 centimeters in each meter, multiply 2 times 100. The common unit (cm) lets you simply multiply length by width: 85 × 200 = 17,000 square centimeters (usually written as 17 000 cm²). To convert this figure to square *meters*, divide by 10,000 (100 cm × 100 cm = 10 000 cm² or 1 square meter). If you divide 17,000 cm² by 10 000 cm², the result is 1.7 m². (Note that commas are not used to separate groups of three digits in the SI Metric system. Instead, a space is used.)

You can achieve the same result, in just one step, by converting 85 cm to its equivalent decimal fraction, 0.85 m, and multiplying 0.85 × 2 = 1.7 m².

7.6.3 Area of a Circle

Measuring and cutting circles are tasks that occur fairly often in heating and cooling work. A circle is measured by a straight line across its widest part. This line is called the ***diameter***. A line from the center of the circle to its perimeter (half the diameter) is called

the ***radius***. See **Figure 7-17**. Radius and diameter might be used, for example, to find the size of a fan blade. Measuring from the center of the hub to the outside edge of one blade, as shown in **Figure 7-18**, will give the radius. Then, just multiply by 2 to find the diameter.

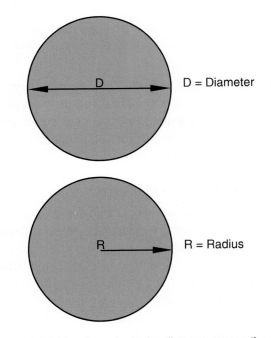

D = Diameter

R = Radius

Figure 7-17. *The diameter is the distance across the widest part of a circle. The radius is half the diameter.*

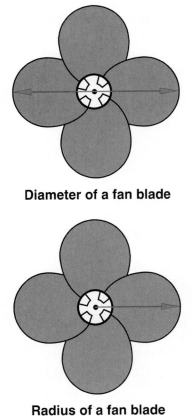

Diameter of a fan blade

Radius of a fan blade

Figure 7-18. *Finding the diameter and the radius of a fan blade.*

Finding the **area of a circle** is sometimes necessary. The most common method or **formula** for finding the area of a circle is:

$$A = \pi r^2, \text{ where:}$$
$$A = \text{area}$$
$$\pi \text{ (pi)} = 3.1416$$
$$r = \text{radius}$$

Reading formulas

Many formulas, like the one for finding the area of a circle, do not show a multiplication sign (×); instead, the × is understood. However, functions other than multiplication are indicated by symbols. The formula $A = \pi r^2$ means:

Area of a circle = 3.1416

(the value of π, pronounced "pi") × radius × radius.

Special rules govern the order (first, second, and so on) in which operations within mathematical formulas are performed:

1. Sections enclosed within parentheses ( ) are performed first. If there are parentheses within parentheses, the operations are performed from the innermost set outward.
2. Base numbers must be multiplied to the specified power.
3. Multiplication and division are performed in the order given, reading left to right.
4. Addition and subtraction are performed in the order given, reading left to right.

Memorizing the phrase, "**P**lease **E**xcuse **M**y **D**ear **A**unt **S**ally" provides a simple aid for remembering the order in which functions are performed. The first letter of each word provides the clue:

- ❑ **P**arentheses
- ❑ **E**xponents
- ❑ **M**ultiply
- ❑ **D**ivide
- ❑ **A**dd
- ❑ **S**ubtract

Now, using the formula $A = \pi r^2$, you can find the area of a circle whose radius is three inches as follows:

$$A = \pi r^2$$
$$A = 3.1416 \times 3^2$$
$$A = 3.1416 \times (3 \times 3)$$
$$A = 3.1416 \times 9$$
$$A = 28.2744 \text{ sq. in.}$$

7.6.4 Cubic (Volume) Measure

Cubic measure, using the three dimensions of length, width, and **height** (or *depth*), makes it possible to determine **volume**. The difference between area and volume is the addition of a third measurement, height. See **Figure 7-19**.

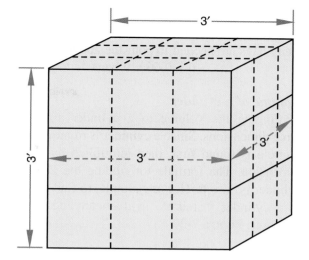

Figure 7-19. *By measuring length and width and height, you can compute the volume of an object, such as this cube. In this example, 3' × 3' × 3' = 27 cu. ft. The volume could also be expressed as 3³ or as "three cubed."*

Volume measurements are necessary, for example, when figuring the number of cubic feet of space in a room or house being air conditioned. Volume measurements are also necessary for such tasks as figuring concrete needed for slabs, determining piston displacement, and finding the capacity of a cylinder.

Volume of a rectangle

To find the volume of a rectangular solid, multiply the three dimensions together. The formula is: volume = length × width × height (or depth), or $V = L \times W \times H$ (or D). Volume is expressed in cubic terms: cubic inches (cu. in.), *cubic* centimeters (cu. cm), cubic feet (cu. ft.), cubic yards (cu. yds.), or cubic meters (cu. m).

$$1 \text{ cu. in.} = 16.387 \text{ cu. cm}$$
$$1 \text{ cu. ft.} = 1728 \text{ cu. in.} = 28\ 316.846 \text{ cu. cm}$$
$$1 \text{ cu. yd.} = 27 \text{ cu. ft.} = 0.7645 \text{ cu. m}$$

A perfect cube would have identical measurements in all three dimensions. For example, a cube measuring 12″ in all three dimensions would have a volume of 1728 cu. in. (12 × 12 × 12 = 1728). This cube can be expressed as 12^3, and is read, *12 cubed or 12 to the third power*. Therefore, 12^3 equals 1728.

In your everyday work, you may have to determine the volume of a rectangle. For example, you must place a concrete pad before you can install an outdoor condensing unit for a residential air conditioning system. What is the volume of a concrete pad measuring 4′ long, 3′ wide, and 6″ deep?

The formula is:

$$V = L \times W \times D$$
$$V = 4 \times 3 \times 0.5 \ (6'' = 0.5')$$
$$V = 12 \times 0.5$$
$$V = 6 \text{ cu. ft.}$$

Knowing that an 80-pound bag of concrete mix will make 2/3 cu. ft. (0.66 cu. ft.) of concrete, how many bags are needed to pour the above pad?

$$6 \div 0.66 = 9 \text{ bags}$$

Volume of a cylinder

Finding the volume of a cylinder also involves three dimensions. Since a cylinder is round, the area of a circle is found first, then multiplied by the height dimension. The formula for finding the volume of a cylinder is: $V = \pi r^2 H$. For example, to find the volume of a cylinder that is 3″ in diameter and 5″ tall, as shown in **Figure 7-20**:

$$V = \pi r^2 H$$
$$V = 3.1416 \times (1.5 \times 1.5) \times 5$$
$$\text{(diameter} = 3″, \text{ so radius} = 1.5″)$$
$$V = 3.1416 \times 2.25 \times 5$$
$$V = 3.1416 \times 11.25$$
$$V = 35.343 \text{ cu. in.}$$

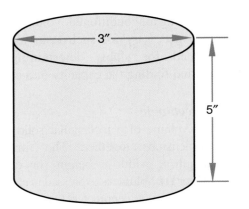

Figure 7-20. *To find the volume of a cylinder, first find the area of a circle; then multiply by the height dimension.*

With this information, it is possible to determine the amount of vapor a compressor can compress in a given time. (Usually, they are rated in cubic feet per minute, or cfm.) Assume the cylinder is on a two-piston compressor, operating at 1800 revolutions per minute (rpm). Since this compressor has two pistons, the cylinder volume must be doubled:

$$35.343 \times 2 = 70.686 \text{ cu. in.}$$

Next, multiply the volume by the rpm. This will result in a cubic *inches* per minute figure:

$$70.686 \times 1800 = 127,234.8 \text{ cu. in./min.}$$

Finally, convert cubic inches per minute to cubic feet per minute. (Since 1728 cu. in. = 1 cu. ft., divide by 1728.)

$$127,234.8 \div 1728 = 73.63 \text{ cu. ft./min.}$$

7.7 Angles and Degrees of a Circle ◆

Angles and degrees of a circle are important when selecting tubing and pipe fittings. The angle of a fan blade is a critical factor for proper airflow.

The circumference (distance around) of a circle is divided into 360 equal angles. The measure of each angle is defined as 1 **degree**. See **Figure 7-21**. A straight line from 360° to 180° would cut the circle in half.

A **protractor** is an instrument used to measure degrees and angles of a circle. A protractor is a half-circle, marked from 0° to 180°, as shown in **Figure 7-22**.

An **angle** is formed by two straight lines drawn from the same starting point. The starting point may be thought of as the exact center of a circle and is called the **vertex**. One line is drawn from zero degrees to the vertex, and the other line extends from the vertex to any degree mark on the circle. The number of degrees between the lines intersecting the circle becomes the *degree of angle.*

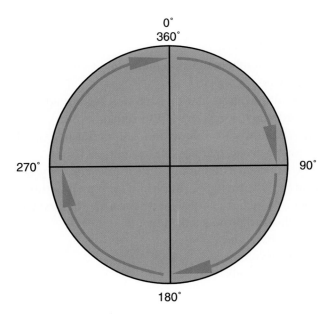

Figure 7-21. *A circle is divided into 360 degrees.*

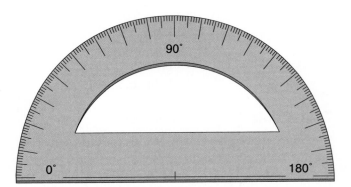

Figure 7-22. *A protractor is used to measure angles and degrees of a circle.*

A 90° angle (often referred to as a right angle) is best illustrated by the hands of a clock reading 3 o'clock. See **Figure 7-23**. The small hand points to the zero (3 on the clock face) and the large hand points to 90 (12 on the clock face). Therefore, the degree of angle is 90.

A 45° angle is halfway between 0 and 90. It takes two 45° elbow fittings to equal one 90° elbow fitting. See **Figure 7-24**. Also, two 90° elbows make a U-bend (also called a U-turn), or 180° turn.

All fan blades are twisted to a particular degree of angle called a *pitch*. Each blade must have the same pitch or angle for efficient operation. The pitch is an important factor in determining the volume of air moved by the fan. All else being equal, a fan with blades at a 20° pitch will not move as much air as one with blades at a 33° pitch. The pitch is usually stamped on one blade.

Figure 7-23. *Ninety degrees is a quarter-circle. Clock hands in the 3 o'clock position, as shown, form a 90° angle.*

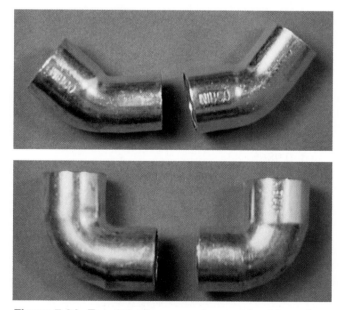

Figure 7-24. *Two 45° elbows can be combined to make a 90° elbow, or "el." Two 90° elbows can be combined to make a U-bend, or 180° turn.*

7.8 Temperature Scales

While the **Fahrenheit** (°F) temperature scale is most common in the United States, the **Celsius** (°C) scale is used in almost all other countries. Technical literature may use either or both. While you are most likely to use the Fahrenheit scale in refrigeration, heating, and air conditioning work, you should also be able to use the Celsius scale. Sometimes, you will have to convert Fahrenheit temperatures to Celsius, or vice versa. See **Figure 7-25** for scale comparisons.

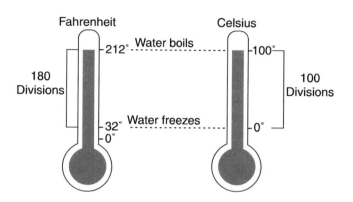

Figure 7-25. *Fahrenheit and Celsius temperature scales compared. Note that the Fahrenheit scale has almost twice as many divisions as the Celsius scale between freezing and boiling points.*

7.8.1 Converting Temperature Readings

There are two methods for converting Fahrenheit to Celsius, or Celsius to Fahrenheit. One uses fractions; the other uses decimals. Either method will yield the correct answer, but the decimal method is easiest to use:

❑ Fahrenheit = (1.8 × Celsius) + 32
❑ Celsius = (Fahrenheit – 32) ÷ 1.8

To convert a temperature of 40°C to Fahrenheit:
$$°F = (1.8 × 40) + 32$$
$$°F = 72 + 32$$
$$°F = 104$$

To convert 104°F to Celsius:
$$°C = (104 – 32) ÷ 1.8$$
$$°C = 72 ÷ 1.8$$
$$°C = 40$$

It is worth noting that the Fahrenheit and Celsius scales read the same at –40°! (You can prove it by using the above formulas.)

7.9 Finding Percentage

One *percent* of a number means 1/100 of that number, six percent means 6/100, and so on. The symbol % stands for the word "percent." Thus, 10 percent is written 10%, 25 percent as 25%, and so on.

The symbol % does the work of two decimal places, or hundredths. For example, 6% = 0.06; 25% = 0.25; 50% = 0.50; 100% = 1.00.

7.9.1 Using Percent for Profit Margin

Parts and supplies used to make repairs are charged to the customer, with a **profit margin** included in the price. The profit margin, or *markup*, is the difference between cost and selling price. The difference is created by increasing cost price by a certain *percentage*.

Since percent means "hundredths," the whole of any number contains 100% of itself. Thus, 100% of 20 is 20. If a 100% profit margin were applied to a part, for example, the customer would be charged $40 for a part that cost the seller $20. With a profit margin of 25%, one-quarter of the cost price is added to obtain selling price. Since 25% of 20 is 5, the $20 item's selling price would be $25.

An easy method of figuring selling price is to consider cost price 100%, and add the percentage of markup. Thus, a 25% profit margin becomes 125% of cost. A 50% profit margin becomes 150% of cost, and 100% profit margin becomes 200% of cost. To change a percent to a decimal number, drop the % sign and move (or add) a decimal point two places to the left. This means 125% = 1.25, 150% = 1.50, and 200% = 2.00.

As an example of how a selling price is determined with this method, consider a part with a base cost of $23.50. The desired profit margin is 25%, so the selling price is figured by multiplying base cost ($23.50) × 1.25. The result is a selling price of $29.38.

Summary

The ability to perform basic math operations is necessary in all phases of the refrigeration, heating, and air conditioning trade. Being able to apply mathematical formulas will prevent mistakes and help lead to early job advancement. The logical, step-by-step thought processes required to perform mathematical operations are invaluable for understanding why and how a system is repaired.

Working with decimals, fractions, and percentages; measuring dimensions; and reading both US Conventional and SI Metric rulers are fundamental to the job of a heating and cooling technician. Other tasks such as completing time sheets and work orders,

pricing parts, and keeping inventory and mileage records will also call on the use of math skills.

This chapter was intended as a refresher to help you remember the skills needed to work with numbers. Later chapters will use mathematics, as needed, to explain and prove how various system components operate.

Test Your Knowledge

Please do not write in this text. Write your answers on a separate sheet of paper.

1. The bottom number of a fraction is the:
   a. Elevator.
   b. Denominator.
   c. Designator.
   d. Numerator.
2. *True or false?* The fraction 9/5 is an improper fraction.
3. Decimal points are used to separate whole numbers from _____.
4. How would you read the number 6.435? (Do not use the word "point.")
5. Inch, foot, and yard are units of length in the _____ system of measurement.
6. There are ____ inches in one foot, and _____ centimeters in one meter.
7. There are ____ sixteenths in one inch.
8. *True or false?* The fractions 24/32, 12/16, 6/8, 3/4 all represent the same value.
9. To find the _____ of a rectangular surface, length and width measurements must be made.
10. How would you read the number 5^2? What does it mean?
11. Half the distance across the widest part of a circle is called the _____.
12. The volume of a rectangular solid is computed by multiplying length times width times _____.
13. The value of π is:
    a. 2.54
    b. 1.666
    c. 3.1416
    d. 0.9865
14. What is the formula for converting Celsius to Fahrenheit?
15. What is the selling price of a part that cost $36.84, if the profit margin (markup) is 50%?

8
Basic Thermodynamic Principles

Objectives

After studying this chapter, you will be able to:
- ❏ Identify the ways heat is transferred.
- ❏ Describe how a material undergoes a change of state.
- ❏ Define and distinguish sensible heat, specific heat, and latent heat.
- ❏ Explain the significance of the British thermal unit.
- ❏ Name the five latent heats.
- ❏ Define saturated conditions, superheat, and subcooling.
- ❏ Explain and calculate a ton of refrigeration.

Important Terms

absolute zero	latent heat of
boiling point	vaporization
British thermal unit	molecule
(Btu)	physical states
change of state	radiation
conduction	refrigeration
convection	saturated conditions
Ice Melting Equivalent	saturation point
(IME)	sensible heat
insulators	specific heat
kilojoule (kJ)	subcooled
latent heat	superheated vapor
latent heat of	thermodynamics
condensation	ton of refrigeration
latent heat of freezing	effect
latent heat of melting	
latent heat of	
sublimation	

8.1 Refrigeration and Heat Movement

Most people think of refrigeration as a cold or cooling process. However, refrigeration actually removes heat. **Refrigeration** is the process of removing unwanted heat and carrying it away to be discarded.

The term "cold" describes a lack of heat, the condition produced by the *removal of heat*. For example, a refrigerator produces "cold" by removing heat from inside the cabinet, then releasing that heat to the outside atmosphere. The refrigeration system simply transfers heat from inside the cabinet to the outside atmosphere, just as you would bail water from a leaking boat and throw it back into the lake. See **Figure 8-1**.

Figure 8-1. *A refrigeration system is similar to bailing out a boat. When bailing, you move water from inside the boat to outside. In a refrigeration system, you move heat from inside a house, refrigerator, or other space to outside.*

Heat constantly leaks into the refrigerator through the appliance's walls and insulation. Warm air enters whenever the door is opened, and still more heat gets in when warm foods are placed inside the refrigerator. To accomplish refrigeration, the heat must be removed faster than it enters, just like bailing water from a leaking boat.

8.1.1 Thermodynamic Laws

Thermodynamics is the science that deals with the mechanical action of heat. There are a number of basic thermodynamic principles, called "the laws of thermodynamics." Two of these laws are very important to the study of refrigeration and air conditioning. Since the HVAC technician constantly works with the controlled movement of heat, an understanding of basic thermodynamic principles is vital.

First Law of Thermodynamics

The First Law of Thermodynamics states that energy cannot be created or destroyed, but can be *converted* from one form to another. This law applies to heat and other forms of energy, such as electrical, mechanical, light, chemical, and atomic.

Heat can be generated by converting another form of energy into heat energy. For example, electrical energy is converted to heat energy by such devices as a toaster, electric range, hair dryer, water heater, or space heater. **Figure 8-2** illustrates an example of energy conversion.

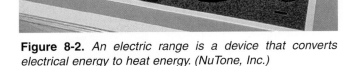

Figure 8-2. *An electric range is a device that converts electrical energy to heat energy. (NuTone, Inc.)*

Second Law of Thermodynamics

In simplest form, the Second Law of Thermodynamics states that heat always travels *from hot to cold*. See **Figure 8-3**. Heat energy is always in motion. For example, a metal spoon placed in hot coffee will absorb heat from the coffee, causing the coffee to begin to cool. How fast heat travels depends upon the temperature difference between the two objects. The greater the temperature difference, the faster heat will travel.

8.1.2 How Heat Travels

Since refrigeration deals with the movement of heat, it is necessary to know *how* heat travels. There are three basic methods by which heat is transferred from one substance to another: radiation, conduction, and convection. Sometimes, a combination of these methods is used.

Figure 8-3. *Heat always travels from hot to cold; the greater the temperature difference, the faster the rate of heat transfer.*

Radiation

Radiation is the transfer of heat by waves similar to light waves or radio waves. The sun's energy is transferred by means of radiant heat waves, which travel through space in a straight path. Radiant heat waves are absorbed by objects, not the air they pass through. If you have ever moved from the shade into direct sunlight on a hot, sunny day, you have felt the effect of these heat waves. You feel the difference because your body immediately absorbs the radiant heat waves.

Dark-colored materials tend to absorb radiant heat waves; light-colored materials tend to reflect them. Clothing manufacturers apply this principle when designing garments for different seasons or climates.

An automobile parked in the hot sun with the windows closed will absorb radiant heat waves, causing the inside of the automobile to become very hot, **Figure 8-4.** For this reason, never leave a child or pet in a parked car on a hot, sunny day.

Figure 8-4. *Radiant heat is transmitted from a heat source to an object without heating the air through which it passes. The interior of a closed car is quickly heated by the sun on a hot, sunny day.*

Conduction

Conduction is the flow of heat through a substance from one end to the other. An iron skillet provides a good example of conduction. If the skillet is placed on a hot fire, heat will travel to the handle until it becomes just as hot as the skillet body. See **Figure 8-5**.

Heat flow by conduction can take place between two substances or objects if they are touching each

Figure 8-5. *If an iron skillet is placed over a fire, the heat will travel from the skillet body to the handle by means of conduction.*

other. Conduction is improved by the amount of physical contact involved. Most metals will conduct heat very well. Copper and aluminum are excellent conductors of heat, so these two metals are used extensively in refrigeration systems.

Substances that are poor heat conductors are called **insulators.** Examples of insulators are cork, wood, fiberglass, mineral wool, and polyurethane foams. Insulators cannot totally stop heat flow but can slow it significantly.

Convection

Convection is the movement of heat by means of a carrier, such as air or water. In a *forced-air* heating system, air is heated by a furnace and discharged into the living areas to warm them. The air is then returned to the furnace to be heated again.

Movement of convection currents can be either *natural* or *forced.* Natural convection involves slow-moving currents of air, with lighter, warm air rising and heavier, cold air falling. This method was once used for furnaces and is still found in refrigerators today.

Forced convection uses fans or blowers to increase the amount of air movement, **Figure 8-6.** This procedure permits the use of small and efficient heat exchangers.

Another method of forced convection uses a liquid to carry heat. In a forced convection cooling unit, **Figure 8-7,** a fan forces warm air over tubes filled with liquid refrigerant. The liquid absorbs the heat and cools the air. The liquid is then transferred by a circulating pump to another location where the heat is released.

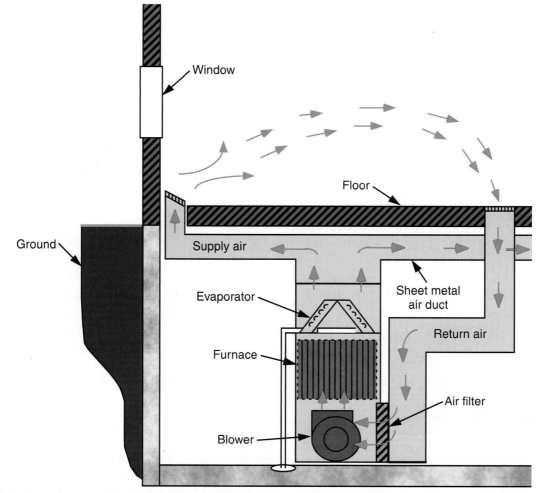

Figure 8-6. *Forced convection transfers heat more efficiently than does natural convection. The A-frame evaporator of a central air conditioning system is installed in the plenum of a household furnace. The system uses forced convection to remove heat and distribute the cooled air throughout the living area of a house.*

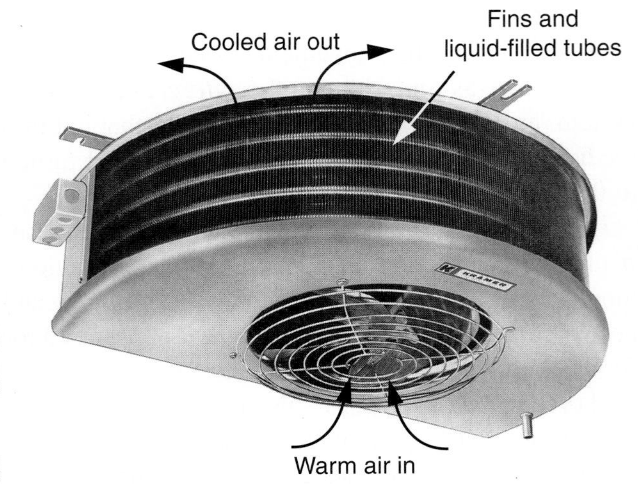

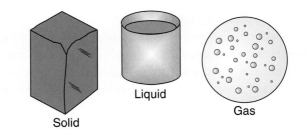

Figure 8-7. *In this forced convection cooling unit, the fan forces warm air through the fins, past tubes filled with liquid refrigerant. The refrigerant absorbs heat, cooling the air. (Kramer)*

8.2 Molecular Theory

All substances are made up of particles known as molecules. A molecule is the smallest physical particle of any substance that can exist by itself and retain its chemical properties. For example, a water molecule (H_2O) consists of two atoms of hydrogen and one atom of oxygen.

Regardless of the substance they make up, molecules are always in rapid motion. As the temperature of a substance increases, the motion of the molecules also increases, **Figure 8-8.** As the temperature drops, the molecules slow down. Even in a solid, such as ice or steel, the molecules are moving unless all heat is removed from the substance. If all heat is removed (at **absolute zero,** –460°F or –273°C), molecular motion will stop completely.

Molecules are like building blocks arranged in certain patterns to form different substances. In a solid, molecules still move or vibrate, but distances are very limited. The force holding the molecules in place is very strong in a solid. Much heat energy is required to overcome this force and permit the molecules to move and form a new pattern.

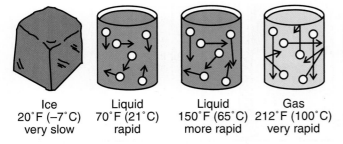

| Ice
20°F (–7°C)
very slow | Liquid
70°F (21°C)
rapid | Liquid
150°F (65°C)
more rapid | Gas
212°F (100°C)
very rapid |

Figure 8-8. *Molecules in a substance speed up or slow down with the rise or fall of temperature. Molecular motion is slowest in a solid, faster in a liquid, and fastest in a gas.*

When heat flows from a warmer to a cooler substance, the faster-moving molecules give up some of their energy to the slower-moving ones. As a result, the faster molecules slow down, and the slower ones move a little faster.

8.3 Physical States

Most substances can exist in three **physical states:** solid, liquid, or gas. As heat is added or removed, the substance may go through a change of state. Water is the substance commonly used to illustrate this process because it can exist as a solid (ice) below 32°F (0°C), a liquid (water) between 32°F and 212°F (0°C and 100°C), and a gas (steam) at 212°F (100°C) or higher. See **Figure 8-9.**

Figure 8-9. *Most substances can exist in any of the three states of matter: solid, liquid, or gas. Water, for example, is a solid below 32°F (0°C), a liquid from 32°F to 212°F (0°C to 100°C), and a gas above 212°F (100°C).*

When a substance is changing from one physical state to another, the temperature level remains *constant* (does not change) until all the molecules are rearranged to the new pattern. This principle is difficult to understand — the question is often asked, "If the temperature doesn't move during a change of state, just where does the heat energy go?" The energy is being used to shift the molecules into a different pattern.

A liquid allows more movement than a solid, but the movement of the molecules is still limited to a specific pattern. When changing from a liquid to a gas (at the **boiling point**), a large amount of heat energy is needed to shift the molecules and permit them to move in total freedom in all directions as a gas.

8.3.1 Change of State

A **change of state** occurs when the temperature and speed of the moving molecules reach a certain level. At this precise level, the molecules will rearrange themselves into a different pattern. The shift in pattern causes the substance to change from a solid to a liquid or from a liquid to a gas.

This rearrangement of molecules is best illustrated by adding heat to ice that has a temperature of 32°F

(0°C). See **Figure 8-10.** The additional heat will not raise the temperature of the ice, but will provide the energy needed to cause the molecules to shift and rearrange themselves into a liquid (water) pattern.

Once the ice becomes water, additional heat will cause the water molecules to increase their speed of motion; temperature will also rise. An increase in temperature and speed will continue until the water is heated to its boiling point (212°F or 100°C). Additional heat at this temperature will provide the energy necessary to cause the molecules to shift their pattern again, becoming steam.

Of course, the reverse is also true. By *removing* heat from a substance, its temperature will decrease. The molecules will slow down, shifting their pattern to a new physical state (gas to liquid, liquid to solid).

8.3.2 Temperature and Change of State

The temperature at which a given substance will change its state is always the *same*. Because the molecules are different for each substance, however, the temperature at which a change of state will occur is *different* for each substance. Water, for example, always boils at 212°F (100°C); ethyl alcohol always boils at 173°F (78°C).

Because they are common and familiar to everyone, the three states of water (ice, liquid water, and water vapor or steam) are used to explain the basic principles of temperature and change of state. The principles are applied to substances that change state at very high or very low temperatures, as well. Try to imagine a substance like liquid ammonia that *boils* at −28°F (−33°C) and *freezes* at −107°F (−77°C)! See **Figure 8-11** for a table of boiling and freezing temperatures for various substances.

Heat energy must be removed to cause a gas to condense back to a liquid. Likewise, heat must be removed to cause a liquid to freeze into a solid. Exactly the same amount of heat energy is involved, regardless of whether heat is being added or removed. It requires the same amount of heat energy to boil one pound of water as it does to condense one pound of steam. This is true of all substances.

8.4 Sensible Heat ◆

Sensible heat is heat that causes a change in temperature but not a change of state. When a substance is heated and the temperature rises as a result of the added heat, the added heat is referred to as sensible heat. Likewise, if heat is removed and the temperature of the substance falls, the heat being removed is sensible heat.

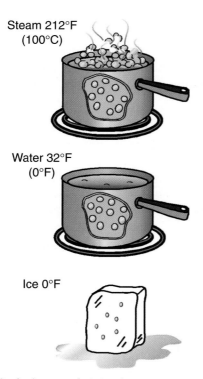

Steam 212°F (100°C)

Water 32°F (0°F)

Ice 0°F

Figure 8-10. *A change of state always occurs at the same temperature for a given substance. For water, the change from solid to liquid comes at 32°F (0°C), and the change from liquid to gas at 212°F (100°C).*

Boiling and Freezing Temperatures

Boiling Temperatures			Freezing Temperatures		
Substance	°F	°C	Substance	°F	°C
Water	212	100	Water	32	0
Ethyl alcohol	173	78	Fruits/ vegetables	30	−1
Chloroform	143	62	Seafood	28	−2
Butane	31	1	Beef and pork	28	−2
Ammonia	−28	−33	Poultry	27	−3
Propane	−43	−42	Carbon tetrachloride	−9	−23
Carbon dioxide	−109	−78	Linseed oil	−11	−24
Acetylene	−118	−83	Chloroform	−81	−63
Oxygen	−287	−177	Ammonia	−107	−77
Nitrogen	−321	−196	Acetone	−139	−95
Hydrogen	−423	−253	Ether	−177	−116
Helium	−452	−269	Ethyl alcohol	−179	−117

Figure 8-11. *Each substance has a specific temperature at which it changes state (freezes or boils). The table lists temperatures for some common foods and chemical substances.*

8.4.1 British Thermal Unit

The ***British thermal unit (Btu)*** is the basic unit used to measure the *quantity* of heat. A thermometer measures only the temperature of a substance. It cannot measure the amount of heat required to reach a certain temperature.

Many years ago, scientists realized a standard method to measure quantities of heat was needed. Since water was such a common substance, they placed a quantity of water that weighed exactly one pound into a container. Next, they measured the precise amount of heat required to raise the temperature of the water one degree Fahrenheit. See **Figure 8-12.** The standard they established was accepted by all scientists and is still used today:

One Btu = The amount of heat that will raise the temperature of one pound of water by one degree Fahrenheit.

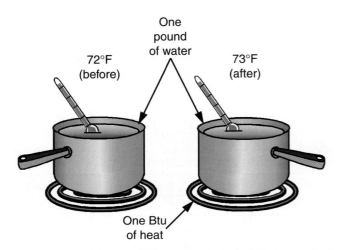

Figure 8-12. *The Btu (British thermal unit) is the standard used to measure quantities of heat. The Btu is the amount of heat needed to raise the temperature of one pound of water by one degree Fahrenheit. The temperature here was raised from 72°F to 73°F.*

If more than one pound of water is involved, it will require one Btu for each pound of water. Likewise, if the temperature is to be raised more than one degree, it will require one Btu for each degree of temperature change. For example:

How many Btu are required to raise the temperature of 10 pounds of water from 72°F to 82°F?

Btu = wt × td (weight × temperature difference)

Btu = 10 × 10

Btu = 100

If a substance is cooled by removing heat, the amount of heat removed is figured the same way. If heat is removed from water, the weight is multiplied by the temperature difference to obtain the number of Btu removed in cooling the water. For example:

How much heat must be removed to cool 50 pounds of water from 75°F to 65°F?

Btu = wt × td

Btu = 50 × 10

Btu = 500

Kilojoule

In the metric system, the unit used to measure quantities of heat is the joule (J). It is, however, a very tiny unit; for practical use in refrigeration work, the ***kilojoule (kJ)*** is used. A kilojoule is 1000 J. To raise the temperature of one kilogram (kg) of water by one degree Celsius (C), the amount of heat required is 4.187 kJ.

8.5 Specific Heat ◆

Specific heat is the amount of heat required to raise the temperature of one pound of *any* substance one degree Fahrenheit. This definition is almost the same as that for a Btu, except it considers *all* substances, not just water. The specific heat of water is one Btu, but the amount of heat required (in Btu/lb.) to cause a temperature change in other substances will vary with each substance. Each substance requires different amounts of Btu per pound. Scientists have already calculated the specific heats for most substances; these figures are readily available in technical manuals or other references in your library. **Figure 8-13** lists specific heats for some common substances.

Computing Btu requirements

Determining the number of Btu needed to raise the temperature of a substance involves three factors:

❑ The number of pounds of the substance involved.

◆ Specific Heats

Substance	Specific Heat (Btu/lb.)	Substance	Specific Heat (Btu/lb.)
Acetone	0.514	Chicken	3.316
Alcohol	0.680	Chloroform	2.340
Ammonia	1.090	Copper	0.095
Bacon	1.474	Fish	3.550
Beef	2.345	Ice	0.487
Beer	3.852	Iron	1.373
Benzene	0.412	Oranges	3.751
Bread	1.993	Peaches	3.818
Butter	1.373	Popcorn	1.172
Cheese	2.077	Water	1.000

Figure 8-13. *Specific heats for a number of foods and other substances.*

❏ The number of degrees the substance's temperature is to be raised.

❏ The specific heat for the substance.

To find the number of Btu required, multiply the *weight (wt)* times the *specific heat (sp ht)* per pound times the desired *temperature change (tc)*.

The formula, *Btu = wt × sp ht × tc*, makes it possible to quickly figure quantities of heat. For example:

How many Btu must be removed to cool 50 pounds of water from 75°F to 55°F?

Btu = wt × sp ht × tc

Btu = 50 × 1 × 20

Btu = 50 × 20

Btu = 1000

How many Btu must be added to 150 pounds of ice to change the temperature from –20°F to 30°F?

Btu = wt × sp ht × tc

Btu = 150 × .487 × 50

Btu = 73.05 × 50

Btu = 3652.5

8.6 Latent Heat

Latent heat is heat energy that causes a change of state but no temperature change. During a change of state, the temperature will remain constant until the process of change is completed. Latent heat is sometimes called "hidden" heat because it does not show on the thermometer. The heat energy is used to rearrange the molecules into a different pattern, and the temperature cannot change until all the molecules have been rearranged.

Each substance has a temperature/pressure point at which a change of state will occur. For example, under normal atmospheric pressure, ice will melt at 32°F (0°C), or water will freeze at 32°F (0°C). Therefore, at 32°F (0°C), it is possible to have all ice, all water, or a combination of water and ice. See **Figure 8-14.** The temperature will remain the same until the change of state is completed. It requires exactly 144 Btu of heat to melt one pound of ice; the same amount of heat must be removed to freeze one pound of water. This is called the latent heat of melting or the latent heat of freezing.

When water is heated, the temperature will stop rising at 212°F (100°C) because the water begins to boil into steam. During this change of state, the temperature will remain at 212°F (100°C) until all the water has turned into steam. It requires 970 Btu to change one pound of water into steam (latent heat of vaporization). If the steam is trapped in a closed container, and 970 Btu are removed by cooling the steam, the steam will condense back to water (latent heat of condensation).

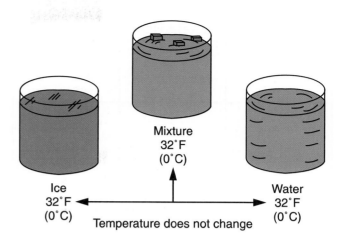

Figure 8-14. *Latent heat causes a change of state without increasing the temperature of the substance. The temperature will remain constant until the change of state is complete. As shown, while ice is changing to water, the temperature remains constant at 32°F (0°C).*

In short, the four latent heats just described are the:

❏ *Latent heat of melting* — the process of changing a solid to a liquid by adding heat.

❏ *Latent heat of freezing* (or fusion) — the process of changing a liquid to a solid by removing heat.

❏ *Latent heat of vaporization* (or evaporation) — the process of changing a liquid to a vapor by adding heat.

❏ *Latent heat of condensation* — the process of changing a gas (or vapor) to a liquid by removing heat.

Figure 8-15 shows latent heat of vaporization/condensation values for some typical refrigerants (water is included as a reference value). Latent heat of vaporization/condensation forms the basis of all refrigeration systems. Refrigerants are the vital working fluids in any refrigeration or air conditioning system. They absorb heat as they change from a liquid to a gas (evaporate). The gas then travels to a place where the heat is removed, causing the gas to return to a liquid state (condense). The refrigerant is then ready for reuse (another cycle).

The formula *Btu = wt × sp ht × tc* can be applied to discover Btu quantities in the heating or cooling process. The following problem illustrates different heat values and the steps required to allow for sensible and latent heats.

PROBLEM: How much heat must be removed to convert 10 pounds of steam at 212°F (100°C) into ice at 12°F (–11°C)?

This problem is performed in four easy steps, with the specific heat and latent heat values obtained from the tables in Figures 8-13 and 8-15.

◆

Latent Heat

Substance	Freezing or Melting (Btu/lb.)	Latent Heat of Vaporization or Condensation (Btu/lb.)
Water	144	970.4
R-134a		67.8
R-12		68.2
R-22		93.2
R-502		68.96
R-717 (Ammonia)		565.0

Figure 8-15. *Latent heat of evaporation/condensation values for water and several common refrigerants. Since the refrigerants are used only in the liquid and gaseous states, no values are provided for the latent heat of freezing or the latent heat of melting.*

1. Condense the steam into water. Multiply weight times specific heat times temperature change:

    Btu = wt × sp ht × tc

    Btu = 10 × 970 (no temperature change — latent heat only)

    Btu = 9700

2. Lower the temperature of the water from 212°F (100°C) to 32°F (0°C). The temperature change will be 180 degrees Fahrenheit (100 degrees Celsius). We must stop at 32°F (0°C), the point where the latent heat of freezing comes into effect.

    Btu = wt × sp ht × tc

    Btu = 10 × 1 × 180 (sensible heat)

    Btu = 1800

3. Freeze the water into ice.

    Btu = wt × sp ht × tc

    Btu = 10 × 144 × 0 (no temperature change — latent heat only)

    Btu = 1440

4. Lower the temperature of the ice from 32°F (0°C) to 12°F (–11°C). The temperature change will be 20 degrees Fahrenheit (11 degrees Celsius) of sensible heat.

    Btu = wt × sp ht × tc

    Btu = 10 × .487 × 20 (sensible heat)

    Btu = 4.87 × 20

    Btu = 97.4

Add the answers to obtain the total amount of Btu removed to change 10 pounds of steam into ice at 12°F (–11°C).

    9700 (condensed steam into water)

    1800 (lowered water temperature to 32°F or 0°C)

    1440 (froze the water into ice)

    97.4 (lowered the temperature of ice to 12°F or –11°C)

    TOTAL = 13,037.4 Btu

Imagine the steam, water, and ice enclosed in a copper container that weighs 15 pounds. The copper must be included in the problem since its temperature also must be lowered. The copper will remain a solid at the temperatures involved, so its temperature can be lowered from 212°F (100°C) directly to 12°F (–11°C). The temperature change will be 200 degrees Fahrenheit (111 degrees Celsius).

    Btu = wt × sp ht × tc

    Btu = 15 × .095 × 200

    Btu = 1.425 × 200

    Btu = 285

    285 + 13,037.4 = 13,322.4 Btu (new total to be removed)

8.6.1 Latent Heat of Sublimation

Latent heat of sublimation describes the process in which a substance changes directly from a solid to a vapor, without passing through the liquid phase. A common example of this process is solid carbon dioxide (CO_2) or "dry ice." This substance will bypass the liquid state and sublime (go directly from a solid to a gas).

Another example of sublimation is a tray of ice cubes left in the freezer for an extended time. The tray may have been full of ice to begin with, but will sublime if not used. When you return from vacation, for example, the ice cube tray may be half full or almost empty.

To review, the five latent heats and their related state changes include:

❑ Sublimation = solid to vapor

❑ Melting = solid to liquid

❑ Freezing (fusion) = liquid to solid

❑ Vaporization (evaporation) = liquid to vapor

❑ Condensation = vapor to liquid

8.6.2 Saturated Condition

Saturated condition is a term that refers to the boiling/condensing point of a substance. When the temperature/pressure combination of a substance is such that the substance can *change its state,* the condition is called the **saturation point.** The boiling/condensing point of a substance is dictated by a specific combination of temperature and pressure.

For water at atmospheric pressure, saturated condition is achieved at 212°F (100°C). See **Figure 8-16.** At this temperature/pressure, it is possible to have all water, a mixture of water and steam, or all steam. Whenever a mixture of vapor and liquid exists, it is *saturated* because it is in the process of changing state.

8.6.3 Superheated Vapor

The term ***superheated vapor*** describes a gas that has been heated to a temperature *above* its boiling point as a liquid at the existing pressure. The air is composed of superheated vapors, since the gases that make up air (nitrogen, oxygen, and carbon dioxide) have boiling points well below zero at atmospheric pressure.

Figure 8-17 shows steam in a superheated condition. Since water vaporizes (becomes steam) at 212°F (100°C), any heat energy added to the vapor will raise its temperature, making it a superheated vapor. The specific heat of steam is 0.46, so it requires 0.46 Btu per pound to superheat steam to any temperature above its saturation (boiling) point. Steam at 217°F (103°C) would be superheated by 5 degrees Fahrenheit (217 – 212 = 5) or 3 degrees Celsius.

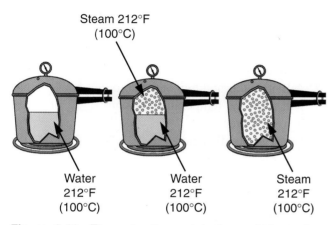

Steam 212°F (100°C)

Water 212°F (100°C)   Water 212°F (100°C)   Steam 212°F (100°C)

Figure 8-16. *The saturation point of a substance is a specific temperature at a specific pressure. For water, it is 212°F (100°C) at atmospheric pressure. Under saturated conditions, the substance can be water, steam, or a mixture of water and steam, all at the same temperature/pressure.*

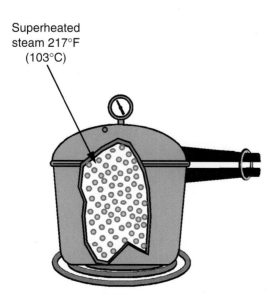

Superheated steam 217°F (103°C)

Figure 8-17 *When a substance is superheated, sensible heat is added after the substance changes its state. Either a liquid or gas can be superheated. The steam (water vapor) shown has been superheated 5 degrees Fahrenheit to a temperature of 217°F (3 degrees Celsius to 103°C).*

Vapors and gases

The terms "gas" and "vapor" are often used interchangeably to describe a substance in the gaseous state. In a strictly scientific sense, there is a difference between the two terms. In practical, everyday applications in the HVAC field, however, there is not. This book follows that trade practice.

8.6.4 Subcooled Substances

Any substance at a temperature *below* its saturation point can be described as **subcooled.** While the term is most often applied to a liquid, it also can apply to substances in the solid state.

An example of subcooling is shown in **Figure 8-18.** Water at 200°F (93°C) is subcooled 12 degrees Fahrenheit (7 degrees Celsius) since the saturation point of water is 212°F (100°C).

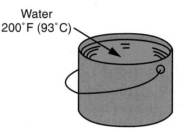

Water 200°F (93°C)

Figure 8-18. *A substance that has a temperature below its saturation point is said to be subcooled. This container of liquid has been subcooled to 200°F (93°C), which is 12 degrees Fahrenheit (7 degrees Celsius) below its saturation point.*

8.7 Ton of Refrigeration Effect

Ice (water in a solid state) was the forerunner of all refrigeration systems and is still used today to compare the refrigerating effect of various systems. The unit of measurement is called a "ton" and refers to the ***Ice Melting Equivalent (IME)*** — the amount of heat absorbed in melting one ton (2000 pounds) of ice at 32°F (0°C) in exactly 24 hours. See **Figure 8-19.**

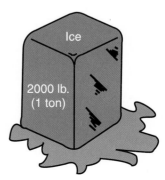

Ice

2000 lb. (1 ton)

Figure 8-19. *A ton of refrigeration effect is equal to the removal of 288,000 Btu from a block of ice in a 24-hour period (12,000 Btu/hr. or 200 Btu/min.).*

By referring to the latent heat table in Figure 8-14, you see that one pound of ice will absorb 144 Btu when it melts. Therefore, 288,000 Btu are needed to melt one ton of ice (2000 pounds × 144 Btu). Any refrigeration system capable of removing 288,000 Btu in 24 hours is, therefore, called a one-ton system.

By dividing 288,000 by 24, you can determine how many Btu are being removed each hour (12,000). Thus, any refrigeration system capable of removing 12,000 Btu per hour is also a one-ton system.

Dividing 12,000 by 60 will identify how many Btu are being removed each minute (200). An absorption rate of 200 Btu per minute is also equivalent to one *ton of refrigeration effect.*

Thus, there are three ways of expressing a Btu removal rate equal to a one-ton system. One of these methods usually appears on the data plate of the system:

❑ 288,000 Btu/24 hrs.
❑ 12,000 Btu/hr.
❑ 200 Btu/min.

An air conditioning system rated at three tons would be capable of removing 36,000 Btu/hr. (600 Btu/min.). A half-ton window unit would be capable of removing 6000 Btu/hr. (100 Btu/min.) Most systems are rated on a per-hour basis, but you may sometimes see a data plate rating the system in Btu/min.

Summary

The fundamental principles explained in this chapter form the foundation for later chapters that will build upon these principles and put them to practical use.

Heat is a form of energy that can be controlled by applying the principles in this chapter. Refrigeration and air conditioning systems are designed to move heat from one area to another. To understand how these systems operate, it is first necessary to know what heat is, how heat travels, and how it is measured.

The refrigerant acts as a vehicle for transporting heat; this transportation process involves regular changes of state. Some of the terms introduced in this chapter explain the conditions of a liquid or vapor when changing its state. These terms will be used frequently in later chapters to explain the condition of the refrigerant as it travels inside the system.

Test Your Knowledge

Please do not write in this text. Place your answers on a separate sheet of paper.

1. Define the term "cold."
2. What is the First Law of Thermodynamics?
3. The Second Law of Thermodynamics states that heat always travels from _____ to _____.
4. Name three methods by which heat travels.
5. How does the color of a material affect radiant heat flow?
6. Name the three physical states of matter.
7. Define sensible heat.
8. A British thermal unit is the amount of _____ that will raise the temperature of one _____ of water by one degree _____.
9. What is specific heat?
10. What is latent heat?
11. Name the five latent heats.
12. Give an example of sublimation.
13. The specific heat of copper is _____.
14. What does the term "saturated" mean?
15. *True or false?* At a temperature of 32°F (0°C), the material in a container can be all water, all ice, or a combination of water and ice.
16. What is superheat?
17. When a substance is at a temperature lower than its saturation point, it is said to be _____.
18. One ton of refrigeration effect is equal to how many Btu/hr.?
19. How many tons of refrigeration effect can be obtained from a window air conditioner rated at 18,000 Btu/hr.?
20. What substance is used to compare the refrigerating effect of various systems?

Temperature and Pressure

Objectives

After studying this chapter, you will be able to:

❏ Identify four types of thermometer scales.

❏ Discuss the importance of Fahrenheit and Celsius absolute temperatures.

❏ Describe atmospheric pressure and how pressures are measured and read.

❏ Describe vacuum measurement and the use of the compound gauge.

❏ Explain and apply Boyle's Law and Charles' Law.

Important Terms

absolute pressure	hydrargyrum
absolute zero	in. Hg
ambient temperature	Kelvin scale
atmospheric pressure	mercury barometer
boiling point	partial vacuum
Boyle's Law	perfect vacuum
Celsius scale	psia
centigrade	psig
Charles' Law	Rankine scale
compound gauge	refrigerant
cryogenic	saturation point
Dalton's Law	temperature
Fahrenheit scale	thermometer
gauge pressure	variable

9.1 Temperature and Pressure

The information on temperature and pressure in this chapter is the foundation for understanding all refrigeration and air conditioning systems. These systems contain a **refrigerant** that readily changes from a liquid to a gas, then is condensed back to a liquid for recirculation. The effect of temperature and pressure changes on the refrigerant is controlled and predictable. A thorough understanding of the principles presented in this chapter is required to help the technician deal with constant temperature and pressure changes within the system.

9.2 What Is Temperature?

Temperature does not reveal the *quantity* of heat in a substance. It merely indicates the *intensity*, or heat level (how hot or cold a substance is). In the molecular theory of heat, temperature indicates the speed of motion of a single molecule. Temperature specifies how hot a substance is but does not indicate how many Btu were needed to reach that temperature level.

For example, consider two containers of water over identical heat sources. In the first container, there is one gallon of water; in the second, five gallons. See **Figure 9-1.** If the same *amount* of heat is applied to each container, the temperature level of the water in the first container will rise more quickly than the level in the second container. If the temperature level of the water in both containers is raised to 180°F (82°C), the second container will require many more Btu of heat than the first to achieve the same temperature level.

9.3 Measuring Temperature

Temperature, or heat level, is measured by a device called a **thermometer.** Thermometers are used only for measuring sensible heat and usually consist of a glass tube calibrated in *degrees.* Inside the tube is a liquid, such as mercury or colored alcohol, that will expand or contract at a known rate with changes in temperature. Expansion or contraction will cause the liquid to rise or fall within the tube, where its level can be compared to the degree marks.

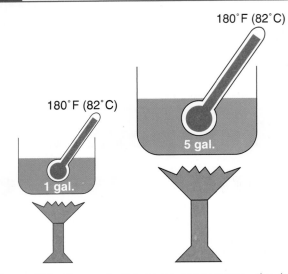

Figure 9-1. *To reach the same temperature, the larger quantity of water in the container at the right would require many more Btu of heat than the smaller quantity in the container at the left.*

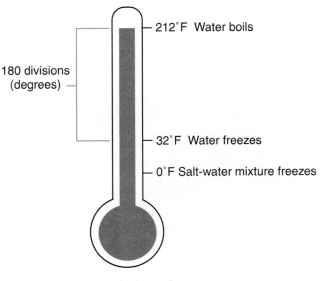

Fahrenheit scale

Figure 9-2. *The Fahrenheit temperature scale has 180 divisions between the freezing and boiling points of water. Zero is set at the point where a salt-water mixture freezes.*

In the United States, temperature is usually measured in degrees Fahrenheit, but in other parts of the world, and for scientific work, the Celsius scale is used.

9.3.1 Fahrenheit Scale

In 1714, the German scientist Gabriel Fahrenheit (1686–1736) became the first person to use mercury as the liquid in a thermometer. Mercury allowed more accurate measurement of temperature changes than other liquids previously used. In order to calibrate his thermometer, Fahrenheit chose the freezing point of a mixture of salt and water as the *zero* point. Ordinary water freezes at 32° on the ***Fahrenheit scale*** and boils at 212°. As shown in **Figure 9-2,** there are exactly 180 divisions *(degrees)* between the freezing and boiling points of water (212 – 32 = 180). Any temperature that occurs below zero is expressed as a negative or "minus." For example, a temperature of 10 degrees below zero is written *–10°F.*

9.3.2 Celsius Scale

The Swedish astronomer Anders Celsius (1701–1744) believed the Fahrenheit scale was impractical for laboratory use because its 180 divisions between the freezing and boiling points of water required troublesome mathematical calculations. To overcome this problem, Celsius developed his own scale in 1742. He called it the ***centigrade*** ("hundred steps") scale because it had exactly 100 divisions between the freezing (0°) and boiling (100°) points of water. See **Figure 9-3.** This new scale was quickly adopted by other scientists. Today, it is used by most countries of the world. Since 1948, the centigrade scale officially has been known as the ***Celsius scale.***

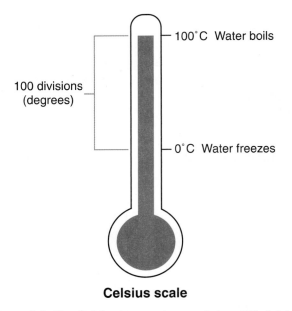

Celsius scale

Figure 9-3. *The Celsius temperature scale has 100 divisions between the freezing and boiling points of water. The scale is easier to use than the Fahrenheit scale, especially in scientific work.*

Absolute zero

When Fahrenheit devised his scale, he set the zero point at the temperature where salt water freezes since this was the lowest temperature he could achieve in the laboratory. However, the *lowest possible* temperature is far below 0°F.

Scientists have established that heat is always present until a temperature of –460°F (–273°C) is achieved. This is the lowest possible temperature because *all* heat has been removed. It is, therefore, called ***absolute zero.***

At absolute zero, all molecular motion stops because there is no heat in the substance. By comparison with absolute zero, the coldest weather we might experience on earth is quite warm. Atmospheric air of 0°F actually contains many Btu of heat!

9.3.3 Rankine and Kelvin Scales

For some types of scientific work, the use of *absolute temperature scales* — scales with no negative numbers — were a great convenience. In the mid-1800s, two Scottish scientists, William Rankine and William Thomson (Lord Kelvin), developed two different approaches to the absolute temperature scale.

In the **Rankine scale,** Fahrenheit divisions are used. Absolute zero is 0°R, the freezing point of water is 492°R, and the boiling point of water is 672°R. The Rankine scale is also known as the Fahrenheit Absolute (F_A) scale.

Kelvin's approach was identical, but he preferred the Celsius scale. Absolute zero is 0°K, water freezes at 273°K, and water boils at 373°K. The **Kelvin scale,** also called the Celsius Absolute (C_A) scale, is widely used by scientists. **Figure 9-4** is a side-by-side comparison of the four temperature scales.

Cryogenics

Cryogenic refrigeration systems (those capable of producing temperatures below –250°F or –157°C are used in laboratories to perform various scientific experiments. For instance, at absolute zero, all substances (even rubber) become perfect conductors of electricity. Research in the cryogenic temperature range has resulted in the discovery of materials that are superconducting (exhibiting no electrical resistance) at temperatures well above absolute zero.

9.3.4 Ambient Temperature

The term **ambient temperature** is used frequently in the refrigeration field. It simply refers to the temperature surrounding an object, typically a motor or condenser. Ambient temperatures can have a great influence on the operating conditions of a system or part of a system. For example, a motor may be rated to give maximum efficiency at any temperature that does not exceed 72°F (40°C) above *ambient*. The word "ambient" makes communication easier when referring to the temperature conditions surrounding an object.

Many times, a refrigeration system will have components located outside the building that are connected to other components located inside the building. The ambient temperature of the inside components is controlled, and thus remains relatively constant. However, the components located outside are exposed to all types of weather conditions, **Figure 9-5**. These ambient temperature conditions can be rather extreme compared to those experienced by indoor components.

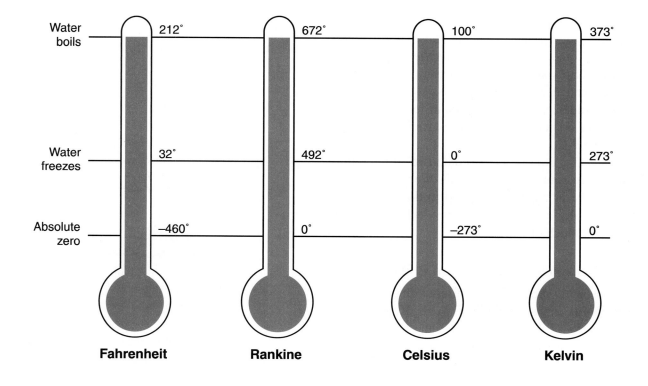

Figure 9-4. *A comparison of four temperature scales. Both the Rankine and Kelvin scales use absolute zero as their zero point. The Rankine scale is based on Fahrenheit degree divisions; Kelvin is based on Celsius divisions.*

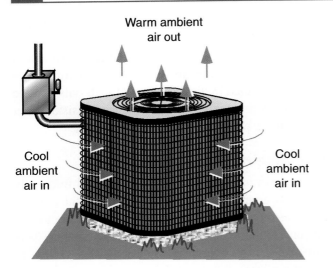

Figure 9-5. *Ambient temperature refers to the temperature surrounding an object. Cool ambient air is drawn through the outdoor condenser of an air conditioning system. As the moving air absorbs heat from the condenser coils, it becomes warm ambient air.*

9.4 Atmospheric Pressure ◆

Our planet is surrounded by a gaseous atmosphere that consists of about 78% nitrogen and 21% oxygen (the remaining 1% is composed of rare gases). The atmosphere extends approximately 50 miles above the earth's surface and is held in place by the force of gravity. The gases that make up the atmosphere are composed of tiny molecules traveling in all directions. Even though they are very small, these molecules are not weightless. The weight of the molecules exerts a downward force, known as *atmospheric pressure,* upon the earth's surface. This pressure is greatest at sea level, and is generally 14.7 psi (101.3 kPa). At higher altitudes, such as a mountain top, the atmospheric layer is much thinner and exerts less pressure. See **Figure 9-6.**

Because the atmosphere is composed of a mixture of gases, pressure is exerted in all directions. At sea level, anyone or anything exposed to the atmosphere will have 14.7 pounds of pressure exerted against every square inch of surface area.

9.4.1 Atmospheric Pressure and Boiling Point

Atmospheric pressure has a direct influence on the *boiling point* of water and other liquids. If the amount of pressure on a liquid is changed, the boiling point will also change. At sea level, where atmospheric pressure is 14.7 psi (101.3 kPa), water boils at 212°F (100°C). At the summit of a mountain 15,000 feet high, however, atmospheric pressure is only 8.3 psi (57.3 kPa), so water boils at 184°F (84°C).

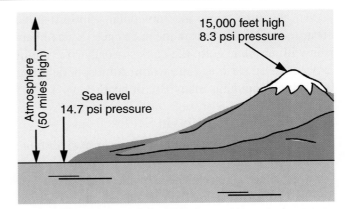

Figure 9-6. *Atmospheric pressure (14.7 psi at sea level) is a result of the weight of gas molecules in the 50-mile-high thick layer of air surrounding the earth. The higher the altitude, the lower the pressure because the blanket of air is thinner (and thus weighs less).*

When a liquid substance, such as water, reaches its boiling point (saturation point), additional heat causes a change of state. The temperature of the substance remains constant until the change is completed. To illustrate this, consider an open pan of water over a flame on a kitchen range, **Figure 9-7.** In this example, the water is at atmospheric pressure, so it will boil when a temperature of 212°F (100°C) is reached. The water cannot get hotter, *regardless of the amount of heat applied;* it will only boil at a faster rate. The temperature will remain constant until the change of state is completed, and all the liquid has changed to a gas.

Atmospheric pressure creates a surface tension on top of the water. The tension opposes the water molecules trying to break through the surface and

Figure 9-7. *At normal atmospheric pressure (14.7 psi), water boils at 212°F (100°C). The water will remain at that temperature until its state is completely changed (turns to steam). Adding heat will merely make the water boil faster.*

become a gas. As the temperature of the water increases, the speed of the water molecules also increases. If the speed of the water molecules increases to the point where surface tension can no longer keep them from escaping, the boiling point has been reached. There is now a mass escape of molecules changing state from liquid to gas (water to steam).

The pressure cooker, **Figure 9-8,** was devised to permit water to be heated to temperatures above its normal boiling point without changing state. Placing a sealed lid on top of the kettle traps the molecules of water vapor in the small space between the lid and the surface of the boiling water. As the space becomes crowded with escaped water molecules, the pressure increases. This increased pressure on the surface of the water prevents the escape of additional molecules. In effect, the water stops boiling. To continue the process of changing the water's state from liquid to gas, the

temperature of the water must be raised to overcome the increased surface pressure. Thus, as the surface pressure is increased, the boiling point is also increased.

9.4.2 Vacuum

While an increase in surface pressure results in an increase in the boiling point, the *reverse* is also true: reducing surface pressure reduces the boiling point. Any pressure less than atmospheric is called a **partial vacuum.** When *all* atmospheric pressure is removed, it is called a **perfect vacuum.** If you were able to achieve a perfect vacuum, water would boil at –90°F (–67.8°C) without a fire or other external heat source. See **Figure 9-9.**

The refrigeration and air conditioning technician typically works with refrigerant pressures that are above and below atmospheric pressure. That is why a good understanding of vacuum principles is a must.

Refrigeration systems are designed to operate with only refrigerant and oil circulating inside them. However, air and moisture do enter the system when the unit is assembled or serviced. Air is a mixture of gases and water vapor. It is noncondensable in a refrigeration system and causes high-discharge pressure (Dalton's Law of Gases, which is discussed later in this chapter). Moisture or water vapor in a refrigeration system forms hydrochloric and hydrofluoric acid, which results in compressor burnout. Proper servicing removes noncondensable gases and moisture from the system prior to charging the system with a single, pure refrigerant.

To "pull a vacuum" means to lower the pressure in a refrigeration system below atmospheric pressure. The procedure is performed with a vacuum pump such

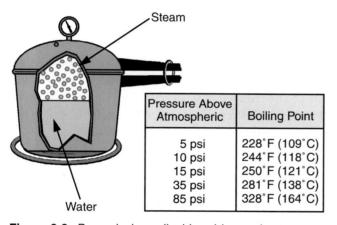

Pressure Above Atmospheric	Boiling Point
5 psi	228°F (109°C)
10 psi	244°F (118°C)
15 psi	250°F (121°C)
35 psi	281°F (138°C)
85 psi	328°F (164°C)

Figure 9-8. *By enclosing a liquid and increasing pressure (as in a pressure cooker), the boiling point of the liquid can be raised. Examples of boiling point increases are shown.*

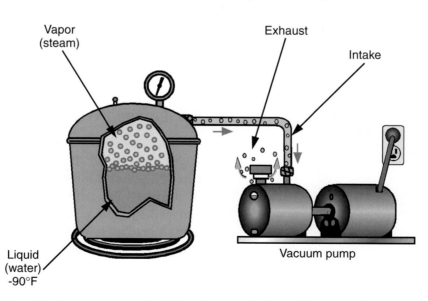

Figure 9-9. *Reducing atmospheric pressure on a liquid (creating a partial or perfect vacuum) lowers the temperature at which it will boil. In this example, a vacuum pump is used to remove all atmospheric pressure from the water in the closed container. In such a perfect vacuum, water will boil at –90°F (–68°C) without an external heat source.*

as the one depicted in Figure 9-9. The vacuum pump is a sucking mechanism that removes any gases and reduces the pressure inside the system to a very low vacuum. The vacuum pump removes moisture by lowering the boiling point of water and then removing the resulting vapor.

Measuring vacuum

There are several methods used to measure vacuum, depending upon the purpose and accuracy required. One instrument used to measure atmospheric pressure is the **mercury barometer.** As shown in **Figure 9-10,** this device consists of a hollow glass tube about 34″ high, sealed on one end and open at the other. The glass tube is filled with mercury, turned upside down, and the open end placed in a dish half-filled with mercury.

The level of mercury in the glass tube will correspond to the atmospheric pressure being exerted on the mercury in the open dish. The normal atmospheric pressure of 14.7 psi (101.3 kPa) exerted on the exposed mercury in the dish will force a 29.92″ column of mercury up into the tube. From this, it can be determined that every pound of atmospheric pressure is equal to 2.035 inches of mercury in the column (29.92 ÷ 14.7 = 2.035).

Mercury was once referred to as "quicksilver" or liquid silver. In Latin, the term "liquid silver" is **hydrargyrum.** The abbreviation of hydrargyrum, Hg, is the chemical symbol for mercury. To indicate inches of mercury in vacuum measurements, the abbreviation **in. Hg** is used. For example, a partial vacuum of 25″ of mercury would be written **25 in. Hg.**

Microns. Deep vacuums are often required before charging special refrigeration systems, such as ultra-low temperature applications or cascade systems. Deep vacuums are measured in microns for extreme accuracy. Atmospheric pressure equals 759,999 microns and a deep vacuum is about 200 microns. Such deep vacuums are seldom required, but they illustrate the need to achieve the best vacuum possible. Pulling a deep vacuum requires a special vacuum pump and electronic vacuum gauge. Microns are considered an *absolute* measurement and only used in special applications. The technician's ordinary vacuum pump is capable of pulling between 500 to 2000 microns.

Millimeters of mercury (mm Hg). Vacuum is sometimes expressed in millimeters of mercury (mm Hg). The atmosphere will support a column of mercury 760 mm (29.92″) high. Millimeters of mercury (mm Hg) is considered an *absolute* measurement and seldom used by service personnel. To achieve a perfect vacuum in a refrigeration system, for example, the pressure inside the system would have to be reduced to 0 psi, 29.92 in. Hg, 0 microns, or 0 mm Hg. This represents a perfect vacuum, which has never been achieved.

9.5 Absolute Pressure ◆

Absolute pressure is any pressure above a perfect vacuum and is expressed in terms of *pounds per square inch absolute*, or **psia.** Absolute pressure measurements allow both vacuums and pressures above atmospheric pressure to be expressed in the same units. A reading of 10 psia would be a partial vacuum (which could also be expressed as 20 in. Hg); a reading of 20 psia would be well above the atmospheric pressure of 14.7 psia.

9.6 Gauge Pressure ◆

Gauge pressure is expressed in *pounds per square inch gauge*, or **psig.** It is different from absolute pressure (psia) in that it totally ignores vacuum. Readings in psig measure only pressures *above* atmospheric; gauges are calibrated to read zero at atmospheric pressure. A reading in psig can easily be converted to the corresponding absolute value (psia) by adding atmospheric pressure (for example: 20 psig + 14.7 = 34.7 psia). A simple way to remember the difference is that psig is pressure *above* atmospheric, and psia *includes* atmospheric pressure. The difference between the two is always 14.7 psi.

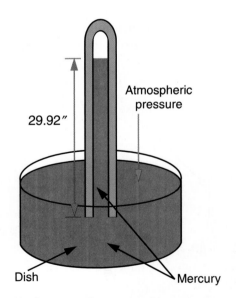

Figure 9-10. *A mercury barometer. Atmospheric pressure on the mercury in the open dish supports a column of mercury inside the tube. By marking the tube in increments of whole and fractional inches, pressure readings can be made. Normal atmospheric pressure will support a 29.92″ column of mercury in such a device. Vacuum is expressed in "inches of mercury" (in. Hg).*

9.6.1 High-pressure Gauge

All refrigeration systems are divided into two sides according to internal pressures. One side of the system operates at high pressure, the other at low pressure. A high-pressure gauge makes readings on the high-pressure side of the system, **Figure 9-11.** The scale on the gauge shown reads from zero (atmospheric pressure) to 500. Readings are in psig because the zero point is calibrated at atmospheric pressure. A high-pressure gauge is usually color-coded red.

9.6.2 Compound Gauge

In refrigeration work, the service technician must frequently check and adjust the operating pressures of the system. As noted above, the system has high-pressure and low-pressure sides. The low-pressure side of the system normally operates at or above atmospheric pressure but may slip into the vacuum area during certain operating conditions. A special gauge that displays two scales makes it possible to read pressures both above and below atmospheric. Such gauges are called **compound gauges** and are usually color-coded blue. See **Figure 9-12.**

Pressure readings

The compound gauge is calibrated with zero at atmospheric pressure and a scale that is accurate up to 120 psig. A special retarded area from 120 psig to 350 psig is provided to protect the gauge against overpressure but is *not* accurate. Pressure readings on the compound gauge are expressed as pounds per square inch gauge (psig).

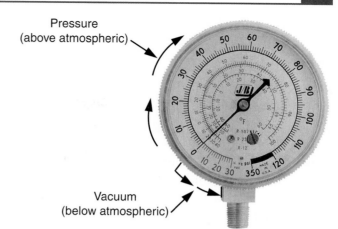

Pressure (above atmospheric)

Vacuum (below atmospheric)

Figure 9-12. *A compound gauge makes possible both pressure and vacuum readings. The zero point is atmospheric pressure; pressure readings are above zero and vacuum readings are below. Vacuum values are the opposite of the scale used with the mercury barometer: normal atmospheric pressure is zero, and a perfect vacuum is 30. (J. B. Industries)*

Vacuum readings

The other scale on the compound gauge is used for vacuum readings. This scale is calibrated downward in inches of mercury (0 in. Hg to 30 in. Hg). This numbering system is the reverse of the one used on the mercury barometer to measure atmospheric pressure changes. On the compound gauge, 30 in. Hg indicates a perfect vacuum.

Rounding off readings

For practical purposes, most technicians "round off" atmospheric pressure from 14.7 psia to 15 psia and perfect vacuum from 29.92 in. Hg to 30 in. Hg. In rounded terms, then, every pound of atmospheric pressure equals two inches of mercury. This is called a 2:1 ratio (2 in. Hg = 1 psia).

As described earlier, when the needle is in the pressure scale, the reading is expressed in pounds (psig). When the compound gauge needle is in the vacuum scale, the reading is expressed in inches (in. Hg). This distinction is important.

9.7 Gas Laws ◆

Refrigerants are the vital working fluids in any system; they are the vehicles that transport heat from one location to another. All refrigeration systems are constantly changing liquid to gas and gas back to liquid for another cycle. To read pressure gauges and diagnose system problems accurately, the technician must fully understand the laws defining the behavior of gases.

Gas laws involve three factors: *temperature*, *pressure*, and *volume*. If one factor changes, the others are affected. A gas law establishes the fact that the gas will *always* behave according to the rule, with no

Figure 9-11. *A high-pressure gauge, used only for reading pressures above atmospheric. The zero point on this gauge actually represents a pressure of 14.7 psi (normal atmospheric pressure). To avoid confusion, pressure readings made with such a gauge are identified as "pounds per square inch gauge" (psig); absolute readings are 14.7 psi higher, and are identified as "psia." (J. B. Industries)*

exceptions. These laws are so exact that the behavior of the temperature-pressure-volume relationship can be predicted and controlled.

9.7.1 Boyle's Law

Robert Boyle (1627–1691) was an English scientist who was among the first to practice chemistry as a true science. He is best known for the gas law that bears his name. **Boyle's Law** explains the relationship between the pressure and the volume of a gas if the temperature does not change (remains constant). Boyle's Law states:

The pressure of a gas varies inversely (opposite) as the volume, provided the temperature remains constant.

Figure 9-13 illustrates this law. If two cubic feet (cu. ft.) of gas at 50 psia are compressed to one cu. ft. (volume reduced by half), the pressure will double to 100 psia, provided the temperature does not change.

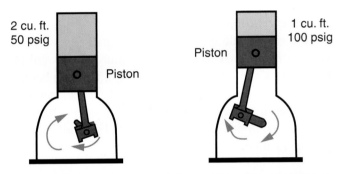

Figure 9-13. *Boyle's Law of Gases describes the inverse relationship of pressure and volume when temperature is constant: if volume is reduced by half, pressure will double. The reverse is true as well: if volume is doubled, pressure will be halved.*

Of course, the opposite is true as well; if the volume is doubled, the pressure is halved. Thus, when either the pressure or volume is changed, the corresponding pressure or volume is changed in the opposite direction (in exact proportion).

Gas laws can be proven mathematically. When working with gas law formulas, remember *absolute* pressures and temperatures must always be used. The formula for Boyle's Law is:

$$Po \times Vo = Pn \times Vn$$

(old pressure × old volume =
new pressure × new volume).

Notice that temperature does not appear in Boyle's formula because the temperature remains constant. Only the pressure and volume change. The old conditions are on the left of the equal sign, and the new conditions are on the right. The equal sign means "the same as." In other words, when you multiply the old conditions, the answer will be the same as when you

multiply the new conditions. For example, in the equation $10 \times 3 = 5 \times 6$, both sides equal 30.

In practice, the problem will always provide three numbers, or *values*, but the unknown fourth value must be discovered by division.

$$10 \times 3 = 5 \times ?$$
$$30 = 5 \times ?$$
$$30 \div 5 = 6$$
$$? = 6$$

Try an actual Boyle's Law problem: What is the new volume if 6 cu. ft. of gas at 35 psig is compressed to 85 psig, providing the temperature stays constant? (Remember to use *absolute* pressures.)

$$Po \times Vo = Pn \times Vn$$
$$50 \times 6 = 100 \times ?$$
$$300 = 100 \times ?$$
$$300 \div 100 = 3 \text{ cu. ft.}$$

Where:

$$Po = 35 \text{ psig} + 15 = 50 \text{ psia}$$
$$Pn = 85 \text{ psig} + 15 = 100 \text{ psia}$$

In this problem, the pressure doubled (50 psia to 100 psia); therefore, the volume should be reduced by half. Is it? Yes, from 6 cu. ft. to 3 cu. ft.

9.7.2 Charles' Law

As with every great invention or discovery, someone will improve upon it or carry the experiment a step further than the originator. In 1787, the French scientist Jacques Charles (1746–1823) expanded upon the work done by Boyle a century earlier. He was aware that temperature would not always stay constant, so he performed various experiments to determine how gases would behave when temperature became a **variable** (a factor that changes).

In these experiments, Charles maintained one constant factor (either pressure or volume). He was mainly concerned with what happened to the pressure or volume when the temperature changed. He did indeed prove that gases behave consistently with temperature changes. **Charles' Law** states:

"At a constant pressure, the volume of a confined gas varies directly as the absolute temperature; at a constant volume, the absolute pressure varies directly as the absolute temperature."

It is actually two laws expressed in one sentence. Charles' Law *first* deals with constant pressure, *then* with constant volume. We will consider the two laws separately.

Constant pressure

The formula for Charles' Law is:

$$Vo \times Tn = Vn \times To$$

(old volume × new temperature =
new volume × old temperature).

Notice that *Tn* and *To* do not follow the form of Boyle's equation, with "old on the left and new on the right" of the equal sign. Pressure and volume do follow this formula but temperature does not. Otherwise, the equation works just like Boyle's.

Always remember to use absolute temperatures whenever you apply gas laws. Fahrenheit is easily converted to Fahrenheit Absolute (Rankine) by adding 460°. Absolute zero is 460° lower than regular Fahrenheit zero. (Refer again to Figure 9-4.) For example, 30°F + 460° = 490°F$_A$ (490°R).

Celsius is just as easily converted to Celsius Absolute (Kelvin) by adding 273° to the Celsius temperature. Absolute zero is 273° lower than regular Celsius zero. (Refer again to Figure 9-4.) For example, 10°C + 273° = 283°C$_A$ (283°K).

Charles proved that, with a constant pressure, a temperature increase will result in a volume increase (in exact proportion). In other words, temperature and volume will change together. If one goes up, the other goes up. If one goes down, the other goes down.

Example 1: You raise 5 cu. ft. of gas at 40°F to 140°F at constant pressure. What is the new volume?

$$Vo \times Tn = Vn \times To$$
$$5 \times 600 = ? \times 500$$
$$3000 = ? \times 500$$
$$3000 \div 500 = 6 \text{ cu. ft.}$$

Where:
$$To = 40°F + 460° = 500°F_A$$
$$Tn = 140°F + 460° = 600°F_A$$

The temperature went up by 100°, from 500° F$_A$ to 600°F$_A$, while the volume went from 5 cu. ft. to 6 cu. ft. The volume and temperature increased in exact proportion, just as Charles' Law said it would.

Example 2: At a constant pressure, what is the new temperature of 2 cu. ft. of gas at 12°C when the volume is increased to 4 cu. ft.?

$$Vo \times Tn = Vn \times To$$
$$2 \times ? = 4 \times 285$$
$$2 \times ? = 1140$$
$$1140 \div 2 = 570°C_A$$

Where:
$$To = 12°C + 273° = 285°C_A$$

In this example, the volume was doubled (2 cu. ft. to 4 cu. ft.), so the temperature should also double. Did it? Yes, from 285°C$_A$ to 570°C$_A$. See **Figure 9-14.**

Constant volume

This part of Charles' Law is *very important* to all technicians since refrigeration systems are completely sealed and, therefore, maintain a *constant volume*. The pressures and temperatures in these systems are easily controlled because the volume remains constant. Charles' Law will prove that when the pressure goes

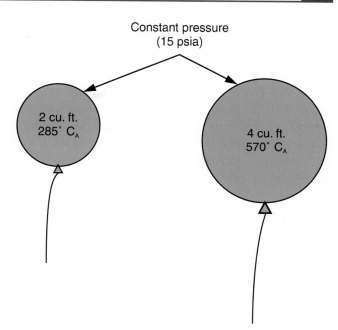

Constant pressure
(15 psia)

2 cu. ft.
285° C$_A$

4 cu. ft.
570° C$_A$

Figure 9-14. *According to Charles' Law of Gases, volume and temperature change in exact proportion to each other if pressure is constant. In this example, doubling the volume of the gas also doubles the temperature.*

up, the temperature also goes up; and when the temperature goes down, the pressure also goes down, in exact proportion. This law illustrates a major principle in refrigeration: *if the pressure is controlled, the temperature will also be controlled, and vice versa.* When solving Charles' Law problems, remember to convert gauge and temperature readings to absolute pressure and temperature.

Example 1: What is the new pressure of a confined gas at 40°F and 35 psig, if the temperature is raised to 90°F?

$$Po \times Tn = Pn \times To$$
$$50 \times 550 = ? \times 500$$
$$27,500 = ? \times 500$$
$$27,500 \div 500 = 55 \text{ psia}$$

Where:
$$Po = 35 \text{ psig} + 15 = 50 \text{ psia}$$
$$To = 40°F + 460° = 500°F_A$$
$$Tn = 90°F + 460° = 550°F_A$$

In this example, the temperature was raised from 500°F$_A$ to 550°F$_A$. Therefore, the pressure should also have increased in exact proportion. Did it? Yes, from 50 psia to 55 psia.

Example 2: What is the new temperature of a confined gas at 20 psig and 4°C, if the pressure is changed to 55 psig?

$$Po \times Tn = Pn \times To$$
$$35 \times ? = 70 \times 277$$
$$35 \times ? = 19,390$$
$$19,390 \div 35 = 554°C_A$$

Where:
$$Po = 20 \text{ psig} + 15 = 35 \text{ psia}$$
$$Pn = 55 \text{ psig} + 15 = 70 \text{ psia}$$
$$To = 4°C + 273° = 277°C_A$$

In this example, the pressure was doubled (from 35 psia to 70 psia), so the temperature should also double (277°C$_A$ to 554°C$_A$). By converting these temperatures back to Celsius (4°C and 281°C), you can readily see that a large increase in pressure can have a drastic effect on temperature.

9.8 Saturation Point

Saturation refers to the boiling (or condensing) point of a substance. The **saturation point** of a substance will vary, depending on temperature and pressure. When a liquid is heated to its boiling point, it has also achieved its saturation point. Likewise, when a gas is cooled to its condensing point, it has achieved its saturation point. When a substance has achieved a saturated condition, it is in the process of boiling or condensing. In a saturated condition, a substance can exist in one of three states: all liquid, a mixture of gas and liquid, or all gas, as shown in **Figure 9-15**. The saturation point of a substance can be changed by changing the temperature or pressure (Charles' Law of Gases).

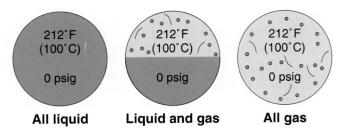

All liquid   **Liquid and gas**   **All gas**

Figure 9-15. *At the saturation point (boiling point), a substance can exist as all liquid, a mixture of liquid and gas, or all gas.*

9.8.1 Saturation Charts

Thanks to the principles discovered years ago by Jacques Charles, saturation charts have been compiled to illustrate the temperature/pressure relationship for refrigerants. A chart like the one shown in **Figure 9-16** is a working tool for all refrigeration and air conditioning technicians and engineers. Pocket-size versions of these charts are usually available free from the local refrigeration parts supplier. (Proper use of this chart is fully explained in Chapter 12, Refrigerants.)

Whenever a pure gas is confined within a constant volume (refrigeration system), the chart will reveal the *saturation temperature and pressure* for each

refrigerant. For example, if refrigerant R-134a is at 10°F, the pressure will be 12 psig. These charts do not reveal superheat or subcooling, which are explained in a later chapter.

9.8.2 True Gas Equation

In some areas of engineering, situations can develop in which there is no constant. All three factors (temperature, pressure, and volume) will change. A "true gas equation" was formulated to handle this situation. While it requires higher mathematics than are used in everyday refrigeration work; it is valuable for making scientific calculations and understanding engineering problems involving the performance of different gases. The formula combines Boyle's Law and Charles' Law:

$$\frac{Po \times Vo}{To} - \frac{Pn \times Vn}{Tn}$$

9.8.3 Dalton's Law of Partial Pressures

From 1803 to 1810, the English schoolteacher John Dalton (1766–1844) conducted a series of experiments that greatly added to the existing knowledge of gases and atoms. Dalton experimented with a *mixture* of gases rather than a single, pure gas. **Dalton's Law** established that in a mixture of gases, each gas behaves as if it occupies the space alone. To obtain the total pressure of a confined mixture of gases, the pressure for each gas in the mixture must be added. For example, if R-134a refrigerant, oxygen, and nitrogen are occupying the same space, you must add the pressure for each gas to obtain the total pressure. See **Figure 9-17.** The formula is:

Total pressure = P1 + P2 + P3

This system of partial pressures is the basis of operation for old absorption-type refrigeration systems. Such systems contain at least two gases, ammonia and hydrogen. Today, however, the vast majority of refrigeration systems are compression-type, designed to use a single, pure gas. If another gas (such as atmospheric air) enters the system, then Charles' Law no longer applies; instead, Dalton's Law becomes effective.

Major problems will develop when another gas is permitted to enter a compression-type system. The pressures will not correspond with the temperature-pressure chart. Atmospheric air contains *moisture*, a violent enemy of the compression refrigeration system. The moisture will quickly form an acid and destroy the motor windings.

Poor service procedures are the primary cause of air entering the system. Various methods can be used to rid the system of air and moisture, but severe cases

— **Saturation Chart**

Temp.		Refrigerant Number and Cylinder Color Code									
		Silver	Orange	White	Green	Purple	Blue-Gray	Light Blue	Yellow	Orchid	Aqua
°F	°C	717	11	12	22	113	R-123	R-134a	500	502	503
−50	−45.6	14.3	28.9	15.4	6.2		29.2	18.6	12.8	0.0	86.1
−45	−42.8	11.7	28.7	13.3	3.0		29.0	16.6	10.0	2.0	95.2
−40	−40.0	8.7	28.4	11.0	0.5		28.8	14.7	7.6	4.3	108.0
−35	−37.2	5.4	28.1	8.4	2.5		28.6	12.3	4.8	6.7	118.8
−30	−34.4	1.6	27.8	5.5	4.8	29.3	28.3	9.7	1.2	9.4	133.0
−25	−31.7	1.3	27.4	2.3	7.3	29.2	28.1	6.8	1.2	12.3	145.4
−20	−28.9	3.6	27.0	0.6	10.1	29.1	27.7	3.6	3.2	15.5	161.0
−18	−27.8	4.6	26.8	1.3	11.3	29.0	27.6	2.2	4.1	16.6	166.5
−16	−26.7	5.6	26.6	2.1	12.5	29.0	27.4	0.7	5.0	18.1	172.9
−14	−25.6	6.7	26.4	2.8	13.8	28.9	27.3	0.4	5.8	19.5	179.4
−12	−24.4	7.9	26.2	3.7	15.1	28.8	27.1	1.2	6.8	21.0	186.1
−10	−23.3	9.0	26.0	4.5	16.5	28.7	26.9	2.0	7.8	22.6	193.0
−08	−22.2	10.3	25.8	5.4	17.9	28.6	26.7	2.8	8.8	24.2	200.1
−06	−21.1	11.6	25.5	6.3	19.3	28.5	26.5	3.7	9.9	25.8	207.3
−04	−20.0	12.9	25.3	7.2	20.8	28.4	26.3	4.6	11.0	27.5	214.7
−02	−18.9	14.3	25.0	8.2	22.4	28.3	26.1	5.5	12.1	29.3	222.3
0	−17.8	15.7	24.7	9.2	24.0	28.2	25.8	6.5	13.3	31.1	230.0
2	−16.7	17.2	24.4	10.2	25.6	28.1	25.6	7.5	14.5	33.0	238.0
4	−15.6	18.8	24.1	11.2	27.3	28.0	25.3	8.6	15.7	34.9	246.2
6	−14.4	20.4	23.8	12.3	29.1	27.9	25.1	9.7	17.0	36.9	254.5
8	−13.3	22.1	23.4	13.5	30.9	27.7	24.8	10.8	18.4	38.9	263.0
10	−12.2	23.8	23.0	14.6	32.8	27.6	24.5	12.0	19.7	41.0	271.8
12	−11.1	25.6	22.7	15.8	34.7	27.5	24.2	13.2	21.1	43.2	280.7
14	−10.0	27.5	22.3	17.1	36.7	27.3	23.9	14.4	22.6	45.4	289.9
16	−8.9	29.4	21.9	18.4	38.7	27.1	23.5	15.7	24.1	47.7	299.2
18	−7.8	31.4	21.5	19.7	40.9	27.0	23.2	17.1	25.7	50.0	308.8
20	−6.7	33.5	21.1	21.0	43.0	26.8	22.8	18.4	27.3	52.5	318.5
22	−5.6	35.7	20.6	22.4	45.3	26.6	22.4	19.9	28.9	55.0	328.5
24	−4.4	37.9	20.1	23.9	47.6	26.4	22.0	21.4	30.6	57.5	338.7
26	−3.3	40.2	19.7	25.4	50.0	26.2	21.6	22.9	32.4	60.1	349.1
28	−2.2	42.6	19.1	26.9	52.4	26.0	21.2	24.5	34.2	62.8	359.7
30	−1.1	45.0	18.6	28.5	54.9	25.8	20.7	26.1	36.0	65.6	370.6
32	0.0	47.6	18.1	30.1	57.5	25.6	20.2	27.8	37.9	68.4	381.7
34	1.1	50.2	17.5	31.7	60.1	25.3	19.7	29.5	39.9	71.3	393.0
36	2.2	52.9	16.9	33.4	62.9	25.1	19.2	31.3	41.9	74.3	404.5
38	3.3	55.7	16.3	35.2	65.6	24.8	18.7	33.1	43.9	77.4	416.2
40	4.4	58.6	15.6	37.0	68.5	24.5	18.1	35.0	46.1	80.5	428.2
42	5.6	61.6	15.0	38.8	71.5	24.2	17.5	37.0	48.2	83.8	440.5
44	6.7	64.7	14.1	40.7	74.5	23.9	16.9	39.0	50.5	87.0	452.9
46	7.8	67.9	13.6	42.7	77.6	23.6	16.3	41.1	52.8	90.4	465.6
48	8.9	71.1	12.8	44.7	80.8	23.3	15.6	43.2	55.1	93.8	478.5
50	10.0	74.5	12.0	46.7	84.0	22.9	15.0	45.4	57.6	97.4	491.7
55	12.8	83.4	10.0	52.0	92.5	22.1	13.1	51.2	64.1	106.6	517.3
60	15.6	92.9	7.8	57.7	101.6	21.0	11.2	57.4	71.0	116.4	551.8
65	18.3	103.1	5.4	63.8	111.2	19.9	9.0	64.0	78.1	125.8	598.7
70	21.1	114.1	2.8	70.2	121.4	18.7	6.6	71.1	85.8	136.6	
75	23.9	125.8	0.0	77.0	132.2	17.3	4.1	78.6	93.9	147.9	
80	26.7	138.3	1.5	84.2	143.6	15.9	1.3	86.7	102.5	159.9	
85	29.4	151.7	3.2	91.8	155.6	14.3	0.9	95.2	111.5	172.5	
90	32.2	165.9	4.9	99.8	168.4	12.5	2.5	104.3	121.2	185.8	
95	35.0	181.1	6.8	108.3	181.8	10.6	4.2	113.9	131.3	199.7	
100	37.8	197.2	8.8	117.2	195.9	8.6	6.1	124.1	141.9	214.4	
105	30.6	214.2	11.1	126.6	210.7	6.4	8.1	143.9	153.1	229.7	
110	43.3	232.3	13.4	136.4	226.3	4.0	10.2	146.3	164.9	245.8	
115	46.1	251.5	15.9	146.8	242.7	1.4	12.6	158.4	177.4	266.1	
120	48.8	271.7	18.5	157.7	259.6	0.7	15.0	171.1	190.3	280.3	
125	51.7	293.1	21.3	169.1	277.9	2.2	17.7	184.5	204.0	298.7	
130	54.4		24.3	181.0	296.8	3.7	20.5	198.7	218.2	318.0	
135	57.2		27.4	193.5	316.5	5.4	23.5	214.5	233.2	338.1	
140	60.0		30.8	206.6	337.2	7.2	26.7	229.2	248.8	359.2	
145	62.8		34.4	220.3	358.8	9.2	30.2	245.6	263.7	381.1	
150	65.6		38.2	234.6	381.5	11.2	33.8	262.8	280.7	404.0	

Italic figures indicate Hg.   Bold figures indicate psig.

Figure 9-16. *Saturation charts show temperature-pressure relationships for various refrigerants. These quick-reference charts are easy to use. For example, at 32°F, the pressure for refrigerant R-22 is 57.5 psig. For refrigerant R-134a at the same temperature, the pressure is only 27.8 psig.*

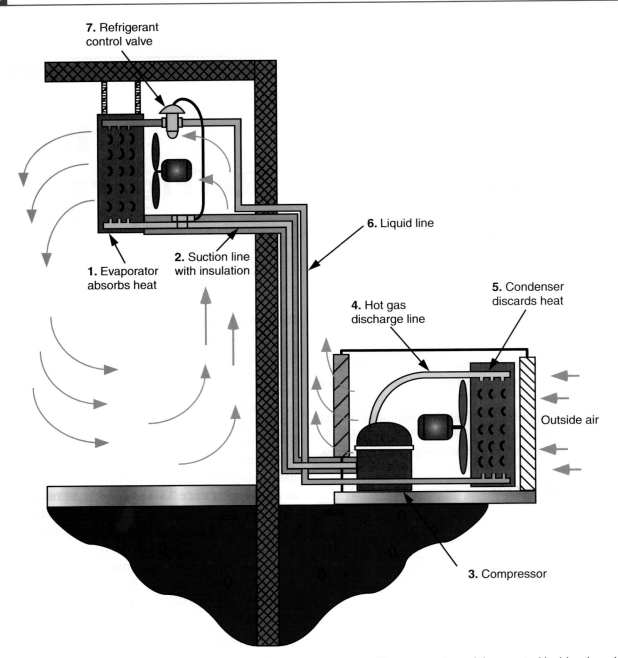

Figure 10-1. *The basic refrigeration system consists of seven components. The evaporator unit is mounted inside where heat is absorbed, and the condensing unit is located outside for disposing heat to the atmosphere. The suction and liquid lines serve to connect the two units.*

hot to cold, so warm air passing through the fins of the evaporator gives up heat to the cooler evaporator. Liquid refrigerant entering the system leaves the evaporator as vapor.

The heat travels through the fins to the copper tubing of the evaporator, **Figure 10-3.** Then, the heat is absorbed by liquid refrigerant traveling through the copper tubing. Heat absorption causes the liquid to boil (evaporate). Liquid refrigerant inside the evaporator boils at a low temperature. A refrigerant with a low boiling point is selected because a temperature difference must exist between the air and the refrigerant for heat to flow.

Saturated conditions (liquid at its boiling point) exist inside the evaporator tubing. A refrigerant metering device ensures that *all* liquid boils off in the evaporator, and only superheated gas (vapor) leaves the evaporator outlet. The normal amount of ***superheat*** (temperature above saturation) at the evaporator outlet is 10 degrees Fahrenheit (5.5 degrees Celsius). This provides maximum cooling effect and prevents liquid from leaving the evaporator.

Common refrigerants have boiling points ranging from –21°F to –50°F (–29°C to –45°C) at atmospheric pressure. Any liquid refrigerant boiling at this low temperature-pressure combination will indeed create

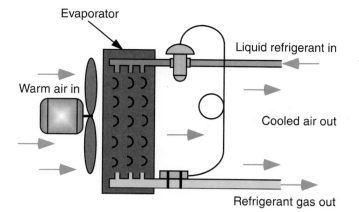

Figure 10-2. *As warm air passes through the fins of the evaporator, the air gives up heat. The heat is conveyed by the fins to the copper tubing and, in turn, to the liquid refrigerant. The added heat causes the refrigerant to boil and evaporate (change to a gas).*

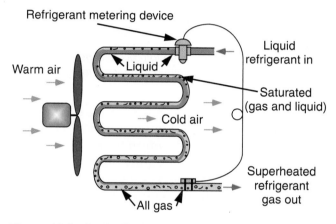

Figure 10-3. *As the liquid refrigerant inside the evaporator tubing absorbs heat from the air, it reaches a saturated condition and evaporates to a gaseous state.*

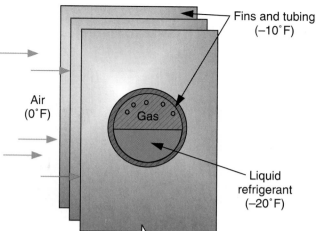

Figure 10-4. *A difference in temperature must exist between the refrigerant and the outside air for proper heat flow from hot to cold. Typically, the difference is about 20 degrees Fahrenheit (11 degrees Celsius): the fins and evaporator tubing are 10 degrees Fahrenheit (5.5 degrees Celsius) colder than the air, and the refrigerant, in turn, is 10 degrees colder than the tubing and fins.*

the temperature difference needed to cause heat to travel from the air, through the evaporator, and into the liquid refrigerant.

Evaporator temperature difference

Refrigerant inside the evaporator tubing is in a low-temperature, low-pressure, saturated condition. The boiling point of the refrigerant is deliberately controlled to provide the temperature difference necessary to remove heat.

The precise function of the refrigeration system is to create a temperature within the evaporator that is about 20 degrees Fahrenheit (11 degrees Celsius) colder than air passing through the evaporator. See **Figure 10-4.** The refrigerant is coldest, about 10 degrees Fahrenheit (5.5 degrees Celsius) colder than the evaporator. The evaporator, in turn, is about 10 degrees Fahrenheit (5.5 degrees Celsius) colder than the air flowing past its fins. The fins increase the surface area of the tubing and aid in the transfer of heat from the air to the tubing.

The evaporator **temperature difference** (td) must exist for heat to transfer from hot to cold, as specified by the Second Law of Thermodynamics. This temperature difference only exists when the system is running (called the "on" cycle). As the temperature of the air over the evaporator becomes colder, the evaporator and the refrigerant also become colder, thus maintaining the 20-degree Fahrenheit (11-degree Celsius) temperature difference.

When the temperature of the incoming air reaches the desired lower temperature level, an air-sensitive thermostat turns the system off and stops the refrigeration process. However, the evaporator fan runs constantly, circulating air through the evaporator even when the other system components are off. The continuous airflow quickly warms the evaporator and refrigerant to the same temperature as the air. Therefore, no temperature difference exists during the "off" cycle. Refrigeration cannot be accomplished during the off cycle because no temperature difference exists.

The system is turned on when the rising air temperature reaches the **cut-in** (turn-on) point of the air-sensitive thermostat. Thus, the thermostat controls the temperature of the air inside the cabinet by controlling the on-and-off cycling of the system. (See Chapter 16 for additional information on evaporator temperature differences.)

10.2.2 Suction Line

The **suction line,** usually made of copper tubing, connects the evaporator outlet to the compressor intake. See **Figure 10-5.** The suction line is normally

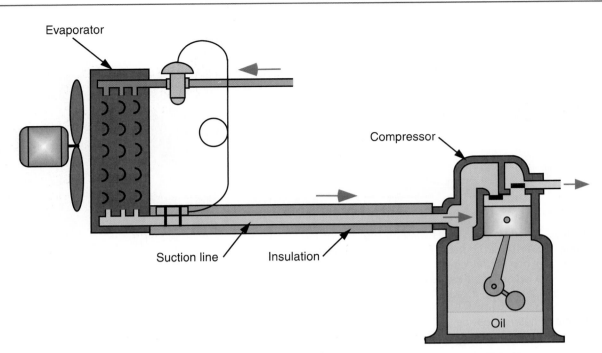

Figure 10-5. *The suction line connects the evaporator and the compressor. The line is insulated to prevent condensation from forming on the line and freezing. The suction line should contain only low-pressure, low-temperature gas; liquid refrigerant could damage the compressor.*

insulated to prevent moisture-laden air from contacting the cold tubing and condensing. Insulation eliminates moisture or frost problems.

The refrigerant inside the suction line is a superheated gas, even at this low temperature. The compressor draws the gaseous refrigerant through the copper suction line, similar to someone sipping soda through a straw. The drawing or sucking action of the compressor creates the low pressure needed in the evaporator to provide a low boiling point for the liquid refrigerant.

Brazing is used to connect the suction line to the evaporator; the compressor connection may be brazed or flared. Regardless of the type of connection used, the suction line must be leakproof.

The condition of the refrigerant inside the suction line is a *low-temperature, low-pressure, superheated gas.* The superheat is obtained in the refrigerant's final pass through the evaporator and all the way through the suction line. Care must be exercised to prevent liquid refrigerant from entering the compressor where it can cause severe damage.

10.2.3 Compressor

The *compressor* is a vital part of the system and probably the least understood. Compressors often are unjustly condemned as being "inefficient" when the problem actually occurs elsewhere. The technician must have a thorough understanding of how the compressor operates. (Descriptions of the operation of various types of compressors are found in Chapters 19 and 20.)

The compressor is one of two division points that separate the *low-pressure side* of the system from the *high-pressure side.* See **Figure 10-6.** The actual division point is located at the *valve plate* that contains the suction and discharge valve reeds.

The purpose of the compressor is twofold. First, it must remove vapor from the evaporator (by the sucking action) to maintain a low boiling-point pressure inside the evaporator. Second, it must compress (squeeze) this low-temperature, low-pressure gas into a small volume, which results in a high-temperature, high-pressure gas.

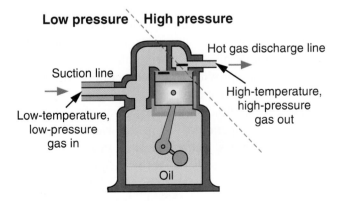

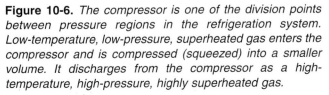

Figure 10-6. *The compressor is one of the division points between pressure regions in the refrigeration system. Low-temperature, low-pressure, superheated gas enters the compressor and is compressed (squeezed) into a smaller volume. It discharges from the compressor as a high-temperature, high-pressure, highly superheated gas.*

Reciprocating compressor

Reciprocating means moving first in one direction, then the opposite (back and forth, or up and down). This is the operating principle of the reciprocating compressor widely used in refrigeration systems, **Figure 10-7.** The reciprocating compressor operates much like an automobile engine. It has two or more pistons driven by a crankshaft that causes the pistons to make alternating suction and compression strokes. At the top of each piston cylinder are two valve reeds: a *suction valve reed* and a *discharge valve reed.*

The downstroke of the piston creates a partial vacuum in the cylinder. This permits the low-pressure

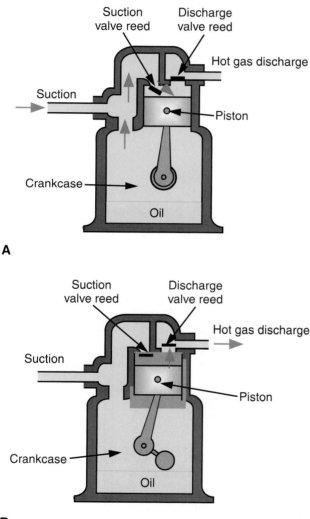

A

B

Figure 10-7. *Compressor operation. A—On the piston downstroke, lowered pressure in the cylinder allows low-pressure, low-temperature refrigerant gas to push open the suction valve reed and flow into the piston cylinder. High pressure in the discharge line holds the discharge valve reed closed. B— On the piston upstroke, increased pressure closes the suction reed valve. At the top of the piston stroke, pressure of the compressed gas is high enough to force open the discharge valve reed. High-pressure, high-temperature gas flows from the cylinder into the hot gas discharge line.*

gas in the suction line to force open the suction valve reed and flow into the cylinder. On the upstroke (compression stroke), increasing pressure in the cylinder allows the suction valve reed to close. The gas is compressed until the pressure within the cylinder overcomes the higher pressure in the hot gas discharge line that holds the discharge valve reed closed. This occurs at the top of the stroke, after the gas has been compressed to a much smaller volume and a much higher pressure and temperature. The discharge valve reed opens, allowing compressed refrigerant to flow into the hot gas discharge line.

The increased pressure created by the compressor pushes the refrigerant around the refrigeration system. Both the vacuum (sucking action) on the low side of the compressor and the pushing action on the high side are important qualities in a compressor. A malfunction on either side will greatly affect the system. The compressor *must* perform both duties simultaneously to have a properly operating system. For this reason, most compressors have at least two pistons. While one piston is on the downstroke, the other is on the upstroke.

The operating pressures of the system are very important. The boiling point (temperature-pressure) inside the evaporator must be controlled to achieve the desired cooling effect. To maintain the correct boiling point, the compressor must be able to remove vapor from the evaporator at the same rate the liquid is boiling off (changing state to a gas). If the temperature-pressure inside the evaporator is permitted to rise, the boiling point also increases (Charles' Law), reducing the cooling effect. Of course, the opposite is also true. Removing the vapor too fast (and lowering its boiling point) causes the evaporator to become too cold.

The compressor is the most expensive and critical component of the system. The compressor must be perfectly matched to other system components to balance the effect of the refrigerant circulating within the system. The refrigerant must be pushed through the high side of the system at the same rate the vapor is being removed from the evaporator by the vacuum action of the compressor. If the refrigerant fails to circulate properly, problems will develop. All system problems can be traced to one of two causes: no refrigerant in the system or poor circulation.

10.2.4 Hot Gas Discharge Line

The refrigerant being discharged by the compressor is in the form of a high-pressure, high-temperature gas. The discharged hot gas travels through copper tubing to the condenser. See **Figure 10-8.** The copper tubing is called the **hot gas discharge line.**

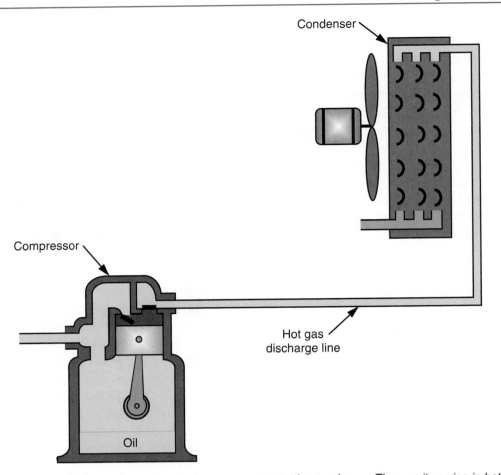

Condenser

Compressor

Hot gas
discharge line

Oil

Figure 10-8. *The hot gas discharge line connects the compressor to the condenser. The gas it carries is both high-temperature and high-pressure, so the connections must be leakproof.*

The hot gas discharge line is smaller than the suction line because the gas has been compressed to a smaller volume and higher pressure. While the suction line is cold or cool, the hot gas discharge line is hot. Due to heat generated by compression, the discharge gas contains about 100 degrees Fahrenheit (55 degrees Celsius) of superheat. Since the line must carry the superheated gas at high pressure (from 100 psig to 350 psig or 690 kPa to 2415 kPa), it must be leakproof.

10.2.5 Condenser

The **condenser** is a heat exchanger, somewhat like the evaporator. While the evaporator is designed to absorb heat, the condenser is designed to release heat. The sole purpose of the condenser is to remove heat from the superheated refrigerant vapor, causing the vapor to condense (change state) back to a liquid, as shown in **Figure 10-9.** Only liquid should leave the condenser. Most condensers are cooled by the flow of ambient (surrounding) air through the fins, but some are water-cooled. The differences between air-cooled and water-cooled condensers is explained later in this chapter.

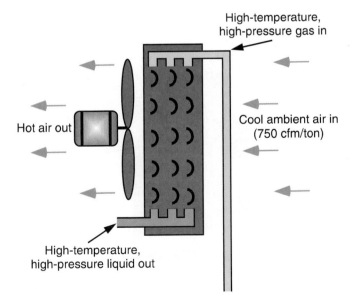

High-temperature,
high-pressure gas in

Hot air out

Cool ambient air in
(750 cfm/ton)

High-temperature,
high-pressure liquid out

Figure 10-9. *The job of the condenser is to remove heat from the refrigerant gas and change it back to a liquid state.*

The high pressure created by the compressor raises the saturation (condensing) point of the refrigerant to the point where ambient air can be used to:

❏ Remove superheat to the saturation point (sensible heat).

❑ Remove enough additional heat to cause the refrigerant to condense back to a liquid (latent heat).

❑ Remove enough additional heat to partially subcool the liquid before it leaves the condenser (sensible heat).

When the refrigerant passes through the first part of the condenser, superheat is quickly removed, lowering the temperature of the high-pressure gas to the saturation point. Continued removal of heat causes a change of state from gas to liquid. In the final passes of the condenser tubing, all the refrigerant is in liquid form, permitting the liquid to fall below saturation point, or **subcool**, before it leaves the condenser. At the condenser outlet, the liquid is still at high pressure, but the temperature is lukewarm (saturation temperature or subcooled).

Condenser operation

The main objective of the operation is to have all liquid refrigerant in the final pass of the condenser tubing. To illustrate, consider ambient air temperature of 70°F (21°C) entering the condenser. On a running system using R-134a, the compressor raises the saturation point about 30 degrees Fahrenheit (17 degrees Celsius) above ambient air temperature, or 100°F (38°C). The saturation pressure for R-134a at 100°F (38°C) is 124 psig (856 kPa). (Refer to the saturation chart, Figure 9-16.)

The temperature of the highly superheated gas at the compressor discharge is about 200°F (93°C). Some superheat is **dissipated** (lost) as the gas travels through the uninsulated hot gas discharge line, so the gas arrives at the condenser at about 150°F (65°C). Ambient air at 70°F (21°C) passing through the condenser first removes all superheat (about 50 degrees Fahrenheit or 28 degrees Celsius), quickly lowering the temperature of the gas to the saturation point of 100°F (38°C).

Ambient air continues to pass through the condenser, removing additional heat (latent heat) and causing the gas to condense back to a liquid. The final lengths of condenser tubing should contain all liquid, which is easily subcooled by at least 10 degrees Fahrenheit or 5.5 degrees Celsius (sensible heat). Including subcooling, the temperature of the liquid leaving the condenser is about 90°F (32°C), lukewarm to the touch. Subcooling in the final condenser passages makes the system more efficient.

During the refrigerant's change of state (and subcooling) in the condenser, the discharge pressure remains 124 psig (856 kPa). Although heat has been removed, pressure gauges continue to indicate saturation pressure. Pressure gauges cannot read superheat or subcooling. Remember, at saturation (boiling point) you can have all gas, a mixture of liquid and gas, or all liquid, depending upon how much heat has been removed.

10.2.6 Liquid Line

The copper-tubing **liquid line** connects the condenser outlet to the metering device (called the refrigerant control), **Figure 10-10.** The liquid line

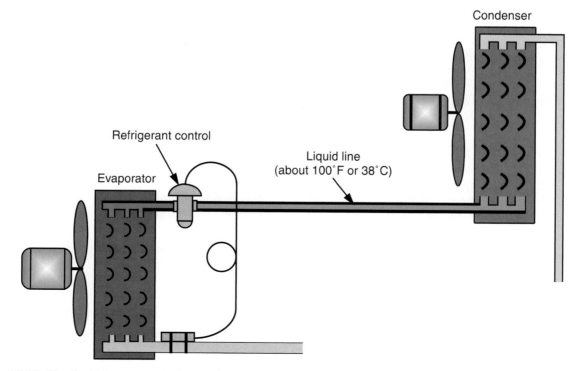

Figure 10-10. *The liquid line connects the condenser with the refrigerant control. Refrigerant leaving the condenser is a lukewarm liquid under high pressure.*

should contain only *liquid* refrigerant subcooled by about 10 degrees Fahrenheit (5.5 degrees Celsius). However, all liquid at the saturation point (with no subcooling) is also acceptable. The liquid refrigerant is still under high pressure (saturation pressure), and the tubing is slightly warm to the touch (about 100°F or 38°C). The refrigerant inside the liquid line is a high-temperature, high-pressure, preferably subcooled liquid.

10.2.7 Refrigerant Control

The *refrigerant control,* **Figure 10-11,** is a device that plays a key role in the efficient and automatic operation of the system. The control meters the flow of liquid refrigerant into the evaporator to ensure that all liquid is boiled off before it enters the suction line. Allowing liquid refrigerant to reach the compressor will result in compressor failure.

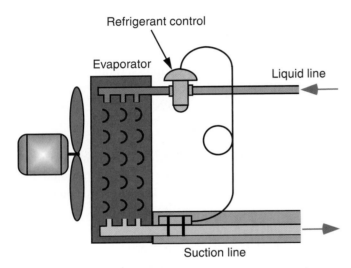

Figure 10-11. *The refrigerant control acts as a valve to meter the flow of liquid refrigerant into the evaporator. Metering is necessary to prevent liquid refrigerant from reaching and damaging the compressor (floodback).*

The refrigerant control is the second division point between the high- and low-pressure sides of the system. (Recall that the compressor is the first division point.) The control works like a valve to automatically regulate the amount of liquid refrigerant entering the evaporator. At the control *inlet,* the liquid is in a high-temperature, high-pressure, subcooled condition. At the control *outlet* (as a result of a pressure drop across the valve), the liquid is in a low-temperature, low-pressure, saturated condition.

The amount of low-pressure liquid entering the evaporator is critical. The liquid must be correctly metered into the evaporator so it boils off before it reaches the compressor. In other words, the amount of liquid entering the evaporator must be controlled to match the speed of evaporation. When the liquid boils rapidly, more liquid must pour into the evaporator. When the boiling process is slow, the valve must restrict the amount of liquid entering the evaporator.

There are different types of refrigerant controls; exactly how each one operates is explained in Chapter 16.

Condition of refrigerant in each component

To avoid confusion and misunderstanding, it is important to understand the basic refrigeration system and the condition of the refrigerant within each of its components. The system illustrated in **Figure 10-12** is the basis for all compression refrigeration systems, regardless of how large or small.

Since refrigerant circulating within a system cannot be seen, thermometers and gauges are used to check system operation. The technician must know where to obtain these readings and how to diagnose problems correctly. Refer to Figure 10-12 to locate the following components:

1. *Evaporator:* Component where heat is absorbed.
   a. *Inlet:* Low-temperature, low-pressure liquid.
   b. *Outlet:* Low-temperature, low-pressure gas.
2. *Suction line:* Low-temperature, low-pressure, slightly superheated gas.
3. *Compressor:* Division point between high and low pressure.
   a. *Inlet:* Low-temperature, low-pressure, slightly superheated gas.
   b. *Outlet:* High-temperature, high-pressure, highly superheated gas.
4. *Hot gas discharge line:* High-temperature, high-pressure, highly superheated gas.
5. *Condenser:* Component where heat is discarded.
   a. *Inlet:* High-temperature, high-pressure, superheated gas.
   b. *Outlet:* Moderate-temperature, high-pressure, saturated liquid.
6. *Liquid line:* High-temperature, high-pressure, (preferably) subcooled liquid.
7. *Refrigerant control:* Division point between high-pressure and low-pressure sections of the system (the first cold spot in the system).
   a. *Inlet:* High-temperature, high-pressure, subcooled liquid.
   b. *Outlet:* Low-temperature, low-pressure, saturated liquid.

10.3 Variations in Basic Refrigeration Components

System components often serve an identical purpose but may vary in size, style, and shape. These differences are necessary to save cost or space, or to

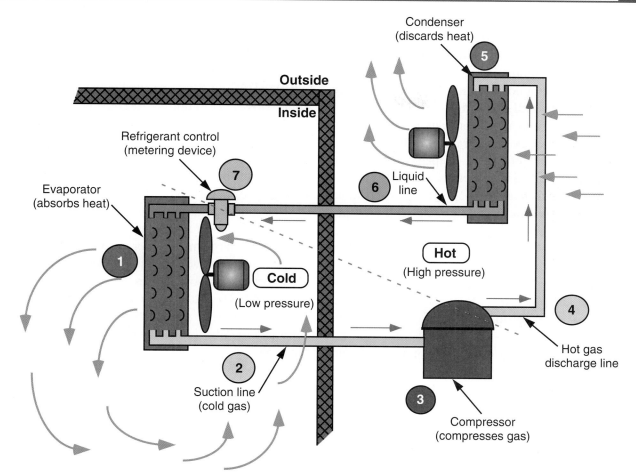

Figure 10-12. *The basic refrigeration system consists of seven components. The diagonal dashed line represents the boundary between the low-pressure (cold) side of the refrigeration system and the high-pressure (hot) side.*

improve efficiency. Variations in a number of major components are described in the following sections.

10.3.1 Evaporators

Evaporators are manufactured in many shapes and sizes to fill specific design or operational needs. They can operate on the principles of **conduction** (flow of heat through metal or another solid) or **convection** (flow of heat through a liquid or gas). Convection can be *natural* or *forced*.

Evaporators are classified into five types:
- ❑ Shell.
- ❑ Shelf.
- ❑ Wall.
- ❑ Plate.
- ❑ Finned-tube.

Shell-type evaporator

The *shell-type evaporator,* **Figure 10-13,** is commonly used as the freezing compartment in domestic refrigerators that do not have automatic defrost. The evaporator takes the form of a metal box that serves as the freezer compartment. Food is placed inside the shell (or box) for freezing.

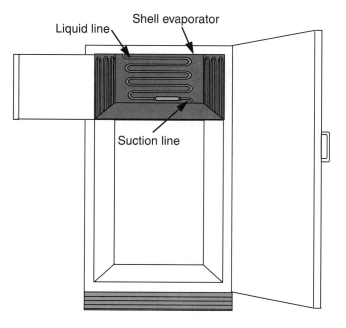

Figure 10-13. *The shell-type evaporator is commonly used for the freezing compartment of manual-defrost domestic refrigerators. The aluminum shell is easily damaged, so care must be exercised when defrosting.*

The shell-type evaporator is made of thin aluminum sheet. The refrigerant travels through passages built into the shell during the manufacturing process. Since aluminum is a good conductor of heat, the shell-type evaporator operates on the principles of conduction and natural convection.

The main problem with this evaporator is its thin aluminum construction. The metal is easily punctured, resulting in loss of refrigerant. Repairs are difficult, time-consuming, and expensive.

Frost accumulates on the evaporator and acts as an insulator, preventing heat from getting to the evaporator. Proper defrosting is important, but the use of a sharp instrument is forbidden because of the danger of puncturing the evaporator. Defrosting is accomplished by turning the refrigerator off and placing a pan of hot water inside the evaporator. After the frost and water are removed, recovery to normal operating temperature is fairly rapid.

The shell-type evaporator includes an **accumulator,** a small reservoir used as a safety device to trap any liquid refrigerant that did not evaporate during passage around the evaporator. Liquid flowing into the accumulator evaporates before being drawn into the suction line.

Shelf-type evaporator

The *shelf-type evaporator,* **Figure 10-14,** is used extensively in domestic upright freezers. The evaporator is made of aluminum tubing that forms the shelves inside the freezer. Additional wires are soldered to the aluminum tubing to increase the strength and evaporator surface area of each shelf. Food containers are placed upon these shelves. Heat transfer is accomplished by conduction and natural convection.

Because warm air rises, one "shelf" of tubing, called the ceiling coil, is located in the uppermost part of the cabinet. Most upright freezers have three or four shelves, not counting the ceiling coil. The refrigerant goes to the ceiling coil first, then down through each of the shelves to the bottom. This is called a **series connection** because the refrigerant must travel through one shelf before going to the next.

The upright freezer is popular because it is easy to select frozen foods from its organized shelving. The drawback is that every time the door is opened, cold air spills out the bottom, and warm, moisture-laden air enters at the top. Frost readily accumulates on the ceiling coil and upper shelf. The excess frost acts as an insulator and prevents heat from getting to the evaporator. Manual defrosting must be done about once a year. Defrosting is accomplished by turning the freezer off, removing any packages of frozen food, placing one or two pans of hot water inside the freezer, and closing the door. Heat from the hot water causes the temperature inside the freezer to rise rapidly, melting or loosening the accumulated frost, which is removed by hand. Like the shell-type evaporator, the use of a sharp instrument on a shelf-type unit is forbidden. The sharp instrument might puncture the aluminum evaporator, permitting the refrigerant to escape. A puncture in the aluminum tubing is expensive to repair.

Wall-type evaporator

The *wall-type evaporator* is commonly used in chest-type domestic freezers. The evaporator tubing is made of steel or aluminum and is firmly attached to the outside surface of the inner cabinet liner.

The arrangement provides a smooth inside surface with uniform cooling throughout the cabinet. The rings of tubing go around the sides only, not across the bottom or lid. The refrigerant travels through the top ring first and continues to the lowest ring. Heat transfer is accomplished by conduction and natural convection.

An advantage of the chest freezer over the upright is that cold air does not spill out when the door is opened. Less frost-accumulation requires less frequent defrosting (once every two or three years) and less running time (making the freezer more energy efficient).

Plate-type evaporator

The commercial *plate-type evaporator,* **Figure 10-15,** is usually made of steel, and the refrigerant travels through passages formed in the process of making the plate. The cold plates are normally hung from the ceiling, with drip pans made from sheet metal located below the plates to catch and drain off the condensate.

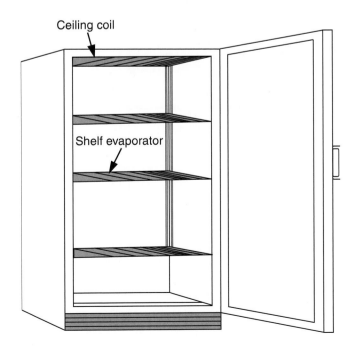

Ceiling coil

Shelf evaporator

Figure 10-14. *The shelf-type evaporator is found in upright domestic freezers. The refrigerant tubing forms the shelves. A ceiling coil cools air that rises to the top of the freezer.*

The plates operate on the principle of natural convection (hot air rises and cold air falls). As warm air rises and encounters the cold plates, it gives up its heat to the plates and sinks toward the floor. This

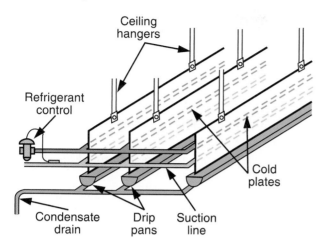

Figure 10-15. *The plate-type evaporator is used in older commercial-type installations, such as walk-in coolers. It usually hangs from the ceiling of the cooler, creating slow, steady air movement by natural convection.*

method creates constant, slow air movement that keeps temperature levels quite uniform.

The cold plate system is not common, but variations of it are used in both domestic and commercial applications. The cold plate was the forerunner of the forced convection evaporator discussed next.

One type of cold plate is used in some domestic refrigerator-freezer combinations. Made of aluminum, it is located in the upper back section of the refrigerator compartment (not the freezer compartment). See **Figure 10-16.** The cold plate is connected "in series" with the freezer evaporator. It cannot receive any liquid refrigerant until the freezer is satisfied; then, the leftover liquid reaches the refrigerator cold plate. In this system, the cold plate functions as a suction accumulator to prevent liquid floodback.

Finned-tube-type with forced convection

The *finned-tube-type* of evaporator is widely used in domestic, commercial, and industrial applications. The introduction of the finned-tube, forced-convection evaporator made possible remarkable advances in the

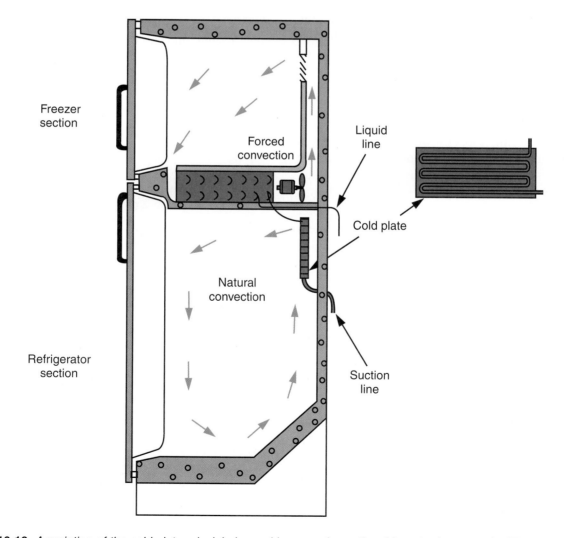

Figure 10-16. *A variation of the cold plate principle is used in some domestic refrigerator-freezer units. The plate is located at the back of the refrigerator section and connected in series with the freezer finned-tube type evaporator.*

field of refrigeration and air conditioning. The *forced convection* evaporator has a mounted electric fan to increase airflow through the evaporator. The system has made it possible to quickly accomplish large amounts of refrigeration (Btu/hr.) with a small evaporator.

The surface area of the evaporator tubing is significantly increased by the addition of fins. Fins are usually made of aluminum and are securely bonded to the tubing. Spacing the fins varies the capacity of the evaporator and compensates for depth. The fins must be kept straight and equally spaced, or airflow is reduced, in turn reducing the capacity of the evaporator. Most evaporators are designed for a specific airflow in cubic feet per minute (cfm); the design airflow should not be changed. Changing the capacity of the evaporator affects operation of other system components which are sized and selected according to the original capacity.

10.3.2 Suction Line

The correct size and proper installation of the suction line is important to the efficient operation of a refrigeration system. The size of the suction line (as well as the liquid line) is determined by several factors involving the condition of the gas it contains. Factors include velocity, pressure, volume, density, and pressure drop. The size of the copper tubing will vary according to the size of the system, ranging from 1/4″ to 6″ (6 mm to 152 mm) OD.

The gas-carrying capacity of the suction line is closely calculated to match the flow of gas from the evaporator to the compressor. A very long suction line, or one with many bends or fittings, will lead to a *pressure drop.* An excessive pressure drop (greater than 2 psi or 14 kPa) in the suction line is equivalent to operating the compressor at a lower pressure, which reduces compressor capacity.

Due to the compressor's lubrication design, some oil tends to circulate through the system along with the refrigerant. It is important that this oil return to the compressor and not become trapped somewhere in the system. Oil return can be improved by slanting the suction line downward from the evaporator so oil drains naturally into the compressor. All horizontal suction lines should slope toward the compressor at the rate of 1/4″ for every 10′ of tubing. A low spot in the suction line functions as an oil trap, **Figure 10-17.** Oil accumulates in such a trap and decreases the efficiency of flow. As the low spot becomes filled with oil, the vapor builds up a pressure against the slug of oil, and the compressor lowers the pressure on the other side. As soon as the pressure difference overcomes the weight of the oil, the oil is sent back to the compressor.

Oil trapping is an advantage in commercial systems where the condensing unit is located above the evaporator section. An oil trap is deliberately installed in the suction line at the base of the *vertical riser,* **Figure 10-18.** The vertical riser is usually reduced-diameter tubing, which increases the velocity of the refrigerant to help sweep the oil up the riser and back to the compressor.

Suction line insulation

Suction lines are insulated for two reasons:

- ❏ To limit superheating of gas that would reduce compressor efficiency.
- ❏ To eliminate condensation that could cause frost and ice problems.

The superheated gas inside the suction line is in a low-temperature, low-pressure state. This cold gas cools the compressor, so the amount of superheat gained through the copper tubing suction line must be limited. High superheat at the compressor inlet reduces compressor efficiency and causes damage due to excessive discharge superheat.

Insulating the suction line eliminates contact between moisture-laden atmospheric air and the cold copper tubing, preventing condensation. If moisture is allowed to condense on the copper tubing, it will drip. If the suction line temperature is below 32°F (0°C), atmospheric moisture will freeze on the tubing, creating frost and ice problems.

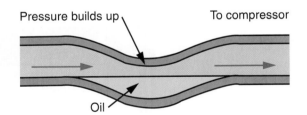

Figure 10-17. *A low spot in the suction line functions as an oil trap. As the oil accumulates, vapor pressure builds up on the evaporator side of the trap. Eventually, it pushes the slug of oil to the compressor.*

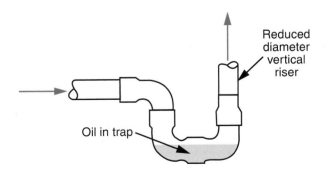

Figure 10-18. *An oil trap in a commercial installation. The vertical riser uses reduced-diameter tubing to increase refrigerant velocity and help sweep oil back to the compressor.*

Suction lines are usually insulated with black, vaporproof flexible tubing. The tubing is available in 4′ lengths, with wall thicknesses of 3/8″, 1/2″, 3/4″, or 1″. Insulation is available with inside diameters (ID) ranging from 3/8″ to 4″ or larger.

In new installations, the insulation is slipped onto copper tubing and over elbows, **Figure 10-19.** For existing systems, insulation is slit vertically with a sharp knife and slipped around tubing and fittings, **Figure 10-20.** A special adhesive is applied to seal all slits, seams, and butt joints.

10.3.3 Reciprocating Compressor Styles

The compressor, **Figure 10-21,** is the "heart" of the refrigeration system and is driven by an electric motor. The electric motor can be mounted outside the compressor or inside the same housing as the compressor.

There are three basic types of compressors:

- ❑ *Open:* This type of compressor is belt-driven from an external electric motor.
- ❑ *Semihermetic:* The electric motor and compressor are enclosed in a forged iron body that is sealed and bolted. The motor shaft is directly connected to the compressor crankshaft.
- ❑ *Hermetic:* The motor and compressor are sealed within a steel body that is welded together. Like the semihermetic type, the hermetic compressor is directly driven from the motor shaft.

Both the open-type and semihermetic compressors are repairable. Since hermetic compressors are sealed within a welded enclosure, they cannot be repaired. (Compressors and motors are fully explained in later chapters.)

10.3.4 Condensers

Condensers are manufactured in a variety of shapes, sizes, and styles. Condensers remove heat that was absorbed by the refrigerant during its passage through the evaporator and compressor. Air-cooled condensers operate on the principle of forced convection or natural (sometimes called *static*) convection.

Forced convection permits the use of smaller condensers because increased airflow removes heat faster. Natural convection condensers remove heat by exposing large metal surfaces to slow-moving air currents. Heat travels from the warm metal surface to the cooler air currents. The disadvantage of natural convection condensers is the need for one or more fans to increase airflow to the required 750 cfm (cubic feet per minute) per ton of refrigeration. The temporary addition of fans, or spraying water on a natural convection condenser, changes the condenser's name. This is a temporary "fix" for a very hot ambient.

The condenser must be located where adequate airflow is available and the discharge of heat is not objectionable. Commercial and industrial refrigeration systems locate the condensing unit outside the building or in equipment rooms. Domestic systems are normally self-contained, and the condensing unit is attached to, but outside, the refrigerated space.

The condenser must be able to remove heat from the refrigerant at the same rate the heat is being absorbed during passage through other system components. The main source of heat to be removed by the

A

B

Figure 10-19. *Installing insulation. A—Insulation is slipped onto tubing and over elbows in new installations. B—For tees and other fittings, the material is cut, then joined with adhesive. (Halstead)*

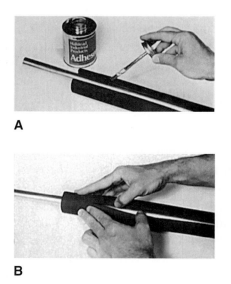

A

B

Figure 10-20. *Insulating existing installations. A—Insulation is slit to fit over tubing and fittings, then adhesive is applied. B—Adhesive bonds seams quickly when they are pressed together. (Halstead)*

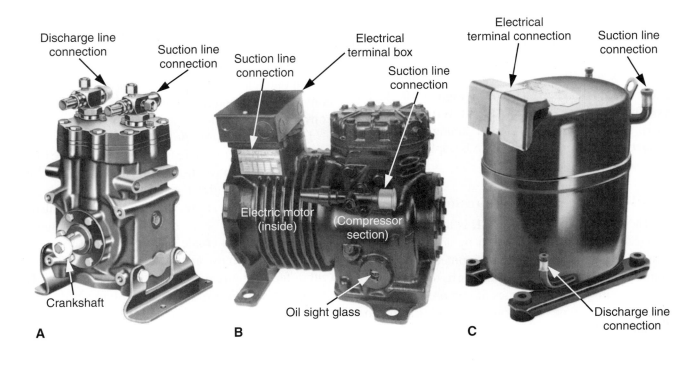

Figure 10-21. *Basic styles of reciprocating compressors. A—Open-type compressor, belt-driven by an external electric motor. (Tecumseh) B—Semihermetic-type compressor, with the motor and compressor inside bolted housing. (Copeland Corp.) C—Hermetic-type compressor, with the motor sealed inside a welded metal shell. (Copeland Corp.)*

condenser comes from the evaporator. Additional heat is absorbed by the cool refrigerant vapor when passing over the compressor motor windings.

To achieve proper condenser operation, the compressor must increase the refrigerant pressure to a saturation temperature 30 to 35 degrees Fahrenheit (17 to 19 degrees Celsius) higher than the ambient air entering the condenser. By raising the temperature-pressure relationship, the saturation point (condensing point) of the refrigerant is increased. This makes it possible to use ambient air or water as a cooling medium to condense the refrigerant to a liquid. The compressor is designed to maintain the 30-degree Fahrenheit (19-degree Celsius) difference during the running cycle, regardless of changing ambient temperature. If the ambient temperature goes up, the discharge pressure also goes up. Likewise, if the ambient temperature goes down, the discharge pressure also goes down.

For example, on a hot summer day, the ambient temperature may reach 95°F (35°C). The condensing temperature (saturation point) would be about 125°F (52°C) and the condensing pressure would be close to 185 psig (1277 kPa).

This method of determining discharge pressure applies to most air-cooled condensers. A high-efficiency condenser is able to reduce the temperature

difference about 20 degrees Fahrenheit (11 degrees Celsius). R-502 systems use high-efficiency condensers to reduce discharge pressure and control pumping ratios. These systems require the use of a 20-degree Fahrenheit (11-degree Celsius) differential for calculating discharge (saturation) pressure. For example, with an ambient of 80°F (27°C), the condensing temperature is 100°F (38°C). Checking the temperature-pressure chart reveals a condensing pressure of about 216 psig (1490 kPa).

These calculated differences are not precise, but they *are* guidelines for evaluating equipment performance. The actual temperature-pressure may vary slightly above (or below) the calculated pressure. The condensing temperature-pressure relationship is dependent upon several factors, but wide differences between actual pressure and calculated pressures should be investigated. Also, remember to use dewpoint pressures when calculating discharge pressures for *zeotropic refrigerants* (fluids with two or more components that have different vapor pressures and boiling points) in the 400-series.

Water-cooled condensers are more efficient than air-cooled condensers, and the discharge temperature-pressure corresponds to a temperature about 20 degrees Fahrenheit (11 degrees Celsius) higher than the temperature of the water exiting the condenser.

Condenser cooling stages

As noted earlier, the refrigerant is cooled in three stages as it circulates through the condenser:

- ❏ First, the superheat of the gas is removed, cooling the gas down to the saturation point. (Sensible heat is removed.)
- ❏ Second, the gas is condensed to a liquid. (Latent heat is removed.)
- ❏ Third, the liquid is subcooled to a point below its saturation point. (Sensible heat is removed.)

As soon as the superheat of the gas is removed, the saturation point is achieved. Further heat removal will cause the gas to condense back to a liquid (hence, the name *condenser*). The condenser's job is to remove enough heat so all the gas condenses, leaving only liquid during the final passages where subcooling occurs.

Air-cooled condensers are easy to install, inexpensive to maintain and, if sized properly, satisfactory in all regions. Water-cooled condensers are very efficient but are expensive to operate and keep clean. Water-cooled condensers are used occasionally where cool ambient air is not available.

10.3.5 Domestic Condensers

Domestic refrigeration commonly uses three types of air-cooled condensers:

- ❏ Finned-tube (forced convection).
- ❏ Wire static (natural convection).
- ❏ Wall static (natural convection).

Finned-tube, forced convection condenser

This is the most commonly used condenser. It is made of steel tubing with external aluminum fins that provide a large and efficient heat transfer surface. Heat transfer is accomplished by forcing large quantities of air through the compact tubing and fin assembly. The amount of air flowing through the condenser should be about 750 cfm per ton of refrigeration. Airflow is critical for proper operation of the condenser. (Refer to Figure 10-9 for typical operation of this type of condenser.)

The forced convection condenser is generally preferred for commercial applications. An adequate supply of fresh air must be available at all times. In order to achieve their compact size, these condensers are normally constructed with a small face (or surface area) and a depth of several rows of tubing. As the air is forced through the condenser, it absorbs heat from the refrigerant inside the tubes.

Most forced convection condensers use a fan that draws, rather than blows, air through the condenser. The draw-type fans result in more uniform airflow, which is preferred because even air distribution increases condenser efficiency.

Air-cooled condensers must be kept clean. The condenser fan draws into the condenser such foreign material as lint, dust, dirt, or leaves. Foreign material acts as an insulator and prevents proper airflow.

A dirty condenser causes the compressor to work harder. The condensing temperature-pressure must be raised to compensate for the reduced condenser capacity. This results in excessive discharge pressures that will seriously affect the operation of the system. Cleaning the condenser is an easy, but often neglected, task. Cleaning is done with a brush, vacuum cleaner, or by other appropriate means.

Wire static condenser

The wire static (natural convection) condenser, **Figure 10-22,** is used on domestic refrigerators. This condenser consists of rows of steel tubing with lengths of wire attached to the tubing. The wire increases the surface area of the tubing, permitting better heat transfer.

The condenser is mounted on the back of the refrigerator, so the appliance should not be placed tightly against a wall. A space of at least 4″ to 6″

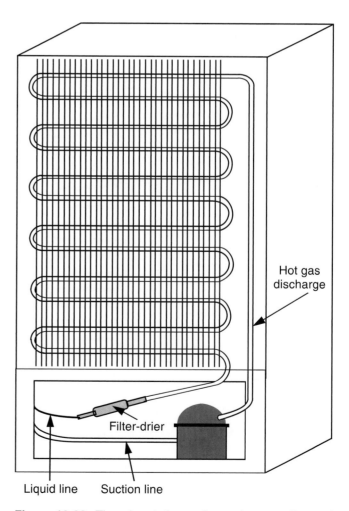

Hot gas discharge

Filter-drier

Liquid line   Suction line

Figure 10-22. *The wire static condenser is generally used for domestic refrigerators. Sufficient air circulation space must be provided around the condenser for efficient cooling.*

(10 cm to 15 cm) must separate the back of the refrigerator from the wall to allow for natural convection airflow over the condenser.

Wall static condenser

The wall static condenser is efficient, hidden from view, and space-saving. It is used on both chest and domestic upright freezers. The condenser tubing is securely fastened to the inside surface of the outer shell, **Figure 10-23.** A layer of insulation separates it from the inner shell (upon which the evaporator is mounted). The unit is actually a box inside a box, with insulation between the two boxes. Of course, the inner box is very cold, while the outside of the freezer is warm to the touch. The insulation between the two boxes retards heat flow between the two temperature levels. When installing these freezers in the home, allowance must be made for proper airflow around the outside of the cabinet.

10.3.6 Commercial Condensers

Forced convection, air-cooled condensers are most common in commercial systems, although water-cooled systems are not unusual. Air-cooled units are cheaper to operate and have fewer service problems than do water-cooled systems. Commercial air-cooled condensers vary greatly in size and style. Some outdoor units are mounted vertically; others are mounted horizontally, with one or more electric fans to draw air through the condenser. See **Figure 10-24.**

Commercial water-cooled condensers

At one time, water-cooled condensers were very common. However, they have become costly to operate because the water is usually discharged into a sewer drain after passing through the condenser. Water and sewage service rates have increased with the growing demand on facilities. In some cases, a water-cooling tower is installed on the roof to cool the water for recirculation through the condenser. As a result of increasing costs, many water-cooled systems are being converted to air-cooled systems.

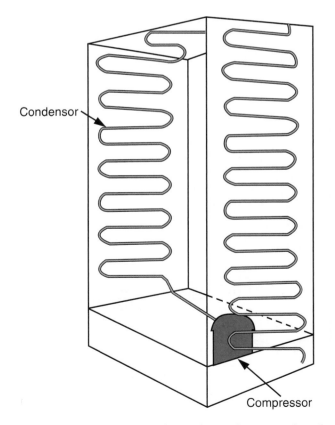

Figure 10-23. *The wall static condenser is mounted on the inner surface of the outer wall of a chest or upright freezer. A layer of insulation separates it from the evaporator mounted on the inner wall. It is an efficient condenser design.*

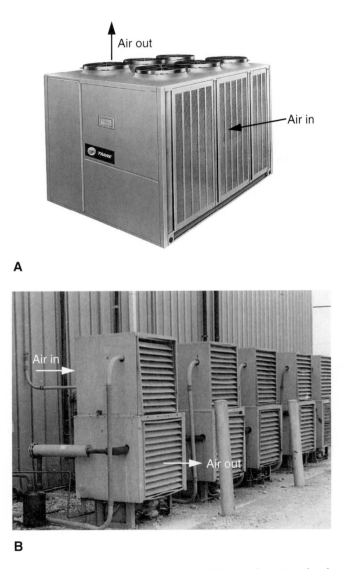

A

B

Figure 10-24. *Air-cooled commercial condensers. A—A typical unit equipped with several electric fans for forced convection. Ambient cool air is drawn through the side of the unit, and hot air is exhausted through the top. (Trane) B—A series of air-cooled condensers at a manufacturing plant. Ambient air is drawn in one side and exhausted out the other.*

Water-cooled condensers are very efficient but also present certain problems related to water conditions. Lime and scale deposits occur inside the condenser tubing, which tends to insulate the tubing and reduce efficiency. Scale deposits can severely restrict or even completely stop waterflow. Special water treatment chemicals are added to the cooling water to reduce deposits and prolong equipment life.

Water-cooled condensers are available in four types, suitably named for their designs:

- ❑ Tube-in-a-tube.
- ❑ Tube-in-a-shell.
- ❑ Tube-in-a-coil.
- ❑ Coil-in-a-shell.

Tube-in-a-tube condenser. The tube-in-a-tube condenser, **Figure 10-25,** has one tube located inside another tube. Cooling water circulates through the inside tube, and hot gas discharges through the space between the tubes. The *counterflow* principle is applied for maximum efficiency, meaning the water flows in one direction while the hot gas flows in the opposite direction. The warmest water is next to the warmest refrigerant, and the coolest refrigerant is next to the coolest water.

Tube-in-a-shell-condenser. This condenser is a cylinder made of steel with copper tubes running from end to end inside it, **Figure 10-26.** Water circulates through the copper tubes and condenses the hot vapors inside the cylinder to a liquid state. The bottom part of the cylinder serves as a storage tank for excess liquid refrigerant.

The tube-in-a-shell condenser has some distinct advantages. It is compact, usually located under the compressor frame, and needs no cooling fan. Manifolds at each end of the condenser control the flow of water through the copper tubing. When these manifold ends are removed, the water tubes are accessible for easy removal of scale deposits.

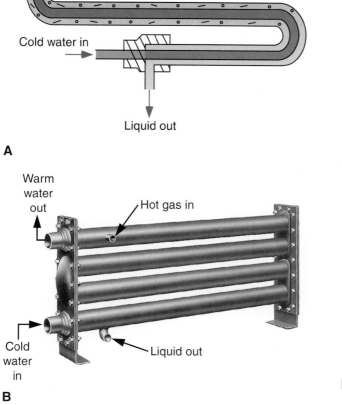

A

B

Figure 10-25. *Tube-in-a-tube condenser. A—Water flows through the inside tube. Water flow is opposite of refrigerant flow (counterflow). B—A typical tube-in-a-tube condenser. (Standard Refrigeration Company)*

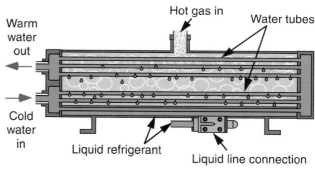

A

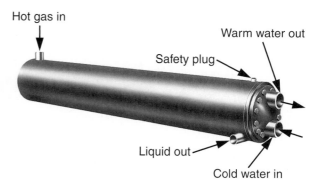

B

Figure 10-26. *Tube-in-a-shell condenser. A—Water flows through the tubes inside the cylinder, absorbing heat from the refrigerant gas and condensing it to a liquid. B—A typical condenser of the tube-in-a-shell type. (Standard Refrigeration Company)*

Cleaning scale deposits from water-cooled condensers is usually done with a wire brush attached to a rod turned by an electric drill. The brush is run back and forth through the water tubing to remove the scale. Removing the scale deposits may reveal small corrosion pits that have penetrated the copper tubing, allowing water to enter the refrigeration system. Cleaning water-cooled condenser tubing is always a gamble.

Tube-in-a-coil and *coil-in-a-shell condensers.* These condensers can only be cleaned with a special acid pump. The pump circulates an acid through the water tubing to dissolve scale deposits. The system is then flushed with fresh water and placed back in service.

10.3.7 Water-cooling Tower

Water-cooling towers, **Figure 10-27,** are commonly used to reduce the cost of operating water-cooled refrigeration systems. The purpose of the ***water-cooling tower*** is to capture the water leaving the condenser and lower its temperature so the water can be recirculated through the condenser for further cooling.

The tower is usually located on a rooftop or other outdoor location where ambient air can cool the water. The hot water leaving the condenser is piped to the top of the tower where a distribution pan or shower heads distribute it evenly, **Figure 10-28.** The falling water droplets flow through a large chamber with many rows of metal fins or wooden slats serving as heat transfer surfaces. The greater the water surface in contact with air flowing through the chamber, the more efficient the cooling action. A large fan forces ambient air between the slats; the water cools as it falls through the chamber. A catch basin, or reservoir, in the bottom of the tower collects the cooled water.

The cooled water in the catch basin is recirculated through the water circuit by a motor-driven water pump. The cooled water first travels to the compressor water jacket, then to the condenser, and back to the tower for recooling.

The catch basin also contains a water makeup valve and float assembly to maintain the proper water level in the reservoir. **Makeup water** is necessary because some water is lost to evaporation. Cooling towers evaporate about two gallons of water every hour for each ton of refrigeration.

Sometimes the water reservoir, water pump, and makeup water valve assembly are located inside the building for protection against freezing. The reservoir must be able to hold all the water in the system. A thermostat prevents the water from freezing by turning the tower fan on and off during cold weather. The sensing element for the thermostat is located in the water reservoir.

10.3.8 Water-regulating Valve

The amount of water traveling through the compressor *water jacket* (casing) and the condenser is controlled by a water-regulating valve located at the inlet to the water jacket, **Figure 10-29.** The water-

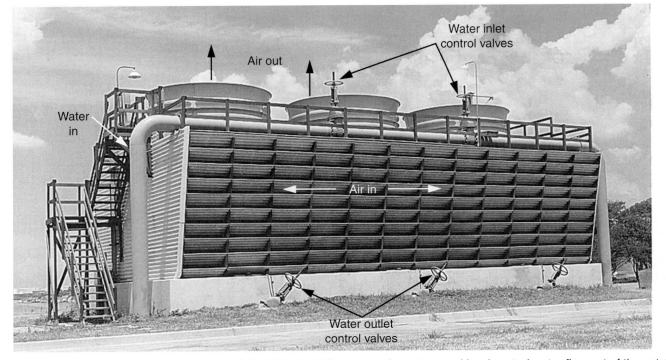

Figure 10-27. *A large cooling tower at an industrial plant. The large valves at ground level control water flow out of the catch basin. (The Marley Cooling Tower Company)*

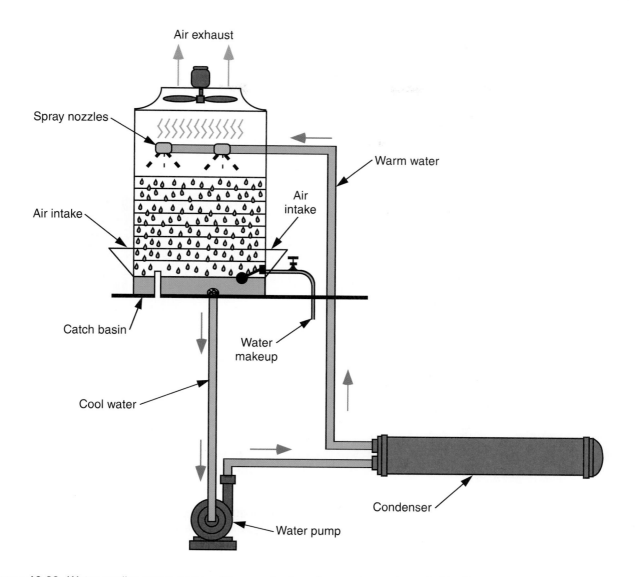

Figure 10-28. *Water-cooling tower system. Warm water from the condenser is cooled in the tower and returned for reuse.*

regulating valve is designed to control the head pressure by governing the flow of cooling water to the condenser. A bellows and diaphragm assembly at the bottom of the valve has a capillary tube connected to the high-pressure side of the compressor head. The capillary tube transfers the head pressure to the bottom of the water-regulating valve. As the head pressure increases, the valve opens and permits more water-flow. Likewise, as the head pressure decreases, the water flow decreases.

The spring tension on top of the water-regulating valve is adjustable and permits the technician to precisely control the head pressure setting. See **Figure 10-30.** The water valve automatically adjusts the amount of water flow according to the head pressure; thus, a constant head pressure is maintained while the system is running. When the compressor stops, the head pressure falls below the valve setting,

and the waterflow is stopped during the off cycle. The water-regulating valve is very reliable, and once set, seldom requires further attention.

On water-cooled systems, the normal head pressure settings should correspond to a saturation temperature of 100°F (38°C). The setting is comparable to an air-cooled system operating at an ideal ambient of 70°F (21°C). To save water, higher head pressure may be desirable.

Dirty tubes can affect the efficiency of a water-cooled condenser. Check the head pressure gauge reading; it should correspond to about 20 degrees Fahrenheit (11 degrees Celsius) above the temperature of the water leaving the condenser. For example, if the water temperature leaving the condenser is 80°F (27°C), the head pressure should correspond to 100°F (38°C). If the saturated head pressure is higher than calculated, the condenser tubes probably need to be cleaned.

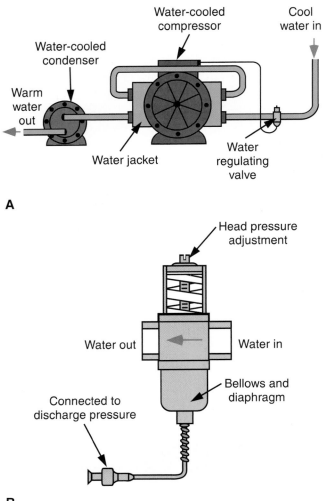

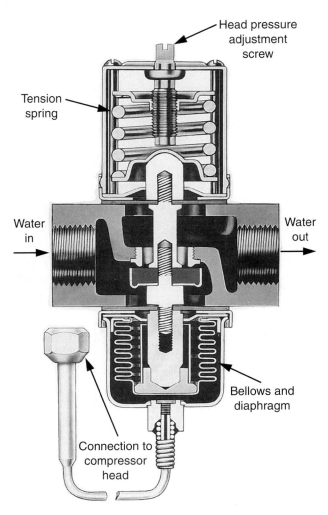

Figure 10-29. *Water-regulating valve. A—The valve is located at the inlet of the compressor water jacket where it controls water flow. B—A capillary tube transmits the compressor discharge pressure to the bellows and diaphragm of the valve to vary the water flow. As head pressure increases, the valve opens further to increase the flow of water.*

Figure 10-30. *A spring-loaded adjustment at the top of the water-regulating valve permits the technician to precisely control head pressure. (Penn)*

Summary

All refrigeration systems are designed to control the movement and condition of the circulating refrigerant. This chapter introduced the basic refrigeration system, the operation of its seven components, and variations in the components. The purpose and operation of each component was fully explained, along with the condition of the circulating refrigerant.

Later chapters will introduce other system components and explain how the entire system is controlled. The technician should be able to draw the basic system from memory and state the condition of the refrigerant as it circulates through the system. This knowledge is necessary for recognizing problems and understanding later areas of study. Troubleshooting and repairs cannot be performed without knowing how and why each component affects the system.

Test Your Knowledge

Please do not write in this text. Write your answers on a separate sheet of paper.

1. Name the seven components of a basic refrigeration cycle in order of refrigerant flow. Begin at any point in the cycle.
2. What is the purpose of the evaporator?
3. What is the condition of the refrigerant when it is in the evaporator?
4. What is the difference in temperature between the air flowing over the evaporator and the refrigerant inside the evaporator?
5. What material is used to make the suction and liquid lines?
6. What is the condition of the refrigerant in the suction line?
7. Name the two division points between the high-pressure and low-pressure sides of the refrigeration system.
8. Briefly describe the function of the compressor.

9. What is the condition of the refrigerant in the hot gas discharge line?
10. True or false? The suction line is larger in diameter than the discharge line.
11. What is the purpose of the condenser?
12. After air has passed through the condenser, is the air hot or cold?
13. After air has passed through the evaporator, is the air hot or cold?
14. In what condition is the refrigerant as it enters the condenser?
    a. Low-temperature, low-pressure, superheated gas.
    b. Medium-temperature, high-pressure liquid.
    c. High-temperature, high-pressure, highly superheated gas.
15. In what condition is the refrigerant as it leaves the condenser?
    a. Low-temperature, low-pressure, superheated gas.
    b. Medium-temperature, high-pressure liquid.
    c. High-temperature, high-pressure, highly superheated gas.
16. Describe the function of the component called the refrigerant control.
17. By controlling the boiling point (pressure) in the evaporator, you also control the _____.
18. Name the five types of evaporators.
19. Which of the following is not a type of reciprocating compressor?
    a. Open.
    b. Semi hermetic.
    c. Hermetic.
    d. Osmotic.
20. Name the three types of air-cooled condensers.

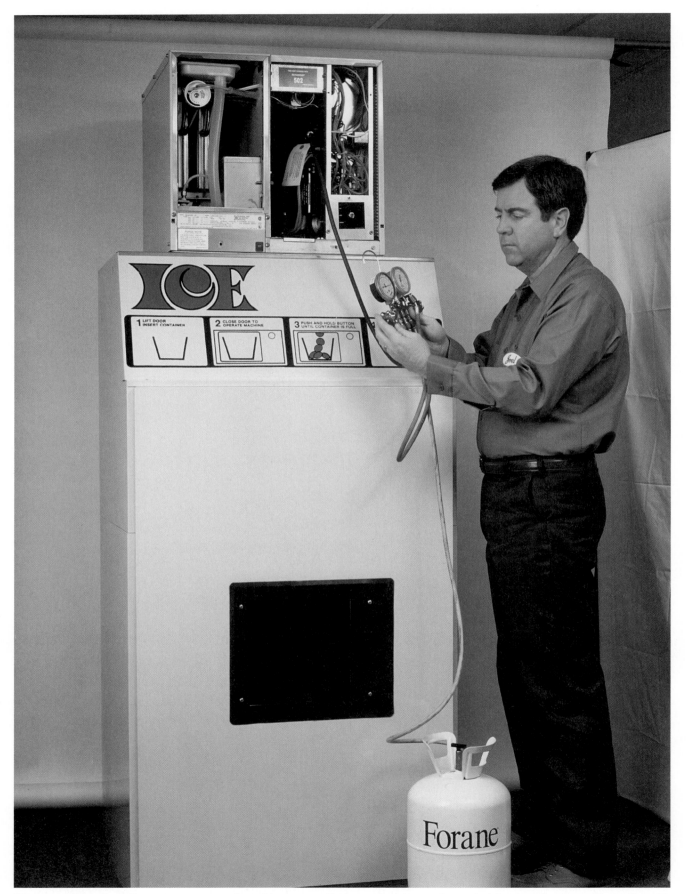

When working with refrigerant systems such as this commercial ice maker, the service technician must be throughly familiar with the functions of each of the components that make up the system.

Other System Components

Objectives

After studying this chapter, you will be able to:
❑ Identify condensing and evaporator units.
❑ Describe the condition of refrigerant in various accessory components.
❑ Describe the purpose of system accessory components.
❑ Identify component variations.
❑ Name accessory components and describe the purpose of each.
❑ Install and use a gauge manifold.
❑ Discriminate between components in domestic and commercial systems.

Important Terms

absorb	liquid receiver
adsorb	liquid receiver service
aspirator hole	valve
backseated	moisture indicator
charging	noncondensibles
condensing unit	overcharge
cracked	pigtail
desiccant	pointer flutter
dip tube	recalibration
discharge service valve	recovery
evacuating	Schrader valve
evaporator unit	sight glass
filter-drier	suction accumulator
frontseated	suction service valve
gauge manifold	undercharged
heat exchanger	vacuum pump
hydrostatic expansion	

11.1 Evaporating and Condensing Units

As described in Chapter 10, the basic refrigeration system consists of seven components: evaporator, suction line, compressor, hot gas discharge line, condenser, liquid line, and refrigerant control.

These components can be further grouped into condensing and evaporating units. The **condensing unit** consists of the equipment necessary to reclaim the refrigerant gas and convert it back to a liquid. The condensing unit, **Figure 11-1,** contains the compressor, condenser, hot gas discharge line, condenser fan, electrical panel box, and some accessory components. The **evaporator unit** consists of the evaporator, refrigerant control, evaporator fan, and some accessory components. The suction and liquid lines connect the evaporator unit with the condensing unit to complete the system.

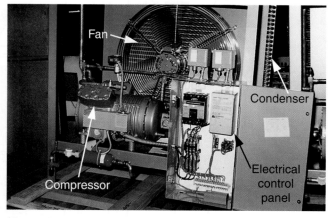

Figure 11-1. *A condensing unit for a commercial installation, consisting of a compressor, condenser, fan, and electrical control panel. The unit may be located far away from the evaporating unit. (Dunham-Bush)*

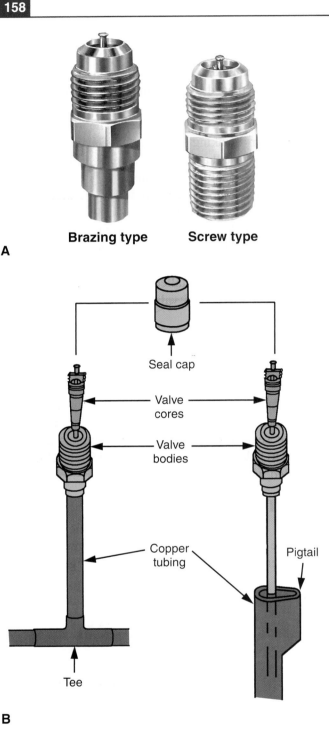

Figure 11-19. *Schrader valves are installed for permanent service access to domestic systems. A—Brazing-type and screw-type Schrader valves. (J.B. Industries) B—Because heat will affect the gasket material, the cores must be removed while the valve body is being brazed to the tubing.*

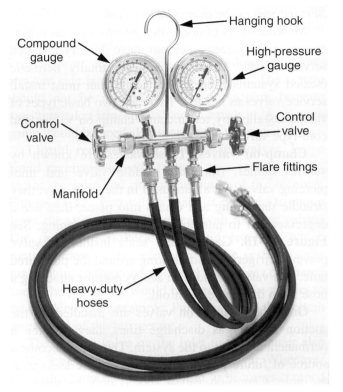

Figure 11-20. *The gauge manifold is the technician's basic tool for determining system pressures. It is equipped with two gauges, two control valves, and three connection hoses. (Uniweld)*

Learning how to use the gauge manifold properly will save countless hours diagnosing field problems, determining their causes, and deciding their solutions. Without the gauge manifold, it is impossible to identify system problems. Therefore, the service technician must become thoroughly familiar with the manifold and its many uses.

11.3.1 Pressure Gauges

The gauge manifold has two gauges for a wide range of readings. See **Figure 11-21.** The compound gauge (blue) is designed for connection to the low-pressure side of the refrigeration system. It has scales to show pressure readings from 0 psig to above 120 psig (828 kPa), and vacuum from 0 in. Hg to 30 in. Hg. The high-pressure gauge (red) is designed for connection to the high-pressure side of the refrigeration system; it typically shows only pressure readings (0 psig to 500 psig or 0 kPa to 3450 kPa).

The rapid compression strokes of the compressor pistons sometimes create pressure pulsations that cause the gauge pointer to swing above and below the actual pressure reading. Called ***pointer flutter,*** it does not indicate problems with the compressor or any other type of defect. The correct pressure reading is obtained at the center of the flutter. Special pressure gauges, **Figure 11-22,** are available with a built-in pulsation dampener to prevent pointer flutter.

as noted earlier in this chapter, is a pressure-checking device with both compound and high-pressure gauges. See **Figure 11-20.** It also has control valves and connectors for hoses to the service valves. The gauges reveal the system's operating pressures which, in turn, help the technician determine the needed repairs. The manifold is also used while performing the repairs.

The saturation temperature-pressure relationship for one or more refrigerants is included on the dial face of refrigeration gauges. The outside scale on the dial face (black numbers) indicates gauge pressure (psig). The three inner scales (red numbers) indicate Fahrenheit saturation temperatures that correspond with the pressure for each of the three refrigerants. To read the scales properly, simply follow the line of the needle to the proper temperature scale to determine the temperature that corresponds with the pressure reading.

Newer gauges include temperature scales for R-134a, while older gauges normally show R-12, R-22, and R-502. The scales are a convenience and do not affect the gauge pressure readings. Consult a saturation temperature-pressure card for other refrigerants.

Remember that refrigeration gauges only reveal *saturation* temperature for a given pressure. The gauges do *not* reveal superheat (temperature above saturation) or subcooling (temperature below saturation). Because the gauges only read saturation, a thermometer is used to determine if the actual temperature is above or below the saturation point (superheat or subcooling).

For efficient operation of the system, the amount of superheat or subcooling at various points must be determined and controlled. For example, a pressure reading at the suction service valve for R-134a may be 12 psig (83 kPa), which corresponds to 10°F (–12°C) saturation temperature. However, if a thermometer is clamped to the suction line at the inlet to the suction service valve, the reading would be about 25°F (–4°C). The difference in temperature readings amounts to 15 degrees Fahrenheit (25 – 10 = 15) or 8 degrees Celsius. Therefore, the refrigerant vapor contains 15 degrees Fahrenheit (8 degrees Celsius) of superheat as it enters the suction service valve.

Recalibrating gauges

Refrigeration gauges are sensitive instruments, initially calibrated to produce accurate readings. While these gauges can withstand some abuse, they should be handled with care. It is not unusual for the gauges to require **recalibration** in the field because of use and handling.

To gain access to the recalibration screw, the clear crystal face on the gauge must be unscrewed, **Figure 11-23.** The recalibration screw is located on the gauge face, just below the needle hub. The hose on

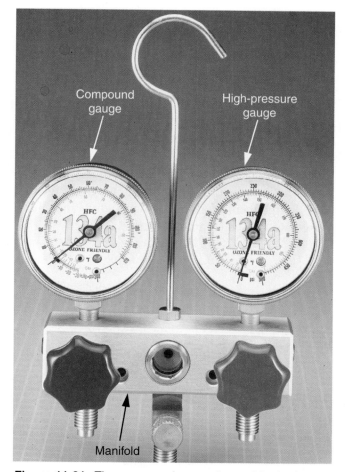

Figure 11-21. *The compound gauge is used for both pressure and vacuum readings, while the high-pressure gauge is used only for pressure readings. (TIF Instruments, Inc.)*

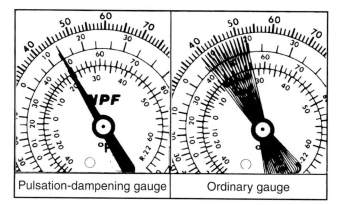

Figure 11-22. *A special gauge with a pulsation dampener eliminates pointer flutter, making more accurate readings possible. (Uniweld)*

Figure 11-23. *Unscrew the clear crystal of the gauge to reach the recalibration screw on the dial face. For accurate readings, gauges must be recalibrated, or zeroed, periodically. (Imperial Eastman)*

the manifold body immediately beneath the gauge is removed to expose the inlet port to atmospheric pressure. A suitable screwdriver is used to slowly turn the recalibration screw until the pointer lines up with zero (atmospheric).

11.3.2 Manifold Body

The compound gauge and the high-pressure gauge are directly connected to the hoses by special bypass gas passages through the manifold body. A hose connection is located directly beneath each gauge. The gas passage between each gauge and its connecting hose is *never closed*. See **Figure 11-24.** The two hand valves located at each end of the manifold control access to the center hose only. These valves *do not* control the bypass gas passages to the gauges.

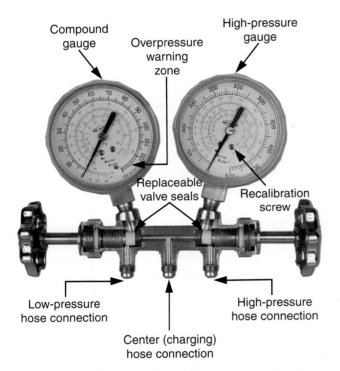

Figure 11-24. *Cross section of the gauge manifold body. The hand valves are used only for access to the center hose. (Uniweld)*

The center hose is for adding refrigerant to a system (**charging**), removing refrigerant from a system (**recovery**), or emptying a system with a vacuum pump (**evacuating**). Each of these procedures requires the hand valves be opened properly to gain access to the center hose. Both valves are normally closed and should be opened only when access to the center hose is needed. The left-hand valve controls access to the center hose via the low-pressure side and the right-hand valve via the high-pressure side. Opening both hand valves opens the center hose to both gauges.

11.3.3 Refrigeration Hoses

Refrigeration hoses are designed to withstand working pressures of 500 psig to 750 psig (3450 kPa to 5175 kPa). The hoses are normally color-coded: *blue* for low pressure, *red* for high pressure, and *yellow* for the center connection. The hoses are also designed to remain flexible under most temperature conditions for easy handling. Each hose has a straight connector at one end and an angled (45°) connector at the other, **Figure 11-25.** The straight ends connect to the 1/4″ flare fittings on the manifold body. The angled ends easily connect to the proper service valve by finger-tightening. Angled ends contain a valve-core depressor for connection to Schrader (core-type) valves.

A special gasket inside the angled hose end provides a leakproof seal on the service valve or Schrader valve. The gasket can become badly worn due to long use or abuse, but it is easily replaced. Replacement gaskets are available from your local supplier. Before fully hand-tightening refrigerant hoses, purge them of air by allowing a tiny amount of refrigerant to escape through them.

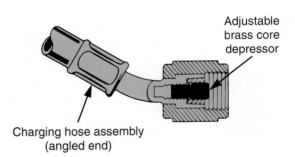

Figure 11-25. *A refrigeration hose is designed to withstand high pressures and temperature extremes. The straight connector attaches to the manifold body; the angled connector attaches to the service valve or Schrader valve.*

11.4 Purging and Venting

"Purging" and "venting" are terms that describe two repair procedures involving the release of refrigerant to the atmosphere.

Purging describes the necessary process of releasing a small quantity of refrigerant from the hose ends after connection to a system. Purging removes atmospheric moisture and air from the service hoses, thus preventing contamination of the system. Small releases of refrigerant due to purging, connecting, or disconnecting hoses is not prohibited by the Clean Air Act.

Some service procedures require the removal of all pressure (refrigerant) from the system prior to performing repairs. Older service procedures simply removed the pressure by *venting* (releasing) the system refrigerant to the atmosphere. Because of ozone depletion, the deliberate release of refrigerant to the atmosphere is prohibited by federal law. New recovery equipment and procedures for saving the refrigerant for reuse are described in Chapter 13.

11.4.1 Evacuating the System

"Evacuation" means to clean a system with a **vacuum pump** to remove all gas and moisture. Evacuation is done before charging a system with refrigerant. **Figure 11-26** shows the arrangement for evacuating a system.

The yellow hose is connected to a two-stage vacuum pump. The vacuum pump removes moisture

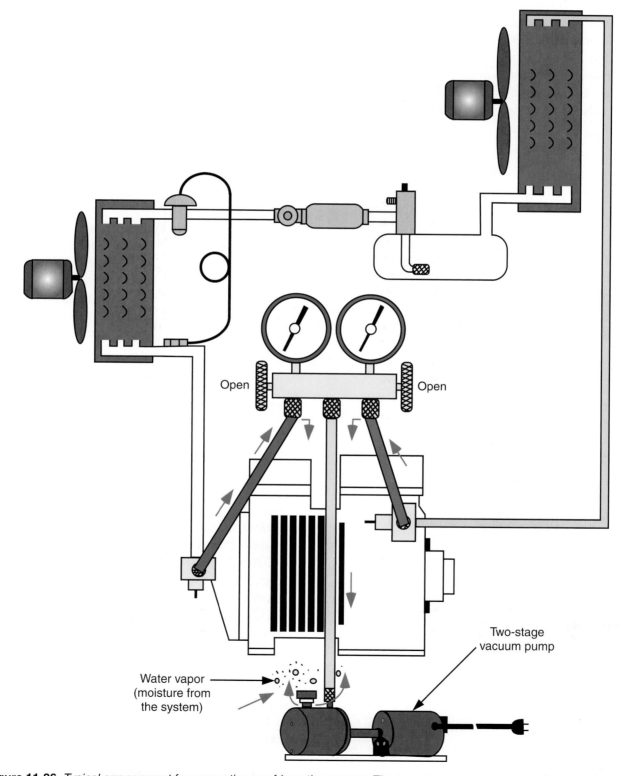

Figure 11-26. *Typical arrangement for evacuating a refrigeration system. The two-stage vacuum pump reduces system pressure low enough to vaporize any moisture and withdraw the vapor from the system.*

from the system by reducing pressure so the moisture boils (vaporizes) at normal atmospheric temperatures. In a perfect vacuum, water changes state from liquid to vapor at –90°F (–68°C). Moisture is removed from the system as a vapor by the vacuum pump.

Reducing pressure too quickly will cause the water to boil rapidly and the remaining water to freeze. Water expands when it freezes, and expansion could cause tubing to burst or leak. Excess moisture should be removed before evacuation by blowing the system out with nitrogen at 150 psig to 300 psig (1035 kPa to 2070 kPa).

Two-stage vacuum pumps, **Figure 11-27,** are necessary to reduce the pressure sufficiently and hold it at a reduced level long enough for the moisture to

A

B

Figure 11-27. *Two-stage vacuum pumps are available in different sizes for varying applications. A—Small 1.6 cfm pump. (Thermal Engineering) B—Cart-mounted 15 cfm unit for use on large systems. (Robinair Mfg. Corp.)*

vaporize and be removed. A second pumping chamber enables the pump to obtain a lower vacuum. In a two-stage pump, the exhaust of the first pumping stage is discharged into the intake of the second pumping stage, rather than to the atmosphere. The second stage pumps at a lower pressure and pulls a deeper vacuum on the system than the first stage could by itself. Two-stage vacuum pumps are capable of pulling down to an extremely low vacuum, but a vacuum seldom goes below 500μ (microns) under field conditions. The two-stage vacuum pumps are able to reach and hold a low vacuum for prolonged periods of time.

Figure 11-28 provides conversion factors for evacuating or dehydrating a system with a vacuum pump.

Evacuation procedure

1. Connect the red and blue hoses on the gauge manifold to the appropriate service valves. Crack open the service valves, and close the manifold hand valves.

2. Connect the center (yellow) hose to a two-stage vacuum pump, and start the pump. Very slowly crack open both manifold hand valves. *Be careful to avoid drawing oil out with the vacuum pump.* After one or two minutes, the compound gauge should read 15 in. Hg (15 inches vacuum). Then, both manifold valves can be fully opened. The vacuum pump should quickly reduce the system pressure to about 29 or 30 in. Hg. If the pump fails to pull an adequate vacuum, there is probably a leak in the system that must be found and corrected. After reaching a vacuum of 29 or 30 in. Hg, permit the vacuum pump to operate for at least *one-half hour.* Adequate time is needed for all moisture to vaporize and all refrigerant gas mixed with oil in the compressor crankcase to be evacuated. Large systems require more time on the vacuum pump.

3. *Before* shutting off the vacuum pump, close both manifold hand valves. Shut down the pump, and remove the yellow hose. At this point, all gases have been removed by the vacuum pump, and the refrigeration system is in a deep vacuum. The system should hold this vacuum, unless a leak permits atmospheric air to enter (or the vacuum pump has not been connected to the system long enough).

Leak detection. Closing the manifold valves and waiting to see if the system loses its vacuum is *not* a proper leak-detection method. The vacuum pump may have been stopped too early for proper evacuation, or the leak may be too small for atmospheric air to enter. Even with a perfect vacuum, the pressure differential between the inside and outside of the tubing is only 15 psi (103 kPa). It is poor

Conversion Factors

Inches of Mercury (Hg)	Pounds Per Square Inch Absolute (psia)	Millimeters of Mercury (mm Hg)	Microns	Boiling Temperature of Water (°F/°C)
0	14.696	760	760,000	212/100
10.24	9.629	500	500,000	192/89
22.05	3.865	200	200,000	151/66
25.98	1.935	100	100,000	124/51
27.95	0.968	50	50,000	101/38
28.94	0.481	25	25,000	78/26
29.53	0.192	10	10,000	52/11
29.67	0.122	6.3	6,300	40/4
29.72	0.099	5	5,000	35/2
29.842	0.039	2	2,000	15/–9
29.882	0.019	1.0	1,000	+1/–17
29.901	0.010	0.5	500	–11/–24
29.917	0.002	0.1	100	–38/–39
29.919	0.001	0.05	50	–50/–46
29.920	0.0002	0.01	10	–70/–57
29.921	0.0000	0	0	–90/–68

One Atmosphere = 14.696 psia (atmospheric pressure at sea leve)
= 760 mm Hg absolute pressure at 32°F (0°C)
= 29.921 in. Hg absolute at 32°F (0°C)

Figure 11-28. *Conversion factors for using a vacuum pump to evacuate or dehydrate a system.*

procedure to leak test with such a low pressure differential. Most leak detection is accomplished with pressures of 200 psi to 300 psi (1380 kPa to 2070 kPa) because the refrigerant molecules are much smaller than air molecules. Various leak detection methods are discussed in Chapter 15.

11.4.2 Charging the System

During compressor operation, the pressure in the high-pressure side of the system is higher than the pressure in the refrigerant cylinder. If the manifold hand valve on the right side (high-pressure) is accidentally opened, refrigerant will be pumped back into the cylinder. This could cause dangerous overpressure in the cylinder and rupture it. *Always feed the refrigerant charge into the low-pressure side of the system, and always feed refrigerant as a gas, not a liquid.*

The gauge manifold arrangement for charging a system is shown in **Figure 11-29.** *Charging is done with the system operating.* Follow this procedure:

1. Close both manifold hand valves. Check gauge pointers for accuracy of zero readings. Recalibrate if necessary.

2. Connect the low-pressure (blue) compound gauge hose to the suction service valve. Finger-tighten it. Crack open the service valve.

3. Connect the high-pressure (red) gauge hose to the discharge service valve; finger-tighten it. Crack open the *service* valve. (Remember the high-pressure *manifold hand valve* must remain closed.)

4. Connect the center (yellow) charging hose to the refrigerant cylinder. Make sure the container is in an upright position so you obtain gas, not liquid. Never charge liquid refrigerant into a system; it will damage compressor valve reeds or remove oil from bearings.

5. Fully open the valve on the refrigerant cylinder to transfer control to the gauge manifold.

6. With the system running, slowly open the left side (low-pressure) manifold hand valve. The pressure inside the refrigerant cylinder is higher than the low-side system pressure, so the gaseous refrigerant will be forced up through the center hose, through the left side of the manifold, into the blue hose, and into the system. The manifold valve on the right side (high-pressure) must remain closed during the charging procedure.

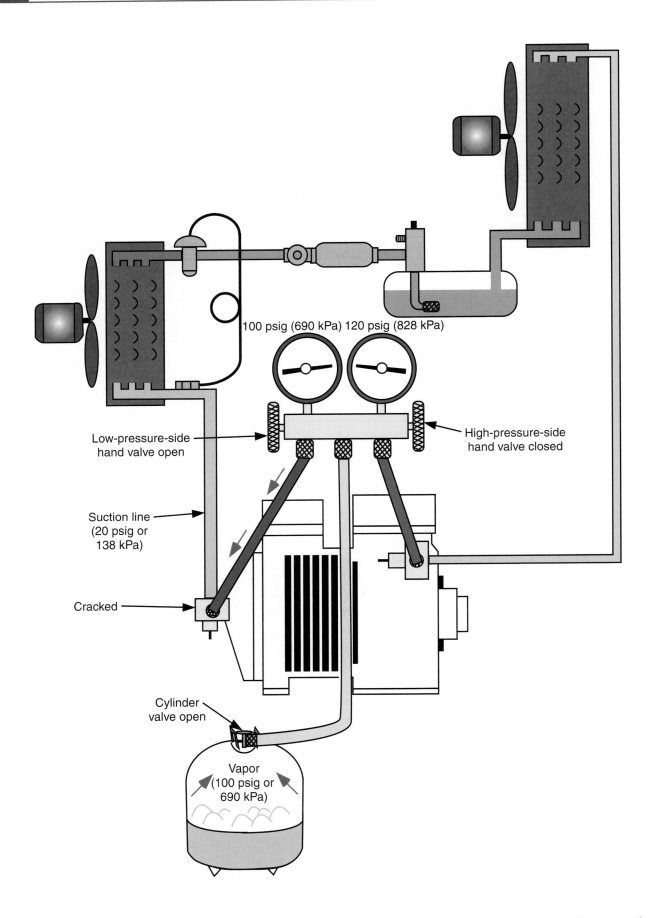

100 psig (690 kPa) 120 psig (828 kPa)

Low-pressure-side
hand valve open

High-pressure-side
hand valve closed

Suction line
(20 psig or
138 kPa)

Cracked

Cylinder
valve open

Vapor
(100 psig or
690 kPa)

Figure 11-29. *Typical arrangement for charging a refrigeration system. The refrigerant cylinder pressure must be greater than the pressure in the suction line so the refrigerant is forced into the system.*

7. To observe the changing conditions of the low-side system pressure, occasionally close the left-side manifold valve, and check the compound gauge. Also, closely watch the high-pressure gauge during charging. Continue adding refrigerant to the system until both high and low pressures achieve normal status.

8. Backseat (close) both service valves, and close the valve on the refrigerant cylinder. Disconnect the hoses from the service valves, and screw the hose ends onto the "dummy" fittings at the manifold. Replace covers or caps on the service valves.

11.5 Filter-driers

Regardless of the care used in evacuating and charging a system, it is safe to assume the system is not completely free of moisture. A *filter-drier* installed in the system, **Figure 11-30,** will absorb the remaining moisture. It will also catch foreign particles circulating with the refrigerant.

On commercial systems, the filter-drier is normally installed in the liquid line, immediately after the liquid receiver. A second unit may be installed in the suction line as well, **Figure 11-31.** On domestic systems, the filter-drier is typically located at the condenser outlet.

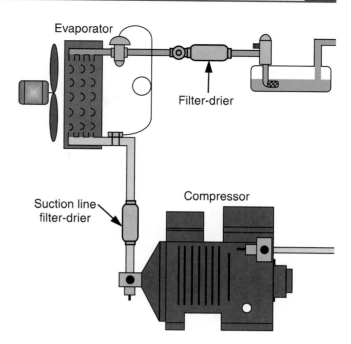

Figure 11-31. *On commercial systems, the filter-drier is usually installed in the liquid line, just after the liquid receiver. For additional protection, technicians sometimes install a second filter-drier, as shown, between the compressor and evaporator.*

Figure 11-30. *Filter-driers remove residual moisture from a system and trap foreign particles circulating with the refrigerant. (Sporlan Valve Co.)*

11.5.1 Moisture Problems

Water or moisture is always present in refrigeration systems and must be kept to an absolute minimum. Acceptable limits vary from one system to another and from one refrigerant to another. Moisture is the primary factor in the formation of acids, sludge, copper plating, and corrosion. The service technician should always be alert to keep the moisture level of the refrigeration system as low as possible.

The main problems caused by moisture are:

❏ *Corrosion* that damages metal parts and adds contaminants to the system.

❏ *Formation of acid* that damages the motor windings. Since the motor windings are exposed to refrigerant,

any acid that forms will cause a breakdown of the insulation and result in a motor burnout.

❏ *Freezing of moisture* in the orifice of the refrigerant control valve or capillary tube. Frozen moisture can block the flow of refrigerant and stop the operation of the system.

Other sources of problems within the system are dirt, sludge, rust, and foreign matter such as flux, copper or brass chips, and solder. These contaminants can damage piston cylinder walls or compressor bearings, or may plug capillary tubes and other refrigerant controls.

11.5.2 Filter-drier Operation

Filter-driers in refrigeration systems work by bringing liquid refrigerant in contact with a substance that absorbs moisture in the refrigerant. The substance is called a drying agent, or *desiccant,* and is usually capable of removing acid as well as moisture.

If properly sized, the filter-drier will not restrict the flow of refrigerant, even when the desiccant is full of moisture and no longer effective. Desiccants are extremely sensitive to moisture and must be protected against it until ready for use. The factory seal on a filter-drier should not be removed until just before installation. Most filter-driers are direction-sensitive, meaning they must be oriented to the direction of refrigerant flow. Direction-sensitive filter-driers have an arrow printed on the body to indicate the proper direction of flow. See **Figure 11-32.**

Desiccants used in filter-driers include activated alumina, silica gel, and activated carbon, which **absorb** (soak up) moisture. They also include molecular sieves, which **adsorb** (collect substances on their surfaces in a condensed layer) contaminants from the refrigeration system. Desiccants are available in granular, bead, and block forms, **Figure 11-33.** Combinations of desiccants can be used in solid cores and have certain advantages over a single desiccant, such as absorption of a greater variety of contaminants.

Do not attempt to reactivate a used filter-drier. It should be discarded when no longer effective. The most common desiccants used today are activated alumina and molecular sieve; occasionally silica gel is used. For desiccants to be returned to their active states, they must be heated for four hours at temperatures ranging from 400°F to 600°F (204°C to 316°C). At these temperatures, all refrigeration oils decompose into sludges or acids. If a drier is reactivated after it has been used, the oil will decompose into sludge and upon installation, will be released into the system. The sludge could clog small ports, such as expansion valves and capillary tubing.

On large commercial systems, a replaceable element filter-drier is used. The filter-drier is bolted together, and replacement cores are available in vacuum-packed metal containers. See **Figure 11-34.**

Special acid-removing filter-driers are available for field installation in the suction line. Such filter-driers are used to clean the system and protect a new compressor following a burnout.

Domestic filter-drier

Domestic systems (refrigerators and freezers) also have a filter-drier, but it is much smaller than those used on commercial systems. The body of the domestic filter-drier is made of copper and is brazed into the system at the outlet of the condenser, **Figure 11-35.** These filter-driers are normally direction-sensitive, but some small-capacity units are nondirectional.

Figure 11-32. *This filter-drier is designed for attachment to the liquid line with flare fittings. Other types are designed for brazing into the line. Note the arrow indicating the proper direction of refrigerant flow through the device. To keep the desiccant at full effectiveness, the seal caps on the fittings should be left in place until just before the filter-drier is installed. (Alco)*

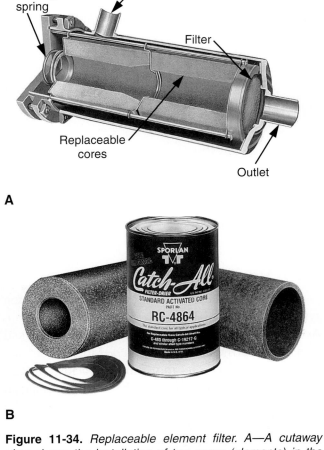

A

B

Figure 11-34. *Replaceable element filter. A—A cutaway view shows the installation of two cores (elements) in the receiver-drier shell. Access is by means of the bolted cover at the left. B—Replacement elements are packed in sealed cans that exclude moisture. Typical cores are shown. (Sporlan Valve Co.)*

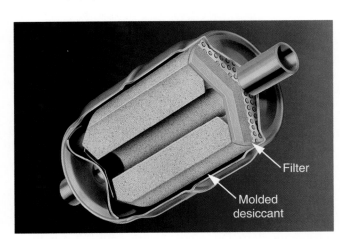

Figure 11-33. *Desiccant granules can be molded into a core for the receiver-drier, as shown in a cutaway view. (Sporlan Valve Co.)*

Domestic systems need a filter-drier to ensure trouble-free operation. Due to the small amount of refrigerant circulating in the system, a single drop of excess moisture will cause a freeze-up. The components in the system are quite small; any foreign material will cause severe problems. When a domestic (hermetic) system is opened for repairs, the filter-drier should always be replaced.

11.6 Sight Glass

The **sight glass, Figure 11-36,** is a small window placed in the liquid line on commercial systems to provide a view of the flowing liquid. The sight glass serves as a valuable service aid: visible bubbles indicate problems within the system. Problems could be low refrigerant charge, low head pressure, insufficient subcooling, restrictions, or poor piping design. The sight glass is usually located close to the liquid receiver and immediately after the filter-drier, **Figure 11-37.**

The sight glass shows clear if the line is full of liquid and shows bubbles if the system is having problems. An occasional bubble is not unusual or harmful, but excessive bubbles indicate trouble in the system.

While the sight glass shows clear when full of liquid, it also shows clear when empty. To determine the correct situation, frontseat the liquid receiver service valve, and pump down the system while looking through the sight glass. An empty system shows no change. A full system, however, appears full, shows bubbles, and then appears empty.

Figure 11-35. *Small filter-drier intended for a domestic unit. The copper tubes at each end are brazed into the system at the condenser outlet. (Parker-Hannifin)*

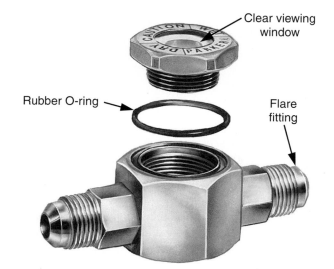

Figure 11-36. *Typical sight glass. The clear window allows the technician to "look inside" the system and diagnose problems by the appearance of the refrigerant. Sight glasses are available with various types of fittings to suit different system configurations. (Parker-Hannifin)*

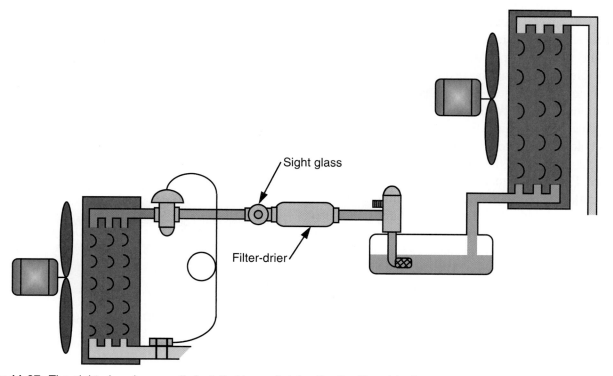

Figure 11-37. *The sight glass is normally installed immediately after the filter-drier in the liquid line.*

11.6.1 Sight Glass with Moisture Indicator

Most sight glasses have a *moisture indicator* centered within the viewing window, **Figure 11-38.** The indicator is highly sensitive to moisture, gradually changing color to reflect the moisture content in the refrigerant. The indicator element is completely reversible and changes color as often as the moisture content of the refrigerant varies.

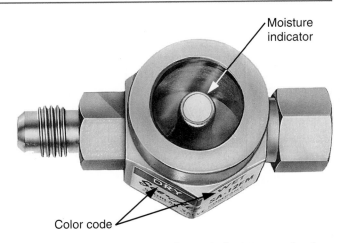

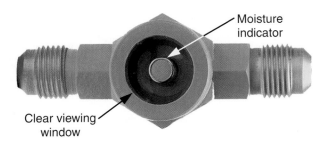

Figure 11-38. *Typical sight glass with a moisture indicator in the center. The indicator gradually changes color to reflect the presence or absence of moisture. (Sporlan Valve Co.)*

Figure 11-39. *Colors the moisture indicator uses to show wet or dry conditions are displayed on the sight glass rim or body. In this example, the color code is on a label attached to the body. Colors for each condition vary among manufacturers. (Sporlan Valve Co.)*

Some change in color takes place rapidly at the start-up of a new system or after the filter-drier is replaced. The system should operate for about 12 hours before you decide (based on the moisture indicator) that another filter-drier change is needed. Drying of the refrigerant should continue until the indicator element changes to the proper color.

All halogenated refrigerants (such as R-12, R-22, or R-502) accept very small amounts of moisture and still function properly. However, when these levels are exceeded, severe problems develop. The amount of moisture in a refrigeration system must be kept to an absolute minimum to provide trouble-free operation. Every precaution must be taken to prevent moisture from entering the system during installation or service operations. Any moisture that *does* enter the system should be removed quickly.

A color reference code is printed around the edge of the sight glass or on the front of the sight glass body, **Figure 11-39.** One manufacturer's color code varies from dark green (dry), to light green (caution), to bright yellow (very wet). Another manufacturer's indicator changes from dark blue (dry), to light blue (caution), to pink (very wet). A plastic or metal cap keeps the glass free of dust, dirt, and grease. The cap should always be replaced. Unlike commercial installations, domestic systems do not use a sight glass or moisture indicator.

The moisture indicator is chemically engineered for long life, accuracy, and reliability. The same indicator can be used for all common refrigerants. The indicator element will show a wet condition before

installation, but that is normal and simply reveals ambient humidity. Most sight glasses are installed with flare connections, but sweat types are also available.

11.7 Heat Exchanger

"Heat exchanger" is a general term to describe any device that transfers heat from one medium to another. However, in the commercial refrigeration industry, *heat exchanger* describes a particular component that transfers heat from the warm liquid line to the cold suction line, **Figure 11-40.** The heat exchanger performs two mutually beneficial tasks:

- ❏ It subcools the refrigerant in the liquid line before it reaches the refrigerant control valve, improving the system's efficiency.
- ❏ It superheats vapor inside the suction line with heat removed from the warm liquid line. This prevents liquid refrigerant from reaching the compressor.

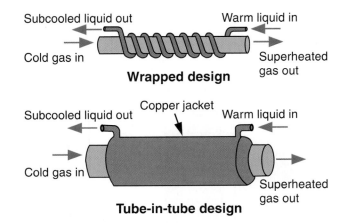

Figure 11-40. *A heat exchanger has a dual role: it subcools liquid refrigerant in one line and superheats refrigerant gas in the other. The two designs shown are common.*

The liquid line is wrapped around the suction line for several turns before traveling to the refrigerant control valve. Some heat exchangers use the highly efficient tube-in-a-tube design. In domestic models, the small capillary tube used for the refrigerant control is soldered to the suction line for almost its entire length.

Heat exchangers are designed to perform a dual function in a system and should not be added to, or removed from, a system without proper engineering information.

11.8 Suction Accumulator

The **suction accumulator** is a device that keeps liquid refrigerant from entering the compressor, **Figure 11-41**. The accumulator is a cylinder that acts as a trap to collect liquid refrigerant and permits only vapor to exit. Liquid entering the accumulator must boil off inside the device before exiting as a vapor.

Figure 11-41. *Commercial condensing unit showing the relationship of the suction accumulator and the compressor. The accumulator keeps liquid refrigerant from reaching the compressor. (Hussmann Corporation)*

Domestic systems locate a small accumulator at the evaporator outlet, with the suction line coming out the top of the accumulator. This prevents liquid from entering the suction line. Liquid entering the bottom of the accumulator remains inside until the liquid boils. Only vapor can exit and enter the suction line at the top of the accumulator.

The accumulator is an upright cylinder with two openings in the top: the *inlet* and the *outlet*, **Figure 11-42.** The suction line is brazed to the inlet opening; any liquid refrigerant entering from the suction line falls to the bottom of the cylinder. The outlet opening has a dip tube that goes to the bottom of the cylinder, makes a 180° bend, and extends upward toward the top of the cylinder where only vapor can enter the tubing to the compressor suction service valve.

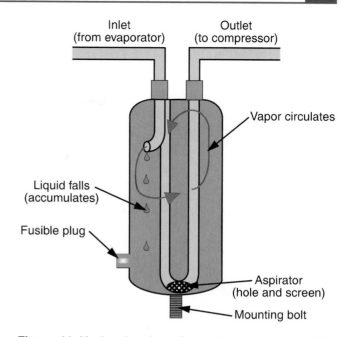

Figure 11-42. *Interior view of a suction accumulator. The 180° bend of the outlet tube permits only vaporized refrigerant to emerge from the accumulator.*

Liquid refrigerant is trapped in the bottom of the accumulator until it evaporates. A small hole (called an **aspirator hole**) is drilled in the side of the 180° bend. The hole permits small quantities of oil to enter the outlet tube and be drawn back to the compressor. Without the aspirator hole, the accumulator would trap oil and deprive the compressor of proper lubrication. Small amounts of liquid refrigerant may enter the aspirator hole but evaporate before reaching the compressor; therefore, they are harmless.

Summary

Figure 11-43 illustrates a refrigeration system with all the components introduced in this chapter. Each component serves a specific purpose. A refrigeration system seldom has all these additional components, but some are found on every system. Many systems require the use of accessory components to properly control the movement and condition of the refrigerant.

This chapter was devoted to explaining how accessory components operate mechanically and *why* they are used on different systems. All systems are designed to control the movement and condition of the refrigerant. Later chapters will explain how the system itself is controlled. Troubleshooting and repairs cannot be performed correctly without understanding the effect of each component on the system. Being able to draw the entire system (as shown in Figure 11-43) from memory will be a plus as you study later chapters.

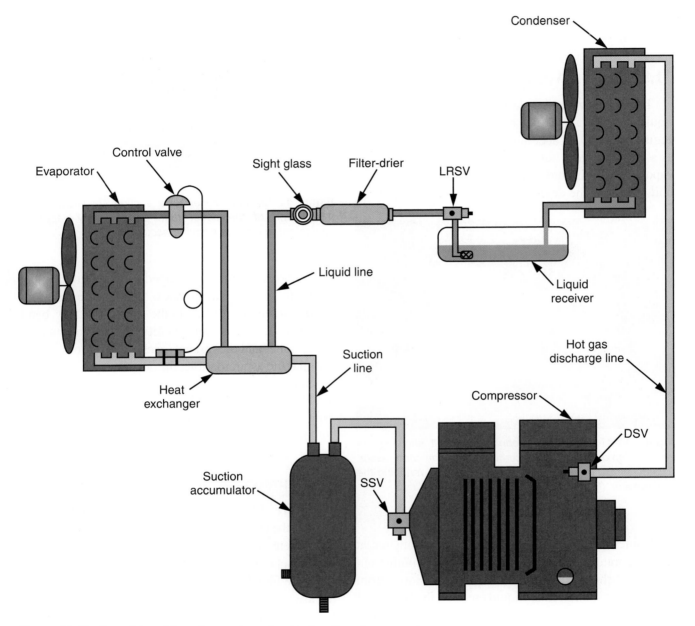

Figure 11-43. *Complete refrigeration system showing location of accessories.*

Test Your Knowledge

Please do not write in this text. Write your answers on a separate sheet of paper.

1. What two copper lines connect the evaporator section with the condensing unit?
2. Where is the condensing unit located on a domestic refrigerator?
3. What is the purpose of the liquid receiver?
4. The liquid receiver should never be more than _____% full when it contains the system's entire refrigerant supply.
5. Name three service valves used on commercial systems.
6. At which two service valves can you obtain high-pressure readings?
7. When a service valve is backseated, which opening is closed?
8. When a service valve is backseated and then cracked, which opening is closed?
9. Frontseating the suction service valve closes which opening?
10. What is the name of the fitting used on service valves to install gauges? What sizes and thread types are the fitting ends?
11. Name two types of service valves installed by technicians on domestic systems.
12. Hand valves on the gauge manifold control access to the _____ hose.
13. Are hand valves on the gauge manifold normally closed or open?
14. Where is the filter-drier located?
15. What is the purpose of the filter-drier?
16. What does a large number of bubbles in the sight glass indicate?
17. Where is the sight glass installed?
18. What is the purpose of the moisture indicator?
19. A heat exchanger has a dual purpose. Describe what it is.
20. Name all 15 possible system components, in order of refrigerant flow, beginning at the evaporator.

Refrigerant is charged into a system using a manifold gauge. (Allied Signal)

12

Refrigerants

Objectives

After studying this chapter, you will be able to:
- ❏ Recognize and identify different refrigerants.
- ❏ Use temperature-pressure charts.
- ❏ Determine superheat at any location in a refrigeration system.
- ❏ Recognize pumping ratio problems.
- ❏ Describe refrigerant applications within specific temperature ranges.
- ❏ Describe and use proper safety precautions with refrigerants.

Important Terms

azeotropic refrigerant	hydrofluorocarbons
CFCs	in-line freezing
chlorofluorocarbons	inorganic refrigerants
cryogenic	latent heat value
dichlorodifluoromethane	monochlorodifluoro-
enthalpy	methane
freezing	ozone depletion
Freon-12®	potential
frozen storage	pumping ratio
halide refrigerants	refrigerant
halogens	saturation
HCFCs	standard conditions
HFCs	subcooled
hydrochlorofluoro-	superheat
carbons	TP card

12.1 What Are Refrigerants?

Refrigerants are the working fluids necessary in every refrigeration system. A **refrigerant** is any fluid or liquid that picks up heat by evaporating at a low temperature/pressure, and gives up heat by condensing at a higher temperature/pressure. The purpose of a refrigerant is to move heat. Refrigerants absorb heat from an unwanted place (boiling process) and carry the heat to a place where it can be discharged (condensing process).

This chapter describes several common refrigerants, and their basic properties and applications. To achieve low temperatures, it is necessary to select a refrigerant with a boiling point lower than the temperature being controlled. Remember, heat always flows from the *warmer* to the *colder* substance.

Factors other than boiling point must be considered when selecting a refrigerant for a specific application. These factors include toxicity, flammability, miscibility with oil, operating pressures, and Btus absorbed per pound. Some refrigerants can be used in a variety of applications, while others have more limited use.

The refrigeration and air conditioning industry is in transition as new refrigerants are replacing older ones in an effort to protect the environment. Many refrigerants are currently available, and others are still being developed. In this textbook, it is not possible to describe all of the available refrigerants. Such information is readily obtainable from manufacturers, supply houses, or the local library. Instead, this chapter concentrates on the composition, selection, identification, and application of the most common refrigerants. You may encounter a new or unusual refrigerant, but the theories and principles described here apply to old and new refrigerants alike.

12.2 Brief History of Refrigerants

The following are key events in the development of refrigerants and the refrigeration and air conditioning industry:
- ❏ In 1834, the first practical compression-cycle refrigeration system was built.

- In 1850, the absorption machine was developed. It used water and sulfuric acid as the refrigerant.
- In 1873, the ammonia compressor was introduced.
- In 1876, the sulfur dioxide compressor was introduced.
- In 1882, thanks to Thomas Edison, the first electric power plant opened in New York.
- By 1890, the demand for small domestic and light commercial systems was increasing rapidly, signaling the demise of the household ice box.
- By 1900, use of the electric motor had become widespread.
- In 1906, Willis Carrier patented his air conditioning system, calling it an "apparatus for treating air." Stuart Cramer coined the term "air conditioning."
- By 1915, sulfur dioxide systems were common in domestic refrigerators. (Note: Sulfur dioxide is a toxic refrigerant and no longer used.)
- In 1922, the first air conditioned movie theater opened in Los Angeles, California (Grauman's Metropolitan Theater).
- In 1928, research to discover a "safe" refrigerant was begun by General Motors Corporation and DuPont Chemical Company. Researchers sought a revolutionary substance that would be nontoxic, nonflammable, odorless, noncorrosive, and nontoxic to humans. Experiments combined halogens, such as chlorine and fluorine, with hydrocarbon compounds (CFCs). The chamber of the U.S. House of Representatives became air conditioned.
- In 1929, the U.S. Senate chamber became air conditioned.
- In 1930, DuPont began manufacturing the first of many new refrigerants using halogen-hydrocarbon compounds. The first compound to be marketed (under the trade name **Freon-12®**) was **dichlorodifluoromethane.** Freon-12 molecules were extremely stable (did not break down) and had many other valuable qualities. This CFC refrigerant revolutionized the refrigeration and air conditioning industry, leading to rapid growth and the discovery of many more refrigerants. Several chemical companies were involved in the research and development of refrigerants.
- In 1930, the White House, the Executive Office Building, and the U.S. Department of Commerce became air conditioned.
- In 1945, the demand for room air conditioners began to increase, with 30,000 produced that year.
- In 1957, the first rotary compressor was introduced, permitting refrigeration and air conditioning units to be smaller, quieter, lighter, and more efficient than reciprocating types.
- In 1977, new technology allowed heat pumps to operate at lower outdoor temperatures when heating on the reverse cycle.
- In 1987, the Montreal Protocol was signed, establishing international cooperation for the phaseout of stratospheric ozone-depleting substances, including CFC refrigerants.
- On December 31, 1995, the United States and many other industrialized countries ended production of CFC refrigerants.

12.3 Halide Refrigerants

For many years, the most widely used refrigerants contained one or more of the chemical substances called **halogens.** Halogens include fluorine, chlorine, iodine, and bromine. Combined with a hydrocarbon compound, such as acetylene, methane, or ethane, halogens produce **halide refrigerants.** Hydrocarbons transfer heat well but are extremely flammable, explosive, and somewhat toxic. However, when hydrocarbons are blended with halogens, the blend is not flammable or explosive.

The manufacturing process for halide refrigerants is complicated. Each mixture is precise and must produce a new substance that has a specific boiling point and acts according to Charles' Law of Gases. The molecule of the new compound also must be stable (not break down spontaneously). The combining of hydrogen with oxygen to form a compound such as water (H_2O) is a naturally-occurring example. The halide refrigerants are classified into three groups according to their chemical makeup: the *chlorofluorocarbons* (CFCs), the *hydrochlorofluorocarbons* (HCFCs), and the *hydrofluorocarbons* (HFCs). These terms refer to manufactured (synthetic) fluids specifically designed for use as refrigerants.

12.3.1 Chlorofluorocarbons (CFCs)

Chlorofluorocarbon refrigerants are composed of chlorine, fluorine, and a hydrocarbon (such as methane, ethane, or propane). They are classified as **CFCs** due to their composition (chlorine-fluorine-carbon). Examples of CFCs are R-11 (trichloromonofluoromethane) and R-12 (dichlorodifluoromethane). The Greek prefixes *mono, di,* and *tri* indicate how many parts of a substance are used in the compound. "Mono" means one part, "di" means two parts, and "tri" means three parts. The name for R-12, dichlorodifluoromethane, describes the molecular structure of the compound. R-12 contains two parts chlorine (*di*chloro) and two parts fluorine (*di*fluoro), combined with methane. Therefore, the number 2 appears twice in the chemical formula: CCl_2F_2. See **Figure 12-1.**

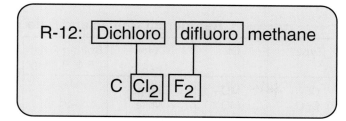

Figure 12-1. *The chemical name and chemical formula of a refrigerant describes its molecular structure. A molecule of R-12 (dichlorodifluoromethane) includes two chlorine atoms and two fluorine atoms combined with a methane molecule.*

Note: Due to their *ozone depletion potential* (ODP), or potential to destroy the protective ozone layer in the stratosphere, all chlorofluorocarbon refrigerants (CFCs) were banned by the 1990 Federal Clean Air Act. Congress designated the Environmental Protection Agency (EPA) to rapidly phase all CFCs out of existence, beginning in 1990. Effective January 1, 1996, the United States and over 70 other industrialized nations stopped manufacturing CFC refrigerants. Chapter 13 provides a thorough discussion of ozone depletion by CFCs.

12.3.2 Hydrochlorofluorocarbons (HCFCs)

Hydrochlorofluorocarbons, known as **HCFCs,** contain hydrogen atoms, causing the compound to be less stable in the atmosphere than CFCs. Therefore, HCFCs are considered less harmful to the ozone layer because the refrigerant molecules break up rather quickly. Only a small percentage of the chlorine in HCFCs is able to reach the ozone layer. The most widely used HCFC is *monochlorodifluoromethane,* designated R-22. R-22 is the most popular refrigerant for air conditioning systems (other than automotive systems). Other examples of HCFCs are R-123, R-124, R-141b, and R-142b. These refrigerants have a very low ozone depletion level and are considered only slightly hazardous to the ozone layer.

12.3.3 Hydrofluorocarbons (HFCs)

Hydrofluorocarbons, or **HFCs,** contain no chlorine atoms. HFC refrigerants are considered environmentally safe, with an ozone depletion level of 0.0 (no chlorine). Examples are R-125, R-152a, and R-134a. These refrigerants, or combinations of HFCs, are rapidly replacing CFCs. R-134a has achieved wide acceptance as the alternative for R-12; R-123 is the acceptable alternative for R-11.

12.4 Naming Refrigerants

Trade terminology is often confusing because each refrigerant can be called by a different name. For example, R-22 can be called by the manufacturer's trade name (Freon-22®), by its chemical name (mono-chlorodifluoromethane), by its chemical formula ($CHClF_2$), by the American Society of Refrigerating and Air Conditioning Engineers "R" number (R-22), or by its composition (HCFC-22). Each refrigerant has its own special characteristics and applications. **Figure 12-2** lists the various names for several refrigerants. A few refrigerants are quite popular, while others are used only in special applications. The service technician must be able to recognize uncommon refrigerants and know where to find information about them.

12.5 Refrigerant Numbering System

DuPont's numbering system for refrigerants came into general use in 1956. By that time, several other chemical companies were manufacturing halogenated refrigerants (those containing chlorine or fluorine). Each separate refrigerant chemical (such as dichlorod-ifluoromethane) was given a number, and each company produced these refrigerants under its own brand name. For example, DuPont's trademark for dichlorodifluoromethane was Freon-12®, while Allied Chemicals used Genetron-12®, and Virginia Chemicals used Isotron-12®.

The American Society of Refrigerating and Air Conditioning Engineers (ASHRAE) has standardized identification of all refrigerants. The system uses the DuPont numbering system, but precedes each number by the letter "R" (for refrigerant), regardless of manufacturer. It appears that the ASHRAE "R" designation, combined with DuPont's numbering system, is most consistent and will continue to be used for years. The DuPont numbering system is explained in **Figure 12-3.**

In response to concern about ozone-depleting compounds, the "R" is sometimes dropped and replaced with letters describing the chemical composition of the refrigerant: CFC-12, HCFC-22, and HFC-134a, for example.

12.6 Azeotropic Mixtures

An *azeotropic refrigerant* is a mixture of two refrigerants that combine to form a third, unique refrigerant having its own individual characteristics. It is mixed in exact proportions, and its composition is always the same in both the liquid and vapor states. An *azeotrope* exhibits a single boiling temperature different from either of the refrigerants of which it is composed. Such a mixture must maintain a constant boiling point and act as a single refrigerant, strictly conforming to Charles' Law of Gases. Azeotropes are

◆———————————————————————————— **Refrigerant Names and Data**

Number	Chemical Name	Formula	Type	Oil	Color	Boiling Point (°F @ 0 psig)
R-11	Trichlorofluoromethane	CCl_3F	CFC	MO	Orange	74.9
R-12	Dichlorodifluoromethane	CCl_2F_2	CFC	MO	White	−21.6
R-13	Chlorotrifluoromethane	$CClF_3$	CFC	MO	Light Blue	−114.6
R-113	Trichlorotrifluoroethane	$C_2Cl_3F_3$	CFC	MO	Purple	117.6
R-114	Dichlorotetrafluoroethane	$C_2Cl_2F_4$	CFC	MO	Dark Blue	38.8
R-115	Chloropentafluoroethane	$CClF_2CF_3$	CFC	MO		−38.4
R-22	Chlorodifluoromethane	$CHClF_2$	HCFC	MO	Green	−41.4
R-123	Dichlorotrifluoroethane	$CHCl_2CF_3$	HCFC	MO/AB	Blue Gray	82.2
R-124	Chlorotetrafluoroethane	$CHClFCF_3$	HCFC	MO/AB	Green	10.3
R-141b	Dichlorofluoroethane	CCl_2FCH_3	HCFC	MO/AB	Sand	89.7
R-142b	Difluorochloroethane	CH_3CClF_2	HCFC	MO/AB	Slate Gray	14.4
R-23	Trifluoromethane	CHF_3	HFC	POE	Med. Gray	−115.7
R-32	Difluoromethane	CH_2F_2	HFC	POE		−61.1
R-125	Pentafluoroethane	CHF_2CF_3	HFC	POE	Lt. Brown	−55.8
R-134a	Tetrafluoroethane	CF_3CH_2F	HFC	POE	Lt. Blue	−15.1
R-143a	Trifluoroethane	CH_3CF_3	HFC	POE		−53.7
R-152a	Difluoroethane	CH_3CHF_2	HFC	POE	Red	−11.3

MO = mineral oil, AB = alkylbenzene, POE = polyol ester

Figure 12-2. *Each refrigerant has its own special characteristics and is identifiable by its ASHRAE "R" number, chemical name, and chemical formula.*

◆——————————————————————

Fluorocarbon Refrigerant Numbering System

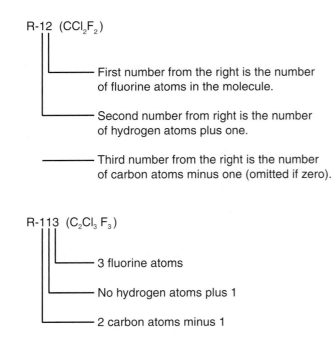

R-12 (CCl_2F_2)

— First number from the right is the number of fluorine atoms in the molecule.

— Second number from right is the number of hydrogen atoms plus one.

— Third number from the right is the number of carbon atoms minus one (omitted if zero).

R-113 ($C_2Cl_3F_3$)

— 3 fluorine atoms

— No hydrogen atoms plus 1

— 2 carbon atoms minus 1

Figure 12-3. *The numbering system developed by DuPont has been adopted by all refrigerant manufacturers. The figure illustrates how the designations for R-12 and R-113 were developed.*

patented refrigerants, mixed in a precise manufacturing process. The service technician should never attempt to make refrigerant mixtures.

ASHRAE designates all azeotropic mixtures with a 500-series number. **Figure 12-4** lists a few of these azeotropes. As each manufacturer introduces a refrigerant within a series, it is assigned a consecutive number. If another manufacturer submits a refrigerant that has exactly the same components, it is assigned the same number. If the components are the same, but the weights by percentages are different, a letter of the alphabet is added at the end of the number. Appropriate R-numbers de-emphasize individual manufacturer's brand names.

The two most common azeotropic mixtures are R-500 and R-502, but these two refrigerants are classified as CFCs and are swiftly being eliminated because of their high ozone depletion and global warming potentials. They are included in the list of azeotropic mixtures in Figure 12-4 to show their history and unique qualities. Similar information is available for each refrigerant, but many have yet to develop a proven history. For example, R-507 is a new, non-ozone-depleting azeotrope designed to replace R-502.

12.7 Inorganic Refrigerants————————◆

The 700-series of numbers is assigned to ***inorganic refrigerants,*** such as ammonia or carbon

Azeotropic Mixtures

Number	Chemical Name	Formula	Type	Oil	Color	Boiling Point (°F @ 0 psig)
R-500	R-12/152a (73.8/26.2)	CCl_2F_2/CH_3CHF_2	CFC	Mineral oil	Yellow	−28.3
R-502	R-22/115 (48.8/51.2)	$CHClF_2/CClF_2CF_3$	CFC	Mineral oil	Orchid	−50.0
R-503	R-23/13 (40.1/59.9)	$CHF_3/CClF_3$	CFC	POE/AB/MO	Aqua	−126.1
R-507	R-125/143a (50/50)	CHF_2CF_3/CH_3CF_3	HFC	Polyol ester	Teal	−52.1

MO = mineral oil, AB = alkylbenzene, POE = polyol ester

Figure 12-4. *Each new azeotrope is assigned a consecutive number in the 500-series.*

dioxide, **Figure 12-5.** The last two numerals represent the atomic number of the substance. For example, the atomic number of ammonia is 17, so the refrigerant number is R-717. The atomic number of carbon dioxide is 44, so the number is R-744. Most inorganic refrigerants are considered "expendable," that is, the vapor is not recovered for reuse. The liquid is vaporized and released to the atmosphere.

Inorganic Refrigerants

Number	Name	Formula	Boiling Point (F° @ 0 psig)
R-702	Hydrogen	H_2	−423.0
R-704	Helium	He	−452.1
R-717	Ammonia	NH_3	−28.0
R-718	Water	H_2O	212.0
R-720	Neon	Ne	−410.9
R-728	Nitrogen	N_2	−320.4
R-732	Oxygen	O_2	−297.3
R-740	Argon	Ar	−302.6
R-744	Carbon dioxide	CO_2	−109.2

Figure 12-5. *Inorganic refrigerants are assigned numbers in the 700-series. These refrigerants are expendable.*

12.8 Refrigerant Cylinder Colors

Most refrigerant cylinders are color-coded for ease of identification and to help prevent accidental mixing of refrigerants. The service technician must be knowledgeable of refrigerants; using the wrong refrigerant or mixing refrigerants will create severe problems in the system. Refrigerants come in cylinders and drums with the refrigerant number clearly labeled. The more common refrigerants are shown in **Figure 12-6.**

The following partial list identifies some of the color-coding for refrigerant cylinders:

R-11	Orange	R-404A	Orange
R-12	White	R-406A	Light gray
R-13	Light blue	R-407A	Lime-green
R-22	Green	R-407C	Chocolate
R-113	Dark purple		brown
R-114	Navy blue	R-408A	Reddish-
R-123	Light gray		purple
R-134a	Light blue	R-409A	Medium brown
R-401A	Pinkish-red	R-500	Yellow
R-401B	Warm orange	R-502	Light purple
R-402A	Whitish green	R-503	Blue-green
R-403B	Light purple	R-507	Aqua blue

Figure 12-6. *Refrigerants come in drums and cylinders of various sizes. To prevent accidental mixing of refrigerants, each container is assigned a specific color (for example, white for R-12). The trade name and refrigerant number are prominently displayed on the label of each container. (Allied Signal)*

Do not rely solely on the cylinder color for identification; shades of colors vary from one manufacturer to another, and some refrigerants have the same or similarly colored cylinder. Always check the R-number in addition to the color code. Sometimes it is necessary to know the name or chemical formula of the refrigerant. All condensing units have a data plate displaying important information about the unit. The data plate may specify the refrigerant and the amount, for example, R-134a: 6.0 lb.; Tetrafluoroethane: 6.0 lb.; or CF₃CH₂F: 6.0 lb. See **Figure 12-7.**

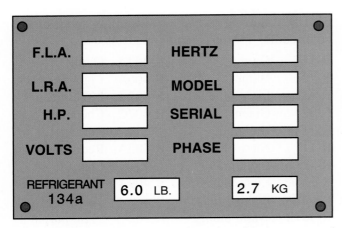

Figure 12-7. *Check the unit data plate for the type and amount of refrigerant.*

12.9 Selecting a Refrigerant

The halide (halogenated) family of refrigerants was responsible for the tremendous growth of the refrigeration and air conditioning industry. The chemical properties of these refrigerants have permitted their use under conditions that would be hazardous with more flammable or toxic chemicals. In many cases, one refrigerant may be used in a number of applications. The choice of refrigerant for a particular application depends upon several factors. A low boiling point is only one feature of a good refrigerant. Other important qualities include toxicity (safety), flammability, latent heat value, suction pressure, condensing pressure, vapor density, oil compatibility, corrosiveness, and cost.

Latent heat value (the number of Btus absorbed for each pound of liquid evaporated) is very important. The latent heat value (Btu/lb.) is different for each refrigerant. If the latent heat value for a refrigerant is fairly high, less refrigerant is required, and system components can be smaller and less expensive. The latent heat of the refrigerant should be as high as possible, but other factors must also be considered. For example, ammonia has high latent heat value (565 Btu/lb.) but is toxic to humans.

Selection of the best refrigerant for a particular system is often a compromise. For example, the pressure in the evaporator must be as high as possible for the boiling temperature desired. A high suction pressure results in a high-density gas that sweeps oil back to the compressor. A refrigerant having low pressure (or a vacuum) in the evaporator results in a thin gas, and oil return could be a problem.

Whereas the evaporator pressure should be as high as possible, the condensing pressure should be as *low* as possible. A low condensing pressure reduces the size of the condensing unit and decreases the load on the compressor. A high condensing pressure requires a high-capacity compressor and may require additional cooling for the compressor.

When selecting a refrigerant, the design engineer tries to achieve high capacity (Btu/lb.) accompanied by low material and energy costs. System components are expensive to purchase and costly to operate. Therefore, the best choice is a refrigerant that has high capacity and low power requirements.

12.9.1. Standard Conditions

Each refrigerant is different, and it is important to select the best one for a given application. Many factors must be considered in the selection process, but the first step is to compare refrigerants under *standard conditions.* Standard conditions have been established as 5°F (–15°C) evaporating temperature and 86°F (30°C) condensing temperature. See **Figure 12-8** for comparisons, under standard conditions, of several common refrigerants. Using standard conditions to compare refrigerants provides a quick method of eliminating unsuitable ones. Final selection is made at actual system operating temperatures. Refrigerant saturation charts provide this important information.

12.9.2. Saturation Point

Saturation point means *boiling point* or *condensing point.* At **saturation,** a refrigerant can exist as a liquid, a mixture of liquid and vapor, or a vapor, at the same pressure and temperature. When a refrigerant is at the saturation point, the temperature-pressure relationship is predictable. See **Figure 12-9.** The saturation point does not include superheat or subcooling. A vapor can be superheated to a temperature above the saturation point without affecting pressure. Likewise, a liquid can be subcooled without affecting pressure. A pressure reading reveals saturation temperature, but actual temperature may be different. For a given pressure, the actual refrigerant temperature may be higher or lower (if superheated or subcooled), **Figure 12-10.**

─── **Thermodynamic Properties**

	R–11	R-12	R-13	R-22	R-113	R-114	R-500	R-502	R-503
Properties at 1 atmosphere (14.7 psia or 101 kPa):									
Freezing point, °F.	–168	–252	–294	–256	–31	–137	–254		
Boiling point, °F.	74.8	–21.6	–114.6	–41.4	117.6	38.4	–28.3	–50.1	–127.6
Condensation at 86°F:									
Specific heat of liquid (Btu/lb./cu.ft.)	0.209	0.235	0.247	0.335	0.218	0.246	0.290	0.305	0.290
Compressor discharge temperature (°F)	111	101	–1	128	86	86	105	99	14
Compressor suction temperature (°F)	5	5	–100	5	10	20	5	5	–100
Compression ratio	6.24	4.08	4.74	4.06	8.02	5.42	4.12	3.75	4.58
Refrigerant circulated per ton (lb./min.)	2.96	4.00	4.30	2.89	3.73	4.64	3.30	4.38	3.72
Horsepower per ton	0.935	1.002	1.12	1.011	0.973	1.045	1.01	1.079	1.15
Coefficient of performance	5.04	4.70	4.20	4.66	4.84	4.64	4.65	4.37	4.23
Evaporation at 5°F:									
Specific volume (cu. ft./lb.) (suction gas)	12.27	1.46	1.55	1.25	27.38	4.34	1.50	0.82	1.32
Net refrigerating effect (Btu/lb.)	66.8	50.0	46.3	70.0	53.7	44.7	60.6	44.9	55.4
Latent heat of vaporization (Btu/lb.)	83.5	68.2	52.1	93.2	70.6	61.1	82.5	68.9	72.1

Figure 12-8. *Standard conditions (specific evaporating and condensing temperatures) are useful when comparing refrigerants. The table presents a side-by-side comparison of a number of refrigerants. (LaRoche Chemicals)*

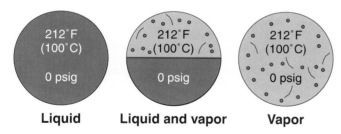

Figure 12-9. *At the saturation point (boiling point), a refrigerant can be a liquid, a mixture of liquid and vapor, or a vapor.*

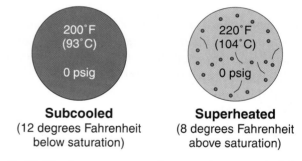

Figure 12-10. *A pressure reading can be converted to saturation temperature, but actual temperature may be different. Saturation charts reveal saturation temperature, not superheat or subcooling.*

12.9.3. Saturation Charts

Highly accurate saturation charts help design engineers determine the best refrigerant for a particular application. Saturation charts are very useful for comparing refrigerants under various operating conditions. At a given temperature, the saturation charts automatically assume the refrigerant is boiling or condensing. This explains why they are called *saturation charts*. At a given temperature, the charts reveal important information regarding the physical properties of the refrigerant: the correct saturation pressure (boiling point), the volume occupied by the saturated vapor (in cubic feet per pound), the density of the saturated liquid (in pounds per cubic feet), and the **enthalpy** (heat content) of both vapor and liquid (in Btu/lb.).

Saturation charts are based on a single, pure gas that behaves according to Charles's Law. If another gas, or group of gases (such as atmospheric air), is permitted to enter the system, Dalton's Law takes effect and the chart becomes useless. The condition is easily recognized because the system pressures greatly exceed the values shown on the saturation chart.

Service and installation technicians are primarily interested in the temperature-pressure relationship, but the charts are also used to select the best refrigerant for a system; to size suction and liquid lines; and to size such system components as the evaporator, condenser, refrigerant control, and compressor. Proper sizing and selection of these components is very important to efficient operation of the system.

12.9.4. Refrigerant Properties

As mentioned, the choice of refrigerant depends upon several factors. Properties of several halide refrigerants and one inorganic refrigerant are discussed next, as well as hazards posed by their use.

R-11

Chemical name: Trichloromonofluoromethane
Chemical formula: CCl_3F
Composition: CFC
Cylinder color: Orange
Boiling point: 75°F (23.9°C) at 0 psig

R-11 is a low-pressure refrigerant often referred to as carrene or methylene chloride. R-11 and R-12 were the first halide refrigerants to receive wide acceptance. R-11 is used almost exclusively in 200-ton to 500-ton industrial centrifugal chiller systems where high volumes of vapor can be handled efficiently. Applications include industrial and commercial air conditioning systems and cooling industrial process water and brine. R-11 is well suited for industrial equipment because it exerts low pressures in the upper temperature ranges, permitting economy in the design of pressure vessels. R-11 liquid refrigerant vaporizes at a temperature of about 40°F (4°C) and a pressure of 15.6 in. Hg. Under these conditions, one pound of liquid produces about 6 cu. ft. of vapor and has a net refrigerating effect of 69 Btu/lb.

R-11 is a synthetic refrigerant that is stable, nonflammable, and nontoxic. The amount of refrigerant in these systems averages about three pounds per ton of capacity. R-11 systems are normally water-cooled and operate in a vacuum. Discharge pressure averages about 10 psig (69 kPa). The latent heat at 5°F (–15°C) is 0.18 Btu/lb. (Latent heat is the difference between the heat content of the liquid and the vapor.) See **Figure 12-11**.

The cylinder color for R-11 is orange, but it is normally supplied in a black, 55-gallon barrel with an orange lid. The safety disc on the barrel ruptures at 15 psig at 115°F (104 kPa at 46°C). Leaks may be detected by using a soap solution, halide torch, or an electronic detector. Due to ozone depletion, CFC-11 systems are rapidly being retrofitted with HCFC-123 (similar to R-11) or redesigned for R-134a.

R-123

Chemical name: Dichlorotrifluoroethane
Chemical formula: $CHCl_2CF_3$

°F	Pressure psig	Volume Vapor cu. ft./lb.	Density Liquid lb./cu. ft.	Heat Content Btu/lb. Liquid	Heat Content Btu/lb. Vapor
–75	29.52	153.32	103.85	–0.017	0.216
–50	28.85	61.14	102.02	–0.004	0.210
–25	27.43	27.57	100.16	0.007	0.204
–20	27.01	23.91	99.78	0.009	0.203
–15	26.48	20.86	99.40	0.011	0.202
–10	26.00	18.07	99.03	0.013	0.202
–5	25.39	15.81	98.64	0.016	0.201
0	24.70	13.86	98.26	0.018	0.200
*5	23.94	12.20	97.88	0.020	0.200
10	23.08	10.77	97.49	0.022	0.199
15	22.20	9.54	97.10	0.025	0.198
20	21.08	8.48	96.72	0.027	0.198
25	19.90	7.56	96.32	0.029	0.197
50	12.03	4.41	94.34	0.039	0.195
75	0.03	2.72	92.31	0.049	0.194
*86	3.4	2.23	91.39	0.053	0.193
100	8.7	1.76	90.20	0.058	0.193
125	20.99	1.19	88.03	0.068	0.192
150	37.66	0.82	85.77	0.076	0.192
175	59.61	0.59	83.40	0.085	0.192

Italics indicate inches of mercury (vacuum).
*Indicates standard conditions:
  5°F evaporating temperature
  86°F condensing temperature

Figure 12-11. *Properties of liquid and saturated vapor for R-11.*

Composition: HCFC
Cylinder color: Light gray
Boiling Point: 82.2°F (27.8°C) at 0 psig

R-123 is a single-component refrigerant with properties similar to R-11. **See Figure 12-12.** Most new low-pressure centrifugal chillers are designed for use with R-123. It is similar to R-11 in many respects and is also used in foam-blowing applications. It is a low-pressure, nonflammable liquid with low chemical reactivity. R-123 is colorless, odorless, and has an ozone depletion level of 0.02 compared to 1.0 for R-11. R-123 has a greater coefficient of performance than R-11, but it has reduced capacity and efficiency in chillers where the only change is from R-11 to R-123. Some equipment modification is usually necessary to obtain maximum performance when retrofitting from R-11 to R-123. R-123 is chemically compatible and *miscible* (able to mix) with the same lubricants currently used in R-11 equipment (mineral oil or alkylbenzene). See **Figure 12-13.**

— R-123

°F	Pressure psig	Volume Vapor cu. ft./lb.	Density Liquid lb./cu. ft.	Heat Content Btu/lb. Liquid	Heat Content Btu/lb. Vapor
−40	28.8	0.0186	101.1	0.00	84.77
−30	28.3	0.0259	100.3	4.35	86.14
−20	27.8	0.0354	99.49	6.51	87.51
−10	26.9	0.0477	98.70	8.69	88.89
0	25.9	0.0632	97.91	10.89	90.28
10	24.5	0.0826	97.11	13.11	91.68
20	22.8	0.1065	96.30	15.35	93.09
30	20.7	0.1358	95.49	17.61	94.50
40	18.1	0.1711	94.67	19.89	95.91
50	14.9	0.2134	93.84	22.20	97.32
60	11.1	0.2636	93.00	24.52	98.74
70	6.5	0.3228	92.15	26.86	100.16
80	1.2	0.3919	91.29	29.22	101.57
90	2.5	0.4723	90.41	31.61	102.98
100	6.2	0.5651	89.52	34.01	104.39
110	10.4	0.6717	88.62	36.44	105.79
120	15.2	0.7935	87.69	38.89	107.18
130	20.8	0.9323	86.75	41.37	108.56
140	27.0	1.0900	85.79	43.87	109.93
150	34.2	1.2670	84.80	46.39	111.29

Italics indicate inches of mercury (vacuum).

Figure 12-12. *Properties of liquid and saturated vapor for R-123.*

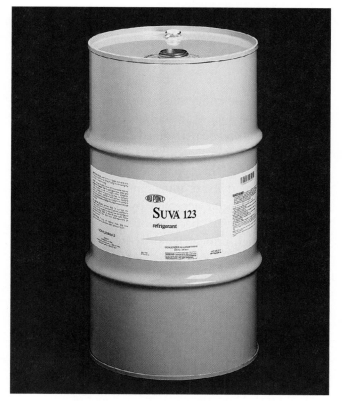

Figure 12-13. *R-123 is normally supplied in 55-gallon drums. (DuPont)*

According to industry standards, refrigerants are classified into three groups, depending on their hazard potential. Group 1 refrigerants are considered the least hazardous; Group 3 refrigerants are the most hazardous.

R-123 is not hazardous when exposure is at or below the Acceptable Exposure Limit (AEL) of 30 ppm. AEL is an airborne-exposure limit with a time-weighted average (TWA) to which workers may be exposed repeatedly during an 8- or 12-hour day, or 40-hour week, without adverse effects.

DuPont has established an Emergency Exposure Limit (EEL) of 1000 ppm for up to one hour, or up to 2500 ppm for one minute. During an emergency, workers may be exposed to these concentrations without harmful effects.

The AEL of R-123 is below its odor threshold. ASHRAE Standard 15 specifies a *compound-specific* monitor and an alarm be located in confined spaces with equipment using R-123. It is recommended that the first alarm level be set for 30 ppm and the second alarm level set for 90 ppm. Be sure to comply with all industry standards when using R-123.

R-12

Chemical name: Dichlorodifluoromethane
Chemical formula: CCl_2F_2
Composition: CFC
Cylinder color: White
Boiling point: −21.7°F (−29.8°C) at 0 psig

R-12 is a colorless, almost odorless liquid that is nontoxic, noncorrosive, nonirritating, nonflammable, and very stable. R-12 is an old and reliable refrigerant that came into general use in 1936. It replaced toxic refrigerants such as sulfur dioxide and methyl chloride. R-12 was often called Freon®, a trade name registered by DuPont. R-12 remains the most common refrigerant in existing household refrigerators and freezers, water-coolers, automotive air conditioners, transport refrigeration units, and commercial refrigeration systems operating in the medium-temperature range. Some large industrial centrifugal chillers use R-12 for comfort cooling and manufacturing process refrigeration.

R-12 has a relatively low latent heat value, an advantage in small refrigeration systems. The large amount of refrigerant circulated permits the use of less sensitive, more accurate, and more positive regulating mechanisms. The suction and discharge pressures are low but not in a vacuum. The latent heat of R-12 at 5°F (−15°C) is 68.2 Btu/lb. Latent heat is the difference between the last two columns of the table in **Figure 12-14.**

An R-12 leak may be detected by using soap bubbles, sniffers (halide torch and electronic), dyes, or ultrasound. Water is only slightly soluble in R-12 and limited to about six parts per million (ppm) by weight.

◆——————————————— R-12

°F	Pressure psig	Volume Vapor cu. ft./lb.	Density Liquid lb./cu. ft.	Heat Content Btu/lb. Liquid	Heat Content Btu/lb. Vapor
–100	27.01	22.16	100.15	–12.47	66.20
–75	23.02	9.92	97.93	–7.31	69.00
–50	15.43	4.97	95.62	–2.10	71.80
–25	2.32	2.73	93.20	3.17	74.56
–20	0.57	2.44	92.70	4.24	75.11
–15	2.45	2.19	92.20	5.30	75.65
–10	4.49	1.97	91.69	6.37	76.20
–5	6.73	1.78	91.18	7.44	76.73
0	9.15	1.61	90.66	8.52	77.27
*5	11.79	1.46	90.14	9.60	77.80
10	14.64	1.32	89.61	10.68	78.34
15	17.22	1.20	89.06	11.77	78.86
20	21.04	1.10	88.53	12.86	79.39
25	24.61	1.00	87.98	13.96	79.90
50	46.70	0.66	85.14	19.51	82.43
75	76.99	0.44	82.09	25.20	84.82
*86	93.34	0.38	80.67	27.77	85.82
100	117.16	0.31	78.79	31.10	87.03
125	169.06	0.22	75.15	37.28	88.97
150	234.61	0.16	71.04	43.85	90.53
175	315.94	0.11	66.20	51.03	91.48

Italics indicate inches of mercury (vacuum).
*Indicates standard conditions:
  5°F evaporating temperature
  86°F condensing temperature

Figure 12-14. *Properties of liquid and saturated vapor for R-12.*

◆——————————————— R-134a

°F	Pressure psig	Volume Vapor cu. ft./lb.	Density Liquid lb./cu. ft.	Heat Content Btu/lb. Liquid	Heat Content Btu/lb. Vapor
–100	28.09	41.52	96.89	–17.93	87.24
–75	25.06	16.56	94.08	–10.47	90.76
–50	18.73	7.55	91.22	–2.99	94.24
–25	6.92	3.83	88.27	4.50	97.72
–20	3.60	3.41	86.47	5.71	98.81
–15	0.05	3.02	85.96	7.17	99.55
–10	1.93	2.69	85.44	8.64	100.28
–5	4.09	2.39	84.91	10.13	101.01
0	6.47	2.14	84.38	11.63	101.74
*5	9.07	1.92	83.85	13.14	102.47
10	11.92	1.72	83.31	14.66	103.19
15	15.11	1.55	82.76	16.19	103.90
20	18.40	1.40	82.21	17.74	104.61
25	22.07	1.26	81.65	19.30	105.31
50	45.62	0.78	78.75	27.27	108.73
75	78.42	0.50	75.63	35.57	111.95
*86	96.60	0.42	74.17	39.33	113.28
100	124.45	0.33	72.22	44.23	114.88
125	185.26	0.22	68.38	53.33	117.41
150	262.80	0.15	63.90	63.06	119.30
175	360.99	0.10	57.60	73.58	120.78

Italics indicate inches of mercury (vacuum).
*Indicates standard conditions:
  5°F evaporating temperature
  86°F condensing temperature

Figure 12-15. *Properties of liquid and saturated vapor for R-134a.*

R-134a

Chemical name: Tetrafluoroethane
Chemical formula: CF_3CH_2F
Composition: HFC
Cylinder color: Light blue
Boiling point: –15°F (–26°C) at 0 psig

R-134a was the first commercially available fluorocarbon refrigerant that was not ozone-depleting. It was developed more than 20 years ago and has characteristics similar to R-12. R-134a is the preferred, long-term alternative refrigerant in most medium- and high-temperature applications where R-12 has been successfully used. As an HFC, it is not scheduled for phaseout under current law. The overall stability of R-134a is very good for virtually all applications where R-12 is suitable. R-134a is compatible with steel, copper, aluminum, and brass. The latent heat value for R-134a is 88.29 Btu/lb. at 5°F (–15°C). See **Figure 12-15.**

Working with 134a requires the same precautions as when working with R-12. Because R-134a has a combustibility potential under pressure, leak-checking should never be done with a mixture of R-134a and air. Leak-checking can be performed safely with a mixture of R-134a and nitrogen. An R-134a leak may be detected by using soap bubbles, sniffers (halide torch and electronic), dyes, or ultrasound.

R-134a has been accepted by the automobile air conditioning industry because of its low hose permeability and high critical temperature. The automobile industry elected to use varying types of polyalkylene glycol (PAG) lubricants. Retrofit kits are available for converting existing automotive air conditioning systems from R-12 to R-134a. R-134a is being used for a number of refrigeration applications including supermarket cases, walk-in coolers, vending machines, watercoolers, and home refrigerators. R-134a is also being used in some centrifugal chillers.

Field tests show the capacity and energy efficiency of R-134a are similar to those of R-12. At evaporating temperatures below –10°F (–23°C), R-134a loses its attractiveness for several reasons:

❑ Capacity and efficiency are significantly lost compared to R-12.

❑ Pumping ratios become very high, reducing compressor reliability.

❑ Low-side pressures are in a vacuum, reducing system reliability.

R-134a is not a "drop-in" replacement for R-12, and it is not miscible with mineral oils (MO) or alkylbenzene (AB) lubricants. R-134a requires the use of polyol ester (POE) or polyalkylene glycol (PAG) lubricants. POE and PAG lubricants are manufactured in several variations. Most automotive original equipment manufacturers have chosen specific PAG lubricants for their systems. For nonautomotive applications, most compressor manufacturers are recommending specific polyol ester lubricants. Check with equipment manufacturers for the recommended lubricants for their systems.

Chlorinated compounds (CFCs) should not be introduced into systems that use R-134a. When performing retrofits, the service technician should make every effort to eliminate residual contamination. The PAG lubricants used with R-134a are not compatible with mineral oil or CFC refrigerants. Tests have shown that up to about 0.5% of R-12 may be present in R-134a before decomposition of the PAG lubricant occurs.

A second concern regarding the mixing of R-12 and R-134a is that these two materials form a compound whose pressure is higher than the pressure of either component. The system pressure may be higher than expected and could result in performance problems. It is troublesome to separate such an azeotropic mixture, and recovery and reclamation efforts become difficult.

Compatibility of R-134a with desiccants can be a problem. Three common types of desiccant materials are used in making driers. They are molecular sieves, alumina, and silica gel. Under some conditions, the molecular sieve XH5, commonly used with R-12, is incompatible with R-134a. Molecular sieves XH7 and XH9 are recommended for use with R-134a. Drier manufacturers have developed driers and filters compatible with R-134a. Such driers can include all three types of drier materials. The solubility of water in R-134a is comparable to the solubility of water in R-22 (19.5 ppm).

Should a large release of R-134a vapor occur, the area should be evacuated immediately. Vapors may concentrate near the floor, displacing available oxygen. Once the area is evacuated, it must be ventilated using blowers or fans to circulate the air at floor level. If vapors are inhaled at high concentrations, death by asphyxia, cardiac irregularities, and possible cardiac arrest could occur. Service technicians have died from entering tunnels and other areas where heavy concentrations of halide refrigerants had displaced the oxygen.

Never expose refrigerants to open flames or glowing hot metal surfaces. At high temperatures, R-12 and R-22 decompose to form hydrochloric acid, hydrofluoric acid, and phosgene gas. Phosgene gas is very toxic to the human system.

R-22

Chemical name: Monochlorodifluoromethane
Chemical formula: $CHClF_2$
Composition: HCFC
Cylinder color: Green
Boiling point: –41°F (–41°C) at 0 psig

R-22 is a synthetic (manufactured) refrigerant that made its debut in the 1940s. HCFC-22 was developed for low-temperature systems operating in the range of –20°F to –40°F (–29°C to –40°C). R-22 has high discharge pressure and high pumping ratios when used for low-temperature applications. (Pumping ratios are discussed later in this chapter.) When used for high-temperature applications (air conditioning), R-22 has higher suction pressure, resulting in low pumping ratios. R-22 is very efficient when applied to high-temperature applications and dominates the air conditioning market.

EPA regulations permit the use of HCFC-22 in new equipment until January 1, 2010. HCFC-22 is available until January 1, 2020, but its use after 2010 is limited to service repair. A total HCFC-22 production ban becomes effective January 1, 2020. In the air conditioning segment of the industry, substantial progress toward phasing out HCFC-22 has been made; however, HFCs are not widely used yet.

R-22 is stable, nontoxic, noncorrosive, nonirritating, and nonflammable. It is more tolerant of moisture than most refrigerants and will absorb 19.5 ppm of water. Desiccants (driers) should be used to remove moisture. Because of its strong attraction to moisture, R-22 requires the use of large amounts of desiccant. The latent heat value of R-22 is 93.2 Btu/lb. at 5°F (–15°C). See **Figure 12-16.** An R-22 leak may be detected by using soap bubbles, sniffers (halide torch and electronic), dyes, or ultrasound.

R-500

Chemical name: Azeotropic mixture
Chemical formula: CCl_2F_2/CH_3CHF_2
Composition: CFC and HFC (73.8% R-12 and 26.2% R-152a)
Cylinder color: Yellow
Boiling point: –28°F (–33.3°C) at 0 psig

◆————————— R-22

°F	Pressure	Volume Vapor	Density Liquid	Heat Content Btu/lb.	
	psig	cu. ft./lb.	lb./cu. ft.	Liquid	Vapor
−100	25.03	18.43	93.77	−14.56	93.37
−75	18.50	8.36	91.43	−8.64	96.29
−50	6.15	4.22	89.02	−2.51	99.14
−25	7.39	2.33	86.48	3.83	101.88
−20	10.15	2.08	85.96	5.13	102.42
−15	13.17	1.87	85.43	6.44	102.94
−10	16.47	1.68	84.90	7.75	103.46
−5	20.06	1.52	84.37	9.08	103.96
0	23.96	1.37	83.83	10.41	104.47
*5	28.19	1.24	83.28	11.75	104.96
10	32.77	1.13	82.72	13.10	105.44
15	35.73	1.02	82.16	14.47	105.91
20	43.03	0.93	81.59	15.84	106.38
25	48.75	0.86	81.02	17.22	106.84
50	84.03	0.56	78.03	24.28	108.95
75	132.22	0.37	74.80	31.61	110.74
*86	158.17	0.32	73.28	34.93	111.40
100	195.91	0.26	71.24	39.27	112.11
125	277.92	0.18	67.20	47.37	112.88
150	381.50	0.12	62.40	56.14	112.73
175	510.72	0.08	56.14	66.19	110.83

Italics indicate inches of mercury (vacuum).
*Indicates standard conditions:
  5°F evaporating temperature
  86°F condensing temperature

Figure 12-16. *Properties of liquid and saturated vapor for R-22.*

◆————————— R-500

°F	Pressure	Volume Vapor	Density Liquid	Heat Content Btu/lb.	
	psig	cu. ft./lb.	lb./cu. ft.	Liquid	Vapor
−100	26.4	22.18	89.64	−13.34	79.78
−75	21.7	10.11	87.45	−7.95	83.13
−50	12.8	5.11	85.20	−2.30	86.44
−25	1.2	2.81	82.87	3.57	89.68
−20	3.2	2.52	82.40	4.79	90.31
−15	5.4	2.26	81.92	6.00	90.94
−10	7.8	2.03	81.44	7.22	91.57
−5	10.4	1.83	80.95	8.46	92.19
0	13.3	1.65	80.46	9.71	92.81
*5	16.4	1.50	79.96	10.96	93.42
10	19.7	1.36	79.46	12.23	94.03
15	23.4	1.23	78.96	13.50	94.63
20	27.3	1.13	78.45	14.79	95.22
25	31.5	1.03	77.93	16.09	95.81
50	57.6	0.66	75.26	22.75	98.64
75	93.4	0.45	72.40	29.68	101.26
*86	113	0.38	71.06	32.85	102.33
100	141	0.31	69.28	36.97	103.60
125	203	0.22	65.82	44.69	105.56
150	281	0.15	61.84	52.98	107.00
175	378	0.10	56.96	62.14	107.55

Italics indicate inches of mercury (vacuum).
*Indicates standard conditions:
  5°F evaporating temperature
  86°F condensing temperature

Figure 12-17. *Properties of liquid and saturated vapor for R-500.*

R-500 is an older refrigerant designed to improve the Btu-per-pound capacity of R-12. It provides about 20% greater refrigerating capacity than R-12 for the same size compressor. Its latent heat value at 5°F (−15°C) evaporating temperature is 82.45 Btu/lb. whereas R-12 is 68.2 Btu/lb. This means R-500 absorbs more heat per pound of refrigerant circulated than does R-12. R-500 is used in some older heat pump applications, domestic dehumidifiers, and commercial or industrial applications where high capacity is needed. Presently, there are replacement refrigerants for R-500, but there is no azeotropic replacement. See **Figure 12-17.**

Servicing refrigeration systems using R-500 is much like servicing units containing R-12. R-500 readily absorbs moisture. It is vital that moisture be kept out of refrigeration systems, or it will cause damage. Moisture is removed by careful dehydration procedures and installation of driers. Leak-detection is accomplished by using soap bubbles, sniffers (halide torch and electronic), dyes, or ultrasound.

R-502

Chemical name: Azeotropic mixture
Chemical formula: $CHClF_2/CClF_2CF_3$
Composition: HCFC and CFC (48.8% R-22 and 51.2% R-115)
Cylinder color: Orchid
Boiling point: −50.1°F (−45.6°C) at 0 psig

R-502 was developed in 1961. It is an excellent refrigerant for low-temperature applications where a range of 0°F to −60°F (−18°C to −51°C) is desired. R-502 has good latent heat capacity and a lower condensing pressure than R-22. The lower condensing pressure increases the life expectancy of the compressor. Furthermore, better lubrication is possible, and the need for liquid injection to cool the compressor is usually eliminated. Leaks are detected by using soap bubbles, sniffers (halide torch and electronic), dyes, or ultrasound. The latent heat of R-502 at 5°F (−15°C) is 67.3 Btu/lb. and 70.8 Btu/lb. at −20°F (−29°C). See **Figure 12-18.**

R-502 will hold 1-1/2 times more moisture than R-12 before the moisture becomes a problem (12 ppm). The higher suction pressure at low temperatures improves oil return to the compressor, but oil separators are frequently used to attain proper oil levels inside the compressor. Most low-temperature systems experience some difficulty with oil return. R-502 has good solubility in mineral oil; and good piping practice ensures proper velocity for sweeping oil back to the compressor.

A long-term replacement solution for R-502 systems is *R-507*, an azeotropic mixture. Both of its components are HFC refrigerants (R-125 and R-143a).

R-503

Chemical name: Azeotropic mixture
Chemical formula: $CHF_3/CClF_3$
Composition: HFC and CFC (40.1% R-23 and 59.9% R-13)
Cylinder color: Aquamarine
Boiling point: –126°F (–87.8°C) at 0 psig

R-503 is an azeotropic mixture that is nonflammable, noncorrosive, and practically nontoxic. R-503 is a low-temperature refrigerant for use in the low section of a cascade system where temperatures range from –100°F to –125°F(–73°C to –87°C). R-503 holds more moisture than some other low-temperature refrigerants. However, all low-temperature applications require extreme dryness. Any moisture not in solution with the refrigerant will form ice at the metering devices. System evacuation requires the use of micron gauges for measuring the vacuum level. The latent heat of R-503 at 5°F (–15°C) is 48.9 Btu/lb. and 77.2 Btu/lb. at –127°F (–88°C). See **Figure 12-19.**

Oil does not circulate well at low temperatures, so oil return to the compressor can be a problem. Cascade systems and other low-temperature units are usually fitted with oil separators and other devices for returning oil to the compressor crankcase. An R-503 leak may be detected by using soap bubbles, sniffers (halide torch and electronic), dyes, or ultrasound.

— R-502

°F	Pressure psig	Volume Vapor cu./ft. lb.	Density Liquid lb./cu. ft.	Heat Content Btu/lb. Liquid	Heat Content Btu/lb. Vapor
–100	*23.28*	10.46	97.86	–12.55	65.89
–75	*15.09*	4.96	95.24	–7.59	68.92
–50	*0.19*	2.59	92.51	–2.25	71.93
–25	12.13	1.47	89.68	3.50	74.87
–20	15.31	1.31	89.08	4.69	75.44
–15	18.80	1.19	88.50	5.91	76.02
–10	22.56	1.07	87.90	7.13	76.58
–5	26.66	0.97	87.29	8.38	77.36
0	31.08	0.88	86.68	9.63	77.69
*5	35.86	0.80	86.06	10.91	78.24
10	41.00	0.73	85.43	12.19	78.78
15	46.53	0.67	84.80	13.49	79.31
20	52.49	0.61	84.15	14.81	79.83
25	58.81	0.56	83.50	16.14	80.35
50	97.42	0.37	80.06	22.98	82.80
75	149.13	0.25	76.22	30.12	84.96
*86	176.59	0.21	74.45	33.36	85.79
100	216.19	0.17	71.97	37.56	86.71
125	301.36	0.12	66.84	45.36	87.84
150	408.35	0.08	60.09	53.85	87.76
175	544.72	0.06	47.55	65.69	83.37

Italics indicate inches of mercury (vacuum).
*Indicates standard conditions:
  5°F evaporating temperature
  86°F condensing temperature

Figure 12-18. *Properties of liquid and saturated vapor for R-502.*

— R-503

°F	Pressure psig	Volume Vapor cu. ft./lb.	Density Liquid lb./cu. ft.	Heat Content Btu/lb. Liquid	Heat Content Btu/lb. Vapor
–150	*16.9*	5.80	94.46	–0.079	0.181
–125	0.56	2.57	91.94	–0.059	0.170
–120	3.13	2.22	91.39	–0.055	0.168
–110	9.28	1.68	90.25	–0.048	0.164
–100	16.9	1.29	89.05	–0.041	0.161
–90	26.3	1.01	87.78	–0.033	0.157
–80	37.7	0.80	86.44	–0.026	0.154
–70	51.3	0.64	85.02	–0.019	0.152
–60	67.4	0.51	83.52	–0.013	0.149
–50	86.1	0.42	81.93	–0.006	0.147
–40	108	0.34	80.25	0.000	0.144
–30	133	0.28	78.46	0.006	0.142
–20	161	0.23	76.56	0.012	0.140
–10	194	0.19	74.52	0.019	0.137
0	231	0.16	72.33	0.025	0.135
*5	251	0.15	71.16	0.029	0.134
10	272	0.13	69.95	0.032	0.132
20	318	0.11	67.35	0.039	0.130
30	369	0.09	64.45	0.046	0.127
40	426	0.07	61.12	0.054	0.123
50	489	0.06	57.09	0.061	0.118

Italics indicate inches of mercury (vacuum).
*Indicates standard condition:
  5°F evaporating temperature

Figure 12-19. *Properties of liquid and saturated vapor for R-503.*

R-507

Chemical name: Azeotropic mixture

Chemical formula: CHF_2CF_3/CH_3CF_3

Composition: HFC-125 and HFC-143a (45% R-125 and 55% R-143a)

Cylinder color: Teal blue

Boiling point: −52°F (−88°C) at 0 psig

R-507 was developed as a long-term substitute for R-502. R-507 is an azeotrope and a non-ozone-depleting refrigerant that possesses characteristics similar to R-502. Few, if any, equipment design changes are necessary to optimize the performance of R-507 in retrofit situations. R-507 is not a "drop-in" replacement for R-502. Mineral oils and alkylbenzene lubricants are incompatible with R-507 and must be replaced with a particular polyol ester lubricant. Consult the original equipment manufacturer for the recommended lubricant.

Polyol ester lubricants are available in a wide range of viscosities (thicknesses) from as low as 15 centistokes (cs) to more than 220 cs at 104°F (40°C). Their range of miscibility (the ability of the refrigerant-lubricant mixture to form a single liquid phase) with R-507 can vary widely.

Avoid mixing R-502 with R-507. Such a mixture would contain four components: R-125, R-143a, R-22, and R-115. CFC-115 and HFC-125 may form an azeotrope and make separation of the four components difficult. Reclamation may be impossible and result in a disposal expense.

The solubility of water in R-507 is comparable to its solubility in R-502 (12 ppm). Leaks are detected by using soap bubbles, sniffers (halide torch and electronic), dyes, or ultrasound. The latent heat of R-507 at 5°F (−15°C) is 75.82 Btu/lb. and 80.22 Btu/lb. at 20°F (−29°C). See **Figure 12-20.**

Respiratory dangers. The hazard presented by low concentrations of fluorocarbon refrigerants is minor, but the possibility of injury (or death) exists in unusual situations or if materials are deliberately misused.

At high temperatures, R-12 and R-22 decompose to form phosgene gas. Phosgene gas is very toxic to the human system.

Refrigerant vapor is almost odorless and is five times heavier than air. High accumulations will displace atmospheric air and prevent access to oxygen. Good ventilation should be provided to eliminate dangerous concentration of vapors.

If a large leak of refrigerant occurs in an enclosed area, immediately vacate and ventilate the area. Inhaling high concentrations of refrigerant vapors or mist may cause cardiac sensitization, heart irregularities, unconsciousness, and oxygen deprivation leading to death by asphyxia.

◆———————————— **R-507**

°F	Pressure	Volume Vapor	Density Liquid	Heat Content Btu/lb.	
	psig	cu. ft./lb.	lb./cu. ft.	Liquid	Vapor
−60	−2.90	3.5367	83.17	−6.61	81.11
−50	0.86	2.7264	82.06	−3.31	82.55
−40	5.48	2.1327	80.93	0.00	83.99
−30	11.10	1.6903	79.78	3.30	85.40
−20	17.84	1.3557	78.62	6.59	86.81
−10	25.83	1.0990	77.43	9.88	88.18
0	35.21	0.8994	76.21	13.15	89.55
*5	39.39	0.8165	75.59	14.78	90.23
10	46.15	0.7423	74.96	16.42	90.89
20	58.78	0.6173	73.68	19.68	92.20
30	73.27	0.5167	72.36	22.93	93.48
40	89.80	0.4349	70.99	26.19	94.73
50	108.56	0.3679	69.57	29.47	95.94
70	153.58	0.2661	66.51	36.12	98.20
*86	197.83	0.2065	63.80	41.63	99.83
100	243.52	0.1665	61.14	46.74	101.05
110	280.64	0.1409	59.02	50.65	101.78
120	321.93	0.1193	56.63	54.89	102.34
130	367.80	0.1001	53.85	59.65	102.67
140	418.74	0.0826	50.42	65.29	102.64
150	475.27	0.0653	45.67	72.77	101.95

Italics indicate inches of mercury (vacuum).

*Indicates standard conditions:
    5°F evaporating temperature
    86°F condensing temperature

Figure 12-20. *Properties of liquid and saturated vapor for R-507.*

Becoming light-headed (dizzy) is a warning that oxygen is not available. Anyone suffering from light-headedness should immediately move, or be moved, to fresh air. The use of epinephrine and similar drugs while working with refrigerants should be avoided because severe heart problems could result. Confined areas should be checked with a leak detector before entering.

R-717

Chemical name: Ammonia

Chemical formula: NH_3

Composition: One part nitrogen, three parts hydrogen

Cylinder color: Silver

Boiling point: −28°F (−33°C) at 0 psig

Freezing point −107°F (−77°C) at 0 psig

Anhydrous ammonia is commonly used in industrial refrigeration systems. "Anhydrous" means free from water. Ammonia is found naturally in the atmosphere and is not harmful to the ozone layer. The latent heat of ammonia is 565 Btu/lb. at 5°F (−15°C). The large latent heat value makes ammonia about eight times more efficient than R-12. Ammonia makes it possible to achieve large refrigerating effects with relatively small-sized equipment. See **Figure 12-21.**

◆━━━━━━━━━━━━━━━━━ R-717

°F	Pressure psig	Volume Vapor cu. ft./lb.	Density Liquid lb./cu. ft.	Heat Content Btu/lb. Liquid	Heat Content Btu/lb. Vapor
−100	*27.4*	182.40	45.52	−63.3	572.5
−75	*23.2*	72.81	44.52	−37.0	583.3
−50	*14.3*	33.08	43.49	−0.6	593.7
−35	*5.4*	21.68	42.86	5.3	599.5
−25	1.3	16.66	42.44	16.0	603.2
−20	3.6	14.68	42.22	21.4	605.0
−15	6.2	12.97	42.00	26.7	606.7
−10	9.0	11.50	41.78	32.1	608.5
−5	12.2	10.23	41.56	37.5	610.1
0	15.7	9.12	41.34	42.9	611.8
*5	19.6	8.15	41.11	48.3	613.3
10	23.8	7.30	40.89	53.8	614.9
15	28.4	6.56	40.66	59.2	616.3
20	33.5	5.91	40.23	64.7	617.8
25	39.0	5.33	40.20	70.2	619.1
35	51.6	4.37	39.72	81.2	621.7
50	74.5	3.29	39.00	97.9	625.2
75	125.8	2.13	37.74	126.2	629.9
*86	154.5	1.77	37.16	138.9	631.5
100	197.2	1.42	36.40	155.2	633.0
125	293.1	0.97	34.96	185.1	634.0

Italics indicate inches of mercury (vacuum).
*Indicates standard conditions:
  5°F evaporating temperature
  86°F condensing temperature

Figure 12-21. *Properties of liquid and saturated vapor for R-717.*

Anhydrous ammonia is becoming more acceptable for replacement of R-12 equipment in large systems. Ammonia is not as hazardous as formerly believed. Small leaks are easily detected by ammonia's sharp, pungent smell, and large leaks are rendered harmless (neutralized) by automatic water sprinklers. R-717 vapor is extremely soluble in water. Accidents involving ammonia are normally caused by the "panic factor." People panic and injure themselves trying to escape from the odor. Ammonia gas can severely irritate the eyes and the mucous membranes of the nose and throat. In high concentrations, it can suffocate a person.

Ammonia (R-717) cannot be used with copper and bronze piping because it attacks these metals in the presence of moisture. Iron and steel pipe is commonly used with R-717, resulting in very few leaks.

Ammonia leak detection

There are several methods of detecting anhydrous ammonia leaks:

❏ *Ammonia test paper (phenolphthalein or litmus paper)* — Moistened test paper changes color when exposed to ammonia.

❏ *Diluted hydrochloric acid* — Fumes from a dilute solution of hydrochloric acid produce a dense white fog when they come in contact with ammonia vapor (ammonium chloride).

❏ *Sulfur tapers* — When sulfur tapers (candles) burn, sulfur dioxide is formed, which reacts with ammonia to produce a dense white cloud (ammonium bisulfite) near the point of escaping fumes.

Ammonia gas is not harmful at low concentrations because the pungent odor of the gas gives adequate warning. It is unlikely an individual would stay in a room with a concentration of ammonia great enough to cause harm, unless trapped. Because of its pungent odor, ammonia is easily identified by smell at 3 ppm to 5 ppm. At 15 ppm, the odor is quite irritating. At 30 ppm, the service technician needs a respirator. Exposure of five minutes at 50 ppm is the maximum allowed by OSHA. Ammonia becomes hazardous to life at 5000 ppm and is flammable at 150,000 ppm to 270,000 ppm. Special training is necessary for proper handling and use of ammonia in refrigeration systems. A recognized safety code is available from the International Institute of Ammonia Refrigeration.

As noted earlier, ammonia represents a possible panic hazard. Because of the discomfort resulting from even traces of ammonia in the air, care should be taken in setting up equipment and in keeping valve packing glands tight. Goggles recommended for anhydrous ammonia service should be worn by all personnel dealing directly with anhydrous ammonia. For users connecting and disconnecting ammonia cylinders regularly, an ammonia gas mask should be kept readily available in case of a leak. Most canister-type ammonia gas masks are limited to a maximum concentration of 3% ammonia in air. For concentrations exceeding 3%, self-contained breathing apparatus should be used.

If a hazardous ammonia leak should develop, spray water around the leaking section until the cylinder can be turned off. Ammonia reacts chemically with water to form ammonium hydroxide (aqua-ammonia). If a mask is not available, a sponge or cloth soaked in water and placed over the nose and mouth offers temporary protection. Since ammonia is lighter than air, the area closest to the floor has the lowest ammonia concentration.

If liquid ammonia has been released, it should be flooded with large volumes of water. Under no circumstances should liquid ammonia be neutralized with an acid.

Liquid ammonia on the skin usually results in severe burns due to a freezing, dehydrating, and caustic effect. For this reason, care should be taken in breaking unions on lines containing liquid ammonia. Before breaking a connection, vent the line to the atmosphere until all frost has left the line.

12.9.5. TP Cards

Service technicians normally use the temperature-pressure relationship for checking system operation and troubleshooting system problems. Special pocket-size temperature-pressure charts, called **TP cards,** show the temperature-pressure relationship for several refrigerants. Most technicians keep a TP card readily available but refer to it less often as they become familiar with the numbers. A typical TP card is shown in **Figure 12-22.** Similar cards are available at no charge from refrigerant manufacturers or from your local parts supplier.

Using temperature-pressure cards

Temperature-pressure cards provide the information necessary to predict and control the behavior of each refrigerant. Service technicians are primarily interested in the temperature-pressure relationship, which differs for each refrigerant. Gauges are used to obtain refrigerant pressure; then, the card is used to convert the pressure reading to a corresponding temperature equivalent.

The service technician can raise or lower the pressure (according to the card) to achieve the desired refrigerant temperature. All refrigeration systems are designed to control the temperature of some substance, usually air. The air temperature is regulated by controlling the refrigerant temperature inside the evaporator. The saturation temperature of the refrigerant inside the evaporator is changed by raising or lowering the suction pressure. A colder evaporator (lower pressure) results in colder air temperature.

The service technician must control the temperature-pressure relationship inside the evaporator. The metal of the evaporator is about 10 degrees Fahrenheit warmer than the refrigerant boiling inside. The temperature of the air entering the evaporator is about 10 degrees Fahrenheit warmer than the metal of the evaporator. Thus, during the run cycle the air temperature remains about 20 degrees Fahrenheit warmer than the refrigerant. See **Figure 12-23.**

There is no universal, all-purpose refrigerant since each refrigerant responds differently to a change in temperature. Knowing exactly how each refrigerant behaves at a particular temperature makes it easy to locate and correct system problems.

Temperature-Pressure Card

	Refrigerant								Refrigerant								Refrigerant						
°F	11	12	22	123	134a	500	502	°F	11	12	22	123	134a	500	502	°F	11	12	22	123	134a	500	502
-50	28.9	15.4	6.2	29.2	18.6	12.8	0.2	10	23.1	14.6	32.8	24.5	12.0	19.7	41.0	55	10.0	52.0	92.6	13.1	51.2	63.9	106.6
-45	28.7	13.3	2.7	29.0	16.7	10.4	1.9	12	22.7	15.8	34.7	24.2	13.2	21.2	43.2	60	7.8	57.7	101.6	11.2	57.4	70.6	116.4
-40	28.4	11.0	0.5	28.8	14.7	7.6	4.1	14	22.3	17.1	36.7	23.9	14.4	22.6	45.4	65	5.4	63.8	111.2	9.0	64.0	77.8	126.7
-35	28.1	8.4	2.6	28.6	12.3	4.6	6.5	16	21.9	18.4	38.7	23.5	15.7	24.1	47.7	70	2.8	70.2	121.4	6.6	71.1	85.4	137.6
-30	27.8	5.5	4.9	28.3	9.7	1.2	9.2	18	21.5	19.7	40.9	23.2	17.1	25.7	50.0	75	0.1	77.0	132.2	4.1	78.6	93.5	149.1
-25	27.4	2.3	7.4	28.1	6.8	1.2	12.1	20	21.1	21.0	43.0	22.8	18.4	27.3	52.5	80	1.5	84.2	143.6	1.3	86.7	102.0	161.2
-20	27.0	0.6	10.1	27.7	3.6	3.2	15.3	22	20.6	22.4	45.3	22.4	19.9	28.9	54.9	85	3.2	91.8	155.7	0.9	95.2	110.0	174.0
-18	26.8	1.3	11.3	27.6	2.2	4.1	16.7	24	20.1	23.9	47.6	22.0	21.4	30.6	57.5	90	4.9	99.8	168.4	2.5	104.3	120.6	187.4
-16	26.5	2.1	12.5	27.4	0.7	5.0	18.1	26	19.7	25.4	49.9	21.6	22.9	32.4	60.1	95	6.8	108.3	181.8	4.2	113.9	130.6	201.4
-14	26.4	2.8	13.8	27.3	0.4	5.9	19.5	28	19.1	26.9	52.4	21.2	24.5	34.2	60.1	100	8.8	117.2	195.9	6.1	124.1	141.2	216.2
-12	26.2	3.7	15.1	27.1	1.2	6.8	21.0	30	18.6	28.5	54.9	20.7	26.1	36.0	65.6	105	11.1	126.6	210.8	8.1	134.9	152.4	231.7
-10	26.0	4.5	16.5	26.9	2.0	7.8	22.6	32	18.1	30.1	57.5	20.2	27.8	37.9	69.4	110	13.4	136.4	226.4	10.2	146.3	164.1	247.9
-8	25.8	5.4	17.9	26.7	2.8	8.8	24.2	34	17.5	31.7	60.1	19.7	29.5	39.9	71.3	115	15.9	146.8	242.7	13.6	158.4	176.5	264.9
-6	25.5	6.3	19.3	26.5	3.7	9.9	25.8	36	16.9	33.4	62.9	18.9	31.3	41.9	74.3	120	18.5	157.7	259.9	15.0	171.1	189.4	282.7
-4	25.5	6.3	19.3	26.5	3.7	9.9	25.8	38	16.3	35.2	65.6	18.7	33.1	43.9	77.4	125	21.3	169.1	277.9	17.7	184.5	203.0	301.4
-2	25.0	8.2	22.4	26.1	5.5	12.1	29.3	40	15.6	37.0	68.5	18.1	35.0	46.1	80.5	130	24.3	181.0	296.8	20.5	198.7	217.2	320.8
0	24.7	9.2	24.0	25.8	6.5	13.3	31.1	42	15.0	38.8	71.5	17.5	37.0	48.2	83.8	135	27.4	193.5	316.6	23.5	213.5	232.1	341.2
2	24.4	10.2	25.6	25.6	7.5	14.5	32.9	44	14.3	40.7	74.5	16.9	39.0	50.5	87.0	140	30.8	206.6	337.3	26.7	229.2	247.7	362.6
4	24.1	11.2	27.3	25.3	8.6	15.7	34.9	46	12.8	44.7	80.8	15.6	43.2	55.1	93.9	145	34.4	220.3	358.9	30.2	245.6	264.0	385.0
6	23.8	12.3	29.1	25.1	9.7	17.0	36.9	48	12.8	44.7	80.8	15.6	43.2	55.1	93.9	150	38.2	234.6	381.5	33.8	262.8	281.1	408.4
8	23.4	13.5	30.9	24.8	10.8	18.4	38.9	50	12.0	46.7	84.0	15.0	45.4	57.6	97.4	155	41.6	295.6	405.1		298.9		432.9

Italic figures = Inches of mercury

Bold figures = Pressure (psig)

Figure 12-22. *Technicians keep a TP card readily available and refer to it often.*

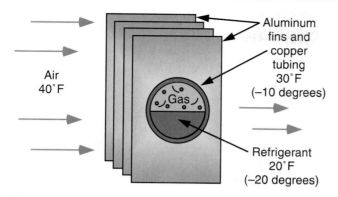

Figure 12-23. *To achieve proper heat flow, a temperature difference of 20 degrees Fahrenheit is created between the airflow and the boiling refrigerant. The metal fins and tubing are 10 degrees colder than the entering air, and the refrigerant is 10 degrees colder than the metal.*

Two pairs of statements will help you remember the relationship of pressure and temperature in a refrigeration system:

❑ Refrigerant *temperature* can be controlled by controlling *pressure.*

 Refrigerant *pressure* can be controlled by controlling *temperature.*

❑ Refrigerant *pressure* is predictable by knowing *temperature.*

 Refrigerant *temperature* is predictable by knowing *pressure.*

12.9.6. Superheat and Subcooling

Superheat describes a gas with a temperature *above* the saturation point. Heat is added *after* the refrigerant changes its state from liquid to a gas. The temperature of superheated gas is higher than the corresponding pressure reading on the saturation table. The gauge manifold does not read superheat or subcooling. When checking system pressures, the pressure reading shown on the gauge manifold will always be *saturation* pressure. If the refrigerant is superheated, the actual temperature of the gas will be above saturation temperature. For troubleshooting purposes, superheat is normally found at three specific places in the refrigeration system:

❑ Evaporator outlet (8 to 12 degrees Fahrenheit or 5 to 7 degrees Celsius) — called *TEV (thermostatic expansion valve) superheat,* or *evaporator superheat.*

❑ Suction inlet to compressor (20 to 30 degrees Fahrenheit or 11 to 17 degrees Celsius) — called *system superheat,* or *suction superheat.*

❑ Hot gas discharge from the compressor (90 to 100 degrees Fahrenheit or 50 to 56 degrees Celsius) — called *discharge superheat.*

Subcooled describes a liquid having a temperature *below* the saturation point. Heat is removed *after* the refrigerant changes its state from a gas to a liquid. The temperature of a subcooled liquid does not correspond to saturation pressure. Refrigerant in the condenser changes from a gas to a liquid and then begins to subcool. Subcooling takes place in the bottom of the condenser and in the liquid line. Subcooling at the condenser outlet is normally 10 degrees Fahrenheit (6 degrees Celsius) but can range from 5 to 15 degrees Fahrenheit (3 to 8 degrees Celsius).

Superheat and subcooling are important because they are required for proper operation of the refrigeration system. In addition to controlling the saturation point in the evaporator and the condenser, the amount of superheat and subcooling within a system must also be controlled. **Figure 12-24** depicts an air conditioning system with gauges and thermometers to illustrate saturation, superheat, and subcooling within an R-22 air conditioning system.

You can easily determine the amount of superheat or subcooling at any point in the system by using a gauge manifold and thermometer. Follow this four-step procedure:

1. Obtain a pressure reading with the gauge manifold.
2. Convert the pressure reading to refrigerant saturation temperature.
3. Using the thermometer, obtain an accurate temperature reading of the copper tubing at the test site.
4. Find the difference in temperature readings (above or below saturation temperature). The difference is the amount of superheat or subcooling, in degrees Fahrenheit.

For example, in Figure 12-24, the pressure reading at the evaporator outlet is 60 psig. By referring to a saturation chart for R-22, you find the saturation temperature at 60 psig is 34°F. A thermometer reading taken at the evaporator outlet shows a temperature of 44°F. The difference in the readings (44 – 34 = 10) is the amount of superheat, 10°F.

12.9.7 Pumping Ratio

Pumping ratio (often called compression ratio) is important when selecting a refrigerant for use in a specific application. The pumping ratio is obtained by dividing the *absolute* suction pressure (psia) into the *absolute* condensing pressure (psia). The pumping ratio should not exceed a certain value for each particular refrigerant. If the pumping ratio is too high, the temperature of the high-pressure vapor going through the exhaust valve will overheat the valve reeds and cause some of the oil in the exhaust pocket to become

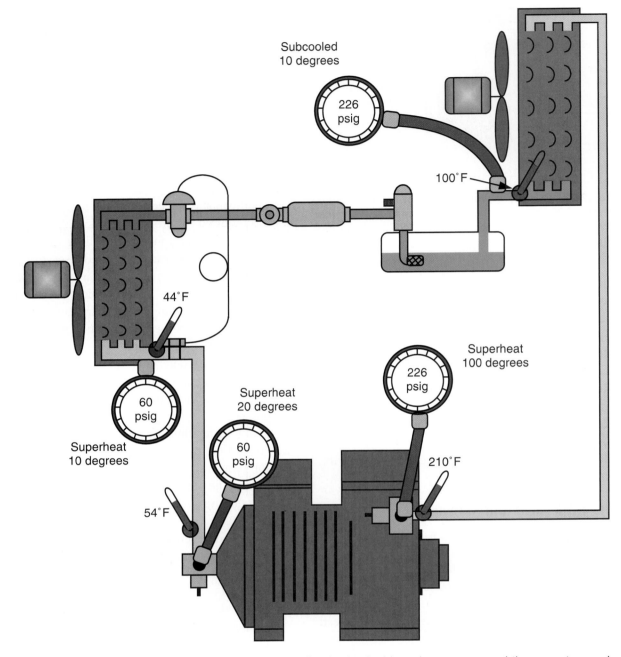

Figure 12-24. *This system uses R-22. Superheat or subcooling is checked by using a gauge and thermometer as shown. The difference between the thermometer and saturation temperature reveals the amount of superheat or subcooling.*

carbonized. A ratio of less than 10:1 makes it possible to use a low-capacity (single-stage) compressor; a pumping ratio of more than 12:1 usually requires a high-capacity (two-stage) compressor. See **Figure 12-25.**

Two-stage compressors are more complicated and costly than single-stage compressors. The two-stage compressor uses a pair of pistons to compress the suction gas to a certain level (first-stage compression), then discharges the gas into a second pair of pistons for compression to a higher level (second-stage compression). The second stage pistons discharge the fully compressed gas through the hot gas discharge line for travel to the condenser. Two-stage compressors usually require special cooling treatment, such as liquid injection, desuperheating TXV, and external fans.

The need for two-stage compressors can usually be avoided by selecting a refrigerant that operates below the 10:1 ratio. For example, a typical residential air conditioner using R-22 on a hot summer day (90°F or 32°C) has a discharge pressure of about 260 psig and a suction pressure of about 60 psig. These gauge readings must be converted to absolute by adding 15 psi for atmospheric pressure:

260 + 15 = 275 psia (absolute discharge pressure)

60 + 15 = 75 psia (absolute suction pressure)

275 ÷ 75 = 3.7 (approximately 4:1 pumping ratio)

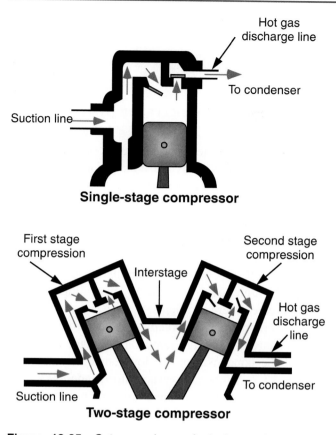

Single-stage compressor

Hot gas
discharge line

To condenser

Suction line

First stage
compression

Second stage
compression

Interstage

Hot gas
discharge
line

To condenser

Suction line

Two-stage compressor

Figure 12-25. *Cutaway views of single- and two-stage compressors.*

Another example uses the same refrigerant (R-22) on a low-temperature application, such as a commercial freezer. Here again, on a hot summer day the discharge pressure is about 260 psig, and the suction pressure is about 0 psig:

260 + 15 = 275 psia (absolute discharge pressure)

0 + 15 = 15 psia (absolute suction pressure)

275 ÷ 15 = 18.3 (18:1 pumping ratio)

In this example, a larger two-stage compressor is needed. Since the compressor becomes very hot at the interstage, additional compressor cooling (such as fans and liquid injection valves) are required. The load on the compressor will be severe due to the double compression needed to achieve the high condensing pressure.

Service technicians often overlook pumping ratio as a possible cause of compressor failure. A dirty condenser can result in high discharge pressure, which seriously affects the pumping ratio. Likewise, a very low suction pressure increases the pumping ratio.

In air conditioning (often called comfort cooling), about one horsepower (hp) is expended for each ton of cooling. In other words, it takes a one-hp compressor to provide one ton of comfort cooling. Therefore, a 10-ton system has a 10-hp compressor motor. This applies to comfort cooling only.

Lower suction pressures, such as those required for refrigeration, have a higher horsepower-per-ton ratio. The lower the suction pressure, the higher the horsepower per ton required.

12.9.8 What Refrigerant Is in a System?

A service technician is often involved with recovering the refrigerant, making system repairs, and recharging the system. Therefore, a service technician must have good working knowledge of the various refrigerants and where they are found. The ozone depletion problem required many systems to have their original refrigerants replaced (retrofitted) with newer refrigerants or refrigerant blends. Identifying the refrigerant in an existing system is becoming an increasingly complex task. Identification is simplified when the technician knows what refrigerants were originally used on various applications and what new refrigerants are used as replacements.

When servicing or repairing a system, you can tell what kind of refrigerant it contains by using one or more of the following methods:

❑ On split systems and packaged equipment, the manufacturer's equipment nameplate includes the refrigerant type and often the refrigerant charge. The refrigerant charge is expressed by weight (in pounds and ounces). When a refrigerant replacement has been performed (retrofitted), all major system components should have a permanent label indicating the new type of refrigerant in the system. If the lubricant was changed along with the refrigerant, the label should also carry that information. If you are replacing refrigerant or labeling a system someone neglected to relabel, you can obtain labels from refrigerant and oil manufacturers and their distributors.

❑ On equipment that uses a thermostatic expansion valve (TEV), the refrigerant type (R-number) is imprinted on top of the valve diaphragm. The valve diaphragm is painted with the proper color code for that particular refrigerant. (Note: The valve manufacturers developed their own color code before the cylinder colors were established. So, for R-12, the *valve* color is yellow, and the cylinder color is white.)

❑ One method of determining the refrigerant in a system without a nameplate or retrofit label is to take a pressure and temperature reading when the system is off and at ambient temperature. Refrigerant temperature-pressure (saturation) charts are consulted to identify the refrigerant. This method must be used with *caution;* many replacement refrigerants possess characteristics similar to the old refrigerants.

12.10 Refrigerant Applications ———◆

For general purposes, refrigerants are categorized for operation within a specific temperature range, or *application*. Some refrigerants operate best in high-temperature applications, others are best in low-temperature applications, and still others are best at ultralow-temperature applications. **Figure 12-26** lists six applications (by temperature), the most popular refrigerants used for each type of application, and their replacement refrigerants.

High-temperature applications primarily refer to air conditioning systems (or comfort cooling) but may also include fresh flower cases and salad bars, for instance. Air conditioning systems are normally designed to maintain building interiors at Fahrenheit temperature levels in the 70s, which is considered high temperature for refrigeration systems. R-22 is a good refrigerant for high-temperature applications due to the low pumping ratio, but it is not good for medium- or low-temperature applications.

Medium-temperature applications refer to domestic refrigerators, dehumidifiers, commercial walk-in coolers, watercoolers, dairy cases, meat cases, and produce cases. R-12 and R-134a are excellent medium-temperature refrigerants but are not practical for low-temperature uses. At low temperatures, R-12 and R-134a must operate with the suction pressure in a vacuum and the Btu per pound (latent heat value) is quite low.

Low-temperature applications normally refer to frozen food cases, walk-in freezers, and ice cream storage. See **Figure 12-27**. R-502 or R-507 are best for low-temperature applications because of such properties as low boiling point, azeotropic mixture, high suction pressure, low pumping ratio, and low latent heat value.

Ultralow and ***cryogenic*** applications are special systems designed for use in research laboratories and industrial applications requiring extremely low temperatures.

12.11.1 Cascade Refrigeration Systems

Cascade systems are composed of two refrigeration systems connected together and operating at the same time. Such systems are mostly found in industrial processes where objects must be cooled to temperatures below –50°F (–46°C). See **Figure 12-28.** One refrigeration system (low-stage) uses a refrigerant with a very low boiling point, such as R-13 or R-503. The suction pressure for R-13 is about 2 psig at –110°F (–79°C), but the condensing pressure is extremely high (over 500 psig). This pumping ratio would create severe problems.

To overcome these problems, a second refrigeration system using R-502 (high-temperature stage) is utilized to cool the condenser of the low-temperature system using R-13. A combination evaporator/condenser serves as a special heat exchanger to interconnect the evaporator of the high stage with the condenser of the low stage. The component is specially manufactured for cascade systems and serves to greatly reduce the discharge pressure for the lower stage, bringing the pumping ratio within normal limits.

Since these systems operate at very low temperatures, the refrigerant must be extremely dry. Such systems require the use of micron gauges with a good vacuum pump during the evacuation process. Special oils are required for lubrication at such low temperatures, and oil separators are installed to help keep oil in the compressors. Compressor cooling is not a problem because the compressor is cooled by the cold suction gas.

———————————————————————————————— **Refrigerant Applications**

Application	Temperature	Refrigerant	Replacement
High temperature	55° to 90°F	R-22 (Air conditioning)	Not required
Medium temperature	32° to 50°F	R-12 (Coolers)	R-134a and R-401A
Low temperature	0° to –50°F	R-502 (Freezers)	R-507 R-404A
Ultralow temperature	–55° to –250°F	R-13 and R-503 (Cascade)	R-403B
Cryogenics	–250° to 0 F_A	R-728 (Expendable)	No ozone depletion
Low-pressure chillers	45° to 55°F	R-11 (Operates in vacuum)	R-123

F_A = Absolute Fahrenheit

Figure 12-26. *The table lists six application areas by temperature, the most popular refrigerants used for each, and their replacements.*

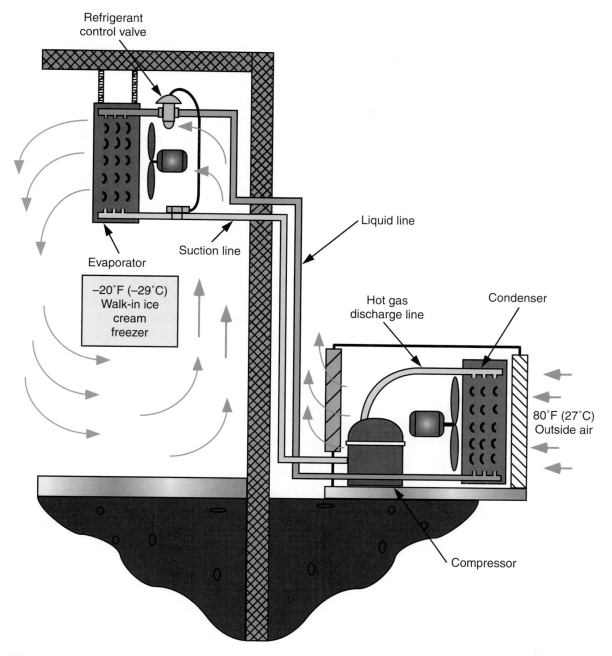

Figure 12-27. *Low-temperature application showing a walk-in ice cream freezer operating at −20°F (−29°C).*

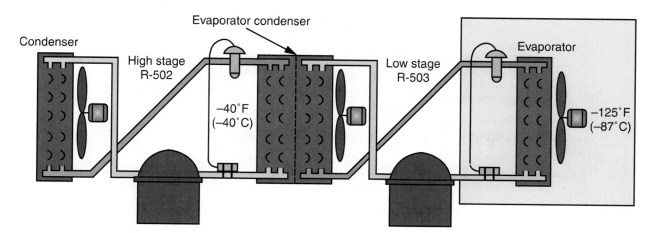

Figure 12-28. *A cascade system uses the high-stage evaporator to cool the low-stage condenser. This method provides normal discharge pressure for the low-stage system.*

12.10.2 Expendable Refrigerants

Expendable refrigerants are used primarily for fast freezing at temperatures below −140°F (−96°C). Generally, expendable refrigerants are classified in the cryogenic range because of their very low boiling points. Refrigerants such as liquid nitrogen or liquid carbon dioxide are sprayed into a freezing chamber, and the resulting vapor is released into the atmosphere. **Figure 12-29** depicts a fast-freezing system using an expendable refrigerant. Expendable refrigerants are only used once. The vapor is not collected and recondensed as in compression systems. Expendable refrigerants have very low boiling points and are relatively cheap when purchased in the liquid state. Their latent heat value (Btu/lb.) is rather high. Low cost and high latent heat values make expendables economically feasible. Systems that use expendable refrigerants are sometimes referred to as *open-cycle* or *chemical* refrigeration systems.

12.10.3 Cryogenic Refrigerants

The use of cryogenic fluids in food manufacturing processes is common. A typical cryogenic fluid is liquid nitrogen (R-728), which has a boiling point of −320°F (−196°C) at atmospheric pressure. Liquid carbon dioxide (R-744) is also acceptable as a cryogen, even though its boiling point is −109°F (−78°C) at atmospheric pressure. Other cryogenic refrigerants are listed in Figure 12-5.

Containers for cryogenic fluids are made of special materials that retain their strength at very low temperatures. These containers are heavily insulated and constructed on the principle of a vacuum bottle. They are never sealed tightly because the vapor must escape. The slow, constant boiling of the liquid to a vapor maintains a very low storage temperature inside the container. (The heat for boiling is extracted from the remaining liquid.)

12.10.4 Modern Freezing Methods

Until recently, commercial and industrial food freezing was not well-advanced. Modern technology has produced methods that have greatly improved the quality of the finished product. Many food products spoil easily and cannot be stored for any significant length of time without suffering loss in quality. Deterioration results from changes within the food caused by bacterial, chemical, or biochemical processes. Modern freezing methods play a vital role in controlling these changes since bacterial and chemical action is greatly retarded by lower temperatures.

It is important to understand the difference between *freezing* and *frozen storage*. **Freezing** is the process by which a product's water content is changed to ice, and its temperature is lowered to the desired level for storage. **Frozen storage** is the storage of an already frozen product at a constant temperature, usually 0°F (−18°C) or lower.

Because water expands when it changes from a liquid state to a solid state, slow freezing of food results in large ice particles. Expansion of the ice damages food fibers, destroys flavor, and decreases shelf life. Fast freezing, however, greatly reduces ice expansion.

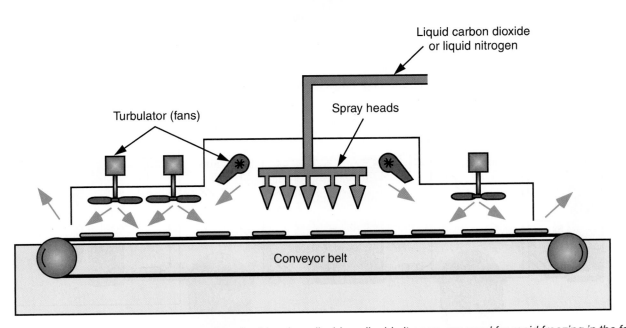

Figure 12-29. *Expendable refrigerants, such as liquid carbon dioxide or liquid nitrogen, are used for rapid freezing in the food processing industry. Since they are actually components of atmospheric air, these refrigerant gases can be released after use without danger to the environment.*

Fast freezing is done at very low temperatures, using cryogenic refrigerants and equipment installed directly in the production line (called *in-line freezing*). The foodstuffs are then packaged and placed in frozen storage which uses common refrigerants and a compression refrigeration system. A storage room should not be considered equipment for freezing even though it is sometimes used for this purpose.

All frozen food placed in storage should be properly wrapped to prevent loss of moisture. Moisture in exposed foods will *sublime* (go directly from a solid to a gas), resulting in a condition known as freezer burn. Freezer burn destroys quality and flavor.

Summary

This chapter has presented the physical properties of refrigerants and discussed how refrigerants are used in different applications. The principles in selecting a refrigerant for a particular system and the properties of each of the common refrigerants were also discussed.

The technician must understand the principles in selecting a refrigerant for a particular system and be aware of the properties of each of the common refrigerants. This knowledge is useful when troubleshooting, repairing, or installing a system. The circulating refrigerant is the key to the entire system. All system problems can be reduced to either:

❑ Poor or nonexistent circulation of refrigerant.
❑ Not enough refrigerant in the system.

Finally, saturation charts and determining super-heat and subcooling were explained.

Test Your Knowledge

Please do not write in this text. Write your answers on a separate sheet of paper.

1. Name some qualities that determine a good refrigerant.
2. What are the two basic problems of refrigeration systems?
3. What was the first halide refrigerant?
4. What are the color codes for containers used for R-12, R-22, R-500, and R-502?
5. What is the chemical name for R-12?
6. Monochlorodifluoromethane is the chemical name for what refrigerant?
7. Write the chemical formula for R-22.
8. What are the boiling points of R-12 and R-22 at atmospheric pressure?
9. The chemical elements chlorine and fluorine are known as _____.
10. The halogenated refrigerants are so stable that at every point in the refrigeration system their condition can be _____ and _____.
11. Name two common azeotropic refrigerants.
12. R-502 can absorb _____ more moisture than R-12 without serious system problems.
13. The service technician should never _____ or attempt to _____ refrigerants.
14. What are the standard conditions used to compare refrigerants?
15. What are the gauge pressures for R-12, R-22, R-500, and R-502 at 10°F?
16. What is the temperature of R-12 at a pressure of 2.32 in. Hg?
17. The best refrigerant choice is one that has _____ capacity and _____ power requirements.
18. How is pumping ratio calculated?
19. Pumping ratios higher than 10:1 usually require a _____ compressor.
20. Give one example each of high, -medium-, and low-temperature applications.

Portable refrigerant recovery and recharging systems like this one are easily wheeled into place on the job site. This unit is designed to recover, recycle, and recharge refrigerant R-134a. (Kent-Moore)

Refrigerant Recovery and Recycling

 Objectives

After studying this chapter, you will be able to:
- ❑ Explain the ozone depletion problem.
- ❑ Recognize CFCs, HCFCs, and HFCs.
- ❑ Comply with the 1990 Federal Clean Air Act.
- ❑ State equipment and technician certification requirements.
- ❑ Use recovery cylinders properly and safely.
- ❑ Recognize recovery and recycle equipment.
- ❑ Properly perform recovery and recycle procedures.

Important Terms

active recovery	ozone depletion
capture	potential (ODP)
certification	ozone depletion
Clean Air Act	passive recovery
Environmental Protection	pumpdown
Agency (EPA)	R&R system
global warming	reclaim
global warming	recover
potential (GWP)	recovery cylinders
hazardous waste	recycle
low-loss fittings	self-contained
Montreal Protocol	system-dependent
motor vehicle air	ultraviolet radiation
conditioning	

13.1 Ozone Depletion

In recent years, much concern has been raised about chlorine-fluorine-carbon compounds (CFCs) and the damage they do to the earth's atmosphere (*ozone depletion*). The stratosphere is the part of the atmosphere extending 10 miles to 30 miles (16 km to 48 km) above the earth's surface. The ozone layer is an essential part of the stratosphere and serves as the earth's security

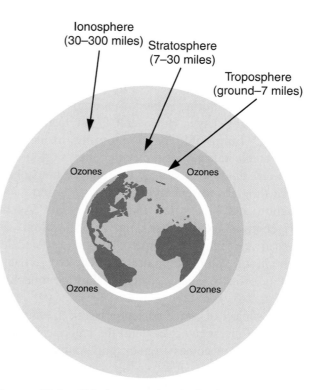

Figure 13-1. *Chlorine, a chemical element in many refrigerants, is the primary culprit in damaging the atmosphere's ozone layer.*

blanket, completely surrounding the planet and blocking harmful solar radiation. See **Figure 13-1.**

Ozone is a very reactive form of oxygen — a blue, irritating gas with a pungent odor. An ozone molecule consists of three oxygen atoms (O_3). While it is a naturally-occurring, beneficial component of the upper atmosphere (keeping most of the sun's harmful radiation from reaching the earth's surface), ozone is a major air pollutant in the lower atmosphere.

Ozone depletion is a serious matter. The ozone layer serves two important functions: it protects life on earth from harmful *ultraviolet radiation* and helps maintain stable earth temperatures. CFC and HCFC refrigerants are known to damage the ozone layer and have come under increasing governmental regulation worldwide. Stratospheric ozone depletion is a global problem that may cause:

❏ Crop loss.
❏ Increased eye diseases.
❏ Skin cancer and other radiation-related diseases.
❏ Reduced marine life.
❏ Increased ground-level ozone.
❏ Deforestation.

Much attention has focused on the destructiveness of widely used CFCs such as R-11 and R-12. *Chlorine,* a chemical element of many refrigerants, is the primary culprit. CFCs are very stable, chlorine-containing compounds that do not break up in the atmosphere as do other, less-stable refrigerants. Unlike other chlorine compounds and naturally-occurring chlorine, the chlorine in CFCs neither dissolves in water nor breaks down into compounds that dissolve in water. This means it does not "rain out" of the atmosphere.

Normally, chlorine in the lower atmosphere cannot reach the ozone layer. However, CFC molecules act as "carriers," lifting chlorine atoms to the ozone layer. These stable refrigerant molecules may float around for more than 40 years in the upper stratosphere, where they absorb energy from the sun's ultraviolet radiation and break into individual atoms. As the refrigerant molecules break up, chlorine atoms are released and attack the protective ozone molecules. Each chlorine atom combines with one oxygen atom from an ozone molecule, forming chlorine monoxide (ClO) and oxygen (O_2).

The subject of ozone depletion has been controversial. Some investigators contend the chlorine found in the stratosphere comes from natural sources, such as volcanic eruptions. However, air samples taken over erupting volcanoes show that volcanoes contribute a small amount of chlorine compared to the amount of chlorine produced by CFCs. Also, the rise in the amount of chlorine measured in the stratosphere over the past two decades matches the rise in the amount of fluorine over the same period; fluorine (also a refrigerant chemical) and chlorine do not originate from the same natural source. Furthermore, the rise in the amount of chlorine measured in the stratosphere over the past 20 years matches the rise in CFC emissions over the same period.

To express the risk to the ozone layer caused by the chlorine in refrigerants, each refrigerant has been assigned a number called its *ozone depletion potential (ODP).* See **Figure 13-2** for a partial list of these numbers. ODP is a measurement of a refrigerant's ability to destroy ozone.

Ozone Depletion Potential (ODP) and Global Warming Potential (GWP)

Compound	ODP	GWP
CO_2	0.0	1.0 (Base)
CFC-11	1.0 (Base)	1.30
CFC-12	0.93	3.70
CFC-113	0.83	1.90
CFC-114	0.71	6.40
CFC-115	0.38	13.80
HCFC-22	0.05	0.57
HCFC-123	0.02	0.28
HCFC-401A	0.03	
HCFC-401B	0.035	
HCFC-402A	0.03	
HCFC-402B	0.02	
HFC-134a	0.0	0.40
HFC-125	0.0	
HFC-507	0.0	

Figure 13-2. *Ozone depletion potential (ODP) is a measurement of a refrigerant's ability to destroy ozone molecules. Global warming potential (GWP) expresses the risk of a refrigerant contributing to global warming.*

The three primary types of refrigerants differ greatly in ODP:

❏ Chlorine-fluorine-carbon (CFC) = high ODP.
❏ Hydrogen-chlorine-fluorine-carbon (HCFC) = low ODP.
❏ Hydrogen-fluorine-carbon (HFC) = no ODP.

R-11, R-12, and R-500 are examples of CFCs; R-22 and R-123 are examples of HCFCs; R-134a is an HFC.

CFCs have the highest ODP. HCFCs have a hydrogen atom in the molecule, making the molecule unstable. Therefore, HCFCs tend to break up quickly when released to the atmosphere and have a low ODP. HFCs contain no chlorine, so they have no ozone depletion potential.

Most scientists agree that certain chemical compounds have another damaging effect on the earth's environment. *Global warming,* also called the *greenhouse effect,* occurs when long-wave (infrared) radiation from the sun reaches the earth but cannot escape. Trapped radiation causes a gradual buildup of heat on the earth's surface. Several refrigerants are suspected of contributing to global warming. To express the risk of a refrigerant contributing to global warming, each one has been assigned a number called its *global warming potential (GWP).* The higher the GWP, the greater the risk. Refer to Figure 13-2.

Refrigerants with high ODP and GWP ratings are systematically being phased out; refrigerants with lower ratings are replacing them. The phaseout is the result of a significant piece of national legislation called the Clean Air Act whose rules largely resulted from an international agreement called the Montreal Protocol. The Clean Air Act is having a profound impact on new equipment manufactured, lubricants used, and procedures required for installation and service of air conditioning and refrigeration systems.

13.2 Montreal Protocol

Scientists became aware of the effect of CFCs on the ozone layer in the 1970s. The United States banned the use of CFCs as aerosol propellants; however, much of the world did not. Finally, it was determined that worldwide cooperation was imperative to solve the ozone depletion problem. In 1987, government leaders from 11 industrialized nations met in Montreal, Canada, and reached an international agreement, called the **Montreal Protocol,** to substantially reduce, and eventually eliminate, production of ozone-depleting CFCs. The Montreal Protocol was signed by the United States and 22 other countries. Since then, more than 90 nations have ratified the Protocol. The Montreal Protocol was amended in 1990 and 1992, and continues to be modified as new information about the effects of refrigerants is uncovered.

Clearly, the regulations governing the production, use, and handling of refrigerants are here to stay and will be enforced. The Montreal Protocol called for production of all CFCs to cease by January 1, 1996. When virgin supplies of CFCs are depleted, all future supplies for service requirements must come from recovered, recycled, or reclaimed CFCs. Developing countries have an additional 10 years to complete the transition to new technologies.

Further regulations in the United States stipulate that equipment produced after 2010 must be HCFC-free, leaving the remaining HCFCs for service requirements only. HCFCs should be available in the United States for new projects until 2010 and for the aftermarket until 2020. The schedule for total phaseout of HCFC production and use is shown in **Figure 13-3.**

A current rush to produce HCFC-free equipment may be economically motivated. One major U.S. refrigerant producer recently announced it would discontinue the sale of HCFC-22 for new equipment by 2005. Comfort-cooling system manufacturers will probably introduce HCFC-free equipment soon and be well on their way to offering complete product lines shortly after the year 2000. The reduced supply of HCFC-22 will most likely raise the price of the

HCFC Phaseout Schedule

Year	Production Levels	EPA Rule (HCFC-22)
1996	2.8% of 1989 use, plus 1989 HCFC use, ODP-weighted	
2004	65% of 1989 use	
2010	35% of 1989 use	Ban on new equipment
2015	10% of 1989 use	
2020	0.5% of 1989 use	Ban on production and consumption
2030	0% of 1989 use	

Figure 13-3. *The schedule for total phaseout of HCFC production and use.*

refrigerant, just as the price of CFC-12 rose when its supply dwindled. The initial reduction in usage is already occurring because of equipment conversions and the early phaseout of HCFCs (by 2003) for solvent cleaning and aerosol use.

13.3 The Clean Air Act

While the Montreal Protocol addressed ozone depletion at the international level, the Clean Air Act dealt with the problem at the national level in the United States. The **Clean Air Act,** a federal law passed in 1990, contains severe restrictions and penalties for **venting** (releasing) *any* refrigerants to the atmosphere. Section 608 of the Clean Air Act directed the U.S. **Environmental Protection Agency (EPA)** to implement and enforce the regulations. The EPA was also granted the authority to establish environmentally safe procedures for using refrigerants. See **Figure 13-4.** The ban on CFC production went into effect January 1, 1996, making it illegal to manufacture new CFC refrigerants; however, existing systems already charged with CFCs could continue in use. Since refrigerant that leaks from a system is permanently lost, the pricing and scarcity of CFCs eventually will make owning and maintaining a CFC system cost-prohibitive.

Failure to comply with the EPA regulations could cost a technician and his or her company as much as $25,000 per day, per violation. A bounty of up to $10,000 lures competitors, customers, and fellow workers to turn in violators. Service technicians who violate the Clean Air Act provisions may be fined, lose their certification, and be required to appear in federal court.

Summary of EPA Compliance Dates

Date	Regulation
July 1, 1992	Illegal to vent CFCs or HCFCs.
August 13, 1992	MVAC technicians must be certified.
July 13, 1993	Evacuation requirements in effect.
November 15, 1993	Newly manufactured R&R equipment must be certified.
November 14, 1994	HVAC technicians must be certified.
November 14, 1994	Sale of CFCs and HCFCs restricted.
November 15, 1995	Illegal to vent HFCs.
January 1, 1996	Ban on CFC production.

Figure 13-4. *Compliance dates established by the EPA under the 1990 Clean Air Act.*

It is a violation of Section 608 of the Clean Air Act to:
❑ Falsify or fail to keep required records.
❑ Fail to reach the required recovery vacuum level prior to opening or disposing of appliances. An "appliance" refers to any refrigeration or air conditioning system, regardless of size.
❑ Knowingly release (vent) CFCs or HCFCs while repairing appliances.
❑ Service, maintain, or dispose of appliances designed to contain refrigerants without being appropriately certified.
❑ Fail to become certified (effective November 14, 1994).
❑ Vent CFCs or HCFCs after July 1, 1992.
❑ Vent HFCs on or after November 15, 1995.
❑ Fail to recover CFCs or HCFCs before opening or disposing of an appliance.
❑ Fail to have an EPA-approved recovery device.
❑ Add nitrogen to a fully charged system for the purpose of leak detection and thereby cause a release of the mixture.
❑ Discard a disposable cylinder without recovering any remaining refrigerant, rendering the cylinder useless, and recycling the metal.
❑ Fail to possess appropriate and approved equipment with low-loss fittings.

Some state and local government regulations may contain additional or stricter regulations than those stipulated in Section 608 of the Clean Air Act. In case of conflict between the Clean Air Act and the Montreal Protocol, the more stringent law regarding the issue takes precedence.

13.4 Venting

As you read, the Clean Air Act prohibits individuals from knowingly venting ozone-depleting compounds used as refrigerants into the atmosphere.

Only four types of release are permitted:
❑ *De minimus* (minor) quantities of refrigerant released in the course of making good-faith attempts to recapture and recycle or safely dispose of refrigerant.
❑ Refrigerants emitted in the course of normal operation of air conditioning and refrigeration equipment, such as mechanical purging and leaks. However, the EPA *is* requiring the repair of substantial leaks.
❑ Mixtures of nitrogen and R-22 used as holding charges or leak test gases. However, a technician may not avoid recovering refrigerant by adding nitrogen to a charged system.
❑ Small releases of refrigerant resulting from purging hoses or from connecting or disconnecting hoses.

13.5 Refrigerant Leaks

Owners of equipment with charges over 50 pounds are required to repair substantial leaks. A 35% annual leak rate is the trigger for requiring repairs in the industrial process and commercial refrigeration sectors. An annual leak rate of 15% is established for comfort cooling chillers and all other equipment with charges over 50 pounds.

13.6 The "Three Rs:" Recover, Recycle, Reclaim

Recover means to remove refrigerant in any condition from a system and to store it in an approved cylinder or external container. It is illegal to purge or vent refrigerants to the atmosphere. Recovery equipment is required to avoid releasing refrigerant to the atmosphere when making repairs or disposing of refrigeration equipment. See **Figure 13-5.** Recovered refrigerant is normally returned to the same system after repairs are finished. Recovery does not address the quality or reuse potential of the refrigerant, so cleaning, processing, and testing are *not* part of the recovery process. Recovered refrigerant may contain acids, moisture, and oil.

Recovery must be performed:
❑ Before the system is opened for repairs.
❑ Before disposing of any system or components containing refrigerant.
❑ When removing excess refrigerant from a system.
❑ Before pressurizing a system for leak detection with an inert gas.

Figure 13-5. *Refrigerant recovery equipment. (Robinair Mfg. Corp.)*

Most HVAC contractors have increased their service prices to help offset the cost of recovery equipment and recovery time. Customers have complained about the increased cost of service. To justify the increase, simply explain you are duty-bound and required by law to recover refrigerants to protect the environment and human health.

13.7 Recycle

Recycle means to clean refrigerant for reuse by removing moisture and contaminants from recovered refrigerant. Recycling is accomplished by repeatedly passing the recovered refrigerant through one or more filter-driers. See **Figure 13-6.** The cleanliness of recycled refrigerant does not have to meet the specifications for new refrigerant, and chemical tests are not required.

EPA regulations limit the use of recycled refrigerant to the same system or another system belonging to the same owner. Recycled refrigerant cannot change ownership by being sold or given away.

13.8 Reclaim

Reclaim means to restore the recovered refrigerant to a level equal to new (virgin) product specifications as determined by chemical analysis. Refrigerant that is *recovered* or *recycled* can be returned to the same system or other systems owned by the same person without restriction. If a refrigerant is to change ownership, however, that refrigerant must be reclaimed. Reclaiming requires cleaning the refrigerant to the Air

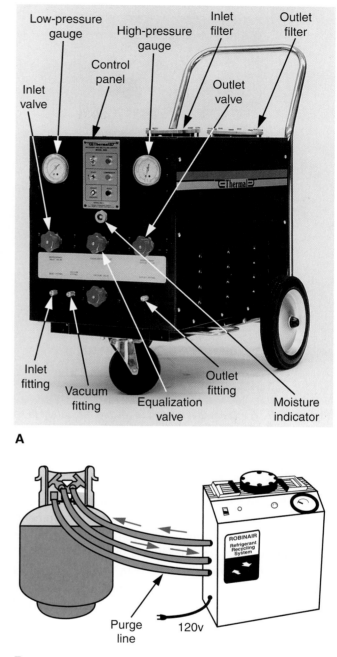

Figure 13-6. *Recycle units. A—Recycle unit removes contaminants from a refrigerant to allow reuse. (Thermal Engineering) B—Inside a recycle unit, a liquid pump (not a compressor) pulls liquid from the bottom of the cylinder via the dip tube (liquid valve). The liquid refrigerant is circulated through a replaceable core filter-drier and returned to the top of the cylinder (vapor valve). Recirculation of the liquid refrigerant through the filter-drier removes moisture and contaminants and may require several hours. (Robinair Mfg. Corp.)*

Conditioning and Refrigeration Institute (ARI) 700 standard of purity and chemically analyzing the restored refrigerant to verify it meets the standard. Reclaiming is a process normally available only at a manufacturing or reprocessing facility. See **Figure 13-7.**

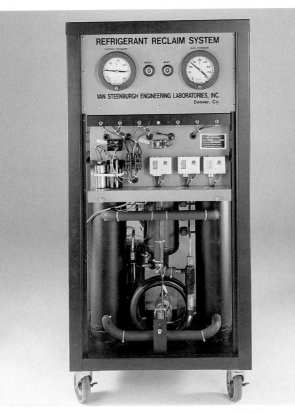

REFRIGERANT RECLAIM SYSTEM

VAN STEENBURGH ENGINEERING LABORATORIES, INC.
Denver, Co.

Figure 13-7. *A reclaim unit is used to restore refrigerant to its original purity. (Van Steenburgh Engineering Laboratories, Inc.)*

13.9 Safe Disposal Requirements

Equipment that is typically dismantled on site before disposal (commercial and industrial equipment) must have the refrigerant recovered in accordance with EPA requirements for servicing. Equipment that typically enters the waste stream with the charge intact (auto air conditioners, domestic refrigerators, room air conditioners) is subject to certain safe disposal requirements.

The final person in the disposal chain (junkyard or landfill owner) is responsible for ensuring refrigerant is recovered from equipment before the equipment's final disposal. However, persons "upstream" may remove the refrigerant and provide documentation of its removal to the final person if it is more cost-effective.

13.10 Record-keeping Requirements

Technicians, appliance owners, refrigerant wholesalers, and reclaimers all must keep accurate records of services and transactions involving refrigerants.

Technicians servicing appliances that contain 50 or more pounds of refrigerant must provide the owner with an invoice indicating the amount of refrigerant added to the system. Technicians must also keep proof of certification at their place of business.

Owners of appliances that contain 50 or more pounds of refrigerant must keep service records documenting the date and type of service, as well as the quantity of refrigerant added.

Wholesalers who sell CFC and HCFC refrigerants must retain invoices that indicate the name of the purchaser, the date of sale, and the quantity of refrigerant purchased. Since 1994, sale of these refrigerants has been restricted to technicians certified in refrigerant recovery.

Reclaimers must maintain records of the names and addresses of persons sending them material for reclamation and the quantity of material sent to them for reclamation. At the end of each calendar year, the reclaimer must submit an annual activity report to the EPA.

13.11 Hazardous Waste Disposal

Recycled or reclaimed refrigerants are not considered **hazardous waste** under federal law. Likewise, used oils contaminated with CFCs are not classified as hazardous provided they are:

❑ Not mixed with other waste.
❑ Subjected to CFC recycling or reclamation.
❑ Not mixed with used oils from other sources.

Individuals with questions regarding the proper handling of used oils should contact the EPA for guidance.

13.12 Certification Requirements

13.12.1 Motor Vehicle Air Conditioning

Section 609 of the Clean Air Act requires all persons who service **motor vehicle air conditioning (MVAC)** units be certified in refrigerant recovery and recycling and proper use of approved equipment when performing service involving refrigerant. Servicing includes repairs, leak testing, and "topping off" air conditioning systems low on refrigerant, as well as other vehicle repairs that require dismantling the air conditioner. MVAC certification can be obtained through a home-study program that includes a booklet and mail-in test. The test is open-book.

13.12.2 Technician Certification

Section 608 of the Clean Air Act mandates that all persons who maintain, service, repair, or dispose of refrigeration and air conditioning appliances, other than motor vehicle systems, must become certified by passing an EPA-approved test. The test is administered by an EPA-approved certifying organization and is monitored by an EPA-approved proctor. Each test contains a different set of questions. A passing score of 70% is required to obtain certification.

If EPA regulations change after a technician is certified, it is the technician's responsibility to comply with any future changes. Failure to comply with EPA regulations can cost a technician and his or her company as much as $25,000 per day, per violation.

There are four categories of EPA technician certification:

Type I. Technicians servicing small appliances must be certified in refrigerant recovery if they perform sealed-system service. The EPA definition of "small appliances" includes products manufactured, charged, and hermetically sealed in a factory with five pounds or less of refrigerant. This includes Packaged Terminal Air Conditioners (PTACs) with five pounds or less of refrigerant. Persons recovering refrigerant during maintenance, service, or repair of small appliances, except motor vehicle air conditioning systems, must be certified as either Type I Technician or Universal Technician. Some EPA-approved organizations are offering Type I certification only, as an open-book, mail-in test. The mail-in test requires a passing score of 84%.

Type II. Technicians maintaining, servicing, repairing, or disposing of high-pressure or extremely high-pressure (R-12, R-134a, R-22, R-502, R-503) appliances, except small appliances and motor vehicle air conditioning systems, must be certified as Type II Technician or Universal Technician.

Type III. Technicians maintaining, servicing, repairing, or disposing of low-pressure (R-11 and R-123) appliances must be certified as Type III Technician or Universal Technician.

Type IV (Universal). Technicians maintaining, servicing, or repairing low- and high-pressure equipment, including small appliances, except motor vehicle air conditioning systems, must be certified as Universal Technician.

13.12.3 Test Format

Most certification tests contain all four sections (A, I, II, III), and each section contains 25 multiple-choice questions. A minimum passing score of 70%, or 18 out of 25 correct, is required in *each section* (or group). A technician is certified in each group where a passing score is achieved. However, a technician *must* achieve a minimum passing score in Section A to receive any type of certification. A minimum passing score in all four sections earns a technician Type IV (Universal) certification.

Section A covers general knowledge about stratospheric ozone depletion, rules and regulations of the Clean Air Act, the Montreal Protocol, the "three Rs," recovery devices, substitute refrigerants and oils, recovery techniques, dehydration, recovery cylinders, safety, and shipping.

Section I contains questions pertaining to Type I certification.

Section II contains questions pertaining to Type II certification.

Section III contains questions pertaining to Type III certification.

The test is closed book, with no time limit for completion. (The average technician completes one section in about 20 minutes and finishes the test in 1 to 1-1/2 hours.) Some personal information is required for the examination. The technician should be prepared to present:

❑ Picture identification.
❑ Social Security number.
❑ Home address, including zip code.
❑ Employer's name, address, zip code.

EPA-approved testing organizations are located throughout the United States. Check with your local air conditioning and refrigeration parts supplier to obtain information about the testing organization(s) in your area. Most EPA-approved testing organizations provide the technician with a preparatory manual before taking the test. Some manuals are quite extensive and elaborate, while others are brief and simple.

EPA *certification* means you have the required knowledge in refrigerant recovery techniques and agree to abide by the EPA regulations prohibiting venting of refrigerants to the atmosphere. EPA certification *does not mean* you are qualified or knowledgeable in the proper repair, service, and installation of refrigeration and air conditioning *equipment.* Job certification skills training is offered by one or two industry organizations, but most contractors and service companies do not require job certification.

13.13 Equipment Certification ——◆

The EPA requires recovery units manufactured on or after November 15, 1993, be tested by an EPA-approved testing organization to ensure the equipment will achieve the required vacuum levels shown in the third column of **Figure 13-8.**

Recovery units manufactured *before* November 15, 1993, including home-made equipment, are exempt (grandfathered) if the equipment meets the standards shown in the second column of Figure 13-8.

The EPA requires that persons servicing or disposing of air conditioning and refrigeration equipment certify to the EPA they have acquired (built, bought, or leased) recovery equipment and are complying with the applicable requirements. The certification must be signed by the owner of the equipment or another responsible officer and sent to the appropriate EPA regional office.

Recovery Requirements

Type of Appliance	Inches Mercury (in. Hg) Vacuum Using Equipment Manufactured:	
	Before 11/15/93	On or After 11/15/93
HCFC-22 appliance normally containing *less* than 200 pounds of refrigerant	0	0
HCFC-22 appliance normally containing *more* than 200 pounds of refrigerant	4	10
Other high-pressure appliance normally containing *less* than 200 pounds of refrigerant (R-12, R-500, R-502, R-134a, R-401A, etc.)	4	10
Other high-pressure appliance normally containing *more* than 200 pounds of refrigerant (R-12, R-500, R-502, R-134a, R-401A, etc.)	4	15
Very high-pressure appliance (CFC-13, CFC-503)	0	0
Low-pressure appliance (CFC-11, HCFC-123)	25 in. Hg vacuum	29.90 in. Hg (500 microns)

Figure 13-8. *If the recovery equipment was manufactured before November 15, 1993, the air conditioning and refrigeration system must be evacuated to the levels indicated in the middle column. If the recovery equipment is manufactured on or after November 15, 1993, the air conditioning and refrigeration system must be evacuated to the levels indicated in the last column. The recovery equipment must be certified by an EPA-approved equipment testing organization.*

13.14 Recovery Requirements

Technicians are required by law to use recovery systems to evacuate air conditioning and refrigeration equipment to established vacuum levels. If the recovery equipment was manufactured *before* November 15, 1993, the air conditioning and refrigeration system must be evacuated to the levels indicated in Figure 13-8. If the recovery equipment is manufactured on or *after* November 15, 1993, the air conditioning and refrigeration system must be evacuated to the levels indicated in Figure 13-8, *and* the recovery equipment must be certified by an EPA-approved equipment testing organization.

Technicians repairing small appliances, such as household refrigerators, household freezers, and water coolers, are required to recover 80% to 90% of the refrigerant in the system, depending on the status of the system's compressor.

13.14.1 Exceptions to Recovery Requirements

If evacuation to the required levels is not attainable due to leaks, or if it would substantially contaminate the refrigerant being recovered, the person opening the system must:

❑ Isolate leaking components from nonleaking components, wherever possible.

❑ Evacuate nonleaking components to the required levels.

❑ Evacuate leaking components to the lowest attainable level without substantially contaminating the refrigerant. The level cannot exceed 0 psig (0 kPa).

13.15 Recovery Cylinders

Recovery cylinders are heavy-duty refillable cylinders that must meet Department of Transportation (DOT) specifications. The color code for recovery cylinders is a gray bottom with yellow shoulders and top. See **Figure 13-9.** Avoid mixing refrigerants in a recovery cylinder. Label your recovery cylinder permanently and clearly. You should have at least one refillable recovery cylinder for each type of refrigerant; otherwise, the cylinder must be emptied and evacuated before each use. Do *not* use dented, rusted, gouged, or damaged cylinders. Examine the cylinder valve assembly for leakage, damage, or tampering.

Refillable cylinders must be retested and recertified every five years from the test date stamped on the cylinder shoulder, in accordance with DOT Title 49 CFR, Section 173.34(e) and 173.31(d). Do not fill a container that is out of date (five years or older); return it empty, and have it retested. As shown in **Figure 13-10,** recovery cylinders have two valves. One valve is marked "liquid," the other "gas." The liquid valve has a dip tube that reaches to the bottom of the cylinder, making it possible to remove liquid without turning the cylinder upside down.

Some manufacturers require special recovery cylinders for use on their recovery units. The cylinders contain built-in safety devices to stop the recovery unit when the cylinder is full. The sensing devices are located inside the steel cylinder and consist of solid state thermistors and float switches. The reliability of these automatic shut-off devices is questionable. All recovery cylinders should be inspected carefully

Figure 13-9. *Three of these refillable cylinders are recovery cylinders, as indicated by the color coding. Always check cylinders for possible damage. (Amtrol)*

Cylinder manufacturers use water instead of refrigerant to determine cylinder capacity by weight. *Never* exceed a cylinder's weight capacity rating for any reason. All liquids expand when heated. The expanding liquid inside a cylinder creates hydrostatic pressure. Hydrostatic pressure in an overfilled refrigerant container will cause the cylinder to burst. Someone could be killed by the explosion.

13.16 Recovery Equipment

Refrigerant recovery equipment is used to *avoid venting refrigerant into the atmosphere* when making repairs to a refrigeration system. The recovery equipment is essentially a refrigeration system designed to remove and store refrigerant in a *refillable* cylinder. See **Figure 13-11.**

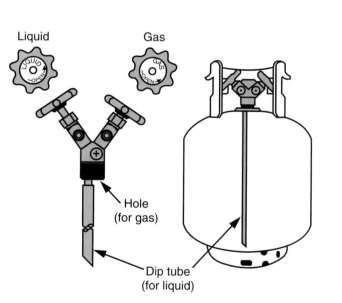

Figure 13-10. *Recovery cylinders have two valves, one marked "liquid" and the other marked "gas."*

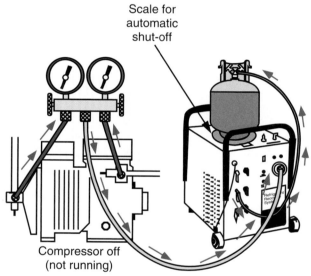

Figure 13-11. *Recovery equipment is a refrigeration system designed to remove and store liquid refrigerant in a refillable cylinder.*

and placed on an accurate scale before being filled. The technician should stop the recovery when the required recovery vacuum has been achieved, when the scale shows the cylinder is full, or when leaks prevent recovery to the proper vacuum level.

Recovery cylinders are considered *full* at 80% of total cylinder capacity. The capacity rating of a recovery cylinder designates the maximum amount of liquid refrigerant (in pounds) the cylinder can safely hold. The capacity rating is stamped on the cylinder and sometimes reads "W.C." for water column.

After recovery, refrigerant may be returned to the system from which it was removed, or to another system owned by the same person or company, without being *recycled* or *reclaimed*. Recovered refrigerant can simply be returned to the system from which it was originally withdrawn, but it should be cleaned of any contamination before use in *another* system owned by the same person.

Many types of recovery and recycling systems are available, and an operator's manual is supplied with each unit for proper connection and operation of the unit. Some recovery machines are capable of removing liquid or vapor, while others remove vapor only. Removal of refrigerant in the liquid state is less time-consuming than removal of vapor. However, after removing all liquid from a system, the remaining vapor must also be recovered. For instance, an average 350-ton R-11 chiller at 0 psig (0 kPa) pressure, with all the liquid removed, still contains 100 pounds of refrigerant in the vapor state.

Refrigerant recovery is accomplished by connecting a gauge manifold to the suction and discharge service valves, then connecting the center (yellow) hose to the recovery unit. The system service valves are opened, the gauge manifold valves are opened, and the hoses are purged of atmospheric air. The center (yellow) hose serves as the suction line for withdrawing vapor from the system. The recovery unit includes an evaporator, compressor, condenser, oil separator, various safety devices, and a separate refillable cylinder for liquid storage.

A good recovery system should provide for oil separation and contain a crankcase oil drain plug, a sight glass, and low-loss hose fittings. As mentioned earlier, most recovery units are able to remove refrigerant in the liquid or vapor state, with liquid removal much faster. However, some recovery units are rated according to pounds of liquid recovered per minute, even when the unit only removes vapor. **Figure 13-12** is a diagram of a recovery system.

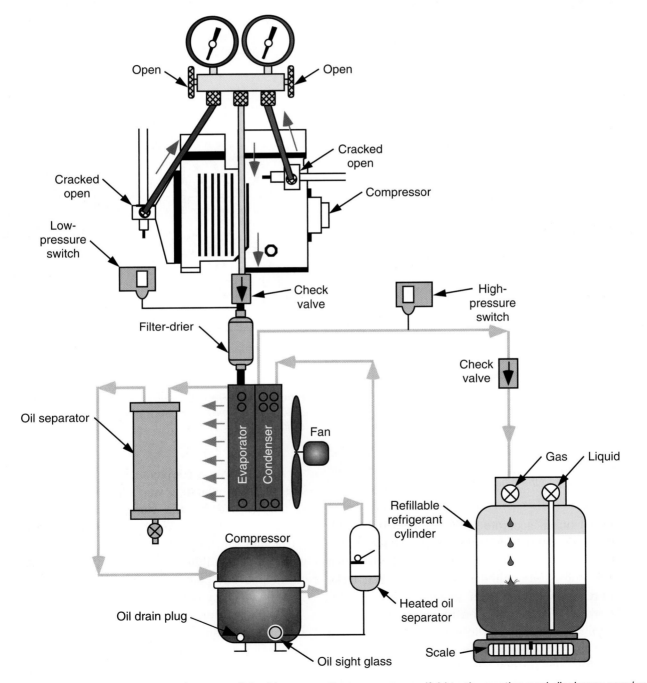

Figure 13-12. *Refrigerant recovery is accomplished by connecting a gauge manifold to the suction and discharge service valves, then connecting the center (yellow) hose to the recovery unit.*

Recovery systems require oil separators because some oil is removed from the system along with the refrigerant. Oil separators are not totally efficient; consequently, some contaminated oil appears in the recovery cylinder. The contaminated oil, in turn, contaminates the recovery compressor oil. The contamination should be removed by replacing the oil in the compressor crankcase.

Oil replacement can only be performed when the compressor has an oil sight glass and drain plug. Before using a recovery unit to process a different refrigerant than the one last processed, the unit's oil may need to be drained and replaced.

The EPA requires the use of *low-loss fittings* on recovery units. These fittings prevent loss of refrigerant to the atmosphere when connecting or disconnecting hoses from the recovery unit to a refrigeration system. The trapped refrigerant inside the recovery unit acts as a "holding charge" and prevents contamination by air. This is not a problem if the recovery unit is used for just one type of refrigerant. However, if it is used for different refrigerants, the recovery unit must be evacuated with a vacuum pump before use. Evacuation prevents contamination in the cylinder by two refrigerants.

13.17 Recovery Procedures

It is important to recover only one type of refrigerant into a recovery cylinder. Each recovery cylinder should be designated and properly marked for a specific refrigerant type. *Do not mix refrigerants in a recovery cylinder;* the mixture may be impossible to reclaim. If a reclaim facility receives a cylinder of mixed refrigerant, it may not be able to reclaim the mixture. The facility may agree to destroy the refrigerant but typically will charge a fee for the service.

The length and diameter of the hose connecting the recovery machine to the system affects the efficiency and time required for the recovery process. Long hoses and small diameters cause excessive pressure drop and increased recovery time. All refrigerants respond to a pressure-temperature relationship. Cold ambients result in low pressure, which results in a thin vapor and slow recovery rate. At low pressure, one pound of refrigerant vapor occupies several cubic feet of space. Likewise, high temperature results in high pressure, which results in a thicker vapor and faster recovery rate. At high pressure, one pound of refrigerant vapor occupies a much smaller volume.

13.17.1 Recovery Methods

The two most common methods of recovering refrigerant from mechanical refrigeration systems are using system-dependent recovery units and self-contained recovery units.

System-dependent recovery units are used with systems that normally contain 15 pounds or less of refrigerant. *Self-contained* recovery units are used with systems that normally contain 15 pounds or more of refrigerant.

System-dependent (passive) recovery method

Nonoperating compressor. In this method, often called *passive recovery,* the gauge manifold is connected to the suction and discharge lines, using service valves or line-piercing valves. The charging hose is connected to a recovery cylinder that has been evacuated to a deep vacuum and placed in a container of ice. Open both manifold valves and purge the hoses of air. Open the recovery cylinder vapor valve. Refrigerant from the system migrates to the cold cylinder and condenses to a liquid inside the cylinder. It is slow and inefficient and recovers about 80% of the system charge. Passive recovery is permitted only on systems containing 15 pounds or less of refrigerant. See **Figure 13-13.**

Operating compressor. The gauge manifold is connected to the suction and discharge lines, using service valves or line-piercing valves. If possible, frontseat the discharge service valve. Do not turn the compressor on until the proper valves are opened. The charging hose is connected to a recovery cylinder that

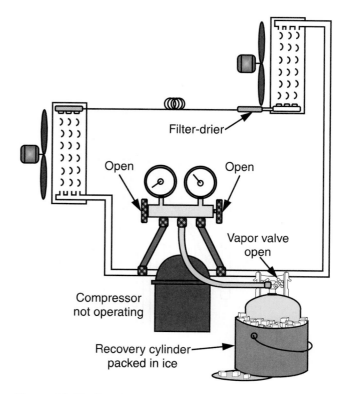

Figure 13-13. *Passive recovery method for a capillary tube system with a compressor that is not operating.*

has been evacuated to a deep vacuum and placed in a container of ice. Open the high-pressure manifold valve only, and purge the hoses. Open the recovery cylinder vapor valve, and turn the compressor on. The system compressor is used to pump high-pressure refrigerant vapor into the recovery cylinder. The recovery cylinder acts as a combination condenser/receiver. This passive recovery method recovers about 90% of the system charge. See **Figure 13-14.**

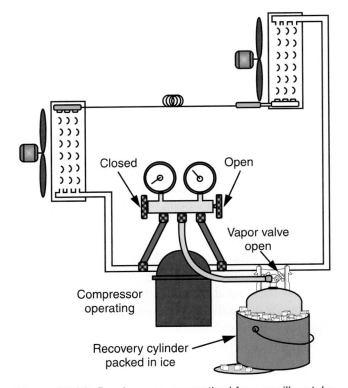

Figure 13-14. *Passive recovery method for a capillary tube system with an operating compressor.*

Self-contained, independent (active) recovery method

In this method, often called **active recovery,** the gauge manifold is connected to the suction and discharge service valves. Crack open the service valves; the manifold hand valves remain closed. The center hose is connected to the recovery unit intake valve. A separate hose connects the recovery unit output valve to the recovery cylinder vapor valve. Carefully purge all hoses. Open the cylinder vapor valve, the recovery unit valves, and the manifold hand valves before starting the recovery unit. The recovery cylinder should be placed on an accurate scale. See **Figure 13-15.**

Recovery connections and start-up procedures may vary among manufacturers. Follow the manufacturer's instructions provided with each unit. Connection procedures will vary if refrigerant is removed in the liquid state. Always use a recovery cylinder designated for use with the type of refrigerant being recovered.

13.18 Recycle Equipment

Recycle units remove contaminants from refrigerant prior to returning the refrigerant to the machinery from which it came, or to other equipment belonging to the same owner. Recycle units contain a liquid pump that withdraws liquid refrigerant from the bottom of the recovery cylinder, circulates it through a large filter-drier, and returns the liquid to the top of the recovery cylinder. Most recycle units use one or more replaceable-core filter-driers for cleaning the refrigerant. It may take four to six hours of recirculation, including filter element changes, to

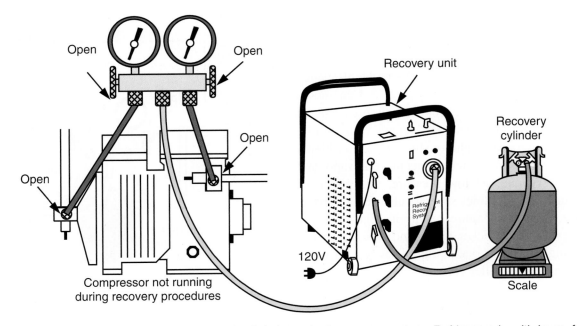

Figure 13-15. *Active recovery using a self-contained, independent recovery system. Refrigerant is withdrawn from the defective refrigeration system as a vapor. It is condensed to a liquid and then stored in the refillable cylinder.*

remove sufficient moisture, acid, and particulate matter before the refrigerant is suitable for reuse. See **Figure 13-16** for a diagram of recycle unit operation.

The cleanliness of recycled refrigerant does not meet the specifications for *new* refrigerant, and chemical tests are not required. Some recycle units have built-in methods to determine the acid and moisture content of the recycled liquid refrigerant. Individual testing kits are available for performing acidity and moisture tests, if desired. Recycling is done at the job site or the local service shop. Recovery and recycle units are often housed in one system, called an **R&R system,** that performs both operations.

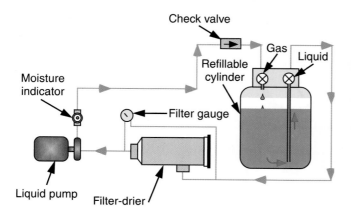

Figure 13-16. *A pump draws liquid refrigerant from the bottom of the cylinder and circulates it through a replaceable-core filter-drier to remove moisture and contaminants from the refrigerant.*

13.19 Capture Methods ◆

In some repair situations, the refrigerant can be stored temporarily in another part of the system not in need of repair. For example, a system that has a liquid receiver with two service valves (inlet and outlet) can store the refrigerant while the area in need of repair is opened. See **Figure 13-17.** The procedure is not recovery because it does not remove the refrigerant to a container *outside* the system. It **captures** (traps) most of the refrigerant in the liquid receiver, with residual vapor residing in the condenser and hot gas discharge line.

13.19.1 Capturing Refrigerant in the Liquid Receiver

Capturing refrigerant in the liquid receiver is called a **pumpdown.** It is performed by closing the receiver outlet valve (frontseated) while the compressor is running. When pumping down, sometimes the compressor shuts off on a low-pressure control before the pumpdown is completed. It may be necessary to use a "jumper" to bypass the low-pressure safety control or otherwise force the compressor to run.

The compressor evacuates (removes) all refrigerant from the low-pressure side and stores it in the liquid receiver. Then, the compressor is turned off and the receiver inlet valve is immediately closed (frontseated). The low-pressure side of the system is in a vacuum. The major portion of the system refrigerant charge is now stored in the liquid receiver, but residual high-pressure vapor remains in the condenser and discharge line. The suction service valve is also frontseated to prevent backflow of the vapor through the compressor valve reeds.

Do not open the low-pressure side when it is in a vacuum, or atmospheric air will enter the system. The air will have to be removed (evacuated) before restarting the system. To prevent atmospheric air from entering the low-pressure side, release a small amount of liquid refrigerant from the receiver outlet valve. (A small amount of liquid makes a large amount of vapor.) The vapor breaks the vacuum by raising the low-pressure side to 1 psig or 2 psig (6.9 kPa or 13.8 kPa). When the suction side is opened for repairs, a *small* quantity of vapor will escape to the atmosphere. Air cannot enter the system, and refrigerant cannot escape because both are at the same pressure (0 psig).

When the system is properly pumped down and the vacuum eliminated, the low-pressure side can be opened for minor, short-term repairs, such as changing a filter-drier. When repairs are completed, the receiver service valves can be opened and the system restored to service. No evacuation or recovery procedures are required.

If the low-pressure side is left open to the atmosphere for too long, atmospheric air will enter the system, and vapor will leak back through the compressor valve reeds. The atmospheric air must be removed with a vacuum pump prior to releasing refrigerant from the receiver, and the lost refrigerant must be replaced. Evacuating air from the low-pressure side is done by connecting the gauge manifold to the receiver outlet service valve, with the valve in the frontseated position. See **Figure 13-18.**

When in the pump-down mode, repairs to the high-pressure side can be accomplished only after recovery of vapor trapped in the hot gas discharge line and condenser. The recovery process is performed at the discharge service valve. See **Figure 13-19.**

13.19.2 Capturing Refrigerant in the Condenser

When the system does not have a liquid receiver, the condenser can be used as a storage area to capture refrigerant. A shut-off (isolation) valve must be available at the condenser outlet. The system is pumped down into the condenser after closing the shut-off

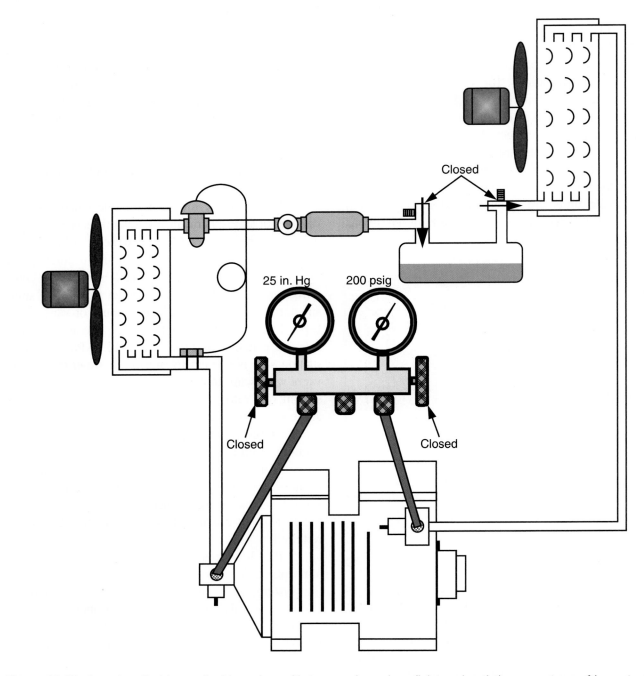

Figure 13-17. *A system that has a liquid receiver with two service valves (inlet and outlet) can capture refrigerant while low-pressure-side repairs are made.*

valve at the condenser outlet. After pumpdown to the condenser, the compressor is shut off, and the discharge service valve is frontseated. This captures all the refrigerant in the condenser and the hot gas discharge line. Using the condenser as a storage area does *not* cause high-discharge pressure (as commonly believed), unless the system is severely overcharged. The condenser will hold the normal system charge and not be completely full. Liquid occupies less space than vapor, and the vapor stops coming into the condenser *before* the condenser becomes full. See **Figure 13-20.**

Summary

This chapter explained ozone depletion and the role refrigerants play, the Montreal Protocol, and consequent laws to protect the atmosphere. Sections 608 and 609 of the Clean Air Act were discussed, as well as penalties for intentionally venting refrigerants to the atmosphere. The Environmental Protection Agency's responsibility for writing and enforcing the rules and regulations to prevent further ozone depletion by refrigerants was also discussed.

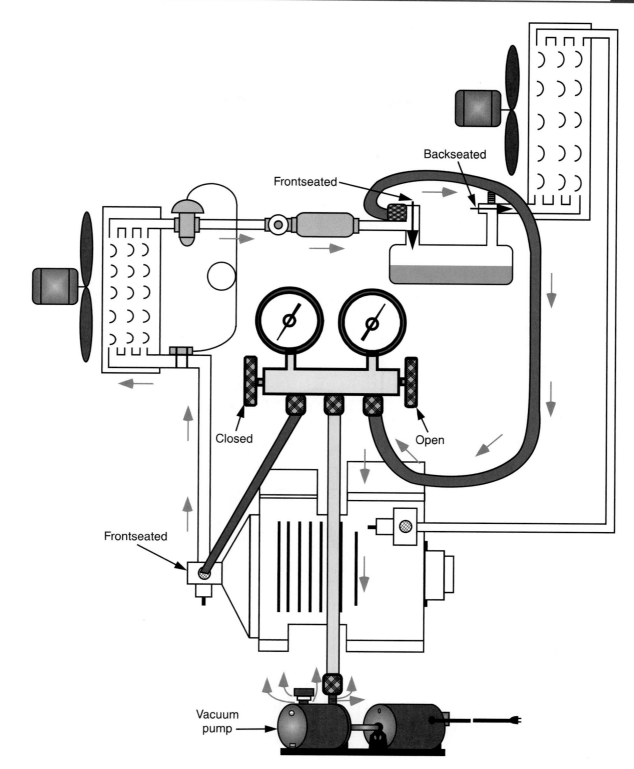

Figure 13-18. *If the system is left open to the atmosphere, air will enter. The evacuation process for removing air from the low-pressure side is shown.*

Proper service procedures when working with both old and new refrigerants have undergone major changes. The new rules and procedures involving recovery, recycle, and reclaim activities were explained. This chapter also covered requirements and testing for technician certification in MVAC, Type I, Type II, Type III, and Universal categories.

Equipment certification and step-by-step procedures for using recovery and recycle equipment were explained. Recovery cylinders are specially designed for use with R&R units. The proper and safe use of these recovery cylinders was fully explained. Finally, recycle equipment, capture methods, and recovery procedures were covered.

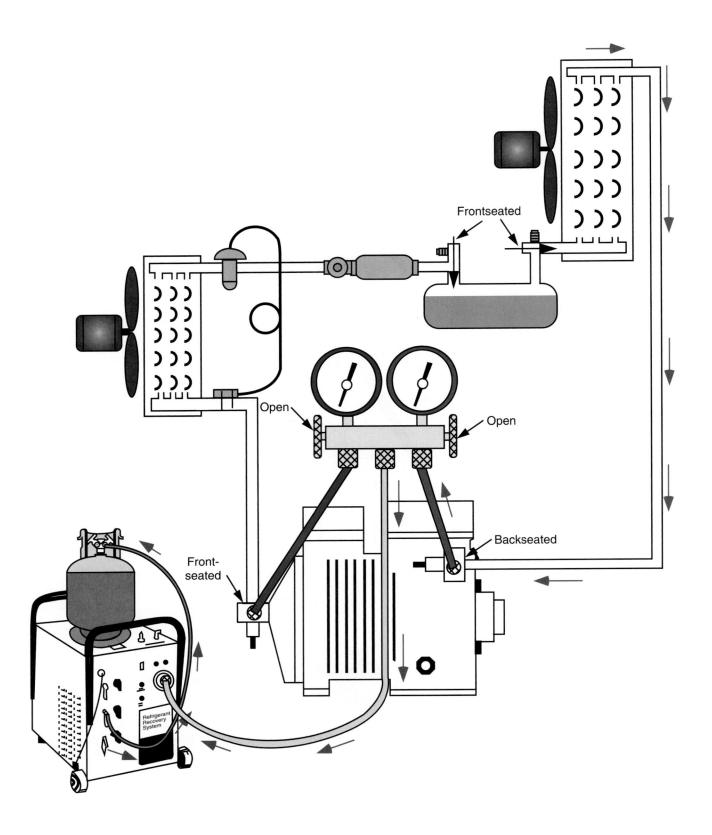

Figure 13-19. *After a pumpdown, vapor trapped in the hot gas discharge line and the condenser must be removed by the recovery method.*

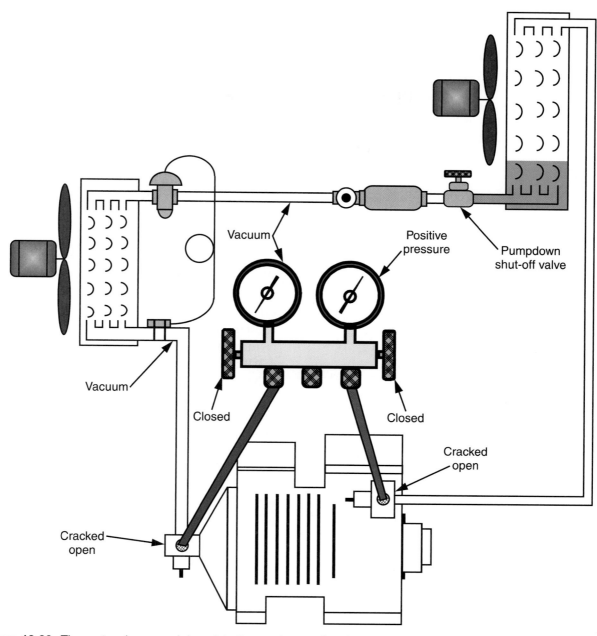

Figure 13-20. *The system is pumped down into the condenser after closing the shut-off valve at the condenser outlet.*

Test Your Knowledge

Please do not write in this text. Write your answers on a separate sheet of paper.

1. The stratosphere is located _____ miles above the earth.
   a. 5 to 10
   b. 10 to 20
   c. 10 to 30
   d. 50

2. The ozone layer provides protection from _____.
   a. UV radiation
   b. skin cancer
   c. deforestation
   d. all of the above

3. Which element in refrigerant destroys ozone?
   a. Chlorine.
   b. Fluorine.
   c. Methane.
   d. All of the above.

4. _____ is a measurement of a refrigerant's ability to destroy ozone.
   a. UFO
   b. ODP
   c. EEC
   d. DOT

5. Which type of refrigerant does not harm the ozone layer?
   a. CFC.
   b. HCFC.
   c. HFC.
   d. All of the above.

6. The Montreal Protocol called for all CFC production to cease by January 1, _____.
   a. 1993
   b. 1994
   c. 1995
   d. 1996
7. Which U.S. government agency was empowered by Congress to regulate and enforce the Clean Air Act?
   a. OSHA.
   b. EPA.
   c. DOE.
   d. ODP.
8. It is a violation of the Clean Air Act to knowingly release _____ refrigerants into the atmosphere.
   a. CFC
   b. HCFC
   c. HFC
   d. all of the above
9. R&R equipment refers to _____.
   a. recycle and reclaim
   b. recover and recycle
   c. reclaim and dispose
   d. all of the above
10. *True or false?* Minor amounts of refrigerants may never be vented to the atmosphere.
11. To "pump down" means to capture refrigerant in the _____.
    a. liquid receiver
    b. evaporator
    c. compressor
    d. suction line
12. Recycle means to clean refrigerant for (reuse, resale).
13. A system containing 50 pounds of R-12 must be recovered to _____ if the recovery unit was manufactured after November 15, 1993.
    a. 0 psig
    b. 4 in. Hg
    c. 10 in. Hg
    d. 500 microns

14. *True or false?* Recovered refrigerant may contain acid, moisture, and oil.
15. The color code for recovery cylinders is _____.
    a. gray
    b. gray bottom, yellow top
    c. yellow bottom, gray top
    d. none of the above
16. *True or false?* Recovery cylinders are considered full at 80% of total cylinder capacity.
17. The expanding liquid inside a cylinder creates _____ pressure.
    a. negative
    b. positive
    c. hydrostatic
    d. all of the above
18. When recycling refrigerant, how long must the refrigerant normally recirculate through a filter-drier?
    a. 30 minutes.
    b. 4 to 6 hours.
    c. All day.
    d. Not required.
19. Recovery cylinders have (one, two) valves.
20. Passive recovery can be used only when the system contains _____ pounds of refrigerant or less.
    a. 10
    b. 15
    c. 20
    d. 25

Zeotropic Blends

Objectives

After studying this chapter, you will be able to:
- ❏ Distinguish between pure component refrigerants, azeotropic mixtures, and zeotropic blends.
- ❏ Explain fractionation.
- ❏ Determine temperature glide.
- ❏ Read and use a temperature-pressure card for blends.
- ❏ Define bubble point and dew point.
- ❏ Appropriately select and use lubricants for blends.
- ❏ Perform system retrofit to a blend.

Important Terms

alkylbenzene (AB)	long-term
azeotropes	miscible
azeotropic refrigerant	NARMs
mixture (ARM)	polyol ester (POE)
baseline data	pure refrigerant
bonded core	retrofit
bubble point	short-term
dew point	temperature glide
flushing procedures	topping off
fractionation	zeotropes
hygroscopic	zeotropic blends

14.1 Introduction

The Clean Air Act resulted in many new non-ozone-depleting refrigerants and the introduction of synthetic lubricants for use with them. From 1990 to 1995, about 20 new refrigerants were introduced. At this writing, many refrigerants are being touted as alternatives to CFCs, but only a few are widely accepted.

A few products have emerged as the most commonly used CFC replacements:
- ❏ R-404A and R-507 replace R-502 for new equipment; R-402A is used to convert existing R-502 equipment.
- ❏ R-134a is the refrigerant of choice for new medium-temperature R-12 equipment; R-401A is used to convert existing R-12 equipment.
- ❏ R-123 is the accepted choice for new and existing R-11 equipment.

Service technicians must be knowledgeable in the use of these new refrigerants, oils, and associated products. New procedures are required for successful use of alternative refrigerants.

Wide-spread conversion from CFC refrigerants is not necessary unless there are strong reasons for doing so. A CFC system that is leak-free and operating properly should be left alone. However, the availability of CFC refrigerants for servicing existing equipment is becoming limited and expensive. Equipment owners are being forced to choose one of the following three options:
- ❏ *Option 1:* Remove the CFC-containing equipment, and replace it with equipment that uses HCFC-22 or new HFC refrigerants.
- ❏ *Option 2:* Convert the equipment to a low-ozone-depleting HCFC refrigerant. Most of the mineral oil is removed and replaced with either alkylbenzene or polyol ester lubricant. Labor and material costs are lower than with Option 3. The system is environmentally acceptable, and the HCFC refrigerant will be available until January 1, 2020. For many equipment owners, this is the preferred option for existing equipment.
- ❏ *Option 3:* Convert the equipment to a non-ozone-depleting HFC. This option is labor-intensive, but the system is environmentally acceptable and serviceable for the remainder of its useful life.

14.2 Pure Refrigerants, Mixtures, and Blends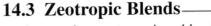

It is important to understand the differences between *pure* component refrigerants, azeotropic *mixtures* and zeotropic *blends*. When a refrigerant's entire composition consists of one molecule type, it is referred to as a single-component or "pure" refrigerant. A **pure refrigerant** is a single-component fluid that does not change composition when boiling or condensing. Examples of pure refrigerants are CFC-11, CFC-12, HCFC-22, and HFC-134a. Pure refrigerants are generally the refrigerants of choice.

Sometimes two pure refrigerants can be combined to form an **azeotropic refrigerant mixture (ARM)** that cannot be separated into its individual components. An azeotropic mixture behaves as a single-component refrigerant with a single boiling point at a particular pressure and temperature. **Azeotropes** do not separate in either the gaseous or liquid state. The American Society of Heating, Refrigeration and Air Conditioning Engineers (ASHRAE) has classified all azeotropic mixtures into a 500-series of refrigerant numbers. Examples of azeotropic mixtures are R-500, R-502, and R-507.

When two or more single-component refrigerants are combined and do not form an azeotropic mixture, the result is a blended refrigerant. Blended refrigerants are commonly known as "blends" or **zeotropes**. Most of the new, non-ozone-depleting refrigerants are classified as zeotropic blends, and ASHRAE has classified all zeotropes into a 400-series of refrigerant numbers. **Figure 14-1** shows several of these new refrigerant blends.

Figure 14-1. *A variety of refrigerant blends from one manufacturer. (DuPont)*

14.3 Zeotropic Blends ◆

As noted, a zeotrope is a *blend* of refrigerants. It usually consists of two or three different refrigerants and may be described as a *binary* (two-part) or *ternary* (three-part) blend. These new refrigerant blends are designed to replace the ozone-depleting CFC and HCFC refrigerants. **Figure 14-2** lists many of the new blends, and the list continues to grow as more are being developed. **Zeotropic blends** are unusual because they do not have a specific pressure for each degree of temperature change. At a given pressure, the temperature may fluctuate by as much as 16 degrees Fahrenheit (9 degrees Celsius). The temperature difference is called *temperature glide* and is the result of *fractionation* (separation of parts).

14.4 Fractionation and Temperature Glide ◆

All zeotropes exhibit fractionation and temperature glide. When both liquid and vapor states are present simultaneously (as in the evaporator or condenser), the composition of the liquid is different from that of the vapor. The components of the blend do not boil and condense as a single refrigerant. Instead, the most volatile component of the refrigerant blend boils first, the next most volatile boils second, and the next most volatile boils third. In other words, **fractionation** (separation of parts) occurs when the most volatile component boils into a vapor more quickly than the less-volatile components. Fractionation also occurs when the system develops a leak or when refrigerant is being charged into the system. Zeotropic refrigerants fractionate inside the charging cylinder when pressure is removed from the liquid. To prevent fractionation in the charging cylinder, zeotropic refrigerants are carefully charged into the system as a *liquid.* Charging the system as a vapor is possible when the entire contents of the charging cylinder are being emptied into the system.

The difference in composition of zeotropic refrigerants leads to a difference in boiling points for the liquid and vapor. There is a boiling *range* because the liquid begins boiling at one temperature and the vapor condenses at a different temperature. The temperature difference caused by fractionation between the evaporator inlet and outlet, or between the condenser inlet and outlet, is called **temperature glide.** The amount of temperature glide varies from blend to blend and with system conditions. The old rules for setting superheat do not apply to zeotropic blends. Temperature glide for most zeotropic blends ranges from 1 to 16 degrees Fahrenheit (0 to 9 degrees Celsius). Some zeotropic blends are referred to as **NARMs** (nearly azeotropic refrigerant mixtures) because the temperature glide is rather small (1 or 2 degrees).

Zeotropic Blends

Refrig. Number	Blend	Trade Name	Type	Lubricant Used*	Replaces	Boiling Point (°F @ 0 psig)
R-401A	R-22/152a/124	MP39	HCFC/HFC/HCFC	AB/POE	R-12/500	−16.5
R-401B	R-22/152a/124	MP66	HCFC/HFC/HCFC	AB/POE	R-12/500	−29.5
R-402A	R-125/290/22	HP80	HFC/HC/HCFC	MO/AB/POE	R-502	−53.6
R-402B	R-125/290/22	HP81	HFC/HC/HCFC	MO/AB/POE	R-502	−52.5
R-403B	R-290/22/218	69L	HC/HCFC/HFC	AB/POE	R-502/R-13	−57.1
R-404A	R-125/143a/134a	HP62/FX70	HFC/HFC/HFC	POE	R-502	−51.8
R-406A	R-22/142b/600a	GHG12	HCFC/HCFC/HC	MO/AB	R-12	−22.0
R-407A,B	R-32/125/134a	KLEA60,61	HFC/HFC/HFC	POE	R-502	−38.0
R-407C	R-32/125/134a	KLEA66	HFC/HFC/HFC	POE	R-22	−34.6
R-408A	R-125/143a/22	FX10	HFC/HFC/HCFC	MO/AB/POE	R-502	−46.3
R-409A	R-22/124/142b	FXS6	HCFC/HCFC/HCFC	MO/AB	R-12	−15.4
R-410A	R-32/125	AZ20	HFC/HFC	POE	R-22	−62.9

*Lubricants: MO = Mineral Oil, AB = Alkylbenzene, POE = Polyol Ester

Figure 14-2. *Some of the new zeotropic refrigerant blends. New blends are continually being developed.*

Temperature glide can affect the system in several ways. First, the evaporator is no longer at a constant temperature, and the "set" temperature is the *average* temperature. The thermostatic expansion valve may sense the temperature at the evaporator outlet as too high. In some systems, temperature glide may cause frosting at the evaporator outlet. Evaporator frosting is a normal characteristic for refrigerants with glide. No problems should occur if glide is recognized and proper adjustments are performed.

14.4.1 Topping Off

Topping off is adding enough refrigerant to bring a system up to full capacity. A leak in a system using a zeotropic blend cannot be topped off in the usual way, because of fractionation. A leak on the liquid side of a system releases proportionate amounts of each refrigerant component; whereas, a leak from the low-pressure side tends to fractionate and leak disproportionate amounts of the lower-boiling components. As a rule of thumb, a leak of less than 50% of the total charge can be topped off with *liquid* without a major effect on system performance.

14.5 Temperature-Pressure Card

The temperature-pressure card for zeotropic blends is unusual. Instead of one pressure for each saturation temperature, it has two. **Figure 14-3** is a partial example of the new temperature-pressure card used for blends.

R-409A Temperature-Pressure

Temp. °F	Temp. °C	Liquid Pressure (Bubble)	Vapor Pressure (Dew)
−30	−34	0.9	7.9
−25	−32	2.9	4.8
−20	−29	5.1	1.4
−15	−26	7.4	1.2
−10	−23	10.0	3.2
−5	−20	12.9	5.5
0	−18	16.0	8.0
5	−15	19.3	10.6

Italics indicate in. Hg vacuum.

Figure 14-3. *A sample of part of a temperature-pressure card used for zeotropic blends.*

The third column of the card shows the *saturated liquid pressure*, called the **bubble point.** These numbers indicate the pressure for a given temperature in which the first bubble of vapor appears in a liquid refrigerant. The last column shows the *saturated vapor pressure*, called the **dew point.** These numbers indicate the pressure at which the first drop of liquid condenses from a vapor. For example, R-409A has a bubble point temperature of −10°F (−23°C) when its pressure is 10 psig (69 kPa). It has a dew point temperature of about

5°F (−15°C) at approximately the same pressure (10.6 psig or 73 kPa).

To determine superheat, use *dew point* values. Superheat is the number of degrees *above* the dew point temperature for a specific pressure. For example, with a suction pressure of 8 psig (55 kPa), the dew point temperature is 0°F (−18°C). An accurate temperature reading at the evaporator outlet reads 10°F (−12°C). The superheat equals 10 degrees Fahrenheit or 6 degrees Celsius *above* dew point.

To determine subcooling, use *bubble point* values. Subcooling is the number of degrees *below* the bubble point temperature for a specific pressure. For example, the discharge pressure for R-409A is 152 psig at 100°F

(1049 kPa at 38°C). An accurate temperature reading at the condenser outlet reads 90°F (32°C). The amount of subcooling is 10 degrees Fahrenheit or 6 degrees Celsius *below* the bubble point.

Temperature glide can be calculated by finding the difference between the dew point and bubble point temperatures at constant pressure. Actual measurements may differ slightly, depending on the state of the refrigerant and pressure loss through the evaporator or condenser.

Figure 14-4 shows a new, complete temperature-pressure card used by technicians when working with blends. These cards, or similar ones, are available free from your local refrigerant supplier or directly from the refrigerant manufacturer.

Temperature-Pressure Card for Zeotropic Blends

		401A		401B		402A		404A		407C		409A	
		MP 39		MP66		HP 80		HP 62		KLEA 66		FX 56	
°F	°C	(Liquid) Bubble	(Vapor) Dew	(Liquid) Bubble	(Vapor) Dew	(Liquid) Bubble	(Vapor) Dew	(Liquid) Bubble	(Vapor) Dew	(Liquid) Bubble	(Vapor) Dew	(Liquid) Bubble	(Vapor) Dew
−40	−40.0	3.8	12.5	7.2	11.3	8.5	7.1	5.1	4.7	3.3	3.2	5.2	13.2
−35	−37.2	2.3	10.1	4.0	8.7	11.3	9.9	7.6	7.2	5.7	0.3	1.9	10.7
−30	−34.4	0.7	7.3	0.4	5.9	14.5	13.0	10.3	9.9	8.3	2.3	0.9	7.9
−25	−31.7	1.1	4.3	1.8	2.7	17.9	16.2	13.3	12.9	11.1	4.6	2.9	4.8
−20	−28.9	3.1	0.9	3.9	0.4	21.6	19.9	16.6	16.2	14.3	7.1	5.1	1.4
−15	−26.1	5.3	1.4	6.3	2.3	25.7	23.8	20.2	19.8	17.7	9.8	7.4	1.2
−10	−23.3	8.9	3.4	8.9	4.4	30.1	28.1	24.1	23.7	21.4	12.9	10.0	3.2
−5	−20.6	10.3	5.6	11.7	6.7	34.8	32.8	26.3	27.9	25.5	16.2	12.9	5.5
0	−17.8	13.2	8.0	14.8	9.3	40.0	37.8	33.0	32.5	29.9	19.8	16.0	8.0
5	−15.0	16.3	10.6	18.2	12.0	45.6	43.3	37.9	37.5	34.7	23.8	19.3	10.6
10	−12.2	19.7	13.5	21.9	15.0	51.6	49.2	43.3	42.9	39.9	28.2	22.9	13.6
15	−9.4	23.3	16.7	23.3	18.3	58.0	55.5	49.1	48.6	45.5	32.9	26.8	16.8
20	−6.7	27.3	20.1	30.2	21.8	65.0	62.3	55.3	54.9	51.6	38.0	31.0	20.0
25	−3.9	31.6	23.8	34.9	25.7	72.4	69.6	62.0	61.6	58.1	43.6	35.5	24.0
30	−1.1	36.2	27.8	39.9	29.8	80.4	77.4	69.2	68.8	65.1	49.6	40.4	28.0
35	1.7	41.1	32.2	45.3	34.3	88.9	85.8	76.1	76.5	72.5	56.0	45.6	32.4
40	4.4	46.4	36.8	51.0	39.1	97.9	94.8	85.1	84.7	80.6	63.0	51.1	37.1
45	7.2	52.1	41.9	57.2	44.3	107.6	104.3	93.9	93.6	89.1	70.6	57.1	42.1
50	10.0	58.2	47.3	63.8	49.9	117.8	114.5	103.2	103.0	98.3	78.6	63.4	47.6
55	12.8	64.7	53.1	70.8	55.9	128.8	125.3	113.2	113.0	108.0	87.3	70.1	53.4
60	15.6	71.6	59.4	78.3	62.3	140.3	136.7	123.7	123.6	118.4	96.6	77.3	59.6
65	18.3	79.0	66.0	86.3	69.1	152.6	148.9	134.9	134.9	129.4	106.5	84.9	66.2
70	21.1	86.9	73.2	94.8	76.4	165.6	161.8	146.8	146.9	141.0	117.1	92.9	73.2
75	23.9	95.2	80.8	103.7	84.2	179.3	175.4	159.4	159.6	153.4	128.4	101.5	80.7
80	26.7	104.0	88.9	113.2	92.5	193.8	189.9	172.7	173.0	166.4	140.4	110.5	88.7
85	29.4	113.4	97.5	123.2	101.3	209.0	204.4	186.7	187.2	180.2	153.2	120.0	97.2
90	32.2	123.3	106.7	133.7	110.6	225.1	221.0	201.5	202.1	194.8	166.8	130.0	106.2
95	35.0	133.7	116.4	144.8	120.6	242.0	237.8	217.1	217.9	210.2	181.2	140.6	115.7
100	37.8	144.7	126.8	156.4	131.1	259.8	255.6	233.5	234.5	226.3	196.5	151.7	125.8
110	43.3	168.5	149.2	181.5	153.9	298.0	293.7	268.8	270.3	261.1	229.7	175.7	147.6
120	48.9	194.8	174.3	209.0	179.4	339.9	335.6	307.1	309.8	299.5	266.7	202.1	171.9
130	54.4	223.7	202.2	238.9	207.6	385.8	331.5	350.3	353.1	341.5	307.7	231.1	198.9
140	60.0	255.3	233.1	271.5	238.8	435.8	431.5	396.9	400.4	387.4	353.1	262.7	228.6
150	65.6	289.8	267.1	306.6	273.2	490.1	485.8	447.5	452.0	437.3	403.1	297.1	261.3

Italic figures indicate Hg.   **Bold figures indicate psig.**

Figure 14-4. *A complete temperature-pressure card for zeotropic blends.*

14.6 Retrofit

Retrofit means to change or upgrade existing equipment to new guidelines. Retrofitting converts existing equipment to use new refrigerants and lubricants. Retrofit normally does not involve major component replacement but serves to extend the "life" of the equipment. Retrofitting existing equipment to new refrigerants and lubricants reduces environmental concerns associated with older refrigerants.

Many zeotropic blends are **short-term** (interim) alternatives intended to function until the equipment needs replacement. When an existing system is retrofitted with an alternative refrigerant, the new refrigerant must be carefully selected for that application. There is no "drop-in" replacement for CFC-12. Service technicians must acquaint themselves with the proper service procedures for working with the new blends. **Figure 14-5** lists the various applications and replacement refrigerants available for retrofit.

When working with zeotropic blends, two key points must be remembered:

❑ Mineral oil lubricants, traditionally used with CFC refrigerants, are much less **miscible** (able to mix) with new refrigerants. To assure proper lubrication and good oil return, the HCFC-based refrigerants use a synthetic alkylbenzene (AB) lubricant. This produces a combination that tolerates some contamination by residual mineral oil. HFC-based refrigerants, such as R-134a and R-507, use polyol ester (POE) oil. Systems using polyol ester lubricants require removal of mineral oil to a residue of not more than 5%.

❑ Zeotropic refrigerants can fractionate (separate). Systems must be charged with *liquid* (not vapor) from the cylinder, unless the entire cylinder content is emptied into the system. Refrigerant cylinders for the 400-series (zeotropic) refrigerants are equipped with a dip tube to permit liquid removal with the cylinder in the upright position.

When retrofitting a system with one of the new refrigerants, it is sometimes necessary to determine the amount of new refrigerant required. The following formula is used to calculate the amount of refrigerant to use:

Pounds of new refrigerant = pounds of original refrigerant × density of new refrigerant ÷ density of original refrigerant (at 80° F).

Example: How many pounds of R-507 are required to replace 1000 lbs. of R-502?

$$\text{Pounds of R-507} = 1000 \times (64.9 \div 75.5)$$
$$= 1000 \times 0.86$$
$$= 860 \text{ lbs.}$$

14.7 Polyol Ester Lubricants

Because the new HFC refrigerants were not compatible with mineral oil, new lubricants were required. The first commercially available synthetic

Refrigerant Applications and Replacements

Application	Temperature	Refrigerant	Replacements	
			Interim	Long-term
High temperature	55°F to 90°F	R-22 (Air conditioning)	R-411A,B	R-410A,B
				R-407C
Medium temperature	32°F to 50°F	R-12 (Coolers)	R-401A,B	R-134a
			R-405A	R-410A,B
			R-406A	R-407C
			R-409A	
Low temperature	0°F to –50°F	R-502 (Freezers)	R-22	R-404A
			R-408A	R-507
			R-402A,B	R-407A,B
			R-403A,B	
Ultralow temperature	–55°F to –250°F	R-13 and R-503 (Cascade)	R-403B	Unknown
Cryogenics	–250°F to 0°F$_A$	R-728 (Expendable)	No ozone depletion	No ozone depletion
Low-pressure chillers	45°F to 55°F	R-11 (Operates in vacuum)	R-123 (HCFC)	Unknown

F_A = Absolute Fahrenheit

Figure 14-5. *Interim (short-term) and long-term replacements for various refrigerant applications.*

shown on the gauge face cannot be used for the new refrigerant. Use the pressure readings and a temperature-pressure card for R-401A. A sight glass in the liquid line can be used with R-401A in most systems. The sight glass should be located downstream of a liquid receiver or near the expansion valve. Cleaning a system after compressor burnout is very similar to the procedure used for R-12 or R-502.

Where feasible, the preferred retrofit for CFC-12 is R-134a with polyol ester lubricant. In some cases, retrofitting with R-134a may be difficult because nearly all of the mineral oil in the system must be removed. In these instances, interim (temporary) blends such as R-401A may be preferred. Leak-detection is accomplished by using soap bubbles, sniffers (halide torch and electronic), dyes, or ultrasound.

Keep in mind that future regulations may restrict the use of HCFC-containing refrigerants and dictate the ultimate use of R-134a for servicing CFC-12 equipment.

14.9.2 R-401B

A blend of HCFC-22, HFC-152a, and HCFC-124, this zeotrope has a boiling point of –29°F (–34°C) at 0 psig (0 kPa). A variation of R-401A, it was designed primarily to replace CFC-12 as a retrofit refrigerant in transport refrigeration. See **Figure 14-12.** However, where high-compression ratios exist, the use of R-401B can result in significant capacity reductions and excessive discharge temperatures. The manufacturer of the equipment *must* be contacted for a recommendation on retrofitting from CFC-12. R-401B is suitable for domestic and commercial freezer applications where the evaporating temperature is below –10°F (–23°C). Leak detection is accomplished by using soap bubbles, sniffers (halide torch and electronic), dyes, or ultrasound. Either alkylbenzene or polyol ester can be used with R-401B, with AB the

recommended lubricant. Working with R-401B is much like working with R-401A. However, the temperature/pressures are different, and the use of a temperature-pressure card is necessary.

14.9.3 R-402A

R-402A is a blend of HFC-125, HCFC-22, and HC-290. Its boiling point is –57.6°F (–49.7°C) at 0 psig (0 kPa). This ternary blend was designed as a short-term alternative to R-502 in low- and medium-temperature commercial refrigeration applications. See **Figure 14-13.** It contains HCFC-22, a federally regulated refrigerant scheduled for elimination by 2020.

Where feasible, the preferred retrofit for R-502 is HFC-507. R-507 is a non-ozone-depleting refrigerant, increasingly used in many OEM (Original Equipment Manufacturer) applications. In some cases, retrofitting with R-507 can be difficult because nearly all mineral oil in the system must be removed. For these instances, an interim refrigerant such as R-402A may be desired.

14.9.4 R-403B

A blend of HCFC-22, CFC-13, and R-290 (propane), this refrigerant has a boiling point of –57.1°F (–49.5°C) at 0 psig (0 kPa). R-403B offers an immediate reduction in ozone depletion potential of nearly 90% compared to the ODPs of R-13 and R-503. R-403B was developed as a "drop-in" replacement for R-13 and R-503 and can be directly charged into systems without replacing existing hardware. Due to the slightly different properties of R-403B, some minor adjustments to the superheat setting of the expansion valve may be necessary. R-403B is a nonflammable, near-azeotropic, ternary blend. Once blended, R-403B cannot separate into its original components, and the mixture behaves as a single substance. All standard mineral oils are compatible with R-403B.

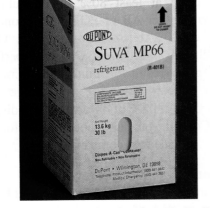

Figure 14-12. *Refrigerant MP66 (R-401B). (DuPont)*

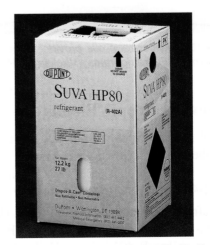

Figure 14-13. *Refrigerant HP80 (R-402A). (DuPont)*

The composition of a near-azeotropic blend may change if amounts of the component refrigerants are removed as vapor. If refrigerant is removed from the cylinder as a liquid, the composition is unchanged. When charging a system, it does not matter whether the refrigerant is put into the system in the vapor or liquid phase, provided the cylinders are emptied.

In the event of a large leak in the compressor room, normal considerations and precautions taken for any escape of refrigerant gas should be applied. Cold gas is more dense than air and initially settles to the bottom of the room before dispersing. A large portion of the charge can be lost without significant change in composition and performance. All liquid and vapor leaks are nonflammable.

14.9.5 R-404A

R-404A is a blend of HFC-125, HFC-143a, and HFC-134a. Its boiling point is –51.7°F (–46.5°C) at 0 psig (0 kPa). This refrigerant was designed to serve as a *long-term* alternative to R-502 and HCFC-22 in low- and medium-temperature commercial refrigeration applications. See **Figure 14-14.** R-404A contains three HFCs and is not scheduled for phaseout under current law. It is a suitable retrofit refrigerant in such applications as supermarket freezer cases, reach-in coolers, display cases, transport refrigeration, and ice machines.

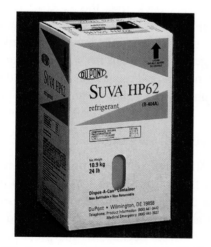

Figure 14-14. *Refrigerant HP62 (R-404A). (DuPont)*

R-404A is not a "drop-in" replacement for R-502 since it is not miscible with existing lubricants used in R-502 systems. The mineral oil used with R-502 must be replaced with polyol ester lubricant to a residual mineral oil level of less than 5%. R-404A requires a polyol ester lubricant to ensure complete miscibility between oil and refrigerant. Miscibility is important for oil return to the compressor, especially in large systems with long runs of piping. Compressor manufacturers provide specific recommendations on the type or brand of lubricant to install in the field. Differences among polyol ester-based lubricants make it difficult to assume they are interchangeable. Always check with the compressor manufacturer prior to retrofit.

14.9.6 R-407C

This zeotrope is a blend of HFC-32, HFC-125, and HFC-134a. Its boiling point is –46.6°F (–43.6°C) at 0 psig (0 kPa). R-407C is a non-ozone-depleting blend of HFC refrigerants. It was designed to closely match the properties of R-22 and can serve as a replacement for R-22 in both new and retrofitted applications. It is nontoxic and nonflammable. R-407C is not a "drop-in" replacement. Due to its high temperature glide and fractionation behavior, R-407C should not be used in centrifugal chillers or other equipment using a flooded evaporator.

Polyol ester lubricants must be used with R-407C; it is not compatible with mineral oil or alkylbenzene lubricants. When retrofitting, a lubricant flush procedure is necessary to lower the original oil content below 5% contamination.

14.9.7 R-409A

A blend of HCFC-22, HCFC-124, and HCFC-142B, the boiling point of R-409A is –29.6°F (–34.2°C) at 0 psig (0 kPa). It is an alternative refrigerant blend for retrofitting R-12 medium- and low-temperature refrigeration systems. R-409A provides a slightly higher capacity than R-12 and R-134a in lower-temperature applications. The approximate R-409A charge for most applications is 85% to 90% of the original (R-12) charge. R-409A is suitable for use with mineral, alkylbenzene, or polyol ester oils. Changing the existing lubricant is not required in most cases. If oil miscibility becomes a problem at lower temperatures (less than 0°F or –18°C), oil return can be improved by using at least 30% alkylbenzene lubricant mixed with mineral oil. R-409A is fully miscible with pure alkylbenzene or polyol ester lubricants.

Controlling evaporator temperature

Starting conditions: From baseline data using R-12, take the desired evaporator temperature using R-12, and add 5 degrees Fahrenheit (2.7 degrees Celsius). This gives the dew point (vapor) temperature for R-409A at the outlet of the evaporator. The additional degrees compensate for the glide across the evaporator when using R-409A. To illustrate:

❑ The evaporator operating temperature using R-12 is 10°F (–12°C). Adding half of R-409A glide, or 5 degrees Fahrenheit (2.7 degrees Celsius), to 10°F (–12°C) = 15°F (–9°C), the dew point (vapor) temperature at the evaporator outlet (if 0 degrees superheat).

❑ Using the temperature-pressure card for R-409A, the dew point pressure at 15°F (–9°C) is 16.8 psig (116 kPa). This pressure is equal to the suction pressure at the evaporator outlet (if 0 degrees superheat).

Figure 14-15 illustrates the process for achieving an average evaporator temperature of 10°F (–12°C) when retrofitting to R-409A.

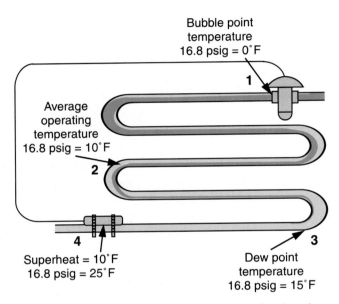

Bubble point
temperature
16.8 psig = 0°F

1

Average
operating
temperature
16.8 psig = 10°F

2

4

Superheat = 10°F
16.8 psig = 25°F

3

Dew point
temperature
16.8 psig = 15°F

Figure 14-15. *Measure temperatures at these four locations to achieve an average evaporator temperature of 10°F when retrofitting to R-409A.*

14.10 System Accessories

The new refrigerants are not identical to the refrigerants they are replacing. The insulating materials in motors may not be compatible with new refrigerants and lubricants. Check with your local compressor wholesaler regarding compatibility. In all cases, the new refrigerants have higher pressures than the original refrigerants. Pressure controls and pressure-operated valves may need to be reset for proper operation. Some controls and valves may need replacing.

14.10.1 Sight Glass with Moisture Indicator

A sight glass/moisture indicator can be used with the new refrigerants and lubricants, but the moisture indicator will be incorrect. The actual moisture level of polyol ester oil will be higher than the sight glass indicates. The color changes in moisture indicators are a function of relative humidity. There is a difference in moisture saturation levels for the new refrigerants, and higher moisture levels are needed to cause a color change. For example, R-134a is able to dissolve about

three times as much moisture as R-12 at the same temperature. Testing of oil samples is required to determine true moisture content in the system.

14.10.2 Filter-driers

Systems retrofitted with alkylbenzene or polyol ester oils tend to circulate solid contaminants to a greater degree than do systems using mineral oil. Systems that run with mineral oil tend to deposit solid contaminants throughout the system. POE and AB lubricants remove these deposits and carry them in suspension throughout the system. The filter plays an important role when used with AB or POE oils.

Activated alumina is used in filter-driers to remove acids from the system. In the presence of excessive moisture, polyol ester oils may hydrolyze and produce acids. Therefore, filter-driers recommended for polyol ester oils contain about 25% alumina. Molecular sieve filter-driers are best for moisture absorption. Given the hygroscopic nature of polyol ester lubricants, the filter-drier should have a high moisture-absorbing capacity. Oversized filter-driers are often suggested. The XH-6 (*bonded core*), XH-7, and XH-9 types of filter-driers are recommended. The XH-6 (*loose fill*) is not recommended.

Solid-core driers made with bauxite tend to absorb polyol ester oil and moisture. If the drier becomes overloaded with excess moisture, it can release the acidic materials back into the system. Driers made with bauxite are not recommended for use with POE lubricants. *Utmost care must be taken to prevent moisture from getting into the refrigeration system.* For this reason, do not leave the compressor or system open to the atmosphere longer than 15 minutes.

14.11 Retrofit Procedures

The following retrofit procedures are general guidelines for converting R-12 systems to the use of alternative refrigerants. These guidelines, along with any manufacturer recommendations, should provide a successful retrofit. Because of the great number and complexity of systems, it is impossible to cover every application. For precise retrofit information regarding a specific system, contact the equipment manufacturer. Some manufacturers require specific procedures and lubricants to maintain their warranties.

14.11.1 Retrofitting from R-12 to R-401A or R-401B

Follow these steps:

1. *Record baseline data.* Before making any changes, be certain the system is operating properly. Record

system performance as a baseline. ***Baseline data*** should include measurements throughout the system, including temperatures and pressures at the evaporator, temperatures and pressures at the condenser, compressor suction and discharge, expansion valve superheat, and compressor amps. These measurements will prove useful when adjusting the system after retrofit.

2. *Recover the R-12 charge.* The R-12 charge should be recovered from the system using a recovery machine capable of meeting or exceeding the required levels of evacuation. The charge must be collected in a recovery cylinder. *Do not vent the refrigerant.* Knowing the recommended R-12 charge (in pounds) for the system is helpful. If the weight is not known, weigh the entire amount of refrigerant removed. The weight can be used as a guide for the initial charge of R-401A or R-401B.

3. *Drain the oil.* Most hermetic compressors do not have oil drains, making it necessary to remove the compressor to drain the lubricant. Remove the compressor, and drain the oil from the suction connection into an approved container. Most of the mineral oil will drain from the system. For large systems, the mineral oil should be drained from any low spot in the system. Most of it can be drained from the compressor crankcase. Mineral oil should also be drained from oil separators and suction accumulators. Residual mineral oil often can be "blown" out with nitrogen pressure. Set the nitrogen regulator for not more than 300 psi (2070 kPa). Do not use nitrogen without a pressure regulator.

4. *Measure the oil removed.* Measure and record the amount of oil removed from the system. Compare this amount with the amount recommended by the manufacturer to ensure the majority of lubricant has been removed. The data will serve as a guide to determine the amount of alkylbenzene lubricant to add in Step 5.

5. *Recharge the compressor with alkylbenzene lubricant.* Add the same amount of alkylbenzene lubricant to the compressor as you removed in Step 4. Use a commercially available alkylbenzene oil of the same viscosity grade as the mineral oil, typically 150 SUS for optimum performance. Follow recommendations by equipment manufacturers.

6. *Reinstall the compressor.* If the compressor was removed to drain the oil, reinstall the compressor using standard service procedures.

7. *Evaluate the liquid-metering device.* Most R-12 systems with expansion valves operate satisfactorily with either R-401A or R-401B. Capillary tube systems need greater restriction to achieve satisfactory performance. A new capillary tube of the

same diameter should be about 50% longer than the original tube. In most cases, the system can be operated with the original capillary tube by undercharging the unit. Operation will be satisfactory if the ambient conditions are relatively constant. Operation over a wide range of condensing temperatures may result in liquid floodback, motor overload at high condensing temperatures, and loss of capillary tube liquid seal at low condensing temperatures.

8. *Replace the filter-drier.* Check with your refrigerant wholesaler to be certain the replacement filter-drier is compatible with R-401A or R-401B.

9. *Reconnect the system, and check for leaks.* Check the system for leaks, using normal leak-testing procedures.

10. *Evacuate the system.* Use normal service procedures to evacuate air and other noncondensables from the system. The system should be evacuated from both sides to a vacuum of 500 microns or less.

11. *Charge the system with the appropriate blend.* Charge the system with liquid refrigerant from the cylinder liquid valve. Initially charge the system with about 75% (by weight) of the original R-12 charge.

12. *Check system operation.* Start the unit, and let the system stabilize. If the system is undercharged, add small amounts of liquid refrigerant until the desired operating conditions are achieved. Suction pressures for R-401A should be within about 1 psi (7 kPa) of normal system operation with R-12 for most medium-temperature systems. Discharge pressures will be about 10 psi to 20 psi (70 kPa to 140 kPa) higher than R-12. Suction pressure for R-401B should be close to that of R-12 when evaporating temperatures are –20°F (–29°C). The suction pressure may be 8 psi or 9 psi (55 kPa to 62 kPa) higher at medium temperatures of 25°F (–4°C). The discharge pressure may be 70 psi (485 kPa) higher when exposed to extreme ambient conditions. It may be necessary to reset system pressure controls and adjust the expansion valve superheat setting.

13. *Label the system.* After retrofitting, label the system to identify the type of refrigerant, its weight, and the alkylbenzene lubricant (by brand name) in the system. Labeling helps ensure the proper refrigerant and lubricant will be used to service the equipment in the future.

14.11.2 Retrofitting from R-12 to R-134a

Follow these steps:

1. *Record baseline data.* This information establishes normal operating conditions for system performance prior to retrofit.

2. *Isolate or recover the R-12 charge.* Perform a system pumpdown, and close both receiver valves. Recover the R-12 in a recovery cylinder. *Do not vent the refrigerant.*

3. *Drain and measure the existing lubricant.* Remove as much mineral oil as possible. R-134a is not miscible with mineral oil. A contamination level of less than 5% of the total oil charge is required.

4. *Recharge the compressor with polyol ester lubricant.* Add the same amount of polyol ester lubricant as the mineral oil drained in Step 3. Polyol ester lubricants differ; check with the compressor manufacturer for the correct viscosity grade and brand of polyol ester lubricant.

5. *Recharge the system with R-12.* If the system was pumped down into the receiver, the receiver valves can be opened and the system purged at the compressor. If the original charge was recovered, the system should be leak-tested, evacuated, and recharged with the original R-12.

6. *Run the compressor.* Operate the compressor with the polyol ester lubricant and R-12 for at least 24 hours. Drain the used polyol ester oil, and recharge with fresh polyol ester oil. Test the drained lubricant for residual contamination of mineral oil. Test kits that check for residual mineral oil content are available from several lubricant suppliers. Normally, it requires about three oil changes to achieve an acceptable contamination level of less than 5%.

7. *Continue the flushing procedure.* Repeat Steps 5 and 6 until residual mineral oil content is below 5%.

8. *Recover the R-12 from the system.* Recover the R-12 into a recovery cylinder. *Do not vent the refrigerant.*

9. *Replace the filter-drier.* Make sure the new filter-drier is compatible with R-134a and polyol ester lubricant.

10. *Check the system for leaks.* Use normal leak-testing procedures to check the system for leaks.

11. *Evacuate the system.* Evacuate the system to 500 microns to remove noncondensables and residual R-12.

12. *Charge the system with R-134a.* Systems being charged with R-134a typically require about 90% (by weight) of the original R-12 charge. It is usually necessary to undercharge a capillary tube system.

13. *Check system operation.* Start the unit, and let the system stabilize. Adjust the R-134a charge until the desired operating conditions are achieved. Suction pressures should be within about 2 psi (14 kPa) of normal operation with R-12 for most medium-temperature systems. Discharge pressure will be about 3 psi to 10 psi (20 kPa to 70 kPa) higher than normal operation with R-12. Check and adjust pressure-control settings as required.

14. *Label the system.* Label the system to identify the type of refrigerant (R-134a), its weight, and the polyol ester lubricant (by brand name) in the system. Labeling helps prevent contamination by other products when the equipment is serviced.

Figure 14-16 can be used as a guideline for making retrofit decisions.

Summary

This chapter described many of the new zeotropic refrigerant blends designed as interim (short-term) alternative refrigerants. Examples illustrated appropriate use of each blend and the refrigerant each blend replaces. Fractionation and temperature glide were explained. New procedures for liquid charging and topping off were introduced. Using bubble point and dew point readings on the temperature-pressure card for blends was also covered.

The new synthetic lubricants were described, and guidelines were given for their proper application and use with the new refrigerant blends. Alkylbenzene (AB) lubricants can be used with the new HCFC refrigerants, but HFC refrigerants require the use of polyol ester (POE) oils. The retrofit procedures for converting from an older refrigerant to a new one were described.

Alternative Refrigerant Reference Guide

Refrigerant Name	Components (weight %)	ODP	GWP	Replaces	Temp. Glide (°F/°C)	Lubricant	Application
R-22	R-22	0.05	0.35	R-12, R-502	0/0	MO, POE, AB	Low- or med.-temp. refrigeration; A/C
R-134a	R-134a	0.0	0.28	R-12	0/0	POE	Med.-temp. refrigeration; automotive A/C
R-401A (MP39)	R-22/152a/124 (53/13/34)	0.03	0.22	R-12	8.9/4.9	AB or POE	Low-or med.-temp. refrigeration
R-401B (MP66)	R-22/152a/124 (61/11/28)	0.035	0.24	R-12	8.7/4.8	AB or POE	Low-or med.-temp. refrigeration
R-406A (GHG12)	R-22/600a/142b (55/4/41)	0.055	0.38	R-12	16/8.9	MO or AB	Stationary refrigeration
R-409A (FX56)	R-22/124/142b (60/25/15)	0.05	0.30	R-12	15.2/8.4	MO, POE, AB	Low- or med.-temp. refrigeration
R-125	R-125	0.0	0.84	R-502	0/0	POE	Low- or med.-temp. refrigeration
R-402A (HP80)	R-125/290/22 (60/2/38)	0.02	0.63	R-502	2.8/1.5	AB or POE	Low- or med.-temp. refrigeration
R-402B (HP81)	R-125/290/22 (38/2/60)	0.03	0.52	R-502	2.9/1.6	AB, POE, MO	Ice machines
R-403B (ISCEON 69-L)	R-290/22/218 (5/56/39)	0.025	3.00	R-502, R-13b1, R-503, R-13	2.1/1.2	MO, POE, AB	Low-temp. refrigeration
R-404A (FX70, HP62)	R-125/143a/134a (44/52/4)	0.0	0.94	R-502	0.9/0.5	POE	Low- or med.-temp. refrigeration
R-407A (KLEA60)	R-32/125/134a (20/40/40)	0.0	0.45	R-502	9/5	POE	Low- or med.-temp. refrigeration
R-407B (KLEA61)	R-32/125/134a (10/70/20)	0.0	0.45	R-502	9/5	POE	Low- or med.-temp. refrigeration
R-408A (FX10)	R-125/143a/22 (7/46/47)	0.023	0.75	R-502	1.3/0.7	MO, POE, AB	Low- or med.-temp. refrigeration
R-507 (AZ50)	R-125/143a (50/50)	0.0	0.98	R-502	0/0	POE	Low- or med.-temp. refrigeration
R-410A (AZ20)	R-32/125 (50/50)	0.0	0.44	R-22	0/0	POE	Med.- or high-temp. refrigeration; A/C
R-407C (SUVA9000)	R-32/125/134a (23/25/52)	0.0	0.22	R-22	11/6.1	POE	Med.- or high-temp. refrigeration; A/C
R-23	R-23	0.0		R-503	0/0	POE	Very low-temp. refrigeration
R-508	R-23/ R-116 (39/61)	0.0		R-503	0/0	MO, AB	Very low-temp. refrigeration
R-123	R-123	0.015	0.02	R-11	0/0	MO or AB	Centrifugal chillers
R-124	R-124	0.020	0.10	R-114	0/0	AB	High-temp. refrig; high ambient
R-236fa	R-235fa	0.0		R-114	0/0	POE	Centrifugal chillers

Figure 14-16. *This table can be used as a guideline for making retrofit decisions. (National Refrigerants)*

Test Your Knowledge

Please do not write in this text. Write your answers on a separate sheet of paper.

1. What is the refrigerant of choice for new medium-temperature equipment?
   a. R-134a.
   b. R-401A.
   c. R-404B.
   d. R-507.
2. What is the refrigerant of choice for new low-temperature equipment?
   a. R-134a.
   b. R-410A.
   c. R-401B.
   d. R-507.
3. What is the alternative refrigerant for R-11?
   a. R-115.
   b. R-717.
   c. R-123.
   d. All of the above.
4. Short-term alternative refrigerants are mainly what type?
   a. CFC.
   b. HCFC.
   c. HFC.
   d. All of the above.
5. R-134a is an example of a(n) ____ refrigerant.
   a. pure
   b. azeotrope
   c. zeotrope
   d. All of the above
6. Which is an example of an azeotropic refrigerant?
   a. R-502.
   b. R-123.
   c. R-401A.
   d. R-13.
7. Which one of the following refrigerants is not a blend?
   a. R-401A.
   b. R-409A.
   c. MP39.
   d. R-22.
8. *True or false?* Fractionation describes the separation of refrigerant components.
9. Temperature glide occurs only with:
   a. CFCs.
   b. Zeotropic blends.
   c. Azeotropic mixtures.
   d. Oils.
10. The saturated liquid pressure is usually referred to as the ____.
    a. dew point
    b. break point
    c. bubble point
    d. condensation point
11. The same gauge manifold can be used for R-12 and ____.
    a. R-502
    b. R-401A
    c. R-134a
    d. All of the above
12. Polyol ester lubricants are compatible with ____.
    a. CFCs
    b. HCFCs
    c. HFCs
    d. all of the above
13. *True or false?* Flushing procedures are used to reduce fractionation.
14. The new HFC refrigerants are not compatible with:
    a. Mineral oil.
    b. AB oil.
    c. POE oil.
    d. PAG oil.
15. When retrofitting to an HFC refrigerant, mineral oil contamination must be reduced to less than ____%.
    a. 10
    b. 5
    c. 1
    d. 0.05
16. Mineral oil is not compatible with ____.
    a. R-134a
    b. HFCs
    c. R-408A
    d. all of the above
17. POE oils are ____ times more hygroscopic than mineral oil.
    a. 10
    b. 20
    c. 50
    d. 100
18. Activated alumina is used in filter-driers to remove ____.
    a. acid
    b. moisture
    c. solids
    d. all of the above
19. *True or false?* Bonded core filter-driers are recommended for use with POE lubricants.
20. After retrofitting, label the system to identify:
    a. Type of refrigerant.
    b. Weight of refrigerant.
    c. Type of lubricant.
    d. All of the above.

Troubleshooting Refrigerant Problems

Objectives

After studying this chapter, you will be able to:
- ❏ Identify and correct contamination problems.
- ❏ Detect refrigerant leaks using approved methods.
- ❏ Use refrigerant cylinders properly.
- ❏ Predict and correct low-side pressures.
- ❏ Recognize and eliminate causes of high head pressure.
- ❏ Use electronic scales and graduated cylinders for charging.

Important Terms

acids	inert
air off	liquid charging
air on	nonflammable
air space	permeation
balance point	pressure control
black oxides	pressure regulator
charging	recovery cylinders
contaminants	red iron oxide
dehydrate	returnable cylinders
disposable cylinders	setpoint
electronic leak detector	sludge
electronic scales	soluble
graduated cylinder	standing pressure test
halide torch	td
head pressure	thermostat
heat load	vapor charging
hydrostatic expansion	working pressure

15.1 Moisture, Air, and Contaminants

Troubleshooting refrigerant problems involves keeping the refrigerant pure and clean, locating and repairing system leaks, and adjusting system pressures.

Moisture, air, and contaminants are major causes of problems in compression refrigeration systems. These systems are designed to operate with a single pure refrigerant. Excess *moisture* in a refrigeration system can cause ice to form, restricting metering devices and preventing proper refrigerant flow. Moisture also causes rusting, corrosion, refrigerant decomposition, oil sludge, and system deterioration. *Air* in the refrigeration system causes high head pressure and high operating temperatures. Sludge, acid, corrosion, and other **contaminants** result in various system problems. Acid, for example, eats the insulation off motor windings, causing a burnout.

15.1.1 How Much Moisture Is Safe?

All refrigerants safely tolerate small amounts of moisture. Nobody knows for sure how much moisture is "safe," but there is general agreement that the less water present the better. Water is more **soluble** (easily dissolved) in some refrigerants than in others. Any excess water will exist as a separate liquid and cause considerable damage. If the temperature is low enough, the water will freeze.

The solubility of water in refrigerants is measured in parts per million (ppm) and varies according to temperature. R-12 is *very* sensitive to moisture — it only holds 6 ppm by weight. R-502 holds twice as much moisture (12 ppm); R-22 absorbs three times more than R-12 (19.5 ppm). The solubility of water in R-134a is comparable to that of R-22.

15.1.2 Moisture and Air Removal

Moisture and air are the primary enemies of a refrigeration system. Proper evacuation and the use of driers are necessary for trouble-free operation.

Thorough evacuation of the system should be performed whenever it has been contaminated or exposed for prolonged periods to moisture-laden atmospheric air. Blowing the system out with an inert gas (nitrogen or carbon dioxide) removes most of the air from the system but does not remove moisture and air from trapped areas. Filter-driers remove small amounts of moisture and contaminants.

Evacuating a system

Before evacuation, the system's internal pressure must be reduced to atmospheric (0 psig). The vacuum pumps used to evacuate systems are not designed to handle pressures above atmospheric.

Proper evacuation of a system is performed with a two-stage, rotary-type vacuum pump. The vacuum pump draws all vapor (gas) out of the system, reducing its internal pressure to a very low vacuum. The vacuum causes any moisture to boil, allowing it to be removed as a gas.

The amount of time required to properly evacuate a system is strictly a matter of judgment; a large system requires more time on the vacuum pump than a small system. Time must be allowed for the pump to pull a vacuum in all parts of the system and for moisture to work its way out of the oil in the compressor crankcase.

A large quantity of moisture in a system cannot boil into a vapor immediately upon lowering the pressure (just as an open pan of water does not immediately flash into a vapor upon reaching the boiling temperature of 212°F or 100°C). The boiling process requires time for the molecules to rearrange themselves in a new pattern. The boiling process must continue until all the molecules are rearranged. Stopping the vacuum pump too soon will leave moisture in the system.

Evacuation of the refrigeration system should be performed using a vacuum pump of the proper size. Vacuum pumps are rated in cfm (cubic feet per minute). The following are suggested minimum vacuum pump ratings for systems of different sizes:
- ❑ Up to 7 tons — 1.2 cfm
- ❑ Up to 21 tons — 3.0 cfm
- ❑ Up to 35 tons — 5.0 cfm
- ❑ Up to 70 tons — 10.0 cfm
- ❑ Up to 105 tons — 15.0 cfm

The time required for a vacuum pump to remove moisture and air from a system depends upon several factors:
- ❑ Size of the system.
- ❑ Amount of water present in the system.
- ❑ Capacity of the vacuum pump being used.
- ❑ Size and length of the connecting lines.

Pulling a good vacuum on the system is not sufficient; the vacuum pump must have time to do its work. The gauge reading reveals pressure at the pump, but it seldom shows actual vacuum at the farthest point of the system. The exact amount of time required to evacuate and **dehydrate** (dry out) a given system cannot be predicted because many factors are involved.

The speed of moisture removal can be increased by applying a heat lamp to the compressor crankcase during the vacuum-drawing process. Severe moisture problems may require the use of heat lamps on other system components as well. See **Figure 15-1**.

15.1.3 Contaminants

In addition to moisture and air, contaminants such as soldering fluxes, metal chips, wax, acid, oil sludge, dirt, black oxides, and rust can harm refrigeration systems. With hermetic and semi-hermetic compressors, the motor windings are exposed to contaminants in the system. Care is required to avoid contamination. Contaminants cause serious damage to the motor windings.

Sludge is caused by a chemical breakdown of oil that combines with other materials, such as carbon, metals, oxides, or salts. These compounds form a dark, gummy mass that blocks valves, oil ports, screens, and filters. Sludge problems can be traced to high compressor discharge temperatures and the presence of contaminants, including air.

Air in the system causes high discharge pressure (and high temperature). *Atmospheric air* is a source of oxygen and moisture. Oxygen and moisture contribute to breakdown of the oil-refrigerant mixture. The breakdown forms **acids**, which are corrosive chemical compounds. Acids are more corrosive in the presence of moisture than in a dry system, reacting quickly with metal parts and contributing (among other effects) to the formation of sludge. Therefore, it is important to keep systems as clean and dry as possible.

When contaminants are in a system, ordinary operating temperatures will produce corrosion. Oxygen from the moisture and air combine to form **red iron oxide** (rust). This leads, in turn, to the formation of iron salts and more water if acid is present.

Soldering fluxes also cause metal salts to form. Methyl alcohol (used as an antifreeze) can react with aluminum and cause corrosion. Moreover, poor brazing practice can introduce flux (acid) into the system.

Contamination during brazing

Brazing procedures can cause **black oxides** to form on the inside and outside of copper tubing. Oxides forming on the outside of tubing cause no harm, but those forming on the *inside* are easily washed from the tubing surface when the system is in operation. The oxides are then free to circulate with the refrigerant and oil. These oil-borne oxides are exposed to high temperatures at the compressor discharge and cause decomposition of the oil and refrigerant (sludge).

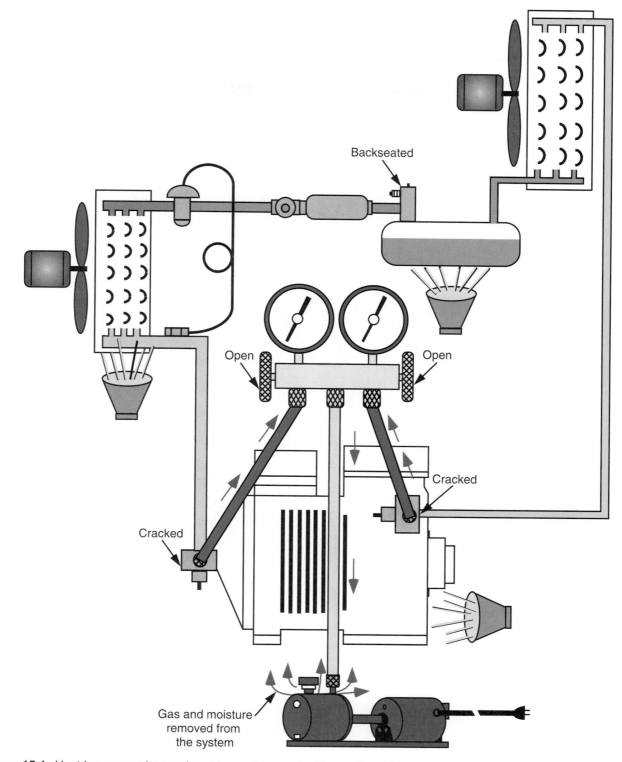

Figure 15-1. *Heat lamps may be used to drive moisture out of the system when evacuating. Moisture in the compressor oil is the last to be removed. Never use a torch or live steam for this purpose.*

Sweeping with inert gas. You can prevent oxide formation on the inside of tubing by sweeping an inert gas through the tubing during the brazing process, **Figure 15-2.** Nitrogen and carbon dioxide are examples of gases that are *inert* (chemically inactive) and *nonflammable* (do not support combustion). The inert gas pushes oxygen and other gases out of the tubing

and thus prevents oxidation (without oxygen, oxides cannot form). The nitrogen should flow very slowly, moving just fast enough to displace the air inside the tubing. A flow of 1 cfm to 3 cfm (a pressure reading of 2 psig to 3 psig) is sufficient.

The inert gas is vented (permitted to escape) after passing the brazing area. Venting eliminates the

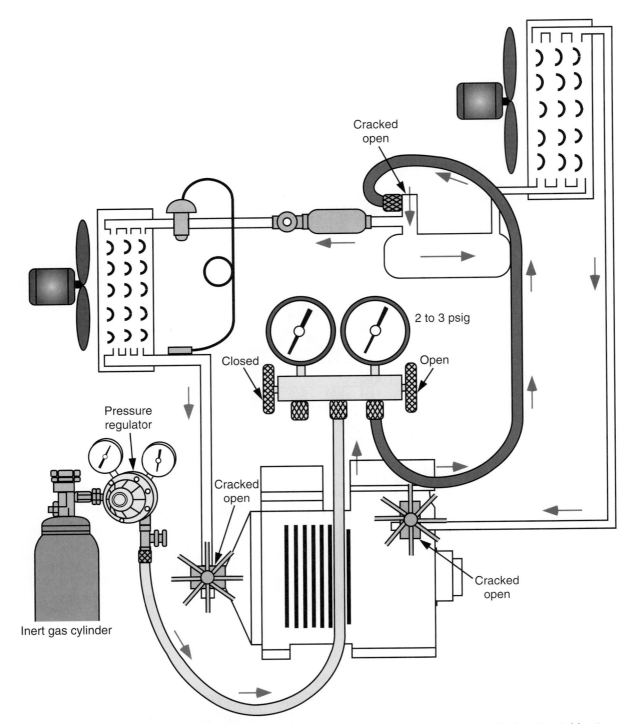

Cracked
open

2 to 3 psig

Closed

Open

Pressure
regulator

Cracked
open

Cracked
open

Inert gas cylinder

Figure 15-2. *An inert gas, supplied at low pressure, pushes ("sweeps") atmospheric oxygen out of system tubing to prevent black oxide formation during brazing. The suction and discharge service valves are cracked open to vent the gas and prevent pressure buildup inside the system.*

pressure buildup that would otherwise result from the heat of the brazing torch acting on the confined gas. An escape or vent *must* be provided. If pressurized gas is allowed to build up in the tubing, it will blow melted brazing alloy out of the connection.

Gas safety. Nitrogen and carbon dioxide are compressed gases supplied in cylinders under high pressure (2350 psig or 16 215 kPa for nitrogen and about 800 psig or 5520 kPa for carbon dioxide). Since

these pressures are much higher than the 2 psig to 3 psig (14 kPa to 21 kPa) needed, you must *always* use pressure-regulating valves on inert gas cylinders. See **Figure 15-3.** A refrigeration hose (or gauge manifold) is used to transfer the inert gas from the outlet of the flow pressure regulator to the system tubing.

WARNING: Never pressurize a system with oxygen. Oxygen will react violently with oil in the compressor crankcase and explode.

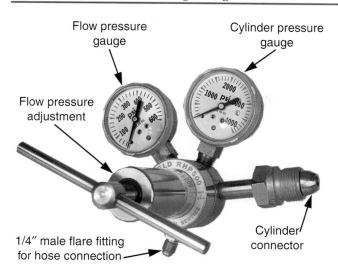

Figure 15-3. *A medium-duty nitrogen regulator capable of regulating output gas flow pressures from 0 psig to 600 psig. It can be fitted on cylinders filled with gases up to 4000 psig. (Uniweld)*

15.2 Leak Detection

Refrigerants are expensive, and a system cannot operate properly without an adequate supply. Therefore, leakage in refrigeration equipment is a major problem for both manufacturers and service technicians. Leak detection, however, should be secondary to leak prevention. Vibration is probably the principal cause of leaks, but improperly made flared connections or poorly brazed joints are high on the list of contributing factors. Eliminating poor brazing techniques, the use of improper fluxes and brazing materials, and poor workmanship will result in fewer leaks.

Sometimes fluxes used for brazing can temporarily prevent a leak by plugging a very tiny hole. When the system is pressurized, however, the flux or residue blows away and a leak develops. Using proper brazing techniques eliminates this problem.

Repairing a leak is not difficult, but *locating* the leak can prove time-consuming and frustrating. Each piece of equipment presents specific problems and limitations. To say the leak cannot be found or to simply add refrigerant is not a solution. Good technicians are highly skilled in preventing leaks and in locating and repairing them. The following approved leak-detection procedures are discussed in this section:

❑ Vacuum method.
❑ Bubble method.
❑ Halide torch.
❑ Electronic "sniffers."
❑ Ultrasound.
❑ Oil/dye.

15.2.1 Vacuum Method

In this method, the system is evacuated with a vacuum pump, and both hand valves on the gauge manifold are closed. When the vacuum pump is removed, the system is left in deep vacuum for several hours. If the system maintains the vacuum, you can assume it is leakproof. If the system loses vacuum, a leak is present and must be located.

The vacuum method is reliable only for detecting *major* leaks that result in quick loss of vacuum. A partial loss of vacuum might indicate a small leak, or it could result from the escape of gases that had been trapped in the oil and were not removed during the vacuum-drawing process. These gases will separate from the oil when left in a vacuum and show on the gauges as a partial loss of vacuum.

Another problem with the vacuum method of leak detection is the relatively small pressure difference between the inside of the tubing and atmospheric pressure. The pressure difference is only 15 pounds (see Chapter 9), which means small leaks cannot be readily detected.

Evacuating a system with a vacuum pump requires considerable time and energy; therefore, evacuation should be performed *after* all leak testing is completed, as necessary preparation for charging the system with refrigerant.

15.2.2 Bubble Method

The use of bubbles to pinpoint leaks is the oldest, least expensive, and most effective leak-detection method available. Testing with the bubble method requires the system be pressurized with refrigerant, nitrogen, carbon dioxide, or a mixture of these gases. A recovery system is often used to remove the refrigerant (to atmospheric pressure) before pressurizing the system with an inert gas. The system should be pressurized to equal the highest pressure it normally produces, usually 150 psig to 300 psig (1034 kPa to 2069 kPa). Leak testing at pressures less than 100 psig is wasting time and effort. Small leaks at low pressure may not appear until the pressure is increased to operating levels. Leaks are always more pronounced and easier to detect when the pressure is high. You should look for leaks on the high-pressure side of the system while the unit is running (if possible) and on the low-pressure side when the system is off.

The bubble method is simple, but certain precautions must be observed for satisfactory results. The soap solution must be of the proper consistency, and the system must be properly pressurized. Apply the soap solution generously to the area being tested; then, wait for escaping gas to cause bubbles to appear. Do not forget to allow time for small leaks to appear. Locating large leaks is not difficult, but locating small leaks requires skill. The first thing most technicians do is apply bubble solution to the most likely spots — flare and solder joints, fittings, and service valves.

Remember, vibration and poor craftsmanship are major causes of system leaks. It is not uncommon to find stress cracks in tubing bends or holes worn in tubing from rubbing together. Most leaks are located in the condensing unit because of the high level of vibration.

Do not forget to check the valve cores with bubbles when the hose connections are removed. It is common for the stem to get bent and stick open slightly, the core to loosen, or the gasket to be bad. Always replace the cap on a core valve. A valve core replacement tool makes it possible to replace a core under pressure without losing refrigerant.

Pressurizing the system

Nitrogen or carbon dioxide are often used to pressurize a system for leak detection. Inert gases like carbon dioxide or nitrogen are not harmful to the system, but the system must *never* be turned on while an inert gas is in it. A **pressure regulator** should always be used with such high-pressure gases to reduce them to a safe **working pressure** (test pressure). See **Figure 15-4.** For domestic systems using R-12, a test pressure of 150 psig (1035 kPa) is correct. A test pressure of 300 psig (2070 kPa) is used for R-22 systems and all commercial systems.

WARNING: Never interconnect a refrigerant cylinder and an inert gas cylinder through a gauge manifold. Accidental opening of the wrong manifold valve would permit high pressure to enter the refrigerant cylinder. High pressure could rupture the cylinder or contaminate the refrigerant.

Detergent solution

A special detergent solution for pressurized leak detection can be purchased from the local refrigeration supply house. This prepared leak indicator is available in plastic bottles with a dauber attached to the cap or a trigger spray. See **Figure 15-5.** Detergent solution is very effective for locating leaks.

An inexpensive detergent solution can be made from concentrated dishwashing liquid. As purchased, these detergents are too thick for proper use in leak detection. They must be thinned by mixing with water in a 1:1 ratio (*equal amounts* of water and detergent). Too much water will make the solution too *thin,* and it will not stick properly or blow bubbles. If the solution is too *thick,* more time will be needed for bubbles to appear (especially on small leaks). The solution must

be of proper consistency to adhere (stick) to the surface but not so thick that spraying or bubble-blowing is difficult.

A handy applicator for detergent solution is a plastic trigger-spray bottle with an adjustable nozzle, like the one shown in Figure 15-5. The adjustable spray makes it easy to apply detergent solution in a fine stream to points up to four feet away. A spray nozzle can eliminate a great deal of bending, climbing, crawling, and reaching to apply solution in hard-to-reach areas.

When leak testing, apply detergent solution *generously* to suspected leak sites to provide plenty of bubble-forming material. (The solution can be easily wiped up with a rag after testing is completed.) A large leak will form large bubbles quickly; small leaks will form small bubbles slowly. See **Figure 15-6.**

Be sure to test the gauge manifold and hoses, as well as all system connections. Ordinary refrigeration hoses may allow gas to seep through their walls (a process called **permeation**). Heavy-duty hoses do not allow refrigerant loss through permeation.

15.2.3 Halide Torch

The **halide torch, Figure 15-7,** has been used for many years as a fast and reliable method for detecting halogenated refrigerant leaks. The torch detects *only* leaking halogenated refrigerants, not other gases such as nitrogen or carbon dioxide.

The halide torch operates on the principle that a flame will change color if it decomposes a halogen in the presence of copper. The essentially colorless flame of the halide torch will change to a color ranging from faint green to bluish green to bright blue, depending upon the size of the leak (and the amount of halogen gas being decomposed in the flame). A halide torch is capable of detecting extremely small leaks, such as a leak of only one ounce of refrigerant per year.

Hydrocarbon fuels, such as propane, butane, acetylene, or alcohol, are burned in the torch to provide an almost colorless flame. As shown in **Figure 15-8,** a copper element is heated by the flame. This alone does not change the flame color — color changes only when the air drawn into the flame contains a halogen. The air is drawn through a "sniffer hose" that is slowly passed over the area of the suspected leak. Leaking gas is drawn into the hose, along with atmospheric air, and carried to the flame. Color changes in the flame are visible through the viewing window of the torch.

Using the halide torch

Since a halogen refrigerant is heavier than an equivalent volume of air, the refrigerant sinks downward from the leak site. Since air contaminated with leaking refrigerant can cause false readings in the

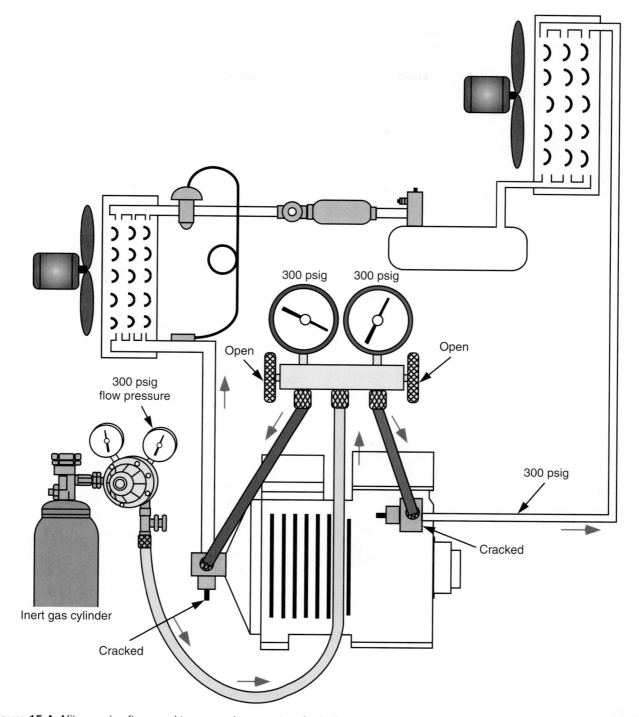

Figure 15-4. *Nitrogen is often used to pressurize a system for leak testing. This illustration is a typical example of leak testing a system with nitrogen and a detergent solution. The pressure causes bubbles to form, which pinpoint the location of the leak.*

halide torch, always begin checking the system at the top and work downward. See **Figure 15-9.** Systematically check each connection or other potential leakage point by slowly passing the sniffer hose over it while closely observing the color of the flame. Sometimes (especially in areas toward the floor), refrigerant contamination of the air makes it impossible to pinpoint the leak site with the halide torch. Apply detergent solution (the "bubble method")

to suspected areas of leakage. Bubble formation will show where the leak is located.

15.2.4 Electronic "Sniffer"

Small, hand-held electronic leak detectors ("sniffers") are extremely sensitive and easy to use. Most are powered by batteries, but some models must be plugged into an electric wall outlet for power. A typical *electronic leak detector,* **Figure 15-10,** consists

Figure 15-5. *Detergent solution for use in the bubble method of leak testing is available commercially, or it can be made from diluted dishwashing liquid. A trigger-spray bottle with an adjustable nozzle permits application in hard-to-reach areas. (Big Blu)*

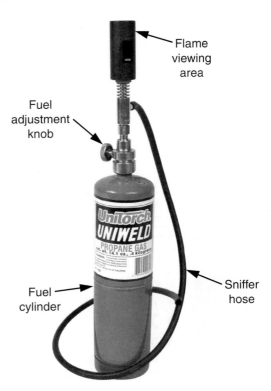

Figure 15-7. *A typical halide torch. The sniffer hose directs air from the suspected refrigerant leak site into the torch flame. Changes in flame color make it possible to detect even very tiny leaks. (Uniweld)*

Figure 15-6. *Bubble testing for leaks. A—When a leak is large, bubbles will quickly appear. B—A froth of small bubbles will slowly form at the site of a small leak. (Big Blu)*

of a probe and a case containing the electronics. The unit may be equipped with visual indicators, in addition to the audible signal used to indicate the presence of halide refrigerant gas.

When in operation, the instrument emits a repeating, short "beep" signal. As the probe is passed over the site of the suspected leak, the presence of even a minute trace of refrigerant causes the sound to change from a repeating tone to a long, steady tone.

Like the halide torch, the electronic detector can give false readings because of refrigerant contamination of the air. Unlike the torch, however, the electronic detector can be recalibrated to compensate for a certain amount of atmospheric contamination. It can be made less sensitive so it responds only to leaks over a certain volume of flow. This means, however, it will not identify smaller leaks. At times, it will be necessary to use the same approach as described for the halide torch: identify the general area of the leak; then, apply detergent solution to pinpoint the leak location.

For accurate readings with a sniffer, it is important to keep the sensing probe clean. Check the probe tip regularly for accumulations of lint, dirt, oil, or moisture. False readings can be caused by a fouled detector tip or from expander gases released from insulating materials. False readings are also caused by soap solutions. An electronic sniffer is the best tool for determining where

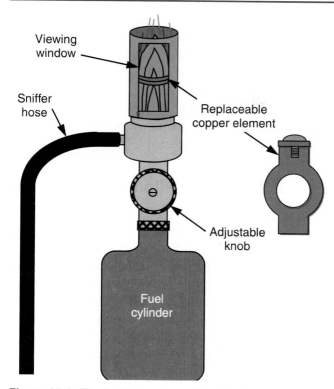

Figure 15-8. *The halide torch consists of a flame source, an air intake, a copper element, and an area for viewing flame changes.*

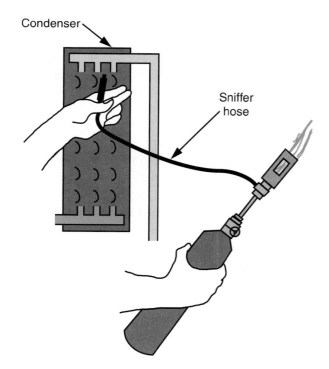

Figure 15-9. *A halide torch is used to check a refrigeration system for leaks. Since refrigerant is heavier than air, begin checking at the top of the system and work downward.*

leaks do *not* exist, such as evaporators, tunnels, and ceilings. Because it is often impossible to pinpoint a leak with an electronic sniffer, they are best used to locate the general area of a leak. The technician should then resort to a more precise method to pinpoint the leak.

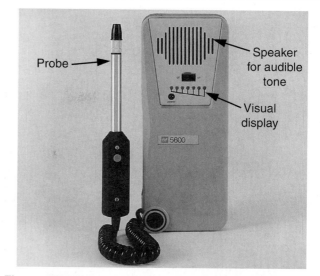

Figure 15-10. *A highly sensitive electronic leak detector. This battery-operated model uses an audible tone to signal leaks and also has a light display to indicate the relative size of the leak. (TIF)*

15.2.5 Ultrasound

Almost every leak produces a noise of some type, usually a high-pitched "hiss." An ultrasonic device detects the "hiss" and responds with audible and visual indications. The ultrasonic leak detector can pinpoint a leak regardless of contamination and works on pressurized systems or a vacuum. The ultrasonic detector is highly directional, an asset in confined spaces.

Some leaks, such as those occurring in valves and other machined devices, do not produce an audible "hissing" noise. You can increase the sound level by spraying the area with plain water. Do not use bubble solution because it will muffle the sound. Be aware that false readings can be obtained from exterior noises (hissing sounds). These noises will adversely affect the ultrasonic detector and render it useless.

15.2.6 Oil As a Leak Indicator

Since some oil is always circulating through the system with the refrigerant, a certain amount escapes at the site of a refrigerant leak. A visual inspection of the system will reveal signs of fresh oil that indicate a leak. Oil spills should always be cleaned up as a safety measure. Cleaning up old oil spills also makes it easier to spot new ones that indicate leaks.

15.2.7 Dyes

Special dyes are available for injecting into the compressor crankcase (oil). The dye mixes with the refrigerant oil and circulates throughout the system with the oil. When a leak occurs, oil and dye escape along with the refrigerant and produce a stain that glows fluorescent green in the presence of ultraviolet light. See **Figure 15-11.**

Figure 15-11. *An aid in troubleshooting refrigeration systems is a leak-detection method that uses a fluorescent tracer dye and an ultraviolet light source. The dye is added to the refrigerant oil, and the system is allowed to operate for a time. Leakage shows up as a bright yellow stain under ultraviolet light. (H.B. Fuller Company)*

Dyes are particularly effective in locating low-level leaks that cannot be found by other detectors. Furthermore, the equipment owner can see evidence of a leak, and the dye remains in the system for future detection.

It is important to allow adequate time between injecting the dye and inspecting for leaks. Time is needed for the dye to circulate and for very small leaks to produce a visible stain. The dye method cannot be used on tubing runs between walls and other enclosed spaces since the tubing cannot be observed.

15.2.8 Standing Pressure Test

Once leaks have been located and repaired, and if time permits, the system should be checked with a *standing pressure test* to determine whether all leaks were found.

If the leak-testing procedure does not involve a refrigerant, an inert gas should be used to raise the system pressure to the proper leak-testing level of 150 psig or 300 psig (1035 kPa or 2070 kPa). Note and record the system pressure and ambient temperature; then, backseat the service valves, and remove the gauge manifold (to prevent leakage). The system is now pressurized and sealed. After several hours (overnight, if possible), reconnect the gauge manifold, open the service valves, and record new pressure readings and ambient temperature. If the pressures remain constant according to ambient, the system has no leaks. Pressure loss indicates an undetected leak remains.

If the system contains a refrigerant rather than an inert gas, it should be fully charged to bring it to full operating pressure. With the system turned off, note and record the pressure and ambient temperature. Seal the system by backseating the service valves and removing the gauge manifold. Recheck the system pressures and ambient after several hours (or overnight) to determine whether all leaks were detected.

15.3 Refrigerant Cylinders ◆

Cylinders used to store and transport refrigerants are made from either steel or aluminum and are available in a number of standard sizes. All but the smallest of cylinders must have a pressure relief device to protect against excessive pressure buildup. The U.S. Department of Transportation (DOT) requires a relief valve or a fusible plug in any refrigerant cylinder more than 12″ (30 cm) tall and 4 1/2″ (11 cm) in diameter. On large steel cylinders, a fusible plug is normally located in the concave cylinder base. Depending upon whether a cylinder is used for a corrosive or a noncorrosive refrigerant, it must be inspected every five or 10 years.

15.3.1 Steel Cylinders

Large cylinders containing 100 lbs. to 150 lbs. (45 kg to 68 kg) of liquid refrigerant are convenient when servicing systems that require large quantities of refrigerant. They also are less costly than smaller cylinders on a per-pound basis. Suppliers usually require a sizable deposit to ensure return of the empty cylinder. A steel protective cap screws down over the valve at the top of the cylinders when they are being moved or are not in use. See **Figure 15-12.**

Steel cylinders are equipped with a dual valve that allows the technician to withdraw refrigerant as a vapor or, by means of a dip tube, as a liquid. There is no need to invert the cylinder (turn it upside down) to withdraw liquid refrigerant.

A refrigerant cylinder should *never* be completely filled with liquid — space must be left for *hydrostatic expansion* (swelling of the liquid) with increases in temperature. A cylinder is considered "full" when liquid occupies 85% of its interior space. The remaining 15% allows for expansion within a normal range of temperatures without danger of rupturing the cylinder. In case of temperatures that might cause expansion too great for the available space (and present the danger of a cylinder bursting), the fuse plug in the base of the cylinder or a special safety valve will open and vent refrigerant to relieve the pressure. See **Figure 15-13.**

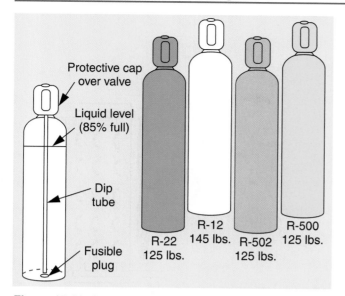

Figure 15-12. *Large steel refrigerant cylinders are returned to the supplier for refilling after use. A dual valve permits the technician to select refrigerant as either liquid (through a dip tube) or vapor while keeping the cylinder upright. To protect against excessive pressure, a fusible plug is installed in the cylinder bottom.*

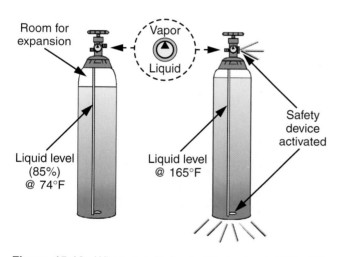

Figure 15-13. *When a cylinder is filled, about 15% of its volume must be left for hydrostatic expansion of the liquid refrigerant. If the temperature of the cylinder increases to 165°F (74°C), fusible plugs will melt or a safety valve will open to relieve pressure and prevent bursting.*

Saturated conditions exist inside the cylinder, so the contents behave according to the temperature-pressure chart for the specific refrigerant involved. The space above the liquid is occupied by a very dense refrigerant gas. A *balance point* is reached in which the pressure exerted on the liquid refrigerant by the gas molecules prevents any further *boiling off,* or changing of liquid to the gaseous state.

When the vapor valve at the top of the cylinder is opened, however, the gas molecules can escape. Pressure on the liquid is reduced, allowing it to boil

and produce a steady flow of refrigerant gas. The flow of gas continues until the valve is closed or all the liquid has changed to a gas.

The heat required for the change of state ("boiling off" of gas) is obtained from the remaining liquid and the metal cylinder. As gas is released, both the temperature and the pressure of the remaining liquid drops. A continuous release of gas actually causes frost to form on the outer cylinder wall. Cylinder pressure may drop so low that it creates a problem when charging refrigerant into a system in vapor form. The cylinder pressure *must* be higher than the system pressure for vapor to flow from the cylinder into the system. See **Figure 15-14.**

To increase cylinder pressure in such a situation, the contents may be carefully warmed. The best and safest way to increase pressure in the cylinder is to place the cylinder in a container of warm water (80°F to 110°F or 27°C to 43°C). *Never* heat a cylinder with a torch or use any other heating method that would produce a cylinder temperature greater than 125°F (52°C).

15.3.2 Disposable Cylinders

At one time, industry practice required technicians to transfer refrigerant from large steel cylinders to smaller, portable steel cylinders that held 25 lbs. to 30 lbs. (11 kg to 14 kg). Not only was this practice time-consuming and dangerous, it often resulted in contaminated refrigerant. The introduction of light-weight *disposable cylinders* has virtually eliminated the need to transfer refrigerants. These throwaway cylinders, **Figure 15-15,** are available in sizes ranging from 25 lbs. to 50 lbs. (11 kg to 23 kg).

Carrying handles on top of the cylinder also serve as protection against damage to the cylinder valve. The valve outlet is a 1/4″ male flare fitting for easy connection to the gauge manifold. To withdraw refrigerant vapor, the cylinder is used in the upright position; to withdraw liquid, the cylinder is inverted. See **Figure 15-16.** Some cylinders are equipped with a dip tube and dual-stem valve and can remain upright for liquid charging.

WARNING: Federal law prohibits refilling disposable cylinders with refrigerant. Violators are subject to fines and jail terms. Although it is not illegal, using empty disposable cylinders to transport compressed air is unwise; moisture in compressed air could cause the thin metal of these cylinders to rust and eventually rupture.

15.3.3 Cylinders for Refrigerant Recovery

Since CFCs may no longer be vented into the atmosphere when installing and servicing refrigeration equipment, the use of recovery (returnable) cylinders with refrigerant recovery equipment has become widespread.

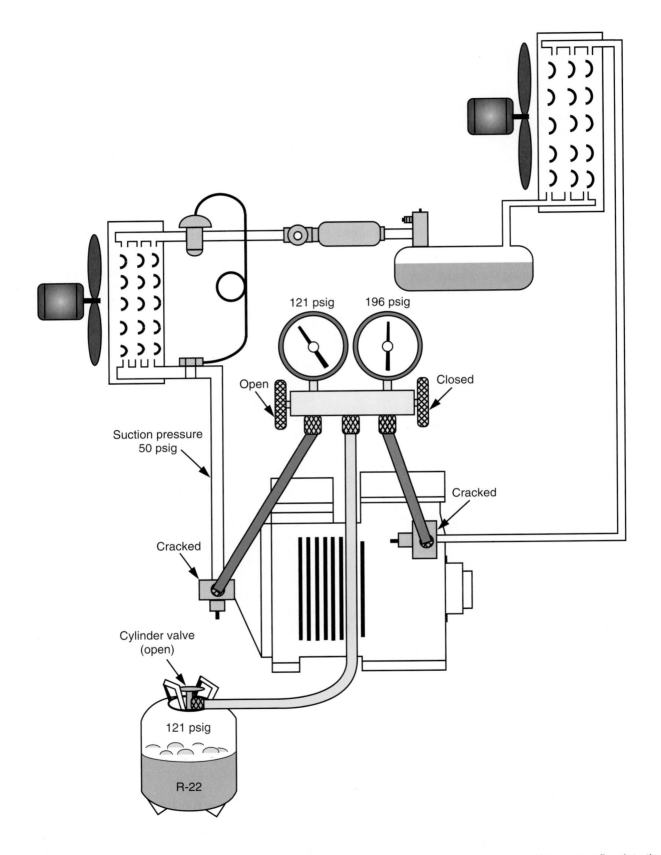

Figure 15-14. *If cylinder pressure is lower than or equal to system pressure, charging will stop. For refrigerant to flow into the system, cylinder pressure must be higher. Pressure can be increased by carefully heating the cylinder.*

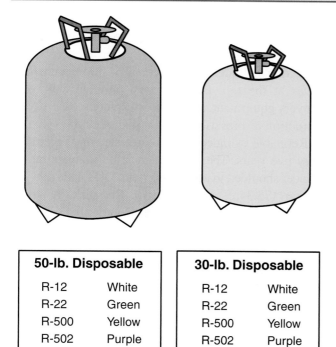

50-lb. Disposable		30-lb. Disposable	
R-12	White	R-12	White
R-22	Green	R-22	Green
R-500	Yellow	R-500	Yellow
R-502	Purple	R-502	Purple
		R-11	Orange

Figure 15-15. *Disposable cylinders are lightweight and intended to be discarded after use. An empty cylinder is considered hazardous waste until a hole is punched in it.*

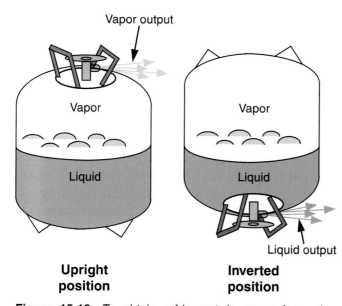

Upright position   **Inverted position**

Figure 15-16. *To obtain refrigerant in vapor form, the cylinder is in the upright position. Inverting it will provide refrigerant in liquid form.*

Recovery cylinders are refillable, heavy-duty, certified pressure vessels meeting DOT specifications. **Figure 15-17** shows three recovery cylinders and one refillable cylinder. Federal law specifies that recovery cylinders must have gray bottoms with yellow shoulders and tops. Also, recovery cylinders must be inspected at regular intervals and repaired or replaced when damaged through handling.

Figure 15-17. *Recovery cylinders have gray bottoms with yellow shoulders and tops. Refillable-type cylinders, like the one in the foreground, may have solid colors. (Amtrol)*

Proper procedures must be followed when using cylinders. The greatest danger to the technician and persons nearby is *overfilling* the cylinder with liquid refrigerant. Even though the cylinders have pressure-relief devices, they can still rupture or violently discharge refrigerant if overfilled. The amount of liquid refrigerant a cylinder can safely hold is measured in pounds, which means quantities vary by the type of refrigerant (some weigh more than others for a given volume). Sufficient space must be allowed in the cylinder for hydrostatic expansion, **Figure 15-18.** Generally, 15% of the cylinder's volume should be left empty for expansion. The EPA is even more cautious, stipulating 20% of a recovery cylinder should be left empty for hydrostatic expansion.

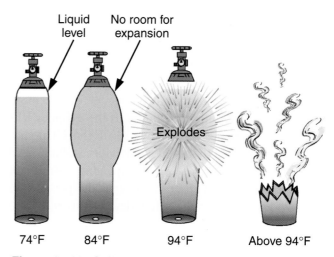

Figure 15-18. *Sufficient room must be left in a cylinder for hydrostatic expansion of the liquid refrigerant. If little or no space is left, an increase in temperature will cause pressure to rise to the point where the cylinder explodes.*

Cylinders must be clearly labeled to indicate the type of refrigerant inside; furthermore, they should always be refilled with the same type of refrigerant. Colored spray paint is often used for marking a cylinder with the number of the refrigerant it contains. Refilling an R-22 cylinder with R-134a could result in charging a system with the wrong refrigerant. Extreme care should be taken to prevent contamination of recovery cylinders.

Before reuse, recovered refrigerant should be recycled to separate oil and remove moisture, acids, and foreign particles. Preferably, the refrigerant should go through the *reclaiming* process, which brings it back to new product specifications.

15.3.4 Cylinder Safety

When working with refrigerants, the technician must practice safety on the job everyday. Safety-consciousness extends to the handling of cylinders that hold and transport refrigerants. Anytime a used refrigerant is transported, clearly label its container with a DOT classification tag to meet the requirements of the Code of Federal Regulation, Title 49. When moving a cylinder, be sure it is firmly strapped onto an appropriate wheeled device. *Never* roll a cylinder on its base or lay it down to roll it.

There are three major concerns when transferring refrigerants:
- ❑ Avoid contamination.
- ❑ Avoid venting to the atmosphere.
- ❑ Handle refrigerants safely.

Refrigerant cylinders must be the correct type and color and be properly marked.

The cylinder used for holding transferred refrigerants must first be evacuated to a vacuum of at least 28 in. Hg. Different refrigerants should not be mixed in a cylinder because no economical method exists to separate refrigerants.

Careless handling of cylinders can result in sudden releases of refrigerant that can cause frostbite, skin damage, or blindness. If skin contact occurs, soak the area in lukewarm water and seek medical attention at once.

A ***returnable cylinder*** must *never* be filled to more than 80% of the container volume. If the cylinder will be exposed to temperatures above 130°F (54°C), do not fill to more than 60% of the container volume. All liquids expand when heated. An expanding refrigerant liquid in an enclosed space determines the pressure in the cylinder. Overpressure safety devices provide some level of safety, but they do not eliminate all risk.

Weigh cylinders on an accurate scale, and inspect them carefully before filling. Do not use dented, rusted, gouged, or damaged cylinders. Examine the valve assembly for leakage, damage, or tampering. Handle cylinders carefully. Do not drop them or bump

them together. If the impact from a fall were to knock off a cylinder valve, the sudden release of refrigerant could turn the cylinder into a projectile. Store refrigerant cylinders in a vertical position with their valves at the top. Become very familiar with your recovery equipment, and apply all safety measures and instruction as prescribed by the manufacturer.

Refillable cylinders must be retested and recertified every five years. The test date must be stamped on the cylinder shoulder, in accordance with DOT Title 49 CFR Section 173.34(e) and 173.31(d). Retesting by visual inspection alone is not permitted. Do not fill a container that is out of date (five years or older). Outdated cylinders should be returned empty for retesting.

Disposable steel cylinders will oxidize and become so weakened by rust that the wall and seams can no longer tolerate pressure or contain gases. Discard disposable containers since they may never be used for recovery or refilling. Refrigerant returned for reprocessing in improper cylinders will not be accepted. The penalty for transporting a refilled disposable cylinder is a fine up to $25,000 and five years imprisonment. Disposable refrigerant cylinders should be stored in a dry location to prevent corrosion. *Never* allow used cylinders with residual refrigerant to sit at a job site; saturated vapor pressure will exist if even the smallest amount of liquid is present. An abandoned cylinder could explode under direct heat from any source. Before discarding a disposable container, recover any refrigerant remaining in the container. Then, for maximum safety, break off the cylinder valve and puncture the container before disposal as scrap metal.

Never use an empty disposable cylinder as a compressed air tank. The inside of the cylinder is untreated steel. When filled with compressed atmospheric air, oxygen and moisture contained in the air will form rust on the inside of the tank. It is not a question of "*will* the cylinder burst?" but rather "*where* and *when*?"

15.4 Temperature-Pressure Relationships ◆

Since pure and azeotropic refrigerants are stable, temperature-pressure relationships are also stable and predictable. For ease of reference and use, these relationships have been compiled into pressure-temperature or *saturation* charts. (Zeotropic blends do not exhibit stable temperature-pressure relationships. Refer to Chapter 14.)

Saturation charts provide a means to quickly convert temperature readings to pressure readings or vice versa. A technician can use a manifold gauge to obtain the pressure of the refrigerant, then, consult a pocket-size chart like the one in **Figure 15-19** to determine the corresponding temperature. Similarly, when

Temperature-Pressure Card

°F	11	12	22	123	134a	500	502
-50	28.9	15.4	6.2	29.2	18.6	12.8	0.2
-45	28.7	13.3	2.7	29.0	16.7	10.4	1.9
-40	28.4	11.0	0.5	28.8	14.7	7.6	4.1
-35	28.1	8.4	2.6	28.6	12.3	4.6	6.5
-30	27.8	5.5	4.9	28.3	9.7	1.2	9.2
-25	27.4	2.3	7.4	28.1	6.8	1.2	12.1
-20	27.0	0.6	10.1	27.7	3.6	3.2	15.3
-18	26.8	1.3	11.3	27.6	2.2	4.1	16.7
-16	26.5	2.1	12.5	27.4	0.7	5.0	18.1
-14	26.4	2.8	13.8	27.3	0.4	5.9	19.5
-12	26.2	3.7	15.1	27.1	1.2	6.8	21.0
-10	26.0	4.5	16.5	26.9	2.0	7.8	22.6
-8	25.8	5.4	17.9	26.7	2.8	8.8	24.2
-6	25.5	6.3	19.3	26.5	3.7	9.9	25.8
-4	25.3	7.2	20.8	26.3	4.6	11.0	27.5
-2	25.0	8.2	22.4	26.1	5.5	12.1	29.3
0	24.7	9.2	24.0	25.8	6.5	13.3	31.1
2	24.4	10.2	25.6	25.6	7.5	14.5	32.9
4	24.1	11.2	27.3	25.3	8.6	15.7	34.9
6	23.8	12.3	29.1	25.1	9.7	17.0	36.9
8	23.4	13.5	30.9	24.8	10.8	18.4	38.9

°F	11	12	22	123	134a	500	502
10	23.1	14.6	32.8	24.5	12.0	19.7	41.0
12	22.7	15.8	34.7	24.2	13.2	21.2	43.2
14	22.3	17.1	36.7	23.9	14.4	22.6	45.4
16	21.9	18.4	38.7	23.5	15.7	24.1	47.7
18	21.5	19.7	40.9	23.2	17.1	25.7	50.0
20	21.1	21.0	43.0	22.8	18.4	27.3	52.5
22	20.6	22.4	45.3	22.4	19.9	28.9	54.9
24	20.1	23.9	47.6	22.0	21.4	30.6	57.5
26	19.7	25.4	49.9	21.6	22.9	32.4	60.1
28	19.1	26.9	52.4	21.2	24.5	34.2	62.8
30	18.6	28.5	54.9	20.7	26.1	36.0	65.6
32	18.1	30.1	57.5	20.2	27.8	37.9	68.4
34	17.5	31.7	60.1	19.7	29.5	39.9	71.3
36	16.9	34.4	62.8	19.2	31.3	41.9	74.3
38	16.3	35.2	65.6	18.7	33.1	43.9	77.4
40	15.6	37.0	68.5	18.1	35.0	46.1	80.5
42	15.0	38.8	71.5	17.5	37.0	48.2	83.8
44	14.3	40.7	74.5	16.9	39.0	50.5	87.0
46	13.6	42.7	77.6	16.3	41.1	52.8	90.4
48	12.8	44.7	80.8	15.6	43.2	55.1	93.9
50	12.0	46.7	84.0	15.0	45.4	57.6	97.4

°F	11	12	22	123	134a	500	502
55	10.0	52.0	92.6	13.1	51.2	63.9	106.6
60	7.8	57.7	101.6	11.2	57.4	70.6	116.4
65	5.4	63.8	111.2	9.0	64.0	77.8	126.7
70	2.8	70.2	121.4	6.6	71.1	85.4	137.6
75	0.1	77.0	132.2	4.1	78.6	93.5	149.1
80	1.5	84.2	143.6	1.3	86.7	102.0	161.2
85	3.2	91.8	155.7	0.9	95.2	110.0	174.0
90	4.9	99.8	168.4	2.5	104.3	120.6	187.4
95	6.8	108.3	181.8	4.2	113.9	130.6	201.4
100	8.8	117.2	195.9	6.1	124.1	141.2	216.2
105	11.1	126.6	210.8	8.1	134.9	152.4	231.7
110	13.4	136.4	226.4	10.2	146.3	164.1	247.9
115	15.9	146.8	242.7	12.6	158.4	176.5	264.9
120	18.5	157.7	259.9	15.0	171.1	189.4	282.7
125	21.3	169.1	277.9	17.7	184.5	203.0	301.4
130	24.3	181.0	296.8	20.5	198.7	217.2	320.8
135	27.4	193.5	316.6	23.5	213.5	232.1	341.2
140	30.8	206.6	337.3	26.7	229.2	247.7	362.6
145	34.4	220.3	358.9	30.2	245.6	264.0	385.0
150	38.2	234.6	381.5	33.8	262.8	281.1	408.4
155	41.6	295.6	405.1			298.9	432.9

Italic figures = Inches of mercury **Bold figures = Pressure (psig)**

Figure 15-19. *A pocket-size temperature-pressure (saturation) chart is a useful reference tool when adjusting system pressures to achieve desired temperatures. Saturation cards show pressures for several popular refrigerants and are available free of charge from supply houses or refrigerant wholesalers.*

temperature is known, the corresponding pressure can be obtained from the chart. When the system requires a particular temperature, the technician can adjust system pressure after consulting the chart to identify the pressure corresponding to the desired temperature.

When working with systems that use one of the most popular refrigerants (R-12, R-22, R-502, or R-134a), the technician does not even have to consult the temperature-pressure chart. Instead, he or she can merely check the face of the gauge manifold pressure gauges. As shown in **Figure 15-20,** the faces of some gauges have several scales: the outer scale shows pressure in psig, while three separate inner scales show the corresponding temperatures (°F) for R-12, R-22, and R-502. Direct-reading gauges are also available for R-134a. The temperature-pressure relationship can be read directly from the gauge face. For other refrigerants, of course, the chart must be used to determine temperature for a given pressure.

15.4.1 Controlling Low-side Pressures

The evaporator of an operating refrigeration unit absorbs heat from the space being cooled. The temperature of a product (such as ice cream in a home freezer) is controlled by the temperature of the air circulating around the product. The product gives up

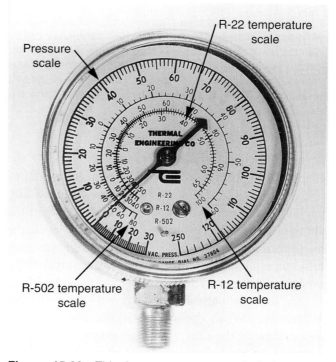

Figure 15-20. *This low-pressure gauge includes corresponding temperature scales for three widely used refrigerants. It allows relationships to be read directly from the gauge face, eliminating the need to consult a chart. A similar direct-reading gauge is available for R-134a.* (Thermal Engineering)

its heat to the air, which in turn gives up its heat to the evaporator. The evaporator temperature is about 10 degrees Fahrenheit (6 degrees Celsius) colder than the temperature of the entering air (called "air on").

The evaporator gives up its heat to the boiling refrigerant inside the tubing. The refrigerant is about 10 degrees Fahrenheit (6 degrees Celsius) colder than the evaporator. These temperature differences (air-to-evaporator and evaporator-to-refrigerant) must be maintained to accomplish heat removal. See **Figure 15-21.**

The amount of heat that must be removed per hour

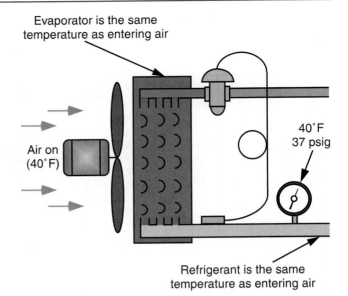

Figure 15-22. *When a refrigeration system is turned off, temperature differences quickly disappear, and heat removal stops. Note the air temperature and the refrigerant temperature at the evaporator outlet. No heat can flow when a temperature difference does not exist.*

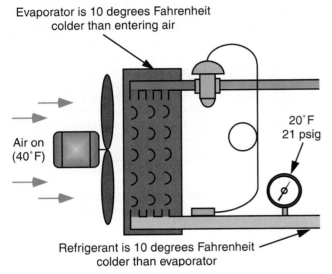

Figure 15-21. *Temperature differences are critical to accomplishing heat removal. The evaporator of an operating refrigeration system will be 10 degrees Fahrenheit colder than the air of the space being cooled, while the refrigerant inside the evaporator will be 10 degrees Fahrenheit colder than the evaporator itself. As shown, if the room air ("air on") temperature is 40°F, the temperature of the refrigerant at the evaporator outlet will be 20°F.*

from an enclosed space determines the size of the evaporator that must be used. The correct amount of refrigerant must be boiled to maintain the necessary temperature inside the evaporator as well. When the air temperature is lowered to the desired level, the refrigeration system is turned off. The temperature differences quickly disappear, and heat removal stops, as shown in **Figure 15-22.**

The low-side pressure always reveals the *refrigerant temperature.* When the system is not operating, the temperature of the refrigerant, the evaporator, and the room air are the same, since no temperature difference (*td*) exists. For example, in a system charged with R-12 that has a low-side pressure reading of 37 psig (255 kPa), the refrigerant temperature at the evaporator outlet is 40°F (4°C). Since the system is not in operation, the temperature of the evaporator is also 40°F, as is the air temperature. Different refrigerants have different

pressure readings at 40°F (68.5 psig for R-22 and 80.2 psig for R-502), but the principle remains the same.

A refrigeration system is designed to quickly *establish* and *maintain* a temperature difference once it begins operating. For example, in a system using R-12 with a low-side pressure of 15.8 psig (109 kPa), the refrigerant temperature is 12°F (–11°C). Thus, the temperature of the evaporator is 22°F (–6°C), and the air temperature is 32°F (0°C). As the entering air becomes cooler, the evaporator and refrigerant temperatures drop accordingly. This relative temperature difference remains constant while the system is running. Pressure readings also decrease as a result of the lowered temperatures.

The temperature difference necessary for cooling is obtained only when the system is running. Once the system is turned off, the evaporator and refrigerant temperatures quickly rise and equal the airflow temperature. To maintain air temperature within a set range, a **thermostat** (temperature-activated switch) may be located within the area being cooled, known as the **air space.**

During the "off" cycle, the air temperature typically is allowed to rise five or six degrees. When the air temperature reaches the highest **setpoint** (top of the range for which the thermostat is set), the system is switched on. When the air temperature drops to the lowest setpoint, the thermostat switches the system off. The air begins to warm again, and the cycle repeats. An example is a walk-in dairy cooler with temperatures maintained in the range of 32°F to 38°F (0°C to 3°C). When air temperature rises to 38°F, the cooler's refrigeration system turns on. It runs until the temperature of the air falls to 32°F, then shuts off.

Since the temperature differences allow low-side *pressures* to be predicted and controlled, many refrigeration systems are turned on and off by a **pressure control,** rather than a thermostat. The pressure control functions as a switch, turning the system on and off as a result of changing low-side refrigerant pressure to maintain temperatures within the desired range.

Evaporator airflow

To maintain proper temperature differences, the amount of airflow across the evaporator must be correct. Medium-temperature evaporators require a flow of about 1350 cfm per ton; low-temperature evaporators require about 2000 cfm per ton. Airflow that is lower in volume than required results in a very cold evaporator with low refrigerant pressure since the **heat load** (air to be cooled) cannot reach the evaporator in sufficient volume. Excessive airflow places too much heat load on the evaporator, resulting in high temperatures and pressures.

Air conditioner temperature differences

The temperature differences designed for air conditioners are different from those designed for refrigeration systems. Because large volumes of very cold air are uncomfortable to people, air conditioning evaporators are designed for an airflow of approximately 400 cfm per ton. The evaporator is designed to produce a temperature difference of 15 to 20 degrees Fahrenheit (9 to 11 degrees Celsius) between the air entering the evaporator (called **air on** or *return air*) and the air leaving the evaporator (called **air off** or *supply air*).

Although temperature differences for most refrigeration systems are calculated from *air on,* temperature differences for air conditioning systems are calculated from *air off.* In other words, the temperature of the evaporator is 10 degrees Fahrenheit (6 degrees Celsius) colder than the air *leaving* it, rather than the air entering it. As shown in **Figure 15-23,** the temperature difference between air off (52°F or 11°C) and the evaporator (42°F or 6°C) is 10 degrees Fahrenheit or 5 degrees Celsius, while the difference between air on (72°F or 22°C) and the evaporator temperature (42°F or 6°C) is 30 degrees Fahrenheit or 16 degrees Celsius.

A supply air (air off) temperature of 52°F (11°C) is considered the low limit for human comfort. As noted, this results in an evaporator temperature of 42°F (6°C) and, in turn, a refrigerant temperature of 32°F or 0°C (at a corresponding pressure of 57.5 psig or 397 kPa

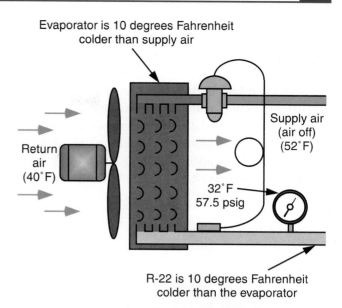

Figure 15-23. *In air conditioning applications, a temperature difference of 10 degrees Fahrenheit is maintained between the cooled air ("air off") and the evaporator. For human comfort, air conditioning systems are usually adjusted to provide a temperature no lower than 52°F.*

for R-22). In air conditioning work, it is common practice to maintain low-side pressure above 57.5 psig.

Allowing the *evaporator* temperature to fall to 32°F (0°C) in an air conditioning situation can create cooling problems. At 32°F, moisture in the return air *freezes on the evaporator surface.* Ice buildup will restrict (or even totally block) airflow through the evaporator.

15.4.2 Controlling High-side Pressures

The condensing unit of a refrigeration system has a single task — to convert gaseous refrigerant back into a liquid form for reuse. The cooling medium is usually outside ambient air. Since heat moves from hotter areas to cooler ones, the gas must be *compressed* to a saturation point higher than the temperature of the ambient air. As the outside ambient air temperature goes up or down, the high-side pressure/temperature follows suit to maintain a temperature difference.

The system is designed so the high-side pressure rises to a saturation point where the refrigerant gas can be converted back to a liquid state. The rate at which the change of state takes place *must* equal the rate at which gas enters the condenser. This is called the **balance point.** The condenser is sized to match system capacity. If the balance point changes, the high-side pressure/temperature will change accordingly.

Air-cooled condensers

The balance point (high-side pressure) of the condenser can be predicted. The compressor must compress the gas to a saturation temperature 30 to 35 degrees Fahrenheit (17 to 20 degrees Celsius) above the temperature of the ambient air. With this information and the ambient temperature reading, a temperature-pressure chart can be used to determine proper **head pressure** (pressure on the high side of the system).

For example, if the ambient temperature is 70°F (21°C), the temperature of the refrigerant (R-12, in this case) in the condenser should be between 100°F and 105°F (38°C and 41°C). See **Figure 15-24.**

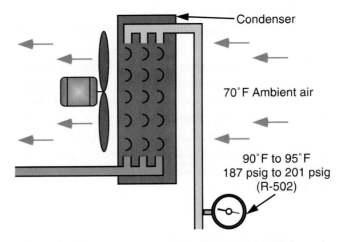

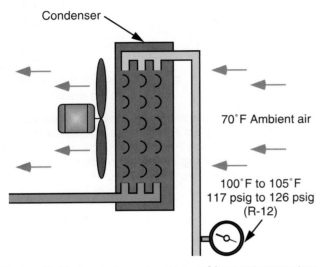

Figure 15-24. *In a balanced system, refrigerant temperature in the condenser will be 30 to 35 degrees Fahrenheit higher than the temperature of the ambient air. This allows use of a temperature-pressure chart to predict head pressure.*

If the ambient temperature increases to 95°F (35°C), refrigerant temperature in the condenser should fall between 125°F and 130°F (52°C and 54°C). Normal head pressure at an ambient temperature of 95°F would be between 168 psig and 181 psig (1159 kPa and 1250 kPa).

This method of predicting head pressure is applicable to most common refrigerants. However, some of today's high-efficiency units use a 20-degree to 25-degree td design. With these systems, the service technician adds 20 to 25 degrees Fahrenheit (11 to 14 degrees Celsius) to the ambient and uses the value to determine the head pressure. See **Figure 15-25.**

Water-cooled condensers

Some refrigeration systems use water-cooled condensers, which are more efficient than air-cooled condensers. Water-cooled condensers remove heat by conduction, providing better heat transfer. As shown in **Figure 15-26,** head pressure should correspond to a temperature of 15 to 20 degrees Fahrenheit (8 to 11 degrees Celsius) higher than the temperature of the water leaving the condenser.

Figure 15-25. *Adding 20 to 25 degrees Fahrenheit to the ambient air temperature allows the technician to determine head pressure in a high-efficiency system.*

Excessive head pressure

As head pressure increases, the compressor works harder to achieve the balance point. Excessive head pressure should be avoided since it reduces system efficiency and causes compressor problems.

Higher-than-normal head pressures can be traced to one of five possible causes:

❑ *Air in the system.* Air is a mixture of noncondensable gases that behave according to Dalton's Law of Gases (pressures are added together to find total pressure). To achieve normal head pressure, the air must be removed.

❑ *Overcharge of refrigerant.* Any excess refrigerant collects in the condenser, reducing condenser capacity and raising the balance point.

❑ *Dirty condenser.* If allowed to accumulate on either an air-cooled or water-cooled condenser, dirt will act as an insulator, reducing condenser capacity and raising the balance point.

❑ *Reduced air or water movement.* If there is no movement of a heat-transfer medium (air or water) past the condenser tubes, head pressure will rapidly rise to an excessive level. Head pressure safety controls prevent damage. Lack of movement in an air-cooled condenser can be caused by a burned-out fan; in a water-cooled compressor, a closed water valve is a possible cause.

❑ *High low-side pressure.* If the low-side pressure is above normal, the head pressure will also rise. High suction pressure and heat content require the compressor to achieve a higher balance point. System pressures correspond — if the high-side pressure goes up, so does the low-side pressure, and vice versa. The only exception is when a compressor is defective — an increase in low-side pressure would not be accompanied by a corresponding increase in high-side pressure.

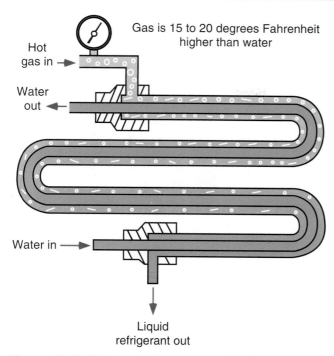

Gas is 15 to 20 degrees Fahrenheit higher than water

Hot gas in →

Water out ←

Water in →

Liquid refrigerant out

Figure 15-26. *Water-cooled systems transfer heat efficiently by means of conduction, so the refrigerant in the condenser is only 15 to 20 degrees Fahrenheit higher than the water leaving the condenser.*

15.5 Charging Refrigerant into a System

The term *charging*, in the context of a refrigeration or air conditioning system, means to add refrigerant. Systems use different kinds and amounts of refrigerant, and the amounts are critical at times. In some domestic systems, for example, a 1/2-ounce overcharge or undercharge can noticeably affect operation. In other systems, there is some room for error. For example, systems that use a liquid receiver can operate with an excess of refrigerant. As a rule, knowing when to *stop* adding refrigerant to a system is as important as knowing *how* to add refrigerant.

Refrigerant can be charged into a system as a *vapor* or as a *liquid*. Before it can be charged, a system must be tested for leaks, dehydrated, and evacuated. Then, a gauge manifold can be connected to the suction service valve and the discharge service valve. Both connections are important since they provide necessary pressure readings.

15.5.1 Vapor Charging

The most common mode of charging a system is to add refrigerant vapor to the low-pressure side (*vapor charging*) using the gauge port on the suction service valve. The center hose from the manifold is connected to a refrigerant cylinder, which must be in the *upright* position to permit vapor charging. The cylinder valve is then opened, transferring control to the gauge manifold.

Cylinder pressure must be *higher* than suction pressure since refrigerant will not flow without a pressure difference. Sometimes, cylinder pressure must be carefully increased by warming the cylinder with hot water. **CAUTION:** Never use a torch to heat a refrigerant cylinder, and never raise cylinder temperature above 125°F (52°C). An explosion could result.

With both service valves in the cracked position, open the left (low-pressure side) manifold hand valve. Refrigerant vapor will travel from the cylinder, through the yellow (center) hose, through the left side of the manifold to the blue hose, and into the compressor through the suction service valve. See **Figure 15-27.**

If the system is turned off, refrigerant can be charged into *both* the high side and the low side by opening both hand valves on the manifold.

WARNING: When the system is running, *never* open the right (high-pressure side) manifold hand valve. Since the system's high-side pressure is greater than the cylinder pressure (refer to Figure 15-27), the system will pump into the cylinder. This could contaminate the cylinder and cause the cylinder to rupture. The high-pressure gauge is connected to the discharge service valve to allow monitoring of high-side pressures while charging.

Charging will take longer if it is done using Schrader valves rather than service valves. The passage for gas or liquid through a Schrader valve is much smaller than the passage through a service valve. Since the smaller passage restricts refrigerant flow, charging takes longer.

However, avoid charging too quickly; fast charging may cause discharge pressure to rise rapidly (it should not be allowed to exceed normal operating pressure). When charging with vapor, the process must be stopped periodically to check low-side operating pressure. The pressure should correspond to the conditions surrounding the evaporator.

Note: System operating pressures will not be normal until the system is fully charged and has achieved normal operating temperatures. Both high-side and low-side pressures will be too low if the system is *undercharged*. If the system is *overcharged*, the high-side pressure will be high. The low-side pressure will be high if the evaporator temperature is high.

Charging of small systems with hermetic compressors should always be done using the vapor method. Liquid refrigerant can easily damage hermetic compressors.

15.5.2 Liquid Charging

Charging a system with refrigerant in a liquid state is faster than vapor charging, since it is done with the

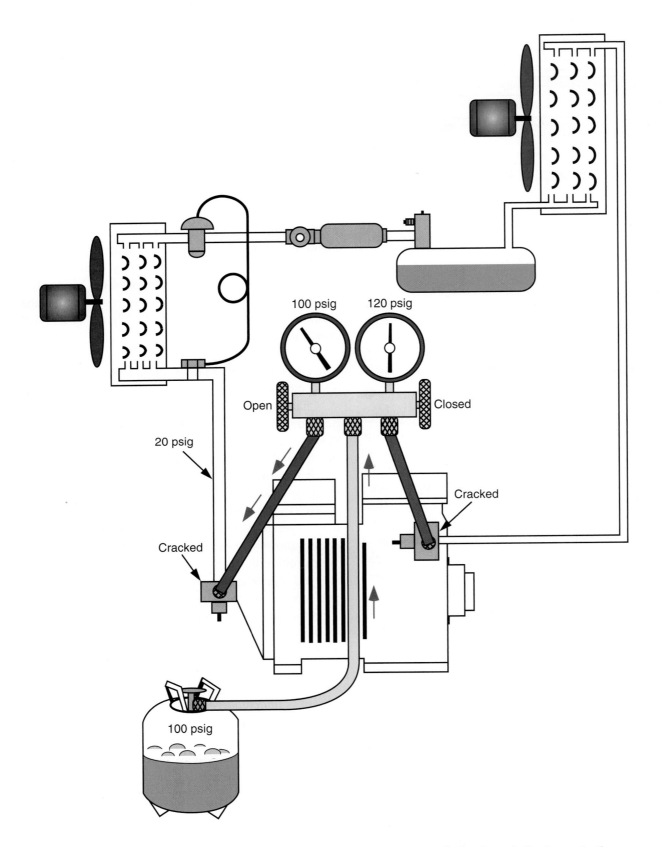

Figure 15-27. *During vapor charging, the refrigerant vapor flows from the cylinder through the hoses to the compressor. Cylinder pressure must be higher than suction pressure so the refrigerant vapor flows into the system.*

system operating; but it does require skill and experience. Since *liquid charging* can easily result in overcharging and cause compressor damage, it should be attempted only by experienced technicians. When liquid charging, the cylinder pressure remains constant since the liquid changes state to a vapor *inside the system*, rather than inside the cylinder. For liquid charging, the cylinder is usually inverted, as shown in **Figure 15-28.** Some cylinders, such as those used for refrigerant recovery, are equipped with a dip tube to allow liquid charging without inverting the cylinder.

Proper procedure for liquid charging entails *metering* the refrigerant into the low-pressure side of the system by adjusting the gauge manifold. The left (low-pressure) manifold valve is cracked open so a pressure 10 psig to 15 psig (69 kPa to 104 kPa) higher than suction pressure is maintained. Frost will form on the manifold body and the hose connected to the suction service valve. Frost should not be allowed to reach the compressor body since it would indicate liquid refrigerant is present.

Because low-side pressure increases as refrigerant is added to the system, the left manifold valve should be closed at regular intervals to check pressure. When the valve is closed, frost disappears. After pressure is checked, the valve should be cracked open once again and the 10 psig to 15 psig (69 kPa to 104 kPa) pressure differential established.

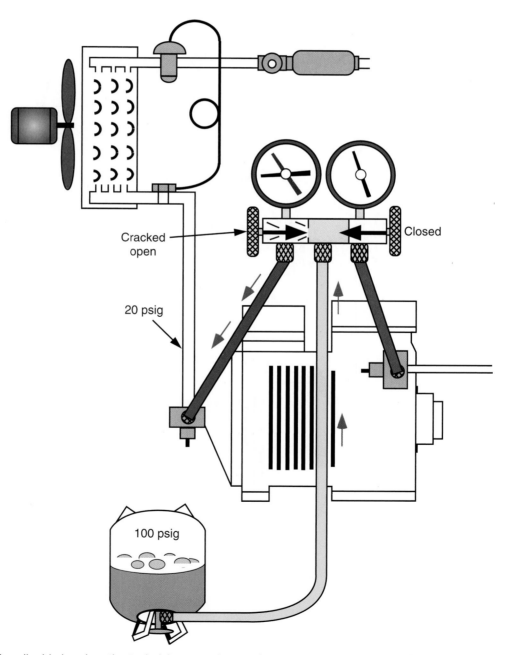

Figure 15-28. *When liquid charging, the technician must be careful to prevent liquid refrigerant from entering and possibly damaging the compressor. Gauge pressure should be 10 psig to 15 psig above actual suction pressure. Some types of cylinders can be left upright for liquid charging.*

Watch the high-side gauge to avoid excessive head pressure. Observe the sight glass for bubbles. When bubbles disappear from the sight glass and the two gauges register the correct pressure, the system is fully charged.

On larger systems, the liquid receiver service valve is usually used for charging. The valve is frontseated, causing the system to pump down (evacuate the low side). When liquid is charged into the system through the LRSV valve, it travels to the expansion valve and the evaporator. The liquid becomes a vapor before it reaches the compressor. See **Figure 15-29.**

Systems with liquid receivers are slightly over-charged on purpose. The technician must determine

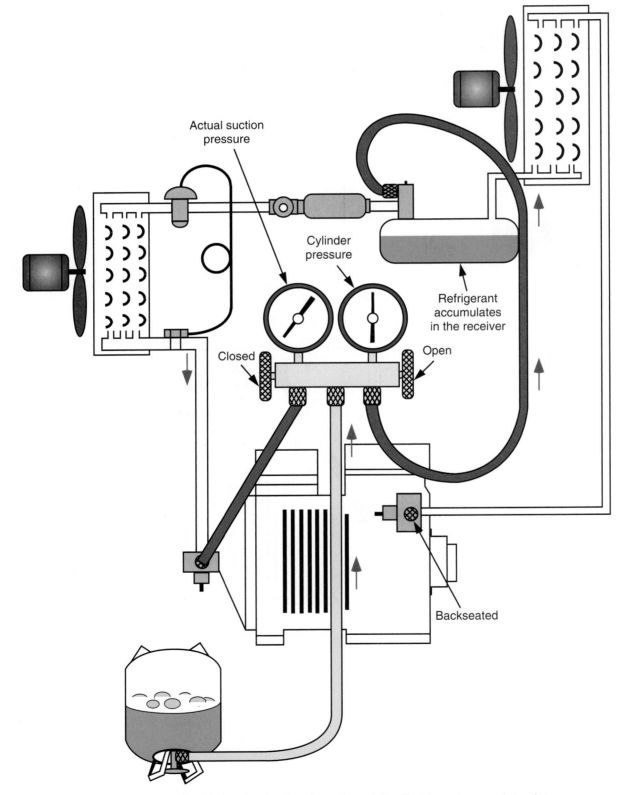

Figure 15-29. *For larger systems, liquid charging is often done through the liquid receiver service valve.*

how much liquid is being stored in the liquid receiver. (On very large systems, it may be big enough to hold the contents of three or four 30-pound cylinders of refrigerant). Periodically, charging should be stopped and the liquid line sight glass checked for bubbles. When bubbles stop appearing, the liquid line is full, and the receiver dip tube (outlet) is covered with liquid. Some additional refrigerant is added to the system, the amount determined by the technician's experience and judgment. The size of the receiver, and how much liquid it already contains, must be taken into account.

Charging by weight

Small, self-contained systems, such as window-mounted air conditioners, domestic refrigerators, and ice-making machines, are usually charged by weight of refrigerant. The recommended charge is stamped on the data plate of the equipment, making it quick and easy to measure an exact amount of refrigerant and charge the system. The system must be thoroughly evacuated before charging.

Electronic scales. Accurate, easy-to-use weighing devices are readily available for use by technicians, **Figure 15-30.** These *electronic scales* have displays that can be set to read zero when a full refrigerant cylinder is placed on them. The display shows the amount of refrigerant (in ounces) being withdrawn from the cylinder and charged into the system. When the desired refrigerant weight is reached, the manifold hand valve is closed to stop the process.

Figure 15-30. *Electronic scales permit very accurate measurement of refrigerant while charging. The cylinder is placed on the scale pad, and the weight readout is adjusted to zero. As refrigerant is charged into the system, the readout will keep a running total in pounds and ounces. (TIF)*

Graduated cylinder. A common method of measuring an exact refrigerant charge is using the *graduated cylinder,* **Figure 15-31.** Such cylinders are very accurate, and have scales for measuring three widely used refrigerants: R-12, R-22, R-502. The graduated scales can be rotated to align the desired one with the external viewing tube that shows the liquid level inside the cylinder.

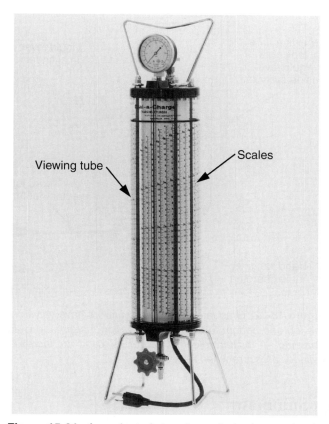

Figure 15-31. *A graduated charging cylinder has scales for the most common refrigerants. The desired scale is rotated into position next to the viewing tube to read the cylinder liquid level. (Robinair Mfg. Corp.)*

The graduated cylinder measures refrigerant by volume rather than weight. Since volume varies with temperature, the cylinder is equipped with a pressure gauge to permit selecting the proper scale. The cylinder is equipped with a top hand valve for vapor charging and a bottom hand valve for liquid charging. See **Figure 15-32.**

Before use, the graduated cylinder is charged with liquid refrigerant from a larger cylinder. The liquid level in the graduated cylinder, as shown by the viewing tube, should be slightly more than needed for the system being charged. The graduated cylinder is connected to the system through a gauge manifold; charging progress is monitored by watching the refrigerant level in the viewing tube. The graduated cylinder method allows precise charging.

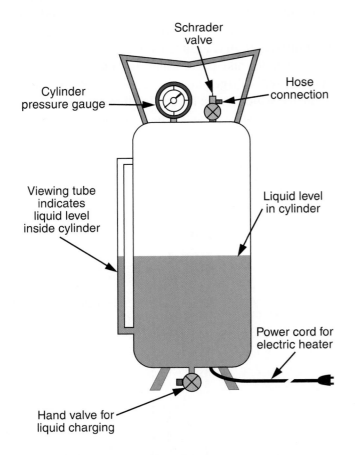

Schrader valve

Cylinder pressure gauge

Hose connection

Viewing tube indicates liquid level inside cylinder

Liquid level in cylinder

Power cord for electric heater

Hand valve for liquid charging

Figure 15-32. *Graduated charging cylinders have valves at the top and bottom to permit either vapor charging or liquid charging. A built-in electric heater is used to increase cylinder pressure when necessary.*

Summary

Knowing how to work safely and efficiently with refrigerants and high-pressure gases is very important. The information presented in this chapter is vital for properly installing, servicing, and troubleshooting refrigeration systems. The chapter explained many servicing and troubleshooting procedures and described several leak-detection methods. Charging methods for both vapor and liquid refrigerant were also described.

Test Your Knowledge

Please do not write in this text. Write your answers on a separate sheet of paper.
1. Why is oxygen never used to pressurize a system?
2. Leak detection using the _____ method is not reliable for detecting small leaks.
3. List three acceptable methods of leak detection.
4. Leak testing should be done with pressures between _____ psig and _____ psig.
5. Name two inert gases commonly used to pressurize systems for leak detection.

6. The halide torch can detect a leak as small as one ounce per _____.
7. Why are refrigerant cylinders filled to only 85% of capacity?
8. _____ cylinders should never be refilled.
9. Temperature charts for three refrigerants are shown on manifold gauge faces. Name the refrigerants.
10. Normally, controlling the temperature of the _____ around a product allows you to control the temperature of the product itself.
11. The evaporator removes heat from the _____, which removed heat from the _____.
12. In a refrigeration system, evaporator temperature should be about _____ degrees Fahrenheit colder than the entering air.
13. Boiling refrigerant inside the evaporator is about _____ degrees Fahrenheit colder than the temperature of the evaporator itself.
14. When a refrigeration system is turned off, is there a temperature difference between the air and the evaporator?
15. In an air conditioning system, what is the temperature difference between the refrigerant and "air off" (the cooled air)?
16. When a refrigeration system (such as one used for a walk-in cooler) is in operation, what is the evaporator temperature when the "air on" temperature is 36°F? What is the refrigerant temperature?
17. The purpose of the condensing unit is to remove _____ from the refrigerant vapor so it changes state back to a _____.
18. High-side pressure depends on the temperature of the _____ air passing through the condenser.
19. How is proper head pressure determined for an air-cooled condenser?
20. On a water-cooled condenser, head pressure should correspond to a temperature _____ to _____ degrees Fahrenheit higher than the temperature of the water leaving the condenser.
21. List five causes of high head pressure.
22. On an air-cooled system using R-502 as a refrigerant, what is the head pressure at an ambient air temperature of 80°F?
23. On an air-cooled system using R-12 as a refrigerant, what is the head pressure at an ambient air temperature of 60°F?
24. What is the most common means of charging a system?
25. For _____ charging, the refrigerant cylinder often must be inverted.

Working with Refrigerant Controls

Objectives

After studying this chapter, you will be able to:
❑ Install and adjust automatic expansion valves.
❑ Install and service capillary tubes.
❑ Install and adjust thermostatic expansion valves.
❑ Determine when and how to use internally and externally equalized valves.
❑ Identify and describe the operation of multiplexed evaporators.

Important Terms

automatic expansion valve (AEV)	metering orifice
	modulates
balanced port valve	multiplexing
bleedover	orifice
boiling point	overcharge
capillary tube	overload
cross-charged	pressure drop
diaphragm	refrigerant control
equalizing	refrigerant distributor
externally equalized	restrict
flash gas	sensing bulb
floodback	starve
hermetic	thermostatic expansion
hunt and surge	valve (TEV)
internally equalized	undercharge

16.1 Refrigerant Controls

A **refrigerant control** is the device that controls the amount of liquid refrigerant entering the evaporator. Since *all* liquid must boil off (vaporize or change state to a gas) inside the evaporator, the refrigerant control has the task of metering into the evaporator the precise amount of liquid refrigerant needed to match the evaporator's boiling capacity. If not enough liquid is metered in, the evaporator will *starve;* if too much is metered in, *floodback* (liquid in the suction line) will result.

Operation of the refrigerant control must be automatic so the correct amount of refrigerant is supplied to the evaporator. If the evaporator starves because too little liquid is supplied, poor cooling will result from reduced system capacity. Also, pressures will be low because the compressor will be able to remove gas from the evaporator at a rate faster than the valve is feeding in liquid. Damage can be caused by too much liquid being fed into the evaporator for the amount of heat available to evaporate it (called *floodback*). Excess liquid in the suction line can be drawn into the compressor and severely damage it.

The refrigerant control acts as the division point between the high-pressure and low-pressure sides of the system. Subcooled liquid at high pressure enters the valve, then exits as a low-pressure, saturated mixture of liquid and gas. This **pressure drop** across the valve is necessary for proper system operation.

Every evaporator must have a refrigerant control; systems with multiple evaporators must have a refrigerant control for each evaporator. Once refrigerant controls are installed and properly adjusted, they are very reliable and virtually trouble-free. Four types of refrigerant controls are discussed in this chapter:
❑ Automatic expansion valve (abbreviated AEV or AXV).
❑ Capillary tube.
❑ Metering orifice.
❑ Thermostatic expansion valve (abbreviated TEV or TXV).

16.1.1 Automatic Expansion Valve

The *automatic expansion valve (AEV)* is a *pressure-type* control installed in the liquid line at the evaporator inlet. See **Figure 16-1.** Most AEVs are installed with flare connections and are clearly marked to reflect refrigerant flow (*in* and *out*). The highly accurate valve will maintain a constant pressure in the evaporator anytime the system is running.

Since the AEV is designed to control pressure at its outlet, it responds to and controls the pressure inside the evaporator as well. When evaporator pressure drops below the selected level (by as little as 1 psig or 7 kPa), the valve opens wider, admitting more liquid refrigerant into the evaporator. If pressure increases above the desired level, the valve closes slightly to reduce the amount of liquid entering the evaporator. The AEV continually **modulates** (adjusts) refrigerant flow in this way to maintain constant evaporator pressure.

When the compressor is off (not operating), the AEV is closed as a result of high evaporator pressure. It remains closed until the compressor cycles back on and reduces evaporator pressure to the valve's setting.

The desired pressure is set by turning an adjustment screw on top of the valve. The screw, protected by a metal cap, permits adjustment over a wide range of pressures.

The technician must know the correct pressure setting for the boiling point of the refrigerant inside the evaporator. To set the pressure, the gauge manifold is installed to obtain a low-side pressure reading at the suction service valve. The automatic expansion valve is then adjusted until the proper pressure is registered on the compound gauge of the gauge manifold.

Determining evaporator pressure

The evaporator pressure is determined by the desired temperature of the *product being cooled.* Since the AEV is a constant pressure valve, the evaporator pressure (and temperature) remain constant. When an AEV is used as the refrigerant control, there is no temperature difference between the evaporator and the refrigerant.

For heat to flow, a temperature difference must exist. When the product (air) temperature is lowered enough to match the evaporator temperature, heat flow stops. As shown in **Figure 16-2,** product temperature

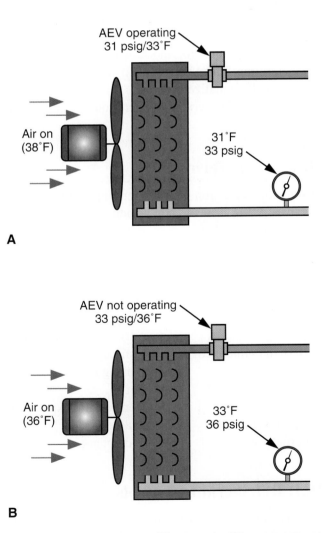

A

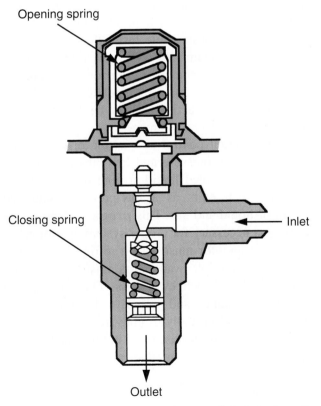

Figure 16-1. *A cutaway view of automatic expansion valve operation. The adjustable opening spring in the top part of the valve forces the rod and ball assembly downward to open the valve. The closing spring moves the ball and rod assembly upward to close the valve.*

B

Figure 16-2. *Temperature difference. A—When a system is running, the air temperature drops until it matches the evaporator temperature. Heat flow stops when the temperatures are the same. B—When a system is not running, all temperatures correspond to the air temperature (no heat flow).*

cannot drop below the evaporator temperature because no temperature difference exists. In the same way, when the system is turned off, the temperature of the evaporator will match that of the ambient air, and no heat will flow.

The temperature-limiting feature of systems equipped with an AEV is very desirable when cooling liquids, such as milk or water. The evaporator is submerged in the liquid, and the AEV pressure is set to keep the evaporator temperature just above the freezing point of the liquid. Since heat flow ceases when there is no temperature difference, the liquid temperature can never drop below the evaporator temperature and cannot freeze.

The feature is illustrated in the milk cooling tank shown in **Figure 16-3.** The wall-type evaporator inside the tank is in direct contact with the milk, allowing excellent heat transfer by conduction. The system uses R-12, so the AEV is set to maintain a pressure of 31 psig (214 kPa). At this pressure, R-12 produces an evaporator temperature of 33°F (0.5°C), one degree above the freezing point of milk. The milk stays cold but cannot freeze.

How the AEV works

The automatic expansion valve functions as a spray nozzle, breaking up the liquid refrigerant into fine droplets as it enters the evaporator. This makes it easier for the refrigerant to *vaporize,* or change to a gas. As the gas is drawn into the compressor, pressure inside the evaporator is decreased. The AEV maintains a constant pressure level in the evaporator by metering

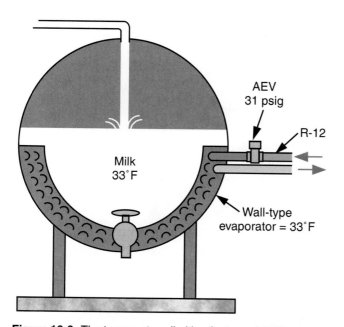

Figure 16-3. *The temperature-limiting feature of AEV systems allows liquids to be kept just above freezing. By holding the pressure of R-12 in the evaporator at 31 psig, the temperature of the milk stabilizes at 33°F.*

in enough liquid to balance the vapor drawn out by the compressor. When the compressor cycles off, pressure in the evaporator rises, and the AEV closes. It remains closed until the compressor cycles back on and lowers pressure in the evaporator.

Working forces in the AEV are:
- ❏ The range adjustment spring pressing downward on the **diaphragm** or bellows, which tends to open the valve.
- ❏ The closing spring exerting an upward pressure on the push rod and ball assembly, which tends to close the valve.
- ❏ Pressure on the diaphragm from the evaporator outlet, which tends to close the valve.

Once set by the technician, spring pressure on the adjustment screw remains constant. As a result, the evaporator pressure determines how much the valve opens or closes. A rise in evaporator pressure will overcome spring pressure and force the valve toward a closed position. A drop in evaporator pressure will allow the spring pressure to open the valve wider. The AEV responds to very slight pressure fluctuations, automatically modulating (partially opening or closing) to maintain a constant evaporator pressure.

Adjustment procedure

A gauge manifold to monitor pressure readings is installed at the suction service valve (evaporator pressure). With the system running, any adjustment of the AEV setting is immediately reflected on the compound gauge.

AEV setting adjustments are performed by removing the metal cap of the valve and turning the adjustment screw in or out. Turning the screw in (clockwise) increases the spring pressure on the diaphragm. More liquid enters the evaporator because higher evaporator pressure is needed to overcome the AEV spring pressure. Increasing the amount of refrigerant in the evaporator causes the evaporator pressure to rise. Turning the adjustment screw out (counterclockwise) has the opposite effect. It reduces spring pressure on the diaphragm, allowing it to open the valve more easily. This reduces the amount of liquid refrigerant entering the evaporator and lowers the evaporator pressure.

Controlling product temperature

A thermostat within the area of the product being cooled controls operation of the AEV system. When product temperature is lowered to the selected ("cut-out") point, the thermostat turns off the system's compressor. With the compressor turned off, pressure in the evaporator rises, closing the AEV.

During the "off" portion of the compressor cycle, the product temperature slowly rises. The evaporator temperature slowly rises, too. Once product temperature rises to the thermostat's cut-in point (typically 5 degrees Fahrenheit or 3 degrees Celsius above the cut-out point), the compressor is turned on again.

The AEV does not open immediately, however. It remains closed until the compressor reduces evaporator pressure to the valve setting. Adjustments to the AEV can be made only while the system is running since it is the only way a constant pressure can be maintained.

Selecting an AEV

The technician should select an automatic expansion valve with a capacity equal to the condensing unit of the system. For example, a system with a one-ton condensing unit should have a one-ton valve. Using a too-small valve will "starve" the evaporator; a valve that is too large will **hunt and surge** (fluctuate from fully open to fully closed) at start-up.

AEV disadvantages

Increased heat load on the evaporator of an AEV-equipped system causes the evaporator pressure to rise which, in turn, causes the AEV to close or "choke down." The evaporator is starved under peak load conditions, and the time needed to bring the product temperature down to the desired level is increased. The AEV is best applied to equipment involving a fairly stable heat load, such as ice-making machines.

To control product temperature, the AEV system requires a thermostat. Since low-side pressure is constant on an AEV system, pressure-type motor controls cannot be used.

Systems using the AEV were once popular but are no longer common. They have largely been replaced by capillary tube or thermostatic expansion valve systems. The *operating principle* of the automatic expansion valve remains in wide use but not as the primary refrigerant control device. Special valves using the AEV principle have different names and uses and are explained later in this book.

16.1.2 Capillary Tube Refrigerant Control

A **capillary tube** is a length of copper tubing with a tiny, accurately sized inner diameter (ID, or hole throughout its length). See **Figure 16-4.** Capillary tubes are available in various sizes for a number of purposes.

The capillary tube is used in small refrigeration systems as a *refrigerant control device*. The tube length and small hole (ID) are accurately calculated to **restrict** liquid refrigerant flow. The capillary tube controls the amount of liquid entering the evaporator.

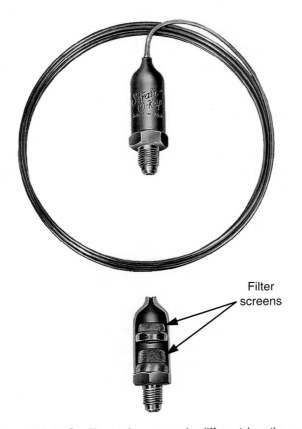

Filter screens

Figure 16-4. *Capillary tubes come in different lengths and inside diameters. This tube has a filter at the inlet. The filter, shown in a cutaway view, has fine screens to trap contaminants that could clog the tiny passage (ID) through the tube. (Watsco)*

In operation, one end of the capillary tube is connected to the condenser outlet and the other end to the evaporator inlet. In this way, the tube serves as both the liquid line and a refrigerant control (metering) device. See **Figure 16-5.**

Capillary tube selection

The capillary tube on a given system must be precisely sized, since the tube is *not* adjustable. Four variables should be taken into account when selecting capillary tubing:
❑ Length.
❑ Inside diameter.
❑ Refrigerant.
❑ Tubing temperature.

The capacity of the tube will vary with the refrigerant used; for example, a capillary tube selected for use with R-22 will have a different capacity if used with R-12 or R-502. Each refrigerant has certain properties that differ with changing temperatures.

On many domestic refrigerators, the temperature of the capillary tube is controlled by soldering it to the suction line. The arrangement functions as a heat exchanger. The suction line cools the liquid refrigerant

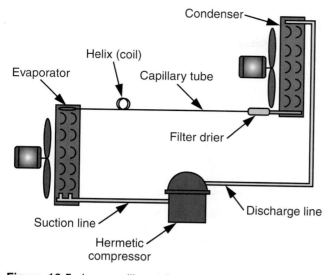

Figure 16-5. *In a capillary tube system, the capillary tube controls the flow of liquid refrigerant into the evaporator. It also serves as the liquid line to connect the condenser and evaporator. Capillary tube systems are often used on domestic and small commercial refrigeration systems.*

inside the capillary tube, while heat from the capillary tube helps keep frost off the suction line. Some refrigerator manufacturers actually run the capillary tube *inside* the suction line for much of its length.

The amount of liquid entering the evaporator must be balanced with the speed of boiling (rate at which the refrigerant changes to a vapor) and the gas-removing ability of the compressor. A perfect match must occur at the desired boiling point (temperature-pressure relationship). If too little liquid enters the evaporator to keep up with the compressor's gas-removing ability, the evaporator will "starve." This results in a lower **boiling point** (suction pressure) and a lower cooling capacity. Excess liquid entering the evaporator raises the boiling point (pressure) and may permit liquid to enter the suction line. Liquid in the suction line (floodback) can cause severe compressor damage.

Capillary tube systems are designed to maintain the same temperature differences explained in the preceding chapter. In a refrigeration system, the evaporator temperature is 10 degrees Fahrenheit (6 degrees Celsius) colder, and the refrigerant temperature is 20 degrees Fahrenheit (11 degrees Celsius) colder than the *air on* temperature. In an air conditioning system, the same temperature differences exist but are measured from the *air off* temperature.

Selection and installation of the proper size of capillary tube is performed by the system manufacturer. The most common size of capillary tube for use with R-12 has an OD of 0.114″ and an ID of 0.049″. Field replacement of capillary tubes is not recommended, although it is possible on some systems. Any replacement tube that

is installed must be an *exact* duplicate; a replacement compressor installed in a capillary tube system also must be an exact replacement. Every component in such a system is critical. Therefore, always follow the manufacturer's recommendations.

Capillary tube advantages

The capillary tube is inexpensive and has no moving parts. In addition, it permits a small amount of high-side pressure to bleed over into the low-pressure side during the off cycle. **Bleedover** helps reduce the high-side pressure (and the starting load) on the compressor. As a result, the system can make use of a less costly compressor with low starting power.

Bleedover process. Sometimes bleedover is referred to as a *balancing* of pressures during the off cycle. This balancing or **equalizing** process does *not* mean the high and low pressures become identical. The pressures do not equalize because the system does not remain off long enough for condenser and evaporator temperatures/pressures to become equal. As the high-side pressure drops to ambient, the amount of flow through the capillary tube is reduced greatly due to bubbles of vapor. The evaporator is *much* colder than the condenser; the pressure in each will behave as described on the temperature-pressure chart for the refrigerant involved.

When a capillary tube system has shut off, a delay of two to five minutes must be allowed before restarting. The delay is necessary because the compressor cannot start under normal head pressure (temperatures 30 to 35 degrees Fahrenheit or 17 to 20 degrees Celsius above ambient). An **overload** (electrical safety device) protects the compressor motor until the head pressure is reduced to ambient. *Cycling on the overload* should be avoided — it is an indication that the motor is having problems. Motors and overloads are explained later in this book.

Capillary tube disadvantages

A capillary tube system has a fixed, nonadjustable capacity, so the amount of refrigerant is critical down to one-half ounce. Overcharging or undercharging the system will greatly affect operation; the amount of refrigerant in the system becomes *the* determining factor for proper control of system pressures.

Capillary systems are self-contained, meaning the condensing unit is located within the cabinet. The cabinets are normally located inside a building where ambient temperatures are controlled (about 70°F or 21°C).

Domestic refrigerator condensers are cooled with ambient air; the head pressure can be predicted by adding 30 to 35 degrees Fahrenheit (17 to 20 degrees

Celsius) to the ambient temperature, then consulting a temperature-pressure chart. A system that uses R-12 should have a head pressure between 117 psig and 127 psig (808 kPa and 873 kPa):

$$70°F + 30 \text{ degrees} = 100°F \text{ (117 psig)}$$
$$70°F + 35 \text{ degrees} = 105°F \text{ (127 psig)}$$

The capillary tube for the system was selected to perform at this head pressure, resulting in an evaporator pressure of 1 psig to 3 psig (7 kPa to 21 kPa). If the head pressure rises above normal, more refrigerant will be pushed through the capillary tube into the evaporator, resulting in a higher temperature-pressure. Likewise, if the head pressure drops below normal, the flow of refrigerant will be reduced, starving the evaporator. Low-side pressure may fall into a vacuum reading.

An **overcharge** causes *high* head pressure because excess liquid accumulates in the condenser. High head pressure, in turn, causes high suction pressure because a greater quantity of liquid is pushed through the capillary tube. An **undercharge** causes *low* head pressure because the condenser is, in effect, "oversized." Since not enough liquid is being pushed through the capillary tube, low head pressure causes low suction pressure. Gas bubbles in the capillary tube compound the low pressure problem because the bubbles tend to increase tube resistance.

A further disadvantage of the capillary tube system is the very small passage through the tubing, which can easily be restricted or plugged by dirt, ice, or wax. To help guard against blockage, a filter-drier is always installed at the capillary tube inlet. When opening a capillary tube system for repairs, thorough evacuation is critical for moisture removal. The filter-drier should always be replaced after service.

Working on capillary tube systems

Capillary tube systems are normally **hermetic** (sealed) systems without service valves. In the manufacturing process, "pigtails" or *process tubes* are used to evacuate and charge the system. When charging is complete, the pigtail is pinched closed and cut off. The closed tubing end is then brazed to totally seal the system. For servicing, the technician must install saddle valves or Schrader (core-type) valves. See **Figure 16-6.**

A restricted (partly blocked) capillary tube or filter-drier causes low suction pressure and low head pressure. When the suction pressure is low, the load on the compressor is reduced and the condenser becomes "oversized." Subcooled liquid accumulates inside the condenser. Attempting to increase suction line pressure by adding refrigerant is a mistake; it will cause excessive head pressure and compressor failure.

A restriction *always* causes a pressure drop, resulting in flash gas and a cold surface at the location of the restriction. When a filter-drier's surface is cold (or actually shows frost), it has a restriction and must be replaced.

Another possible cause of low suction and head pressures is an *undercharge*. An undercharge indicates the system is losing refrigerant through a leak. Adding refrigerant will temporarily alleviate the problem but will not fix it. The only way to remedy a leak is to find and repair it. See **Figure 16-7.**

Charging a capillary tube system

A capillary tube system can be charged by slowly adding small quantities of refrigerant vapor to the low-pressure side. This must be done carefully because capillary tube systems are easily overcharged. Charging is done slowly to permit the system to stabilize after each addition of refrigerant. The evaporator

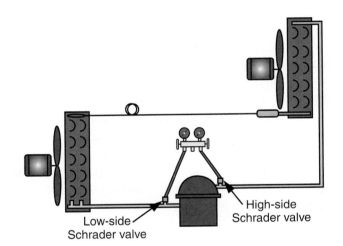

Figure 16-6. *Service valves are not provided on capillary tube systems. Schrader (core-type) valves or saddle valves must be installed at the points shown to obtain pressure readings.*

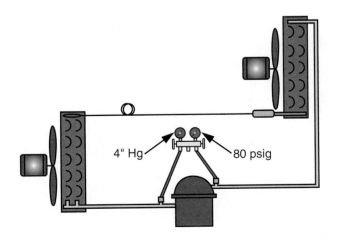

Figure 16-7. *Low pressure on both sides of the compressor indicates either a restriction or a leak. The technician must discover the problem causing the low pressure and fix it.*

temperature, suction pressure, and discharge pressure will not attain normal status until the system reaches normal operating temperatures.

An overcharge can be recognized by high discharge and suction pressures and by the presence of frost on the suction line. Correct the overcharge by slowly purging gas from the system using recovery equipment. The frost line will slowly retract until it disappears inside the cabinet.

Proper charging cannot be checked until the cabinet temperature is reduced to normal. During charging, the head pressure quickly rises to an above-normal point, while the suction pressure reading is in vacuum. Head pressure must not be allowed to exceed 150 psig (1034 kPa) when using R-12.

Vapor charging occurs faster than the condenser can liquefy incoming gas. Gaseous refrigerant in the capillary tube does not flow like liquid. System pressures should be allowed to stabilize after each addition of refrigerant. Cabinet temperature must be reduced to almost normal before the pressures become normal.

Weighing the charge. A graduated cylinder or electronic scale can be used, as explained in the previous chapter, to quickly and easily charge a capillary tube system. The correct amount of refrigerant is measured and quickly charged into the system as a vapor. The procedure saves time, but the precise amount of refrigerant to be charged into the system must be known. The specified amount can usually be found on the data plate of the refrigeration unit. If necessary, check the service manual, or contact the manufacturer to obtain the information.

16.1.3 Metering Orifice

The refrigerant **metering orifice,** like the capillary tube, is designed to operate on the principle of *restriction.* While the capillary tube is several feet long, the metering orifice is very small. See **Figure 16-8.** The unit has flare connections at both ends and is installed in the liquid line at the evaporator inlet.

The brass body of the metering device contains a small hole (**orifice**) sized to match the equipment exactly. The manufacturer's recommendations must be followed; the orifice controls the amount of liquid entering the evaporator, and a mismatch could cause problems. The orifice is easily removed for cleaning or replacement.

The operating characteristics of a metering orifice and a capillary tube are identical, but the orifice has the advantages of being accessible and easily replaceable. These advantages account for its popularity among manufacturers of residential air-conditioning systems and heat pump units.

16.1.4 Thermostatic Expansion Valve (TEV)

The **thermostatic expansion valve (TEV)** is the most common refrigerant control, especially in commercial and industrial refrigeration and air conditioning applications. The TEV, **Figure 16-9,** can be mounted in any position at the evaporator inlet and is reliable, efficient, and virtually trouble-free. The operating principle and characteristics of the thermostatic expansion valve are often misunderstood. As a result, the valve is frequently and mistakenly blamed for other system troubles.

Controlling superheat

Veteran technicians often refer to the TEV as a "superheat valve" because it maintains a constant level of superheat at the evaporator outlet. A constant level provides maximum efficiency under all types of load conditions. Too much superheat at the evaporator outlet starves the evaporator of liquid and reduces capacity. No superheat at the outlet permits liquid to enter the suction line.

The amount of superheat at the evaporator outlet varies according to the speed at which vaporization (boiling) takes place. The speed at which the liquid boils depends on the heat load placed on the evaporator. A high heat load causes the liquid to evaporate

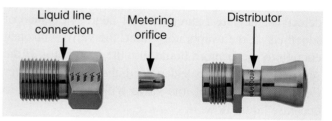

Figure 16-8. *The metering orifice is easily replaced. The assembly is located at the evaporator inlet. (Aeroquip Corp.)*

Figure 16-9. *Thermostatic expansion valves come in a variety of sizes and connections. (Parker-Hannifin)*

rapidly. A low heat load results in a slower change of liquid to gas.

Ideally, the evaporator should be well supplied with liquid refrigerant under all types of heat load conditions. All liquid refrigerant entering the evaporator *must* be converted to a superheated gas before leaving the evaporator.

How the TEV works

The thermostatic expansion valve is designed to maintain a precise superheat at the evaporator outlet. This provides proper flooding of the evaporator with liquid refrigerant under all load conditions. The valve controls superheat by metering varying amounts of liquid into the evaporator. High superheat at the evaporator outlet causes the valve to open, while low superheat causes the valve to close.

As shown in **Figure 16-10,** the valve can be mounted in any position at the evaporator inlet, but the *sensing bulb* must be located on the suction line at the evaporator outlet.

Superheat at the location of the sensing bulb controls the amount of liquid flowing through the valve at the evaporator inlet. The desired amount of superheat (typically 10 degrees Fahrenheit or 5.6 degrees Celsius) is set with an adjustable control on the valve body. The valve opens and closes slightly (modulates) in response to superheat changes. If, for example, the sensing bulb detects 11 degrees Fahrenheit (6.1 degrees Celsius) of superheat at the evaporator outlet, the valve will open enough to meter more liquid into the evaporator. If the sensing bulb registers 9 degrees Fahrenheit (5 degrees Celsius) of superheat, the valve will close slightly to decrease liquid flow.

The advantage of the valve's modulating action is a properly flooded evaporator at all times, despite changing heat load conditions.

The TEV operates by moving the valve stem (pin) away from the valve seat. As shown in **Figure 16-11,** valve stem movement is controlled by four pressures exerted on working parts:
- ❏ Sensing bulb pressure (opens valve).
- ❏ Evaporator pressure (closes valve).
- ❏ Spring pressure (closes valve).
- ❏ High-side pressure (opens valve).

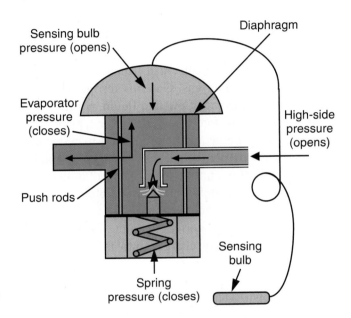

Figure 16-11. *The action of four pressures on the TEV while a system is operating. Pressures tend to either open or close the valve by moving the stem (pin) into or out of its seat.*

Sensing bulb pressure

The sensing bulb contains a special charge of refrigerant that is extremely sensitive to changes in temperature. As the suction line temperature changes, the bulb pressure increases or decreases accordingly. For example, a rise in temperature would cause an increase in bulb pressure that is transferred by means of a capillary tube to the diaphragm of the valve. The diaphragm is pressed downward, acting on push rods. The push rods exert a downward force on the pin carrier, moving the pin out of the seat to open the valve. See **Figure 16-12.**

Sensing bulb charges. To provide accurate superheat control for different applications, different types of charges are used in the sensing bulb. In most cases, the bulb is charged with the same refrigerant used in the system. Different operating characteristics are obtained by varying the sensing bulb volume and, consequently, the amount of charge. Each valve is designed and charged for a particular

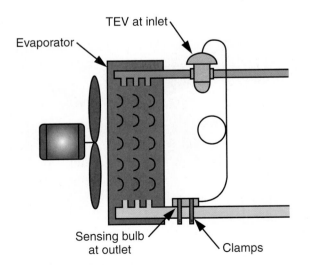

Figure 16-10. *The TEV is mounted at the evaporator inlet where it controls the flow of liquid refrigerant. The sensing bulb is mounted at the evaporator outlet to monitor changes in superheat.*

Valves designed for high-temperature applications have sensing bulbs charged with very little liquid. This provides a *pressure-limiting* feature; that is, when all the liquid has vaporized, a further increase in temperature produces almost no additional bulb pressure. The amount of pressure exerted upon the valve diaphragm is limited. Such valves are normally used in air conditioning applications.

Sensing bulb mounting. Proper mounting of the sensing bulb is essential for good valve control. As shown in **Figure 16-13,** the bulb is clamped to the suction line at the evaporator outlet. It should be located on a horizontal section of the line, upstream from any trap, and make good thermal contact along its entire length with the suction line. The bulb is normally secured to the suction line with two copper straps.

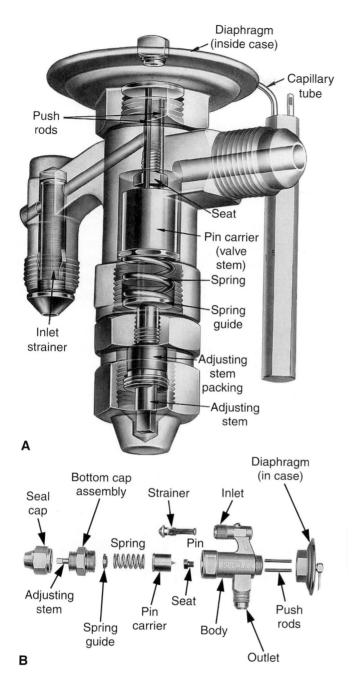

Figure 16-12. *Thermostatic expansion valve. A—Cutaway view of a typical TEV. Varying pressures on the diaphragm and spring modulate operation of the valve. B—Exploded view shows the relationships between parts. (Sporlan Valve Co.)*

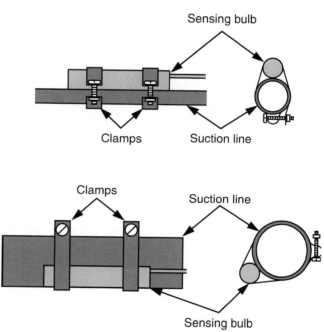

Figure 16-13. *Proper sensing bulb location. A—On suction lines 7/8″ (22 mm) or smaller, the bulb should be clamped to the top or side of the line. Avoid bottom-mounting. B—On lines larger than 7/8″ (22 mm), mount the bulb at the 4 o'clock or 8 o'clock position.*

application — a valve designed for low pressure cannot be used for high-pressure applications. Sufficient refrigerant is charged into the bulb to ensure it contains some liquid under all temperature conditions.

Valves designed for use in low-temperature conditions are usually ***cross-charged,*** meaning the bulb is charged with a refrigerant different from the one used in the system. The refrigerant used in the bulb is more sensitive to low temperatures.

On suction lines 7/8″ (22 mm) or smaller in diameter, the bulb can be mounted on the top or on either side of the line. The bottom should be avoided. Any oil flow in the line will be along the bottom, which could prevent the bulb from sensing temperatures correctly. On suction lines larger than 7/8″ (22 mm), the sensor should be mounted at the 4 o'clock or 8 o'clock position for the most accurate temperature readings.

Evaporator pressure

Evaporator pressure acts on the bottom of the valve diaphragm, providing an upward movement that tends to close the valve (refer to Figure 16-11). An increase in evaporator pressure causes the valve to close until the compressor causes a reduction in evaporator pressure.

Slight changes in either evaporator pressure or bulb pressure make the diaphragm move up or down in small increments, causing corresponding changes in the size of the valve opening (modulation) and the flow of liquid refrigerant.

Internally equalized valve. In an *internally equalized* valve, a special passage built into the valve body transfers evaporator pressure to the underside of the diaphragm. This type of valve is used when there is a pressure drop of less than 2 psig (14 kPa) through the evaporator. See **Figure 16-14.**

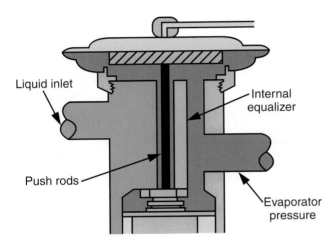

Figure 16-14. *Internally equalized valve. A special passage transfers evaporator pressure to the underside of the diaphragm.*

Evaporator pressure drop. The length of tubing comprising the evaporator, return bends in the tubing, and the vacuum (sucking) action of the compressor combine to develop a *pressure drop* through the evaporator. As a result, pressure at the evaporator inlet is higher than pressure at the outlet. A large evaporator may have a pressure drop of 6 psig (41 kPa) or more. For proper operation of the TEV, correct evaporator pressure on the underside of the diaphragm is essential. Evaporators that have a pressure drop greater than 2 psig (14 kPa) must use an externally equalized TEV.

Externally equalized valve. An *externally equalized* valve is like the internally equalized type, but evaporator pressure on the underside of the diaphragm is supplied from the evaporator *outlet* rather than the *inlet.* As shown in **Figure 16-15,** a separate passageway to the diaphragm is connected to a 1/4″ male flare fitting on the valve body. Tubing connected to this fitting is run to a point just downstream of the sensing bulb and inserted into the suction

line, **Figure 16-16.** This transfers the correct evaporator pressure to the valve diaphragm.

An externally equalized TEV must *always* be connected to the suction line. It should *never be capped,* or the valve may flood, starve, or operate erratically. An externally equalized valve may be used with an evaporator that has very little pressure drop. When the pressure drop exceeds the values shown in **Figure 16-17,** an externally equalized TEV *must* be used.

Spring pressure

The valve pin is mounted on a metal pin carrier that rides on top of a spring. Two push rods extend from the bottom side of the diaphragm to the top of the pin carrier. Upward pressure from the spring creates a closing force, pressing the pin into the valve seat. The force is also exerted on the push rods, which transmit it to the underside of the diaphragm.

Pressure *on top* of the diaphragm (exerted by expansion of refrigerant in the sensing bulb) opposes pressure on the bottom of the diaphragm resulting from a

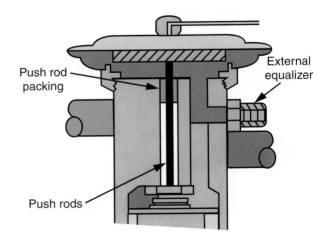

Figure 16-15. *Externally equalized valve. The external equalizer must be connected to the evaporator outlet through 1/4″ copper tubing to compensate for pressure drop through the evaporator.*

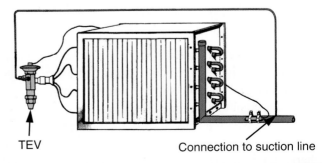

Figure 16-16. *The tubing from an externally equalized TEV is connected to the suction line at a point immediately downstream of the sensing bulb. (Sporlan Valve Co.)*

Refrigerant	Evaporating Temperature (degrees Fahrenheit)				
	40	20	0	−20	−40
	Pressure Drop (psig)				
R-12 & R-500	2	1.5	1	0.75	0.5
R-22	3	2	1.5	1.0	0.75
R-502	3	2.5	1.75	1.25	1.0

Figure 16-17. *The table gives pressure drops for various refrigerants at different evaporating temperatures. If the pressure drop across the evaporator exceeds these values, an externally equalized TEV must be used.*

combination of spring pressure and evaporator pressure. To open the valve, the bulb pressure must be greater than the combined evaporator and spring pressures.

Liquid refrigerant boiling to a vapor at the evaporator outlet will cool the sensing bulb and reduce bulb pressure. Temperatures above saturation (superheat) will warm the sensing bulb and increase bulb pressure. The amount of superheat required to open the valve is determined by the closing force exerted by the spring. When the spring pressure is increased, higher bulb pressure is needed to overcome the spring pressure. More superheat is needed to warm the bulb and increase bulb pressure. Decreased spring pressure has the opposite effect — less superheat is needed.

Spring pressure is changed by turning the valve stem. Turning the stem clockwise increases pressure; counterclockwise rotation decreases pressure.

High-side pressure

High-pressure liquid refrigerant coming into the TEV tends to open the valve by pushing the pin away from the valve seat. This opening force can be offset by spring pressure. However, spring pressure is fixed, and wide changes can occur in head pressure due to changing ambient conditions at the condenser. Severe head pressure changes can result in an imbalance among the valve's operating forces. Preventing wide changes in head pressure ensures more uniform operation of the TEV.

All thermostatic expansion valves are designed to operate at a certain *minimum* head pressure. For example, valves for use with R-12 require a minimum head pressure of 100 psig (690 kPa). Valves for use with R-22 require 200 psig (1379 kPa); those designed for R-502 need a minimum of 180 psig (1241 kPa). The TEV cannot operate properly when head pressures drop below these minimums. Various methods used to control head pressures are discussed later in this book.

One way of dealing with varying head pressure is to install a **balanced port valve.** Construction of the valve distributes head pressure between the valve pin and the push rods. By equalizing opening and closing forces, the valve design compensates for a wide range of head pressure changes.

16.1.5 TEV Superheat Adjustment

Basically, a TEV operates like any other valve: turning the stem in (clockwise) *closes* it; turning it out (counterclockwise) opens it. Closing and opening the valve permits different amounts of liquid refrigerant to flow into the evaporator. While the valve stem physically changes the opening of the valve, it actually is — in operating terms — a *superheat* adjustment.

Normal superheat settings

Normally, the TEV superheat setting is 10 degrees Fahrenheit (6 degrees Celsius). High superheat settings reduce evaporator capacity since more evaporator surface is needed to produce the superheat. *Some* superheat is needed, however, for proper operation of the valve. In air-conditioning applications, the superheat setting can be as high as 15 degrees Fahrenheit (8 degrees Celsius) before there is a noticeable loss of evaporator capacity. Many low-temperature systems that use R-502 for a refrigerant require 5 to 6 degrees Fahrenheit (2.8 to 3.3 degrees Celsius) of superheat. The lower superheat setting provides maximum capacity from the evaporator and makes use of the colder suction line vapor to cool the compressor.

If you have questions about the correct superheat setting for a particular system, consult the manufacturer. Recommendations from a manufacturer are normally the result of extensive laboratory testing.

When a valve is first installed, superheat adjustment is often necessary. Once properly installed and adjusted for superheat control, the TEV is virtually trouble-free.

Some thermostatic expansion valves are nonadjustable; they are factory set for a specific superheat value determined by the manufacturer. If symptoms indicate a superheat adjustment is needed, carefully check for other system causes of incorrect superheat.

Measuring a TEV superheat setting

How well a TEV is performing is easily determined by measuring superheat at the sensing bulb location. For accurate temperature and pressure readings, the system must be running and in a stabilized condition. Measurement is made as follows:

1. Install the gauge manifold on the compressor service valves, and take a pressure reading.

2. Obtain an accurate reading of the suction line temperature at the sensing bulb location.

3. Convert the pressure reading to a temperature value by consulting the temperature-pressure chart for the refrigerant in the system. This is the *saturation* temperature.

4. Subtract the saturation temperature from the actual temperature read at the sensing bulb. The difference is *superheat*.

Allowing for pressure drop. A pressure reading at the suction service valve will be lower than a pressure reading at the evaporator outlet. Called *suction line pressure drop*, it is limited to about 2 psig (14 kPa) if proper line sizing and good piping practice were followed when installing the system. For an accurate superheat reading, the pressure drop must be taken into account. To do so, add 2 psig (14 kPa) to the pressure at the suction service valve *before* converting that pressure to a temperature reading.

Figure 16-18 illustrates how superheat is measured on a system using R-22. As shown, the temperature of the suction line at the sensing bulb is 52°F (11°C). Pressure at the suction service valve is 66 psig (455 kPa). To allow for pressure drop, 2 psi (14 kPa) is added for a total of 68 psig (469 kPa).

Converting 68 psig to a temperature reveals a saturation temperature of 40°F (4°C). Subtracting the saturation temperature from the actual temperature of 52°F (11°C) yields superheat of 12 degrees Fahrenheit (7 degrees Celsius).

Changing the superheat setting. To change the superheat setting of the TEV, the adjusting stem is turned clockwise to increase superheat or counterclockwise to reduce superheat. When the stem is turned in (clockwise), the valve opening is reduced in size, permitting less liquid to enter the evaporator.

Figure 16-18. *Allowing for suction line pressure drop when calculating superheat. (Sporlan Valve Co.)*

When the stem is turned out (counterclockwise), the valve opens wider so more liquid can flow into the evaporator.

Adjustment should be made gradually to avoid overshooting the desired setting. Make no more than one turn of the valve stem at a time; then, allow enough time for the system to stabilize (as much as 30 minutes may be needed) before making another adjustment.

Evaporator temperature

The TEV does not control evaporator temperature or low-side pressure; it maintains *a constant superheat*, regardless of temperature or pressure. It is sized to match the capacity of the evaporator and compressor.

During operation, there is a difference of approximately 20 degrees Fahrenheit (11 degrees Celsius) between the temperature of the product being cooled and the temperature of the refrigerant. The *evaporator* temperature falls midway between the two — 10 degrees Fahrenheit (6 degrees Celsius) colder than the product but an equal amount warmer than the refrigerant.

As the system runs, the product temperature is reduced. The heat load on the evaporator is reduced as well. As the heat load is reduced, less liquid flows into the evaporator.

Since the compressor removes vapor from the evaporator at a fixed rate, less liquid flowing into the evaporator results in lower pressure inside the evaporator. Lower pressure means a lower boiling point for the refrigerant and a colder evaporator. Therefore, the difference of 20 degrees Fahrenheit (11 degrees Celsius) between the product and the refrigerant is maintained. The superheat setting is maintained as well, regardless of evaporator temperature.

When product temperature reaches the desired level, the system is turned off by a thermostat or a low-pressure control. When the system is turned off, the compressor stops running and cannot remove vapor from the evaporator. Evaporator pressure rises quickly as the refrigerant remaining inside the evaporator warms up to match the product temperature. The increase in evaporator pressure is transferred to the bottom of the diaphragm in the TEV, closing the valve. The valve remains closed during the "off" portion of the cycle because the evaporator (closing) pressure is sufficient to overcome the bulb (opening) pressure.

Flash gas

The result of liquid refrigerant passing through the TEV orifice is known as **flash gas.** The sudden pressure drop (from high to low) causes some of the liquid to instantly boil (flash) into a gaseous state, or vapor. See **Figure 16-19.** The heat needed to cause this boiling action is drawn from the remaining liquid, which is cooled to the lower pressure. The remaining liquid then boils at the proper evaporator saturation temperature.

The amount of flash gas is determined by the amount of pressure drop across the valve. This "loss" of liquid refrigerant is necessary as the refrigerant moves from a higher pressure, higher temperature part of the system to a lower pressure, lower temperature section (the evaporator). The expansion valve is the division point between the high-pressure and low-pressure sides of the system.

The amount of liquid that flashes into a gas can be reduced by subcooling the liquid before it reaches the valve orifice. Subcooling the liquid makes the system more efficient. A heat exchanger can be installed to subcool the liquid, but many systems accomplish subcooling by simply having long, uninsulated liquid lines between the condenser and the TEV.

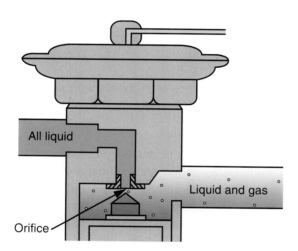

Figure 16-19. *A portion of the liquid refrigerant will "flash" into a gaseous state as a result of the pressure drop when it passes through the TEV orifice.*

16.2 Refrigerant Distributor ──────◆

On large evaporators with multiple circuits, a **refrigerant distributor** controls the amount of liquid refrigerant entering the evaporator. The distributor, **Figure 16-20,** receives refrigerant directly from the thermostatic expansion valve and divides it equally among all evaporator circuits.

In any refrigerant system, the refrigerant control is a division point where high pressure is changed to low (evaporator) pressure. On systems that do not include a distributor, all pressure drop is taken across the valve. When a distributor is used, some pressure drop occurs *across the distributor.* Pressure drops across the distributor for common refrigerants are shown in

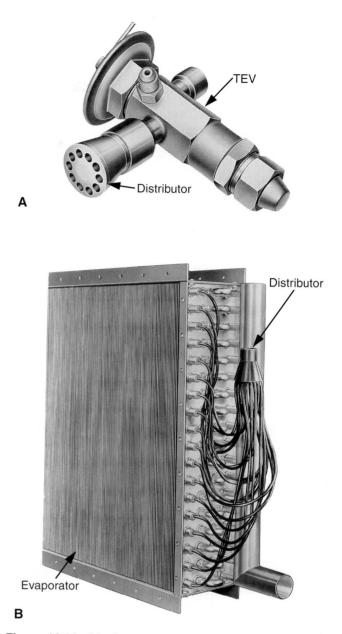

Figure 16-20. *Distributors are used in large systems with multiple evaporator circuits. A—A distributor mounted directly to a TEV. B—A distributor mounted in a vertical downflow position to direct refrigerant to 18 evaporator circuits. (Sporlan Valve Co.)*

Figure 16-21. Any system using a distributor *must* use an externally equalized TEV.

When brazing a distributor to a TEV, be careful to direct the torch flame away from the valve body, as shown in **Figure 16-22.** To prevent excessive heat from damaging the valve diaphragm while brazing, wrap the valve body in a wet cloth.

16.2.1 Valve Mounting

The thermostatic expansion valve may be mounted in any position, but it should be installed as close to the evaporator inlet as possible. If a distributor is used, it is mounted directly to the outlet of the TEV. Most valves are installed with a flare connection or mechanical flanges, but some make use of brazed connections. See **Figure 16-23.**

16.2.2 TEV Capacities

Thermostatic expansion valve capacities are rated in tons of refrigeration. Capacity varies based on these factors:

- ❑ Orifice size.
- ❑ Pressure drop across the valve (liquid line pressure to evaporator pressure).

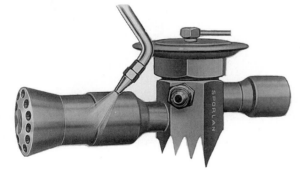

Refrigerant	Average Pressure Drop Across Distributor
R-12	25 psig
R-22	35 psig
R-500	25 psig
R-502	35 psig

Figure 16-21. *Pressure drops across the distributor for several common refrigerants.*

Figure 16-22. *Direct the flame of the brazing torch away from the valve body to prevent heat damage to the diaphragm. Technicians often wrap the valve body in a wet cloth to protect it from heat. (Sporlan Valve Co.)*

- ❑ Amount of flash gas (liquid temperature versus evaporator temperature).
- ❑ Refrigerant used.

It is vital to use a valve of the correct capacity. If an undersized valve is used, the evaporator will starve, despite superheat adjustment. If the valve is oversized, it will hunt and surge (alternately starve and flood the evaporator).

Color codes

When selecting a TEV, a major consideration is the type of refrigerant used in the system since expansion valves are not interchangeable among refrigerants. To help identify valves, the tops of the diaphragms are color-coded. The colors are not *always* the same as those used to identify refrigerant cylinders. The TEV codes do, however, provide another method to identify the refrigerant being used in a system. The American Refrigeration Institute (ARI) has established a standard (750-81) for the color coding of thermostatic expansion valves:

    Yellow — R-12
    Green — R-22
    Orange — R-500
    Purple — R-502
    Blue — Uncommon refrigerants with no
             designated color

16.2.3 Number Codes

Printed on the body or diaphragm of each valve is a set of numbers (or letters and numbers) displaying the manufacturer's coded information about the valve. The codes are not secret, and information on how to read them is freely available from manufacturers. The code numbers are usually needed when ordering replacement parts.

16.3 Multiple Evaporator Systems ⬥

The operating principle of the thermostatic expansion valve makes it possible to connect two or more evaporators to a single compressor, called *multiplexing.*

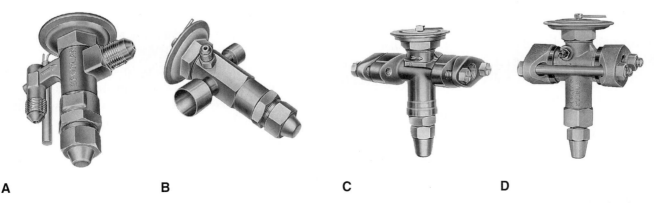

A   **B**   **C**   **D**

Figure 16-23. *TEV mounting variations. A—Flare connection. B—Brazed connection. C—Pipe thread/mechanical flange connection. D—Brazed/mechanical flange connection. (Sporlan Valve Co.)*

Each of the evaporators operates at the same temperature level since all are connected to a common liquid line and a common suction line. See **Figure 16-24.**

In a multiplexed setup, evaporator capacities are added together to determine the size of the compressor needed. An example of multiplexing is found in supermarkets: a line of four or five frozen food cases might be served by a single compressor located in an equipment room. Each case has its own TEV, evaporator, fans, and other components. Common liquid and suction lines connect the evaporators in the cases.

It is very important, in multiplexed installation, to adjust each TEV for the proper superheat. Adjusting the TEVs individually for proper superheat control provides maximum efficiency from each evaporator and prevents floodback.

16.3.1 Multiplexing Advantages and Disadvantages

The major advantage of multiplexing is the elimination of duplicated components, which saves money on materials. Multiplexing also saves installation time, energy costs, and space.

The only disadvantage of multiplexing is the use of a single compressor. If the compressor burns out or malfunctions, the freezer cases may have to be emptied and the products moved to another freezer until repairs are made.

Summary

To become a successful technician, you must understand the operating principles of the various refrigerant controls. One or more of these controls is found on every refrigeration system. Each control operates differently to meter the amount of refrigerant entering the evaporator. As a result, service and repair procedures differ for each type of control or system.

The temperature-pressure at which the refrigerant boils (vaporizes) in the evaporator is controlled by the amount of liquid being metered into the evaporator. The flow of liquid into the evaporator must be balanced with the amount of vapor leaving the evaporator through the suction line.

This chapter described four types of refrigerant controls: the automatic expansion valve, the capillary tube, the metering orifice, and the thermostatic expansion valve. Use of a refrigerant distributor on large evaporators with multiple circuits was also described.

Test Your Knowledge

Please do not write in this text. Write your answers on a separate sheet of paper.
1. What is the purpose of the refrigerant control?
2. Name four common types of refrigerant controls.
3. Refrigerant controls are located at the inlet of the _____.
4. What is the purpose of the automatic expansion valve?
5. Does the AEV open or close on a rise in the evaporator pressure?
6. What are the two forces that operate the AEV?
7. The capillary tube works on the principle of _____.

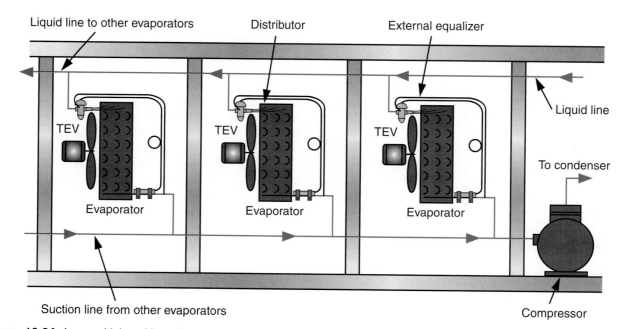

Figure 16-24. *In a multiplexed installation, a number of evaporators are connected to a single compressor, sharing common liquid and suction lines.*

8. *True or false?* The four variables of a capillary tube are length, ID, refrigerant, and tubing temperature.

9. Name three advantages of the capillary tube as a refrigerant control.

10. In capillary tube systems, what device controls product temperature?

11. Name two disadvantages of the capillary tube.

12. The refrigerant metering orifice is used in place of the _____.

13. What is the purpose of the TEV?

14. Does low superheat cause the TEV to open or close?

15. Name the four pressures that affect operation of the TEV.

16. Proper mounting of the _____ is essential to correct operation of the TEV.

17. On suction lines less than 7/8″ in diameter, the sensing bulb should not be mounted on the _____ of the tubing.

18. Is the external equalizer line of a TEV connected to the suction line upstream or downstream of the sensing bulb?

19. If you increase the spring pressure on a TEV, will you increase or decrease the superheat?

20. The normal superheat setting is _____ degrees Fahrenheit.

Special-Purpose Valves

Objectives

After studying this chapter, you will be able to:
❑ Identify hand valves and describe their installation and use.
❑ Identify check valves and describe their installation.
❑ Install solenoid valves.
❑ Install and adjust EPR and CPR valves.
❑ Install and adjust head pressure control valves.
❑ Install and adjust hot gas bypass valves.

Important Terms

capacity control
check valves
clockwise
counterclockwise
crankcase pressure
  regulator
desuperheating TEV
differential valve
discharge bypass valve
evaporator pressure
  regulators
hand shut-off valves

head pressure control
  valves
hot gas bypass valve
humidity
multiplexed evaporators
normally closed (NC)
normally open (NO)
pumpdown cycle
short cycle
solenoid valve
two-temperature
  system

17.1 Accessory Flow Control Valves

Many refrigeration systems use special accessory flow control valves to increase system efficiency, provide safeguards, facilitate repairs, or add system features. These special flow control valves are installed *in addition to* regular refrigerant controls, such as the TEV. It is quite common to see one or more special purpose valves on a system. Each flow control valve serves a particular purpose, and a malfunction in

a valve greatly affects operation of other system components. The operation and purpose of these valves are simple, but they can be misleading. This chapter explains these special purpose valves and the functions they serve in the system. Knowledge of these valves will prevent time-consuming mistakes.

17.2 Hand Shut-off Valves

Hand shut-off valves are installed on commercial and industrial systems as a convenience for service technicians. Hand valves make it possible to quickly isolate sections of a system to perform a particular task, such as changing a filter-drier. For safety, local codes often specify the installation of hand valves in commercial and industrial refrigeration and air conditioning systems.

Manually-operated valves must be sturdy and able to withstand frequent opening and closing without leaking. Pressure drop across the valve must be minimal. Diaphragm-type, globe-type, and ball-type valves are designed for such applications. Each may be soldered or flare-connected to copper tubing. See **Figure 17-1.**

For proper operation, a hand valve must be installed according to the designed direction of flow through the valve body. An arrow on the valve body indicates correct refrigerant flow direction. Diaphragm-type hand valves are used for smaller lines, while ball and globe valves are used for larger lines.

The turning knob (handle) of a *diaphragm valve* is connected directly to the valve stem. *Ball valves* have a metal cap nut over the valve stem. The cap nut must be removed for access to the valve stem. *Globe valves* also have a cap nut over the valve stem, but the cap nut has extension handles for easy turning. The cap nut also has a recessed opening so it can be turned upside down and used as a wrench to turn the valve stem.

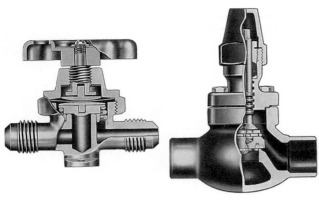
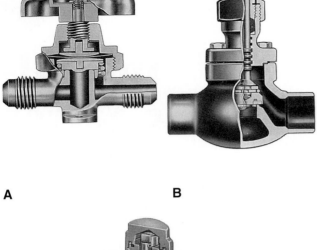

A

B

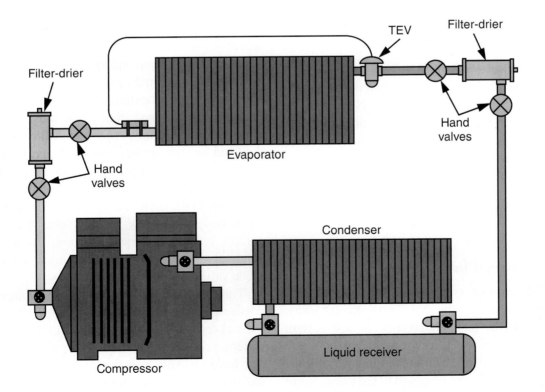

C

Figure 17-1. *Manually-operated (hand) valves. A—Diaphragm-type valve. B—Globe-type valve. C—Ball-type valve. (Henry Valve Co.)*

Hand valves can be installed in any line to evacuate and isolate specific sections of the system, **Figure 17-2.** Hand valves make it possible to perform repairs without loss of refrigerant.

17.3 Check Valves

Check valves are used in some refrigeration systems to prevent the flow of refrigerant in the wrong direction. Proper direction of flow is indicated on the body of these nonadjustable valves. If the liquid or vapor is flowing in the proper direction, the check valve will be open. A change of direction causes the valve to close tight, preventing reverse flow. See **Figure 17-3.**

17.4 Evaporator Pressure Regulator (EPR) Valve

Evaporator pressure regulators are special valves used to prevent evaporator pressure from falling below a set limit, **Figure 17-4.** EPRs make it possible to connect multiple evaporators with different temperatures to a single compressor.

An EPR valve is installed in the suction line and operates much like an automatic expansion valve. While the AEV controls pressure at the valve *outlet*, the EPR controls pressure at the valve *inlet*. A TEV ensures proper evaporator flooding. The EPR is

Figure 17-2. *Hand valves can be used to isolate and evacuate portions of the refrigeration system without loss of refrigerant. They are often used to permit easy changing of filter-driers.*

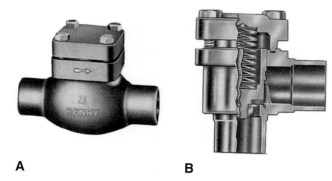

A   **B**

Figure 17-3. *Check valves. A—The arrow on the body of this globe-type check valve shows proper direction for flow. B—A cutaway view of an angle-type check valve shows the spring that holds the seat closed. Flow into the valve from the bottom is allowed to pass, but backward flow (from the right) is blocked. (Henry Valve Co.)*

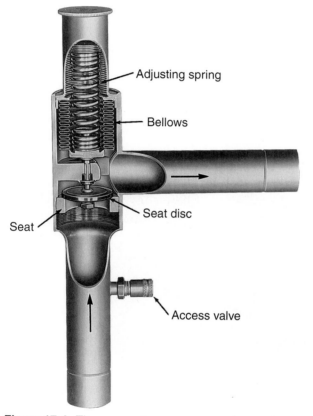

Figure 17-4. *The evaporator pressure regulator valve (EPR) keeps pressure at the evaporator inlet from dropping below a specified limit. A cutaway view shows the major components of a typical EPR. (Sporlan Valve Co.)*

installed in the suction line to provide a low temperature-pressure limit. This prevents evaporator pressure (and temperature) from dropping below the EPR setting. The combination of valves ensures proper flooding of the evaporator *and* a low pressure limit. The EPR can be installed anywhere in the suction line and still perform properly. See **Figure 17-5.**

As noted, the primary function of the EPR is to prevent evaporator pressure from falling below a

certain point. If the evaporator pressure rises above the valve setting, the EPR will open and release the pressure. When the evaporator pressure drops to the valve setting, the EPR will close, preventing further pressure drop. The EPR is a "holdback" type of valve that automatically throttles vapor flow from the evaporator to maintain its setting. The valve modulates very closely to the actual setting, constantly making slight adjustments to control evaporator pressure.

To adjust an EPR valve, remove the cap, and turn the adjustment screw with the proper size hex wrench (1/4″ or 5/16″). A ***clockwise*** (CW) rotation increases the valve setting; a ***counterclockwise*** (CCW) rotation decreases the setting. To obtain the desired setting, a pressure gauge should be attached to the inlet side of the valve so the effects of any adjustments can be observed. Make small adjustments, and allow adequate time between adjustments for the system to stabilize at the new setting.

EPRs are normally available in two pressure-adjustment ranges: 0 psig to 50 psig (0 kPa to 345 kPa) and 30 psig to 100 psig (207 kPa to 690 kPa). The 0 psig to 50 psig range is suitable for most applications. The 30 psig to 100 psig range is for applications using R-22 and R-502, with evaporating temperatures of 20°F (–7°C) and above. An access port or Schrader valve is provided on the EPR inlet side for connecting a pressure gauge. Inlet pressure readings are necessary when making pressure adjustments. Adjustments to the EPR should be performed *before* checking TEV superheat and making any necessary adjustments.

17.4.1 Application of EPRs

A typical EPR application is a single evaporator system, such as a water chiller, where the EPR is used to prevent freezing of the water. See **Figure 17-6.** The valve's pressure adjustment is set to keep the evaporator pressure (and the refrigerant temperature) above the freezing point of water. The EPR prevents evaporator pressure from dropping below this setting.

Another EPR application is the ***two-temperature system.*** The system in **Figure 17-7** has two evaporators operating at different temperatures but connected to a single compressor. The evaporator for the walk-in meat cooler is operating at 32°F (0°C), while the evaporator for the cutting room is operating at 40°F (4°C). An EPR is installed on each evaporator to prevent pressure from dropping too low. This system explains why EPRs are referred to as "two-temperature valves." EPRs are a necessity when one compressor is connected to multiple evaporators.

An EPR is used when one evaporator operates at low temperature (a freezer) and the other at medium temperature (a cooler), see **Figure 17-8.** With two

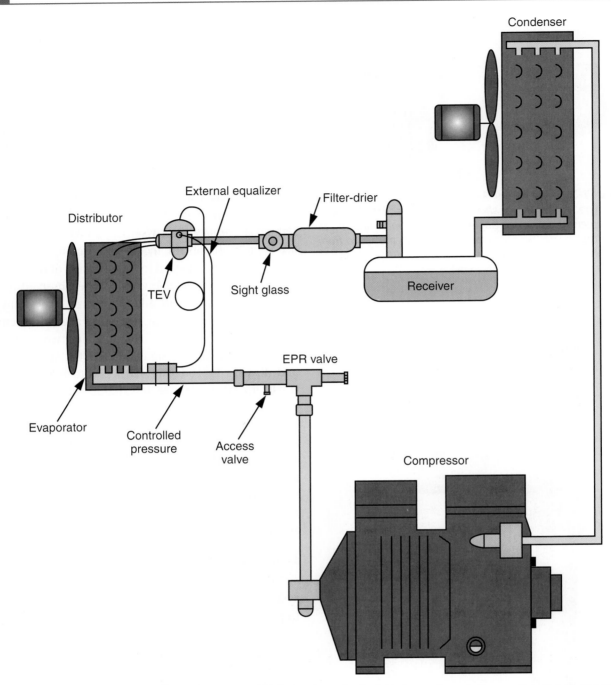

Figure 17-5. *An EPR valve is an inlet pressure regulator and can be installed anywhere in the suction line to prevent the evaporator pressure from going too low.*

evaporators, it is unnecessary to use an EPR for the freezer. The operating controls (thermostat or pressure control) are set for controlling the *freezer*, while the *cooler* temperature is controlled by the EPR. With a two-temperature system, the warmer evaporator becomes a "parasite," feeding off the colder system.

Check valves are used on two-temperature applications in which two evaporators operating at different temperatures are connected to one compressor. A check valve is installed in the suction line of the coldest evaporator, as shown in Figure 17-8. The check valve prevents warm suction gas from flowing into the

freezer evaporator during the off cycle. Warm suction gas entering the freezer evaporator would create problems in the colder evaporator, possibly causing the system to ***short cycle*** (turn on and off rapidly).

17.4.2 Multiplexed Evaporators

EPRs make it possible to operate multiplexed evaporators at different temperature levels. ***Multiplexed evaporators*** refer to a system in which several evaporators are connected to a single compressor. See **Figure 17-9.** Normally, such evaporators operate at nearly the same temperature. However, EPRs make it

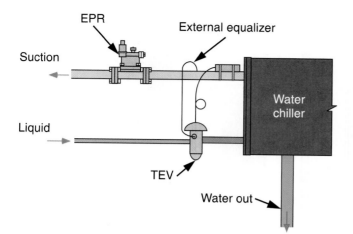

Figure 17-6. *An EPR valve is often used to prevent evaporator freeze-up. The EPR serves to prevent reduction of evaporator pressure below the EPR setting.*

possible to operate each evaporator at a different temperature. There is no limit to the number of evaporators that may be connected to a single compressor.

17.4.3 Disadvantages of EPR Valves

Using several evaporators means a large compressor is needed, leading to the danger of system failure. Replacing a large compressor is time-consuming and expensive, and considerable loss of product could occur during downtime.

Compressor capacity under reduced load conditions often must be controlled. Under low-load conditions, the large compressor reduces suction pressure too rapidly, causing short cycling to occur. Compressor cylinder unloaders may be required, or a hot gas bypass valve may be installed to meter small amounts of hot gas into the suction line during low-load periods.

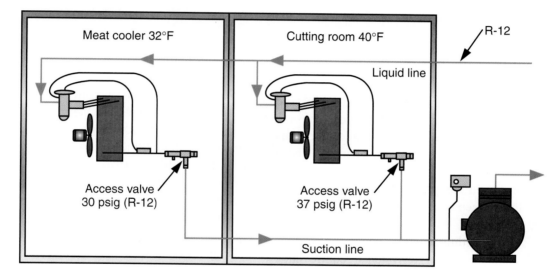

Figure 17-7. *A two-temperature system using two EPR valves. Using R-12 refrigerant, the meat cooler has a temperature of 32°F and a pressure at the access valve of 30 psig. The cutting room has a temperature of 40°F and a pressure of 37 psig.*

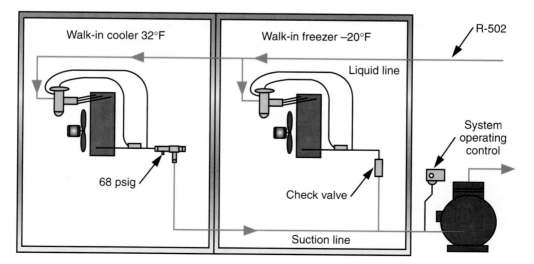

Figure 17-8. *A two-temperature system using one EPR and the system operating control to regulate temperature in two areas. An EPR keeps the walk-in cooler at 32°F (a pressure of 68 psig), while the system operating control keeps the freezer at –20°F. The refrigerant is R-502.*

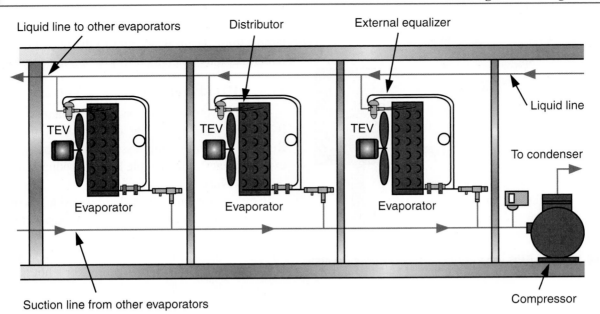

Figure 17-9. *Multiplexed evaporators, each with its own EPR valve to permit individual control of pressure (and temperature). Any number of evaporators can be connected to a single compressor and individually controlled in this way.*

17.4.4 Multiple Compressors

The problems caused by using multiplexed evaporators are easily overcome by using multiplexed *compressors.* Rather than a single 60-hp compressor, three 20-hp units are mounted side-by-side. The units are connected to a common suction manifold and a common discharge manifold. See **Figure 17-10.**

With multiplexed compressors, if one compressor fails, the others keep the system operational until repairs are completed. All compressors will operate under a full-load condition, which occurs when most of the EPRs are open. Large systems require large EPRs, such as the one in **Figure 17-11.**

The load is reduced when the EPRs throttle down, causing a reduction in suction line pressure. When the suction line pressure is reduced to a predetermined point, one of the compressors is turned off. Each compressor is cycled off in rotation, according to

reducing suction line pressure. The individual pressure switches restart each compressor in response to increasing suction line pressure.

Many supermarkets use the multiplex system with EPRs and multiplexed compressors, such as the one in **Figure 17-12.** One rack of four compressors operates on R-502 and controls all the low-temperature evaporators, such as frozen food cases, walk-in freezers, and ice cream cases.

Another rack of four compressors operates on R-12 or R-22 and controls all medium-temperature evaporators, such as walk-in coolers and dairy, produce, and fresh meat cases. On four-compressor systems like these, one compressor operates constantly because the load is never reduced to the point the compressor can shut off. The other three compressors cycle on and off in response to changing load conditions signaled by pressure changes in the suction manifold.

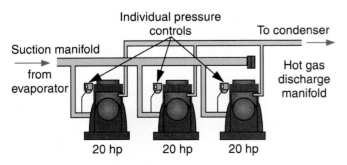

Figure 17-10. *A multiplexed compressor system overcomes problems associated with multiplexed evaporators. Individual pressure controls cycle the compressors on as needed. If a compressor fails, the system will continue to operate with the remaining units.*

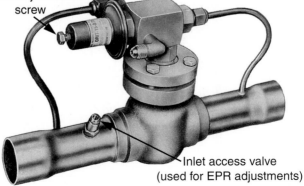

Figure 17-11. *A large evaporator-pressure-regulating valve of the type used with large multiple-compressor installations. (Sporlan Valve Co.)*

17.5 Solenoid Valves

Solenoid valves, **Figure 17-13,** are commonly used in refrigeration and air conditioning systems. The valves serve a vital role in the automatic flow control of refrigerant, water, or other gases and liquids involved in the cooling process.

17.5.1 Solenoid Operating Principles

A **solenoid valve** is an electrically-operated valve that is either fully open or fully closed, depending upon whether electricity is on or off. Solenoid valves contain an electromagnet surrounding a plunger connected to a short valve stem. The electromagnet, **Figure 17-14,** is a coil of insulated copper wire wrapped around a soft iron core with a round hole through the middle. Electricity flowing through the coil produces magnetism in the iron core. When the electrical flow is stopped, magnetism also stops. The magnetic coil is replaceable and available in different voltages.

The valve body section contains the valve seat, tubing connections, and a free-floating plunger connected to the valve stem, **Figure 17-15.** The plunger and valve stem are enclosed in a metal tube extending upward from the

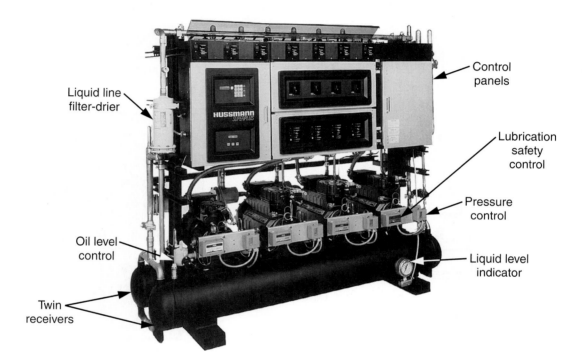

Figure 17-12. *A multiple-compressor system of the type used to meet the varied refrigeration needs of modern supermarkets. (Hussmann Corporation)*

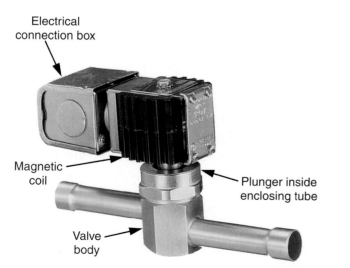

Figure 17-13. *A typical solenoid valve. A magnetically-operated plunger opens or closes the valve to regulate the flow of fluid. Valves are supplied with either flare or sweat-type connections. (Sporlan Valve Co.)*

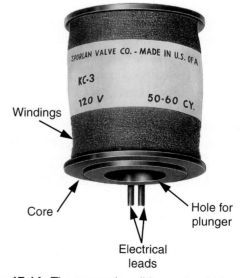

Figure 17-14. *The magnetic coil in a solenoid has a hole in the middle for the plunger. Coils are replaceable. (Sporlan Valve Co.)*

valve body. A special lock nut holds the enclosing tube to the valve body for a leaktight seal. The heavy plunger is free to move up and down in the enclosing tube, which fits into the hole in the electromagnet.

When the coil is magnetized by the flow of electricity, an opposing magnetic field is induced in the plunger. The plunger is pulled upward into the center of the magnetic field, **Figure 17-16.** Lifting of the plunger opens the valve. When the coil is deenergized, the plunger drops to the closed position. Sometimes, a small spring is located on top of the plunger to assist in closing.

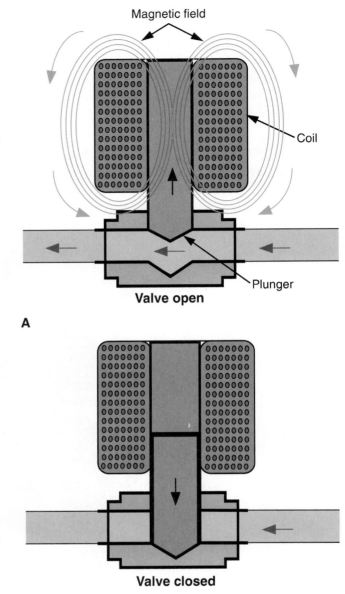

Valve open

A

B

Valve closed

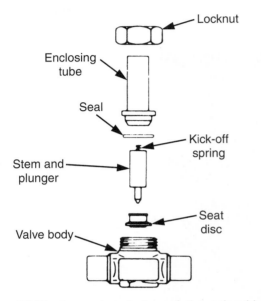

Figure 17-15. *An exploded view of the solenoid valve assembly. The magnetic coil, which surrounds the enclosing tube, has been removed for clarity. (Sporlan Valve Co.)*

Figure 17-16. *Operation of the solenoid. A—Electricity energizes the coil, creating a magnetic field that pulls the plunger up into the enclosing tube. The valve is open. B—With no electricity flowing to the coil, the plunger is in the "normally closed" position, blocking flow. The valve is closed.*

17.5.2 Solenoid Installation

For proper plunger operation, a solenoid valve must be mounted in an upright position or at no more than a 45° angle. Special solenoid valves that use a closing spring, instead of relying on gravity, can be mounted horizontally (sideways). Most solenoid valves are **normally closed (NC)** until energized, but valves that are **normally open (NO)** and close when they are energized are also available. Direction of flow is indicated by an arrow or the word "IN" on the valve body.

Before brazing, solenoid valves should be disassembled to protect internal components from heat damage. Merely removing the coil is not sufficient. Remove all the internal solenoid valve parts illustrated in **Figure 17-17.** It is important to avoid overheating the valve body. Brazing should be performed quickly, and the flame should be directed away from the body. Avoid allowing excess alloy to flow into the valve.

17.5.3 Solenoid Valve and Pumpdown Cycle

A common application of a solenoid valve is the automatic pumpdown cycle. See **Figure 17-18.** A solenoid valve is installed in the liquid line between the sight glass and the TEV. Location may be just before the TEV (evaporator area) or just after the sight glass (condensing unit area).

A temperature control (thermostat) is mounted within the cooled space and electrically connected to the solenoid coil. When temperature within the space rises to the cut-in point on the thermostat, a switch closes and completes an electrical circuit to the

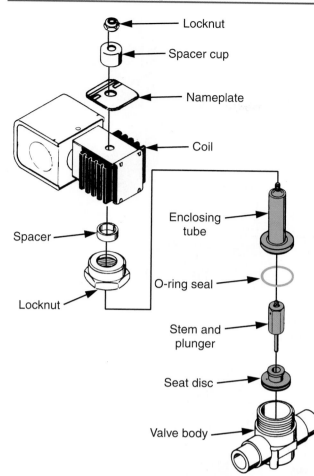

Figure 17-17. *Internal parts of the solenoid, shown in this exploded view, should be removed before brazing. This will protect them from heat damage when the body is brazed to the tubing. (Sporlan Valve Co.)*

solenoid coil. The solenoid coil is energized, opening the valve and allowing refrigerant flow.

The system continues to operate until the temperature in the space to be cooled is lowered to the cut-out point on the thermostat. At the cut-out point, the switch opens, stopping the flow of electricity to the solenoid coil. The solenoid plunger drops to the closed position. Refrigerant flow stops, but the compressor continues running.

On a ***pumpdown cycle,*** the compressor is controlled by a pressure-sensitive switch that senses suction line pressure. The pressure control is adjusted to stop the compressor just before suction pressure reaches a state of vacuum. The switch also turns the compressor on again when suction pressure rises to the proper setpoint.

Pumpdown run cycle

When the solenoid coil is energized by the thermostat, the valve opens and permits liquid refrigerant flow to the TEV. Refrigerant flowing into the evaporator causes the suction pressure to rise. The pressure-sensitive motor control turns the compressor on. Normal refrigeration continues until product temperature is lowered to the cut-out point on the thermostat.

Pumpdown off cycle

When the thermostat switch opens, electrical flow to the solenoid coil stops, and the valve closes. Refrigerant flow to the TEV stops. See **Figure 17-19.** The compressor continues to run until the low-side pressure is reduced to the cut-out point on the pressure-sensitive motor control (about 1 psig to 2 psig or 7 kPa to 14 kPa). The compressor should stop before suction pressure reaches a vacuum, resulting in removal of all refrigerant from the low-pressure side following each run cycle. Some refrigeration and air conditioning systems *require* such a pumpdown cycle to avoid high suction pressure at start-up. Pumpdown also ensures good oil return to the compressor and evacuates the evaporator prior to defrost. Otherwise, heat from electric defrost heaters would create very high suction pressure at start-up.

17.6 Crankcase Pressure Regulators ◆

Crankcase pressure regulator (CPR) valves are used to protect the compressor from excessive suction pressure, which most often occurs at start-up. Like the EPR valve, the CPR is a pressure-limiting device. The CPR, however, is constructed to limit *outlet* pressure. See **Figure 17-20.**

The CPR is a holdback-type valve. As shown in **Figure 17-21,** it is mounted in the suction line very close to the compressor. CPR valves control or *limit* pressure leaving the valve, protecting the compressor from excessive pressure. When suction pressure exceeds the CPR setting, the valve throttles the pressure down to the setting. When suction pressure is reduced to the valve setting, the CPR is fully open.

High suction pressure typically occurs after a defrost cycle when the evaporator is very warm. If such high suction pressure is allowed to enter the compressor, it will cause very high condensing pressure. This condition overloads the compressor, causing it to short cycle on the overload until the suction pressure is lowered to normal conditions. Short cycling increases the amount of oil pumped out of the crankcase. Severe oil pumping causes broken piston rods, dry bearings, and damaged valve reeds.

17.6.1 CPR Adjustments

The CPR is adjusted by turning a screw located under the cover cap. When adjusting the valve, the gauge manifold is installed on the suction service valve (SSV) to obtain accurate readings of pressure *entering the compressor.* Most CPRs are set to limit entering suction pressure at 18 psig to 20 psig (124 kPa to 138 kPa). Pressure adjustments can only be made when suction pressure (on the inlet side of the

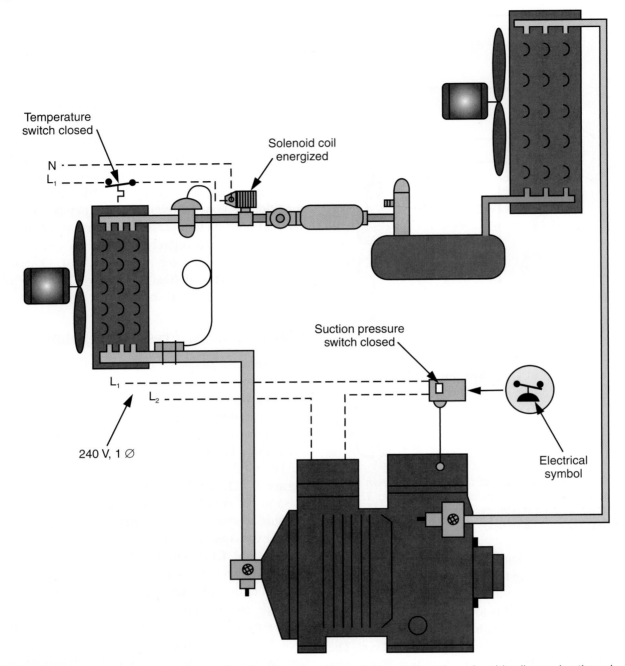

Figure 17-18. *In the pumpdown "run" cycle, the closed temperature switch energizes the solenoid coil, opening the valve and permitting refrigerant to flow. The suction pressure switch controls the compressor.*

valve) *exceeds* the valve setting. Accurate adjustments cannot be made when the valve is fully open, a condition that occurs whenever low-side pressures are below the valve setting.

17.7 Discharge Bypass Valves ◆

Refrigeration and air conditioning systems are usually designed to provide a specific capacity at maximum load conditions. Large systems must make provision for operation at *reduced* load conditions to maintain temperature and **humidity** (air moisture) control. Discharge bypass valves are one method of

controlling humidity and load variations. Some systems control capacity by compressor cylinder unloading or by using two or more compressors (multiplexing). On multiple compressor systems, individual compressors are turned off as the load and suction pressure is reduced.

It is often necessary to limit minimum evaporator pressure during periods of low load. This prevents frost or ice formation on the evaporator and avoids operating the compressor at low suction pressures. Without such **capacity control,** the compressor would cycle on and off and rapidly reduce its lifespan. Such on-off control of air conditioning systems permits

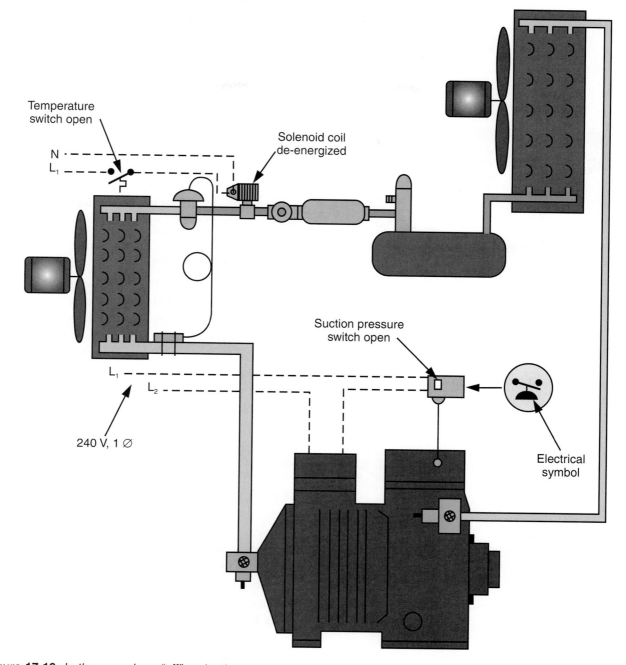

Figure 17-19. *In the pumpdown "off" cycle, the open temperature switch cuts current to the solenoid coil, closing the valve and stopping refrigerant flow. The compressor keeps running until the pressure switch opens (just before suction line pressure reaches a vacuum). This results in all refrigerant being removed from the low side of the system after each run cycle.*

wide temperature variations and does a poor job of controlling humidity. Human comfort requires control of both temperature and humidity.

17.7.1 Valve Application

A practical and economical solution to the problem is to *bypass* a small portion of hot discharge gas directly into the low-pressure side. The amount of hot gas bypassed is controlled by installing a modulating control valve, called a **discharge bypass valve** or **hot gas bypass valve.** See **Figure 17-22.** The valve opens on a decrease in suction pressure and can be set to automati-

cally maintain a desired *minimum* evaporator pressure. Hot gas only bypasses when suction pressure is reduced to the valve setting. This method places a "false load" on the evaporator to maintain a minimum suction pressure.

17.7.2 Valve Location

The discharge bypass valve is installed in a branch line off the hot gas discharge line as close to the compressor as possible. Refer to Figure 17-22. The hot gas is piped to the low-pressure side at the evaporator inlet or directly into the suction line. The evaporator inlet is the preferred location because the

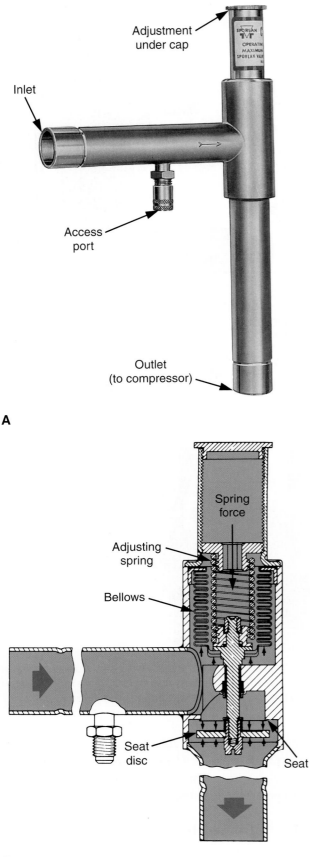

A

B

Figure 17-20. *CPR valve. A—The valve spring adjustment is under the cap. Note the direction-of-flow arrow on the body. B—A cutaway view shows component locations. (Sporlan Valve Co.)*

thermostatic expansion valve responds to the increased superheat leaving the evaporator and provides the liquid required for desuperheating. The evaporator also serves as an excellent mixing chamber for the bypassed hot gas and the liquid-vapor mixture from the expansion valve. Oil return from the evaporator is improved since the velocity in the evaporator is increased by the hot gas.

A solenoid valve should be installed ahead of the bypass valve. The solenoid valve is energized only during the run cycle. This permits the system to operate with automatic pumpdown and guards against leakage during the "off" cycle.

17.7.3 Valve Operation

Discharge bypass valves respond to changes in suction pressure. When suction pressure is above the valve setting, the valve remains closed. As suction pressure drops below the valve setting, the valve begins to open. As with all modulating valves, the amount of the opening depends upon what is being controlled (in this case, the suction pressure). As the suction pressure continues to drop, the valve opens wider until the limit of the valve piston stroke is reached. However, in normal applications, the amount of pressure change is seldom sufficient to open the valve to the limit of its stroke.

Discharge bypass valves are usually rated at 6 degrees Fahrenheit (3 degrees Celsius) temperature change from closed position to rated opening. A typical application would be a low-temperature compressor designed to operate at a minimum suction gas temperature on R-22 of –40°F (0.5 psig). The required suction temperature at normal load conditions is –30°F (5 psig). A discharge bypass valve would be selected to start opening at 3 psig (the pressure equivalent to –34°F) and bypass enough hot gas at 0.5 psig (–40°F) to prevent further decrease in suction pressure.

Many large air conditioning systems use hot gas bypass to prevent evaporator freeze-up and short cycling of the compressor. The hot gas bypass valve begins opening at 60 psig to 61 psig (using R-22). See **Figure 17-23** for a valve used in this type of application.

17.8 Desuperheating Thermostatic Expansion Valve

In applications where hot gas must be bypassed directly into the suction line, an extra thermostatic expansion valve is required. The valve is called a *desuperheating TEV,* or liquid injection valve. See **Figure 17-24.**

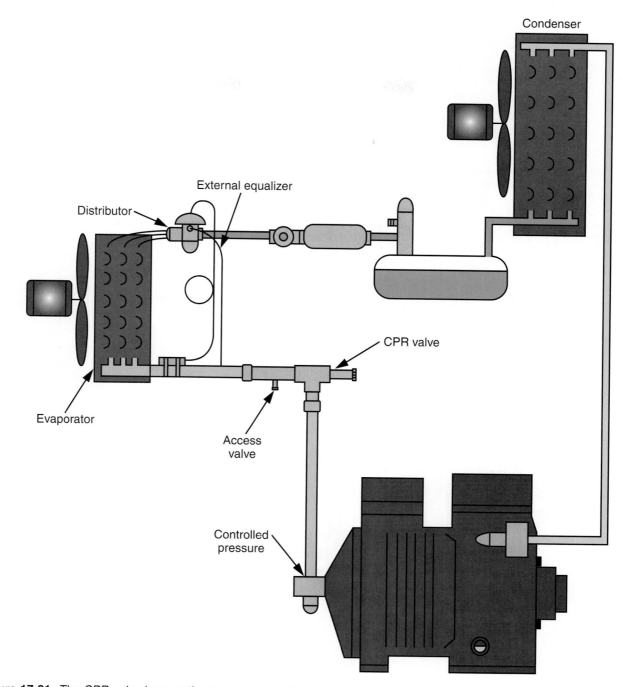

Figure 17-21. *The CPR valve is an outlet pressure controller and may be installed anywhere in the suction line just before the compressor. It protects the compressor by limiting the pressure entering the compressor.*

The purpose of this valve is to supply enough liquid refrigerant to cool the hot discharge gas to the recommended suction temperature. Cooling protects the compressor from becoming overheated by high suction gas temperatures. Hot gas entering the suction line must be desuperheated to remain within the suction temperature limits specified by the compressor manufacturer. Again, a solenoid valve is installed in the liquid injection line to permit an automatic pumpdown cycle and prevent leakage during the off cycle.

Proper mixing of the hot gas, liquid injection, and suction gas is important. The compressor must be protected at all times against high suction temperatures due to superheat and liquid floodback. Proper mixing of these "additives" to the suction line must be obtained. Several mixing methods are available, but with the generally recommended method, the discharge gas (and liquid injection) enters the suction line at an angle of flow against the direction of gas flow in the suction line. See **Figure 17-25.**

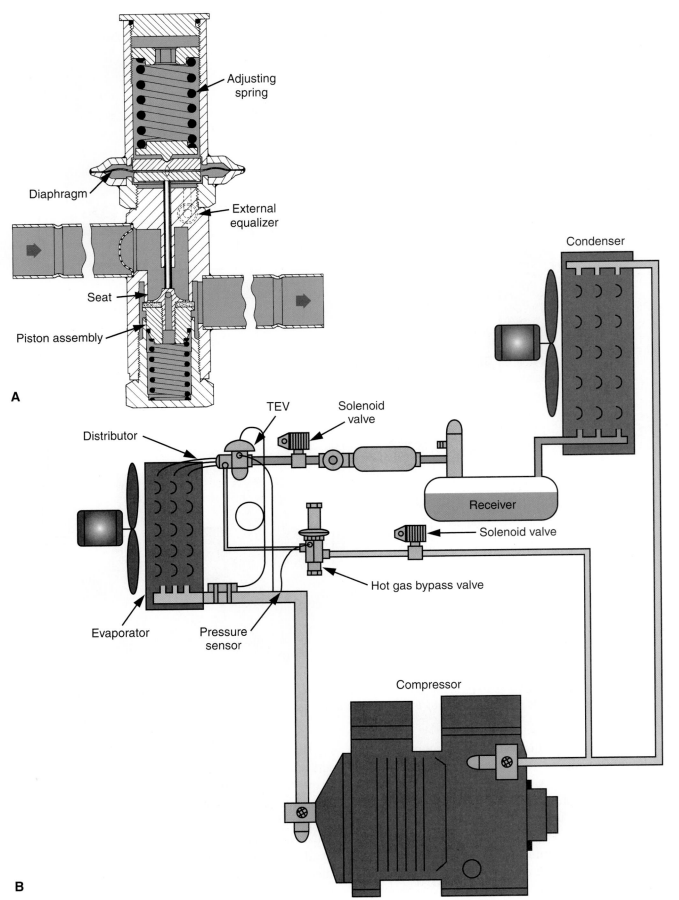

Figure 17-22. *Hot gas (discharge) bypass valve. A—A cutaway view of an adjustable hot gas bypass valve. (Sporlan Valve Co.) B—Location of a hot gas bypass valve.*

Arranging a bypass directly into a suction accumulator is often a convenient way to obtain proper desuperheating of suction gas. In any event, introducing the hot gas and liquid into the suction line with separate connections is generally not recommended.

17.9 Controlling Head Pressure ──◆

Many commercial systems place the air-cooled condensing unit outdoors, often on the roof. Placing the unit outdoors conserves space inside the building and utilizes outdoor ambient air for condenser cooling. Outdoor units perform very well, but cold winter temperatures present problems because of low head pressure. Many air conditioning systems must operate during the winter months to provide controlled environments. In such applications, provision must be

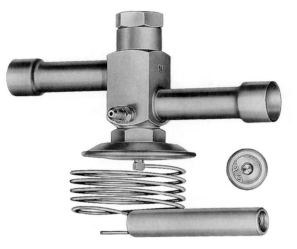

Figure 17-23. *A hot gas bypass valve is used in large systems to prevent evaporator freeze-up. (Sporlan Valve Co.)*

Figure 17-24. *A desuperheating thermostatic expansion valve is used for hot gas bypass systems. (Sporlan Valve Co.)*

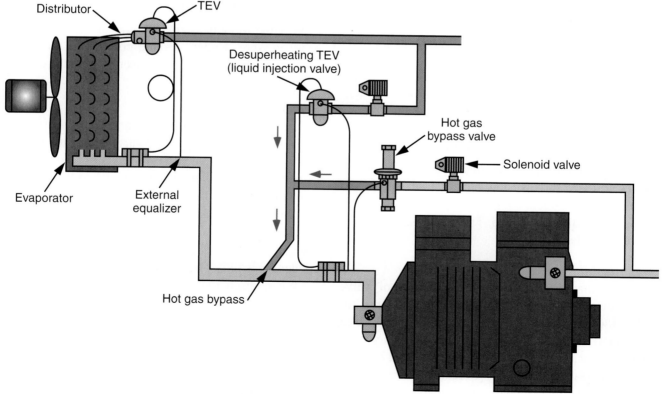

Figure 17-25. *An arrangement for liquid injection into the suction line to desuperheat the hot gas entering the line.*

made to maintain proper head pressure during periods when ambient temperatures are low.

The high-side pressure must be sufficient to maintain proper pressure drop across the thermostatic expansion valve. For proper operation of the TEV, certain minimum head pressures are required. Valves used with R-12 require a head pressure of 100 psig (690 kPa); 180 psig (1242 kPa) is required for R-502, and 200 psig (1380 kPa) for R-22. When head pressure drops too low, the expansion valve cannot feed properly. When head pressure goes down, suction pressure also goes down, resulting in poor refrigeration or an iced evaporator.

Low ambient temperatures cause low head pressure on air-cooled condensing units, observable with a gauge manifold. The gauges reveal low pressures on each side, and the sight glass shows many bubbles (flash gas). The system contains the proper amount of refrigerant for summer operation but not for winter operation.

Air-cooled condenser capacity is selected to provide for efficient operation at 90°F (32°C) ambient airflow. Cold ambient airflow greatly increases condenser capacity, resulting in low head pressure. The low head pressure is not adequate to push liquid out of the receiver and maintain correct pressure drop across the expansion valve. Liquid backs up in the receiver and lies there.

A *temporary* solution to the problem is adding enough refrigerant to partially fill the condenser with liquid. Filling the condenser with liquid reduces its capacity, causing head pressure to rise. When the outdoor ambient temperature rises, however, head pressure will rise drastically as a result of decreased condenser capacity. The system will cycle on the overload (or high-pressure safety control). The excess refrigerant must be removed from the system to return the condenser to normal capacity. Head pressure will return to normal — until winter arrives again.

17.9.1 Solving Low Head Pressure Problems

Two common types of **head pressure control valves** are used to solve low-head-pressure problems caused by low ambient temperatures. One method uses a single nonadjustable valve. The other uses two fully adjustable valves.

Nonadjustable valve

The nonadjustable head pressure control valve will maintain 100 psig (690 kPa); head pressure for R-12, 180 psig (1242 kPa) for R-502, and 200 psig (1380 kPa) for R-22. It is a three-way modulating valve controlled by discharge pressure. See **Figure 17-26.**

When head pressure drops to the valve setting, the valve will direct some (or all) hot gas to bypass the condenser and travel directly to the receiver,

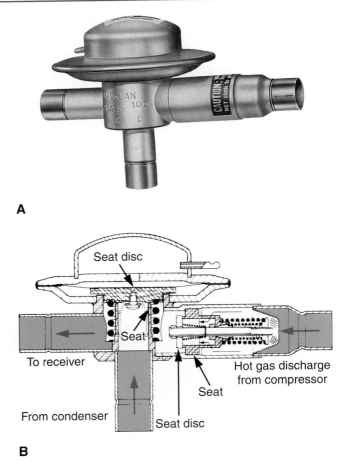

A

B

Figure 17-26. *Head pressure control. A—A nonadjustable head pressure control valve. B—A cutaway view of the valve shows the valve components. (Sporlan Valve Co.)*

Figure 17-27. The valve modulates in any position to control the amount of bypass that maintains proper head pressure. Under extreme conditions, the liquid receiver will act as the condenser as a result of cold ambient temperatures.

Principle of valve operation

The charge in the valve dome exerts pressure on top of the diaphragm, which controls the seat disc. The downward diaphragm pressure tends to open the seat disc, permitting gas to bypass directly to the receiver. The discharge (or condensing) pressure pushes upward on the diaphragm. The upward pressure tends to close the valve and prevent bypassing. Proper discharge pressure keeps the valve closed so the condenser is fully operational.

When discharge pressure falls below diaphragm pressure, the valve opens and allows discharge (bypass) gas to be metered into the receiver. This creates a higher pressure at the condenser outlet. When the valve opens to permit hot gas to bypass, the condenser outlet line is restricted, causing liquid to back up inside the condenser. When the valve is fully open, the condenser outlet is fully closed. The beauty

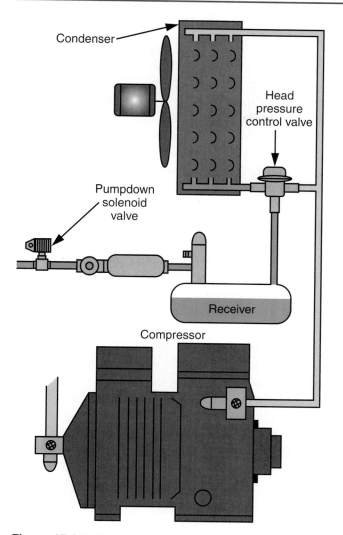

Figure 17-27. *When head pressure drops sufficiently, the valve allows hot gas to bypass the condenser and go directly to the receiver.*

of this valve is that it can modulate and, thus, control the amount of liquid in the condenser, according to ambient temperature. Head pressure is maintained by controlling condenser capacity.

All head pressure control valves require the use of a liquid receiver large enough to hold the total charge for both summer and winter operation. Excess refrigerant is needed during cold weather to permit proper flooding of the condenser. These valves are only used on systems having a thermostatic expansion valve.

Manufacturers of outdoor condensing units usually include a head pressure regulating valve and an oversized liquid receiver. All other components are sized for normal operation. The liquid receiver must be over-sized because good refrigeration practice requires the total system charge not exceed 75% of receiver capacity. The excess charge needed for winter operation must be included in the receiver capacity.

Head pressure control valves should not be used on systems without a liquid receiver nor on systems

where the receiver is too small. A lack of receiver storage space will cause liquid to back up in the condenser. When ambient temperatures begin to rise, discharge pressure will become excessively high due to liquid refrigerant reducing condenser capacity.

Adjustable valves

The adjustable head pressure control valve is combined with a differential valve to provide an improved system of head pressure control. See **Figure 17-28.** These valves provide another method of maintaining constant receiver pressure during all types of low ambient conditions. These head pressure valves can be used for common refrigerants such as R-12, R-22, R-134a, or R-502 because they are adjustable over a range of 65 psig to 225 psig (4485 kPa to 15 525 kPa).

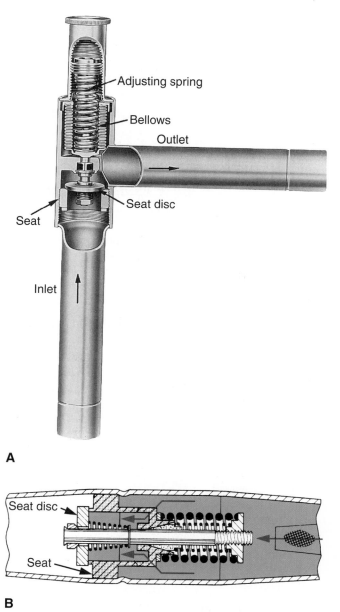

Figure 17-28. *Head pressure control. A—A cutaway view of an adjustable-type head pressure control valve. B—A cutaway view of a differential valve. (Sporlan Valve Co.)*

As shown in **Figure 17-29,** the adjustable valve is located in the liquid line between the condenser and the liquid receiver. The differential valve is located in a hot gas line bypassing the condenser.

Valve operation

During periods of low ambient temperature, head pressure falls until it approaches the setting on the adjustable valve. The valve then throttles down, restricting flow of liquid from the condenser. Liquid backs up in the condenser, reducing condenser capacity and raising condensing pressure.

Since it is actually *receiver* pressure that must be maintained, the bypass line with the differential valve is needed. The **differential valve** is preset for 20 psig and opens when the pressure difference between the receiver and hot gas discharge exceeds that value. Hot gas flowing through the bypass line heats up the cold liquid being passed by the adjustable valve. As a result, the liquid is warm when it reaches the receiver and has sufficient pressure to properly feed the expansion valve.

With proper refrigerant charge, the two valves modulate refrigerant flow automatically to maintain proper receiver pressure regardless of outside ambient temperatures.

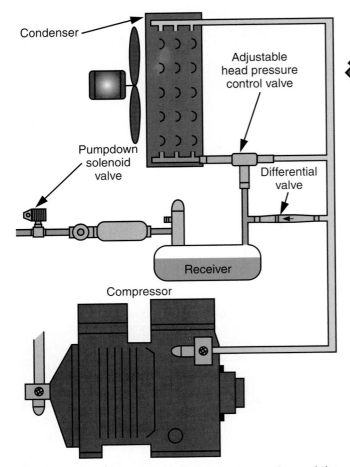

Figure 17-29. *The adjustable head pressure valve and the differential valve work together to control receiver pressure.*

17.10 Brazing Valves to Tubing ◆

When brazing special purpose valves into the system, wrap the valve body with a *wet* cloth to keep the body temperature below 250°F (121°C) and prevent heat damage. The torch tip should be large enough to avoid prolonged heating of the copper connections. Always direct the flame away from the valve body, and perform the brazing operation quickly and carefully. Be careful not to allow melted alloy to flow into the valve body.

◆ Summary

Service technicians frequently encounter special purpose valves, so it is important that you understand the intent and operation of such valves. Each valve serves a special purpose in controlling the circulating refrigerant. Any malfunction by the special purpose valves greatly affects system operation. Installing and adjusting these valves is quite easy, but you must understand their purpose and how they affect the system.

Hand valves are installed for the benefit of the service technician and are useful for isolating sections of the system. Electrically actuated solenoid valves are frequently used to control refrigerant flow. Special pressure regulating valves serve to protect system components and improve system efficiency.

◆ Test Your Knowledge

Please do not write in this text. Write your answers on a separate sheet of paper.

1. Name three types of hand valves.
2. Check valves are used to prevent _____ flow.
3. EPRs are used to keep _____ pressure from falling below a setpoint.
4. EPRs are also referred to as _____ valves.
5. EPRs are installed anywhere in the _____ line.
6. *True or false?* EPRs are adjusted before adjusting TEV superheat.
7. Where are multiplex systems used?
8. A solenoid valve operates on _____.
9. When energized, are solenoid valves normally open or closed?
10. What is a common use for a solenoid valve?
11. CPR valves limit pressure entering the _____.
12. Name two methods to control reduced load conditions.
13. What is the purpose of a desuperheating TEV?
14. What is minimum head pressure for R-12? R-22? R-502?
15. Will cold ambient airflow over the condenser increase or decrease condenser capacity?

Troubleshooting Flow Controls

Objectives

After studying this chapter, you will be able to:
- ❑ Diagnose and correct problems in AEV or TEV systems.
- ❑ Identify and correct capillary tube problems.
- ❑ Recognize and properly use various types of hand valves.
- ❑ Diagnose EPR and CPR valve problems and make necessary adjustments.
- ❑ Identify problems and make repairs needed for correct operation of solenoid valves.
- ❑ Diagnose and correct problems in head pressure control and discharge bypass valves.

Important Terms

amperage	low evaporator load
case	migrate
condensate water drain	overcharging
erratic	slugging
evaporator freeze-up	troubleshooting
isolated	undercharged

18.1 The Troubleshooting Process

Troubleshooting is the logical gathering of information to make an intelligent decision about a system problem. Once the decision has been made, the technician can take the necessary steps to correct the problem. When troubleshooting the various flow controls used in refrigeration systems, the service technician *must* know how the devices operate and how to make corrective adjustments. Such knowledge eliminates mistakes and avoids *creating* problems where none previously existed. Flow controls are very important to the efficient operation of the system and only rarely need adjustment or replacement. Problems with refrigerant flow controls most often can be traced to faulty installation or poor service techniques.

Ultimately, every system problem can be reduced to one of two basic causes:
- ❑ Refrigerant is not moving properly.
- ❑ Refrigerant quantity is insufficient (the system is undercharged).

18.2 Troubleshooting the AEV

The automatic expansion valve (AEV) prevents evaporator pressure from dropping below a specified level. The compressor must be able to remove excess pressure from the evaporator. The AEV is a very reliable system component. It cannot control evaporator pressure during the off cycle.

18.2.1 Undercharged System

An *undercharged* system is one that suffers from a lack of refrigerant. Proper system pressures cannot be maintained without sufficient refrigerant in the system. Many AEV systems do not have a sight glass for determining refrigerant charge. In such systems, troubleshooting the AEV is best performed with the gauge manifold, taking pressure readings from both sides of the system. Trouble with the AEV is usually indicated when evaporator pressure becomes too low. The valve may be defective, the valve setting may be wrong, or the system may be low on refrigerant.

The valve cannot feed properly unless sufficient liquid is present. When the system is undercharged, a low-side pressure reading will be deceptive because the valve will try to maintain a constant pressure. The valve may be wide open, but sufficient liquid is not available to increase evaporator pressure. See **Figure 18-1.**

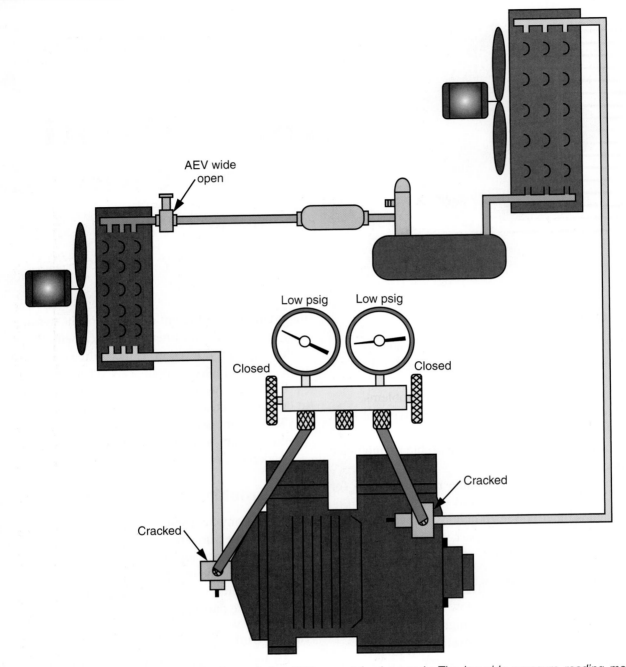

Figure 18-1. *When the system is undercharged, the AEV cannot feed properly. The low-side pressure reading may be deceptive because the valve will try to maintain a constant pressure.*

An undercharged system must be checked for leaks. When found, the leaks must be repaired and the repairs tested. It will be necessary to change the filter-drier, then evacuate and recharge the system.

18.2.2 Erratic Valve Settings

A valve that is *erratic* does not maintain a particular setting for any length of time. If the refrigerant charge is correct, but the AEV requires periodic readjustment, then the valve should be replaced. An erratic valve is difficult to identify; a thorough examination of other possibilities is required before determining the valve is erratic.

18.2.3 Contaminants

Erratic valve operation can also be caused by a foreign substance (ice, sludge, or wax) fouling the valve seat. Ice, sludge, or wax can be detected by turning off the system and warming the valve with hot water. *Do not use a torch!* Heat will loosen or melt the foreign matter so it passes through the valve. Then, the AEV will work properly until the contaminants once again foul the valve seat. A contaminated system can be cleared up by changing the filter-drier as often as necessary to restore cleanliness.

18.2.4 Poor Cooling, Low Head, Normal Suction

The customer may complain, "The system is not cooling right. The product temperature is high and the system runs all the time." The gauges reveal low head pressure, but suction line pressure is acceptable. Is the system low on refrigerant? Maybe. Check for bubbles in the sight glass. The sight glass may show a full charge. Adding more refrigerant (**overcharging**) will raise the head pressure but will not correct the problem. See **Figure 18-2.**

The compressor may have defective internal valve reeds or a broken piston rod. The compressor will run but is unable to draw gas and compress it properly; it is performing at half its normal capacity. The low-side pressure will be normal because the AEV is choked down to match the lower capacity of the compressor. The low rate of refrigerant removal from the evaporator will result in slow lowering of product temperature.

A defective compressor is usually indicated by *lower-than-normal* discharge pressure and *higher-than-normal* suction-line pressure. Troubleshooting the customer's problem was somewhat difficult because the AEV maintained a normal suction pressure.

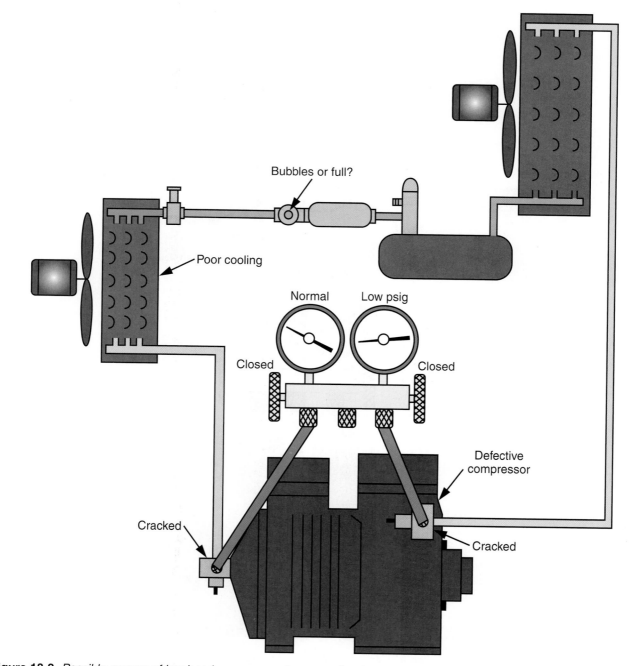

Figure 18-2. *Possible causes of low head pressure and poor cooling include undercharging and a defective compressor.*

18.2.5 Poor Cooling, High Head, Normal Suction

Poor cooling with long running time is frequently caused by high head pressure. See **Figure 18-3.** Pressure drop across the AEV is excessive. It results in extra "flash gas," which increases evaporator pressure. High head pressure causes the AEV to throttle down, resulting in reduced cooling capacity and long running time. The AEV will maintain constant suction pressure, but head pressure will be higher than normal. Correcting head pressure will alleviate the problem, permitting the valve to feed subcooled liquid.

18.3 Troubleshooting the Capillary Tube System

Troubleshooting the capillary tube system requires keen use of your eyes, ears, and hands since the system does not have service valves. Saddle valves are a last resort. It is normal for a capillary tube ("cap tube") to make a gurgling sound as liquid enters the evaporator. It pays to become familiar with this sound — a gurgle indicates liquid, and a whistle indicates vapor.

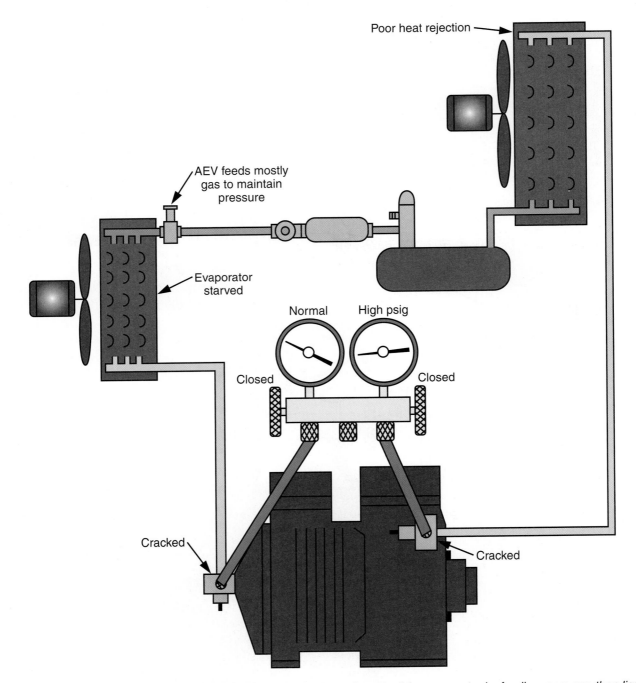

Figure 18-3. *High head pressure causes the AEV to throttle down, "starving" the evaporator by feeding more gas than liquid. The cooling capacity of the system is reduced significantly.*

18.3.1 Poor Cooling, High Head, High Suction

The most frequent problems encountered with a capillary tube system are a dirty condenser or a burned-out condenser fan. Each of these situations causes high head pressure which, in turn, causes high suction pressure (and temperature) and excessive running time. When normal head pressure is restored, suction line pressure is also corrected, and the evaporator pressure/temperature returns to normal. This permits proper cooling and shorter running time. See **Figure 18-4.**

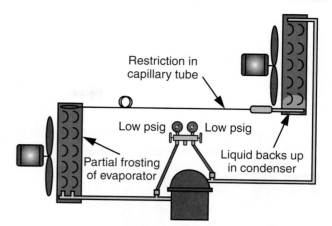

Figure 18-5. *A restriction in the capillary tube causes undercharging of the evaporator. Adding refrigerant may overcharge the system, leading to a burned-out compressor motor.*

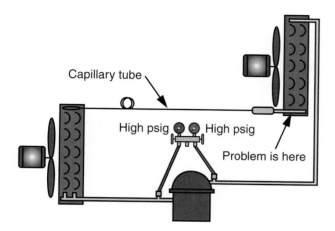

Figure 18-4. *A dirty condenser or a defective condenser fan are major causes of high head pressure in a capillary tube system. High head pressure can also be caused by a refrigerant overcharge, air in the system, or high suction line pressure.*

18.3.2 Undercharged Evaporator

In a capillary tube system, *low* head pressure can be caused by an undercharged evaporator. The condition is revealed by partial frosting of the evaporator. The frost line is incomplete due to improper flooding of the evaporator (and low suction pressure).

The most common (but *incorrect*) method of overcoming an undercharged evaporator is to add refrigerant; the result is usually compressor motor burnout due to system overcharge.

The fact that a *partially restricted capillary tube* can cause an evaporator to be undercharged is a surprise to most technicians. The restriction causes liquid to back up in the condenser and become subcooled. The subsequent low suction pressure results in low head pressure. See **Figure 18-5.**

Overcharging the system increases head pressure because more liquid backs up inside the condenser, reducing its capacity. The higher head pressure overcomes the partial restriction, producing higher suction line pressure. However, because condenser capacity has been reduced by the overcharge, higher suction pressure produces even higher head pressure.

Overcharging a system only compounds the problem of a restricted capillary tube. Product temperature can be reduced to the proper level, although head pressure will be excessive. When the system shuts off, more time is needed to reduce the excessive head pressure before restarting.

The problem occurs when the thermostat requires a system restart before head pressure is reduced. The compressor shortcycles on the overload until head pressure is reduced. Continued shortcycling leads to compressor motor burnout.

An undercharged evaporator can be very misleading. To rely on appearance and a low-pressure reading only adds to the problem. The most reliable method of diagnosing a restriction is to include a high-pressure reading and check **amperage** (current flow) to the compressor. Low amperage indicates the compressor is loafing; high amperage indicates an overload. (Electricity and amperage readings are explained in Chapter 22.)

18.3.3 Poor Cooling, Low Head, High Suction

Suction pressure on capillary tube systems changes along with discharge pressure. When one goes up, the other goes up. When one goes down, the other goes down. The *only* exception is the effect of a defective compressor — high suction pressure with low discharge pressure. See **Figure 18-6.** Defective valve reeds or a broken piston rod render the compressor inefficient and unable to compress refrigerant properly. Compressor replacement is indicated.

Defective compressor symptoms are a cool condenser (low head pressure) and a warm suction line (high suction pressure), combined with low power usage and a hot gas discharge line that is very hot. The compressor is loafing (indicated by a low amperage draw), and very little refrigerant is moving through the

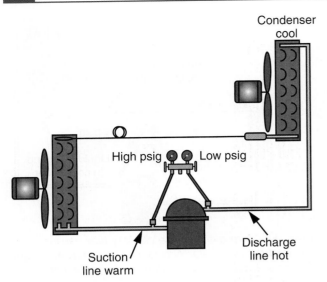

Figure 18-6. *A defective valve reed or a broken piston rod in the compressor can cause poor cooling performance. Suction line pressure will be high and discharge pressure low.*

system. The hot gas is mostly moving back and forth between the compressor and the hot gas discharge line. The compressor discharge valve reed is supposed to prevent backflow of gas into the compressor after being discharged. Broken or damaged compressor valve reeds result in an inefficient compressor. Compressor replacement is indicated. (Compressor operation is explained in more detail in Chapter 19.)

18.3.4 Working with Capillary Tubes

Capillary tubes are easily damaged. Kinks, sharp bends, and vibration should be avoided. It is sometimes necessary to cut off one or two inches where the tube was brazed into the filter-drier. When cutting a capillary tube, care must be exercised to avoid closing the hole. See **Figure 18-7.** Proper procedure is to use a file or knife blade to make a notch (cut) in the tubing. Then, the tubing is bent back and forth at the notch until the tubing breaks off.

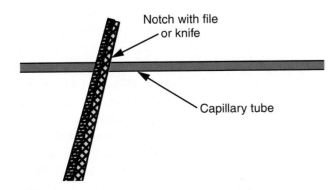

Figure 18-7. *The proper method for cutting a capillary tube, without closing the hole, is to first notch the tube with a knife or file. Then, bend the tubing back and forth until it breaks at the notch.*

Extreme care must be exercised when brazing the capillary tube to a filter-drier. As shown in **Figure 18-8,** insert several inches of the capillary tube into the filter-drier; then, carefully crimp the copper tube onto the capillary tube. Be careful not to crimp the capillary tube. Perform brazing quickly, and use very little alloy. The alloy must not be permitted to follow the capillary tube inside the filter-drier and plug the capillary opening. The possibility of the alloy plugging the capillary tube is increased by excessive or prolonged heating of the joint or by using excess alloy.

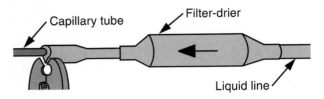

Figure 18-8. *Before brazing, carefully crimp the copper tube of the filter-drier over the capillary tube. Do not pinch the capillary tube shut. Use only a small amount of alloy when brazing.*

18.3.5 Moisture

Moisture cannot be tolerated in a capillary tube system. One drop of moisture freezing inside the capillary tube will completely plug the tiny passage through the tubing. The system will continue to run, but liquid refrigerant will not move past the ice plug. The evaporator will become very warm, and low side pressure will drop to a vacuum. Head pressure will go down as a result of low suction pressure. If the system is turned off until the ice plug melts, it will operate perfectly for a period of time (until the drop of moisture freezes inside the capillary tube again).

Proper service technique will prevent contamination when working on capillary tube systems. Any time a capillary tube system is opened for repairs (such as fixing a leak), it *must* be thoroughly evacuated and the filter-drier replaced prior to recharging. Capillary tube systems allow no margin for error.

18.4 Troubleshooting the TEV———◆

Proper thermostatic expansion valve (TEV) performance is easily determined by measuring superheat at the evaporator outlet. So, checking superheat is the first step in diagnosing TEV performance. A starved evaporator results in high superheat. Excessive liquid flooding of the evaporator causes low superheat. Although both these symptoms point to improper TEV superheat control, the actual cause is usually elsewhere. Always check for proper installation of the thermal bulb and external equalizer.

Water, or a mixture of water and sludge, freezing at the valve seat can lock the TEV closed, open, or partly open. This problem greatly affects proper flooding of the evaporator and can cause high superheat, low superheat, or no superheat.

Shut the system off, and use warm water (*never* a torch) to warm the TEV valve body and melt the obstruction. Restart the system, and recheck superheat. The valve should return to normal operation. Check the moisture indicator, and change the filter-drier.

18.4.1 High Superheat

An undercharged system prevents the TEV from feeding properly and results in high superheat. The situation is signaled by a hissing sound at the valve, bubbles in the sight glass, low suction pressure, and low head pressure. Gas bubbles in the sight glass can also be caused by noncondensables (air) in the system or by a partially restricted filter-drier.

18.4.2 Restricted Filter-drier

A partially restricted filter-drier becomes cold and possibly frost -covered. A restricted filter-drier is caused by a screen plugged with contaminants, *not* by moisture. When a filter-drier has absorbed all the moisture it can hold, additional moisture passes through the filter-drier and freezes at the valve. Moisture cannot plug or restrict a filter-drier.

18.4.3 Improperly Seated Valve

Dirt, sludge, or foreign material lodged in the valve seat may prevent the TEV from seating properly. Liquid refrigerant may pass through the valve during the off cycle and fill the evaporator with refrigerant. (If the seat leak is severe, the valve will feed too much refrigerant during the run cycle.) A valve seat leak is detected by a gurgling or hissing sound during the off cycle. Also, the sight glass may indicate continued flow for a long period after shutdown. Disassemble the valve to determine if dirt or foreign material is responsible for the leaky seat. Replace the TEV if the seat is damaged.

18.4.4 Dead Thermal Bulb

If the thermal bulb loses its charge, the TEV cannot open. The capillary tube connecting the thermal bulb to the valve diaphragm is often a cause of trouble. For example, excess vibration causes the tube to become work-hardened and break off. Vibration is normally caused by airflow from the evaporator fan. Vibration of the coils against one another, or against the evaporator, can wear a hole in the soft capillary tube. The TEV must be replaced when the thermal bulb loses its charge.

18.4.5 Low Evaporator Load

Low evaporator load describes a condition in which the heat load cannot reach the evaporator surface. Low evaporator load causes the evaporator to become too cold. See **Figure 18-9.** Under a low-evaporator-load condition, a lesser amount of liquid is needed to maintain proper superheat, so the TEV restricts flow. Restricted flow causes low suction line pressure, a colder evaporator, and normal superheat. Low evaporator load can be caused by insufficient air over the evaporator, undersized blowers, dirty air filters, frost formation on the evaporator, or low temperature of the entering air. Correct the condition responsible for low heat load.

18.4.6 Liquid Migration

If a condensing unit is located outside, and the ambient outdoor temperature is colder than the indoor evaporator, gaseous refrigerant will **migrate** (travel) to the compressor during the off cycle (since the compressor is colder than the evaporator, and heat travels from hot to cold).

The cold ambient temperature also causes vapor inside the compressor to condense to a liquid. Upon restarting, the compressor must pump *liquid* rather than *vapor* (a condition called **slugging**). Liquid slugging can cause severe compressor damage. This particular situation is not caused by the TEV.

Crankcase heater

Any compressor located in cold ambient temperatures should be equipped with a crankcase heater, **Figure 18-10.** The device is an electric heater, fastened to the crankcase, that keeps the oil warm under cold conditions. Warm oil keeps condensation from forming in the crankcase and prevents liquid migration.

Most condensing units located in areas where cold ambient temperatures occur are equipped with a liquid line solenoid valve for automatic pumpdown. The pumpdown removes refrigerant from the low-pressure side before system shutdown.

18.5 Troubleshooting Multiplexed Evaporators

Troubleshooting a multiplexed evaporator system is fairly easy; the problem can usually be isolated to a particular area. For example, if one **case** (individual cooling system) is having problems, and other cases on the system are functioning normally, the problem is inside the malfunctioning case. See **Figure 18-11.** Each case contains an evaporator, TEV, fans, condensate water drain, defrost heaters, and other components. A malfunction of any of these components will create problems within one case only.

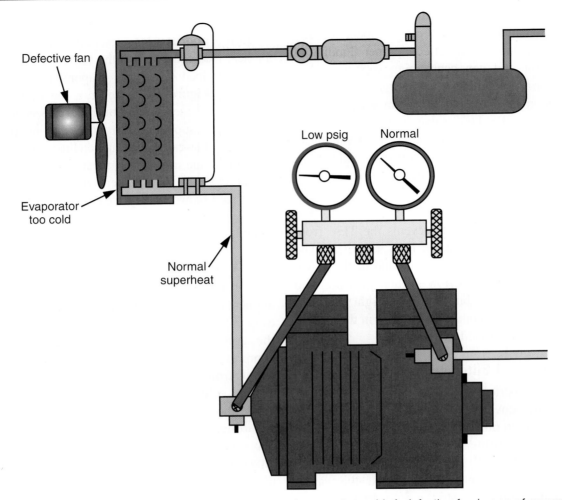

Figure 18-9. *A low evaporator load causes the evaporator to become too cold. A defective fan is one of several possible causes of insufficient airflow past the evaporator.*

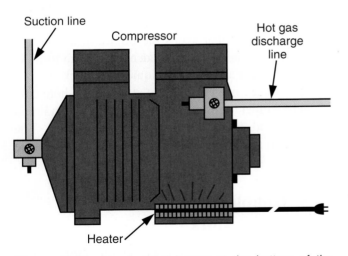

Figure 18-10. *An electrical heater at the bottom of the compressor warms oil in the crankcase under cold ambient conditions. This keeps condensation from forming and eliminates damaging slugging by preventing liquid migration.*

Proper airflow over the evaporator and product must be maintained at all times. Improper airflow will cause the product temperature to rise because the heat load is not reaching the evaporator. Reduced heat load causes the TEV to choke down and create a very cold evaporator. A colder evaporator accumulates more frost, further restricting airflow.

Each case is supplied with a ***condensate water drain,*** which can become plugged. Accumulated water inside the case can cause several problems: it can leak onto the floor, cause fan blades to break off, or burn out fan motors. Accumulated water freezing in frozen food cases can cause damage to equipment and product. Failure of the *defrost heater* will permit the evaporator to accumulate excessive frost. (Defrost systems are explained in Chapter 27.)

18.5.1 Evaporator Freeze-up

The term ***evaporator freeze-up*** describes an evaporator that has become totally restricted by frost and ice buildup. The TEV cannot recognize freeze-up and continues to maintain a constant superheat, regardless of evaporator condition or temperature. Severe freeze-ups may result in floodback problems due to a very cold evaporator.

Freeze-ups are normally caused by poor loading of the evaporator resulting from inadequate airflow. The

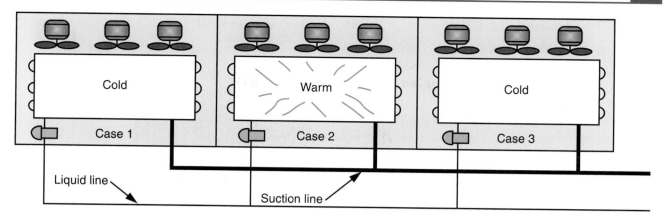

Figure 18-11. *In a multiplexed evaporator system, problems can usually be isolated to one case. In this example, there is a problem in Case 2, but Cases 1 and 3 are operating normally.*

airflow problem may result from a burned-out fan, dirty filters, or other causes. Remove all frost and ice from the evaporator, and correct the loading problem. The TEV will function properly when the evaporator is clean and airflow is sufficient.

18.5.2 Condensing Unit Problems

If all cases in a multiplexed evaporator system have the same problem, a condensing unit malfunction should be suspected. A malfunction could involve high head pressure, loss of refrigerant, a blown fuse or circuit breaker, compressor problems, a defrost timer problem, or shutdown by a safety control. (Motor controls and safety controls are explained in Chapter 26.)

18.6 Using Hand Valves ⬥

The use of hand valves is illustrated by a hand shut-off valve installed immediately after the sight glass. The filter-drier and sight glass are normally located immediately after the liquid receiver outlet service valve (LRSV). See **Figure 18-12.**

When the LRSV is frontseated (closed), and the hand valve is closed, the filter-drier is *isolated* (blocked off) and easily changed. Pumpdown and time-consuming service procedures are eliminated.

18.6.1 System Pumpdown

Another method of changing the filter-drier is to manually pump down the system. The high-pressure gauge (red) hose is connected to the liquid receiver service valve port. With the system running, frontseat the LRSV. The compressor removes all refrigerant from the low-pressure side, all the way back to the LRSV. All refrigerant is stored inside the receiver, condenser, and hot gas discharge line.

The compressor should be stopped just before the system enters a vacuum state. If the compressor is not stopped before a vacuum is achieved, "kill" the

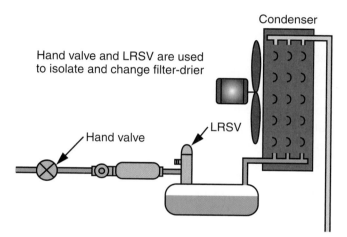

Hand valve and LRSV are used to isolate and change filter-drier

Figure 18-12. *Installing a hand shut-off valve immediately after the sight glass allows easy isolation and changing of the filter-drier.*

vacuum by briefly cracking the LRSV. Release just enough liquid refrigerant from the receiver into the low side of the system to raise its pressure to 1 psig or 2 psig (7 kPa or 14 kPa).

Close the hand valve only when the liquid line pressure is slightly above a vacuum. The filter-drier is isolated so it can be removed and replaced. The small amount of refrigerant vapor remaining inside the filter-drier escapes, preventing atmospheric air from entering. A replacement filter-drier should be at hand so removal and replacement can be done quickly. After replacement, the LRSV is opened but the hand valve is not. Before opening the hand valve, the filter-drier and sight glass are pressurized with liquid and checked for leaks using a soap bubble solution. When connections are leak-free, both valves are fully opened, and the gauge hose is removed. The system is then turned on.

Any hand valve that does not open or close properly should be replaced. Repairs are not recommended. Refrigerant leaking from around the valve stem is rare but indicates a defective valve.

18.7 Troubleshooting EPRs

When installing EPRs with brazed connections, wrap the valve with a wet cloth to keep the valve body temperature below 250°F (121°C). The torch tip should be large enough to avoid prolonged heating of the connections. Overheating can be reduced by directing the torch flame away from the valve body. Be careful to avoid allowing molten alloy into the internal parts of the valve.

Some EPRs are hermetic and cannot be disassembled for inspection and cleaning. They must be replaced if they become inoperative. If an EPR fails to open or close properly, or if it will not adjust, solder or other foreign material lodged in the port may be at fault. It is sometimes possible to dislodge these materials by turning the adjustment nut all the way out with the system running. Once cleared, the valve can be reset.

If the EPR develops a refrigerant leak around the spring housing, it probably has been overheated during installation, or the bellows has failed due to severe compressor pulsation. When the bellows fails, the EPR closes until the inlet pressure becomes greater than the outlet pressure *plus* the spring pressure. The EPR then becomes a *pressure differential* valve, and the evaporator pressure changes according to varying suction pressures. This produces an evaporator pressure higher than desired and cannot be corrected. The valve must be replaced.

18.8 Troubleshooting Solenoid Valves

Troubleshooting a solenoid valve is primarily limited to determining whether or not the coil is energized. The coil cannot be energized unless proper voltage is applied to it — no voltage, no magnetism. (Electricity and methods for checking voltage are explained in later chapters.)

To quickly determine if the coil is energized, hold the tip of a small steel screwdriver close to the top center of the coil. See **Figure 18-13.** If the coil is energized, magnetism will pull the screwdriver tip downward. If the coil is not energized, the screwdriver will not be pulled toward the coil. The magnetic field is not very strong, so a *small* screwdriver is required for proper detection.

Solenoid valves are noiseless during normal operation. A steady humming sound (an "alternating current" hum) indicates the coil is not correctly positioned or not properly anchored to the valve. Solenoid valves normally make a sharp "snap" sound when energized, a sound caused by the plunger striking the top of the enclosing cylinder. To check solenoid operation, energize and deenergize the coil while listening for the snapping sound.

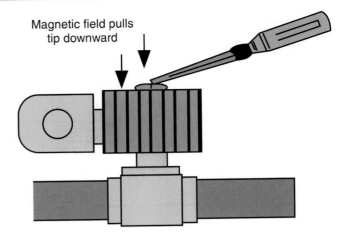

Magnetic field pulls tip downward

Figure 18-13. *A small steel screwdriver can be used to test for an energized solenoid coil. Magnetic force will pull the screwdriver tip downward if the coil is energized.*

Coil burnouts are rare and normally caused by improper wiring or improper voltage. A burnout can occur, however, if the plunger is not free to be pulled upward when magnetic force is exerted. The problem usually is caused by a damaged enclosing tube. A burnout also occurs if the coil is removed from the valve while energized.

Failure of the valve to open or close properly is usually caused by foreign material lodged in the valve seat. Failure can also be caused by a valve body that is deformed by overheating. A defective solenoid coil is easily replaced, but a defective valve body requires replacement of the entire valve.

18.9 Troubleshooting Head Pressure Controls

There are several possible causes for malfunctions in systems with head-pressure controls. These causes can be difficult to isolate from each other. Troubleshooting is not a "guessing game." Detailed information must be obtained to narrow the problem to a specific area or component. Operating temperatures and pressures must be known before system problems can be pinpointed.

Figure 18-14 lists the causes of the most common head-pressure-control malfunctions and recommended solutions to those problems.

Troubleshooting head-pressure-control valves to determine how the valve is feeding can sometimes be accomplished with your hands as tools. During very cold ambient temperatures, the valve should completely bypass the condenser; the bypass line to the liquid receiver should feel warm. When ambient temperatures are high, the valve should be fully closed, and the bypass line should feel cold. Also, with

Low Head Pressure

Possible Cause	Remedy
Low charge, cannot flood condenser	Add refrigerant
Low-pressure setting on valve	Increase setting
Valve fails to close due to foreign material stuck in valve	Open valve wide to pass material. If unsuccessful, replace valve
Valve will not adjust	Replace valve
Valve fails to close due to loss of element charge	Replace valve
Valve fails to open	Replace valve

High Head Pressure

Possible Cause	Remedy
Dirty condenser	Clean condenser
System overcharged	Purge until proper head pressure is obtained
Undersized receiver	Check receiver capacity
Air in system	Purge air from system
Valve fails to adjust or open	Replace valve
Bypassing hot gas when not required	Replace valve

Figure 18-14. *Possible causes and remedies for head pressure problems.*

a proper charge in the system, the head pressure should not drop below 100 psig (690 kPa) for R-12, 200 psig (1380 kPa) for R-22, and 180 psig (1242 kPa) for R-502. In addition to checking the temperature of the copper tubing, always check pressures, and observe the liquid level in the sight glass.

A system low on refrigerant normally exhibits low head pressure. However, the head-pressure-control valve will bypass the condenser in its attempt to maintain proper head pressure. This condition is revealed by a very warm receiver and an empty or low liquid level in the sight glass. The valve is working properly, but the system suffers from an undercharge.

18.10 Troubleshooting Discharge Bypass Valves

There are two possible problems with a discharge bypass valve (DBV): failure to open and failure to close. Failure to open is revealed by low suction pressure. Failure to close is revealed by high suction pressure.

Problems with the DBV are often caused by foreign material lodged inside the valve. The valves usually can be disassembled for cleaning. Defective sensing elements are replaceable without changing the entire valve.

Summary

Troubleshooting refrigerant flow controls involves systematic gathering of information to make a decision. Hasty decisions usually result in a wrong diagnosis and a recall. For proper troubleshooting, take the time to analyze the problem. When correct information is available, the decision is obvious and guesswork is eliminated. Hands, ears, eyes, thermometers, and a gauge manifold are all useful tools for gathering information.

The technician must know how each valve operates and what purpose it serves in the system. A malfunction by one particular valve can cause problems throughout the system. It is important to narrow the problem to the valve responsible. When the source of the problem is corrected, other system components usually function properly.

This chapter has presented many methods of troubleshooting system components. Each type of system requires a certain logic and reasoning. Most common problems associated with refrigerant flow controls were covered. Methods for diagnosing and isolating system problems were explained as well.

Test Your Knowledge

Please do not write in this text. Write your answers on a separate sheet of paper.
1. Name the two basic causes of system problems.
2. List the symptoms that indicate an undercharged AEV system.
3. An inefficient compressor is usually indicated by _____ head pressure and _____ suction pressure.
4. Poor cooling with long running time usually is caused by _____ pressure.
5. *True or false?* On capillary tube systems, high head pressure causes high suction pressure.

6. Name three possible causes of an undercharged evaporator on a capillary tube system.

7. Describe the proper way to cut a capillary tube.

8. What is the first step in diagnosing TEV performance?

9. Evaporator freeze-ups are normally caused by poor _____.

10. Manual pumpdown is performed when changing the _____.

11. EPRs are easily damaged by _____ from torches.

12. Whether a solenoid coil is energized can easily be determined by using a small _____.

13. A sharp _____ sound signals a solenoid valve has been energized.

14. If the _____ is not free to move when a solenoid coil is energized, the coil can burn out.

15. Name two basic problems with discharge bypass valves.

An open-type compressor, like this large six-cylinder model, is more serviceable in the field than semihermetic types. (Vilter Mfg. Co.)

19

Compressors

Objectives

After studying this chapter, you will be able to:
- ❏ Identify five types of compressors.
- ❏ List the advantages of the compliant scroll compressor.
- ❏ Identify the three basic compressor designs.
- ❏ Discuss the operation of each type of compressor.
- ❏ Properly align pulleys and flywheels.
- ❏ Select, size, and replace V-belts.
- ❏ Identify the components of a reciprocating compressor.
- ❏ Describe the advantages of the disc-type valve design.

Important Terms

access opening	piston rings
augers	reciprocating
centrifugal	rotary
clearance pocket	rotor
compliant	rpm
crankshaft	scroll
direct drive	semihermetic
disc-type compressor	stubs
valve	valve plate
flywheel	valve reed
full-floating	vanes
hermetic	volume
orbiting	volute
piston	

19.1 What Is a Compressor?

The compressor is a mechanical pump that circulates refrigerant. It is the heart of any refrigeration system and is the most expensive and vital component of the system. A compressor malfunction will greatly affect the operation of other components. Compressor problems are diagnosed without breaking into the compressor to examine individual parts. Problems involving the compressor are considered major because of the time and expense of replacement or repair. The technician must fully understand compressor performance to eliminate confusion and wrong diagnoses.

The purpose of the compressor is to remove heat-laden, low-temperature vapor from the evaporator and compress (squeeze) it into a much smaller volume. As the *volume* of the vapor is reduced, heat is concentrated in a smaller area, resulting in a much higher temperature.

19.2 Types of Compressors

The compressor is a mechanical device driven by an electric motor. It is available in five basic types:
- ❏ Reciprocating.
- ❏ Rotary.
- ❏ Screw.
- ❏ Centrifugal.
- ❏ Scroll.

19.2.1 Reciprocating Compressor

The reciprocating compressor is the workhorse of the industry. It is used for domestic and commercial applications ranging in size from fractional horsepower to the 100-ton to 150-ton range. The *reciprocating* compressor operates much like an automobile engine. It has one or more pistons driven from a crankshaft that is turned by an electric motor. These pistons make alternate suction and compression strokes in a cylinder equipped with suction and discharge valve reeds. See **Figure 19-1**.

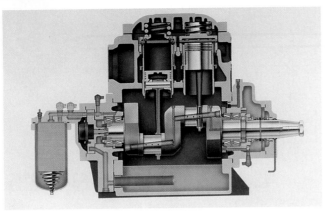

Figure 19-1. *This reciprocating compressor, shown in a cutaway view, is typical of those used widely in refrigeration applications. (Vilter Mfg. Co.)*

19.2.2 Rotary Compressor

There are two types of rotary compressors. One has blades that rotate with the eccentric (rotor); the other has stationary blades (**vanes**). Rotary compressors are so named because they operate in a rotating or circular motion. The rotary compressor is a positive displacement unit and more efficient than a reciprocating compressor. See **Figure 19-2.**

In the domestic field, rotary compressors occasionally are used in fractional horsepower sizes. Recently, the rotary has gained some use in the 1.5-ton to 5-ton residential air conditioning market. Large rotary compressors are sometimes used in low-temperature industrial systems.

19.2.3 Screw Compressor

The screw compressor uses a pair of matched **augers** (or "screws") to compress refrigerant gas. See **Figure 19-3.** The vapor enters one end of the paired augers and is compressed as it travels to the other end where it is discharged. Screw-type compressors currently in use for chilled-water systems range in size from 100 tons to 700 tons. Screw machines also are used in large-tonnage refrigeration and air conditioning systems of 20 tons and up.

19.2.4 Centrifugal Compressor

The **centrifugal** compressor is sometimes called a "turbo compressor" because it belongs to a group of turbo machines that includes fans, propellers, and turbines. See **Figure 19-4.** Refrigerant vapor enters at the center of a rapidly spinning disc with radial blades (impeller vanes) on its surface. The disc is located inside a case called a **volute.** The impellers force the vapor outward, using centrifugal force to compress the gas against the walls of the volute. A continuous flow of compressed gas exits through the discharge port. Because refrigerant flow is continuous, centrifugal compressors can handle greater volume than can other types of compressors. Centrifugal compressors currently start at the 80-ton to 100-ton range and extend to more than 8000 tons. The design and cost of centrifugal compressors make them impractical for systems of 50 tons or less.

19.2.5 Scroll Compressor

The **scroll** compressor was first patented in 1905 but only recently achieved widespread popularity. See **Figure 19-5.** The operation of this compressor involves the mating of two spiral coils (scrolls) to form a series of crescent-shaped pockets. During compression, one scroll (the *fixed* scroll) remains stationary while the other (**orbiting**) scroll *orbits* (but does not rotate) around the fixed scroll.

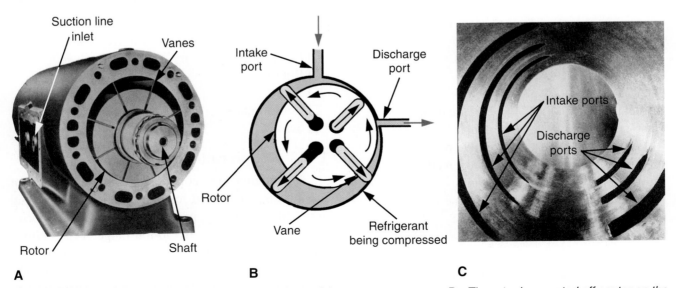

Figure 19-2. *Rotary compressor. A—Major components of the rotary compressor. B—The rotor is mounted off center on the shaft so that, as it rotates, refrigerant in the space between the vanes is compressed into a smaller and smaller volume before being discharged. C—With rotor and shaft removed, the intake and discharge ports of the compressor are visible. Note that the intake ports are much larger than the discharge ports. (Fuller Co.)*

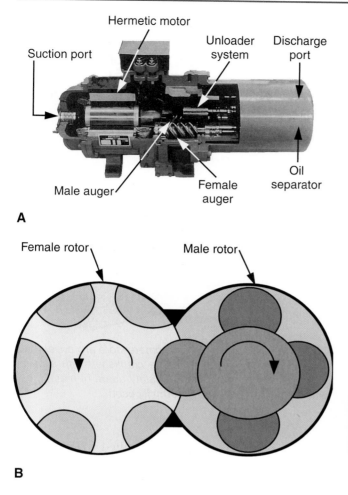

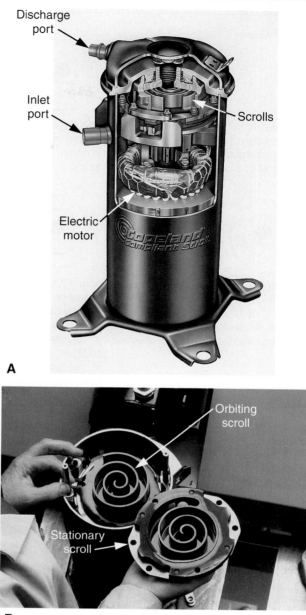

A

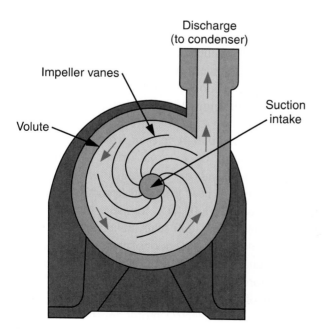

Figure 19-3. *Screw-type compressor. A—Cutaway view of typical screw compressor. It has fewer moving parts than a reciprocating compressor and operates more quietly. (Bohn Heat Transfer) B—End view of a matched pair of augers in a screw-type compressor.*

Figure 19-4. *In a centrifugal compressor, refrigerant gas is introduced at the center of the spinning spiral and forced outward by the impeller vanes. The centrifugal force exerted on the gas compresses it against the walls of the volute.*

Figure 19-5. *Scroll compressor. A—Cutaway view of a typical scroll compressor. B—Compressor partly disassembled to show the two scrolls. (Copeland Corp.)*

As shown in **Figure 19-6,** the orbiting motion traps gas in the pockets between the two scrolls. As the orbit continues, the pockets are reduced in size, compressing the gas and forcing it toward the center, where it is discharged. Several pockets of gas are compressed simultaneously during each orbit. Intake and discharge are continuous, providing smooth operation.

Advantages of the scroll compressor

Only two components, a fixed scroll and an orbiting scroll, are required to compress gas. These two components replace about 15 components in a typical piston-type compressor. Scroll compressors require no valves, while piston-type compressors require

Scroll Gas Flow

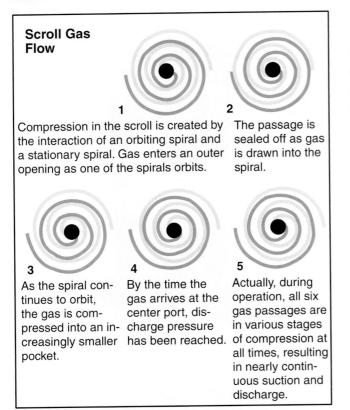

1 Compression in the scroll is created by the interaction of an orbiting spiral and a stationary spiral. Gas enters an outer opening as one of the spirals orbits.

2 The passage is sealed off as gas is drawn into the spiral.

3 As the spiral continues to orbit, the gas is compressed into an increasingly smaller pocket.

4 By the time the gas arrives at the center port, discharge pressure has been reached.

5 Actually, during operation, all six gas passages are in various stages of compression at all times, resulting in nearly continuous suction and discharge.

Figure 19-6. *Operation of the scroll compressor. (Copeland Corp.)*

a discharge and suction valve for each piston. The scroll compressor is 10% more efficient than a reciprocating compressor of the same size due to elimination of the valves. Scroll compressors also eliminate the need for suction accumulators and crankcase heaters, and can be started under any load without special start kits.

Unlike reciprocating compressors, which can be badly damaged by liquid slugging, scroll compressors have superior tolerance to liquid refrigerant. The scrolls are designed to be *compliant* (yielding); they will "give" or separate from each other in the presence of liquid. The pocket of liquid is simply pushed to the discharge opening, and the scrolls return to their original position. Compliant scroll compressors also can handle debris (foreign matter) without being damaged.

Compliant scroll technology offers many advantages to meet the needs of air conditioning and heat pump systems. At least seven major manufacturers, both domestic and international, are developing and marketing scroll compressors.

19.3 Compressor Designs

There are three basic compressor designs:
- ❑ Open-type (externally driven).
- ❑ Hermetic (totally sealed).
- ❑ Semihermetic (accessible hermetic).

19.3.1 Open-type Compressor

An open-type compressor refers to a compressor that is driven by an external source of power, usually an electric motor. The motor may be direct-connected or may use a V-belt to transmit power to the compressor. See **Figure 19-7.**

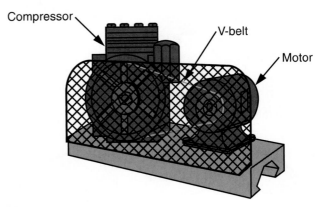

Figure 19-7. *Many open-type compressors are driven by a V-belt connecting a pulley on the crankshaft with a pulley mounted on the motor shaft. A safety guard of metal mesh over the belt and pulleys is normally used.*

Open-type compressors are bolted together with the crankshaft extending through the crankcase. A source of power is connected to the external portion of the crankshaft. The most common method is to attach a *flywheel* (pulley) to the external crankshaft. An electric motor is connected to the compressor crankshaft pulley by one or more V-belts. Another, less-common method eliminates pulleys and belts by connecting an electric motor directly to the external crankshaft.

Although open-type compressors represent a small part of the total number of compressors sold today, a great number of them are still operating in commercial refrigeration applications. New open-type designs are currently being developed and marketed.

Disadvantages of open-type compressors

The open-type, or external-drive, compressors have advantages and disadvantages. The primary *disadvantage* is the need for a leakproof seal where the compressor crankshaft comes through the crankcase. These crankshaft seals are a common source of leaks — it is not unusual to observe traces of leaking oil directly under the crankshaft seal. A small leak is usually ignored, but a large leak requires field replacement of the crankshaft seal. See **Figure 19-8.** The bearings of the electric motor must be oiled (or greased) every six months.

To avoid unusual wear and tear of V-belts and to obtain good belt grip, the motor pulley and the

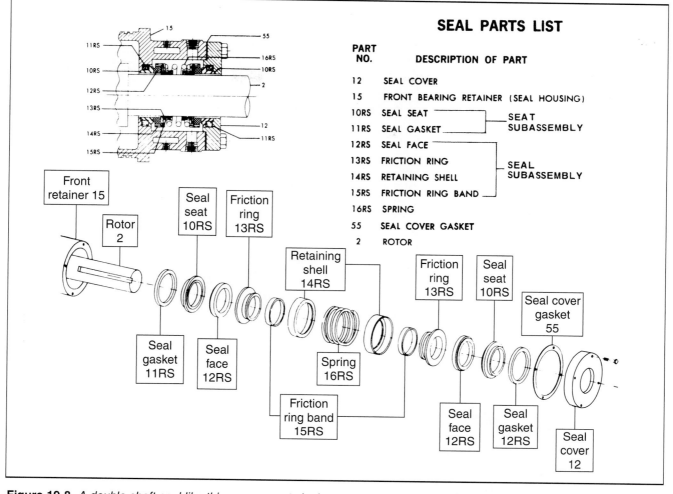

SEAL PARTS LIST

PART NO.	DESCRIPTION OF PART	
12	SEAL COVER	
15	FRONT BEARING RETAINER (SEAL HOUSING)	
10RS	SEAL SEAT	SEAT SUBASSEMBLY
11RS	SEAL GASKET	
12RS	SEAL FACE	
13RS	FRICTION RING	SEAL SUBASSEMBLY
14RS	RETAINING SHELL	
15RS	FRICTION RING BAND	
16RS	SPRING	
55	SEAL COVER GASKET	
2	ROTOR	

Figure 19-8. *A double shaft seal like this one prevents leakage, when properly installed. (Fuller Co.)*

compressor flywheel must be properly aligned. Alignment of the flywheel and pulley is checked by using a straightedge or a length of string. As shown in **Figure 19-9,** both edges of the flywheel and both edges of the pulley must touch the straightedge or string when held in a straight line. Proper alignment is obtained by moving the electric motor to conform with alignment of the flywheel.

Check alignment using a string in this way: catch one end of a string under the belt near the top of the flywheel. Turn the wheel by hand until the string crosses the rim at the 3 o'clock position, as shown in Figure 19-9A. Stretch the string across the flywheel and the motor pulley in a straight line. If the flywheel and the pulley are properly aligned, both edges of the flywheel and both edges of the pulley should barely touch the string. In the same way, a straightedge can be held against the flywheel and pulley to check alignment, if they are close enough together.

V-belt tension

Proper belt tension is important. V-belts that are too tight cause severe wear of the front motor bearing. Loose

belts slip, squeal, and wear out quickly. Proper tension is measured by using a thumb to depress the slack area between the flywheel and the pulley. The belt should deflect about one inch, as shown in **Figure 19-10.**

The 42° angle on V-belts, **Figure 19-11A,** is designed to give more gripping surface in the pulley V-grooves. The *sides* of the belt provide the gripping surface against the pulley grooves. Some types of belts are also notched on the top or bottom to improve gripping power and fit properly on smaller-diameter pulleys. The top of the belt should ride flush with the top of the pulley, not down in the groove. See **Figure 19-11B.** A belt that rides down in the groove is usually loose as a result of wearing down of the belt sides. Check belt width.

When installing V-belts, always place the belt on the motor pulley first. Systems using multiple V-belts require matched belts. Minor differences exist between individual V-belts, but matched belts are identical and wear evenly.

V-belt sizes. V-belts are available in standard widths of 3L, 4L, and 5L. A 3L belt is 3/8″ wide, 4L is 1/2″ wide, and 5L is 5/8″ wide. The numbers imprinted

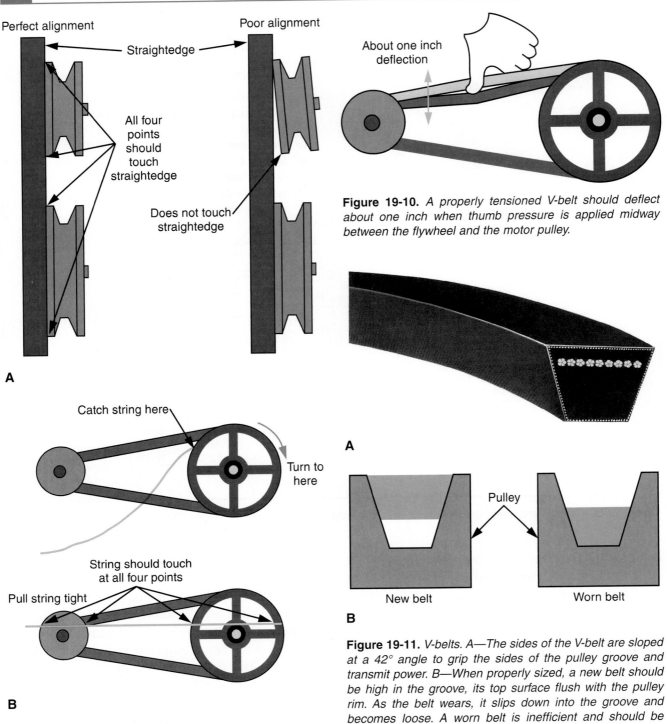

Perfect alignment · Poor alignment

Straightedge

All four points should touch straightedge

Does not touch straightedge

A

Catch string here

Turn to here

String should touch at all four points

Pull string tight

B

Figure 19-9. *A straightedge or a string can be used to check proper alignment of the flywheel and motor pulley. A—The straightedge method. B—The string method.*

About one inch deflection

Figure 19-10. *A properly tensioned V-belt should deflect about one inch when thumb pressure is applied midway between the flywheel and the motor pulley.*

A

Pulley

New belt · Worn belt

B

Figure 19-11. *V-belts. A—The sides of the V-belt are sloped at a 42° angle to grip the sides of the pulley groove and transmit power. B—When properly sized, a new belt should be high in the groove, its top surface flush with the pulley rim. As the belt wears, it slips down into the groove and becomes loose. A worn belt is inefficient and should be replaced.*

on each belt indicate the width and length of the belt. See **Figure 19-12A.** Thus, a 4L360 belt is 1/2″ wide and 36″ long. The ending zero has no meaning but appears on all belt sizes. A 5L600 belt is 5/8″ wide and 60″ long.

Some V-belts are labeled "A" or "B." These are width designations. An "A" belt is equal to a 4L (1/2″ wide). A "B" belt is equal to a 5L (5/8″ wide). See

Figure 19-12B. Length numbers follow the A or B designation. All "A" belts are two inches longer than the comparable 4L belt. For example, an A36 belt is equal to a 4L380. All "B" belts are three inches longer than the comparable 5L belt. For example, a B55 belt is equal to a 5L580.

Knowing the differences in belt sizes makes it easier to convert one numbering system to the other. A 4L420 belt is replaced with an A40. A B52 belt is replaced with a 5L550.

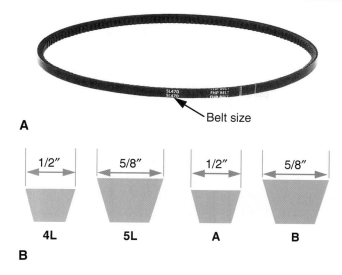

A

B

Figure 19-12. *V-belt sizes. A—Belt size is printed on the belt for easy reference. The designation 5L470 specifies that this belt has a 5/8″ (5L) width and a 47″ length. (Browning) B—Widths of 4L and "A" belts are 1/2″; 5L and "B" belts are 5/8″.*

Advantages of open-type compressors

Open-type compressors can be used with electric motors that have odd voltages or frequencies, an important factor in the world market. Open compressors are always field-serviceable. In case of a motor burnout, the motor is easily replaced.

The primary advantage of open-type compressors is flexibility. The speed can be changed, allowing a single compressor to operate with units of two or three horsepower sizes. By changing the size of the motor pulley or flywheel, the speed can be adjusted for use with different sizes of motors. By changing the speed, the same compressor can be used for high-, medium-, or low-temperature applications.

For belt-driven compressors, speed is determined by the size of the motor pulley since the flywheel is normally fixed. The speeds of the motor and compressor (in *rpm*, or revolutions per minute) are directly related to the diameters of the pulley and flywheel. The following formula is used to solve speed questions:

Motor rpm × Pulley size =
　　　　　Compressor rpm × Flywheel size

For example, what size motor pulley is needed to provide 500 rpm on a compressor with an 8″ flywheel? Motor rpm is fixed at 1750 rpm.

Motor rpm × Pulley size =
　　　　　Compressor rpm × Flywheel size

1750 rpm × Pulley size = 500 rpm × 8″
　　　　1750 × ? = 4000
　　　4000 ÷ 1750 = 2.2857
　　　　　　(approximately 2-1/4″ pulley)

Question: What size pulley is needed to increase compressor rpm to 650?

Motor rpm × Pulley size =
　　　　　Compressor rpm × Flywheel size
1750 rpm × Pulley size = 650 rpm × 8″
　　　1750 × ? = 5200
　　5200 ÷ 1750 = 2.9714 (3″ pulley)

19.3.2 Hermetic Compressor

The word **hermetic** refers to being airtight, or totally sealed against all external forces. The hermetic compressor unit contains an electric motor that is connected directly to the compressor. The entire assembly is enclosed within a welded steel housing. See **Figure 19-13.**

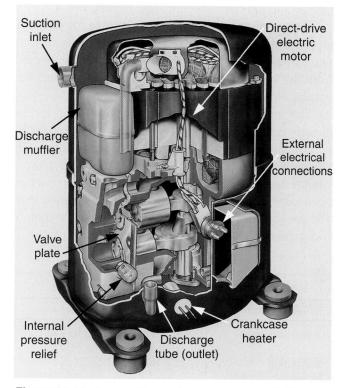

Figure 19-13. *A typical hermetic compressor unit. Since the unit is sealed, it is replaced, rather than repaired, in the field. (Tecumseh)*

The hermetic compressor has several advantages: it is small, compact, almost vibration- and noise-free, has a refrigerant-cooled motor, and has positive lubrication. There is no crankshaft seal and no belts requiring adjustment or replacement.

Since the first welded hermetic compressor was introduced by Tecumseh in 1937, the company has produced over 160 million of them. Other manufacturers have produced millions more. Hermetic compressors include many types and styles, and differ widely from each other in appearance, application, and service.

Welded hermetics dominate the residential and light commercial market because of their performance characteristics and high reliability. In size, they range from fractional horsepower units for domestic refrigerators and freezers to 20-hp units for commercial air conditioning systems.

Hermetics are engineered to perform a specific air conditioning or refrigeration task. Using hermetic compressors within their design limits will yield favorable results. Asking them to perform outside their design limits will result in poor pumping efficiency and other problems. The key to selecting the right hermetic for a particular job is the evaporator temperature of the system. Each hermetic compressor is designed for a specific evaporator temperature range. Each compressor is also designed for use with a specific refrigerant. Do not substitute. See **Figure 19-14.**

Hermetics are not field-serviceable. In the event of a motor burnout, the entire hermetic compressor unit must be replaced. The refrigerant must be recovered from the system and a filter-drier installed to clean the system of contaminants.

The motor-compressor combination (the motor directly connected to the compressor) is usually used in a vertical position and mounted on springs to absorb vibration. A small space exists between the motor-compressor and the housing shell. This interior space is open to refrigerant gas drawn in through the suction line. The gas entering the interior space helps cool the motor-compressor. It also ensures oil return to the crankcase.

The factory welds short copper-coated steel *stubs* to openings in the housing. See **Figure 19-15.** One of these is for brazing of the suction line to the compressor. Another, smaller opening in the shell, called an **access opening,** is provided with a 1/4″ or 3/8″ stub. A Schrader valve is brazed onto the stub for evacuation and charging procedures during installation. The valve permits easy connection of the low-pressure gauge from the gauge manifold. After evacuation and charging are performed, the valve is closed off with a cap nut that has a rubber seal inside. If a valve is not installed on the access stub, the stub must be brazed shut.

The *discharge stub* is welded where it passes through the housing and is directly connected to the compressor discharge. This procedure prevents the interior suction pressure from mixing with discharge pressure. No provision is made for installing gauges or obtaining access to discharge pressure.

On some low-temperature applications, the hermetic unit has a special coil of tubing located inside the base (crankcase) of the shell. The ends of this coil are copper-coated and extend through the shell as stubs. The coil arrangement, which is not open inside the compressor shell, cools the crankcase oil. See **Figure 19-16.**

On such low-temperature applications, the hot gas discharge line is run to a small air-cooled condenser (called a precooler). See **Figure 19-17.** The temperature of the hot gas is lowered in this small "condenser"

Hermetic Compressor Applications

Application	Evaporator temperature
Air conditioning	+32°F to +55°F (0°C to +13°C)
Heat pump (approved models)	–15°F to +55°F (–26°C to +13C)
High evaporator temperature	+20°F to +55°F (–7°C to +13°C)
Medium evaporator temperature	–10°F to +30°F (–23°C to –1°C)
Low evaporator temperature (normal torque motor)	–30°F to +10°F (–34°C to –12°C)
Low evaporator temperature (high torque motor)	–40°F to +10°F (–40°C to –12°C)

Figure 19-14. *Evaporator temperature ranges for hermetic compressors used in different applications. A hermetic compressor is designed for use with a specific range of evaporator temperatures.*

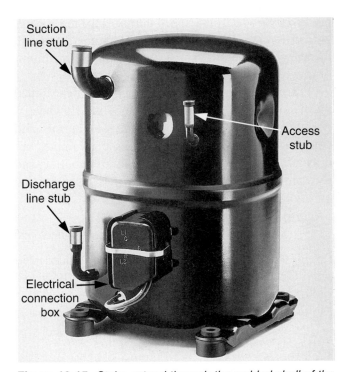

Figure 19-15. *Stubs extend through the welded shell of the hermetic compressor unit to permit necessary connections. The stubs are copper-coated for ease of brazing the connections. (Tecumseh)*

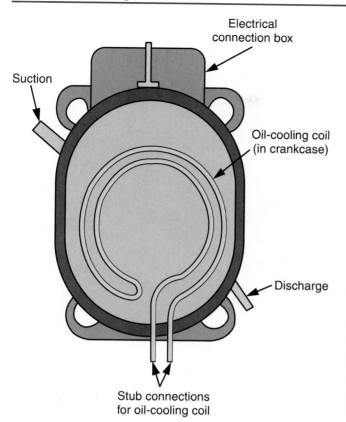

Figure 19-16. *An oil-cooling coil is used in the crankcase of some hermetic units to lower the temperature of the compressor oil. The tubing is not open inside the shell, so pressures are separately maintained.*

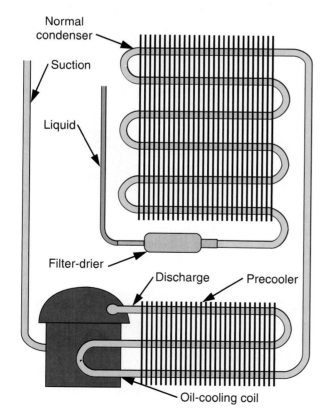

Figure 19-17. *A precooler (small, air-cooled condenser) lowers the temperature of the hot refrigerant gas sufficiently so it can, in turn, be used to cool the compressor oil. From the oil-cooling coil, refrigerant is piped to the normal condenser.*

but not enough to cause condensation to occur. Then, the warm gas is run through the oil-cooling coil in the bottom of the hermetic compressor. Since the gas is cooler than the oil, the oil gives up its heat to the gas. From the exit of the oil-cooling coil, the refrigerant gas travels to the normal condenser for actual condensation of refrigerant.

19.3.3 Semihermetic Motor-Compressor

The accessible, or *semihermetic,* motor-compressor design was pioneered by Copeland Corporation and is widely used throughout the industry for light, medium, and heavy commercial systems. See **Figure 19-18.** The semihermetic type combines all the desirable features normally associated with the open-type or the hermetic compressor. The crankshaft seal that creates problems in open-type compressors is eliminated, however. The compressor is driven by an electric motor mounted directly on the horizontal compressor crankshaft. The working parts of both the motor and the compressor are hermetically sealed within a common enclosure. The enclosure is bolted together, and gaskets are used to obtain leakproof seals between the mating parts.

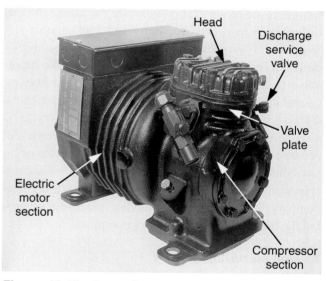

Figure 19-18. *A semihermetic compressor. Unlike the welded hermetic type, these motor-compressor units can be serviced in the field. (Copeland Corp.)*

Such a design is compact, economical, efficient, and maintenance-free. The removable heads, end bells, and bottom plates allow access for field repairs in the event of compressor damage.

Bolted semihermetics can be totally disassembled in the field and fitted with new parts as needed. This is a major advantage when dealing with large-tonnage machines where size and weight make total replacement undesirable. In most cases, however, a defective semihermetic is simply replaced and *not* repaired in the field. There are not enough skilled technicians in the industry to handle the field repair of millions of compressor installations, nor can manufacturers and dealers stock all the parts needed for field repairs. Fortunately, mass production, standardization, and improved quality have eliminated the need for field repairs in most cases. System reliability and compressor life expectancy have improved to the point where the failure rate for these devices is minimal. Occasional repairs are needed to replace a valve plate or an oil pump, but extensive overhaul is seldom required.

Reciprocating compressor operation

Refrigerant vapor enters the compressor through the suction service valve, which is bolted onto an opening in the compressor body. The suction gas is permitted to scatter throughout the compressor body. The suction service valve is often located at the electric motor end bell to take advantage of the incoming cool suction gas for motor cooling.

Compressor construction

Although variations between manufacturers do exist, the general principles of construction and operation of semihermetic motor-compressors are the same. The real difference is in quality, durability, and application.

The semihermetic motor-compressor was specifically designed for field repairs, if required. Removable heads, valve plates, stator covers, bottom plates, and housing covers allow repair access in the event of compressor damage. Understanding how these units are constructed is vital to effective troubleshooting and repair work. An accurate diagnosis must be performed *before* removing any bolts. The exploded view in **Figure 19-19** illustrates typical compressor construction details. Individual components may vary among models and manufacturers, but the basic method of assembly is similar.

Compressor head

The compressor head is divided into two chambers: one side is suction pressure, and the other is discharge pressure. See **Figure 19-20.** Openings inside the compressor body create a pathway for suction gas to reach the suction chamber in the head. The compressor head rests on top of the valve plate. The head chamber for discharge pressure is the *only* part of the compressor under high pressure. All other parts of the compressor are at low-side pressure.

A special gasket is placed between the head and the valve plate, and another between the valve plate and the compressor body. Cap screws hold these parts together. The bolts extend through the head, gaskets, and valve plate and screw directly into the compressor

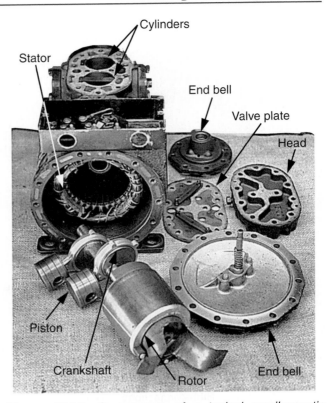

Figure 19-19. *Components of a typical semihermetic compressor, which can be disassembled for field service.*

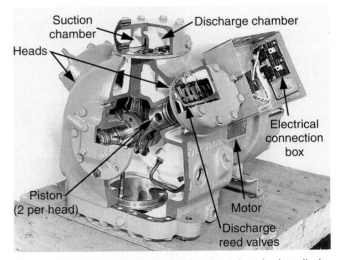

Figure 19-20. *A cutaway view of a three-head, six-cylinder compressor shows the basic construction of a semihermetic compressor. The motor is behind the cutaway, below the electrical connection box. (Carrier)*

body. The cap screws must be tightened slowly and evenly to achieve a leakproof connection.

Compressor valve plate

The **valve plate** is a specially milled steel plate between the head and the piston cylinders. See **Figure 19-21.** The valve plate is a division point between the high-side pressure and the low-side pressure. The suction and discharge valve reeds are considered part of the valve plate.

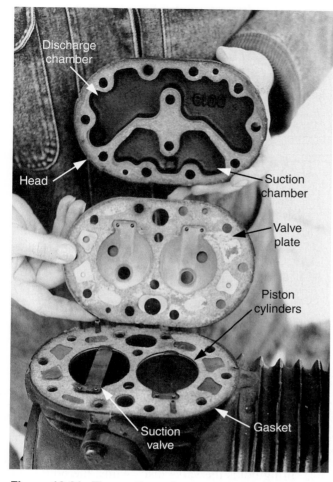

Figure 19-21. *The suction valve reed is mounted beneath the valve plate or on the cylinder wall, and moves downward to admit refrigerant gas on the piston downstroke. Gaskets above and below the valve plate provide a tight seal.*

Suction valve reed

As shown in Figure 19-21, the suction **valve reed** consists of a thin piece of steel located on top of the cylinder and under the valve plate. This thin reed is anchored at one end by two steel pins and the valve plate itself. The other end of the reed is permitted to move up and down slightly, due to a small depression cut into the cylinder wall. The piston is not permitted to rise above this small depression (or strike the valve plate), so a **clearance pocket** is formed at the top of the stroke. See **Figure 19-22.**

The downstroke of the piston causes an area of lowered pressure (vacuum) below the reed. Since pressure of the refrigerant gas in the suction chamber is higher, the end of the reed is pushed down, opening the hole in the valve plate.

Refrigerant vapor flows through the hole in the valve plate, around the suction reed, and fills the cylinder cavity. On the upstroke of the piston, pressure inside the cylinder becomes higher than the pressure in the suction chamber of the head. This presses the suction valve reed firmly against the valve plate hole,

trapping the refrigerant gas in the cylinder. Pressure inside the cylinder continues to rise as the piston moves upward, compressing the trapped gas into a smaller volume.

Compressor discharge valve reed

The discharge valve reed is mounted on top of the valve plate, inside the discharge chamber of the head. It covers one or more holes through the valve plate, opening into the cylinder. See **Figure 19-23.** High pressure inside the head chamber keeps the discharge valve reed closed. When the increasing cylinder pressure exceeds pressure in the head, the discharge reed is forced open, and the compressed gas flows into the head chamber. From the head chamber, the high pressure/high temperature gas travels through the discharge service valve and the hot gas discharge line into the condenser.

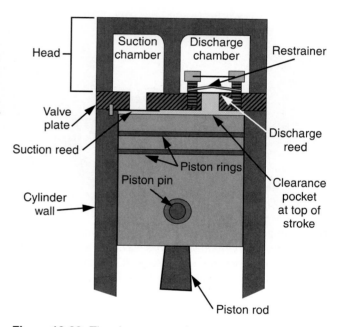

Figure 19-22. *The clearance pocket at the top of the cylinder avoids contact between the piston and the suction valve reed. Some compressed refrigerant gas remains in the clearance pocket after the discharge reed closes and is re-expanded as the piston moves downward.*

Figure 19-23. *Valve plate with discharge valve reeds and restrainers. The actual valve reed is a thin metal plate held in place by the bolts and restrainers (the thicker metal, three-fingered pieces shown). The restrainers prevent excessive flexing and damage to the reed from liquid slugging. (Carrier)*

The discharge reed usually has restrainers that help prevent the reed from bowing too far. These devices protect the discharge valve reed when the compressor is slugging oil or liquid refrigerant.

Each piston cylinder must be equipped with suction and discharge reeds. On larger cylinders, two or more of each may be used.

Disc-type compressor valve

The *disc-type compressor valve* was designed by Copeland Corporation and marketed under the trade name Discus®. It has been very successful in new commercial refrigeration and air conditioning applications. Other than the new design of the valve plate and head area, the Copeland Discus is a conventional semi-hermetic compressor. See **Figure 19-24.**

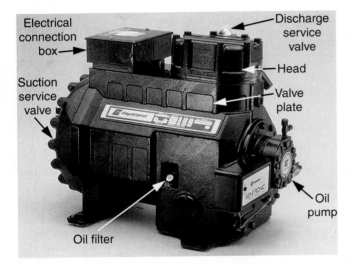

Figure 19-24. *The Discus® compressor is similar in appearance to conventional reed-valve units but has a radically different valve plate and discharge valve design. (Copeland Corp.)*

The disc-type valve design increases compressor capacity up to 25% by improving volumetric efficiency. The valve plate virtually eliminates the clearance pocket at the top of the piston stroke, minimizing the re-expansion of gas. Improved volumetric efficiency provides the compressor with more capacity and saves up to 16% in operating energy costs.

The Discus valve concept. As shown in Figure 19-22, ordinary reed-valve compressors have a clearance volume built into the valve plate. High-pressure refrigerant gas is trapped in the discharge port area and re-expands during the piston downstroke. The volumetric efficiency of the compressor depends on the number and size of the discharge ports as well as the compression ratio.

Figure 19-25 shows the disc-type valve design. The valve plate is hollow and serves as a passage for the low-pressure suction gas. Conventional suction

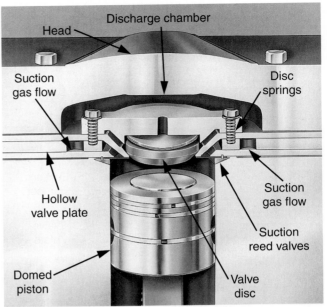

Figure 19-25. *The disc-type valve design is shown in this artist's view. The hollow valve plate provides passage for suction gas and has conventional reed valves for gas flow into the cylinder. The large discharge valve disc is held in place by disc springs. The design improves volumetric efficiency up to 25%. (Copeland Corp.)*

valve reeds control the flow of gas into the cylinder. The large discharge port is closed by the valve disc. A disc spring holds the valve disc against the bottom of the discharge port. This minimizes the clearance volume and improves the compressor's volumetric efficiency. In addition, disc-valve compressors have domed pistons, nearly eliminating the clearance volume above the piston.

Compressors with disc-type valves are extremely rugged and dependable, but they can be damaged by abuse in the same manner that a reed-valve compressor can be damaged. Proper application and installation procedures must be followed whenever a compressor replacement is made. Do not expect a compressor with disc-type valves to correct system problems.

The disc-valve design results in increased capacity for a given displacement. When replacing a conventional reed-type compressor with a compressor that has disc-type valves, it may be necessary to select a unit with a smaller displacement.

Compressor cylinder arrangement

Compressors with one, two, or three pistons have just one valve plate and one head. Larger compressors with four or more pistons use two valve plates and have two heads in a "V" arrangement. See **Figure 19-26.** Six-cylinder compressors normally have three heads, and eight-cylinder compressors have four heads. It does not matter how many cylinders are involved —

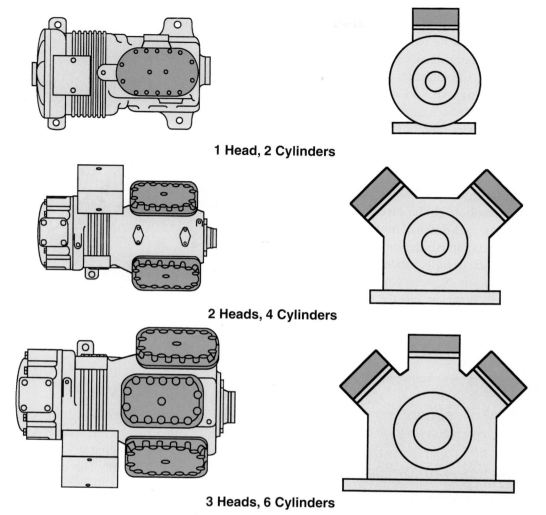

1 Head, 2 Cylinders

2 Heads, 4 Cylinders

3 Heads, 6 Cylinders

Figure 19-26. *Typical cylinder head arrangements for semihermetic compressors are shown in top and end views. Cylinders are usually grouped two per head. (Copeland Corp.)*

the compressors all work the same. The only difference is that the larger compressors have more capacity. A four-head, V-type compressor performs the same task as two, separate, single-head compressors that have the same-sized pistons as the four-head machine.

Compressor pistons

A **piston** is the device that moves up and down inside the cylinder to compress refrigerant gas. Most pistons are made from cast iron, but the small, high-speed, hermetic compressor pistons are made of die-cast aluminum. **Piston rings** are not used on compressors of less than 10 hp. See **Figure 19-27.**

In a compressor, the temperature of the piston rarely exceeds 250°F (121°C), so expansion of the piston or the cylinder wall is minimal. Minimal expansion permits the piston to be fitted with as little as 0.0002″ (0.005 mm) clearance for each inch in diameter. A film of oil provides the necessary seal between the piston and cylinder wall.

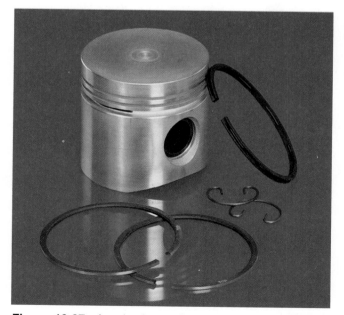

Figure 19-27. *An aluminum piston for use in a hermetic compressor. Rings are not used on most smaller-capacity compressors. (Vilter Mfg. Co.)*

A piston pin fastens the piston to the connecting rod, **Figure 19-28.** The piston pin is accurately machined and *full-floating* (free to rotate within the connecting rod and the piston).

The connecting rod attaches the piston to the crankshaft. Connecting rods vary in design, depending upon the type of crankshaft. Some are clamped to the crankshaft, while others merely slide over the crankshaft eccentric.

Crankshaft

A *crankshaft*, **Figure 19-29,** is used to change the rotary (circular) motion of the electric motor into a reciprocating (up-and-down) motion. There are various ways to accomplish the motion change using different crankshaft designs, such as the eccentric shaft, crank-throw, or Scotch yoke. Each of these crankshaft designs, however, has the same purpose — to cause the piston to move up and down.

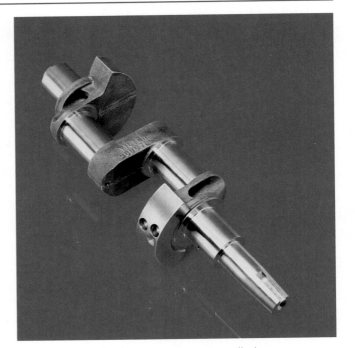

Figure 19-29. *A crankshaft for a two-cylinder compressor. (Vilter Mfg. Co.)*

19.3.4 Compressor Motor

On all types of reciprocating compressors, an electric motor turns the crankshaft. Open-type compressors use a separate motor to drive the compressor with a V-belt. The hermetic and semihermetic compressors have the electric motor inside the compressor shell or housing. The rotating part of the motor (the *rotor*) is fastened directly to the crankshaft. This method is called *direct drive.*

◆ Summary

The service technician must know how compressors work and be able to determine when repair or replacement is necessary. Compressor manufacturers report that many compressors returned as defective are victims of wrong diagnoses — either there is nothing wrong with them, or they failed because of a problem with another system component.

To accurately diagnose compressor problems, the technician must know exactly how the compressor operates. Such knowledge eliminates guesswork and wrong diagnoses. This chapter has explained how each compressor operates to perform its designed task. The reciprocating compressor dominates the industry and, therefore, was covered most thoroughly. Later chapters will further discuss compressor malfunctions, lubrication, and electrical characteristics.

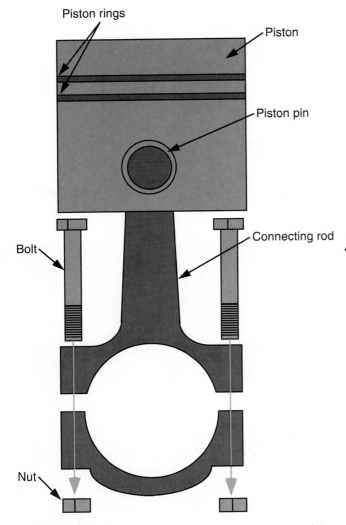

Figure 19-28. *Components of a typical compressor piston assembly.*

◈ Test Your Knowledge

Please do not write in this text. Write your answers on a separate sheet of paper.

1. Name five basic types of compressors.
2. The two types of rotary compressors are the _____ and the _____.
3. The screw compressor has two matched _____.
4. Name the two scrolls used in a scroll compressor.
5. Which type of compressor is not likely to be damaged by liquid slugging or debris in the refrigerant?
6. Name three basic compressor designs.
7. Power for an open-type compressor is normally transmitted by a _____.
8. Which type of compressor is not field-serviceable?
9. *True or false?* The semihermetic compressor has a direct-drive electric motor mounted vertically to the crankshaft.
10. Which type of compressor does not have a fixed capacity?
11. In a reciprocating compressor, what area is under high pressure?
12. The division point between high and low pressure on a reciprocating compressor is the _____.
13. _____ valve reeds are located under the valve plate and _____ valve reeds are above the valve plate.
14. The disc-type valve design increases compressor capacity up to _____%.
15. Compressors of less than _____ hp do not use piston rings.

All refrigerant equipment, such as this vapor compression-type chiller, rely on compressors to change the volume of refrigerant vapor and aid in the heat removal process. (York)

20

Compressor Lubrication and Accessories

Objectives

After studying this chapter, you will be able to:
- ❏ Select the proper oil for each system.
- ❏ Identify the proper oil level in a compressor.
- ❏ Describe the operation of oil separators.
- ❏ Add or remove oil from a compressor.
- ❏ Discuss the factors that can affect compressor capacity.
- ❏ Describe the means to control liquid migration in a system.
- ❏ Identify and install vibration eliminators.
- ❏ Interpret and use information from compressor data plates.

Important Terms

alkylbenzene oils	oil separator
clearance pocket	polyalkaline glycols
compression	(PAGs)
compression ratio	polyol ester oils
crankcase heater	(POEs)
data plate	radial
density	retrofitted
discharge port	slugging
eccentric	splash system
extension	stable
forced-feed system	suction pressure
interstage pressure	two-stage compressors
lubrication	unloaders
miscibility	vibration eliminators
napthenic	viscosity
oil passages	volumetric efficiency
oil pressure safety control	

20.1 Refrigeration Oil

Proper **lubrication** is vital to the efficient operation and long life of the compressor. *Some* oil normally circulates along with the refrigerant, but precautions must be taken to ensure an adequate supply of oil in the compressor at all times. Today's refrigeration oils belong to a specially refined napthenic group of mineral (petroleum) oils very different from the oil used to lubricate automobile engines and other motors.

Requirements for a good refrigeration oil are stringent. The oil must effectively lubricate and cool the compressor's moving parts. In a properly operating system, the oil will not decompose, wear out, or need replacement. In addition, the oil must be **stable** (not break down) and remain fluid in all parts of the system, even when in direct contact with hot motor windings, discharge valves, or cold evaporators.

Refrigeration oil is compatible with any system using common CFC and HCFC refrigerants. Some systems, however, may require special oil additives to properly lubricate moving parts. Manufacturers' recommendations should be followed in these special situations.

20.1.1 Contamination

Store refrigeration oil in sealed containers. Never leave oil exposed to atmospheric air; it will absorb moisture and other contaminants. Refrigeration oils are available in barrels or in one- or five-gallon containers. Most service companies prefer the one-gallon plastic containers, **Figure 20-1,** for easier handling and contamination control.

Figure 20-1. *One-gallon containers of refrigeration oil are preferred for ease of handling and storage. They also are less subject to contamination than larger containers with contents that might be only partly used for a given installation. The container shown holds 150 viscosity oil; 300 and 500 viscosity oils are also available. (Calumet Refining Co.)*

20.2 Oil Viscosity

Viscosity is the ability of an oil to maintain good lubrication despite temperature changes. Choosing the proper viscosity depends on the temperature range to which the oil is exposed. *Higher* viscosity numbers indicate an oil will change very little over an extreme range of temperatures. Oils with *lower* viscosity numbers may not flow properly or may even become solid at very low temperatures.

Three viscosities are available for older refrigerants using napthenic (mineral) oil: 150, 300, and 500. When choosing an oil viscosity for a specific application, *always follow the equipment manufacturer's recommendation.* If no recommendation is available, use these guidelines:

❏ For R-11, R-12, and R-113 with evaporating temperatures above –20°F (–29°C), use 150 or 300. Below –20°F (–29°C), use only 300.

❏ For R-13, R-22, R-114, and R-502, use 300.

❏ For automotive air conditioning compressors, use 500.

A good refrigeration oil must have these properties:

❏ *Low wax content.* Wax separating from the oil will clog valves and capillary tubes.

❏ *Good thermal stability.* The oil must withstand temperature extremes.

❏ *Low pour point.* The oil must remain fluid at low temperatures.

❏ *High dielectric strength.* The oil must resist electrical current flow. (Moisture reduces dielectric strength.)

❏ *Good miscibility.* The oil must readily mix with refrigerant.

20.2.1 Oil-refrigerant Mixture

Refrigeration oils are highly soluble in liquid refrigerant, and at normal room temperatures, they mix completely. Oil and refrigerant vapor do not mix as readily, however. The ability of oil to mix with refrigerant is called **miscibility.**

Oil in the refrigeration system is never a pure oil; it is really an oil-refrigerant mixture. The amount of refrigerant dissolved in oil depends upon pressure and temperature in each part of the system. The oil's lubricating ability and viscosity rating are reduced when it mixes with refrigerant.

20.2.2 Synthetic Oils

The mineral group of refrigerant oils is classified as **napthenic;** these oils are *not compatible* with most of the new HFC refrigerants developed to lessen ozone depletion. Research has produced new synthetic oils that are compatible with new refrigerants. Three of the most popular families of synthetic oils are polyalkaline glycol (PAG), polyol ester (POE), and alkylbenzene (AB).

Polyalkaline glycols (PAGs) were the first synthetic oils developed specifically for use with the new refrigerants, such as HFC-134a, a chlorine-free refrigerant that does not harm the ozone layer.

PAGs have some disadvantages. They attract moisture and are not fully soluble in R-134a. They also have poor lubricating ability in aluminum-on-steel situations, an important consideration in compressors with aluminum pistons in steel cylinders. The biggest problem with PAGs, however, is incompatibility with chlorine. Consequently, a system formerly charged with R-12 or another CFC cannot be **retrofitted** (changed) to the use of PAGs. Systems using PAGs with R-134a must be new and charged at the factory.

Alkylbenzene oils are manufactured from propylene and benzene. The molecule is a *hydrocarbon* (contains hydrogen and carbon atoms). Testing has shown the new HCFC refrigerants perform better with alkylbenzene oils than with existing napthenic mineral oils. The mineral oils are not completely soluble in the new refrigerants. The HCFC refrigerants are soluble, however, in a mixture of alkylbenzene and mineral oil if the mineral oil concentration is 20% or less. As a result, the older refrigerant and air conditioning systems being retrofitted (updated) to use the new refrigerants will not require extensive flushing to rid them of mineral oil residues.

Polyol ester oils (POEs) are a family of synthetic lubricants rapidly gaining popularity. They are compatible with all refrigerants (CFCs, HCFCs, and HFCs) and are miscible with mineral oil and alkylbenzene oils. A mixture of at least 50% polyol ester oil in mineral oil provides excellent lubrication when retrofitting to HCFC refrigerants. Polyol ester oil is made from expensive base stock materials and, therefore, costs more than other lubricants. However, some characteristics of polyol ester oil help offset its higher cost. For instance, polyol ester oil is backward-compatible with mineral oil, which means a compressor containing polyol ester oil can be installed in a refrigeration system that contains mineral oil.

Another advantage of polyol ester oil is that it can be designed to meet lubrication requirements equal to those of mineral oil. The viscosity of polyol ester oil varies less with temperature than does mineral oil. Polyol ester oil is miscible with all refrigerants, and it has oil-return characteristics similar to mineral oil when used with conventional chlorine-containing refrigerants.

Mineral oil is not compatible with HFC (R-134a) refrigerants. When retrofitting a system to an HFC refrigerant and POE oil, the mineral oil contamination must be reduced to less than 5%. Because there are many types and grades of polyol ester oils, it is important to consult the compressor manufacturer for detailed instructions before retrofitting. See Chapter 14 for retrofit procedures when converting from R-12 to an alternative refrigerant.

Oil Additives

Oil additives are sometimes used for better performance. Such additives can lower floc and pour points, improve stability, prevent foaming, improve viscosity, and prevent rust. Additives can be combined to provide special properties.

The new refrigerants and oils have complicated the selection procedure. The old guidelines, using viscosity and temperature to select the proper oil, no longer apply. Mistakes in oil selection can be costly, so it is important to consult the manufacturer's recommendations before adding oil to a system.

20.3 Compressor Lubrication ◆

Two lubrication methods are utilized in reciprocating compressors: the *splash system* and the *forced-feed (oil pump) system.*

20.3.1 Splash System

In the **splash system,** the crankcase is filled with enough oil to bring the level up to the bearings. As the crankshaft revolves, the crank throw (**eccentric**) dips into the oil and splashes it around inside the compressor. Small dippers or scoops are sometimes added to help sling oil around to other parts. See **Figure 20-2.**

The splash system is most common in small compressors of less than 3 hp. Hermetic compressors for use in domestic refrigerators, freezers, window air conditioners, and light commercial systems are factory-assembled under controlled conditions. Although the amount of oil in these systems is critical, no method is provided to observe or check the oil level. When repairing hermetic systems, make every effort to prevent loss of oil. Any oil that is lost should be measured and replaced. If necessary, remaining oil must be removed and a new factory charge installed.

Most welded hermetic compressors are installed in systems without service fittings. Adding oil to a welded hermetic unit usually involves cutting the suction line to allow the oil to be poured into the compressor. (The suction connection on a welded hermetic opens directly into the shell.)

20.3.2 Forced-feed (Oil Pump) System

The **forced-feed system** uses an oil pump to deliver oil under pressure to all bearing surfaces and other critical parts. The oil pump delivers oil through special **oil passages** drilled in the crankshaft, connecting rods, and other components. With this pressure system, the

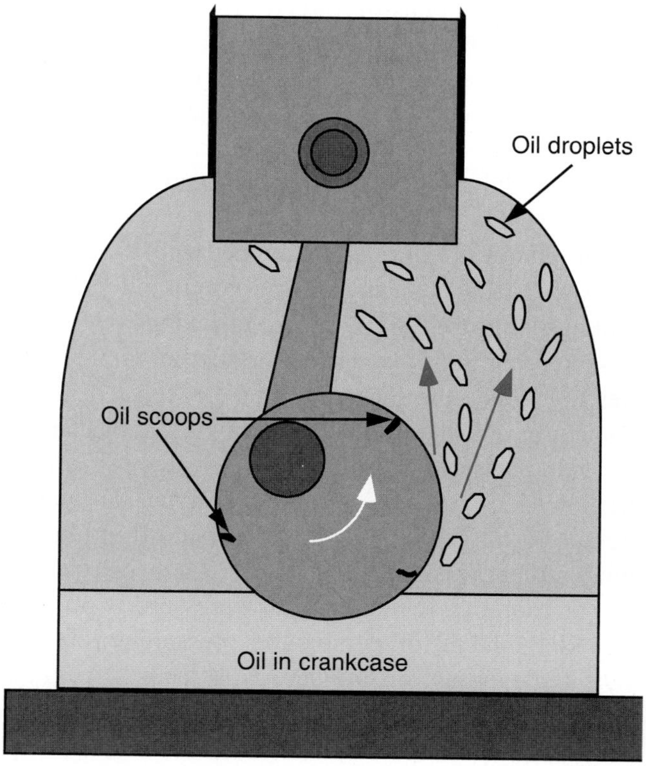

Figure 20-2. *Small oil scoops on the eccentric of the crankshaft splash oil onto compressor parts for lubrication and cooling.*

compressor gets better lubrication, permitting smaller bearing clearances. Oil pumps are usually provided on compressors of 3 hp or more.

The oil pump is normally mounted on the end bell at the crankshaft bearing housing. As shown in **Figure 20-3,** the oil pump has a shaft with a flat end that fits into a slot cut into the crankshaft. The oil pump is driven by the turning of the compressor crankshaft.

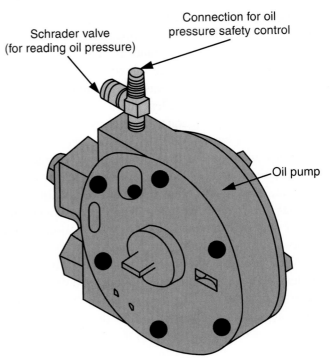

Figure 20-4. *A Schrader valve threaded into a tee fitting permits use of a gauge manifold for reading oil pressure at the pump.*

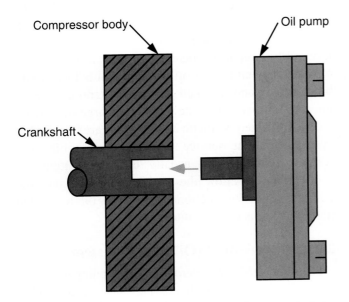

Figure 20-3. *The oil pump is bolted to the end bell of the compressor and driven by the crankshaft.*

The pump obtains oil directly from the crankcase and forces it through a hole in the crankshaft to the compressor bearings and connecting rods. The pump has a spring-loaded ball check valve that acts as a pressure-relief device. The valve allows oil to bypass directly to the crankcase if the oil pressure rises above its setting.

Oil pumps are normally equipped with a Schrader valve to attach the gauge manifold for oil pressure readings, **Figure 20-4.** When checking oil pump pressure, you must also check the suction (crankcase) pressure. Since the oil pump intake is located in the crankcase, and the crankcase is always subjected to low-side pressure, the oil pump inlet pressure equals the low-side pressure. To obtain *actual* oil pump pressure, subtract the crankcase pressure from the oil pump outlet pressure.

For example, if the crankcase pressure is 60 psig (414 kPa), and the oil pump outlet pressure is 100 psig (690 kPa), the net oil pressure is 40 psig (276 kPa). Hence, 100 – 60 = 40 psig (net oil pressure).

In normal operation, actual oil pressure will vary depending on compressor size, temperature and viscosity of the oil, and the amount of clearance in the bearings. Acceptable oil pressures vary from one

compressor manufacturer to another, but *minimum* oil pressures are fairly standard. (See the Oil Pressure Safety Control heading in this chapter.)

20.3.3 Oil Level

For proper lubrication, an adequate supply of oil must be maintained in the crankcase at all times. Most semi-hermetic units have an oil sight glass (small window) on the side of the crankcase for determining the oil level in the crankcase.

Figure 20-5 illustrates proper oil levels in the widely used Copeland and Carlyle (Carrier) compressors. In Copeland units, the window should be half-full when the system is running and stabilized. For Carlyle "D" compressors, the oil level should also be half-full. For "E" compressors, the oil level should be from 1/8″ to 3/8″ up the sight glass. An excessive amount of oil could result in slugging and possible damage to the compressor valves. On field-installed systems, it may be necessary to add or remove oil after the system stabilizes at its normal operating condition to maintain the proper level.

20.3.4 Oil Pressure Safety Control

A high percentage of all compressor failures is caused by lack of proper lubrication. Lack of lubrication can result from several factors:

❏ Shortage of oil in the system.
❏ Improper oil return from the evaporator ("logging").

❑ Shortage of refrigerant.
❑ Refrigerant migration or floodback to the crankcase.
❑ Failure of the oil pump.
❑ Faulty operation of refrigerant flow controls.

A special oil pressure safety control, **Figure 20-6,** protects the compressor against loss of oil pressure. The majority of compressor failures caused by loss of lubrication can be prevented by the safety control.

Operation of the **oil pressure safety control** depends upon the difference in pressure between the oil pump outlet and the crankcase. The top of the control contains a low-pressure bellows (diaphragm). A capillary tube from the bellows is connected to the compressor crankcase, transferring crankcase pressure

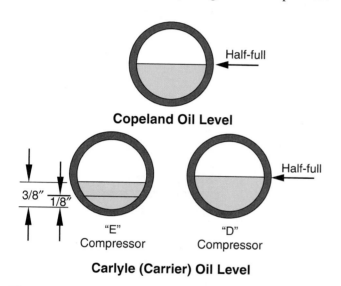

Copeland Oil Level

3/8"   1/8"

"E"
Compressor

"D"
Compressor

Carlyle (Carrier) Oil Level

Figure 20-5. *Sight glass oil levels for two popular brands of compressors.*

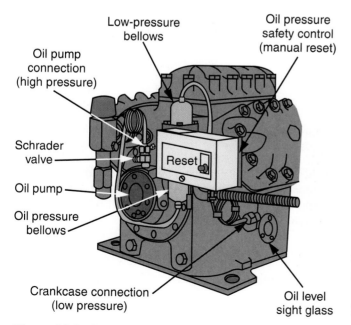

Low-pressure bellows

Oil pressure safety control (manual reset)

Oil pump connection (high pressure)

Schrader valve

Reset

Oil pump

Oil pressure bellows

Crankcase connection (low pressure)

Oil level sight glass

Figure 20-6. *An oil pressure safety control installed on a typical Copeland compressor.*

to the top of the control. The bellows on the *bottom* of the control has a capillary tube that must be attached to the oil pump. The capillary tube transfers oil pump pressure to the bottom of the control.

If oil pressure falls below safe limits, a switch inside the control opens, stopping the compressor. The compressor cannot restart until the manual reset button on the front of the control is pressed. Oil pressure safety controls have a time-delay feature. Obtainable with a 30-, 60-, or 120-second delay, the feature helps avoid nuisance shutdowns caused by brief fluctuations in oil pressure during start-up.

The delay prevents a tripout for the specified number of seconds at start-up, regardless of oil pressure. When the time delay expires, net oil pressure *must* be at least 12 psig to 14 psig (83 kPa to 96 kPa), or the control will stop the compressor. During a normal run cycle, the control prevents net oil pressure from dropping below 9 psig (62 kPa).

Tripping of the oil pressure safety switch is a warning that the system has been without proper lubrication for the time-delay period. It is a clear indication that a lubrication problem exists and needs to be corrected. A well-designed system should not trip the oil pressure safety control. Once tripped, the safety control cannot be reset until it cools (about two minutes).

Compressor manufacturers often *require* the use of an oil pressure safety control. Eliminating or bypassing the safety control will void the compressor warranty. Operation of the safety control and proper wiring to the compressor electrical circuit are fully explained in Chapter 26.

20.4 Oil Separators

An **oil separator** removes oil from the refrigerant and returns the oil to the compressor crankcase. The oil separator is installed in the hot gas discharge line between the compressor and the condenser, **Figure 20-7.**

When the compressor is operating, some oil is pumped out along with the hot compressed gas. A small amount of oil circulating in the system is normal. The system is designed to provide gas velocities that sweep circulating oil through the system and back to the compressor. Some low-temperature systems require an oil separator because the density of the refrigerant vapor is not sufficient to sweep oil back. The oil tends to accumulate (log) in the evaporator, condenser, and lines, depriving the crankcase of sufficient oil.

20.4.1 How the Oil Separator Works

Hot discharge gas containing oil in the form of a fog enters the oil separator inlet and passes through the

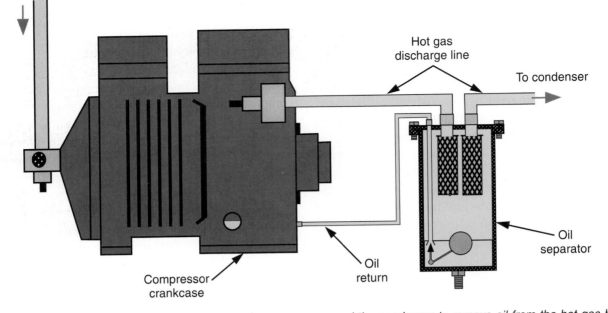

Figure 20-7. *The oil separator is placed between the compressor and the condenser to remove oil from the hot gas being discharged from the compressor. The oil is returned to the compressor crankcase.*

inlet baffling and screens. The baffles and screens force small oil particles to collide and combine into larger, heavier particles. The heavier oil particles drip from the screens to the bottom of the separator, which acts as a sump to collect oil, sludge, and foreign matter. See **Figure 20-8.** A magnet in the sump collects metallic particles and keeps them from being circulated in the system.

Oil collects in the sump area until the level is high enough to raise a float and open the oil return needle valve. Only a small quantity of oil is required to activate the float mechanism, so only a small percentage of the system's oil is absent from the crankcase at any time. Oil returns quickly to the crankcase because the separator contains high-pressure gas that pushes oil back through the return line.

20.4.2 Oil Separator Application

Most compressor manufacturers require an oil separator on two-stage compressors, ultra-low temperature systems, or any system where oil return is critical. A separator is almost always used on large air conditioning units up to 150 tons. Removing oil from the refrigerant improves the efficiency of the circulating refrigerant. In fact, overall efficiency of the system is greatly improved when an oil separator is used.

An oil separator is a system aid, *not* a cure-all or substitute for good system design. Oil separators are never 100% efficient. On a system where the piping design promotes oil logging in the evaporator, an oil separator only delays lubrication-related problems.

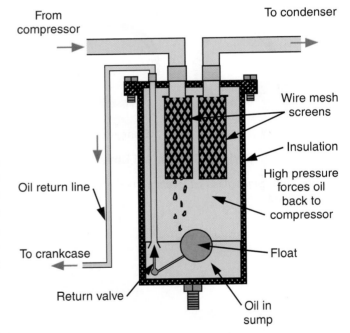

Figure 20-8. *Fine particles of oil in the hot discharge gas collide and combine as the gas passes through the wire mesh screens and baffles of the separator. The larger, heavier drops of oil collect on the screens and drip down to the bottom of the separator. A float opens the return needle valve at a preset oil level, and high gas pressure forces the oil back to the compressor crankcase.*

If the separator is exposed to low ambient temperatures, it must be insulated to keep refrigerant from condensing in the sump during the off cycle. Liquid refrigerant in the separator would be returned to the compressor crankcase.

20.5 Adding Oil to a Compressor

There are three methods for adding oil to a compressor: *the open system method, the oil pump method,* and *the closed system method.* The third method, however, is not recommended; it can introduce air to the system and cause damage to the compressor from slugging.

20.5.1 Open System Method

Open-type and semi-hermetic compressors are equipped with a removable crankcase plug for adding oil. If the system contains refrigerant, the crankcase can be isolated by stopping the compressor, frontseating the suction service valve, and reducing crankcase pressure to about 1 psig or 2 psig (7 kPa or 14 kPa). Remove the crankcase plug, and add the required amount of oil, **Figure 20-9.**

Refrigerant vapor in the crankcase produces a slight positive pressure that prevents entry of air and moisture during the oil-adding procedure. If desired, the crankcase can be purged by cracking the suction service valve off its seat for one or two seconds. After the oil has been added, replace the crankcase plug, backseat the service valve, and restore the system to operation.

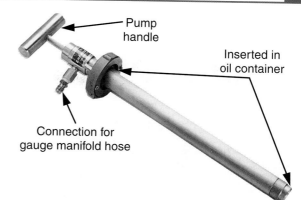

Figure 20-10. *A small hand pump of the type used to add oil to the compressor while the system is running. The pump is connected to the center hose of the gauge manifold. (Thermal Engineering)*

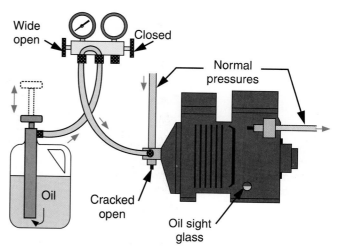

Figure 20-11. *While the system is running, oil is added using a hand pump connected to the suction service valve through the gauge manifold.*

overcome the operating suction pressure and add oil as needed. Adding oil while the system is *operating* is desirable because the oil level can be observed through the oil sight glass. An accurate oil level cannot be determined when the system is off.

20.5.3 Closed System Method

Although the method is not recommended, oil may be drawn into the compressor through the suction service valve using the gauge manifold. Connect the low-pressure gauge hose (blue) to the SSV, and crack the valve open. Crack open the left manifold valve, and vent a small amount of refrigerant through the hoses to purge them of air. Submerge the end of the center charging hose (yellow) in a container of refrigeration oil while venting it. See **Figure 20-12.** Close the manifold valve after venting.

Frontseat the SSV, and pull a vacuum in the compressor crankcase by forcing the unit to run. Open

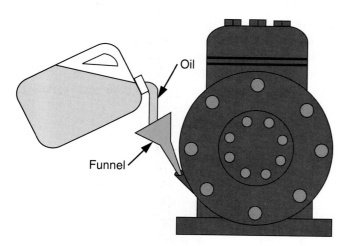

Figure 20-9. *The open system method of adding oil is used when the system is not running. Oil is added through the crankcase fill plug.*

20.5.2 Oil Pump Method

Most technicians purchase a small, hand-operated oil pump for adding oil to compressors. The hand pump is similar to a small bicycle tire pump, **Figure 20-10.** The pump permits the technician to add oil to an operating compressor through the gauge manifold and suction service valve (SSV), **Figure 20-11.** The hand pump contains a check valve to prevent backflow. The technician can easily develop sufficient pressure to

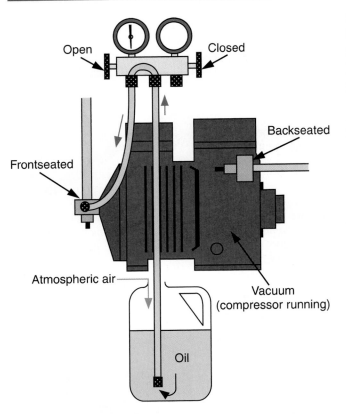

Figure 20-12. *Connections used to add oil to the compressor by the closed system method. This method is not recommended because of the danger of damaging the compressor or introducing air into the system.*

the left manifold valve to permit the compressor to draw oil from the container through the gauge manifold and into the SSV.

IMPORTANT: Be careful to keep the center hose end submerged in oil at all times. Otherwise, air will be drawn into the compressor. On smaller compressors, oil must be added very slowly to prevent slugging. Continue adding oil until the proper amount has been drawn into the compressor.

20.6 Removing Oil from a Compressor

Sometimes excess or severely contaminated oil must be removed from the compressor. Most compressors are equipped with a drain plug. Draining is performed with the compressor turned off.

20.6.1 Removing Excess Oil Using the Drain Plug

Install a gauge manifold, and frontseat the SSV to reduce crankcase pressure to 1 psig to 2 psig (7 kPa to 14 kPa). Stop the compressor, and frontseat the discharge service valve (DSV) to isolate the compressor. Carefully loosen the oil drain plug, but do not completely disen-

gage the threads. Allow pressure to bleed off. Drain oil to the desired level by seepage around the threads (do not totally remove the plug). The oil seal at the drain hole and crankcase pressure prevent entry of air and moisture into the system.

When draining the excess oil is completed, tighten the plug, open the service valves, and restore the compressor to operation. If the crankcase is totally drained because of oil contamination, tighten the plug, and refill with oil using one of the methods described earlier.

20.6.2 Removing Excess Oil through the Oil Fill Hole

Install a gauge manifold, and reduce crankcase pressure to 1 psig or 2 psig by frontseating the SSV. Stop the compressor, and frontseat the DSV to isolate the compressor.

Carefully loosen the oil fill plug, and bleed off pressure before the threads are completely disengaged. Remove the oil fill plug, and insert a length of 1/4″ soft copper tubing so its end is near the bottom of the crankcase. Use tubing that is long enough to bend the other end down below the level of the crankcase, forming a siphon arrangement as shown in **Figure 20-13.** Wrap a clean rag tightly around the copper tubing to seal the oil fill opening. Crack open the SSV slightly to pressurize the crankcase to 4 psig or 5 psig (28 kPa or 34 kPa).

Crankcase pressure forces oil out through the copper tubing and into a container. The oil will continue to drain until the crankcase is empty, if desired. Refrigerant pressure in the crankcase also serves to prevent entry of air and moisture.

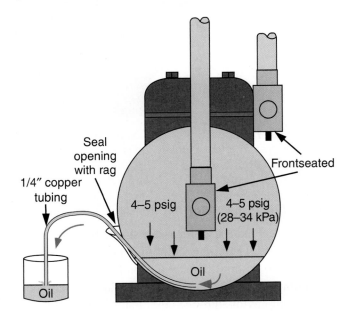

Figure 20-13. *Procedure for removing oil from the compressor through the crankcase opening. Sealing the opening around the tubing with a rag maintains the pressure difference necessary for removing the oil.*

When oil removal is completed, remove the copper tubing and quickly reinstall the oil fill plug to prevent loss of refrigerant. Both service valves can be reopened and the compressor restored to operation. If the crankcase is totally drained, refill before operating the compressor using one of the methods described earlier.

20.7 Testing Oil for Acid

Acid in a refrigerant system is picked up by the refrigeration oil, which acts as a scavenger. Acid test kits, **Figure 20-14,** are available to check the oil for acid contamination. Each kit includes complete instructions for use.

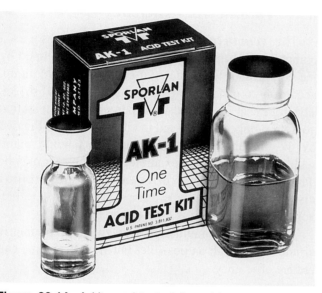

Figure 20-14. *A kit used to test for acid contamination in refrigerant oil. (Sporlan Valve Co.)*

The test is performed by obtaining a small (usually 1/2 oz.) oil sample from the system being tested The oil sample and a test solution are combined in specified amounts in a test tube or similar container, and the contents are thoroughly shaken to mix the ingredients. After the ingredients have stood for a few minutes, the color is observed to determine whether acid is present and, if so, what percentage. The kit contains information on interpreting the color changes.

Periodic oil checks during a cleanup procedure will indicate when acid contamination has been reduced to a safe level. Cleanup is performed by installing (and changing) acid-removing filter-driers until the test color reveals an acceptable acid level.

Carrier Corporation markets a kit called Totaltest™ that permits accurate readings of acid and moisture levels *without* shutting down or opening the system. The test instrument accepts disposable tubes containing indicator chemicals, **Figure 20-15.**

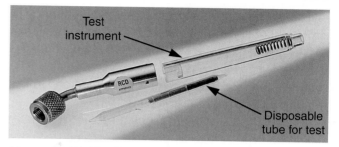

Figure 20-15. *The Totaltest™ system consists of an instrument that screws onto a service or Schrader valve, and a disposable tube filled with indicator chemicals that fits inside the instrument. (Carrier Corporation)*

The instrument is attached to a service valve or Schrader fitting where a minimum pressure of 60 psig (414 kPa) is available. The chemical crystals in the disposable tube react to corrosive acid or any acid-causing moisture that might be present in the system. After 10 minutes, the tube is removed from the instrument, and the crystal color is compared to a chart enclosed in the kit.

20.8 Determining Compressor Capacity

The reciprocating compressor is designed as a positive displacement unit — its capacity depends upon the number and speed of the pistons, and the volume of the cylinders. Since these compressors are normally driven by an electric motor, the speed is fixed at either 1725 rpm or 3450 rpm. Therefore, the amount of refrigerant pumped is calculated by the number of strokes per minute times the total cylinder volume. The capacity for any given compressor is limited.

Compressor capacity is determined by four factors:
- ❑ The *diameter* of the piston.
- ❑ The *number* of pistons.
- ❑ The *length* of stroke.
- ❑ The *speed* (strokes per minute).

Capacity plays a very large role in compressor selection for a particular refrigeration system. An oversized compressor (too much capacity) would rapidly remove refrigerant vapor from the evaporator and suction pressure would be too low. Likewise, an undersized compressor (low capacity) would not remove refrigerant from the evaporator fast enough; suction pressure would remain high for a prolonged length of time.

Other factors influencing compressor capacity are the type of refrigerant being pumped and the temperature of the suction gas. The **density** of a gas (how closely the molecules are packed) varies with temperature. A low-temperature gas is "thin" and occupies more space (volume) per pound of refrigerant than does a denser, high-temperature gas. The compressor piston may have to make several strokes before it can move one pound of refrigerant. A high temperature gas is more dense; thus, a pound occupies less space. The same compressor would require fewer piston strokes to move one pound of refrigerant.

The boiling point (*suction pressure*) in the evaporator should be kept as high as possible to require the least amount of compressor capacity. As suction pressure goes down, compressor capacity must go up. Low-temperature compressors require more capacity than medium- or high-temperature applications. For example, R-12 is *not* a good refrigerant for low-temperature applications because suction pressure is very low (possibly in a vacuum), and the gas is too thin, requiring a high-capacity compressor. Using R-502 on a low-temperature application is preferable. It has a higher suction pressure and much thicker gas, permitting use of a low-capacity compressor.

20.8.1 Clearance Pocket

The volumetric efficiency of a compressor varies with compressor design. If the valve reeds seat properly, the most important factor affecting compressor efficiency is the *clearance pocket.* A clearance space (or pocket) must exist at the top of the piston stroke to prevent the piston from striking the valve plate, **Figure 20-16.** Since the discharge valve reed is located on top of the valve plate, the *discharge port* (hole through the plate) adds some volume to the clearance space as well. At the top of the piston stroke, the clearance pocket remains filled with hot, compressed gas that does not exit.

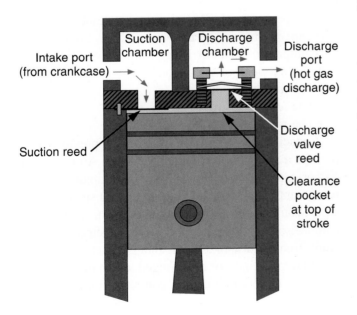

Figure 20-16. *The clearance pocket at the top of the piston stroke affects compressor efficiency. The hot gas remaining in the pocket after the discharge valve reed closes will re-expand as the piston moves down on the intake stroke, reducing the volume of new vapor that can be drawn in.*

As the piston starts downward on the suction stroke, the residue of high-pressure gas re-expands. As it expands, its pressure is reduced. Vapor from the suction line cannot enter the cylinder until the pressure of the re-expanding gas drops below suction pressure. From the standpoint of capacity, the first part of the suction stroke is actually lost. For this reason, the clearance pocket must be kept as small as possible to improve volumetric efficiency of the compressor.

On low-temperature applications, it is often necessary to reduce the clearance pocket to obtain the desired capacity. Low-temperature valve plates usually have smaller discharge ports.

20.9 Volumetric Efficiency ◆

The *volumetric efficiency* of a compressor is a comparison of the amount of gas *actually* pumped by the compressor to the amount of gas it *should* pump, according to piston displacement and length of stroke. The figure is expressed as a percentage. The actual volume pumped (expressed in cubic inches) is divided by the amount of gas the compressor should pump without any losses. For example, a compressor may be designed to pump 20 cu. in. of vapor per stroke (piston displacement) but actually pumps 12 cu. in. Therefore, volumetric efficiency is 60% (12 ÷ 20 = 0.60).

Compressor manufacturers try to keep volumetric efficiency as high as possible, but field problems can reduce efficiency.

20.9.1 High Head Pressure = Decreased Efficiency

As head pressure increases, efficiency decreases due to high-pressure gas left in the clearance pocket at the top of the stroke. Also, as suction pressure decreases, efficiency decreases since the pressure of the gas left in the clearance pocket must be reduced further to drop below the lower suction pressure. This reduces the amount of incoming suction gas and lowers volumetric efficiency. Every effort should be made to operate the system within its designed capacity.

Compressor efficiency also depends upon the size of ports or openings inside the compressor. Anything reducing the flow of gas through these ports will reduce compressor efficiency.

Each semihermetic compressor is designed to be most efficient for a particular evaporator temperature (high, medium, or low) and a particular refrigerant. However, a given compressor may be approved for two operating ranges involving different refrigerants. For example, a compressor may be approved for R-12 at medium temperature or R-502 at low temperature.

20.9.2 Compression Ratio

Compression ratio refers to the relationship between the low-side pressure and the high-side pressure. In other words, exactly how much compression is taking place? The compression ratio is figured by dividing the low-side absolute pressure (psia) into the high-side absolute pressure (psia).

Single-stage compressors

The majority of compressors in use today are the single-stage type. The compression ratio for single-stage compressors goes as high as 12:1. If the ratio becomes higher than 12:1, a two-stage compressor normally is used. Lower compression ratios place less demand upon the compressor, making it more energy efficient.

Two-stage compressor operation

Because of the high compression ratios needed in ultralow-temperature applications, *two-stage compressors* were developed. They provide increased efficiency when evaporating temperatures are in the range of –30°F to –80°F (–34°C to –62°C). **Figure 20-17** illustrates a typical two-stage system.

Two-stage compressors are divided internally into low (or first) and high (or second) stages. The three-cylinder models have two cylinders on low stage and one on high, while the six-cylinder models have four cylinders on low and two on high.

Refrigerant gas enters the low-stage cylinders directly from the suction line. The gas is compressed and discharged into the interstage manifold at *interstage pressure.* The interstage pressure enters the motor chamber and crankcase, so the crankcase is at interstage pressure as well.

Interstage discharge gas is at high temperature (highly superheated). Liquid refrigerant is metered into the interstage manifold area by a desuperheating expansion valve (TEV type). The liquid refrigerant immediately boils to desuperheat the discharge gas. Desuperheating cools the motor and prevents excessive temperatures during second-stage compression.

Desuperheated refrigerant vapor at interstage pressure enters the suction ports of the high-stage cylinders, where it is compressed and discharged to the condenser.

Unloaders

Capacity can be controlled with unloaders. When large compressors must operate at reduced load, *unloaders* are used to reduce compressor capacity,

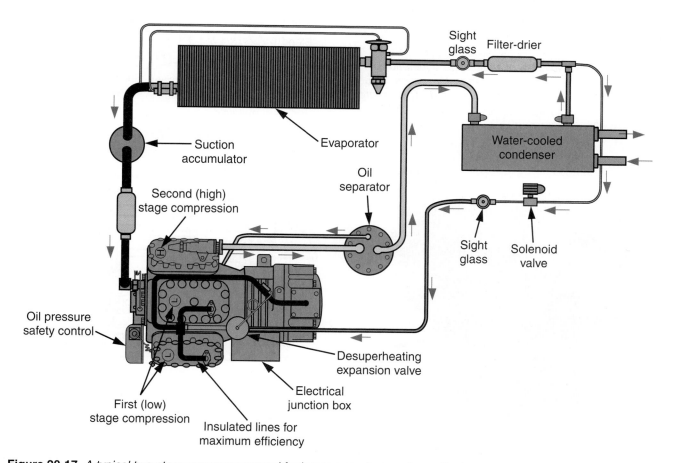

Figure 20-17. *A typical two-stage compressor used for low-temperature systems. Such compressors have high compression ratios and provide increased efficiency.*

Figure 20-18. Suction valves on one or more cylinders are held open or closed mechanically in response to reduced suction pressure. With the suction intake blocked off, the cylinder performs no pumping action.

The unloader is controlled by an electric solenoid valve mounted directly on the head and mechanically connected to the unloader. See **Figure 20-19.** Power supply to the solenoid is controlled by a pressure-sensitive switch operating on suction pressure. **Figure 20-20** illustrates refrigerant flow through the unloader in both loaded and unloaded positions.

Tandem motor-compressors

Sometimes it is desirable to combine two motor-compressors into a single refrigeration system as a means of changing compressor capacity according to reduced load. See **Figure 20-21.** Combining two compressors, however, presents lubrication problems. Unless the pressures in the two crankcases are constantly equal, oil will leave the crankcase with the higher pressure.

Tandem compressors were developed to overcome oil and vibration problems and still offer capacity control. The tandem unit consists of two individual compressors with an interconnecting housing that replaces the individual end bells. The compressors can operate individually or together. The tandem offers greater safety than a single compressor; if one

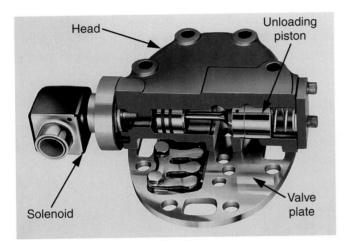

Figure 20-18. *Cutaway view of an unloader. Operation of the solenoid valve controls the unloading piston, which, in turn, blocks or opens the suction manifold port. This controls capacity by switching the cylinder "on-line" or "off-line," as necessary. (Carrier Corporation)*

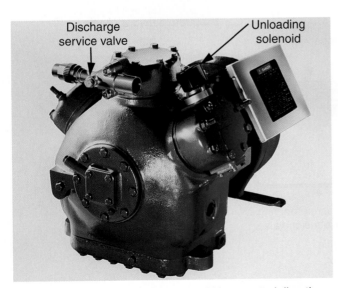

Figure 20-19. *The unloading solenoid is mounted directly on the head of the cylinder it controls. This six-cylinder compressor has unloading solenoids on two cylinders for capacity control. (Carrier Corporation)*

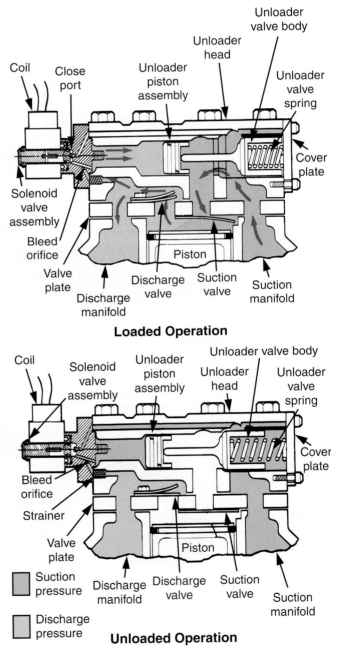

Figure 20-20. *Operation of an unloading system in the loaded and unloaded states. (Carrier Corporation)*

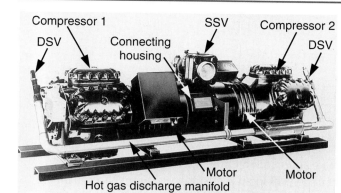

Figure 20-21. *A typical tandem compressor consists of two individual units joined with a connecting housing. (Copeland Corp.)*

compressor fails, the remaining one will operate until repairs are completed. Furthermore, staggered starting reduces electrical requirements. Tandem compressors provide simple, foolproof capacity control with maximum power savings. The tandem arrangement greatly simplifies system control.

20.10 Compressor Cooling

The hot discharge temperatures created at the top of the piston stroke (heat of compression) greatly affect efficient operation of the compressor. The highly superheated discharge gas temperature is directly related to the temperature of the incoming suction gas. Discharge temperatures above 325°F to 350°F (163°C to 177°C) will cause oil breakdown, resulting in sludge and acid. Discharge temperature must be kept below this range. Peak temperatures occur at the discharge valves; the temperature of the discharge line will be from 50 to 100 degrees Fahrenheit (28 to 55 degrees Celsius) *lower* than the temperature at the valve plate. The maximum allowable discharge line temperature is 250°F (121°C), measured six inches from the discharge service valve exit.

The temperature difference between gas at the discharge valve and the discharge line is absorbed by the metal parts of the compressor. This heat cannot be permitted to accumulate, so various methods are used to remove it.

Air cooling

Air-cooled compressors must have a sufficient quantity of air blowing directly on the compressor body to cool the motor. Proper cooling can normally be accomplished by locating the compressor directly in the stream of discharge air from the condenser fan. If the compressor cannot be located in the condenser discharge airstream, cooling must be provided by another fan discharging air directly against the compressor. See **Figure 20-22.** On compressors with

multiple heads, a fan is often mounted on top of the compressor with airflow directed downward over the heads, as shown in **Figure 20-23.**

Refrigerant cooling

Refrigerant-cooled motor-compressors are designed to provide a flow of cool suction gas around and through the motor. At temperatures below 0°F (–18°C), the incoming refrigerant gas is too thin (reduced in density) to provide sufficient cooling ability. Additional motor cooling by means of airflow is usually necessary.

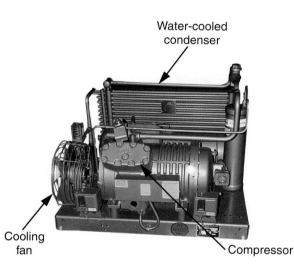

Figure 20-22. *A separate fan blowing air directly onto the compressor provides cooling. (Dunham-Bush)*

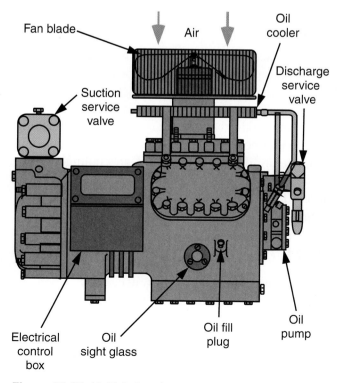

Figure 20-23. *Multiple-head compressors are often cooled by a fan blowing air directly down on the heads.*

Water cooling

Water-cooled compressors are provided with a special water jacket that allows water to be circulated around the compressor body before going to the condenser. See **Figure 20-24.** On smaller compressors, a coil of soft copper tubing is wrapped around the compressor body to act as a water-circulating jacket.

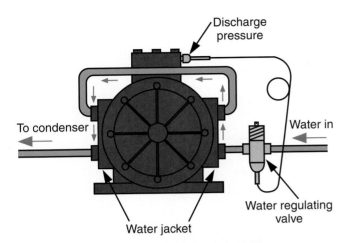

Figure 20-24. *On a water-cooled compressor, cool water circulates through a water jacket that surrounds the compressor body. After exiting the jacket, the water is piped to the condenser to provide further cooling action.*

Cooling for equipment rooms

If compressors or condensing units are located in an equipment room, proper ventilation must be provided to control room temperature. Equipment rooms are normally equipped with one or more large exhaust fans individually controlled by room thermostats. See **Figure 20-25.** As the equipment room temperature rises, one fan comes on and pulls fresh air into the room while exhausting hot air. If room temperature continues to rise, another fan comes on, and then another, as necessary, according to settings on the thermostats. Likewise, as the equipment room temperature drops as a result of air cooling, the fans cycle off and maintain the temperature at the desired level. These thermostats are usually set to maintain room temperature at 68°F to 72°F (20°C to 22°C).

20.11 Crankcase Heater

A **crankcase heater** is necessary when the compressor is surrounded by an ambient temperature that may become lower than the evaporating temperature. Due to low ambient temperature conditions, migration of liquid refrigerant (or condensation in the crankcase) occurs during the off cycle when the compressor is cold. Electric crankcase heaters keep the crankcase oil warm. Warming the oil helps prevent condensing of refrigerant vapor in the crankcase. Under mild conditions, the crankcase heater vaporizes liquid refrigerant and forces it out of the compressor.

Heat always travels toward colder-temperature areas, so if the compressor becomes colder than the evaporator, refrigerant will migrate (travel) to the compressor. Migration is assisted by the pressure difference between the evaporator and compressor crankcase.

Liquid refrigerant in the crankcase readily mixes with oil. Upon start-up, the crankcase pressure is rapidly lowered, and some liquid boils. An oil-and-liquid-refrigerant mixture results and is pulled into the cylinders, a condition called *slugging,* which can cause severe damage to the discharge valve reeds.

Slugging is prevented by eliminating liquid refrigerant in the crankcase. Elimination may require the use of a crankcase heater, suction accumulator, or pump-down cycle.

The crankcase heater in some systems is energized continuously (never shut off). Other systems, however, turn the crankcase heater off during the run cycle. An oversized heater may overheat the oil, while an undersized heater will be ineffective. Consult the compressor manufacturer or local dealer to determine the correct crankcase heater type and size. Some heat pumps and air conditioners *require* a crankcase heater to maintain the compressor warranty.

20.12 Compressor Mounting

The compressor is usually mounted on four mounting bolts equipped with spring and washer assemblies. After installation, the top mounting nut is loosened to allow the compressor to float on the mounting springs. See **Figure 20-26.** A space of about 1/16″ (1.6 mm) between the top nut and the rubber spacer is normal. This method of mounting reduces the transmission of noise and excessive movement during starting or stopping, when vibration of the motor-compressor is greatest.

20.13 Vibration Eliminators

Vibration eliminators prevent transmission of noise and vibration to refrigerant piping. Vibration of the piping causes leaks. Vibration eliminators are installed close to the compressor in both the suction and discharge lines. See **Figure 20-27.** On small units, the soft copper tubing is wrapped in a coil at the compressor to provide vibration protection.

20.13.1 Installing Vibration Eliminators

Unless the vibration eliminator is properly installed, stress will cause it to fail or defeat its

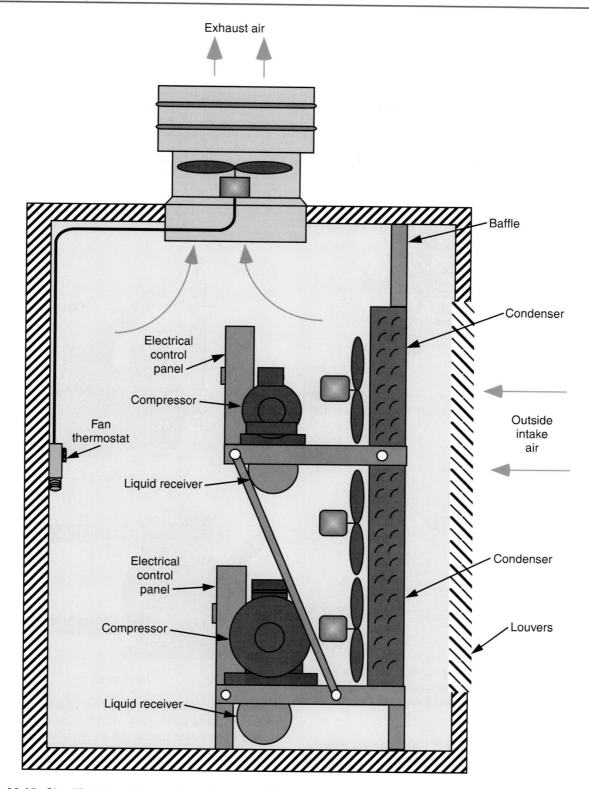

Figure 20-25. *Simplified view of an equipment room cooling system. One or more thermostatically controlled fans cycle on and off to draw outside air through the room and keep room temperature at the desired level.*

purpose. The metallic vibration eliminator is designed to adjust to movement in a ***radial*** (circular) direction. It must not be subjected to stress that causes ***compression*** or ***extension*** (push or pull). Vibration eliminators should be installed as close to the compressor as possible, parallel to the crankshaft. See **Figure 20-28.**

The starting torque (turning power) of the motor tends to rock the compressor from side to side when starting. Parallel mounting to the crankshaft allows the vibration eliminators to easily adjust to the movement.

Internal connections of the metallic vibration absorbers are made with a brazing compound with a

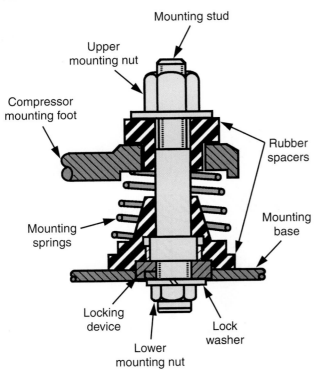

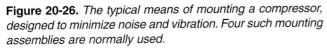

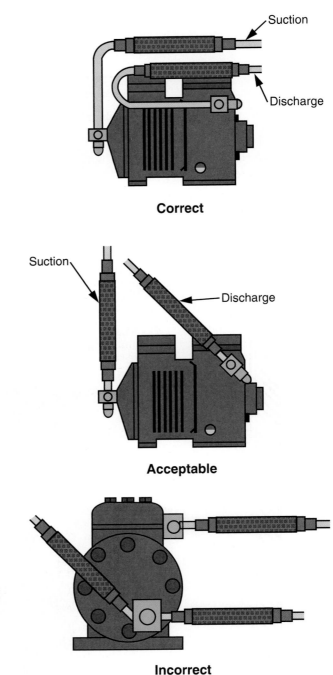

Correct

Acceptable

Incorrect

Figure 20-26. *The typical means of mounting a compressor, designed to minimize noise and vibration. Four such mounting assemblies are normally used.*

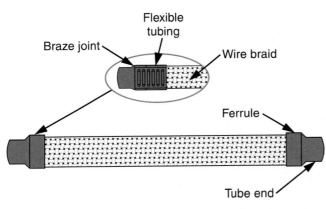

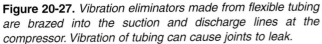

Figure 20-27. *Vibration eliminators made from flexible tubing are brazed into the suction and discharge lines at the compressor. Vibration of tubing can cause joints to leak.*

Figure 20-28. *Methods of installing vibration eliminators. A—Correct installation, with vibration eliminators parallel with the compressor crankshaft. B—An acceptable installation. C—An incorrect installation, with vibration eliminators at a right angle to the crankshaft.*

melting point of about 1300°F (704°C). To prevent damage to these internal joints, line connections should be made with a silver alloy with a melting temperature below 1200°F (649°C). Wrapping a wet rag around the vibration absorber before brazing is recommended. Brazing should be done quickly. Overheating may cause damage to the internal brazed connections.

20.14 Compressor Data Plates

The *data plate* mounted on each compressor contains coded information that reveals construction details of the unit. Knowing how to interpret the

information (letters and numbers) can be useful to the technician. The data plate always contains the model number and serial number of the unit. These numbers *must* be used when ordering parts or obtaining an exact replacement. With these two numbers, the supplier or manufacturer has access to precise information concerning every detail of the unit. The data plate often includes useful electrical information pertaining to the motor compressor as well.

IMPORTANT: The data plate is normally spot-welded or riveted to the unit. The manufacturer will usually *void* (not honor) the warranty if the data plate has been removed.

Data plates from two popular compressor brands (Tecumseh and Copeland) are explained next.

20.14.1 Tecumseh Data Plates

Tecumseh does not indicate the model and serial numbers as such on the data plate. The technician must know where to look for these numbers on the data plate. See **Figure 20-29** to identify these and other important numbers.

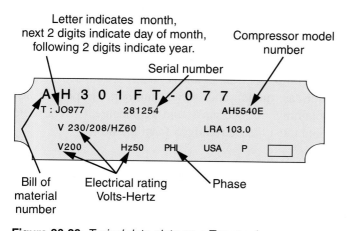

Letter indicates month, next 2 digits indicate day of month, following 2 digits indicate year.

Compressor model number

Serial number

A H 3 0 1 F T - 0 7 7

T : JO977   281254   AH5540E

V 230/208/HZ60   LRA 103.0

V200   Hz50   PHI   USA   P

Bill of material number

Electrical rating Volts-Hertz

Phase

Figure 20-29. *Typical data plate on a Tecumseh compressor. The serial number and the model number are not labeled as such on the plate. The technician must know where to look. (Tecumseh)*

Tecumseh model number codes

Model numbers are coded to reveal a great deal of important information in a very small space. Being able to decipher the coded numbers is not important to the service technician; however, the code *is* important for obtaining a replacement or finding application information. As shown in **Figure 20-30,** the Tecumseh model number usually consists of two letters, four numbers, and another letter.

For example, AH5540E contains the following information:

❏ The first two letters indicate the compressor family, which describes the body shape and style.

❏ The first number indicates the application (high-, medium-, or low-temperature) and the starting torque: 5 = air conditioning; 45°F evaporating temperature; PSC (permanent split capacitor) motor.

❏ The second number indicates the number of digits in the Btu capacity: 5 = a 5-digit number (00,000).

❏ The last two numbers indicate the first two digits of the Btu capacity: 40 = 40,000.

❏ The letter following the four numbers (E, in this case) indicates the type of refrigerant to be used:

A,B,C,D = R-12
E,F,G,H = R-22
J,K,L = R-502

❏ Another letter may be added to the end of the model number to indicate specific data concerning the entire condensing unit and its accessories.

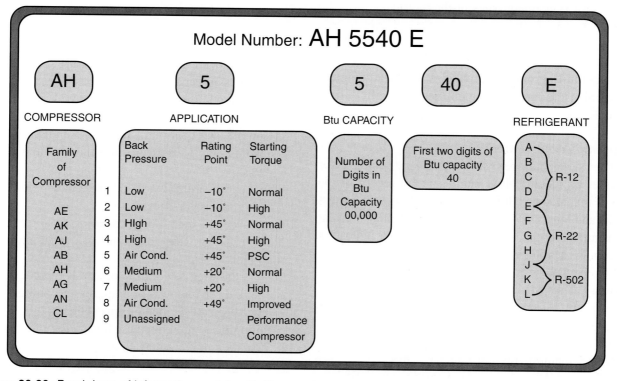

Figure 20-30. *Breakdown of information contained in Tecumseh compressor model number.*

20.14.2 Copeland Data Plates

The model number on compressors manufactured by Copeland provides a great deal of information. Unlike Tecumseh, Copeland clearly labels the model and serial numbers on its data plates. See **Figure 20-31.**

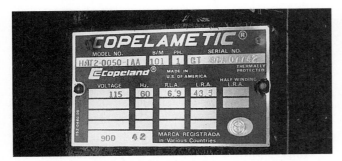

Figure 20-31. *Typical data plate for a Copeland-manufactured compressor. Model and serial numbers are easy to identify.*

A typical model number might be 4RA1-1000-TFC-200. The number identifies a Copeland motor-compressor as follows:
- ❏ The first group of numbers and letters (4RA1) identifies the characteristics of the compressor:
    - 4 = compressor family (body shape and style)
    - R = refrigerant-cooled
    - A = 2380 cubic feet per minute displacement
    - 1 = basic physical characteristics
- ❏ The second group of numbers (1000) identifies the horsepower (1000 = 10 hp).
- ❏ The third group letters (TFC) identifies the electrical characteristics of the motor:
    - T = three-phase
    - F = internal motor protection
    - C = 208/230 volts, three-phase, 60 hertz
- ❏ The fourth group of numbers is used for complete condensing units (200 identifies a specific bill of materials and product variations).

Copeland serial numbers

The serial number on a Copeland compressor provides an identification number and a record of the date of manufacture. It is composed of eight characters (digits and letters). The first two digits identify the year of manufacture; the third character is a letter indicating the month. Months of the year are represented by the first twelve letters of the alphabet (A for January, B for February, C for March). The last five digits of the serial number are assigned in sequence for each month's production. For example, the serial number shown in Figure 20-31 (87A-07742) would indicate:

87A = manufactured in January, 1987.

07742 = the 7742nd unit built in January, 1987.

Summary

Properly lubricating and cooling the compressor leads to its efficient operation and long life. Refrigeration oils have special qualities enabling them to withstand extreme temperature changes while maintaining good lubrication. Some oil circulates with the refrigerant, so provisions must be made for proper oil return to the crankcase. Good piping practice, oil separators, and oil pressure safety controls ensure proper compressor lubrication.

Liquid migration causes severe slugging problems. Several methods are used to eliminate slugging, including crankcase heaters, suction accumulators, and pumpdown cycles.

Vibration causes copper tubing to become work-hardened and crack open, permitting loss of refrigerant. Vibration eliminators absorb vibration that results from starting and stopping the compressor.

Compressor data plates contain coded information, such as model and serial numbers, important for obtaining correct parts and replacements. The data plate must not be removed from the unit.

Test Your Knowledge

Please do not write in this text. Write your answers on a separate sheet of paper.
1. Refrigeration oil is a specially prepared _____ oil.
2. Oil kept in open containers will absorb _____.
3. Name two methods of compressor lubrication.
4. The oil pressure safety control operates on the difference in pressure between the _____ and the _____.
5. The oil separator is designed to separate oil from _____ and return the oil to the _____.
6. *True or false?* As suction pressure goes down, compressor capacity goes down.
7. Why should the clearance pocket be kept as small as possible?
8. Compression ratio is figured by dividing the absolute _____ pressure into the absolute _____ pressure.
9. As head pressure increases, does the compression ratio increase or decrease?
10. Because of high compression ratios, two-stage compressors are used in _____ temperature applications.
11. _____ are used to control compressor capacity.
12. Name three methods of compressor cooling.
13. The crankcase heater prevents _____ from migrating to the crankcase during the off cycle.
14. *True or false?* Liquid slugging can cause severe damage to the suction valve reed.
15. Where are the model and serial numbers found on a compressor?

Water Chillers

Objectives

After studying this chapter, you will be able to:
- ❑ Explain the operating principles of vapor-compression-type chillers.
- ❑ Describe the operation of a centrifugal chiller.
- ❑ Recognize open-loop and closed-loop systems.
- ❑ Explain temperature difference in a closed loop.
- ❑ Describe the operation of a low-pressure chiller.
- ❑ Explain the operating principles of absorption-type chillers.
- ❑ Explain the operating principles of water-ammonia absorption-type refrigerators.

Important Terms

absorber shell	four-pipe system
absorption	generator shell
ammonia	impeller
booster pumps	intercooler
capacity control	lithium bromide
chiller	open loop
chiller barrel	prerotation vanes
closed loop	production process
comfort cooling	cooling
compression stages	purge unit
delta tee	standby
diffuser	tubing bundle
economizer	two-pipe system
eliminators	volute
evaporator	water box
float chamber	

21.1 What Are Chillers?

A **chiller** is a packaged refrigeration system whose sole purpose is to cool (chill) a liquid; the cooled liquid is then used to cool something else. There are two types of chillers. The most common is the *vapor-compression* type that uses a compressor. The other is the *absorption* type that cools by chemical means, instead of using a compressor. Chiller capacities normally range from 25 tons to 1500 tons of refrigeration, but smaller and larger versions are available. The large capacity units are typically vapor-compression units using reciprocating, screw, or centrifugal compressors. They are normally found in hospitals, colleges, large office buildings, and industrial applications. Do not be overwhelmed by the size of the large-capacity machines; they are simply large versions of smaller machines. See **Figure 21-1.**

The absorption-type chiller does not use a mechanical compressor; it relies on the combination and separation of chemicals to absorb and give up heat. These special machines use very little energy. The operating principles of absorption chillers are explained later in this chapter.

Chillers are normally located in an equipment room where noise, maintenance, and access to utilities are not problematic. They are used to chill large quantities of a circulating liquid (usually water) throughout a closed-loop piping system. The cool liquid is then used to cool building occupants (**comfort cooling**), equipment (oil), or a product (**production process cooling**). Selecting and matching the components of a system require the skills of professional engineers.

In chillers, unlike refrigeration and air conditioning systems, lubricating oil is not permitted to circulate with the refrigerant. Instead, an oil pump is used to lubricate moving parts. The process is similar to an automobile lubricating system where an oil pump and oil filter are used for proper lubrication of moving parts.

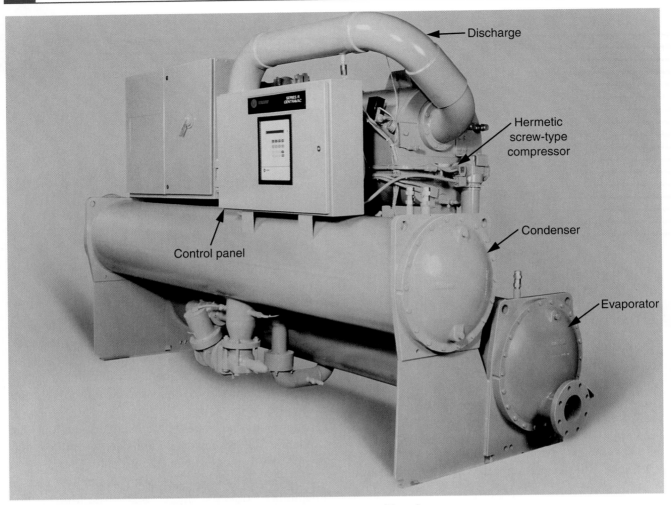

Figure 21-1. *A typical 250-ton chiller using a screw-type compressor. (Trane)*

21.2 Vapor-compression Chillers ———◆

All large chiller systems, and most of the smaller ones, are the vapor-compression type. These chillers make use of the same basic principles and components found in other refrigeration systems: an evaporator in which circulating refrigerant absorbs heat and vaporizes, a compressor that places the gas under pressure, and a condenser in which heat is transferred out of the refrigerant gas, allowing it to return to a liquid state. In refrigeration systems, the refrigerant absorbs heat from and transfers heat to *ambient air.* In a chiller system, heat is absorbed from and transferred to a *liquid* (usually water). See **Figure 21-2.**

21.2.1 Condenser

Water-cooled (tube-in-a-shell) condensers are common on both compression and absorption chillers. The tube-in-a-shell condenser is normally located beside or above the chiller's evaporator. The water tubes inside the condenser must be kept clean and free of debris. A continuous circuit (loop) of insulated steel piping carries cooling tower water to and from the

chiller's water-cooled condenser. Water pumps recirculate water from the tower to the condenser and return it for reuse. The continuous loop of condenser water pipe is called an **open loop** because the water circuit is open to the atmosphere inside the tower. See **Figure 21-3.**

Hot vapor from the compressor enters the condenser, while pumps circulate water from the cooling tower through the water-cooled condenser. The temperature of the water entering the condenser is about 85°F (29°C). As the water passes through tubes in the condenser, it picks up heat from the refrigerant vapor. Water leaving the condenser is about 95°F (35°C); then, it goes back to the cooling tower to be cooled for reuse in the condenser. See **Figure 21-4.**

21.2.2 Evaporator

The **evaporator** is often called the **water box** or **chiller barrel** because the water tubes are submerged in liquid refrigerant. This arrangement of multiple water tubes is known as the **tubing bundle.** Water circulates inside the tubes, transferring heat from the water to the refrigerant surrounding the tubes. The

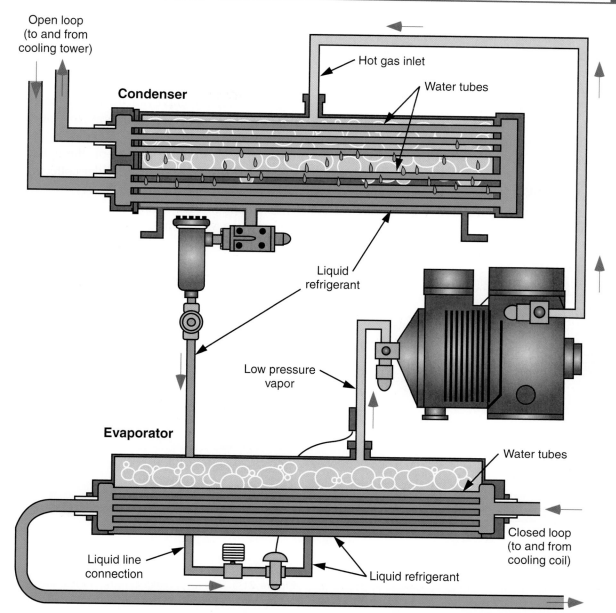

Figure 21-2. *This cutaway view of a water-cooled, reciprocating compressor chiller illustrates the major components of vapor-compression-type units.*

transfer of heat chills the water and vaporizes the refrigerant. The liquid refrigerant level in the evaporator is maintained so it just covers the tubing bundle. See **Figure 21-5.**

Pumps recirculate water from the evaporator to a cooling coil (heat exchanger). The cooling coil can be located inside a metal air duct (air handler) or inside production machinery where it absorbs heat. The warm water leaving the cooling coil is returned to the evaporator for heat removal and recirculation.

The cold water pipe is a continuous loop of insulated steel pipe. It is called a ***closed loop*** because it is not open to the atmosphere. The circulating cold liquid is usually water, but brine (saltwater) is used for applications where plain water would freeze. A glycol

additive (antifreeze) is used for even colder applications. The cooling coil inside the air handler looks like a forced-air evaporator. However, cold water is circulating through this coil, not refrigerant. Sometimes the cold water is used for comfort cooling or for process cooling on a production line. The cold water absorbs heat and is returned to the chiller tubes for heat removal and recirculation through the closed-loop piping system. See **Figure 21-6.**

The temperature of the liquid refrigerant in the flooded evaporator is maintained at about 35°F (2°C), providing a "water out" temperature of 45°F (4°C). A large collection chamber above the tubing bundle moderates the velocity of the exiting vapor. ***Eliminators*** (baffles) are used to limit vapor velocity and prevent liquid drops from entering the impeller of the centrifugal compressor. The temperature of the water entering the evaporator tubes is about 55°F (13°C). As the water flows through the tubes, it is cooled to 45°F (7°C). The temperature difference between the water entering and exiting the evaporator is normally about 10 degrees Fahrenheit (6 degrees Celsius). The temperature difference is expressed as Δ_t (*delta tee*).

21.2.3 Centrifugal Chiller

Although vapor-compression chillers are also manufactured with reciprocating or screw-type compressors, the centrifugal compressor is most common, especially in installations where system capacity is greater than 500 tons. For this reason, the following discussion will focus on the centrifugal compressor.

A centrifugal compressor uses a device called an **impeller** to convert kinetic energy to static energy (pressure). The impeller rotates at high speed, capturing and slinging the vapor through a small passage (**diffuser**) and into a circular collection space called a **volute.** See **Figure 21-7.**

To maintain the centrifugal force needed to compress the vapor, the impeller must rotate at very

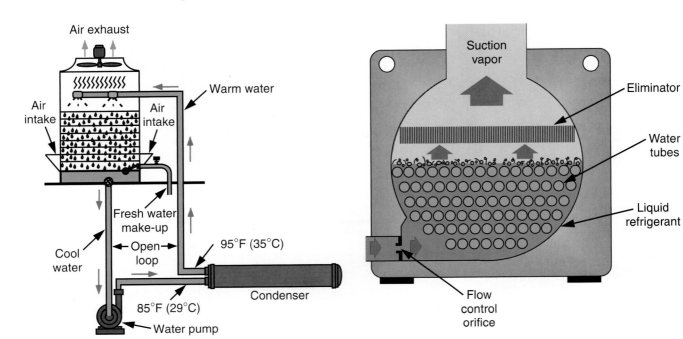

Figure 21-3. *Warm water from the condenser is cooled in the tower and returned to the condenser for reuse. The loop of condenser water pipe is called an open loop.*

Figure 21-5. *The level of liquid refrigerant in the evaporator is maintained so as to just cover the tubes in the bundle.*

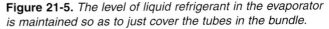

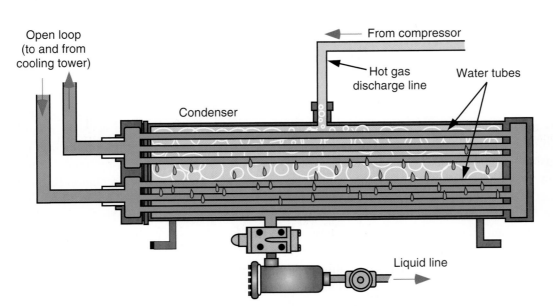

Figure 21-4. *Water in the open loop travels from the condenser to the cooling tower to be cooled for reuse in the condenser.*

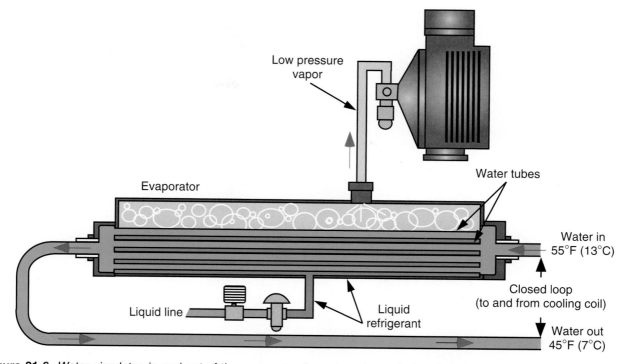

Figure 21-6. *Water circulates in and out of the evaporator through a pipe called a closed loop. Chilled water travels to the cooling coil to absorb heat.*

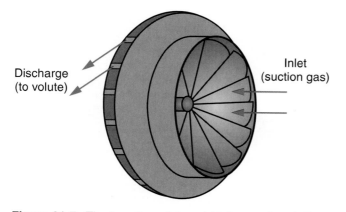

Figure 21-7. *The impeller rotates at high speed and slings the vapor into a collection space called a volute.*

high speed — 25,000 rpm is common. Centrifugal chillers are high-pressure, high-speed, gear-driven machines operating with refrigerants R-12 or R-134a in the 80-ton to 1600-ton range. The machines remain under positive pressure at all times. Centrifugal machines cannot build up pressure like reciprocating and rotary (positive displacement) machines. Therefore, impellers are placed in series to increase vapor pressure. After the vapor leaves an impeller, it enters another impeller, or it enters the discharge line. Centrifugal compressors may have two, three, or four impellers, each one providing some compression. The size and application of the system dictates the capacity of the compressor, the number of *compression stages,* and compressor speed. See **Figure 21-8.**

Low-pressure, low-temperature refrigerant vapor is drawn from the evaporator through *prerotation* (inlet) *vanes* and into the eye of the impeller. Prerotation vanes control the amount of vapor entering the impeller, stabilizing compressor performance over a range of load conditions (gas flow rates). The vanes are closed at startup to permit the machine to start under *no load* conditions. If the evaporator becomes too cold, the prerotation vanes adjust (close slightly) to maintain proper evaporator pressure. The prerotation vanes provide *capacity control* during start-up and low-load conditions. As the impeller spins, the vapor is discharged at high velocities into the volute. The volute directs the relatively high-temperature, high-pressure vapor to another impeller or to the discharge line. See **Figure 21-9.**

21.2.4 Economizer or Intercooler

Various methods are employed to maintain the proper level of liquid refrigerant in the evaporator. A thermostatic expansion valve is often sufficient, while large units may use an orifice (large hole). Another metering device, called an *economizer* or *intercooler,* is a type of liquid receiver with an adjustable float assembly. The unique float assembly maintains a constant liquid level in the evaporator. See **Figure 21-10.**

Liquid refrigerant drains from the condenser to the upper *float chamber* of the economizer where a float valve maintains a liquid seal. As refrigerant level in the upper chamber rises, the float also rises, opening a

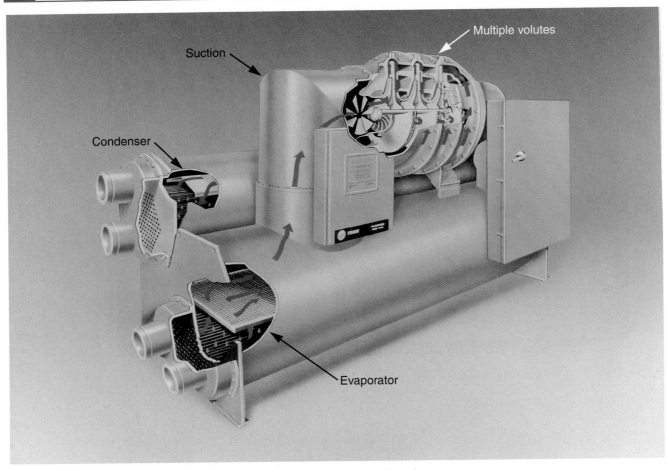

Figure 21-8. *Cutaway view of centrifugal chiller with multiple volutes. (Trane)*

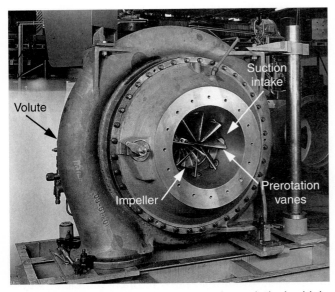

Figure 21-9. *The volute directs the relatively high-temperature/pressure vapor to another impeller or to the discharge line. (York)*

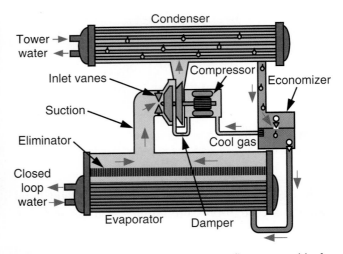

Figure 21-10. *The economizer has a float assembly for maintaining a constant liquid level in the evaporator.*

valve to allow liquid refrigerant to enter the lower chamber. The float valve separates the high-pressure side (upper chamber) from the low-pressure side (lower chamber). Due to low pressure, a portion of the liquid flashes into vapor when it enters the lower

chamber. The heat required for vaporization is taken from the remaining liquid, thus cooling the liquid to the saturated temperature of the lower chamber.

The compressor motor is cooled by taking cool vapor from the economizer's lower chamber and passing it through the motor windings before entering the second-stage compression intake. Refer to Figure 21-10.

21.2.5 Low-pressure Chillers

CFC-11 and HCFC-123 are the most commonly used refrigerants in *low-pressure chillers*. See **Figure 21-11** for a typical low-pressure chiller. R-113 is sometimes used in the 60-ton to 400-ton range and R-11 in the 200-ton to 1600-ton range. The low-pressure CFC refrigerants are rapidly being replaced with HCFC-123. The discharge pressure on low-pressure chillers is normally 7 psig to 9 psig. A rupture disc installed between the condenser and evaporator opens at 10 psig. The amount of refrigerant in a low-pressure, centrifugal chiller amounts to about 3 pounds per ton (lbs./ton) of refrigeration.

Low-pressure chillers operate with evaporator pressure that is less than atmospheric pressure. Because these chillers operate in a vacuum, any leaks in the evaporator section permit air and moisture to enter the system. Air and moisture must be removed with a purge unit.

21.2.6 Purge Unit

A ***purge unit*** removes water vapor and noncondensable gases (air) from the system. The purge unit is simply a small condensing unit attached to the chiller. The purge unit takes its suction from the top of the chiller's condenser where the noncondensable gases are trapped. A mixture of refrigerant vapor, noncondensable gases, and water vapor is drawn from the top of the chiller condenser and compressed by the purge compressor. The compressed gases are then discharged into a heated oil-separator tank. Heat is needed to separate any gases from the oil. The oil collects in the bottom of the separator tank, and a float valve opens to return the oil to the chiller compressor. See **Figure 21-12**.

The high-pressure gases leave the oil separator and enter the purge unit drum. Cold water circulates through a coil inside the drum. The condensable portion of the gaseous mixture is condensed on the surface of the water coil and falls to the bottom of the drum. Condensed liquid refrigerant is released by a float valve and re-enters the system at the evaporator. Any water vapor that condenses on the water coil accumulates and floats on the surface of the refrigerant within the purge drum. A manual drain is provided for removing water from the drum. A sight glass shows when water needs to be drained. See **Figure 21-13**.

Noncondensable gases (air) are trapped in the top of the drum. When sufficient pressure has developed, the relief valve opens and releases the gases to the atmosphere

Figure 21-11. *A hermetic, centrifugal, low-pressure chiller using R-123. It is high-efficiency and environmentally safe. (Trane)*

until pressure is reduced. When the relief valve opens, a small quantity of refrigerant is also lost to the atmosphere. Frequent purging of air to the atmosphere indicates the chiller is leaking and repairs are needed.

21.2.7 Iron Pipe Loops

Iron pipe is used to build the loops for circulating liquid to and from the chiller. Unlike copper tubing, iron pipe seldom leaks. In addition, the iron pipe is supported to eliminate vibration and is properly insulated. Insulation is required to prevent heat loss or heat gain when the liquid is in transit. The closed loop (to and from the evaporator) is connected to a heat

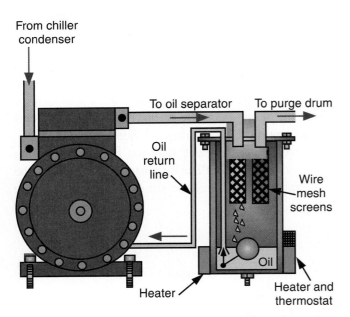

Figure 21-12. *An oil separator returns oil to the purge compressor.*

exchanger or air handler located where cooling is desired. The open loop (to and from the condenser) is connected to the cooling tower.

Centrifugal liquid pumps produce the force to circulate liquid through the iron pipe. The size and number of pumps required to provide sufficient flow depend upon the size of the system. The amount of flow is critical for proper operation and is determined by engineers. An extra liquid pump is often installed as a *standby.* The standby pump is completely installed but remains off (not running), and the valves are closed. The standby pump is immediately available if an operating pump needs repair or replacement.

A large chiller may be located in the basement and comfort cool an entire building. Cold water from the chiller is piped to and from many individual heat exchangers located throughout the building. *Booster pumps* are often required on long (or high) liquid lines to maintain sufficient flow and pressure to air handlers located far from the chiller.

21.2.8 Heat Exchangers

Some type of heat exchanger is required to provide heat flow. A large cooling coil (heat exchanger) can be located inside an air handler. The heat exchanger is not an evaporator even though it may look like one. The copper tubing is made of Type M (thin) copper with aluminum fins. A cold liquid (water) from the chiller's closed loop circulates through the heat exchanger, and a fan forces air through the cold coil. **Figure 21-14** illustrates an air-cooled chiller condensing unit, located on a rooftop, with an air handler on the first floor.

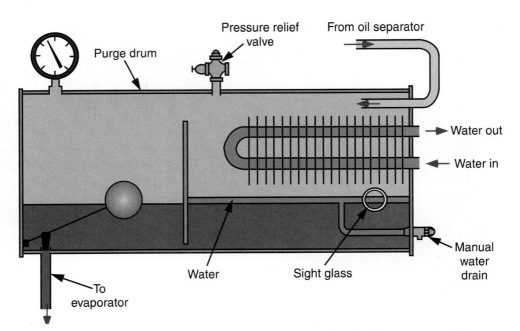

Figure 21-13. *A purge drum removes water vapor and noncondensable gases from the low-pressure chiller.*

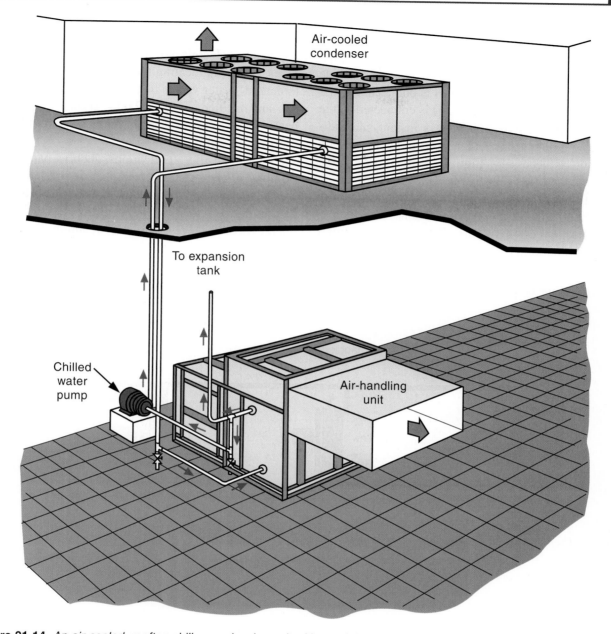

Figure 21-14. *An air-cooled, rooftop chiller condensing unit with an air handler on the first floor.*

Two-pipe system

The system shown in Figure 21-14 is called a ***two-pipe system*** because only two pipes (in and out) serve the coil. For comfort cooling, one pipe delivers cold water to the coil, and the other pipe returns warm water to the chiller. During the heating season, the chiller is shut down, and the closed-loop liquid is directed to a boiler for heating. Hot water is supplied to the coil and used for heating. Thus, a two-pipe system provides either heating or cooling.

The two-pipe systems can be used with multiple coils rather than one large coil. Individual air handlers with coils are located in specific areas of the building. The two main pipes travel the length (or height) of the building; branch circuits to the individual coils are attached to the main feeder pipes. Multiple heat exchangers provide individual temperature control. A thermostat controls an electrical (solenoid) water valve installed in the supply line to the air handler. Booster pumps are frequently used with this system. **Figure 21-15** illustrates the principle of multiple coils on a two-pipe system.

The two-pipe system is inconvenient during the spring and fall since changeover from heating to cooling (or cooling to heating) requires time and effort and is not performed until necessary. Once the system is changed over, it normally remains in position until a permanent seasonal change occurs.

Four-pipe system

A four-pipe system provides a better opportunity for controlling temperature than does a two-pipe system. The *four-pipe system* uses two coils inside the air handler. One coil (cooling) is connected to the chiller, and the other coil (heating) is connected to the boiler. The use of two coils allows instant and individual changeover from heating to cooling, or cooling to heating, as necessary. **Figure 21-16** illustrates the use of two separate coils on a four-pipe system.

A four-pipe system can be used with cooling/heating coils located throughout the building. These special coils are manufactured as one unit but are divided in half; half the coil is for cooling and the other half for heating. The four main pipes travel the length (or height) of the building; branch circuits to the individual coils are attached to the main feeder pipes. A cooling/heating thermostat controls electrical (solenoid) water valves installed in the supply lines to the air handler. **Figure 21-17** shows combination heating and cooling heat exchangers operating on a four-pipe system.

Some industrial applications, such as paper, pharmaceutical, and candy manufacturing are more concerned with *humidity* than temperature. Humidity control may require that both cooling and heating coils operate together. The air is cooled, and moisture drips off the coil into a condensate pan and into a drain.

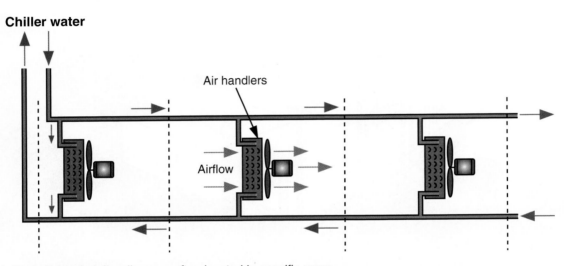

Figure 21-15. *Individual air handlers are often located in specific areas.*

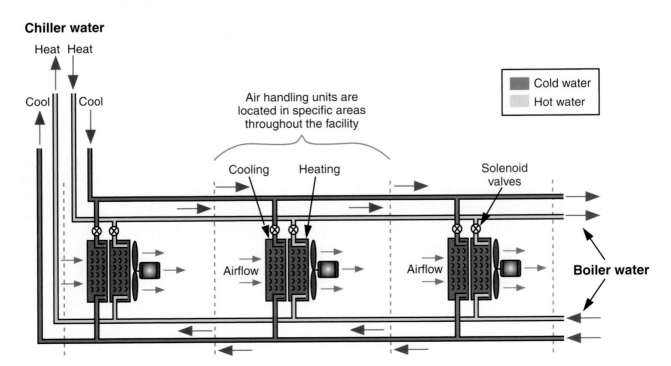

Figure 21-16. *Air-handling units are located in areas that require individual heating, cooling, or humidity control. Thermostats control solenoid valves in the incoming water pipes.*

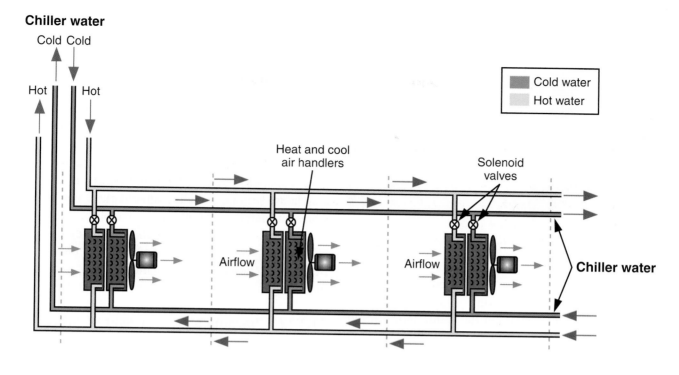

Figure 21-17. *Combination (heat and cool) air handlers operate on a four-pipe system.*

Moisture is removed because cold air cannot hold as much moisture as hot air. The cold air exiting the cooling coil has high relative humidity (100% rh) but low moisture content. The cold air is heated to bring the relative humidity down to the proper level (30% rh to 40% rh) before discharge into the controlled area.

21.2.9 Safety

R-11 and R-113 are classified as A1 refrigerants, and R-123 is classified as B1. ASHRAE Standard 15 requires all equipment rooms containing equipment using an A-1 refrigerant (such as R-11, R-12, R-113, R-134a) be fitted with an *oxygen-deprivation sensor* and alarm. ASHRAE Standard 34 requires a *refrigerant sensor* and alarm in the equipment room when using B1 refrigerants. The refrigerant sensor measures contamination of air in the room by refrigerant and sounds an alarm when contamination exceeds the safe level for humans measured in parts per million (ppm).

In the event of a large release of refrigerant in a confined area, *immediately vacate and ventilate* the area. In large quantities, A1 refrigerants are heavier than air and will displace oxygen, leading to death by suffocation. Inhaling refrigerant vapors or mist may cause heart irregularities, unconsciousness, and oxygen deprivation, leading to death by asphyxia. When A1 refrigerants are exposed to open flames or hot metal surfaces, they decompose and form hydrochloric acid, hydrofluoric acid, and phosgene gas.

21.2.10 Refrigerant Leaks

When leaks occur in the evaporator tubing bundle, water enters the system and causes excessive operation of the purge unit. Water flowing inside the tubes has a positive pressure, while the refrigerant surrounding the water tubes is in a vacuum. Thus, any leaks will permit water to enter the system. Excessive air and moisture entering the system cause the purge unit to operate more frequently. Therefore, frequent operation of the purge unit indicates leaks in the evaporator tubing bundle.

Leaks in the tubing bundle can be located by shutting the chiller down, stopping the water pumps, and closing the *incoming water valve* (closed loop) at the chiller. Water remaining inside the tubes is permitted to drain. Open the drain pipe, and insert the leak detector probe through the water drain valve. The leak detector will indicate if refrigerant is present inside the tubing.

The closed-loop end plates can be removed from the chiller to expose the ends of the water tubes. Removing the end plates is required when cleaning the water tubes. See **Figure 21-18.** A leak detector will determine which tube (or tubes) is leaking. If only one or two tubes are leaking, those tubes can be plugged at each end. Plugging the tubes removes them from active status but does not affect the other tubes. Complete retubing is required when several tubes are leaking or are already plugged. Remember that plugged water tubes result in reduced chiller capacity.

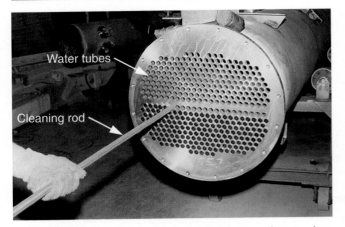

Figure 21-18. *A procedure for cleaning water tubes to clear them of debris and corrosion. (Dunham-Bush)*

Controlled hot water is often used for detecting exterior gasket leaks or making minor repairs. The chiller must be shut down, but the evaporator (closed loop) water pumps remain on. Controlled hot water is permitted to enter the closed-loop system until the water temperature reaches 100°F (37.8°C). The refrigerant pressure increases to a corresponding level of about 9 psig or 62 kPa (no vacuum). *Do not exceed 10 psig (69 kPa), or the rupture disc will collapse and release the refrigerant.* Pressure of no more than 10 psig is sufficient for leak detection.

21.2.11 Recovery and Evacuation

Recovery and evacuation operations for low-pressure chillers require special procedures. The chiller must be shut down, but the evaporator (closed loop) water pumps remain on. Water must be circulated through a chiller when evacuating refrigerant to keep the water from freezing inside the tubes. Recovery begins by removing all liquid refrigerant from the evaporator. You *must* recover the refrigerant vapor after liquid recovery is performed. For instance, once all the liquid has been removed, and the pressure is reduced to 0 psig, an average 350-ton, R-11 chiller still contains 100 lbs. of vapor. When removing vapor from an R-11 chiller, the system water pumps, recovery compressor, and recovery condenser water should all be on.

When using recovery equipment manufactured *before* November 15, 1993, the low-pressure chiller should be evacuated to 25 in. Hg. If the recovery equipment was manufactured *after* November 15, 1993, evacuation is required to 25 mm Hg absolute (29 in. Hg). When leaks in the chiller make evacuation to the prescribed level unattainable, the chiller should be evacuated to the lowest attainable level (or 0 psig). A recovery unit's high-pressure safety cut-out is set for 10 psig when evacuating refrigerant from a low-pressure

chiller. The rupture disc on the recovery vessel (barrel) relieves at 15 psig.

EPA regulations require all systems containing more than 50 lbs. of refrigerant (except commercial and industrial process refrigeration systems) *must* be repaired when the annual leak rate exceeds 15%. Commercial and industrial process refrigeration systems *must* be repaired when the annual leak rate exceeds 35%.

Commercial refrigeration includes systems used to store perishable goods in the retail food and cold storage warehouse sectors, including supermarkets, convenience stores, restaurants, and other food establishments.

Industrial process refrigeration includes complex, customized systems used in the chemical, pharmaceutical, petrochemical, and manufacturing industries, including industrial ice machines and ice rinks.

21.2.12 Charging a Low-pressure Chiller

Charging liquid into a deep vacuum will cause the refrigerant to boil and will freeze water in the tubes. When charging refrigerant, the evaporator (closed loop) water pumps should remain on to circulate water through the chiller. Circulation helps keep the water from freezing inside the tubes. Initial charging should be performed in the vapor phase. A vapor pressure of 17 in. Hg (36° F) should be achieved in the system before charging the chiller with liquid refrigerant.

21.2.13 Safety

ASHRAE Standard 15 requires all equipment rooms containing equipment using an Al refrigerant (such as R-12, R-11, R-134a, R-22, R-502) be fitted with an oxygen-deprivation sensor and alarm. R-123 is classified as B1 and requires a refrigerant sensor and alarm in the equipment room.

21.3 Lithium Bromide Absorption Chiller

The lithium bromide absorption chiller is a large industrial machine used to chill (cool) large quantities of water. The chilled water is used for comfort cooling (air conditioning) or for industrial process cooling of manufactured products. **Figure 21-19** shows a typical lithium bromide absorption chiller. The absorption cycle differs from the compression cycle in that heat energy, instead of mechanical energy, is used to complete the refrigeration cycle. A heater or generator is used instead of a compressor. Gas, steam, or an electric heating element may be the source of heat. Lithium bromide units operate continuously on low power without a compressor.

Figure 21-19. *A lithium bromide absorption chiller with a single-shell design. (Trane)*

The operation of the absorption chiller depends upon an absorbent to which the refrigerant is strongly attracted. The refrigerant is water. **Lithium bromide,** a salt, is the absorbent and typically used in large-capacity absorption chillers. The evaporator is maintained in a vacuum of about 0.248 in. Hg (0.12 psia or 0.83 kPa). The temperature of the refrigerant corresponding to this pressure is approximately 40°F (4°C).

When it is dry, lithium bromide is in crystal form. The concentration of the lithium bromide solution is stated in terms of the amount of lithium bromide (by weight) in the solution. For example, a 100-pound solution may have 65 pounds of lithium bromide and 35 pounds of water. It is called a *65% concentration* because 65% of the total weight is lithium bromide.

As stated, the lithium bromide unit is an absorption machine that uses water as a refrigerant and lithium bromide as the absorbent. The cycle of the machine is based upon two principles:

❑ Lithium bromide solution has the ability to absorb great quantities of water vapor.
❑ Water will boil (flash cool itself) at low temperatures when in a deep vacuum.

These two principles function in the absorption chiller to produce refrigeration. Water is sprayed into a shell that is maintained at a high vacuum. A portion of the water boils to a vapor due to the reduced pressure. Heat required for the boiling process is obtained from the remaining water, reducing its temperature (called *flash cooling*). The water vapor is then absorbed by the lithium bromide solution.

21.3.1 Evaporator and Absorber

Figure 21-20 illustrates the two principles just described. One shell is partially filled with lithium bromide solution, and the other shell contains water. A pipe connects the two shells. Air is removed so only water vapor and lithium bromide are present. The lithium bromide absorbs the water vapor by way of the connecting pipe, reducing the pressure on the remaining water. As the pressure is reduced, water boils to generate more vapor. Heat for the boiling process is obtained from the remaining water.

As shown in **Figure 21-21,** a pump circulates water in the evaporator through a spray header to facilitate the absorbing process. Liquids tend to vaporize easily if sprayed from a nozzle. A chilled water coil (closed loop) located inside the evaporator carries return water from air conditioning coils or manufacturing process cooling coils. The chilled water coil is cooled by the flash cooling of evaporator water on the outside of the tubes.

Another liquid pump is used by the absorber to circulate lithium bromide solution through a spray header. The lithium bromide spray header is an efficient means of facilitating water vapor absorption.

21.3.2 Generator

As the lithium bromide absorbs water, it becomes diluted, and its absorption ability decreases. Another pump circulates the weak solution to a generator where heat is applied to separate water vapor from the lithium bromide, **Figure 21-22.** The solution becomes concentrated again before returning to the absorber where it absorbs more water vapor. The heat is supplied by steam condensing in the generator tubing bundle. The weak solution going to the generator must be heated, while the strong solution coming from the generator must be cooled. A heat exchanger in the solution circuit transfers heat and conserves energy.

21.3.3 Condenser

Water vapor released from the weak solution in the generator passes to the condenser, **Figure 21-23.** The water vapor condenses on the cool condenser tubes and returns to the evaporator. The cycle is completed without loss of water or solution.

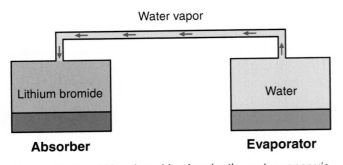

Figure 21-20. *Lithium bromide absorbs the water vapor via the connecting pipe, reducing pressure on the remaining water. Reduced pressure allows the water to boil, generating more vapor.*

Before the water travels through the condenser tubes, it must travel through the absorber tubing bundle. The condenser water absorbs the "heat of dilution" generated by the solution as it absorbs water vapor in the absorber.

21.3.4 Lithium Bromide Absorption Cycle

Figure 21-24 shows the arrangement of the absorption chiller components. The lower shell, called the **absorber shell,** contains the absorber and the evaporator. The absorber and the evaporator are combined in a single shell because they operate at nearly the same temperature and pressure. The upper shell, called the **generator shell,** contains the generator and the condenser. Refer to Figure 21-24 as you read the following description of the absorption cycle:

1. The evaporator water is pumped from the bottom of the evaporator to the evaporator spray header and is sprayed over the chilled water tubing bundle (closed loop). The chilled water is cooled by the flash-cooling process occurring in the evaporator.
2. The water vapor formed in the evaporator is absorbed by the lithium bromide solution sprayed in the absorber. The absorber pump delivers solution from the bottom of the absorber to the solution spray header.
3. The generator pump removes the weak solution from the absorber and pumps it through the heat exchanger to the generator.
4. Steam circulates inside the generator tubing bundle to boil the solution.
5. The solution returns to the absorber through the heat exchanger.
6. The cooling tower water passes through the absorber coil where it picks up the "heat of dilution;" then, it travels to the condenser where it picks up the "heat of condensation."
7. Water vapor boiled off in the generator is condensed in the condenser, and the water returns to the evaporator.

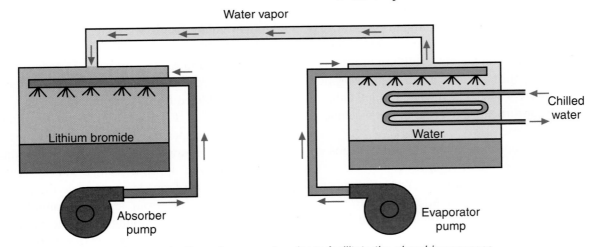

Figure 21-21. *A pump circulates water through a spray header to facilitate the absorbing process.*

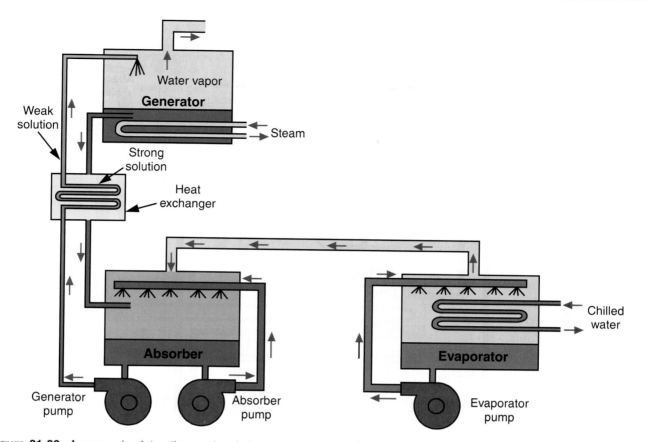

Figure 21-22. *A pump circulates the weak solution to a generator where heat is applied to separate water vapor from the lithium bromide. A heat exchanger transfers heat and conserves energy.*

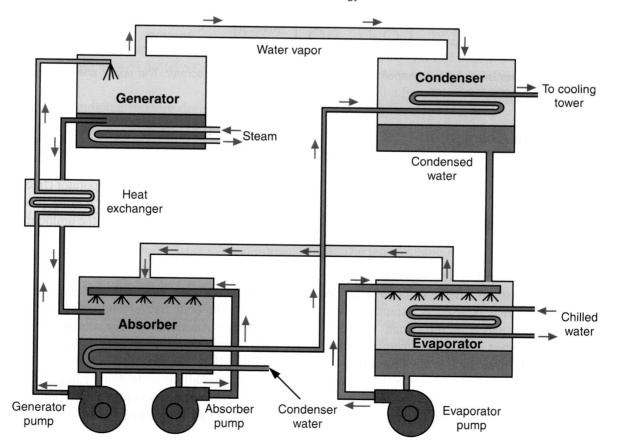

Figure 21-23. *Water vapor released from the weak solution in the generator passes to the condenser where it condenses on the cool tubes and returns to the evaporator.*

13. What type of pipe is used to build the loops for circulating liquid?
    a. Copper.
    b. Iron.
    c. Plastic.
    d. All of the above.
14. How is the liquid forced through the loops?
15. Which system provides better temperature control, two-pipe or four-pipe?
16. What type of alarm is required for A1 refrigerants?

17. When A1 refrigerants decompose, they form:
    a. Hydrochloric acid.
    b. Hydrofluoric acid.
    c. Phosgene gas.
    d. All of the above.
18. What type of alarm is required for HCFC-123?
19. What type of chiller uses water and lithium bromide for cooling?
20. *True or false?* Controlled hot water is used to pressurize a low-pressure chiller.

Service technician is using a clamp-on ammeter. (Fluke)

What Is Electricity?

 **Objectives**

After studying this chapter, you will be able to:
- ❑ Define electricity.
- ❑ Demonstrate proper use of electrical test meters.
- ❑ Describe the application of Ohm's Law.
- ❑ Identify open circuits and short circuits.
- ❑ Determine proper wire size for various uses.
- ❑ Select and use proper insulation on conductors.
- ❑ Describe how to connect loads and switches in electrical circuits.
- ❑ Describe the functions of various types of plugs, receptacles, and switches.
- ❑ Demonstrate how to safely connect portable electric power tools.

Important Terms

alternating current	direct current
ammeter	disconnects
ampacity	DPDT switch
amperage	DPST switch
ampere	duplexes
atoms	electrical power
busbar	electricity
circuits	electromotive force (emf)
circular mils	electrons
conductance	frequency
conductors	fuses
conduit	ground fault
contacts	grounded
continuity	insulators
coulomb	intensity
counter-electromotive	line voltage
force (c-emf)	load
current	lockout
dead short	multimeter

National Electrical Code (NEC)	short circuit
neutral	siemens
neutrons	single-phase loads
normally closed	SPDT switch
normally open	SPST switch
nucleus	stranded
ohm	switches
ohmmeter	tagout
Ohm's Law	three-phase loads
open circuit	throw
parallel	valence electrons
phases	ventricular fibrillation
pole	volt
polyphase generation	voltage
potential difference	voltage drop
protons	voltmeter
reactive	VOM
resistance	watt-hour meter
resistive	wattmeter
safety ground	watts
schematic	wires
series-parallel	zero potential
	zero resistance

22.1 Understanding Electricity

In industrialized nations such as the United States, electricity is an integral part of nearly everyone's life. So essential is it that a single power-station failure can "immobilize" thousands of individuals and businesses until power is restored. Electricity is taken for granted until problems occur. People simultaneously rely on electricity and fear it. Most people are afraid of electricity because they know it can cause severe burns, shock, or even death. Knowledge eliminates fear, but you must never lose *respect* for electrical energy. The potential for injury or death is very real.

About 70% of heating and cooling service calls are for electrical problems. Understanding how electricity controls and operates heating and cooling equipment is essential for troubleshooting system problems and performing routine repairs.

This chapter explains the release of electrical energy by electron movement between atoms. Electromotive force and proper use of voltmeters are described as well. These basic principles lay the foundation for a discussion of electron movement and the use of ammeters. Explanations of resistance, Ohm's Law, and the use of ohmmeters clarify how electrical energy is converted to another form of energy. Watt meters and kilowatt-hour meters as measurements of electrical power are also explained.

22.1.1 Energy Conversion

Electricity is a form of energy. Other forms of energy are heat, light, chemical (battery), atomic (power plant), and mechanical (motor). Energy cannot be destroyed, but it can be converted from one form to another. For example, heat is used to produce electricity, and the electricity is used to produce light. The reverse is also possible. Photoelectric cells convert light into heat, and the heat is used to make hot water. Electrical appliances and devices *convert electrical energy to another form of energy* to perform useful tasks.

22.1.2 What Is Electrical Energy?

Electricity is a form of energy in which electrons move from one atom to another atom. All matter is composed of **atoms,** the smallest particle of any element. Although atoms are tiny, they are a vast source of potential energy. The center of an atom is called the nucleus. The **nucleus** contains protons and neutrons. **Protons** have a positive charge (+) and **neutrons** have no charge (neutral). The atomic number of an element indicates the number of protons in its nucleus. Orbiting around the nucleus are **electrons** carrying negative charges (–). The number of orbiting electrons normally equals the number of protons in the nucleus. The positive charges in the nucleus exert a strong magnetic force that holds the negatively charged electrons in orbit around the nucleus. See **Figure 22-1.**

The simplest atom is the hydrogen atom; it contains one proton and one electron. The proton has a positive charge (+) that exactly balances the negative charge (–) of the electron. Other elements have more than one proton and one electron in their atoms. All electrons possess the same *negative charge*, while all protons possess an equal *positive charge*. Their charges are equal and opposite; hence, they neutralize

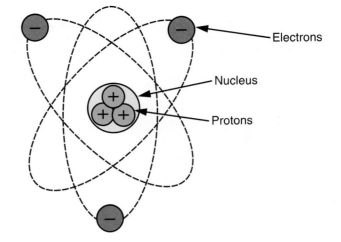

Figure 22-1. *Negatively charged electrons orbit around the nucleus of an atom. The nucleus contains neutrons (which have no electrical charge) and positively charged protons.*

each other. Any stable or electrically balanced atom has an equal number of protons and electrons. A stable atom is said to have a **neutral** charge.

An atom can have from one to over 100 electrons, depending on the element. Oxygen has an atomic number of 8; therefore, it has 8 electrons orbiting around the nucleus. Copper has an atomic number of 29; it has 29 electrons. Uranium has an atomic number of 92, so it has 92 electrons. Electrons orbit around the nucleus much like the planets orbit around the sun. Electrons travel in pathways called shells, and an atom can have as many as seven shells. Electrons traveling in shells located close to the nucleus are more tightly bound to the nucleus than those in the outer shells. The outer shell can hold no more than eight electrons. Electrons located in the outer shell are called free, orbital, planetary, or **valence electrons.** See **Figure 22-2.**

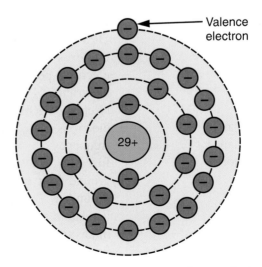

Figure 22-2. *Copper is a good conductor of electricity because it has one valence electron. All the energy in the valence ring is contained in this one electron.*

22.1.3 Electron Movement

The number of valence electrons in an atom's outer shell is meaningful in the study of electricity. Electrons orbiting close to the nucleus contain less energy than those orbiting farther away. The farther an electron orbits from the nucleus, the greater the electron's energy. If enough energy is added to an electron, it will move out of its orbit to the next outer shell. If enough energy is added to a valence electron, it will move out of its atom to another atom because no higher shell exists.

The energy released by the movement of free electrons from one atom to another is called *electricity.* Since valence electrons have the highest energy level, they are most easily set free. The amount of energy supplied to the valence shell is limited and is distributed equally among the electrons orbiting in that shell. The more valence electrons there are, the less energy each electron receives.

Electrons in close orbit around the nucleus are tightly bound to the nucleus and can leave only with great difficulty. As stated, free (valence) electrons traveling farthest from the nucleus are lightly held. Free electrons can be forced to pass from one atom to another. When an atom loses an electron, it becomes positively charged (+) because it has an excess proton. If an atom has an extra electron, it becomes negatively charged (–). See **Figure 22-3.**

Atoms try to maintain an equal number of positive and negative charges (protons and electrons). The Law of Electric Charges states: *Like charges repel and opposites attract.* When a substance with excess electrons contacts a substance lacking electrons, the electrons are attracted and flow to the atoms lacking electrons. To make electricity perform useful work, a constant and steady movement of electrons must be produced.

22.1.4 Potential Difference

When the electrical charge (positive versus negative) between two points is not in balance, the imbalance of electrons is called a *potential difference.* With a potential difference, excess electrons have accumulated and are waiting for an opportunity to connect with atoms lacking electrons. See **Figure 22-4.** The unit of measurement for potential difference is the *volt.* Potential difference (or *voltage*) is another way of describing the strength of the electromotive force (emf) that causes electrons to move. To *perform work* (carry out a useful function), a potential difference must be maintained while electrons are moving.

A variety of energy sources can be harnessed to create a potential difference (electromotive force) between two points. Some of these sources are friction

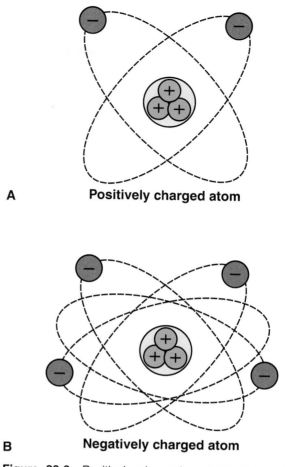

A   **Positively charged atom**

B   **Negatively charged atom**

Figure 22-3. *Positively charged and negatively charged atoms. A—If an atom has three protons in its nucleus but only two electrons, it has three positive charges and two negative charges. This makes the atom positively charged overall. B—With three positively charged protons and four negatively charged electrons, this atom has an overall negative charge.*

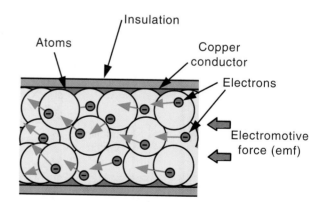

Figure 22-4. *Electrons already exist inside the copper atoms. Electromotive force (voltage) causes the electrons to travel to the next atom, creating a domino effect throughout the length of the wire.*

(static electricity), chemical energy (battery), thermoelectric energy (heat), photoelectric energy (light), and magnetic energy (generator). Typical measurements of electromotive force (voltages) include: 1.5 volts (V) for a flashlight battery, 6 V for lantern batteries, 9 V for radio batteries, 12 V for automobile batteries, 120 V for a house, and 240 V for commercial power.

The terms *potential difference, electromotive force (emf),* and *voltage* mean the same thing and are used interchangeably.

22.1.5 Electromotive force

Electromotive force (emf) is the stimulus that causes free electrons to reel out of orbit and travel to an adjoining atom. In commercial power production, the local power plant generates the electromotive force that causes electron movement. The power plant produces enough force to maintain high voltage through wires over long distances. The high voltage is reduced to normal levels before it reaches consumers. The basic unit of voltage, the volt, is often too large (or too small) for practical use. The following information shows how fractional or multiple values may be used instead:

Units of Voltage

1 microvolt (μV) = 1/1,000,000 V
1 millivolt (mV) = 1/1000 V
1 kilovolt (kV) = 1000 V
1 megavolt (MV) = 1,000,000 V
1 volt = 1000 mV
1 volt = 1,000,000 mV
1 volt = 0.001 kV
1 volt = 0.000001 MV

In 1798, Alessandro Volta, an Italian physicist, discovered and named the principle of *electromotive force (emf).* Mathematicians further abbreviate emf with the letter E; most people, however, refer to electromotive force as *volts.* Electromotive force is *not* electricity. It is the driving force that causes electrons to move from one point to the other. To obtain electron flow (current), a potential difference must exist between two points. The flow of electrical current is much like the flow of water through pipes. The pipes serve as a "circuit" (pathway) for water traveling to a lower pressure. The force (voltage) is comparable to water pressure.

The quantity of emf (volts) is measured with an instrument called a **voltmeter, Figure 22-5.** Voltmeters measure the potential difference between two specific points. The voltmeter has two probes and detects a difference of potential between the probes. Some voltmeters provide very accurate readings, while others make approximate readings. Voltmeters are used to check electrical circuits for problems or to verify proper operation. To successfully troubleshoot electrical problems, the probes must be properly placed and the meter readings correctly interpreted.

Electrical energy conversion devices (called *loads*) are designed for connection between a specific potential difference. Proper voltage must be supplied to force electron movement through the device. The load converts the electrical energy to another form of energy. The connecting wires supply the necessary electrons and provide a complete circuit (pathway) for electron flow. See **Figure 22-6.**

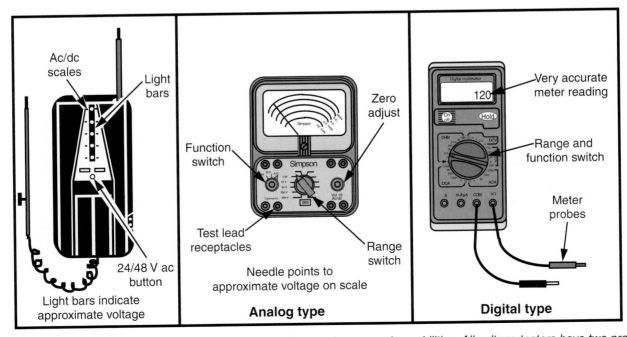

Figure 22-5. *Voltage testers are available in a variety of types, shapes, and capabilities. All voltage testers have two probes and a meter that indicates the potential difference (volts) between the two probes. A zero reading means there is no difference between the probes.*

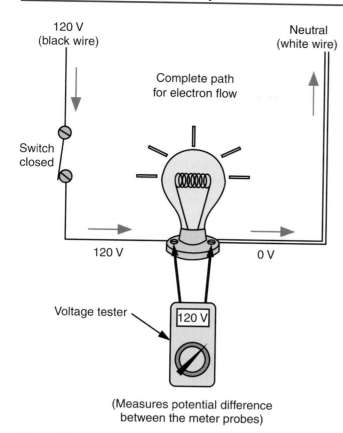

(Measures potential difference
between the meter probes)

Figure 22-6. *An energy-conversion device (such as a light-bulb) must be part of a complete circuit, or pathway, for electron flow. Voltage (electrical pressure) pushes the electrons through the filament (resistance) inside the lightbulb. A potential difference must exist across the bulb for electrons to flow.*

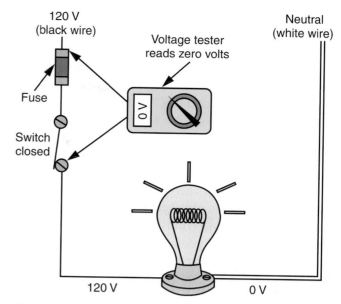

Figure 22-7. *The voltmeter reads zero when the potential is the same at both probes. No potential difference exists across a closed switch or an intact fuse.*

Voltage is the force that causes electrons to move. Electrons cannot move unless they have a place to go. They cannot flow through an open switch or a broken wire. An opening in the circuit (pathway) acts like an open drawbridge. Traffic can flow through a closed drawbridge, but traffic stops when the drawbridge is open. Switches in electrical circuits act as "drawbridges" to start and stop the flow of electrons.

A potential difference between the two probes must exist for a voltmeter to register a reading. The voltage tester reads zero when voltage and polarity are the same at both probe locations. A voltmeter can also be used to check fuses and switches. **Figure 22-7** illustrates how a voltage tester can be used to discover an open switch or a blown fuse.

When a load is connected to both sides of a potential difference, the load should operate. If the device (such as a lightbulb) does not operate, it is defective. A voltage tester can quickly detect a defective device, **Figure 22-8.**

22.1.6 Electrical Safety

Electrical **circuits** (pathways) are composed of wires and devices that control the flow of electrons

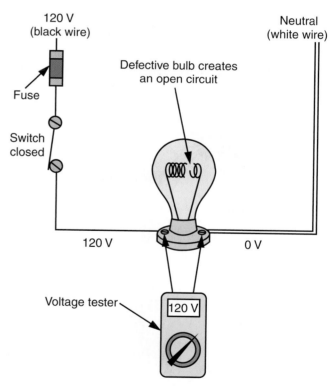

Figure 22-8. *A defective bulb causes an opening in the complete path between points of potential difference. Electrons cannot flow in an open circuit because a complete circuit, or pathway, does not exist.*

(**current**). Electrons should travel within the electrical circuit; electrons on any other pathway are dangerous. Your body is zero voltage and could serve as a pathway for current to flow from a voltage source (black wire), through your body, and to the ground. Never permit your body to become part of an electrical circuit!

Electricity is energy looking for someplace to go, and if your body provides a pathway for a current to travel, it will indeed travel through your body. Thus, your body becomes an energy conversion device generating heat.

You have heard it said that "electricity travels to the ground." The earth (ground) is an excellent conductor of electricity. Electrons can use the ground as a pathway (circuit) back to the power source. The human body can easily complete an electrical pathway to the ground, becoming an electrical energy conversion device if you make contact with an electrically live (hot) wire at the same time you make firm contact with the earth. Damp concrete is a good conductor to the ground, especially if the concrete is reinforced with steel rods. Wet shoes make a very good connection to the ground. See **Figure 22-9.**

Approximately 1000 people are killed by electricity every year. About one-half of these electrocutions occur on low voltage; ordinary household current can kill in a fraction of a second. These deaths usually occur because people take chances.

Never let your body become a pathway for electron flow. When electricity is contained within the conductors and devices where it is supposed to function, there is nothing to fear. When damage occurs within the device, however, you are likely to get shocked. An electrical shock may not be fatal, but it can cause you to jump or fall off a ladder or scaffold and injure yourself. Never become overconfident or take unnecessary risks. Your life may depend on it!

WARNING: Never work on an electrical circuit unless it has been de-energized (turned off). Use a voltage tester to check the circuit *before* and *after* it is turned off. Always recheck the voltage tester on a live circuit to be certain the meter is working properly. (Meters and probes can become damaged and give false readings.) Be certain the circuit cannot be energized while repairs are being made. Lock the main switch in the "open" position (called *lockout*), **Figure 22-10.** Also, tag all operator switches to inform others that repairs are in process (called *tagout*). See **Figure 22-11.**

22.1.7 How Electrons Move

A power plant does not *produce* electrons; electrons are already present inside the copper conductors (**wires**). The copper conductors provide the necessary free electrons and a pathway for electron movement. The electromotive force produced by the power plant generator forces free electrons inside the copper conductor to travel to the next atom within the conductor. Electron movement (from one atom to another) occurs throughout the length of the conductor (refer to Figure 22-4). The movement of electrons in a conductor is very rapid, creating a domino effect. An electrical impulse travels through a conductor at the speed of light (186,000 miles per second). It is this electron movement that produces electricity.

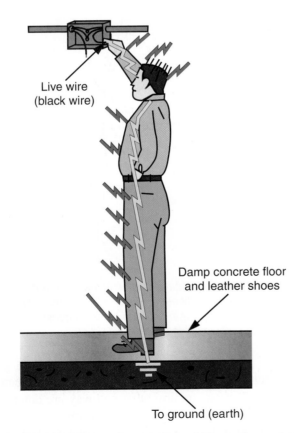

Live wire (black wire)

Damp concrete floor and leather shoes

To ground (earth)

Figure 22-9. *Always exercise care to avoid becoming part of an electrical circuit. Your body could provide a path for electron flow, with painful (or even fatal) results.*

DANGER

DO NOT OPERATE

EQUIPMENT LOCKED OUT BY

Name: *T.J. Martin*

Dept.: *MAINTENANCE*

Date: *6/10*   Time: *12:30*

Figure 22-10. *Lockout tags like this one help prevent accidents.*

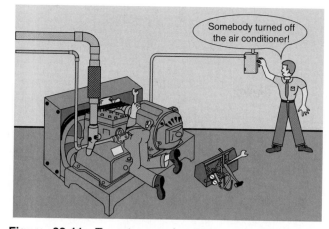

Figure 22-11. *Tagout procedures prevent the circuit or equipment from being energized while you are working on it. Failure to follow tagout procedures could have serious, even fatal, results.*

To illustrate the rapid movement from one atom to another, picture a long plastic tube filled with ping-pong balls reaching more than 800 miles from New York City to Chicago. See **Figure 22-12.**

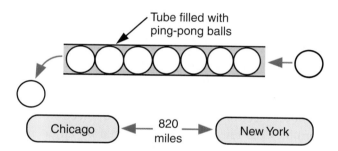

Figure 22-12. *Ping-pong balls illustrate how electrical impulse travels at high speed over long distances. If a ball is inserted at one end of the tube, another ball immediately emerges from the other end.*

If someone in New York City inserts another ball, a ball *immediately* pops out in Chicago. Each ball only moved a short distance, but the *impulse* of bumping each ball traveled at extremely high speed. If additional balls are rapidly inserted at New York City and permitted to fall out in Chicago, it would be an example of *direct current.* Electron flow would be in one direction. If additional balls were alternately inserted at each end, it would be an example of *alternating current:* movement would alternate first in one direction, then in the other direction.

22.1.8 Amperage

In 1836, Andre Ampere, a French physicist, measured the number of electrons flowing past a given point in one second. The unit of measurement was called a *coulomb* of electricity. A **coulomb** equals

6.25×10^{18}, or 6,250,000,000,000,000,000 electrons flowing past a given point in one second. The term "coulomb" is now used only in textbooks as a measure of electron flow. The unit **ampere,** named for the scientist, is the term most often used to describe the flow of electrical current.

The following list gives fractional values for ampere:

Units of Current

1 ampere (A) = 1 coulomb/sec.
1 milliampere (mA) = 1/1000 A
1 microampere (μA) = 1/1,000,000 A
1 ampere = 1000 mA = 1,000,000 μA

The words *ampere, amperage, amps, current,* and *flow* all describe the movement of electrons. Because of the vast amount of energy generated by electrons, their movement must be controlled to produce the desired amount of energy conversion. Toasters, light-bulbs, and motors are examples of energy-conversion devices (loads). An **ammeter** measures the amount of electron flow. The clamp-on ammeter is most common and is available with a needle pointer and scale (analog type) or a digital readout. See **Figure 22-13.**

When electrons flow through a conductor, a magnetic field is produced around the conductor. The strength of the magnetic field is determined by the amount of electrons flowing. High current flow produces a strong magnetic field, while low current flow produces a weak magnetic field. When electrons are not flowing, no magnetic field exists around the conductor. See **Figure 22-14.**

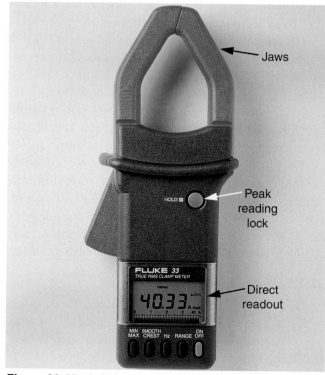

Figure 22-13. *A digital clamp-on ammeter. (Fluke)*

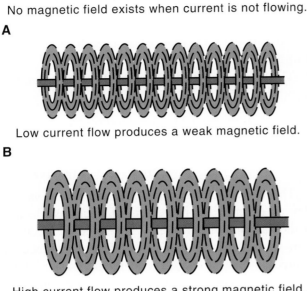

No magnetic field exists when current is not flowing.

A

Low current flow produces a weak magnetic field.

B

High current flow produces a strong magnetic field.

C

Figure 22-14. *The strength of the magnetic field around a conductor is determined by current flow. A—When no current is flowing, no magnetic field is generated. B—When current flow is low, a weak field is generated. C—When current flow is high, a strong magnetic field results.*

Figure 22-15. *When the ammeter's jaws are clamped around a conductor, the meter reads the intensity of the magnetic field. The meter converts the information to an amperage reading.*

When the jaws of an ammeter are clamped around a single conductor, the meter measures the ***intensity*** (strength) of the magnetic field. The ammeter converts the information to an amperage reading. See **Figure 22-15.** Because the ammeter measures intensity, the letter I is used in mathematical calculations to indicate intensity of electron flow.

The ammeter is designed to read the magnetic field around *one* wire. Do not clamp the jaws of the meter around two conductors at the same time. The two wires will be of opposite polarity (one positive and one negative) because the currents are flowing in opposite directions. The two magnetic fields will cancel one another and give a false meter reading of zero.

The clamp-on ammeter is a valuable tool for troubleshooting circuits. It is generally a multimeter, capable of reading volts, amperes, and resistance on several scales. Adapters and probes are included for easy conversion to another type of reading, and complete instructions are included with each ammeter.

22.1.9 Resistance

Electron flow is energy in motion. In electricity, ***resistance*** refers to anything slowing the flow of electrons. All materials offer some resistance to the flow of electrons, some more than others. Resistance converts electrical energy to another form of energy.

The amount and type of resistance determine the outcome of the energy-conversion process. Resistance performs the energy conversion and controls the amount of electron flow. The amount of electron flow determines the rate of work performed by an energy conversion device (load). Loads convert a specific amount of electricity into another form of energy. A load cannot function properly if the electron flow (***amperage***) is too high or too low.

Types of resistance

There are two types of resistance: *resistive* and *reactive*. A resistance that is ***resistive*** remains constant (does not change). Examples of devices with fixed resistance are incandescent lightbulbs, toasters, and electric heaters. Each has a fixed (constant) resistance to controlled electron flow during operation.

With a fixed resistance, higher voltage increases current flow and lower voltage decreases current flow. If resistance is increased, however, current flow decreases; if resistance is decreased, current flow increases.

The ***reactive*** form of resistance is found in devices that have a magnetic coil. The coil produces a magnetic field that creates a voltage acting in direct opposition to the supply voltage. The *counter-voltage* causes an increased resistance that decreases current flow. Counter-voltage is generally described as ***counter-electromotive force (c-emf).*** Motors, transformers (primary side), and solenoid coils are some devices that produce magnetic fields. The amount of counter-emf generated is determined by the strength of the magnetic fields. The c-emf acts as additional resistance, created whenever the device is operating. Counter-emf is further explained in Chapter 24.

22.1.10 Human Beings as Loads

Human skin differs in resistance, depending upon whether the skin is damp or suffering from a cut. Damp skin normally has a resistance of about 1000 ohms. Dry skin has more resistance, and certain types of clothing or shoes can provide some additional insulation. This may explain why some electrical shocks do not result in death.

Three factors determine whether or not an electrical shock is fatal:

❑ How much current (amperage) passes through your body.

❑ How long (duration) the current passes through your body.

❑ Whether the current passes through your heart.

A milliampere (ma) is 1/1000 of an ampere. About 100 ma to 200 ma of current will kill a person. This is about 1/10 of the current required to operate a 100 W bulb.

Safe current values

❑ *1 ma* — Causes no sensation, nothing felt.

❑ *1 ma to 8 ma* — Shock sensation, but not painful. No loss of muscle control.

Unsafe current values

❑ *8 ma to 15 ma* — Painful shock. No loss of muscle control.

❑ *15 ma to 20 ma* — Painful shock. Hand muscles contract. Cannot let go.

❑ *20 ma to 75 ma* — Painful shock. Severe muscle contractions. Breathing is extremely difficult.

❑ *100 ma to 200 ma* — Painful shock and **ventricular fibrillation** of the heart. This is like a pacemaker telling the heart to beat 3600 times per minute! It is a fatal heart condition for which there is no known remedy or resuscitation. It usually means *death* because sophisticated equipment is required to quickly restore normal heartbeat.

❑ *Over 200 ma* — Severe burns. Muscular contractions so severe that chest muscles clamp the heart and stop it for the duration of the shock, preventing ventricular fibrillation. Artificial respiration should be administered immediately. In most cases, the victim can be revived if removed from the circuit quickly.

22.2 Ohm's Law

The relationship between voltage (E), amperage (I), and resistance (R) was discovered by the German scientist, George Ohm, at the beginning of the 19th century. The relationship, called **Ohm's Law,** is used to predict and control the activity of electricity. The unit of resistance is called the **ohm.**

Ohm's Law proves that E = I × R (voltage equals amperage times resistance). The mathematical formula can be depicted as a pie cut into three pieces, **Figure 22-16.** The horizontal line cut across the middle indicates *division,* and the vertical line cut from the middle indicates *multiplication.*

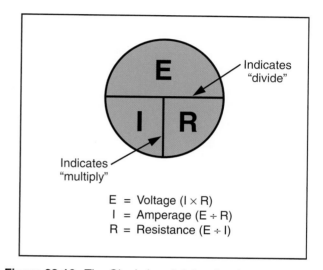

E = Voltage (I × R)
I = Amperage (E ÷ R)
R = Resistance (E ÷ I)

Figure 22-16. *The Ohm's Law "pie" makes it easy to determine the unknown value of an electrical circuit when the other two values are known. The horizontal line indicates division; the vertical line indicates multiplication.*

To use the pie, cover the item to be determined (the unknown). Then, perform the math operation indicated by the horizontal or vertical line. For example, to discover E, multiply I × R; to discover I, divide R into E. The English letter R or the Greek letter Ω (omega) indicate resistance (ohms). Ohm, the basic unit of resistance, is often too large or too small for practical use. The following information gives fractional or multiple values that avoid the use of large (or very small) numbers.

Units of Resistance

R = Resistance = Ohms = Ω
1 kilohm (kΩ) = 1000 Ω
1 megohm (MΩ) = 1,000,000 Ω
1000 ohms = 1 kΩ = 0.001 MΩ

22.2.1 Ohmmeter

An **ohmmeter** is an instrument for accurately checking resistance. Ohmmeters are available in analog or digital type; however, the digital meters provide the most accurate readings. Ohmmeters measure resistance according to the voltage supplied by a battery inside the meter. Analog meters provide an adjustment knob for recalibrating the meter to compensate for a weak battery. Digital meters have automatic recalibration. See **Figure 22-17.**

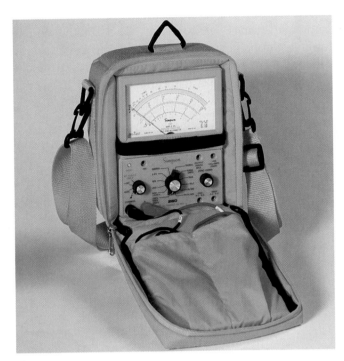

A

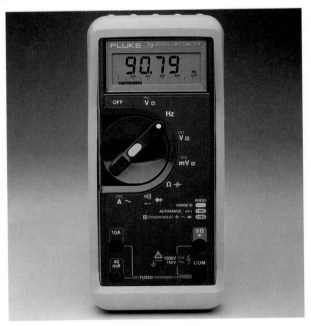

B

Figure 22-17. *An ohmmeter contains its own power supply (a small battery) and reads resistance to the battery voltage. Most technicians prefer a multimeter, like those shown, that can read resistance, voltage, or amperage. A—Analog multimeter. Resistance values (in ohms) are read from the uppermost scale. (Simpson Electric Co.) B—Digital multimeter. (Fluke)*

Continuity

An ohmmeter is used to check **continuity.** Continuity registers **zero resistance,** indicating a complete path for electron movement from one point to another. Continuity proves there are no broken wires, blown fuses, or open switches in the electrical circuit (path of flow). See **Figure 22-18.**

When checking continuity, the adjustable-type ohmmeter is set to the lowest resistance scale (R × 1). Do not use the meter multipliers (R × 10, R × 100). Electrical circuits normally contain minimal resistance to electron flow. Very low resistance cannot be determined unless the meter is calibrated to multiply the reading. However, most digital type (LED) ohmmeters automatically multiply until a small decimal (fractional) resistance reading is obtained.

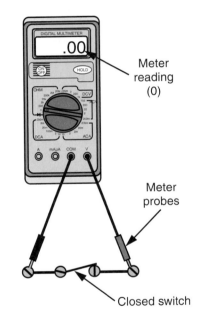

Figure 22-18. *A reading of zero resistance (0 Ω) shows continuity in the component or circuit. A complete path exists for electron movement. Continuity indicates no broken wires, open switches, or blown fuses.*

Open circuit

Electrons cannot flow in an **open circuit.** The ohmmeter indicates an open circuit by registering unlimited resistance, because resistance is too high for the meter to measure it. Unlimited resistance is indicated by the symbol for infinity (∞). An analog meter shows unlimited resistance when the needle goes all the way over to the peg (called *pegged out*). A digital meter shows unlimited resistance by displaying the letters OL, or "overloaded," because the meter cannot read such a large number (infinity).

An open circuit indicates an open switch, blown fuse, or broken wire, among other possible failures.

A failure *disconnects* the circuit, preventing electrons from flowing across the opening. A drawbridge is a good analogy: when the bridge opens (infinite resistance), traffic (electron flow) stops. The electrons must wait until the drawbridge recloses (zero resistance) before continuing their flow. See **Figure 22-19.**

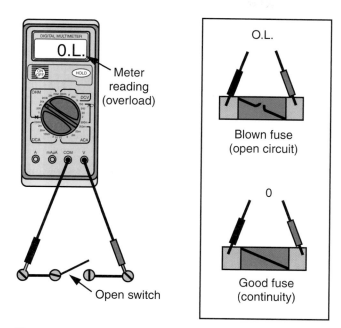

Figure 22-19. *Electrons cannot flow in an open circuit (a break in the electrical path). The ohmmeter indicates total resistance to electron flow by showing large numbers (999999), infinity (∞), or overload (OL).*

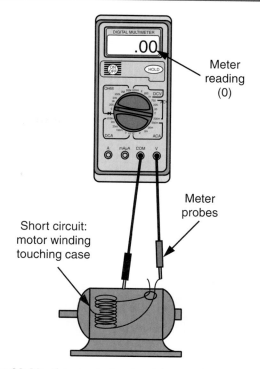

Figure 22-20. *If a short circuit exists, an ohmmeter shows no resistance between points where resistance should exist. A short circuit is a dangerous condition.*

Short circuit

The ohmmeter detects a **short circuit** by indicating zero resistance between two points that *should* have resistance. In other words, a complete circuit (continuity) exists where none should. See **Figure 22-20.**

A short circuit provides a "short cut" for unlimited electron movement (current flow). Some type of resistance is required to control current flow. However, in a short circuit there is none (zero resistance). Too much current flow causes wires to become hot, fuses to blow, fires to occur, and devices to explode. A short circuit is a dangerous situation because current flow is unrestricted and *unlimited*.

22.2.2 Multimeters

Electrical test meters can be combined into one instrument that performs multiple functions. The instrument is called a **multimeter** or **VOM** *(volt-ohm-milliameter)*. A milliampere is 1/1000th of an ampere, so the milliampere range on the meter permits the detection of very small currents. To achieve proper results with a multimeter, probes must be carefully

placed and the dial settings correctly chosen. Always consult the manufacturer's instruction manual if you are unsure of proper probe hookup. Mistakes cause false readings and could result in meter damage.

22.3 Power and Watts

An English scientist, James Watt (1763–1819), determined how to measure electrical power. **Electrical power** is the rate at which electrons are harnessed to perform a useful function. The function (or task) is measured in units called **watts.** Watts are calculated by multiplying amperage (I) times voltage (E), expressed as $W = I \times E$, or $P = I \times E$. You can use either the letter P (for power) or the letter W (for watts) because both represent the same thing. The basic unit of power, the watt, is often too large or too small for practical use. Instead, fractional or multiple values are used as follows:

Units of Power

Power = Watts (W)

1 watt (W) = 0.00134 horsepower (hp)

1 horsepower (hp) = 746 W

1 kilowatt (kW) = 1000 W

1 megawatt (MW) = 1,000,000 W

1 milliwatt (mW) = 1/1000 W

1 microwatt (μW) = 1/1,000,000 W

1000 kW = 1 MW

1000 W = 1 kW = 0.001 MW

22.3.1 Wattmeter

A *wattmeter* measures the power (wattage) used by an appliance or house. The wattmeter is plugged into a power outlet, and the appliance is plugged into the wattmeter. Many electrical devices, such as lightbulbs, window air conditioners, and resistance heaters indicate a wattage number on the data plate or schematic. The wattage number tells the amount of power needed to operate the device. The wattage rating is often converted to terms more useful for troubleshooting. See **Figure 22-21.**

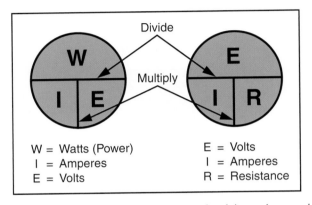

W = Watts (Power)   E = Volts
I = Amperes   I = Amperes
E = Volts   R = Resistance

Figure 22-21. *Power can be determined by using a pie diagram like the one on the left. It is used in the same way as the Ohm's Law pie on the right.*

The main purpose of any electric circuit is to carry electricity to energy conversion devices (loads). The amount of energy converted by the loads (the rate of work) depends upon three factors: voltage, amperage, and elapsed time. A *watt-hour meter* measures the electrical energy used by a household or business. **Figure 22-22** shows the meter normally located where the electrical power supply wires enter the building. The wattmeter measures how many thousands (kilo) of watts are used per hour (kWh). The watt-hour meter performs several functions at the same time. It measures voltage, amperage, and elapsed time. Then, the meter automatically multiplies the three quantities and records the number of kilowatt-hours used on its circular dials. The utility company sends a representative to read the meter dials every one or two months to record the number of kilowatt-hours used since the last reading.

22.4 Conductors

Electrical wires (*conductors*) provide the necessary free electrons and serve as the pathway for electron flow. Metals are conductors of electricity, but not all metals conduct with equal ability. All metal conductors have an internal resistance that opposes the flow of electrons. While *resistance* is the opposition to

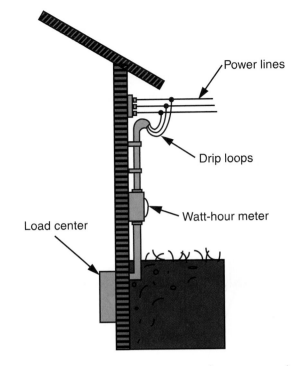

Figure 22-22. *The electric meter on a house or apartment building is a watt-hour meter that records a household's power consumption.*

flow, *conductance* is the ease with which the current flows. Conductance is the number of amperes flowing in a conductor per volt of applied emf and is measured in units called *siemens.*

A conductor that has high internal resistance automatically converts electrical energy to heat energy, so the conductor becomes hot. Internal resistance also causes a drop in voltage. The amount of internal resistance in a conductor depends upon the type of metal from which it is made, the wire size and length, and the ambient temperature. Materials such as tungsten alloys, nichrome, and steel have high internal resistance. They heat up quickly, so they are often used for electric heating elements.

To be a good conductor, a material must have high conductance and low internal resistance. **Figure 22-23** lists the conductance and resistance values of different metals. Silver, copper, and aluminum are good conductors because they have high conductance and low resistance. Copper is the most common conductor. Copper wire bends easily, is strong, resists corrosion, and is easily joined. Aluminum is sometimes used because of its low cost. However, aluminum corrodes easily, and aluminum connections have a tendency to loosen. Loose connections become hot and can destroy insulation and cause fires. Consult your local electrical inspector before using aluminum conductors. The high cost of silver keeps it from being widely used. Silver is limited to such applications as contacts in switching devices.

Conductance and Resistance of Metals

Metal	Conductance	Resistance
Silver	2207	1.00
Copper	2030	1.09
Gold	1471	1.50
Aluminum	1277	1.73
Tungsten	634	3.48
Brass	461	4.79
Iron	304	7.26
Nickel	263	8.39
Steel	235	9.39
Nichrome	34	64.91

Figure 22-23. *Conductance and resistance values of various metals. The higher the conductance value (in siemens), and the lower the resistance number (in ohms), the better the conductor. Note the very high resistance value for nichrome, which is used as a heating element for toasters and portable heaters.*

22.4.1 Voltage Drop

Voltage drop is a term that describes the loss of emf (voltage) between the source and the equipment (load). Voltage drop can be caused by another load, a loose connection, or an undersized conductor. All conductors have some resistance to the flow of electrons. The type, size, and length of wire selected for a conductor must minimize voltage drop through the length of the conductor. If the wires are too small, voltage drop within a conductor will cause the conductor to get hot and prevent the load from receiving its designed operating voltage. Any voltage lost between the power supply and the equipment is voltage unavailable to the equipment.

All loads contain a specific resistance and are designed to operate within a specific voltage range. Most loads are designed to operate properly within ±5% of the rated voltage. Voltage drop should occur across the resistance of the load. Excess voltage drop is easily detected by an accurate voltage tester. See **Figure 22-24.**

22.4.2 Wire Sizes

The amount of current that a conductor can safely carry without becoming overheated is limited. Limited current-carrying capacity is called **ampacity.** The ampacity of a conductor depends upon the wire size, length, location, and type, quality, and quantity of insulation. An electrical conductor can be a single strand of solid wire or a group of wires twisted together (**stranded**). Large conductors are stranded mainly to increase their flexibility.

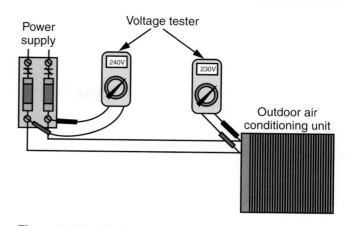

Figure 22-24. *Check voltage at the source and at the load. The difference between the two readings indicates the amount of voltage drop.*

Wire sizes in the United States are established by the American Wire Gauge (AWG) system. The AWG lists wire sizes by number. See **Figure 22-25.** The largest wire is #0000 (4/0); #50 is the smallest. Larger and smaller sizes are available but are not commonly used. The most commonly used wire sizes range from #4 to #18; #12 copper wire is the most popular. Number 12 is often used when a smaller wire would be approved. Other than cost, there is no problem with oversizing a wire. Undersizing, however, causes severe problems from voltage drop and overheating. **Figure 22-26** lists wire sizes, resistance, and ampacity for standard sizes of copper wire. Consult the **National Electrical Code (NEC)** for complete, accurate, and up-to-date information regarding ampacity and application of various types and sizes of electrical conductors approved for use in your area.

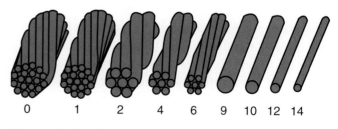

Figure 22-25. *Common sizes of solid and stranded wire. Conductors are stranded to improve their flexibility.*

The area of an electrical conductor is often measured in **circular mils** (CM) or *1000 circular mils* (MCM). A circular mil is the cross-sectional area of a wire 1 mil or 0.001″ in diameter. For electrical purposes, a wire's diameter is changed to mils. For example, a wire with a diameter of 0.325″ would equal 325 mils (0.325 ÷ 0.001). Circular mils are obtained by squaring the diameter in mils (circular mils = diameter in mils²): $325^2 = 105,625$ ($325 \times 325 = 105,625$). The circular mil system is

Dimensions, Typical Resistances, and Ampacity of Commercial Wire

Gauge No. (AWG)	Diameter Bare wire (inches)	Ohms per 1000 ft.		Current Capacity (Amperes)			
				Copper		Aluminum	
		70°F	167°F	TW UF	RH, RHW, THHW, THW, THWN	TW UF	RH, RHW, THHW, THW, THWN
0000 (4/0)	0.460	0.050	0.060	195	230	150	180
000 (3/0)	0.410	0.062	0.075	165	200	130	155
00 (2/0)	0.365	0.800	0.095	145	175	115	135
0 (1/0)	0.325	0.100	0.119	125	150	100	120
1	0.289	0.127	0.150	110	130	85	100
2	0.258	0.159	0.190	95	115	75	90
3	0.229	0.202	0.240	85	100	65	75
4	0.204	.254	0.302	70	85	55	65
6	0.162	0.40	0.480	55	65	40	50
8	0.128	0.645	0.764	40	50	30	40
10	0.102	1.02	1.216	30*	30*	20	25
12	0.081	1.62	1.931	20*	20*	16	18
14	0.064	2.57	3.071	15*	15*		
16	0.051	4.10	4.884	10*	10*		
18	0.040	6.51	7.765	5*	5*		

*Load current rating and overcurrent protection must not exceed these figures.

Figure 22-26. *Wire size, resistance, and ampacity for the most common sizes of conductors.*

primarily used for wire sizes larger than 4/0 AWG. Circular mil sizes run from 250 MCM (about 0.5″ in diameter) to 750 MCM (about 1.0″ in diameter).

22.4.3 Insulation for Wire Coverings

Insulators are materials that make it difficult for electrons to set themselves free. Insulators have stable atoms, and their outer valence shells have eight electrons. Atoms with less than eight valence electrons try to achieve the stable state by releasing their electrons to join with others. Conversely, those atoms that are more than half filled (the insulators) try to collect electrons to fill up the valence shell and achieve the stable state. Examples of good insulators are glass, rubber, mica, Bakelite™, asbestos, paper, and silk. Insulation acts as a barrier to electron flow and prevents electrons from traveling an unwanted path. When insulation is wrapped around conductors, the insulation forces electrons to remain inside the conductor.

All insulators can break down under moisture, heat, excess current flow, or chemical attack. Problems associated with breakdowns can be overcome by increasing the thickness and quality of insulation.

Heat-resistant, moisture-resistant, and oil-resistant types of insulation are available. The type of insulation and covering determines where the conductor can be safely used. Always use care to avoid damaging the insulated covering on a wire. When pulling conductors through metal conduit, use a special grease that makes pulling easier and helps prevent damage to insulation on sharp metal edges or turns.

Insulation markings

The NEC requires all electrical conductors be properly marked for easy identification. The markings appear on the outside surface of the wire's insulated covering. Markings contain:
- ❏ AWG size.
- ❏ Type of insulation (coded).
- ❏ Voltage rating.
- ❏ Testing agency (for example, UL or CSA).
- ❏ Number of conductors (for cable).

The most commonly used insulation types are TW, THHN, THW, and THWN. The more unusual types are reserved for special applications. **Figure 22-27** illustrates how insulated coverings are marked and lists codes and their applications.

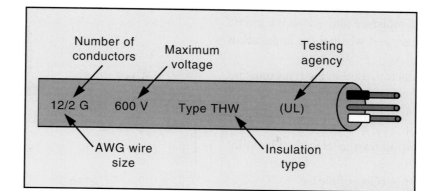

Type of Insulation	Letter Code	Maximum Temperature	Application
Asbestos	A	392°F	Dry locations only
Asbestos and varnished cambric	AVA	230°F	Dry locations only
Rubber	RH	167°F	Dry or wet locations
Heat-and moisture-resistant rubber	RHW	167°F	Dry and wet locations
Thermoplastic	T	140°F	Dry locations
Moisture-resistant thermoplastic	TW	140°F	Dry and wet locations
Heat-resistant thermoplastic	THHN	194°F	Dry locations only
Heat- and moisture-resistant thermoplastic	THW	167°F	Dry and wet locations
Underground feeder	UF	140°F	Moisture resistant
Varnished cambric	V	185°F	Dry locations only

Figure 22-27. *Manufacturers of electrical conductors print coded information on the outside of the insulation. The codes describe where the conductor can be safely used. Consult the National Electrical Code for a complete list of insulation types and approved uses.*

Insulation color

Insulation on conductors is color-coded to identify wiring. The NEC specifies the *neutral wire* be white or natural gray. The *grounding* conductor or safety ground wire must be bare copper, green, or green with yellow stripes. The *current-carrying conductors* (hot wires) are normally black or red, but hot wires can be any color except those designated for the neutral and ground wires.

22.4.4 Protective Electrical Enclosures

For customer protection and electrical safety, wires and connecting circuits must be suitably protected from exposure and abuse. The type of protective enclosure required for different applications is mandated by local, county, and state codes and by the NEC. All electrical connections must be performed inside an approved metal box. Each type of box has a removable cover for easy access. The size of the enclosure varies according to the application. Some considerations for enclosures include the presence of moisture and explosive gases; the voltage and amperage; and whether the application is indoors or outdoors.

In addition, the conductors used in a circuit must be enclosed within an approved protective covering or pipe. The types of protective systems most used today are:
- ❏ Rigid metal conduit.
- ❏ Thin-walled conduit, known as electrical metallic tubing (EMT).
- ❏ Flexible metal conduit (Greenfield).
- ❏ Liquid-tight flexible metal conduit (LTFMC).
- ❏ Armored cable (BX or AC).
- ❏ Nonmetallic sheathed cable (NM or Romex, NMC, UF, and USE).

These protective systems are adequate for most normal installations. Some cannot be used for unusual hazards, such as extreme moisture or explosive gases. Many types of wiring systems are available, and each system provides adequate safety when used properly and for the purpose designed. Special connectors are required for each type of system when connecting it to an approved protective metal box.

22.4.5 Electrical Loads

A *load* is an energy-conversion device that converts electrical energy to another form of energy. Examples of loads are lightbulbs, electric drills, and motors. The applied *voltage and resistance* determine the amount of electron flow (as described by Ohm's Law). *Wattage* measures the rate of energy conversion.

Manufacturers of electrical loads install the correct resistance to perform a particular job. Knowing how to connect these loads to the required voltage is critical. The voltage forces electrons through the resistance at a specific rate. If voltage is too high, the work is performed too fast. If voltage is too low, the work is performed too slowly. The amount of current flowing through the load is determined by the combination of voltage and resistance.

When connecting a load to a voltage source, a minimum of two conductors must be used. One conductor is connected to each end of the resistance. One of these wires contains excess electrons under pressure (voltage); the other wire has zero pressure. This dual setup is typical of a 120 V system used for lighting. The black (or hot) wire supplies the necessary free electrons at 120 V of pressure. The white (neutral) wire has zero voltage and completes the necessary pathway for electrons to return to their source. Both wires are necessary to complete an electrical circuit. The two wires (black and white) provide the required pathway and the necessary potential difference for electrons to flow through the resistance. See **Figure 22-28.**

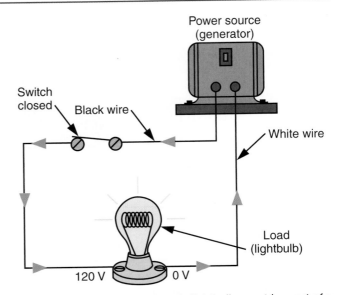

Figure 22-28. *A load, such as a lightbulb, must be part of a complete circuit to function. The black wire has voltage and carries electrical power from the source to the load. The white wire has no voltage and carries electrical power from the load back to the source. The two wires provide the necessary pathway for electron flow.*

All loads *must* be connected from one side of the power source (L_1) to the other (L_2 or neutral). Electrons must be able to flow into and out of the load. The same amount of current flowing into the black wire also flows into the white wire. An amperage reading can be taken on either wire. The resistance serves to convert the electrical energy to another form of energy. Voltage drops across the resistance of the load will be complete. Incoming voltage remains at 120 V, but exiting voltage remains at zero. This potential difference must be maintained for the load to operate properly. The amount of electrons flowing within the circuit (amperage) is determined by Ohm's Law ($E = I \times R$). With a fixed amount of resistance, an increase in voltage results in an increase in amperage. Likewise, a decrease in voltage results in a decrease in amperage.

As noted, the equipment manufacturer determines the amount and type of resistance to be installed in a particular appliance. The resistance must be connected to the proper voltage as designated by the manufacturer. When connected and operating properly, the appliance will draw a specific amount of amperage. The rate of energy conversion is expressed in watts. Voltage and wattage information is located on the unit data plate or stamped on the device.

22.4.6 Switches

Electrical switches offer no resistance to the flow of electrons. Rather, *switches* stop and start the flow of electrons to an energy conversion device (load). When a switch is open, the pathway is broken, and electrons cannot flow to the load. (Remember, when the drawbridge is open, all traffic stops until the draw-bridge is closed.) See **Figure 22-29.**

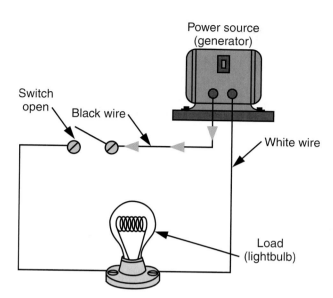

Figure 22-29. *A switch functions like a drawbridge to control the flow of electrons. When it is closed, the circuit is complete, and electrons can flow. When it is open, electron flow ceases because there is no longer a complete circuit.*

Series connections

An electrical load cannot operate unless the circuit is completed for electrons to flow into and out of the load. For proper operation, a switch is connected in series with the load. "In series" means the electrons must travel through the switch before they reach the load. Switches are never connected from one power leg to the other. Such a connection permits unrestricted flow of electrons and is a dead short. See Figure 22-30. **Fuses** are installed to protect against unrestricted current flow.

Switches are always connected in series with a load, but often more than one switch controls a load. With multiple switches, all switches must be connected in series with each other. Thus, electrons must travel through one switch before reaching the next switch and the next. All switches must be closed for the load to operate, but any switch can open and stop the flow. See **Figure 22-31.**

Parallel connections

Energy conversion devices (loads) are connected in **parallel,** or independent from all others. Loads are designed for connection to a specific potential difference (voltage). The ultimate test of a parallel connection is when one load can be removed without affecting any others. It is common practice to connect more than one load to a supply circuit, but each load must be connected in parallel. See **Figure 22-32.**

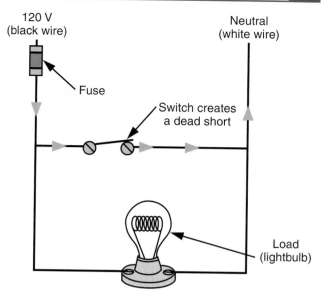

Figure 22-30. *A switch should never be connected in parallel with a load. Closing the switch creates a dead short, permitting unrestricted current flow between power supply wires. Electrons will not travel through the lightbulb (which has resistance) when they can take an easier path.*

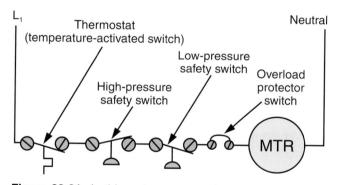

Figure 22-31. *In this series-connected motor circuit, all four switches must be closed for the motor to run. If any switch opens, the motor will stop because electron flow has ceased.*

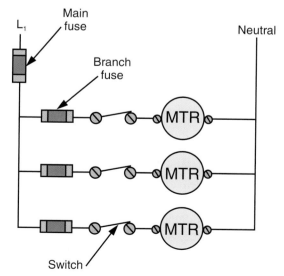

Figure 22-32. *These three loads are connected in parallel so each branch line is independent of the others. If the switch is opened or a fuse blows on any branch line, it will affect only the motor on that branch. The others will continue to operate.*

Series-parallel connections

When more than one load is connected to a circuit, switches are required to control the individual loads. A separate switch can be installed to control each load, but a main switch is often installed in the main supply wire. A main switch serves to control power supply to all loads. The individual switches are connected in series with each load, and each load is connected in parallel. The combination of circuits is called a **series-parallel** connection.

Safety

Switches should be located in the black wire supplying power (voltage) *to the load.* When the switch opens, voltage stops at the switch, and the load is disconnected from the potential difference. If the switch is located in the neutral wire, electricity is available through the load. If the load becomes defective (ground fault), anyone touching the load could complete the circuit and be shocked. See **Figure 22-33**.

22.4.6 Direct Current (dc)

There are two basic types of electricity: *direct current* (which flows in only one direction) and *alternating current* (which reverses direction many times each second). Current continues to flow when a potential difference is maintained. If the polarity (+ vs –) of the potential difference never changes, the current will flow in one direction; this is called **direct current** (dc).

A battery is the most common source of dc power. By converting chemical energy to electrical energy, it becomes a self-contained power source. To obtain continuous current flow, a constant voltage difference must exist between the negative and positive battery terminals. A "dead" battery exists when the chemical reaction stops (no voltage difference exists).

Two theories describe how electrons flow from one point to another. The most widely accepted theory is the *electron flow theory,* which states that electrons travel from negative to positive (magnetic attraction). The electron flow theory is used by most electricians and people working with ac current. The *current flow theory* states that current flows from positive to negative (difference of potential). The current flow theory is used by automobile technicians and people working with solid state devices. Regardless of the theory involved, a *complete circuit* begins at the power source (positive or negative), typically travels through a switch and a load device, and returns to the other side of the power source. See **Figure 22-34.**

22.4.7 Alternating Current (ac)

Alternating current flows first in one direction, then reverses and flows in the other. Since current normally flows from negative (–) to positive (+), the polarity of the power source must alternate from – to +, and then + to –. The rapid alterations in magnetic polarity cause the electrons to move back and forth rapidly inside the conductor. The *movement* of electrons, not the direction of flow, produces electrical energy.

Alternating current, or ac, is the name for a reversing current. Power plants use alternators to produce a rapidly alternating polarity by rotating a conducting loop through a strong, stationary magnetic field. See **Figure 22-35.**

As the sides of the rotating conducting loop cut through the stationary magnetic lines of flux, a voltage (emf) is applied to the free electrons in the conductor. As the conducting loop rotates, each side cuts the magnetic field in opposite directions. Polarity changes as each side of the loop approaches the other magnetic pole (north and south). The result is alternating polarity and alternating current.

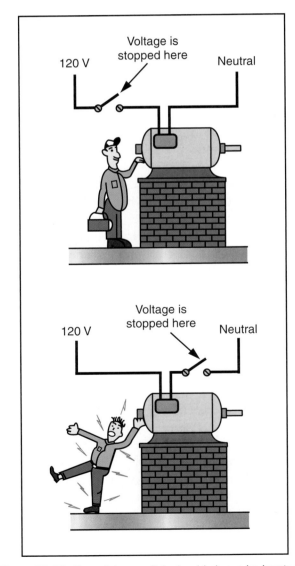

Figure 22-33. *For safety, a switch should always be located on the line side of the load rather than on the neutral side. When a switch is wired into the neutral side, a shock hazard exists.*

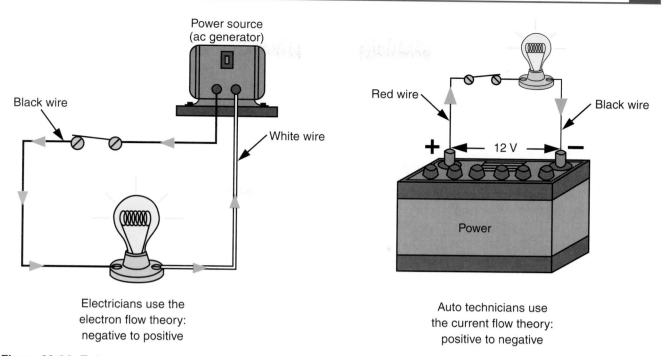

Figure 22-34. *To have a complete circuit in either ac or dc, current must begin at the power source, travel through a switch and a load device, and return to the other side of the power source.*

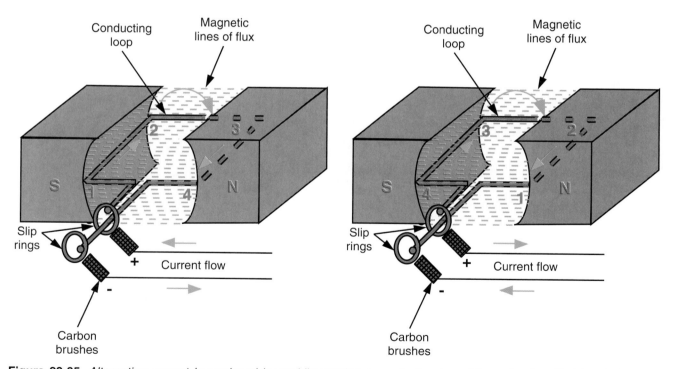

Figure 22-35. *Alternating current is produced by rapidly rotating a conducting loop (shown in simplified form) through a stationary magnetic field. Rotation of the loop causes the generated current to rapidly and regularly reverse polarity and direction. Polarity changes when the conducting loop rotates from one magnet to the other.*

Ac sine wave

The sine wave, **Figure 22-36,** illustrates the principle of alternating current. Voltage builds up at the magnetic north pole, falls back to zero, builds up again at the magnetic south pole, and again falls back to zero.

Electromotive force (voltage) drops to zero when the conducting loop is straight up and down and not cutting magnetic lines of flux. Voltage increases as more and more lines of flux are cut. Voltage is highest when the conductor is closest to the magnetic pole and is cutting lines of flux at right angles. Voltage drops off as the conductor travels away from the magnetic pole.

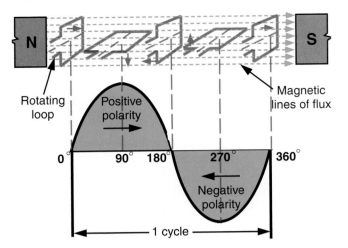

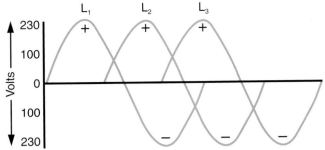

Figure 22-37. *In polyphase generation, the three loops (and the three phases) are 120° apart, as shown by the sine wave.*

Figure 22-36. *As alternating current is generated by the rotating loop, the voltage builds to a peak, then declines to zero. Next, it reverses polarity and repeats the pattern in the opposite direction. The entire cycle is shown in a sine wave. In North America, the frequency of alternating current is 60 cycles per second.*

Cycles or hertz (Hz)

To complete one cycle, or revolution, the loop passes one magnetic pole and then the other. Complete revolutions occur at the rate of 60 times (cycles) per second. Flow reversal occurs twice per cycle (1/2 cycle = north and 1/2 cycle = south). Flow reversal (120 times per second) is so rapid that a lightbulb will not flicker. The voltage and current flowing through the conductor is constant because of the speed of the cycles.

The *frequency* of alternating current is the number of complete cycles that occur in one second. In the United States and Canada, alternating current is produced at a frequency of 60 cycles per second. Most other countries use 50 cycles per second. The unit "cycles per second" is referred to as *hertz (Hz)* and is named for the German physicist Heinrich Hertz, one of the pioneers of alternating current.

22.5 Polyphase Generation ◆

In a power plant, a generator rotates three conducting loops at the same time. This means of producing electrical power is known as *polyphase generation.* The three conducting loops are spaced apart exactly 120° inside the generator. The loops are called *phases* or legs. The three phases are "out of step" along the sine wave; while one phase is positive, the second is negative, and the third is at zero. The three-way positioning is continuous as each loop takes its turn changing polarity from positive to negative to zero. Polyphase generation produces alternating current in three phases that are out of step with each other. See **Figure 22-37.**

The Greek letter phi (ϕ) is an abbreviation for phase. On a *schematic* (electrical circuit diagram) the letter L (line) plus a number identify each phase (L_1, L_2, L_3). See **Figure 22-38.** A potential difference exists between any two "hot" wires because of the difference in polarity. As stated earlier, the normal color for hot wires is black or red. However, they can be any color except white or green.

Each hot wire has the same voltage but is different in polarity. When one wire is pulling (+), the other is pushing (–). Electrons flow according to polarity, but a voltage (120 V) exists on each wire. Therefore, the potential difference between any two hot wires (+ and –) is additive, or the sum of their voltages (120 + 120 = 240).

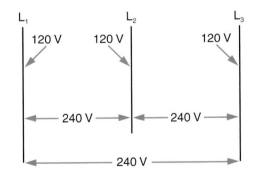

Figure 22-38. *In a three-phase system, voltage between any two wires is additive because the wires have different polarities. While one wire is negative, the other is positive. While one wire is pushing, the other is pulling, which doubles the voltage differential.*

22.5.1 Three-phase vs Single-phase Loads

Many electrical devices are designed for connection to all *three* hot wires; such devices are called **three-phase loads.** Other devices are designed to operate with just *two* hot wires and are called **single-phase loads.** Again, the voltage between the two hot wires is additive. The term "two-phase" is never used because it refers to an older method of power generation no longer used. Voltage higher than 120 V can be obtained only by using two of the three phases.

Many electrical devices operate with just one hot wire from a three-phase system and a second wire called the neutral. This method is also called single-phase because only one hot wire is used. A potential difference exists because the hot wire has voltage and polarity, while the neutral wire has zero voltage and no polarity.

22.5.2 Neutral Wire

The earth is a gigantic mass of elements and compounds that serves as an excellent conductor of electricity. Damp soil is a better conductor than dry soil. The earth is always at *zero potential* (no voltage) and can complete an electrical circuit.

The neutral conductor in a circuit has white or gray insulation and is connected to the earth (*grounded*). The connection to the earth is typically made by the utility company. A bare copper wire travels from the transformer on the utility pole to a solid copper rod driven eight feet into the ground. The copper rod is called a grounding electrode. See **Figure 22-39.**

Grounding the neutral wire provides a pathway for electrons traveling to and from the earth. The neutral wire has zero voltage because it connects directly to ground. The neutral wire should not be broken and is

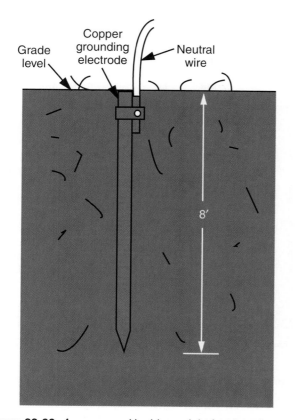

Figure 22-39. *A copper rod is driven eight feet into the earth by the power company and used to ground the neutral wire of the electrical supply system.*

never fused. The electrical symbol for the neutral wire is the letter N. The neutral wire is used as a current-carrying conductor, but it has no voltage or polarity. A potential difference exists between the neutral wire (no voltage) and any black or red (hot) wire with 120 V. See **Figure 22-40.**

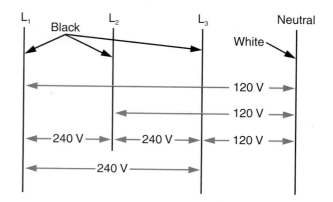

Figure 22-40. *A potential difference of 120 V exists between the neutral electrode and any of the hot wires. When the hot wire is negative, current flows to the neutral electrode. When the hot wire is positive, current returns from the electrode to the hot wire.*

The NEC refers to the neutral conductor as the *grounded* conductor. Another wire (green) is reserved as the *grounding* conductor. The two wires serve entirely different purposes and should never be confused. The purpose and proper use of the grounding conductor (green wire) will be explained shortly. Single-phase loads designed to operate on 120 V require the use of one hot (black) wire and the grounded neutral (white) conductor. The black conductor supplies electrons, voltage, and magnetism, while the white conductor supplies electrons at zero voltage and no magnetism. This provides a potential difference so electrons are able to flow back and forth between the black and white wires.

Current flowing in the hot leg also flows in the neutral. When the load is operating, an amperage reading can be obtained from either conductor. See **Figure 22-41.** The voltage and energy are used up by the energy conversion device (load), so the neutral wire remains at zero voltage. If either wire is disconnected from the load, the potential difference is removed and electrons cannot flow.

Some single-phase loads are designed to operate on 240 V. Such loads require the use of *two* hot wires containing 120 V each (no neutral). The voltage between the two hot wires is additive (120 V + 120 V = 240 V). The two hot wires have the *same* voltage but opposite polarity. *Opposite* polarity (negative/positive) provides the necessary potential difference. Even though it is called single phase (1ϕ), the *higher voltage* can only be obtained by using two hot wires. See **Figure 22-42.**

22.5.3 Equipment Grounding Conductor

The equipment-grounding conductor, or *safety ground,* is a wire added for safety. The wire is *required* by the National Electrical Code on all new electrical systems. The color code for this wire is green, or it can be bare (uninsulated) copper. The symbol for an equipment-grounding electrode is shown in **Figure 22-43.**

Safety-grounding conductors (green wires) from the local devices are connected to a busbar located inside the distribution panel. A *busbar* is a metal bar that serves as a common connecting device. A busbar has multiple screws for connecting wires, and it is securely bonded to the distribution box to obtain good metal-to-metal contact.

For residential wiring systems, the equipment-grounding conductor (green) and the neutral conductor (white) share the same grounding busbar. The two busbars, located inside the distribution center (breaker box), are connected as shown in **Figure 22-44.** The breaker box is the only place where the neutral and safety grounds are joined. The NEC recently ruled that all new installations must keep the neutral and safety wires separated at all times.

The grounding busbar *must* be connected to a grounding electrode (also called a grounding rod). The

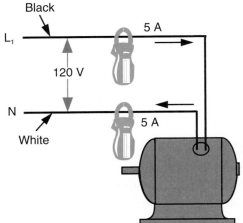

Figure 22-41. *Current flowing in the black wire also flows in the white wire. When the load is active, an amperage reading can be obtained from either wire.*

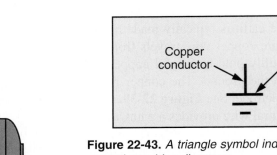

Figure 22-43. *A triangle symbol indicates the safety (earth) ground on wiring diagrams.*

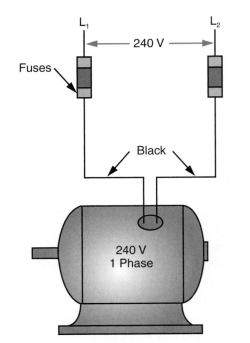

Figure 22-42. *A 240 V single-phase motor like this one uses two hot wires (but no neutral).*

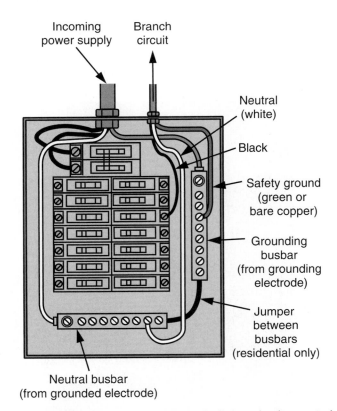

Figure 22-44. *For residential installations only, the neutral and equipment-grounding busbars are connected to each other via a jumper, and both are connected to a grounding electrode.*

grounding rod is a solid copper rod driven eight feet into the ground to make good contact with moist earth. See **Figure 22-45.** Consult your local electrical inspector or the NEC regarding approved grounding techniques.

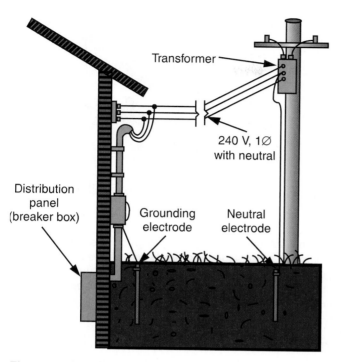

Figure 22-45. *The copper grounding rod, or electrode, is driven at least eight feet into the earth to make contact with moist soil.*

Commercial and industrial electrical systems require complete separation of the grounding electrode from the neutral electrode. As shown in Figure 22-39, the neutral is grounded by the utility company at the pole. A separate grounding electrode is required for the equipment-grounding conductor and must be installed at the site.

The distribution panels of commercial installations are equipped with separate busbars, **Figure 22-46.** The equipment-grounding busbar is often connected to the steel beams used in constructing the building. The beams serve as a grounding electrode.

Purpose of the safety ground

The safety ground conductor (green or bare wire) is *never* connected to the normal electrical circuit. The safety ground conductor serves as a "safety valve" when a malfunction occurs in the electrical equipment. The safety ground wire is connected to the frame of a motor or appliance, **Figure 22-47.** Under normal conditions, the frame of a motor or appliance is safe to touch. However, an internal electrical problem may occur, permitting electron flow to the frame. Called a **ground fault,** the flow of electrons makes the frame electrically live. Anyone touching the frame will receive a severe electrical shock; the body becomes a "load" by completing a circuit from the frame to ground. See **Figure 22-48.**

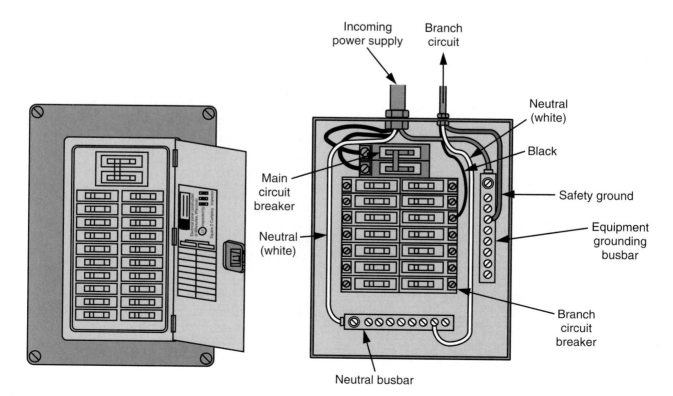

Figure 22-46. *In commercial and industrial applications, the neutral and equipment-grounding busbars must remain separated at all times. The equipment ground is often connected to the steel framework of the building rather than to an electrode. The busbars are never connected to each other, as permitted in residential installations.*

Figure 22-47. *An equipment ground provides a path for electron flow to ground in the event of a ground fault, preventing an electrical shock if the frame is touched.*

Figure 22-48. *Without a safety ground, if the frame of a motor accidentally becomes electrically "live," touching the frame will make your body a load and complete the circuit to ground.*

When a ground fault occurs, the safety ground wire provides an unbroken pathway for electrons to travel from the frame to the grounding electrode. The activity amounts to a "dead short," resulting in excessive current flow that will blow a fuse or trip a circuit breaker. So, the green wire provides an escape path for electrons when a ground fault occurs.

Because the safety ground wire has proven very effective in preventing electrical injuries, it is required by the NEC. The third (rounded) prong on plugs and the matching hole in receptacles are reserved for the safety ground connection. See **Figure 22-49.**

Using portable power tools

Caution must be exercised when using portable power tools. These handheld electrical devices can be dangerous if a ground fault occurs. Power tools are often constructed with a metal frame. Such tools should have a three-prong plug with the metal frame connected to the safety ground. If the frame becomes electrically "live" due to a ground fault, the third wire (safety ground) will safely carry the current to ground.

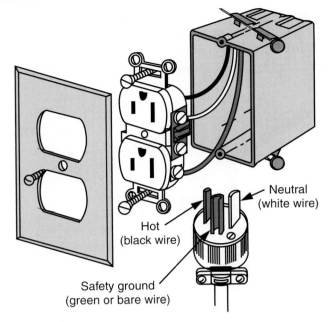

Figure 22-49. *The rounded prong on the plug and the matching socket in the receptacle are reserved for the safety ground. The receptacle should be installed with the ground uppermost to help prevent accidental shorting across the hot and neutral prongs.*

The third wire protects the operator from becoming a pathway to ground. Excess current flow will blow a fuse or trip a circuit breaker. Always use properly grounded power tools, and connect them to properly grounded circuits.

Double-insulated tools

Many power tools do not have the third (safety ground) prong on the male plug. Instead, the devices are supposed to be "double insulated" to prevent a ground fault from the hot wire to the tool case. The insulating plastic is intended to protect the operator. A crack in the plastic case, however, can permit moisture to enter the insulation and defeat its purpose. Always examine such tools for cracks or damage. Never use a tool with a damaged case or one with a plug that has had the third prong removed.

Adapters

Many old-style wall receptacles are two-prong and do not accept a three-prong plug because they are not equipped with the safety ground feature. A three- to two-prong adapter makes it possible to connect a hand tool to these receptacles. For proper protection from a ground fault, the green safety ground wire on the adapter must be connected to a suitable ground, such as a steel beam, a metal stake driven into the ground, or a similar good connection to earth. See **Figure 22-50.** Failure to connect the safety ground can result in death for the person whose body becomes a ground.

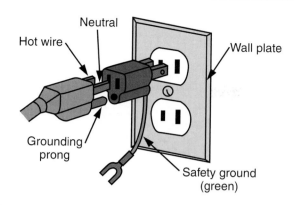

Figure 22-50. *When a three-to-two-hole adapter is used with a portable power tool, the green safety ground wire must be connected to a good ground. These adapters are sometimes called "suicide plugs" because of the fatal consequence of not connecting the safety ground wire to a good ground.*

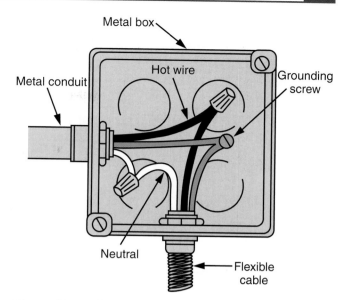

Figure 22-51. *Good metal-to-metal contact must be maintained for effective operation of the safety ground when metal conduit and boxes are used as the path to ground.*

Most two-hole receptacles are *not* grounded, so ground fault protection is lost. Always use a voltage tester to determine if the receptacle is properly grounded. Zero voltage between the hot wire and a metal cover reveals no safety ground. A 120 V reading between the hot wire and the metal cover indicates a good safety ground.

If the receptacle box is not grounded, use another piece of wire to connect the adapter's green wire to a good ground. Taking the extra precaution of properly connecting the safety ground may save your life if the power tool becomes "shorted" to the metal case.

Receptacles and conduit

Receptacles (called **duplexes**) should be installed with the ground connection at the *top*. If mounted sideways, the *neutral* (wide slot) should be on top. This prevents metal objects from falling between the plug and receptacle and contacting the hot wire. The neutral wire and the safety ground wire each have continuous zero voltage yet serve entirely different purposes. Never substitute the safety ground for the neutral wire; dangerous conditions will result.

Commercial and industrial applications, and some local codes for residential installations, require electrical wires be installed inside metal piping (**conduit**). The metal pipes and metal enclosures are securely bonded together at each connection. The continuous run of conduit and enclosures must be properly connected to the equipment-grounding electrode. This method of providing a safety ground is effective and mandated by the NEC. See **Figure 22-51.**

22.6 Power Circuit Devices

Circuits supplying electrical power to residential or commercial installations include *receptacles*, which allow easy connection and disconnection of equipment, and *switches* and *disconnects*, which control part or all of a circuit.

22.6.1 Receptacle Types

Various female outlet receptacles are available for equipment that has a flexible cord and a male plug. As a safety feature, each plug and receptacle is sized and styled for a particular voltage and amperage. For example, a 250 V plug cannot be inserted in a 120 V receptacle. Likewise, a plug rated for 20 A cannot be plugged into a 15 A receptacle.

The shapes and styles of plugs and receptacles are standardized by amperage and voltage. **Figure 22-52** shows some of the many types of receptacles.

22.6.2 Switch Types

As stated earlier in this chapter, the purpose of a switch is to control the flow of electrons to one or more electrical devices. All switches contain one or more sets of **contacts** that are opened or closed by the movement of a **pole**. A switch may contain one or more poles. Movement of the pole is called the **throw**.

Switches are rated to operate with a specific voltage and amperage. The specified ratings *must not* be exceeded. Switches for use in residential circuits are commonly rated 125 V at 10 A or 250 V at 5 A. Special-purpose switches may be rated for smaller or larger voltages and amperages.

The common household light switch is a **single-pole, single-throw (SPST) switch.** It has one pole and makes (closes) or breaks (opens) one contact. The operation of an SPST switch, in its simplest form, is illustrated in **Figure 22-53.**

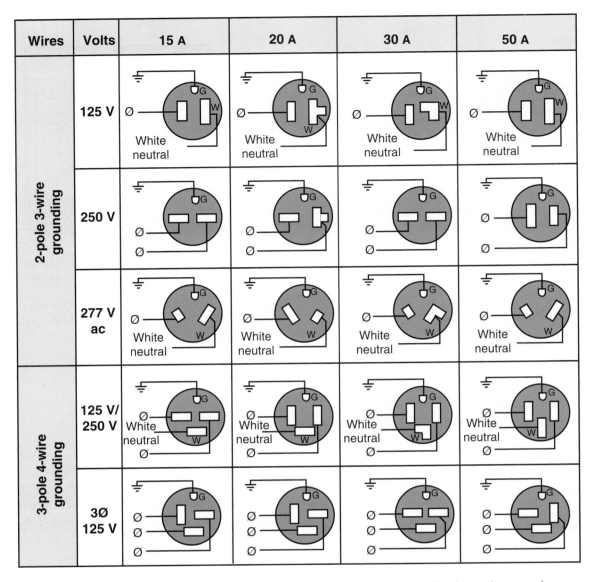

Wires	Volts	15 A	20 A	30 A	50 A
2-pole 3-wire grounding	125 V	White neutral	White neutral	White neutral	White neutral
	250 V				
	277 V ac	White neutral	White neutral	White neutral	White neutral
3-pole 4-wire grounding	125 V/ 250 V	White neutral	White neutral	White neutral	White neutral
	3Ø 125 V				

Figure 22-52. *Each type of plug and receptacle is sized and styled for a particular application, voltage, and amperage.*

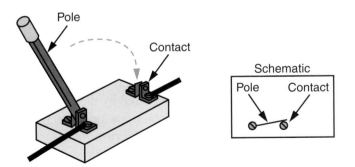

Figure 22-53. *The single-pole, single-throw switch (like the knife switch shown here) is the simplest type of switch. It is either on or off. The common household light switch uses a toggle mechanism but operates on the same principle: it either makes or breaks a contact.*

WARNING: Wire color codes are not mandated, and residential wiring procedures may ignore color codes. The *white* wire is often used as the "hot" wire for connecting switches in a circuit. Indoor residential wiring is typically done with nonmetallic cable (usually called Romex), which has an insulated covering around two insulated #12 solid wires (one black and one white). The safety ground wire is bare copper. Because the cable has only two colors, the black and white wires are used to connect a switch and load in the circuit as shown in **Figure 22-54.** Never assume that the white wire is a neutral. When a black wire and a white wire are connected, the white wire becomes a "hot" supply wire. For safety, always check the circuit with a voltage tester before working on any wiring.

A *single-pole, double-throw (SPDT) switch* has one pole and operates two contacts, **Figure 22-55.** One contact is *normally open* (NO), and the other contact is *normally closed* (NC); there is no "off" position on this switch. Movement of the pole causes the NC

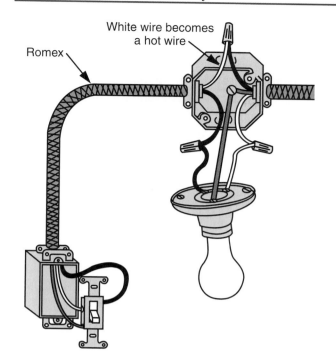

Figure 22-54. *When nonmetallic, two-conductor (Romex) cable is used in residential wiring, a white wire can sometimes be "hot." Never assume that a white wire is neutral. Always use a voltage tester to check any wires before working on them.*

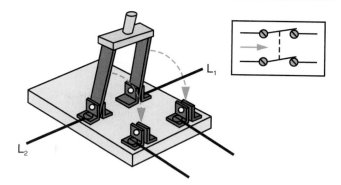

Figure 22-56. *A double-pole, single-throw switch can control two circuits or both hot wires in a 230 V, 1φ circuit.*

Disconnects

Disconnects are heavy-duty, manually operated contacts for disconnecting large loads from the power supply. Disconnect switches are normally located close to the load for quick and easy access. A

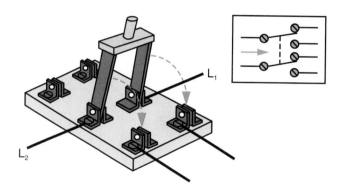

Figure 22-57. *A double-pole, double-throw switch can control a variety of loads. Two of the contacts are normally open (NO), and two are normally closed (NC).*

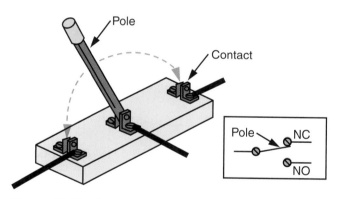

Figure 22-55. *On a single-pole, double-throw switch, the pole moves from one contact to another. One contact is normally open (NO), and the other is normally closed (NC).*

contact to open and the NO contact to close. Power supply is normally connected to the pole.

A *double-pole, single-throw (DPST) switch* has two poles and a contact for each pole. Both contacts are either open or closed, making it possible to open or close two circuits at the same time. This switch is often used to control both hot wires on a 230 V, single-phase circuit. See **Figure 22-56.**

A *double-pole, double throw (DPDT) switch* has two poles and two contacts for each pole. Power supply must be connected to the poles. The DPDT switch makes it possible to control a variety of loads from one location. Two contacts are NC, and two contacts are NO. See **Figure 22-57.**

disconnect normally has a three-pole switch mounted inside a metal enclosure. All three switches are simultaneously controlled by the manually operated handle, **Figure 22-58.**

When a disconnect includes fuses, it is called a fused disconnect. The power supply (*line voltage*) is connected to the terminals located in the top of the unit, and lines are labeled L_1, L_2, and L_3. These power supply terminal connections remain electrically live at all times, even when the disconnect is turned off. The switches are located between the power supply and the fuses. The *load* is connected to lugs provided at the bottom of the fuses labeled T_1, T_2, and T_3. The arrangement permits safe replacement of fuses when the disconnect is off. See **Figure 22-59.** Always use a voltage tester to be certain all switches open properly. Even though the switches are open, use an insulated fuse puller to remove or replace fuses.

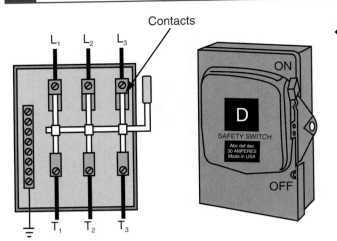

Figure 22-58. *Disconnects are manually operated, heavy-duty contacts that allow large loads to be quickly and easily disconnected from the power source.*

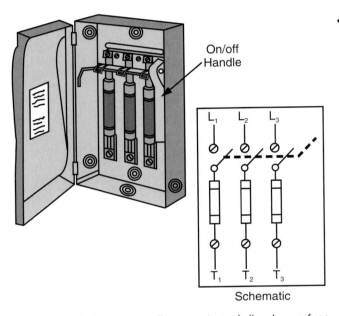

Figure 22-59. *In a fused disconnect, each line has a fuse located below the switch. The dotted line on the schematic indicates that all three switches open and close at the same time. An insulated fuse puller should be used to remove or replace fuses.*

Disconnects are rated and sized for a maximum amperage (such as 30, 60, or 100) and voltage. Each disconnect size requires a different fuse size. The fuse holders are designed to prevent using a too-large fuse. For instance, a 60 A fuse will not fit a disconnect designed for 30 A service.

Do not open the disconnect switches when the load is operating. Always stop the load before operating the disconnect switches. Under certain conditions (such as a short circuit), a sudden inrush of extremely high current could cause the disconnect box to explode. Always operate the disconnect handle with the left hand, and stand to the side and face away from the box in case of malfunction.

Summary

Electricity is a form of energy produced by the movement of electrons from one atom to another. Electrical energy can be converted to other forms of energy by forcing electrons through a resistance.

Conductors provide the necessary supply of electrons and serve as a pathway for electron flow. Electrons cannot flow in a circuit unless the circuit is complete and a difference in potential exists. Various types of switches are available to control the flow of electrons by opening and closing one or more circuits. A safety ground prevents electrical shock by providing a path to ground. Portable electric power tools must be properly grounded for safety.

Test Your Knowledge

Please do not write in this text. Write your answers on a separate sheet of paper.

1. The movement of free electrons is called _____.
2. Electromotive force (emf) is measured with a _____.
3. Amperage is measured with a(n) _____.
4. *True or false?* Resistance is used to convert electrical energy to another form of energy.
5. What type of meter measures resistance?
6. An ohmmeter is used to check for _____ and _____ circuits.
7. When connected to an open circuit, an ohmmeter reads _____.
8. In a short circuit, _____ current flows through a wire.
9. Electrical power (the ability to do work) is measured in _____.
10. Name three good conductors of electricity.
11. What five items determine the ampacity of a conductor?
12. What is the most commonly used wire size?
13. What color is the neutral wire?
14. What is a load?
15. When connecting a load to a voltage source, a minimum of _____ conductors must be used.
16. Are multiple switches connected in series or in parallel?
17. *True or false?* Loads in a circuit are always connected in parallel.
18. Name two devices used for protection against a dead short.
19. A voltmeter connected between the hot contact of a receptacle and the metal cover plate shows a reading of zero volts. Is the receptacle grounded?
20. *True or false?* All four contacts can be closed at the same time in a double-pole, double-throw switch.

Power Transmission and Circuits

Objectives

After studying this chapter, you will be able to:
- ❑ Select and connect transformers.
- ❑ Identify common voltages and their uses.
- ❑ Size and use circuit protectors properly.
- ❑ Make good terminal connections.

Important Terms

amperage interrupting capacity (AIC)	magnetic field
	overcurrent
cartridge fuses	overload
circuit protection devices	oversizing
	power factor
conductors	primary winding
crimp	reset
dead leg	secondary winding
delta	short circuit
double-pole breaker	single-pole breaker
dual element	step-down transformer
full load amperage (FLA)	step-up transformer
	stinger leg
fuse	terminal connectors
fusible element	terminals
ground fault circuit interrupter (GFCI)	three-pole breaker
	transformers
induced voltage	undersized
insulation	wire nuts
load center	wye

23.1 Power Transmission

Electrical power is normally generated as three-phase, 60-Hz, alternating current at about 26,000 V. A *step-up transformer* increases the voltage to 120,000 V (or more). In transmission lines, high voltage reduces current flow (P = I × E). Reduced current flow permits the use of smaller-diameter wire over long distances. Power is often transmitted for hundreds of miles using steel towers to support the wires. The three-phase, four-wire system (with grounded neutral) is used; the neutral line grounds the steel towers and protects them from lightning. See **Figure 23-1.**

Substations are located at various points along the transmission circuit. At the substations, *step-down transformers* reduce voltage to 40,000 V for distribution through a particular region. Transformers at other substations *within* a region further step down voltage to 13,200 V or 4800 V. Transformers for reducing voltage to individual homes and businesses are often located on utility poles but may be at ground level in protective enclosures. The final voltage varies from 120 V, single-phase to 480 V, three-phase, depending upon the needs of the customer.

23.1.1 Meeting Various Electrical Needs

The distribution system can supply electrical power in different configurations to meet the requirements of customers from large industrial plants to individual households.

For commercial and industrial applications, three-phase alternating current is most common. Three-phase power is cheaper than single-phase and provides other advantages. For example, three-phase motors do not require starting components, and they offer better start and run characteristics. Three-phase is easier to understand and control. Easy access to 120 V, single-phase for lighting and receptacles is provided by including the neutral wire. **Figure 23-2** illustrates a typical 230 V, three-phase, four-wire system.

Three-phase motors use all three "hot" wires (no neutral). All three hot wires (L$_1$, L$_2$, and L$_3$) are required to obtain a three-phase supply; the voltage is measured between any two hot wires. Single-phase power is obtained from a three-phase supply by using just two hot wires, or one hot wire and the neutral. All voltmeters have two test leads that check the potential difference between any two hot wires or between one hot wire and the neutral.

An industrial customer, such as a factory, is supplied with a 480 V, three-phase, four-wire system (three hot and one neutral). The safety ground (grounding rod) is installed at the site. Much equipment inside the factory is operated by 480 V, three-phase motors. The fluorescent lighting is operated on 277 V, single-phase (one hot leg and the neutral). See **Figure 23-3.** A step-down transformer can be used within the plant to reduce the 480 V, three-phase to 240 V, three-phase. Voltage reduction provides additional options for use with smaller motors, incandescent lighting, or wall outlets.

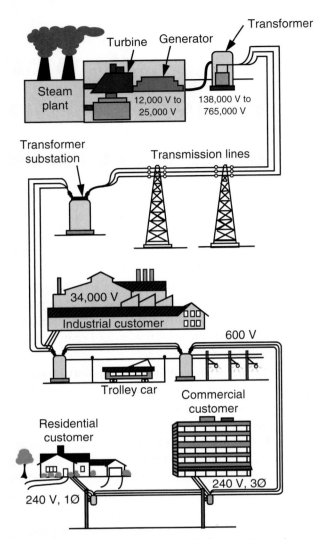

Figure 23-1. *The power transmission system steps up voltage to send electricity over long distances. Then, voltage is stepped down to meet the various needs of users connected to the distribution system.*

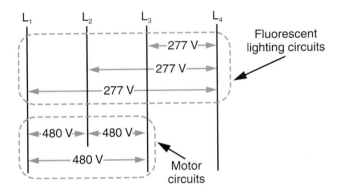

Figure 23-3. *A 480 V, three-phase, four-wire system is generally provided for customers with large power needs, such as industrial plants.*

A commercial customer, such as a supermarket, is supplied with a 240 V, three-phase, four-wire system (three hot and one neutral). The safety ground (grounding rod) is added at the site. The three hot wires power refrigeration and air conditioning equipment. Some equipment in the bakery, meat department, and deli operate on 240 V, single-phase, using two hot wires (and a safety ground). By using one hot wire and a neutral (with a safety ground), 120 V, single-phase is obtained for lighting circuits and receptacles. The single-phase loads must be equally divided among the three hot wires to balance the load on all three phases.

Most residential units are provided with two hot wires and a neutral wire (240 V, single-phase). The safety ground is installed at the site. See **Figure 23-4.**

Some home appliances, such as the water heater and central air conditioning unit, require 240 V for operation. These appliances require two hot wires and the safety ground. Other appliances, such as the clothes dryer and electric range, require both 240 V and 120 V. *Dual voltage* appliances require two hot wires, the neutral, and the safety ground.

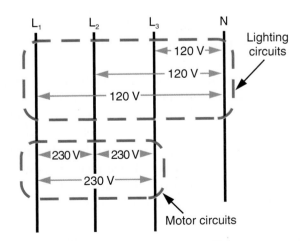

Figure 23-2. *Three-phase power is used for many commercial and industrial applications. Various hot and neutral wire combinations provide different voltages.*

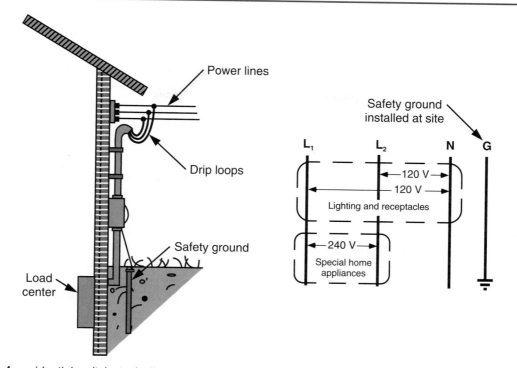

Figure 23-4. *A residential unit is typically supplied with a 240 V, one-phase, three-wire system. The safety ground is connected to a rod driven into the earth at least eight feet.*

Residential lighting and receptacle circuits call for 120 V, single-phase connections. The connection is obtained by using one hot wire, the neutral, and the safety ground. The various 120 V circuits should be equally balanced between the two hot supply wires.

23.2 Transformers

A chief advantage of alternating current is that it can be generated at one voltage, transmitted at a higher voltage, and reduced to a lower voltage at the point of use. *Transformers* make it possible to increase (step up) or decrease (step down) the voltage. See **Figure 23-5.** This flexibility is highly desirable and used throughout the world for residential, commercial, and industrial electrical systems.

Transformers are 96% to 98% efficient. They have no moving parts and require very little maintenance because of their simple, rugged, and durable construction. A transformer has two copper windings (coils) that are electrically separated from each other. The coils are wound around a core of laminated soft iron sheets. The laminated core provides a circuit for the magnetic field. See **Figure 23-6.**

A transformer operates on the principle that electrical energy can be efficiently transferred from one winding to another by magnetic induction. When an

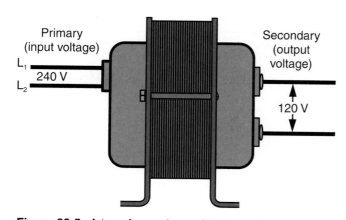

Figure 23-5. *A transformer is used to increase or decrease voltage. An electric power transmission and distribution system uses many transformers.*

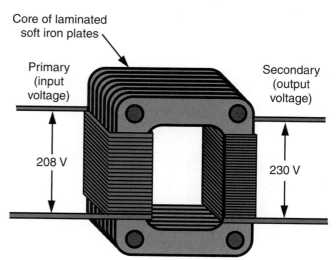

Figure 23-6. *Electrically separated windings transfer electrical energy through the iron core by magnetic induction. The differing number of turns of wire in the windings is responsible for the increase or decrease in voltage.*

alternating current flows in the ***primary winding,*** an alternating ***magnetic field*** is established in the laminated soft iron core. This induces an alternating current in the ***secondary winding.*** The magnetic field cuts across the turns of both windings; therefore, the same voltage is induced in each turn of the two windings. The ***induced voltage*** is proportional to the number of turns in each winding.

Voltage at the secondary (output) winding is determined by the number of coils or wraps versus the number of coils in the primary (input) winding. Fewer coils in the secondary winding produce a lower output voltage (a ***step-down transformer***). More coils in the secondary winding increases output voltage (a ***step-up transformer***). See **Figure 23-7.**

Electrical energy conversion devices (loads) are manufactured to tolerate a voltage of ±10%. However, many motors and other devices are limited to a 5% variation. A transformer is required to increase or decrease the supply voltage to match the load voltage specification. For example, a single-phase motor designed to operate on 230 V cannot be connected to 208 V. A transformer is needed to change the

power supply from 208 V to 230 V. As shown in **Figure 23-8,** the 208 V power supply is connected to the transformer's primary winding, and the motor is connected to the 230 V secondary winding.

Some single-phase transformers offer a choice of voltages and are called *multi-tap transformers.* The windings may have three or more connections (*taps*), which make it possible to use different input voltages or obtain different output voltages. Only *one connection* can be made on each winding at any time. The unused wires must be capped and sealed. Multi-tap transformers are convenient because one transformer can serve several applications. A wiring diagram or instructions are included on each transformer to ensure proper connections. See **Figure 23-9.**

23.2.1 Transformer Symbols and Terminals

Electrical schematics use the symbol in **Figure 23-10A,** or some variation, to indicate a single-phase transformer. Three-phase transformers consists of separate insulated windings for the three phases.

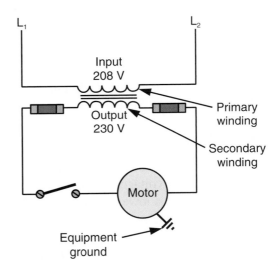

Figure 23-8. *A step-up transformer increases output voltage to the correct level for a motor.*

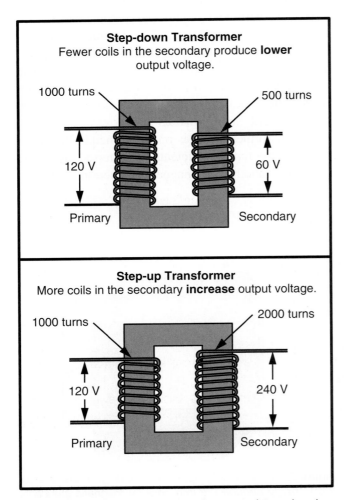

Figure 23-7. *Transformers can decrease (step down) or increase (step up) voltage.*

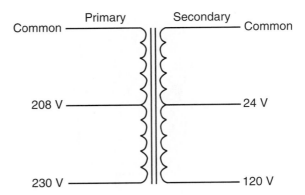

Figure 23-9. *Multi-tap, single-phase transformers provide several possible connections on both the input and output sides. Only one voltage connection per side can be made at a time: 208 V or 230 V input; 24 V or 120 V output.*

The windings are on a three-legged core so three magnetic fields may be spaced 120° apart. Three-phase transformers can be either delta- or wye-type. **Delta** is similar to a triangle (Δ), while **wye** (sometimes referred to as "star") is similar to the letter Y. See **Figures 23-10B** and **23-10C**.

Transformer **terminals** are labeled and tagged with letter and number combinations. Standard practice is to tag the transformer's primary terminals with H_1, H_2, H_3, and so on. The secondary terminals are tagged X_1, X_2, X_3, etc. (X_0 indicates the neutral terminal.) See **Figure 23-11.**

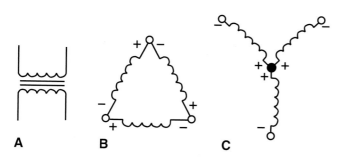

Figure 23-10. *Transformer symbols for use on electrical schematics. A—Single-phase. B—Three-phase delta. C—Three-phase wye.*

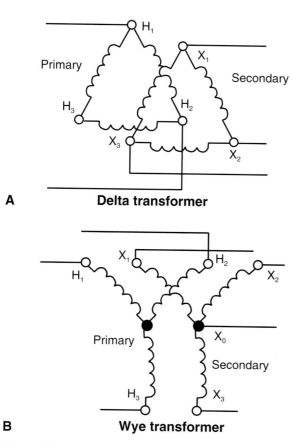

Figure 23-11. *Transformer terminals are identified by the letter H on the primary side and the letter X on the secondary side. A—Delta transformer. B—Wye transformer.*

23.2.2 Transformer Ratings

In addition to the desired primary and secondary voltages, transformers are selected by the *amperage* that the secondary winding can safely carry. The secondary winding must be able to carry more than the minimum required amperage to accommodate the load(s) connected to it. **Oversizing** the transformer by 125% helps protect it from becoming overloaded.

Single-phase transformers are rated by VA (volt-amperes) at the secondary. VA = Volts × Amps = E × I. For example, a 48 VA transformer that has a 24 V secondary can safely carry two amperes of current (24 V × 2 A = 48 VA).

What size transformer should be selected if the secondary is to carry 20 A at 230 V? At minimum, 4600 VA (230 V × 20 A = 4600 VA). To protect the transformer from overload, a 125% oversizing is recommended; therefore, a transformer rated at approximately 6000 VA should be chosen.

Transformer ratings of over 1000 VA are normally given in kVA (k = 1000). Therefore, the rating for a 6000 VA transformer is 6 kVA. Transformers with 1 kVA or higher ratings are always grounded by a separate cable or bus to a grounding electrode. The grounding electrode may be the steel beams supporting the building or a rod driven into the earth. A transformer rated under 1 kVA may be grounded to the cabinet or the conduit serving it. The cabinet or conduit is connected to the grounding electrode.

An overloaded or **undersized** transformer burns out because the secondary coil cannot carry the current. A minor overload causes a slow burnout, and a large overload causes a quick burnout. Transformers should be protected with fuses. *Oversizing* a transformer does no harm but is more expensive. When replacing a burned-out transformer, always check amperage in the secondary circuit to be sure there is adequate capacity.

23.2.3 Three-phase Transformers

As noted earlier, three-phase transformers can be either wye- or delta-type, although the wye (star) is most popular. Three-phase transformers have a better **power factor** (computed by multiplying VA × 1.73 and converting the answer to kVA) than single-phase transformers.

Because electricity is produced in three-phases (*polyphase generation*), a variety of transformer types (wye, delta, wye-delta, or delta-wye) must be used to produce the secondary voltages that match customer needs. Loads and other electrical equipment connected to the transformer secondary *must* match the voltage supplied by the transformer. **Figure 23-12** shows some commonly used voltages.

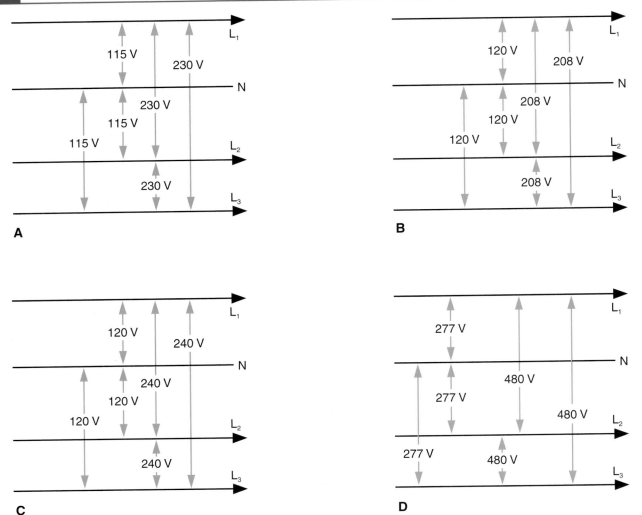

Figure 23-12. *Transformers can provide a variety of voltages to meet different customer needs. A—230 V, three-phase, four-wire. B—208 V, three-phase, four-wire. C—240 V, three-phase, four-wire. D—480 V, three-phase, four-wire.*

Some older installations use a "high-leg" system from a three-phase, delta transformer. Voltage readings from two of the hot legs to the neutral read 115 V. However, a reading from *one* of the hot legs to neutral registers 208 V. The higher-voltage wire is called the ***stinger leg*** and cannot be used for 115 V circuits. The stinger leg is normally color-coded orange and should be tagged or marked for easy identification. See **Figure 23-13.**

Another basic system is the ***dead leg*** system from a three-phase, delta transformer. One corner of the delta transformer, (the dead leg) is grounded. A reading of 240 V can be obtained between any two phases (including the dead leg). A voltage reading from the dead leg to ground is 0 V. See **Figure 23-14.**

23.3 Overcurrent

All conductors (wires) are sized to carry a limited amount of current without overheating. ***Overcurrent*** occurs when too much current flows through a wire. Overcurrent causes the wire to become hot, damaging

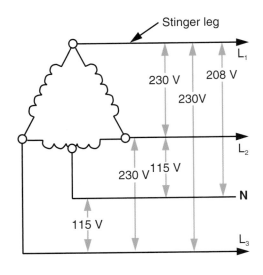

Figure 23-13. *In a 230 V, three-phase, four-wire system with a "high leg," one leg provides higher voltage than the other legs. In this case, the L₁ to neutral reading is 208 V; readings from the other legs to neutral are 115 V. The high or "stinger" leg should be marked for easy identification.*

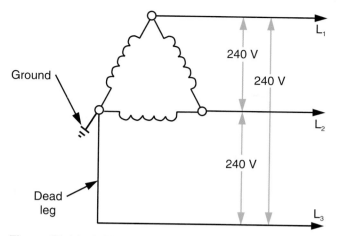

Figure 23-14. *A "dead-leg" system has one leg grounded. Any two legs provide 240 V. The reading from the dead leg to ground is 0 V.*

or destroying the insulated covering. Insulation damaged by overcurrent presents a serious hazard; exposed (bare) wire is dangerous to equipment and people. Overcurrent can be caused by such electrical problems as loose connections, ground faults (short circuits), defective resistance, or excessive loads.

Overcurrent in the form of an **overload** or short circuit is described as "current in excess of the normal flow for a given circuit." Overloads can range from two to 10 times the normal current. Overloads are typically caused by excessive loads in the circuit, loose connections, and dry motor bearings. An overload is usually confined to the circuit wiring. A ground fault or **short circuit** occurs when electrons are permitted to travel unrestricted through a path with very low, or no, resistance. See **Figure 23-15.** The resulting overcurrent may exceed the normal current hundreds of times. High heat is generated rapidly and explosions may occur. A ground fault or short circuit is very dangerous.

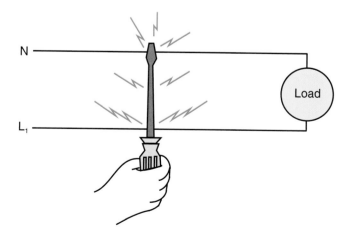

Figure 23-15. *Never work on a live circuit; a slip of a screwdriver or other tool can easily cause a short circuit. The resulting high overcurrent can have painful or even fatal results.*

23.4 Circuit Protection

Every electrical circuit must have some type of safety device, such as a fuse or a circuit breaker, to protect against overcurrent. Fuses and circuit breakers are manufactured in various shapes and sizes. They are designed to stop the flow of current when it exceeds safe limits. These **circuit protection devices** are rated in amperes and volts. Their amperage rating must not be greater than the ampacity of the wires being protected.

Fuses and circuit breakers must be able to interrupt the extreme overcurrent created by a short circuit. However, each device has an *overcurrent interrupting limit* beyond which it can no longer interrupt current flow. If the overcurrent limit is exceeded, the device may violently arc or even explode and start a fire. The limit is called the **amperage interrupting capacity (AIC)** of the device. The AIC is much larger than the load current rating of a fuse or circuit breaker. The *load rating* is the current carrying-capacity during normal operating conditions.

All overcurrent protective devices are labeled with their normal load current rating and AIC for a given voltage. The amperage interrupting capacity of most circuit breakers is between 10,000 AIC and 20,000 AIC. Most fuses are rated about 200,000 AIC.

23.4.1 Fuses

The purpose of a **fuse** is to detect excessive load current and open the circuit before danger arises. Fuses are typically located in the main power supply and in each branch circuit. A blown fuse in a branch circuit helps confine the problem to a specific area. Fuses and circuit breakers are used to protect wires and equipment, not people. See **Figure 23-16.**

How a fuse operates

Zinc is a metal that has moderate internal resistance to the flow of electrons and a rather low melting point. For these reasons, it is often used as the connecting link (**fusible element**) in a fuse. See **Figure 23-17.** When excess current flows through the fuse, the link becomes overheated and melts (blows). The circuit opens and stops current flow.

The *amount* of excess current flow determines how fast the link melts. On a short circuit, the link is instantly vaporized. Overloads cause a slow buildup of heat, and the link melts slowly. A fusible link opens in one to 15 minutes on a 50% circuit overload. However, most fuses can withstand a 10% overload indefinitely.

Fuses protect wires and equipment against overloads and short circuits. A blown fuse indicates a serious circuit problem that must be corrected. The

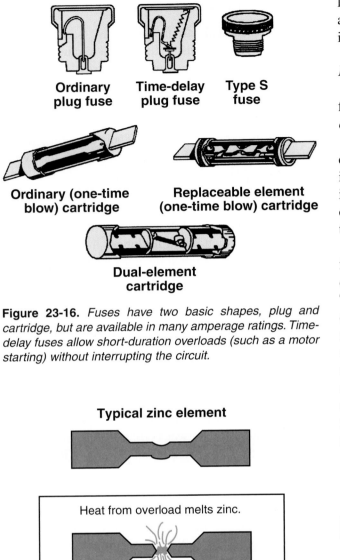

Ordinary plug fuse　　**Time-delay plug fuse**　　**Type S fuse**

Ordinary (one-time blow) cartridge　　**Replaceable element (one-time blow) cartridge**

Dual-element cartridge

Figure 23-16. *Fuses have two basic shapes, plug and cartridge, but are available in many amperage ratings. Time-delay fuses allow short-duration overloads (such as a motor starting) without interrupting the circuit.*

Typical zinc element

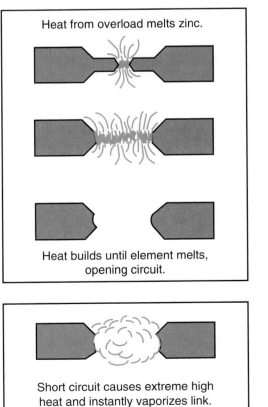

Heat from overload melts zinc.

Heat builds until element melts, opening circuit.

Short circuit causes extreme high heat and instantly vaporizes link.

Figure 23-17. *The zinc element in a fuse reacts progressively to an overload and instantly to a short circuit.*

load amperage rating of a fuse must not exceed the ampacity of the feeder wire. Oversizing fuses is an invitation to disaster.

Plug-type fuses

Plug-type fuses (10 A to 30 A) are sometimes used for 120 V circuits. Three types of fuses are available: *ordinary, time-delay,* and *Type S.* See **Figure 23-18.**

Ordinary plug fuse. This type of fuse has no time delay; the link melts (blows) when the amperage rating is exceeded. The fuse prevents overheating of wiring in the circuit due to excess current flow, which could cause a fire. The fuse should open the circuit before the wire becomes hot.

Time-delay fuse. The time-delay (***dual element***) fuse is designed to permit an overload of short duration but to blow instantly if a short circuit occurs. Time-delay fuses are necessary when fusing a circuit containing an electric motor. They permit the momentary high starting current of the motor (about six times the normal running current) without opening the circuit. The high current is very brief but occurs every time the motor starts. As the motor achieves full speed, counter-emf (increased resistance) fully develops, and the current drops to the normal running level.

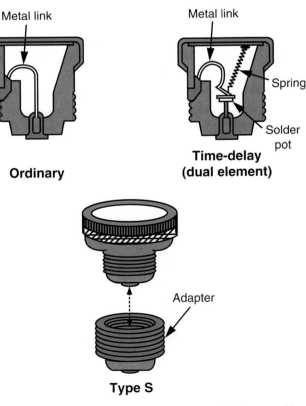

Metal link　　　　Metal link

Spring

Solder pot

Ordinary　　**Time-delay (dual element)**

Adapter

Type S

Figure 23-18. *Plug-type fuses are available as either ordinary or time-delay types. Type S is a special time-delay fuse with threads sized differently for each ampere rating. Adapters ensure a properly sized Type S fuse is used.*

An ordinary fuse used with a motor must be over-sized to permit high starting current. However, an oversized fuse is too large to protect the wires during normal operation. If the fuse is sized for normal running current, it will blow every time the motor starts. The time-delay feature is the solution.

As shown in **Figure 23-19,** time-delay is accomplished by adding a spring and attaching one end of the link to a heat sink with a low-melting-point solder. When an overload occurs, heat is generated. If the condition continues long enough, the heat will melt the solder. The spring pulls the link away from the heat sink, opening the circuit. (If a short circuit occurs, the metal link will vaporize instantly and open the circuit.)

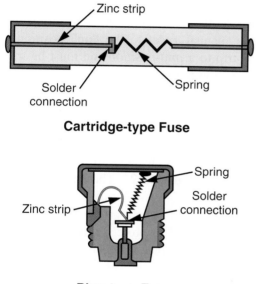

Cartridge-type Fuse

Plug-type Fuse

Figure 23-19. *Time-delay fuses permit brief overloads, such as the short period of high current flow when a motor starts. Prolonged overloads melt the solder connection, causing the spring to open the circuit.*

No heat is generated if the motor starts quickly, so the solder does not melt. If the motor binds or does not start promptly, excess current flow generates enough heat to melt the solder. The delay feature permits the fuse to be sized according to the motor's *running* current, not the starting amps.

Time-delay fuses are sized according to the **full load amperage (FLA)** of the motor. FLA information is located on the motor data plate. A time-delay fuse is selected with a rating 25% greater than the motor FLA (*oversized* by 25%). For example, a motor with an FLA of 8 A should use a 10 A fuse.

Type S fuse. This fuse has the time-delay feature but is made with threaded sections. The size of the threaded section varies according to the load amperage rating of the fuse. Type S fuses are used with an adapter that

screws into the socket on the fuse panel and locks in place. The adapter is almost impossible to remove and prevents installing the wrong size fuse in the circuit. Type S fuses are commonly used in mobile homes because of the greater risk of fire in such installations.

Cartridge fuses

Cartridge fuses are available as ordinary fuses or the time-delay, dual-element type, **Figure 23-20.** Round-end cartridge fuses are suitable up to 60 A. Knife-blade contacts are used for fuses over 60 A. It is important that the fuse ends make good contact in the fuse holder, **Figure 23-21.** Poor connections, or high air temperature around the fuse, reduce its amperage rating and cause needless blows and shutdown.

One-time-blow

Replaceable element

Dual-element

Figure 23-20. *Cartridge fuses are available in standard or time-delay (dual-element) types. In standard cartridges, either disposable (one-time-blow) fuses or those with replaceable elements can be used.*

Fuse holding devices

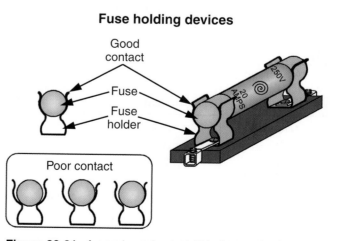

Figure 23-21. *A good mechanical fit between the fuse and fuse holder is important. Poor connections reduce the fuse's amperage rating.*

Amperage and voltage ratings determine the physical size of the cartridge fuse. Fuse-holding devices, **Figure 23-22,** are also sized so fuses and holders match. The fuse holders limit the maximum fuse size. Such limitations helps prevent oversizing fuses on circuits designed for a certain maximum amperage. For example, a cartridge fuse rated at 250 V with a rating from 1/10 A to 30 A only fits in a receptacle designed for a maximum of 30 A service at 250 V. See **Figure 23-23.**

23.4.2 Circuit Breakers

Circuit breakers are available as single-pole (one switch), double-pole (two switch), and three-pole (three switch) types. See **Figure 23-24.** The *single-pole breaker* is used to disconnect the black (hot) wire on 120 V, single-phase branch circuits. The *double-pole breaker* is used to disconnect both hot wires on 230 V, single-phase branch circuits. A *three-pole breaker* is used to disconnect all three hot wires on a three-phase circuit.

Figure 23-22. *Fuse-holding devices for cartridge fuses come in various forms. This fuse holder kit has a screw-on cap that makes a good mechanical connection. (Cutler-Hammer)*

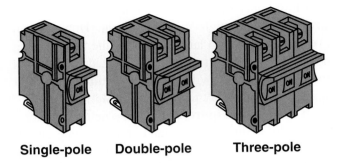

| Single-pole | Double-pole | Three-pole |

Figure 23-24. *Circuit breakers have one, two, or three switches, depending upon the application.*

Circuit breakers perform the same job as fuses but can be reset after they open the circuit. When its load amperage rating is exceeded, the breaker *trips* (switches to the "off" position). All circuit breakers require a manual *reset* to the "on" position. Some breakers trip to a *mid-position* and must be turned to the off position before resetting to the on position. A tripped circuit breaker indicates an overcurrent in the circuit. Failure to correct the problem will only result in another trip-out.

Circuit breakers serving a building or portion of a building are usually located in a *load center,* or breaker panel, **Figure 23-25.** The load center supplies electrical power to several branch circuits. The load center normally has a main circuit breaker located at the top of the panel to protect and disconnect power supply to the entire panel. Other circuit breakers are located below the main breaker and protect each of the branch circuits that obtain power from the panel. See **Figure 23-26.**

Amperage Range	250-Volt Fuse Length (inches)	600-Volt Fuse Length (inches)
1/10 to 30	2	5
35 to 60	3	5 1/2
70 to 100	5 7/8	7 7/8
110 to 200	7 1/8	9 5/8
225 to 400	8 5/8	11 5/8
450 to 600	10 3/8	13 3/8

Figure 23-23. *Ampere ranges and lengths for 250 V and 600 V cartridge fuses. Fuse holders are sized to limit the maximum amperage fuse they can hold.*

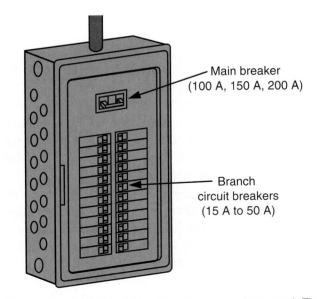

Main breaker (100 A, 150 A, 200 A)

Branch circuit breakers (15 A to 50 A)

Figure 23-25. *A typical load center or breaker panel. The main circuit breaker disconnects the incoming electrical supply from the branch circuits. Branch circuit breakers protect individual circuits.*

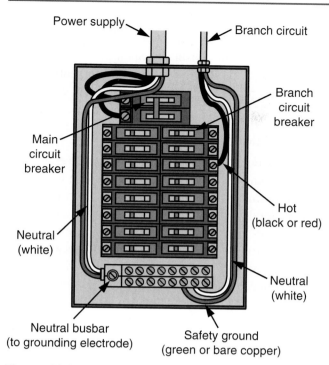

Figure 23-26. *The hot wire for each branch circuit is connected to the appropriate breaker, while the neutral wire is connected to a grounded neutral busbar. In some residential installations, the safety ground wire is also connected to the neutral busbar. For commercial installations, a separate safety ground busbar must be used.*

The hot wires (black or red) for the branch circuit are connected to a circuit breaker. All white (neutral) wires are connected to a neutral busbar which, in turn, is directly connected to a grounding electrode (earth ground). The safety ground busbar is directly connected to another grounding electrode. In some residential systems, however, the safety ground wires are connected to the neutral busbar for direct access to a single grounding electrode. This is the *only* place where the neutral and safety ground may be connected together.

23.4.3 Ground Fault Circuit Interrupters (GFCI)

The importance of a good *safety ground system* cannot be overemphasized. Many people have lost their lives because of no safety ground or poor grounding methods. A ground fault can occur as a result of defective, worn, or misused equipment, especially when the equipment is operated in damp or wet areas. Examples include portable power tools, hair dryers, electric shavers, and kitchen appliances. Such equipment *does* fail; when it fails, it presents a dangerous situation to the operator.

Fuses and circuit breakers do not protect people using such equipment. The **ground fault circuit interrupter (GFCI)** is designed to perform a dual role: it protects the circuit from overcurrent *and*

protects people from potentially hazardous ground faults arising from the use of defective appliances or portable tools. See **Figure 23-27**. The National Electrical Code requires the use of GFCIs outdoors and in bathrooms, swimming pool areas, attached garages, motel guest rooms, and all countertop receptacles within six feet of the kitchen sink. GFCIs are available as circuit breakers and receptacles. Extension cords with built-in GFCI protection are available for use with portable tools.

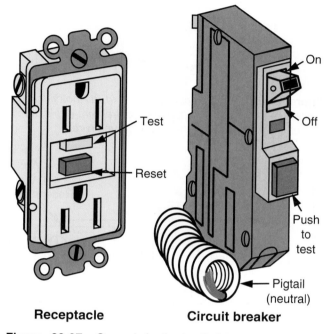

Receptacle   **Circuit breaker**

Figure 23-27. *Ground fault circuit interrupters protect people from potentially hazardous ground faults. GFCIs are available as receptacles or circuit breakers.*

How the GFCI works

As shown in **Figure 23-28**, current normally travels to an appliance along the black (hot) wire and returns along the white (neutral) wire. The amperage in each conductor is the same. No current flows through the safety ground.

When a ground fault occurs, however, the normal current flow pattern (through hot and neutral wires) changes: some current travels along the safety ground wire (or through a person's body) to ground. The flow creates an amperage *imbalance* between the hot and neutral wires.

The GFCI monitors current flow in both the hot wire and the neutral wire and is very sensitive to any imbalance. Since an imbalance can only occur during a dangerous fault to ground, the GFCI opens the circuit whenever the imbalance reaches about 4 mA to 6 mA (milliamperes). See **Figure 23-29**.

As noted earlier, the GFCI serves a dual purpose: people *and* circuit protection. If the normal load amperage rating (15 A, 20 A, or 30 A) is exceeded, it

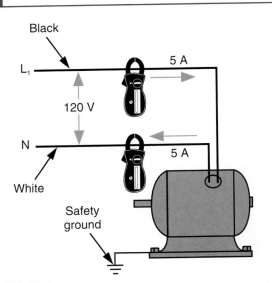

Figure 23-28. *Normal current flow in a motor circuit. Amperage in the hot wire (L₁) and neutral wire (N) is equal. No current flows through the safety ground wire.*

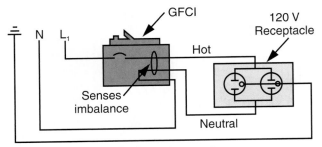

Figure 23-29. *A GFCI senses an amperage imbalance between hot and neutral wires as a result of a ground fault. It quickly opens the circuit to prevent injury to a person who might become part of the circuit.*

will open the circuit. In this manner, it operates much like an ordinary circuit breaker.

23.5 Electrical Connections

Wires are the **conductors** or pathways that carry electrical energy from place to place. Wires connect to each other, terminating at various electrical devices, and are covered with **insulation** to prevent energy loss. Where connections must be made, 1″ of insulation is removed from the wire ends, allowing enough bare wire to make proper metal-to-metal connections.

Properly made connections require a clean, tight contact between the conductor and the device terminals. The National Electrical Code does not permit splicing or connecting wires except inside a proper housing. Connections to an electrical device also must be made inside an approved housing or box. Connections between copper and aluminum can be made *only* with use of an approved connector.

Poor connections are a constant source of electrical problems. A loose connection permits arcing and becomes a built-in resistance. Such poor connections cause voltage drop, energy loss, and overheating. The importance of good electrical connections cannot be overemphasized.

23.5.1 Making Terminal Connections

When connecting a solid conductor to a screw-type terminal, bend the wire end into a loop. Use long-nose pliers to form a circular loop to the right, **Figure 23-30.**

For good contact, the loop should encircle the screw shaft in a *clockwise* direction. The screw is inserted through the loop and turned clockwise to tighten it; this causes the loop to be pulled inward for a tight connection. If the loop faces left (counterclockwise), tightening the screw will cause the loop to open outward, resulting in a poor, and possibly unsafe, connection. See **Figure 23-31.**

Terminal connectors

Quick-connect ***terminal connectors*** that crimp onto wire ends are popular and easy to use. Types and sizes are available to fit most applications. See **Figure 23-32.** When installed properly, terminal connectors avoid most problems arising from loose connections.

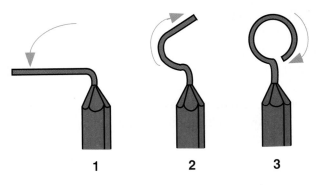

Figure 23-30. *A long-nose pliers is used to make a proper loop for wire connection to a screw terminal. The loop must be made in a clockwise (right-hand) direction.*

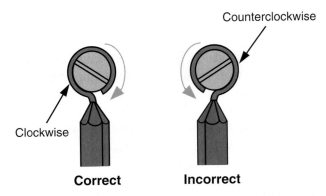

Figure 23-31. *The loop must encircle the screw in a clockwise direction so it is pulled inward as the screw is tightened. A counterclockwise loop tends to open as the screw is tightened.*

To install a connector on a wire, remove about 1/2″ of insulation from the wire end. Then, insert the bare end into the connector. A special tool is used to **crimp** (squeeze) the connector onto the wire end, **Figure 23-33.** Two crimps usually achieve a secure bond between the wire and the connector. After crimping, make sure the connector does not pull off the wire. The connection is covered with a plastic insulator.

Solderless screw-on connectors

Solderless screw-on connectors are commonly called **wire nuts.** They eliminate the need for soldering connections and are NEC-approved. Wire nuts are available in several sizes and made of plastic or Bakelite® insulating material. To install a wire nut, remove about 1/2″ of insulation from the wire ends. Twist the bare ends together, and screw the nut onto the twisted wires. See **Figure 23-34.** Be certain the nut is tight and cannot fall off. If wire nuts become loose, the wires inside will separate. To prevent loosening, use electrical tape to secure the nut on the wire ends.

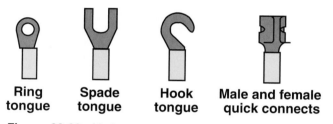

Ring tongue   Spade tongue   Hook tongue   Male and female quick connects

Figure 23-32. *Various types of terminal connectors are available for various applications. Terminal connectors almost eliminate loose connection problems.*

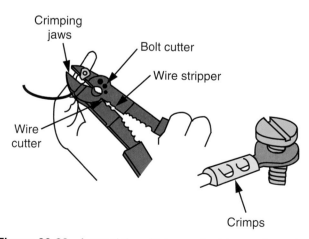

Crimping jaws
Bolt cutter
Wire stripper
Wire cutter
Crimps

Figure 23-33. *A special multiple-use tool is used to crimp the connector to the wire end. As shown on the right, two crimps are usually made in the connector to ensure a good bond to the wire.*

Figure 23-34. *A wire nut is installed by twisting the bare wire ends together, then screwing the nut onto the twisted wires. For a sound mechanical and electrical connection, the wire nut must be tight.*

Summary

Troubleshooting and repairing motor control and circuit protection devices are routine activities for the HVAC technician. This chapter described the power transmission system and the various types of transformers used to reduce voltage to usable levels in residential and commercial installations. Also described were the different types of circuit protection devices and their operation. Finally, the importance of making good terminal connections and types of connectors was discussed. Skills acquired in these areas will improve your ability to diagnose and correct system problems.

Test Your Knowledge

Please do not write in this text. Write your answers on a separate sheet of paper.

1. *True or false?* Single-phase power can be obtained from a three-phase system.
2. Most residential units have two hot wires and a neutral wire. What is the supplied voltage? How many phases are supplied?
3. An advantage of alternating current is that it can be _____ at one voltage, transmitted at _____ voltage, and reduced to a lower voltage at the _____.
4. What is the operating principle for a transformer?
5. Does a step-up transformer increase or decrease voltage?
6. Ampacity is the amount of _____ a wire can safely carry.
7. Name the two devices that protect against over-current.
8. *True or false?* A dual-element plug fuse should be used in circuits serving electric motors.
9. What two ratings determine the physical size of a cartridge fuse?
10. A GFCI protects people against a _____ fault.

toward the shaft end. Direction of rotation should always take into account which end of the motor is being viewed.

Disconnecting the start winding

The sole purpose of the start winding is to get the rotor moving (start the electric motor). Because of its high resistance, the start winding will *burn out* if it is not disconnected from the circuit once the motor reaches about 75% of normal speed. Burnout can happen very quickly, usually in less than one second.

One of three methods is used to disconnect the start winding from the circuit, depending upon the type of motor. Open-type motors use a *centrifugal switch;* small hermetics use an *amperage relay;* larger hermetics and semihermetics use a *potential relay.* Amperage and potential relays are fully explained in Chapter 25.

The centrifugal switch used to disconnect the start winding of an open-type motor is inside the motor frame. The switching device (amperage or potential relay) used by hermetic and semihermetic compressors is outside the metal housing. *Exterior* relays must be used because of the **arcing** (sparking) that occurs when switch contacts open or close. Arcing would create damaging acids inside the hermetic unit by causing a breakdown of refrigerant and compressor oil.

The run winding and the start winding are two separate loads and must be connected in parallel. See **Figure 24-10.** A switch is connected in series with the

start winding and disconnects that winding after start-up. A main switch controls the power supply to the motor and is connected in series with both windings. The main switch can be installed at any remote location.

Locked rotor amps (LRA)

At start-up, little or no counter-emf is generated by the motor. Amperage flow is determined by the resistance of the motor windings. Starting current flow is typically six times higher than normal running amperage. The high current flow during start-up is called **locked rotor amps (LRA).** As the motor picks up speed, counter-emf is generated and rapidly increases the motor's resistance to current flow. As noted, the start winding is disconnected at about 75% of normal speed. Consequently, amperage flow is substantially reduced; the motor operates on the run winding alone. Amperage flow is now determined by the resistance of the run winding combined with the counter-emf generated by the motor.

Full load amps (FLA)

Immediately after the start winding is disconnected, the resistance of the run winding is increased by the counter-emf generated by the motor. The operating ("running") amperage is called **full load amps (FLA).** Induction motors usually operate at less than FLA because the motor rarely works at fully loaded condition. An overload occurs when the amperage flow exceeds the FLA rating on the motor data plate. Except for the temporary high start-up amperage, *any* amperage that exceeds the FLA rating is converted to heat energy.

24.2.2 Overload Protectors

The induction motor is designed to convert electrical energy to mechanical energy. If the rotor does not turn normally, excess amperage flow will result. The motor will convert the excess amperage to heat energy, which will burn out the motor windings. The two most common causes of motor problems are:

❑ A locked rotor (rotor unable to turn).
❑ An overload (rotor turning too slowly).

The **motor overload protector** (usually referred to simply as an "overload") is a device that protects the motor windings from damage caused by overheating due to overloaded conditions or poor ventilation.

The most common form of motor protection uses a "snap-acting" **bimetallic** disc to make and break a set of contacts. See **Figure 24-11.** The snap action is obtained by fusing together two thin, circular discs of different metals, usually steel on top and copper on the bottom. The bimetallic disc is anchored at the center.

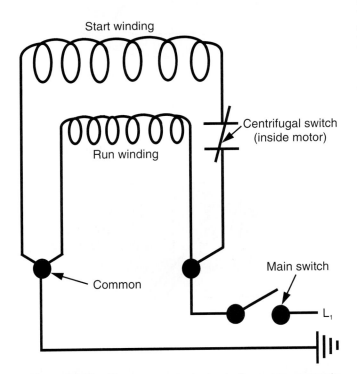

Figure 24-10. *The run and start windings are separate loads and connected in parallel. A centrifugal switch is connected in series with the start winding.*

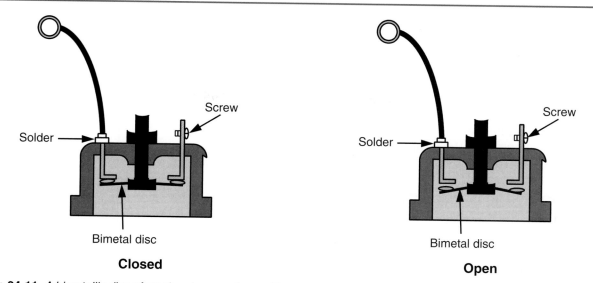

Closed

Open

Figure 24-11. *A bimetallic disc of steel and copper is used in many overloads. In normal operation, the disc holds the contacts in the closed position. When sufficiently heated by an overload, the differing expansion rates of the metals cause the disc to curl in one direction. This opens the contacts and breaks the motor circuit. When the motor cools sufficiently, the disc returns to its normal shape, closing the contacts again.*

Since copper expands faster than steel, a rise in temperature causes the disc to expand. The outer edge curls (bows) upward.

The movement of the bowed edge opens a set of stationary contacts. When the bimetallic disc cools, it snaps back to its original (closed) position. Excessive heat from any source will affect the overload and cause it to open the circuit supplying power to the motor.

Cycling on the overload

If the overload condition is not corrected, the bimetallic disc will trip repeatedly, a malfunction known as **cycling on the overload.** After each trip, more and more time is required for the bimetallic disc to cool. Eventually, the bimetallic disc could require up to an hour to cool and reset. Cycling on the overload is easy to hear; the motor makes a noise described as "hmmm...click." The humming noise is made by the motor trying to start, and the sharp "click" is made when the bimetallic disc snaps open.

Cycling on the overload indicates a condition that *must* be corrected. The overload protector is sensitive to three sources of heat:

❏ Excess amperage flow (usually an electrical problem related to the start winding and start components).
❏ Excess motor heat (usually resulting from dry or badly worn bearings).
❏ Excess compressor heat (low on oil or compressor cooling is not working).

Tripping the overload protector is a warning that the motor is in danger of a burnout. It is important to locate and repair the cause of an overload.

Two-wire overload connections

The terminals on the overload protector are numbered 1, 2, and 3. The contacts are located between numbers 1 and 2; a small pure resistance heater wire is located between numbers 2 and 3. The resistance heater makes the overload react more quickly to overheating. See **Figure 24-12.** Connections to the overload are limited to Terminals 1 and 3, which simplifies making electrical connections and prevents bypassing the heater. The heater is internally connected in series with the contacts.

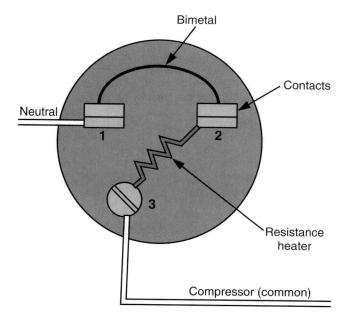

Figure 24-12. *A resistance heater wire, connected in series with the bimetallic element, makes the overload react more quickly.*

One side of the supply voltage is connected to Terminal 1 so the current must flow through the contacts *and* the small heater to reach Terminal 3. A wire from Terminal 3 is connected to the *common* on the motor windings. **Common** describes the wire that connects one end of the run winding to one end of the start winding. Current flowing through the common supplies power to both windings; hence, it is "common" to both windings. See **Figure 24-13.**

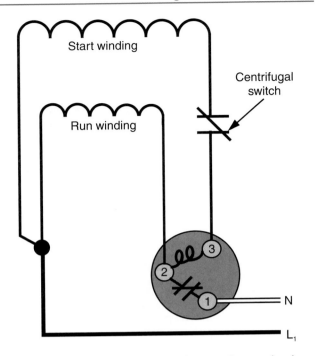

Figure 24-14. *Three-wire connections to the overload use all three terminals. The heater is connected in series with the start winding so the overload reacts quickly to any problem with the winding.*

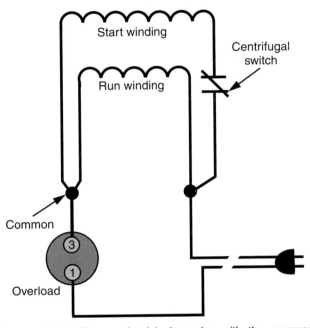

Figure 24-13. *The overload is in series with the common connection to the run and start windings, allowing the overload to protect against excess current flow through either winding.*

Three-wire overload connections

On open-type motors, a three-wire connection is often used for the overload protector. Power supply is connected to Terminal 1, the run winding to Terminal 2, and the start winding to Terminal 3. See **Figure 24-14.**

The heater inside the overload is connected in series with the start winding. The connection makes the overload respond quickly to any fault in the start winding, the source of many motor problems. During normal operation, the *run* winding is protected by the bimetallic disc only. However, if the bimetallic disc opens the contacts between Terminals 1 and 2, the power supply to both windings will be interrupted.

Internal overload protection

Many large hermetics and semihermetics have an overload protector buried inside the motor windings. The overload can sense the current draw of the motor and the temperature of the windings much better than can external overloads. Internal overload protection is installed at the factory and is not accessible to the technician. See **Figure 24-15.**

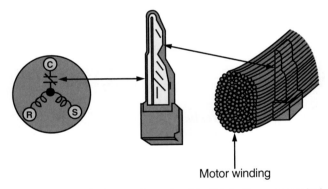

Figure 24-15. *An internal overload built into the motor windings is not accessible to the technician.*

The internal overload is a bimetallic element that will open the circuit and stop the motor if the winding temperature reaches 200°F to 250°F (93°C to 121°C). It will close again when the winding temperature drops to 150°F to 175°F (66°C to 79°C). Depending on ambient temperature conditions, it may take one or two hours before the protector cools enough to close the contacts. Cooling the unit with forced air or ice speeds up the process. Unless an ohmmeter test confirms an open motor winding, do not condemn a motor until sufficient time has passed for the overload to reset.

24.3 Types of Induction Motors ◆

Five types of single-phase induction motors, plus the three-phase induction motor, are in general use for heating, cooling, and refrigeration applications. The type of motor for a particular application is determined by the location and type of load. Each motor is available in various horsepower ratings. Single-phase induction motors are described in the order of turning power (torque), from the least to the greatest. The three-phase motor is described later in this chapter.

The types of single-phase induction motors are:
❏ Shaded-pole.
❏ Split-phase.
❏ Permanent split capacitor (PSC).
❏ Capacitor-start, induction-run (CSIR).
❏ Capacitor-start, capacitor-run (CSCR).

24.3.1 Shaded-pole Motor

The **shaded-pole motor** is a uniquely designed, low-cost motor used to operate small fans, such as the evaporator and condenser fans on domestic and light commercial refrigeration units. See **Figure 24-16.** The shaded pole motor has very low starting torque, but it can carry a full load once it is up to speed. The capability of carrying a full load makes the shaded-pole motor ideal for operating small fans. At start-up, the only load on the motor is the weight of the aluminum fan blades. Full load is not reached until the blades are turning at operating speed and moving volumes of air.

Figure 24-16. *A shaded-pole motor used to operate a condenser fan on an air conditioning unit.*

Construction of the shaded-pole motor is different from construction of other induction motors. As shown in **Figure 24-17,** a slit is cut at the edge of each stator pole, producing a small prong. A single closed loop of wire, called a **shading coil,** is placed over the prong. The shaded pole produces a tiny magnetic field out of step with the main poles. It is *not* a start winding in the usual sense because the shading coil is not individually connected to a power source.

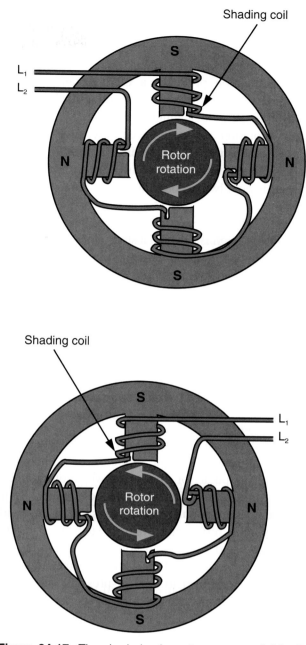

Figure 24-17. *The shaded-pole motor uses small "shading coils" on each pole to provide the necessary impetus for starting. Direction of rotation is determined by location of the shading coils on the poles.*

The shaded pole provides just enough magnetism to start the rotor turning; then, the rotating magnetic field in the main poles provides the necessary torque to bring the rotor up to full speed. The shaded pole is constantly energized when the motor is running, but such a small amount of energy is only a factor at start-up.

Shaded-pole motors are available from 1/100 hp to 1/6 hp. If a shaded-pole motor malfunctions, it is usually cheaper to replace it than repair it. Making electrical connections is simple because only two wires are involved — a power supply is connected to each end of the motor winding.

Figure 24-27. *Start capacitors are available in various sizes, depending upon motor requirements. They are less-durably constructed than run capacitors because start capacitors are used only briefly, whenever the motor is started.*

is disconnected along with the start winding. Typically, a start capacitor can withstand about 20 starts per hour before suffering internal damage.

The start capacitor absorbs (stores) excess electrons during one current alternation, then discharges the electrons into the start winding when the current alternates again. Current flow and the strength of the magnetic field generated are increased in the start winding. The result is the necessary torque, or turning power, to start the motor.

The size (capacity rating) of the start capacitor determines the amount of excess electrons discharged into the start winding. Capacitor size varies according to the horsepower rating of the motor and the load placed upon it. Be sure to use an *exact* replacement for a start capacitor.

Open-type CSIR

Electrical connections for the CSIR motor are similar to electrical connections for the split-phase motor. The only difference is the addition of a start capacitor. See **Figure 24-28.** On this type of motor, the start capacitor, centrifugal switch, and start winding are connected in series. The *order* in which the

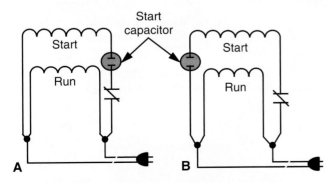

Figure 24-28. *The order of components in the start circuit does not matter as long as they are wired in series. A—The start capacitor is between the centrifugal switch and the start winding. B—The start winding is between the switch and the capacitor.*

components are connected makes no difference; however, the components must be in series so the switch can disconnect them from the circuit.

Reversing rotation. On the open-type CSIR motor, rotation is reversed by swapping the two wires that control current flow to the start winding. Information for identifying these wires is located on the motor's data plate or inside the cover plate.

Hermetic CSIR motor

The hermetic CSIR motor is used when the compressor is required to start under high head pressure and high load conditions, such as in systems using automatic and thermostatic expansion valves.

Hermetic electrical connections. The CSIR hermetic looks like other hermetics, except the start capacitor is mounted immediately above (or beside) the electrical terminal box to simplify wiring connections. See **Figure 24-29.**

One leg of the power supply (white wire) is connected to Terminal 1 of the overload. Terminal 3 of the overload is connected to the common terminal on the hermetic. The other leg of the power supply (black wire) is connected to the relay and travels directly to the run terminal. A wire from the relay switch is

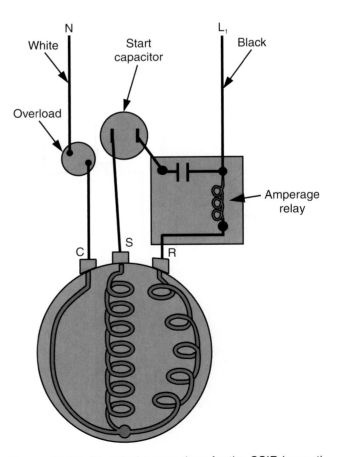

Figure 24-29. *Electrical connections for the CSIR hermetic motor-compressor. The start winding, start capacitor, and relay switch are in series.*

connected to the start capacitor. Another wire, from the other side of the start capacitor, is connected to the start terminal.

These connections place the overload in series with the common terminal, and the run winding is connected parallel. The start winding is parallel, but the switch and start capacitor are in series with the start winding.

24.3.5 Capacitor-start, Capacitor-run Motor

The *capacitor-start, capacitor-run (CSCR) motor* is another improvement on the split-phase motor. The most powerful of the single-phase induction motors, it provides both high starting torque and high running torque. The CSCR motor is used when the motor must start under a heavy load and carry a heavy load during the run cycle. The motor is typically used in small commercial systems with a thermostatic expansion valve.

The CSCR motor is essentially a split-phase motor that combines the operating principles of the PSC and CSIR motors, using both start and run capacitors. At start-up, it operates as a CSIR motor because of the start capacitor. After start-up, a relay switch disconnects the start capacitor. The run capacitor *remains in the circuit* and keeps the start winding energized. The motor continues to run as a PSC.

Hermetic CSCR electrical connections

One leg of the power supply (white wire) is connected to the overload at Terminal 1. See **Figure 24-30.** A wire from Terminal 3 of the overload is connected to the common terminal. The other leg of the power supply (black wire) is connected to the relay and directly to the run terminal. Overall, the run winding is connected from one side of the line to the other (parallel), and the overload is in series with common.

Connections from the relay and capacitors to the start winding are critical but really quite simple. First, connect the start capacitor in series with the relay switch and start terminal by connecting a wire from the switch to one side of the start capacitor. Another wire is connected from the other side of the start capacitor to the start terminal. These connections are exactly like those on the CSIR motor.

The run capacitor is connected to provide another circuit for the electrons to reach the start winding. This circuit must *bypass* the switch and start capacitor. The connections are called *series-parallel* because each capacitor is *in series* with the start winding but *parallel to* each other. Such a connection permits each capacitor to act independently and perform its duty to the start winding. At start-up, both capacitors are

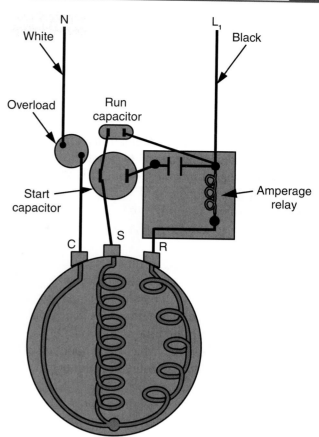

Figure 24-30. *Electrical connections for the CSCR hermetic motor-compressor. The start winding, start capacitor, and relay switch are in series. The run capacitor is wired in parallel with the start capacitor and relay switch.*

energized. When the relay switch opens, only the start capacitor is disconnected. The circuit for the run capacitor is still available for controlled amounts of electrons to reach the start winding.

24.4 Capacitors

A capacitor is an electrical device used to improve the operating characteristics of single-phase induction motors by increasing or decreasing the strength of the magnetic field produced by the start winding. **Figure 24-31** shows the two types of capacitors:

❏ *Start capacitor.* This is a dry-type capacitor intended for intermittent operation. It is fragile and only withstands 20 starts per hour.

❏ *Run capacitor.* This is an oil-filled capacitor intended for continuous operation. The oil serves to dissipate any heat buildup.

24.4.1 Start Capacitor

The start capacitor consists of two layers of aluminum foil, called **electrodes** or **plates,** separated by an insulating layer of specially-treated paper. The foil and paper layers are about three inches wide and marking, **Figure 24-35.** The power supply should always be connected to the identified terminal, which is the terminal most likely to short to the metal case and be grounded in the event of a capacitor breakdown. If the capacitor should break down and become grounded, the fuse in the power supply line will blow before the motor is damaged.

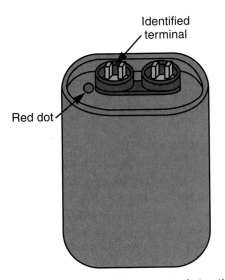

Figure 24-35. *The terminal connected to the outer aluminum strip is marked on a run capacitor. The power supply connects to this lead to help protect the motor. Some manufacturers use a dot; others prefer an arrow or a dash.*

24.4.5 Testing Capacitors

One method of testing a capacitor is to replace the possibly faulty device with a known good one. If the motor operates properly, the old capacitor is faulty.

Another method of testing is to use an ohmmeter. Before testing, the capacitor must be disconnected from the circuit and discharged by shorting across the terminals. This can be done by using a screwdriver with an insulated handle.

Touch the two leads from the ohmmeter to the capacitor terminals. If the capacitor is good, the meter battery (9 V dc) will slightly charge one side of the capacitor. When the leads are reversed from one terminal to the other, one side of the capacitor discharges into the meter and the other absorbs electrons from the battery. Each time the meter leads are reversed, the meter needle deflects slightly and immediately returns to zero. (A digital ohmmeter shows a slight resistance reading, then returns to zero.) Deflection of the needle is slight because the meter battery is weak compared to normal line voltage. Increasing meter resistance to ×10,000 increases needle deflection for easier reading. See **Figure 24-36.**

If the ohmmeter needle deflects to infinity and stays deflected, the capacitor is open. If the needle

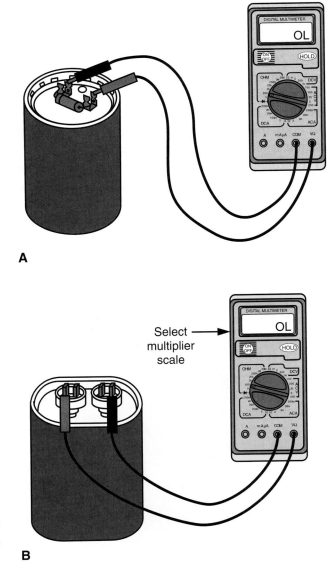

Figure 24-36. *Checking capacitors with an ohmmeter. A—Reversing leads should cause the needle of an analog meter to deflect slightly, then return to zero. A digital meter will briefly show a low resistance reading, then return to OL (overload). B—The very slight needle deflection (or digital meter reading) can be increased for easier viewing by selecting the ×10,000 resistance scale on the meter.*

does not deflect at all, the capacitor is shorted. See **Figure 24-37.** An open or shorted capacitor is faulty and must be replaced.

An ohmmeter test will not reveal a weak capacitor. Low mfd capacity can only be discovered by using a tester specifically designed to test ac capacitors. See **Figure 24-38.** These meters are accurate, inexpensive, and simple to operate.

24.4.6 Changing Motor Types

It is possible to add start components to a PSC motor and convert it to a CSCR motor. The procedure is sometimes necessary when the PSC

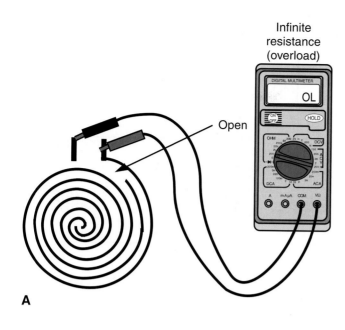

Infinite resistance (overload)

Open

A

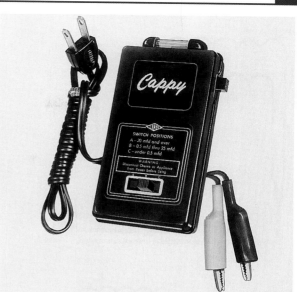

Figure 24-38. *A tester made specifically for use with capacitors helps the technician quickly diagnose problems. (Watsco)*

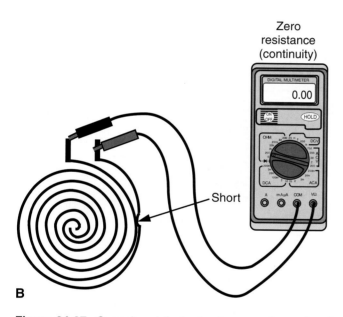

Zero resistance (continuity)

Short

B

Figure 24-37. *Capacitor defects. A—An open (burned out) capacitor causes the meter to show an infinite resistance reading. B—A shorted capacitor shows zero resistance (continuity).*

motor-compressor has difficulty starting, which is not unusual with residential central air conditioning systems.

Selecting a properly sized capacitor and relay is critical for proper operation. Sizing information is available from the local dealer or compressor manufacturer. Most manufacturers of PSC motor-compressors furnish a *hard start kit* for their units' motors. The kit consists of a start capacitor and relay, along with complete instructions for installation. The model and serial numbers of the motor-compressor are needed to obtain the correct kit.

24.5 Dual-voltage Motors

Dual-voltage motors are manufactured to operate on either of two supplied voltages. The voltage rating on the motor data plate may read, for example, *115/230V*, indicating the motor can be connected to either a 115 V supply or a 230 V supply. Dual-voltage motors have two run windings that must be connected properly, as shown in **Figure 24-39.** For lower voltage, the run windings are connected in parallel (providing lower resistance). For higher voltage, the windings are connected in series, doubling the resistance.

The magnetic fields of the two run windings *must* rotate in the same direction. If one field is "left" and the other is "right," opposing polarities will cause the rotor to lock. The factory connects the motor windings to terminals located in the terminal box.

Most dual-voltage motors are equipped with flat copper *jumper bars* for ease of change from one voltage to the other. Instructions for placement of these jumper bars are inside the terminal box cover. **Figure 24-40** depicts the proper placement of jumper bars on an open-type dual-voltage motor.

24.6 Multi-speed Motors

Multi-speed motors are very common in heating, cooling, and refrigeration systems. In contrast to the two wires used in single-speed motors, multi-speed motors have several wires. A *multi-speed motor* is necessary in applications that require variable-speed operation, such as a window air conditioner with a fan that operates at high, medium, or low speeds. Many furnace blower motors are two-speed. Variable speeds allow

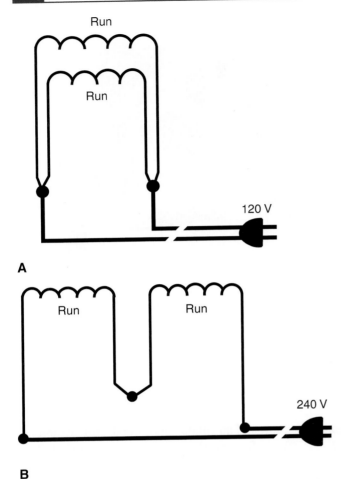

A

B

Figure 24-39. *Dual-voltage motor run winding connections. A—For the lower voltage, the windings are connected in parallel (for lower resistance). B—For high-voltage operation, the windings are connected in series (for high resistance).*

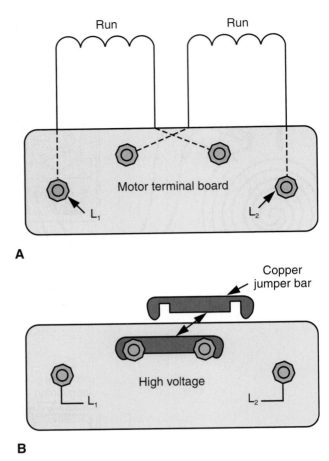

A

B

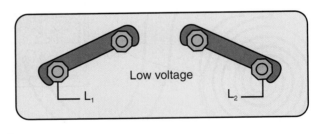

C

Figure 24-40. *Changing a dual-voltage motor from one voltage to another. A—Factory-installed internal connections at the terminal board. B—For high voltage, the copper jumper bar connects the run windings in series. C—Using jumper bars in this configuration connects the windings in parallel for lower resistance.*

more air to be moved during the cooling season (high speed) than during the heating season (low speed).

Motor speed is determined by the number of poles in the run winding. The multi-speed motor has different *taps* (wires) that make it possible to select the number of poles being used. See **Figure 24-41.** A high-speed tap uses two poles (3600 rpm), medium speed uses four poles (1800 rpm), and low speed uses all six poles (1200 rpm). One leg of the power supply is connected to the common terminal and the capacitor. The other side of the power supply is connected to the tap for the desired speed. Unless a selector-type switch is used, each unused tap is capped with a wire nut and sealed with electrical tape to prevent the motor windings from becoming grounded by the unused wires.

Only one end of the start winding is available for connection to a capacitor by the technician. The other end of the start winding is wired at the factory. A power supply that is parallel to the start winding is provided, regardless of the speed (tap) selected. Most manufacturers color-code the external wires for

making connections to multi-speed motors. See **Figure 24-42** for the color codes.

24.7 Three-phase Motors

Three-phase induction motors are common in commercial applications because they are smaller and more powerful than single-phase motors of equal horsepower. Furthermore, they are cheaper to operate, have fewer problems, and are easier to connect and control.

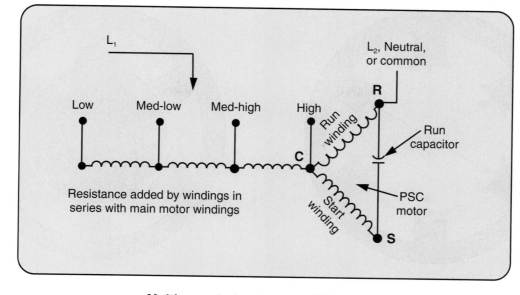

Multi-speed, single-phase PSC motor

Figure 24-41. *Different numbers of poles are used for each speed on a multi-speed motor. Connections are made to the common terminal and the proper tap for a desired speed. A selector switch may be installed for ease of changing from one speed to another.*

Coding for Multi-speed Motors

Tap	Color
Common	White
High	Black
Medium	Yellow
Low	Red
Capacitor	Purple (2 wires)

Figure 24-42. *Colors used by manufacturers to code wires used for making external connections to multi-speed motors.*

Three-phase alternating current induction motors offer high starting torque and high running torque without the need for a start winding, capacitors, or relays. Only the three "hot" wires (L_1, L_2, and L_3) are needed for proper operation, but the ground (green) wire is included for safety. The green wire is connected to the motor frame to provide an escape for electrons if the motor becomes **shorted to ground** (grounded).

24.7.1 Method of Operation

Three-phase motors have three pairs of stator poles (windings), one pair for each supply wire. Each winding shows identical resistance. As shown in **Figure 24-43,** each pair of windings has a north pole and a south pole directly opposite each other and called "one pole per phase". The stationary poles are equally spaced around a circle, exactly 60° apart ($6 \times 60 = 360°$).

With three-phase alternating current, the three power-supply wires take turns changing polarity from north to south to zero. See **Figure 24-44.** Changing of polarity in the supply wires produces a strong rotating magnetic field. The alternating zero does not produce polarity in the stator poles, which permits the other two poles to produce the rotating push-pull effect on the rotor. Because the zero pole is constantly rotating around the stator poles, the rotor chases the rotating magnetic fields.

The alternation of north, south, and zero in a three-phase motor produces a strong starting torque and a strong running torque. If the motor is running, and one wire is disconnected from the circuit, the motor may continue to run as a single-phase motor (called "single-phasing"). However, when the motor shuts off, it cannot restart.

Changing rotation

Direction of rotation is determined by direction of the rotating zero, where no polarity is generated. A three-phase motor is easily reversed by changing any two supply wires, causing the zero to rotate in the opposite direction.

Resistance of windings

All the windings in a three-phase motor are identical in resistance. One end of each winding is factory-connected inside the motor, providing an ideal location for the internal overload. When the overload opens, all three wires are disconnected at the same time. See **Figure 24-45.** When reading

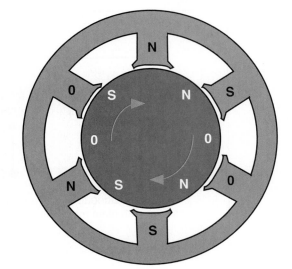

Figure 24-44. *The three poles change polarity from north to south to zero, producing a rotating magnetic field for the rotor to "chase."*

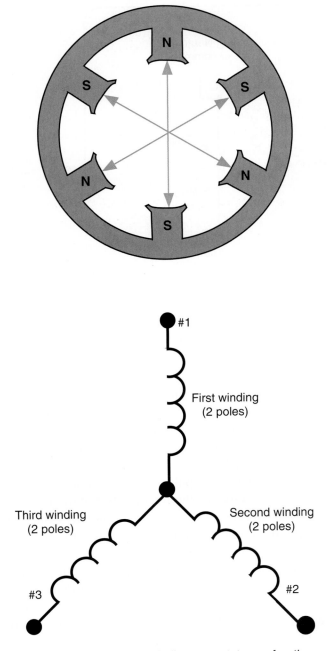

Figure 24-43. *The three windings, or stators, of a three-phase motor each have a north pole and a south pole. The stators are equally spaced around a circle.*

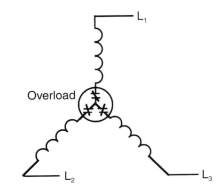

Figure 24-45. *All three windings are connected to the internal overload so when the overload opens, all the windings are disconnected.*

resistance between any two motor leads, the ohmmeter is actually reading through two windings. The resistance reading between any two wires should be the same.

24.7.2 Dual-voltage Three-phase Motors

Many three-phase motors are of the dual-voltage type. The motor data plate specifies the voltages (220/440 V, for example) that can be applied. Instead of three wires to connect, however, this motor has *nine* wires that are tagged and numbered for identification. *Never remove these numbers!*

A dual-voltage three-phase motor contains an extra set of three poles. In addition to the three original wires, two wires for each of these extra poles are brought outside the motor. See **Figure 24-46.** All nine wires must be connected. The motor windings are connected in parallel for low voltage and in series for high voltage.

Instructions for connecting the nine wires for each voltage are included on the motor data plate. The instructions are accurate, but each manufacturer uses different methods to explain the connections.

A simple Y-type diagram can be constructed for connecting the motor for either voltage, **Figure 24-47.** Draw a "Y," and put extensions on each of the branches, allowing a small space between the extensions. Each line represents one of the stator windings. Next, number all the ends. Start by numbering 1 on any outside branch, and continue numbering around the "Y" clockwise, working constantly toward the inside, until all ends are numbered.

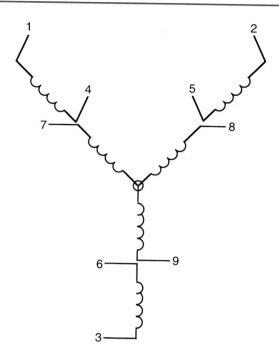

Figure 24-46. *A schematic of wye-connected, dual-voltage, three-phase motor windings shows the numbering method for all nine wires.*

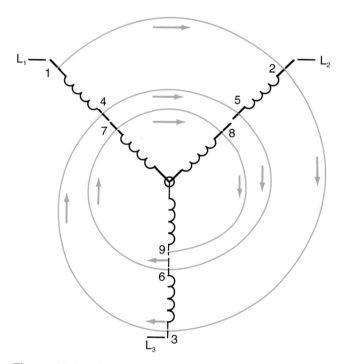

Figure 24-47. *Constructing a "Y" diagram and numbering the wires on a dual-voltage, three-phase motor. The spiral arrow shows the proper sequence for numbering the ends of the windings.*

The three power supply wires are always connected to motor terminals 1, 2, and 3. Except for direction of rotation, it does not make any difference which supply wire is connected to which terminal since the supply wires are the same.

High-voltage connection

Making the high-voltage connection is easy because the windings are wired in series, and the branches in the diagram are simply connected to form a larger "Y." See **Figure 24-48.** Wires 4 and 7 are connected, 5 and 8 are connected, and 6 and 9 are connected. The power supply wires are *always* connected to wires 1, 2, and 3. For high voltage, there are two wires at each connection, and the current flows through four windings before exiting through another supply wire. Each wire connection should be made with a wire nut and secured with electrical tape.

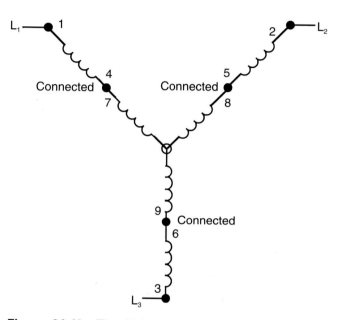

Figure 24-48. *The high-voltage connection places the windings in series. Wires 4 and 7 are connected, as are wires 5 and 8, and wires 6 and 9. Power supply always goes to 1, 2, and 3.*

Low-voltage connection

The parallel hook up for low voltage is more complicated than for high voltage, but the "Y" principle is still used. As shown in **Figure 24-49,** L_1 is connected to 1 and 7; L_2 is connected to 2 and 8; L_3 is connected to 3 and 9; 4, 5, and 6 are connected to each other.

For low voltage, there are three wires at each connection, and the current flows through two windings before exiting through another supply wire.

Resistance check of windings

The motor windings on a three-phase motor can be checked with an ohmmeter. If a resistance reading of zero occurs, the motor is shorted. If a reading is obtained from any motor lead to ground, the motor is grounded. A reading of infinite resistance (full scale

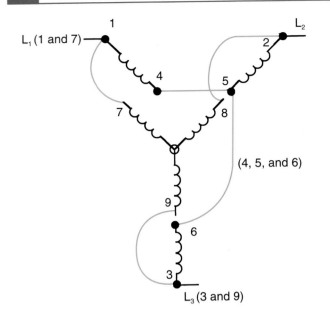

Figure 24-49. *The low-voltage connection places the windings in parallel. There are three wires at each connection, compared to two wires in the series connection for high voltage.*

deflection on an analog meter) means the winding is open. In each of these cases, the motor must be rewound or replaced.

Depending upon the size of the motor, resistance readings on three-phase motor windings range from 1 ohm to 50 ohms (the larger the motor, the smaller the resistance). On a single-voltage, three-phase motor, all windings have the same resistance. On a dual-voltage three-phase motor, the three "extra" windings have one-half the resistance of the main windings.

24.7.3 Identifying Compressor Terminals

Most manufacturers of single-phase, hermetic motor-compressors mark the three electrical terminals C, S, and R, or Common, Start, and Run. If these terminals are not marked, they can be identified by checking the resistance of the windings.

Electrical power supply to the motor must be disconnected and the connections to the three motor terminals opened. This isolates the motor windings and prevents backfeed from other electrical circuits that would cause a false ohmmeter reading.

The first test with the ohmmeter is to determine which two terminals have the highest resistance. Three separate readings are required to cover all the possible combinations. The two terminals that produce the highest reading (18 ohms in **Figure 24-50**) are the start and run terminals because the meter is reading through both windings (in series). By the process of elimination, the high reading means the other terminal is the common terminal.

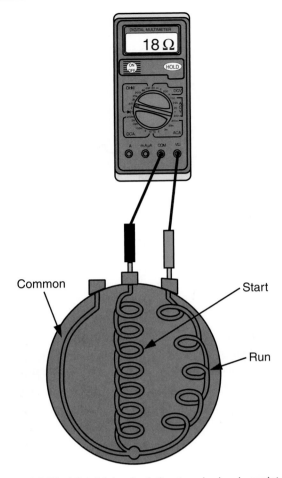

Figure 24-50. *The highest of the terminal-pair resistance readings is for the start and run windings. By the process of elimination, the remaining terminal is the Common.*

With the common terminal identified, read from common terminal to each of the other two terminals. Common-to-Start shows more resistance than Common-to-Run. See **Figure 24-51.** The approximate start winding and run winding resistance readings for fractional horsepower single-phase motors are presented in **Figure 24-52.**

24.8 Identifying Motor Problems

The ohmmeter is used to check motor windings, a process called "ringing out" a motor. All electrical connections to the motor must be removed to prevent false readings and possible damage to the meter.

24.8.1 Grounded Windings

Each motor terminal must be checked to determine whether it is *shorted to ground.* Use one of the ohmmeter probes to scratch through paint or oxidation on the steel shell of the hermetic unit to provide a good electrical connection. Connect the other ohmmeter lead to one of the motor terminals (C, S, or R). The ohmmeter should read infinity (unlimited resistance).

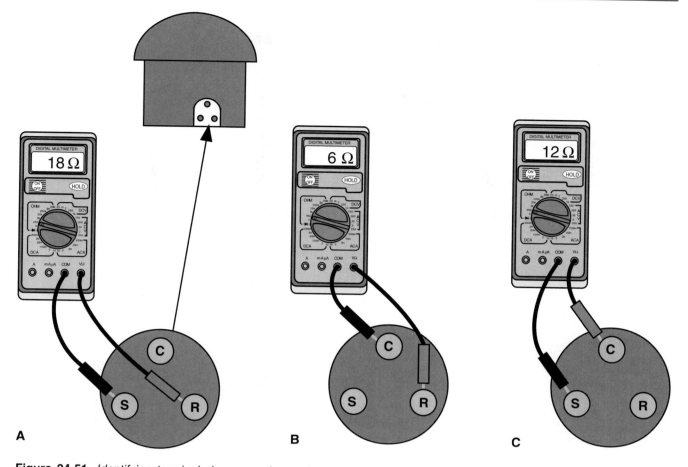

Figure 24-51. *Identifying terminals by measuring resistance. A—The common terminal is the one remaining when the highest terminal pair (run/start) reading is obtained. B—A reading from Common to Run yields the lowest reading. C—The Common-to-Start reading falls between the other two.*

Winding Resistances

Motor hp	Run	Start
1/8	4.5	16
1/6	4.0	16
1/5	2.5	13
1/4	2.0	17

Figure 24-52. *Run and start winding resistances, in ohms, for common sizes of fractional horsepower single-phase motors.*

A reading of zero resistance, or less than infinite resistance, indicates a winding has become shorted to ground. Repeat the test for each motor terminal. A "grounded" hermetic is not repairable and must be replaced. See **Figure 24-53.**

A motor with a winding shorted to ground is indicated by a blown fuse or by cycling on the overload. As described earlier, cycling on the overload occurs when the motor attempts to start but trips the overload each time. Because the overload is more temperature-sensitive than a fuse, the overload may open before the fuse blows.

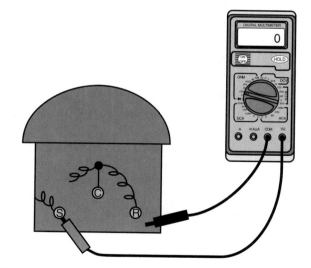

Figure 24-53. *A zero resistance reading when the ohmmeter probes are touching the hermetic case and a motor terminal indicates a short to ground. The hermetic must be replaced.*

24.8.2 Open Windings

An **open winding** refers to a broken wire in the motor winding that is *not* touching the motor shell. To locate or determine an open winding, the motor must be checked with an ohmmeter. Check each winding by touching the meter probes to two terminals, and note the reading. Repeat for each pair of terminals until all combinations have been checked. Since there is no complete path (circuit) for current flow if the circuit is open, the resistance indicated by the meter would be infinity. See **Figure 24-54.** A broken wire can occur in either winding or in the common connection. The condition is not repairable in the field, so the hermetic must be replaced.

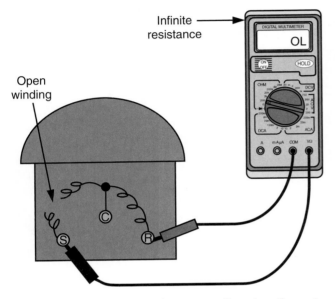

Figure 24-54. *An infinite resistance reading when the meter probes are touching two motor terminals indicates an open winding. The hermetic must be replaced.*

24.9 Motor Bearings and Lubrication

All motors have a **bearing** (lubricated support for a rotating shaft) inside the end bell (cover) at each end of the rotor. The rotor shaft is inserted through the bearing, which holds the shaft in perfect alignment. Alignment is *critical* because the air gap between the rotor and the stator is only a few thousandths of an inch, and the gap must be equal on all sides. The end bells are held in position by four bolts that extend through the stator shell from end to end.

The primary cause of motor failure is dry bearings, which cause the rotor to drag, lock up, or wear out. Dry bearings also cause excess amperage draw, while worn bearings cause a noisy motor.

24.9.1 Permanently Oiled Bushings

Many bearings for fractional horsepower motors are permanently lubricated. The bearing is actually a **bushing** (sleeve around a shaft) made of porous bronze that is saturated with oil at the factory. The

bushing is considered permanently lubricated and is used in fan motors and other low-horsepower applications. No additional oiling of the bearing is required for the life of the motor.

24.9.2 Oil-type Bearings

Oil-type bearings should receive three to six drops of oil in each oil port every six months. Avoid over-oiling, since excess oil may accumulate in the stator and collect dirt. The oil ports are located at each end bell of the motor for direct access to the bearing. See **Figure 24-55.** The oil ports are normally capped with rubber or plastic plugs to keep dirt out of the bearing. The plugs are easily removed for adding oil and should always be replaced. Regular 10W30, nondetergent automotive oil is used to oil motor bearings.

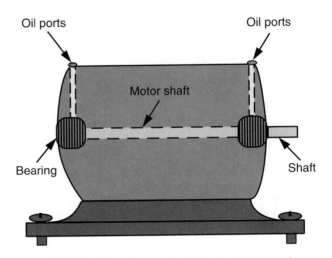

Figure 24-55. *Oil ports are located atop the bearings at each end of the motor. Periodic lubrication is needed to prevent motor damage.*

24.9.3 Grease-type Bearings

Most large horsepower motors have ball bearings lubricated with grease rather than oil. A grease fitting (called a **zerk fitting**) is installed on top of each bearing, and a removable plug is located at the bottom of each bearing. See **Figure 24-56.** An ordinary grease gun is used to pump high-grade, medium grease to the bearings every three months. The bottom plug must be removed before pumping grease into the bearing. If it is not removed, pressure can rupture the diaphragm that holds grease in the bearing. A ruptured diaphragm will permit grease to enter the rotor area, resulting in a dry bearing.

Removing the bottom plug prevents any pressure on the diaphragm because new grease forces old grease (if any) out the bottom port. Continue pumping until new grease appears at the bottom port. Usually,

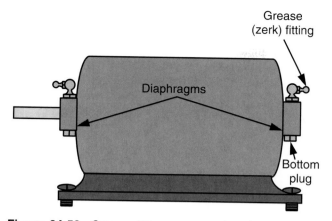

Figure 24-56. *Grease fittings are used to force lubricant under pressure into a bearing. The bottom plug must be removed when adding grease to prevent rupturing the diaphragm and allowing grease to leak out of the bearing.*

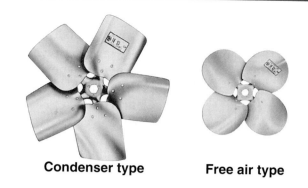

Condenser type   **Free air type**

Figure 24-57. *Square blades and round blades are used for different applications. Square blade fans are used in most air conditioning applications.*

only two or three pumps on the grease gun are required. If the bearing requires more pumps on the gun, the diaphragm has probably been ruptured, and grease is entering the motor. A ruptured diaphragm is revealed by a grease stain or puddle immediately under the motor. The bottom plug should always be reinstalled after greasing the bearing. A motor with a ruptured diaphragm must be greased frequently to delay or prevent bearing failure.

24.10 Fans and Blowers

Fans and blowers are essential parts of most refrigeration and air conditioning systems. Two types of blades are used to move air:

❑ *Axial* (propeller) blades are used for fans.
❑ *Centrifugal* (squirrel cage) blades are used for blowers.

24.10.1 Fans

The purpose of any fan is to move a specific volume of air at a specified *velocity* (speed). The amount of air moved (volume) is measured in **cubic feet per minute (cfm)**. The volume of air a fan can move is determined by several factors. If any one of these factors changes, the volume of air moved also changes:

❑ Speed of rotation.
❑ Number of fan blades.
❑ Blade pitch (angle, in degrees).
❑ Blade length.
❑ Blade width and shape.

Axial blade types

There are two types of propeller (axial) blades, *round* and *square*. See **Figure 24-57.** Each type serves a definite purpose. The round blade is best suited for

free-air applications, such as pedestal fans. It does not perform well operating against pressure. To obtain proper airflow, the movement of inlet air and discharge air must be unrestricted. The round blade has excellent efficiency in free-air conditions, and is almost noiseless because the air enters and exits at different angles along the blade length.

The square-type propeller blade is designed to operate against medium and high pressures in such applications as evaporators and condensers. The discharge blade tip is at full diameter, providing an air seal at the corners. The square blade fan is very efficient when placed in a cowling or orifice that surrounds the blade tips. A gap of about 1/4" (6 mm) is permitted between the blade tips and the orifice.

Fan blade rotation

Proper rotation of a fan blade is determined at the factory and cannot be changed in the field. Most propeller-type blades are mounted directly onto the motor shaft, so the rotation of the fan blades must match the rotation of the motor.

Rotation of fan blades is said to be clockwise (CW) or counterclockwise (CCW) as viewed from the *discharge* air side. Rotation for draw-type fan blades (those that pull air back over the motor) must be viewed from the motor side. Rotation for blow-type fan blades must be determined from the front of the blades, with the motor behind.

Most fan blades are not designed to be reversed (mounted backward). Reversing a fan blade causes severe reduction in the volume of air moved, increases motor and fan blade fatigue, and causes system problems. It is always best to replace a fan blade with an exact duplicate, but some allowance is permitted. The pitch can vary two degrees either way without trouble. Substituting a four-fan blade for a three-blade fan adds very little to the volume of air moved.

A motor and fan must be properly matched to obtained desired results. Motor speed and direction of rotation are important. The motor shaft must precisely fit the hole in the fan hub to prevent slipping or wobbling. The fan is usually anchored to the motor shaft by one or more setscrews through the hub between the blades and the motor.

When purchasing fan blades, the supplier needs to know the direction of rotation, number of blades, degree of pitch, hub hole size, blade type (square or round), and diameter (from blade tip to blade tip).

24.10.2 Centrifugal Blowers

Centrifugal (squirrel cage) blade blowers are used in applications where airflow is controlled and directed by ducts, such as furnaces and central air conditioning units. Excess airflow causes air pressure (called *static pressure*) to build up inside the duct. Static pressure ensures even distribution of air to all vents. As duct pressure increases, the volume of air delivered by the fan is reduced. The centrifugal blower is sized to deliver enough airflow to maintain proper static pressure levels.

Centrifugal blowers are measured by blade length and the diameter across the blade wheel. Some blowers are mounted directly on the motor shaft (direct drive), while others are belt-driven, **Figure 24-58.** The direction of rotation is determined by the curvature of the blades. Forward-curved blades are designed to cup incoming air and throw it forward by centrifugal force.

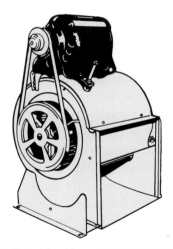

Figure 24-58. *Belt-drive blowers are common in heating and air conditioning installations. (Lau)*

Summary

This chapter explained the six types of induction motors commonly used by the industry. Induction motors have many similarities, but the operating principles differ for each type. Single-phase and three-phase motors require different connections to accomplish specific purposes. Automatic start-up involves a start winding and start components. Troubleshooting procedures require sufficient knowledge of each type of motor. Information on reversing rotation and connecting dual-voltage and multi-speed motors was also provided.

A discussion on the proper lubrication of motor bearings was included. The purpose, styles, and types of fans and blowers were explained, along with the importance of proper air movement to heating, refrigeration, and air conditioning systems.

Test Your Knowledge

Please do not write in this text. Write your answers on a separate sheet of paper.

1. Name two main parts of an electric motor.
2. The motor windings are located in the _____.
3. *True or false?* An opposed magnetic field is produced in the rotor by a process called induction.
4. Motor speed is determined by the number of stator _____.
5. Which of the motor windings determines direction of rotation?
6. *True or false?* The start winding has no other purpose than to provide automatic starting of an electric motor.
7. The start winding must be disconnected at about _____% of normal motor speed.
8. On a hermetic compressor, why is the switch to disconnect the start winding located outside the shell?
9. What do the abbreviations LRA and FLA stand for?
10. Name the two most common motor problems.
11. Cycling on the overload can be caused by:
    a. Excess amperage draw.
    b. Excess motor heat.
    c. Excess compressor heat.
    d. All of the above.
12. Name five types of single-phase induction motors.
13. The PSC motor uses a_____.
14. The _____ is the only difference between a split-phase motor and a CSIR motor.
15. Capacitors are always connected in _____ to the start winding.
16. Name three methods used to check a capacitor.
17. *True or false?* Dual-voltage motors have two separate run windings.
18. How is the direction of rotation changed on a three-phase motor?
19. Name the three electrical terminals located on a hermetic motor-compressor.
20. What test meter is used to check a motor for a short to ground or open winding?

Electromagnetic Control Devices

Objectives

After studying this chapter, you will be able to:
- ❑ Use and troubleshoot solenoid valves.
- ❑ Recognize and troubleshoot relay circuits.
- ❑ Connect and troubleshoot contactors and line starters.
- ❑ Read and use both pictorial and schematic diagrams.

Important Terms

armature
contactor
control circuit
cycling on the overload
dummy terminals
electrical symbols
electromagnets
electromagnetism
ferrous metals
hard-start kit
induced magnetism
jumper wire
ladder diagram
line starter

lockout
magnetism
pickup voltage
pictorial diagram
positive temperature
  coefficient (PTC)
  ceramic thermistor
relay
schematic diagram
solenoid valve
start capacitor
thermal relay
voltage relay

25.1 Magnetism

An important magnetic principle is that a magnetized object can induce **magnetism** in certain other metals. **Ferrous metals**, which are metals containing iron, will readily accept magnetism. This means a ferrous metal will become a magnet if placed within a magnetic field (lines of flux). This is called **induced magnetism.** Soft iron is easily magnetized, but quickly loses its magnetism when the magnetic field is removed. This feature makes soft iron ideal for making **electromagnets** (magnets that can be turned on and off). Electromagnets are used in many electrical devices, such as motors, relays, and solenoids.

25.1.1 Electromagnetism

Electromagnetism is the phenomenon of generating a magnetic field around a conductor (wire) as a result of current flowing through the wire. When the wire is wrapped into the form of a coil, the magnetic field becomes stronger. If the wire is coiled around a soft iron rod (core), the wire's magnetic field will induce magnetism in the iron, **Figure 25-1.** Magnetism in the core is controlled by starting or stopping current flow in the coil. Turning off the electrical supply to the coil kills the magnetic field in both coil and core.

As the electric current alternates, polarity of the magnetic field around the wire and coil also alternates. With 60-cycle current, polarity alternates from north to

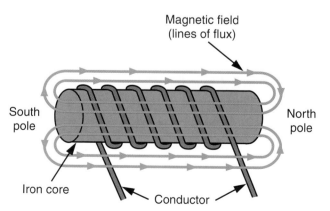

Figure 25-1. *When current flows through a wire coil wrapped around an iron core, the iron is magnetized. The magnetic field can be turned on or off by turning the electric current on or off.*

south 120 times per second. The induced magnetism in the core is always opposite in polarity to that of the coil. This electromagnetic principle is used in operating solenoid valves, relays, contactors, and line starters.

25.2 Electromagnetic Control Devices

Electromagnetism is used to operate a number of control devices used in various applications. These applications include opening and closing of valves, motor starting, and circuit control.

25.2.1 Solenoid Valves

Solenoid valves use electromagnetism to electrically control the opening and closing of valves. See **Figure 25-2.** A solenoid valve uses an electrical coil to control a movable iron core (plunger). The magnetism generated by the coil induces opposite polarity in the iron plunger. Because opposite magnetic poles attract, the plunger is lifted into the coil's magnetic field and the valve is opened. The valve remains open as long as the coil is energized. When the coil is de-energized, magnetism disappears and the plunger returns to the closed position. See **Figure 25-3.**

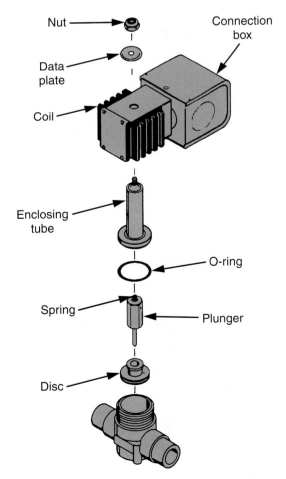

Figure 25-2. *A typical solenoid valve has the coil mounted on top of the valve body. (Sporlan Valve Co.)*

Solenoid valves are frequently used on applications where automatic control of liquids is desired. Some type of switch is required to control power supply to the solenoid coil. **Figure 25-4** shows how proper liquid level is automatically maintained in a tank by using a float switch to control a solenoid valve. When the liquid level falls, the float switch closes. This energizes the solenoid coil, opening the valve and permitting liquid to flow into the tank. When the proper liquid level is reached, the float switch opens, cutting off power to the solenoid valve and allowing the valve to close.

25.2.2 Relays

An internal centrifugal switch cannot be used for hermetic motor-compressors, due to arcing of the switch

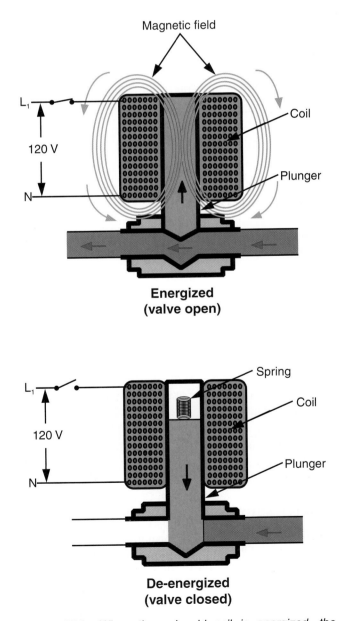

Figure 25-3. *When the solenoid coil is energized, the opposing polarities pull the plunger upward, opening the valve. The valve will remain open until the coil is de-energized.*

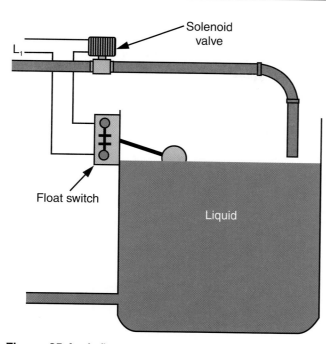

Figure 25-4. *A float switch controls current flow to the solenoid valve. The valve is opened and closed as needed to automatically maintain the liquid level in the tank.*

contacts. The arcing would cause breakdown of oil and refrigerant into acid and destroy the motor windings. Exterior starting relays are used to disconnect the start winding on single-phase motors; contactors are used for starting and stopping three-phase motors.

A *relay* is an electrically operated switch that can be used to automatically disconnect the start winding and/or the start capacitor of a motor. There are just two types of relays, amperage and potential. Each type is easily recognized. The method of operation and the electrical connections of the two are very different.

A relay opens or closes switch contacts using the same electromagnetic principle as the solenoid: a coil of wire is wrapped around an iron core located near a movable arm called an *armature.* An electrical contact is installed at one end of the movable arm, and a matching fixed contact below the arm.

Relay operation

As shown in **Figure 25-5,** when the relay coil is energized, magnetism is induced in the core and the armature is pulled downward. This causes the contacts to close. When the relay coil is de-energized, a spring returns the arm to its original position, opening the contacts.

The electrical circuit to the coil is entirely separate from the circuit through the contacts, **Figure 25-6.** This separation permits use of a low-voltage (24 V) circuit to open or close a line-voltage (120 V) circuit. In such an application, the low-voltage circuit is referred to as a *control circuit.*

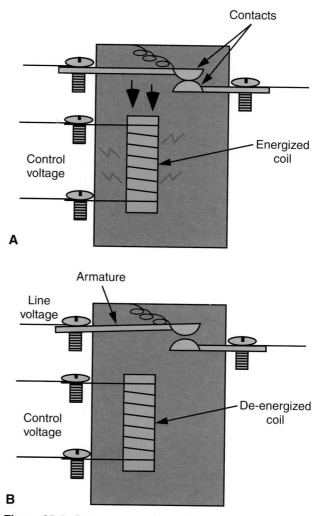

Figure 25-5. *Relay operation. A—When the coil is energized, the armature is pulled downward. This closes the contacts and completes the circuit. B—When the coil is de-energized, a spring pulls the armature upward again. This opens the contacts.*

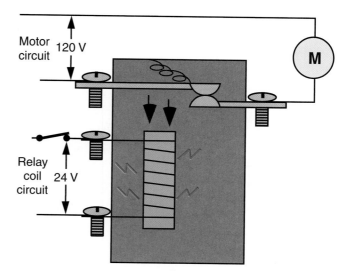

Figure 25-6. *The 24 V circuit for the relay coil is electrically separate from the 120 V motor circuit that it controls. The use of a low-voltage circuit to control a higher-voltage circuit is common in appliances and in heating and cooling equipment.*

The relay contacts have low-current ratings, with a maximum of 10 A being considered normal. Current flow in the *relay coil circuit* is even lower (often less than 0.25 A). The low voltage (24 V) used in the control circuit is safer for the technician, permits the use of small wire, and is less likely to cause a fire. Also, switches last longer due to less arcing of the contacts.

The relay case often has a schematic (electrical drawing) or symbols for electrical components that reveal the internal operation of contacts and coil terminals. See **Figure 25-7**. The relay data plate reveals coil voltage and amperage rating of the contacts. Relays often contain more than one set of contacts. For example, it is not uncommon to find relays used in applications requiring 10 or 12 poles per device. On a schematic, relay contacts are always shown in the state they assume when the coil is de-energized—normally open (NO) or normally closed (NC). All sets of contacts will change position at the same time when the coil is energized. See **Figure 25-8**.

A relay is often used to control an electrical load from a remote location. For example, a 120 V motor located in a basement could be controlled from the building's second floor by using a relay. The relay coil could be 24 V, permitting the use of small wires in the control circuit. **Figure 25-9** shows both pictorial and

schematic views of such a system. Schematic drawings are often used to illustrate electrical circuits. Schematics do not show the *size* or *length* of wires.

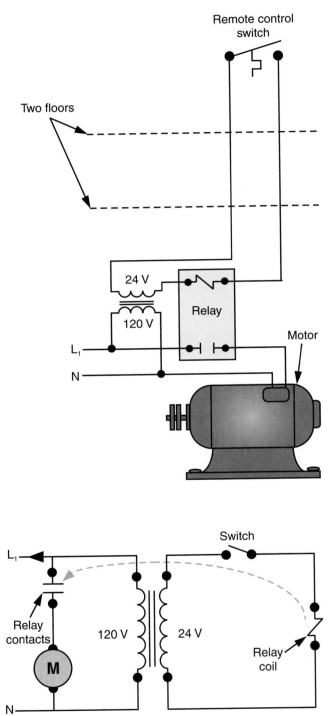

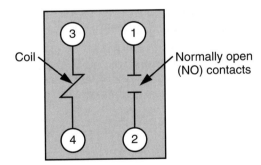

Figure 25-7. *Symbols like these are often used to depict relay components on schematics.*

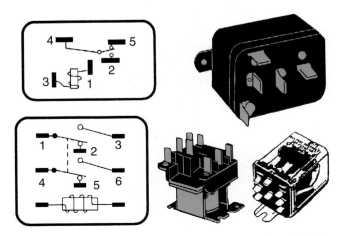

Figure 25-8. *A relay can have a number of contacts to control different devices simultaneously.*

Figure 25-9. *A relay can be used in a low-voltage control circuit that operates a distant device. As shown in the pictorial view at top, a motor is located two floors below the remote control switch. When the switch is closed, it energizes the relay coil. This closes the contacts in the 120 V motor circuit. The circuit is shown in schematic form at bottom. Note that the schematic shows voltages and symbols for all devices and conductors, but does not show size or length of the conductors.*

Amperage relays

When a single-phase motor starts, current flow to the run winding is very high. The rotor is not moving and the resistance of the run winding is very low. As the rotor begins to turn, the amperage (current flow) will decrease as a result of increased resistance created by counter-emf. This principle of temporary high-current flow to the run winding at start-up can be used to operate a relay.

Relay contacts. The contacts on amperage relays are *normally open (NO)*, with the plunger located below them. The plunger must move upward to close the contacts and complete the circuit to the motor's start winding. Once the start winding is energized, the rotor quickly speeds up. As it does so, counter-emf is generated and current flow through the run winding decreases. When the rotor speed reaches 75% of normal, amperage through the run winding drops and the magnetic field generated by the relay coil weakens. The plunger is pulled downward by gravity, opening the contacts to the start winding. See **Figure 25-10.**

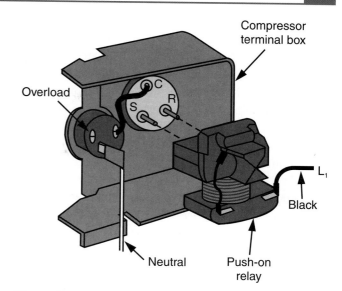

Figure 25-11. *Installation of the amperage relay is done by simply pushing it onto the S and R terminals of the compressor. This mounting method helps assure correct positioning.*

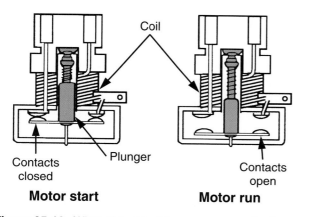

Figure 25-10. *When the coil of the amperage relay is energized, the plunger is pulled upward and closes the contacts to energize the motor's start winding. As motor speed increases, current flow to the coil decreases and the magnetic field weakens. This allows the plunger to drop down, which opens the contacts.*

The contacts used on amperage relays must be durable, because of the arcing that takes place as they open and close under high-current-flow conditions. Arcing can burn out contacts or weld them together.

Push-on amperage relays. Most amperage relays are designed for mounting directly onto the compressor terminals at S and R. See **Figure 25-11.** These are push-on type relays, with the mounting procedure assuring an upright position. Electrical connections are simplified because the relay makes direct contact with the terminals.

Normally, this type of relay is used on the split-phase hermetic compressor. Connection is easy, with the relay simply being pushed onto the compressor's

S (Start) and R (run) terminals. The relay contacts are between S and R, while the coil is connected to the R terminal on the compressor and the L (line) terminal on the relay housing. The technician merely connects the black (hot) wire to the L terminal. The overload is mounted inside the terminal box and is already connected by a short wire to the compressor's Common terminal. The technician connects the *neutral* (white wire) side of the power supply to the overload.

Other amperage relays. A *thermal relay* is an amperage relay that is totally enclosed in a plastic case and is not mounted directly on the compressor. If the relay is position-sensitive, the word "UP" and a directional arrow often will be printed on the case. See **Figure 25-12.** Also printed on the case will be letters designating the terminal screws—L for line or power supply, S for start winding, and M for main or run winding. The contacts inside the relay are always between M and S, and the coil is connected between L and M.

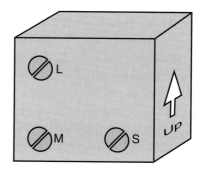

Figure 25-12. *An amperage relay in a plastic case that is not mounted on the compressor. The letters designate terminals.*

Some amperage relays have **dummy terminals**, which are extra terminals designated by *number*. These terminals have no effect on the operation of the relay. Instead, they serve as a convenience to the technician by providing connection points for splicing wires. They eliminate the need for wire nuts and electrical tape. Use of the dummy terminals is optional.

A typical use for dummy terminals is to connect devices such as the thermostat or condenser fan at the relay. See **Figure 25-13.** The system thermostat must be connected in series with one side of the power supply before the power reaches the relay. This permits it to cycle the condensing unit in response to temperature changes. The condenser fan is connected in parallel, *after* the thermostat, so it can be cycled on and off along with the compressor.

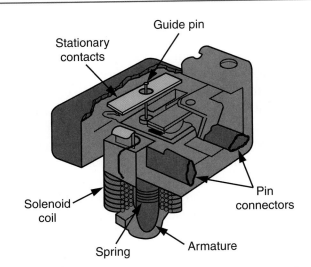

Figure 25-14. *In a manner similar to other amperage relays, solenoid-type amperage relays use high amperage in the run winding to close the contacts.*

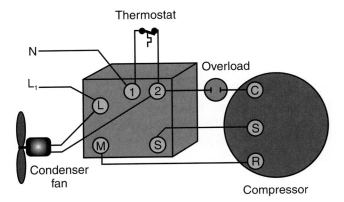

Figure 25-13. *Dummy terminals on the relay case are useful for connecting devices such as the thermostat and the condenser fan.*

Solenoid-type amperage relay. The solenoid-type amperage relay operates on the principle of excess amperage flow to the motor's *run* winding at start-up. Therefore, the current-sensing device must be connected in series with the run winding.

The sensing device is usually a small coil of heavy-gauge copper wire that operates like an electromagnet or solenoid. See **Figure 25-14.** The large wire used for the coil will require large amperage flow to create the necessary magnetism to lift the core plunger to close the contacts. When the coil is de-energized, gravity (assisted by a spring) returns the contacts to the open position.

The solenoid-type amperage relay is position-sensitive. It must be mounted both upright *and* right-side-up. If the relay is not upright, the plunger will not be able to rise and fall. If mounted upside-down, the contacts will remain closed because the plunger cannot fall back to the open position.

Amperage relay replacement. Fractional horsepower single-phase motors rated at less than 1/2 hp use amperage relays. Single-phase motors with ratings of 1/2 hp and above must use *potential-type relays*, which differ considerably from amperage relays.

Amperage relays are selected according to motor horsepower, and replacement must be exact for proper operation. Manufacturers place an identification number on each relay, so exact replacement is merely a matter of specifying the manufacturer's name and the relay identification number. If the number is missing or unreadable, you should provide the electrical supplier with the model number and serial number of the motor-compressor. A cross-reference listing will allow the supplier to identify the correct replacement relay.

Potential relays

The **potential relay** (also known as a **voltage relay**) does the same job as an amperage relay, but depends upon the counter-emf (potential) that is generated by the start winding. The speed of the rotor determines the amount of counter-emf generated by the start winding—as speed increases, so does the counter-emf.

Terminals on potential relays are identified by numbers, instead of the letters used on amperage relays. **Figure 25-15** shows the active terminals, which are conventionally labeled as 1, 2, and 5. The contacts are located between Terminals 1 and 2; the coil between 2 and 5. This information is important to remember, since not all potential relays show it on the case. Most potential relays have dummy terminals labeled 4 and 6, which are included as a convenience for splicing wires. Use of these terminals to connect a thermostat in series with the run winding is shown in **Figure 25-16.**

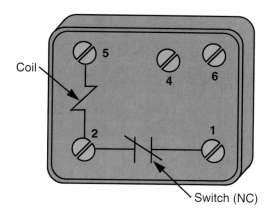

Figure 25-15. *Active terminals on the potential relay are labeled as 1, 2, and 5. Terminals 4 and 6 are dummies that can be used for wire splicing.*

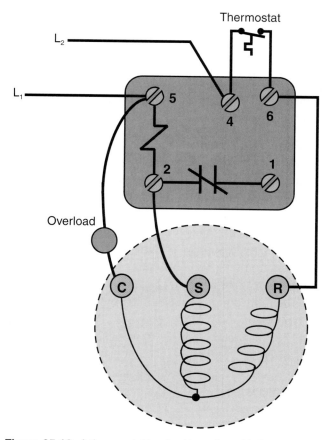

Figure 25-16. *A thermostat is wired in series with the run windings, using dummy Terminals 4 and 6 of the potential relay.*

Some potential relays are position-sensitive and others are not. A relay that *is* position-sensitive will have some indication (usually an upward-pointing arrow) printed on the case. The difference between the two types is in how the armature is moved to return the contacts to their normally closed (NC) position. Relays that are not position-sensitive use spring tension to move the armature and reclose the contacts. The position-sensitive types have a weight on the contact end of the arm. When the coil is de-energized,

the weight pulls the contacts down to the closed position. The relay must be mounted upright to permit proper closing of the contacts.

Testing relays. When a motor fails to start, the relay, start capacitor, and start winding are checked to identify the source of the problem. Components usually fail in this order:

1. Start capacitor.
2. Relay.
3. Start winding.

Testing of start capacitors and start windings is covered later in this chapter.

The amperage *relay* is checked for proper operation by using an *multimeter*. For safety and proper testing, the relay must be disconnected from the circuit and removed from its mounting. To check the contacts, first hold the relay in the upright position and place the test leads at the M and S terminals. The relay contacts should be *open* in this position, giving a resistance reading of infinity. If a zero resistance reading is obtained, the contacts are welded together and the relay must be replaced. If an infinity reading was obtained in the upright position, turn the relay upside down. The multimeter should now read zero resistance, indicating that the contacts have closed. Another means of checking a relay is with the use of a shop-made test cord. This procedure is described later in this chapter in the *Troubleshooting* section.

To check the relay coil, place the multimeter test leads at M and L. A reading that shows a *small* resistance, such as 1 ohm, means the coil has continuity. If an infinity reading is registered, the coil is open and the relay should be replaced.

Potential relay operation. The potential relay uses a normally closed (NC) switch that is *opened* by the energizing of the coil. See **Figure 25-17.** The contacts control the power supply to the start winding of the motor. Since resistance of the coil is high, it will take a moderately high voltage to overcome resistance and energize it. Coil resistance is precisely measured so that, when the rotor speed reaches 75% of normal, counter-emf will be sufficiently high to allow energizing of the coil. This point is the *pickup voltage.* Coils are available with pickup voltages in different ranges. See **Figure 25-18.**

When the coil is energized, it pulls downward on one end of the armature. The other end of the armature pushes the electrical contacts open, cutting off current flow to the start winding. The relay coil will hold the contacts open as long as motor speed is high enough to generate sufficient counter-emf.

Terminal connections. Since the potential relay functions on counter-emf generated by the start winding, the coil must be connected to both ends of

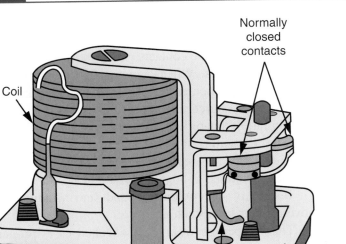

Figure 25-17. *The contacts of a potential relay are normally closed. The coil is energized when speed of the motor is fast enough to generate a predetermined amount of counter-emf. The magnetic field generated by the coil moves the armature, which opens the contacts and takes the start windings out of the circuit.*

Coil No.	Pickup Voltage	Drop Out Voltage	Continuous Voltage	Coil Ohms (Approx.)
1	139–153	15–55	130	760
2	140–153	20–45	170	1400
3	159–172	35–77	256	3320
4	261–290	50–100	336	5180
5	280–310	50–100	395	7150
6	299–327	50–100	420	10000
7	323–352	60–135	495	11950

Figure 25-18. *This table shows the pickup voltage ranges and other information on different relay coils that are available.*

the start winding. One end of the coil is connected to the winding at Terminal 2, and the other end through the Common at Terminal 5. See **Figure 25-19.**

The potential relay is the center of electrical activity, with many devices connected to or through it. These usually include such devices as the overload, system thermostat, condenser fan, and start capacitor. When connecting these devices, you *must be correct—* electricity does not permit mistakes. When you are making a number of connections, the best course is to consider and connect one wire at a time, while double-checking each circuit as you work on it.

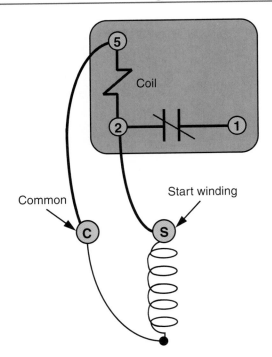

Figure 25-19. *The coil must be connected to both ends of the start winding. It is connected through the Common at Terminal 5, and (on the other end) at Terminal 2.*

Connecting hermetics. There are a number of methods used to make connections to the relay and compressor terminals. If extra terminals are available for multiple connections, a fair amount of wire can be saved. No matter what method is used, the basic principles remain the same:

❑ The power supply is always connected to the Common and Run terminals.
❑ The overload is wired in series with Common.
❑ The relay coil is connected to both ends of the start winding.
❑ The start capacitor is in series with the relay switch.
❑ The run capacitor must be connected to bypass the relay switch.

Only three terminals (*Common, Start, Run*) are available on the hermetic compressor itself. Most wiring connections must be made on the relay. **Figure 25-20** shows how the power supply is typically connected to Common from Terminal 5. The counter-emf wire from Terminal 5 to Common serves a dual purpose. The power supply traveling to Common will not interfere with the counter-emf circuit. The voltage also will not influence the overload (OL), since the current flow through the counter-emf circuit is extremely small.

The other side of the power supply must be connected to the run winding. This is done through the dummy Terminal 4 to allow connection of the system thermostat in series with the run winding. The thermostat is connected to Terminals 4 and 6. See **Figure 25-21.** This temperature-activated switch will control power supply to the motor/compressor at run.

If the motor-compressor is a split-phase type (one without capacitors), a jumper wire must be installed

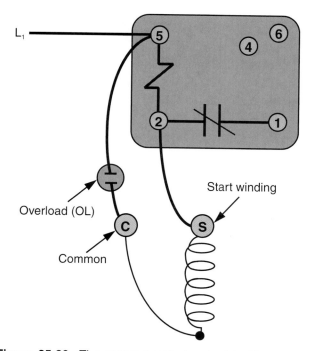

Figure 25-20. *The power supply is wired to Common at Terminal 5. Note the overload wired in series with Common.*

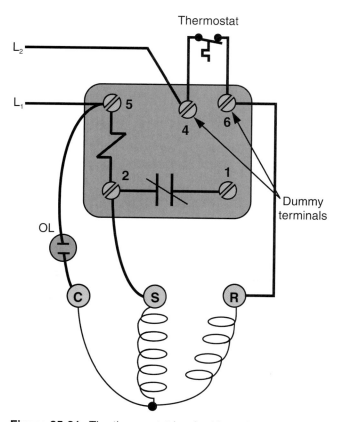

Figure 25-21. *The thermostat is wired in series with the run winding and one side of the power supply. It is wired using dummy Terminals 4 and 6.*

from Terminal 6 (power) to Terminal 1 (switch) on the relay. See **Figure 25-22.** A **jumper wire** is a short wire installed to complete a connection from terminal to terminal. The jumper wire completes the parallel connection from one end of the start winding to the other side of the power supply. The relay switch is in series with the start winding. The relay switch is closed (NC) at start-up, and will not open until the motor reaches 75% of normal speed. When the switch opens, the start winding is disconnected from the power supply. The counter-emf circuit remains intact whether the relay switch is open or closed.

If the hermetic motor-compressor is a *capacitor-start, induction-run (CSIR)* type, the jumper wire is used to install a *start capacitor* between Terminals 6 and 1. See **Figure 25-23.** This places the capacitor in series with the switch, so both the capacitor and the start winding are disconnected from the power supply when the switch opens.

A *capacitor-start, capacitor-run (CSCR)* motor has a run capacitor wired so it bypasses the start capacitor and switch. A typical method of wiring the start capacitor into the circuit is shown in **Figure 25-24.** When both start and run capacitors are wired as shown, opening the relay switch will take just the start capacitor out of the circuit. The run capacitor will keep the start winding slightly energized to assist the run winding under loaded conditions.

Condenser fan connections. Since the condenser fan is usually cycled on and off with the compressor, it is wired to the relay. See **Figure 25-25.** The condenser fan is connected to Terminals 5 and 6 (controlled by

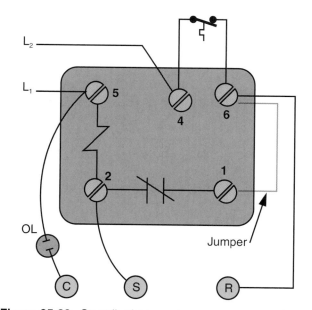

Figure 25-22. *On split-phase motor-compressors, a jumper wire connects Terminals 6 and 1 on the relay. This provides a parallel connection between the power supply and the relay switch.*

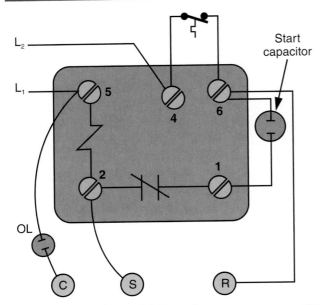

Figure 25-23. *On a CSIR motor-compressor, a start capacitor is connected between Terminals 6 and 1 so that it is in series with the relay switch.*

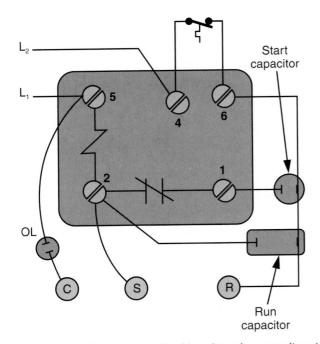

Figure 25-24. *This is the method used to wire capacitors to the relay. When the relay switch opens, the start capacitor will be taken out of the circuit.*

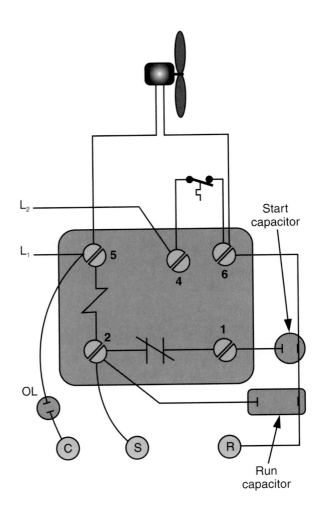

Figure 25-25. *When the condenser fan is wired to Terminals 5 and 6, the opening and closing of the thermostat contacts will cycle it on and off along with the compressor.*

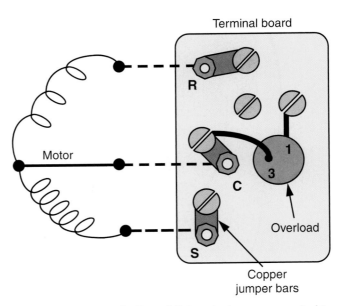

Figure 25-26. *The C, S, and R terminals are connected to dummy terminals with copper jumper bars. The dummy terminals should be used for all connections.*

the thermostat) to achieve this. If the fan were connected to Terminals 4 and 5, ahead of the thermostat, it would run continuously.

Connecting semihermetics. To prevent possible damage to the actual C, S, and R terminals, most semihermetic compressors have a terminal board with dummy terminals. These dummy terminals are connected to the nut-and-bolt type C, S, and R terminals with copper jumper bars. See **Figure 25-26.** The overload is normally installed at the factory. The relay and capacitors are housed in a separate metal box, usually called a *power pack*. The wires connecting the terminal board and power pack are carried in a flexible metal conduit.

Figure 25-27 shows the common method for wiring a potential relay and capacitors to the terminal board. The wiring is summarized as follows:

❑ The overload is in series with Common.
❑ The relay coil is connected between Common and Start.

❑ The relay switch and start capacitor are in series with Start.
❑ The run capacitor is in series with Start, but bypasses the relay switch.
❑ The system thermostat is in series with one power supply line.
❑ The condenser fan is connected in parallel with the compressor, after the thermostat.

Solid-state relays

Solid-state relays, which are used primarily in residential air conditioning, are convenient and reliable.

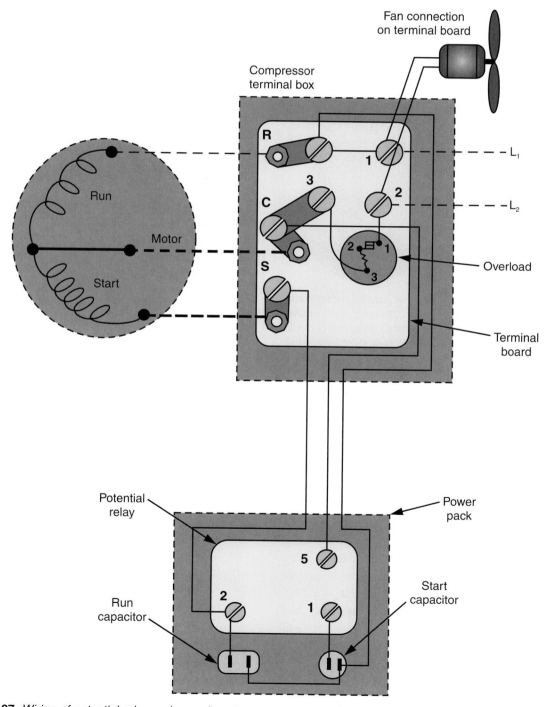

Figure 25-27. *Wiring of potential relay and capacitors to compressor terminal board.*

They can be used to turn any permanent split capacitor (PSC) motor into a capacitor-start induction-run (CSIR) motor to resolve hard start problems. The *hard-start kit* consists of a solid-state relay and a capacitor in a single package. The two terminals on the kit are simply connected in parallel with the run capacitor terminals. See **Figure 25-28.**

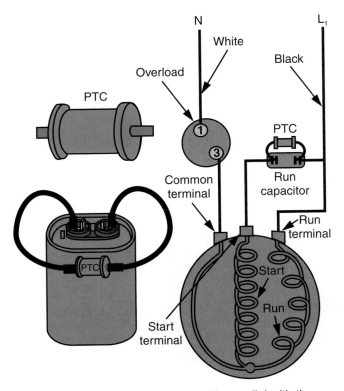

Figure 25-28. *A hard-start kit, wired in parallel with the run capacitor, turns a PSC motor into a CSIR motor to resolve starting problems. A positive temperature coefficient ceramic thermistor in the hard-start relay controls current flow to the windings.*

The solid-state relay contains a *positive temperature coefficient (PTC) ceramic thermistor* that reacts rapidly to temperature change (has a "steep slope") by greatly increasing its resistance to current flow. As used in a hard start kit, the PTC will initially allow the large current needed for starting to flow to the motor's start winding. However, in a period of less than one second, heat from that large current flow will cause the PTC resistance to increase from about 50 ohms to approximately 80,000 ohms. This high resistance effectively stops current flow though the relay, allowing normal current flow through the run capacitor and motor winding. Advantages of the hard start kit (solid-state relay and capacitor) include:

❏ Low cost.
❏ Simple two-connection installation.
❏ Wide applicability (usable on all single-phase PSC compressors up to 340 V and motors up to 5 hp).
❏ Reliability and long life.

❏ Mounting flexibility (not position-sensitive).
❏ Increased torque (up to 300%).
❏ Capability of further increasing torque by wiring second kit in parallel to first.

Troubleshooting

If the system is not running, always check the incoming power supply before making other electrical tests. Power supply to a potential relay can be checked by touching the leads of a multimeter to Terminals 4 and 5. If no voltage reading is obtained, check the fuse or breaker for the circuit serving the system. The fuse may be blown or the breaker tripped. If correct voltage is present at Terminals 4 and 5, move the test lead from Terminal 4 to Terminal 6. If there is no voltage present at Terminals 5 and 6, the system thermostat is open.

Shop-made test cord. A simple test cord with two switches can be constructed by the technician. See **Figure 25-29.** This test cord can be used along with a clamp-on ammeter to perform many troubleshooting procedures on small hermetic and semihermetic compressors. It is especially useful for bypassing start components and manually operating hermetics. Among other uses, it can also be used to safely check the switch on potential relays. **Figure 25-30** lists the materials and shows the schematic needed to construct the test cord. The test cord can easily be converted to operate on 230 V by constructing another length of SJ cord with a 120 V female plug on one end and a 230 V male plug on the other end. The test cord will operate the same way with either voltage.

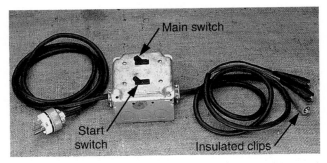

Figure 25-29. *When servicing hermetic and semihermetic units, the shop-made test cord is useful for many troubleshooting procedures.*

WARNING: Exercise extreme care and obey all electrical safety precautions when using this test device, since it contains no built-in safety devices. Always connect a safety ground wire to the device being tested. Install a ground fault circuit interrupter (GFCI) and an additional safety ground wire from the switch box to the device being tested. Illustrations are intended to show principles of use, and do not include recommended safety features.

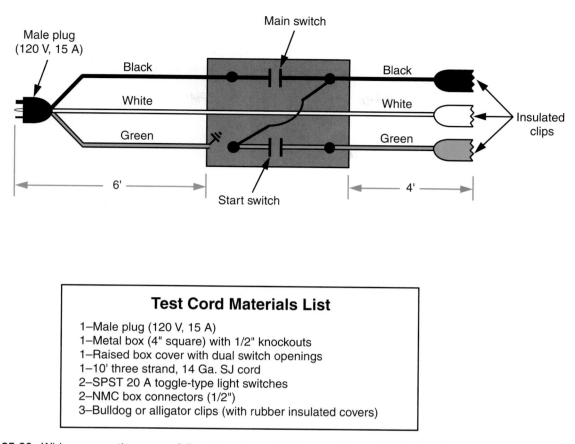

Test Cord Materials List

1–Male plug (120 V, 15 A)
1–Metal box (4" square) with 1/2" knockouts
1–Raised box cover with dual switch openings
1–10' three strand, 14 Ga. SJ cord
2–SPST 20 A toggle-type light switches
2–NMC box connectors (1/2")
3–Bulldog or alligator clips (with rubber insulated covers)

Figure 25-30. *Wiring connections must follow the schematic diagram shown. Label all switches and clips to prevent confusion. If desired, components of different colors (such as brown and ivory) can be used for the* main *and* start *toggle switches. Power supply to the box is controlled by the* main *switch; the* start *switch controls power supply to the green wire, which is attached to a bulldog clip. The green wire is used for the start winding connection. It becomes a power supply wire only when* both *switches are closed.*

Manual compressor test-cord start. Turn off both test cord switches and insert the plug into a power receptacle. Disconnect wires from the compressor's C, S, and R terminals and replace them with the test cord clips—white to Common, black to Run, and green to Start. Clamp an ammeter around the wire to Common. To attempt start-up, turn on both test cord switches at the same time to supply power to both the start and run windings. After about one-half second to one second, open the test-cord "start" switch to cut off the power being supplied to the start winding. The motor should continue to run and pull normal amperage.

Use a clamp-on ammeter to observe amperage flow while operating the switches. If necessary, a start capacitor can be easily connected in series with the green clip and the start winding.

Checking potential relays. When the test cord is used to check a potential relay, both switches are placed in the off position and the plug is *not* connected to a power source. The *system* power supply is turned off, as well. The wire from relay Terminal 1 is removed and replaced by the green clip. See **Figure 25-31.** The black clip is connected to relay Terminal 2.

The test-cord "start" switch is then closed, providing a circuit from the "start" clip to the "run" clip, with the switch in series between them. A clamp-on ammeter is then placed around the wire to Common for observing amperage flow during the start-up procedure.

Observe the ammeter while turning on the *system* power supply and (after about one second) moving the test-cord start switch to the off position. If the compressor starts and runs properly, the relay is defective and should be replaced.

An alternate test method uses just an *ammeter* when troubleshooting problems with solid-state relays. The meter is clamped around one of the wires connecting the relay to the run capacitor and used to read amperage (current) flow. At start-up, the amperage should be high, then quickly drop to normal operating level as the PTC opens the relay contacts and disconnects the start winding. If there is no amperage flow through the relay, or if the amperage does not drop (indicating that the relay contacts have not opened), the relay is defective and should be replaced.

Faulty run capacitor. If the motor-compressor starts okay, runs for a short time, and then shuts off on

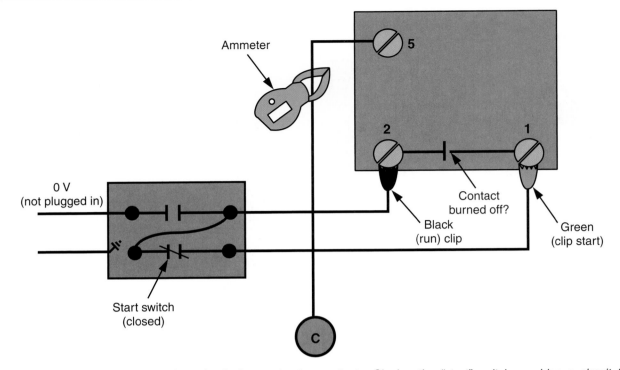

Figure 25-31. *Using the test cord to check the start relay contacts. Closing the "start" switch provides a circuit from Terminal 1 to Terminal 2, with the switch in series between them.*

the overload, the run capacitor may be suspected. However, the run capacitor seldom proves faulty and high head pressure can cause the same symptoms. Install a gauge manifold and check for high head pressure before checking the run capacitor.

If the relay checks out properly, but the compressor fails to start, the problem is in the *compressor*. It may have a faulty start winding, be physically binding, or locked up.

Open start winding. The start winding may be open, and can be checked for continuity with a multimeter. A reading of infinite resistance will be obtained if the circuit is open. **CAUTION:** To protect the meter from damage, all power to the circuit being tested *must* be off. For safety, remove all wires to the compressor terminals before testing for continuity.

Starting a locked-up hermetic. Sometimes a hermetic or semihermetic will not start when the start components and motor windings prove okay. This condition indicates a mechanical problem in the compressor: a bearing may have seized or a small piece of solder may have become wedged between the piston and cylinder wall. Sometimes the unit can be restarted. The effort is worth the trouble if successful; if not, the unit must be replaced.

There are three methods commonly used to try breaking loose a locked-up compressor. **CAUTION:** These methods are to be tried only briefly (about one second) to avoid possible electrical damage to the system.

❑ **Reversing rotation** by switching the wires from the start and run windings. This reverses the magnetic field and makes the motor run backward. A high-capacity start capacitor (rated at 400 mfd to 500 mfd) should be connected in series with the new start winding (actually the run winding). For a stuck *three-phase compressor*, reverse rotation by exchanging connections of any two supply voltage wires. It makes no difference which two wires are changed. The magnetic fields are automatically reversed.

❑ Using a **high-capacity start capacitor** (400 mfd to 500 mfd) will create a very strong magnetic field in the start winding. This method does not reverse rotation, but greatly increases starting torque.

❑ **Using higher voltage** to increase starting torque. If the stuck motor-compressor is 115 V, install a high-capacity start capacitor, then apply 230 V to the compressor terminals. Reversing rotation can be combined with the higher voltage, as well.

If the procedure is successful and the compressor restarts, reconnect all wiring to its original terminals. Remove the extra capacitor.

It is not always possible to successfully start a locked-up compressor. Even if it is restarted, there is no guarantee the compressor will not become stuck again. Sometimes, a restarted compressor runs for years, and then again, it may lock up again in just a day or two. Always advise the customer that this procedure is worth trying, but may not succeed or be a permanent fix. This can avoid trouble and any argument regarding a "recall." The customer may choose not to gamble and authorize immediate compressor replacement.

Cycling on the overload. It is quite possible for the motor-compressor to be *cycling on the overload* while the condenser fan is operating. This condition is marked by operation of the overload each time the motor attempts to start (resulting in a "hmmmmm...click, hmmmmm...click" sound pattern). This provides the following items of information:

❏ The overload is doing its job of protecting the compressor from damage.

❏ The power supply is adequate and the thermostat is closed, or the fan would not be running.

If the head pressure is normal, the problem is in the start components or the compressor motor windings.

To pinpoint the problem's location, use a clamp-on ammeter to check each circuit connected from the relay to the compressor. The *order* in which checks are performed is a matter of experience and the component locations.

Motor shorted to ground. This is difficult to check, because there is an extremely short period in which to measure current flow before the overload contacts reopen. Clamp the ammeter around either the Run or Common wire from the relay to the compressor, and wait for the overload to cool. When the overload contacts close, the extreme current flow caused by the short will generate sufficient heat to reopen them almost instantly.

The amperage reading, if you can obtain one, will be very high (as much as six times the current flow for normal operation of the motor). This indicates the motor windings are shorted to ground. Use a multimeter reading ohms to check the windings and conform the diagnosis. If there is a short-to-ground (ground fault) situation, the motor-compressor must be replaced.

Making bench tests. The test cord described earlier in this section can be used for bench testing many electrical devices for proper operation. The test cord is used to supply power to the device for the test. For use in bench testing, both test cord switches are turned off and the plug is connected to a power receptacle. The Common and Run clips are attached to the device being tested. An ammeter is then clamped around either wire to permit the technician to observe the amperage flow during start-up, **Figure 25-32.** For safety, connect a ground wire to the device being tested.

While observing the ammeter, close only the "main" switch on the box. This procedure supplies power to the device being tested. If the device is defective, the switch can be quickly turned off.

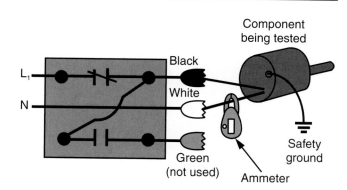

Figure 25-32. *Using the test cord for bench-testing an electrical component. The cord provides a 120 V (or 230 V with adapter cord) power supply.*

25.2.3 Contactors

A *contactor* is an electromagnet used to control multiple *heavy duty* contacts that are opened or closed at the same time. The contacts are normally open (NO), and close when the relay coil is energized. A 2-pole contactor has two separate contacts and a 3-pole contactor has three contacts. A three-phase motor uses a 3-pole contactor to simultaneously control all three hot legs. See **Figure 25-33.** The major difference between a relay and a contactor is the size of the contacts. Contactors are used to control power to large loads. Higher amperage flow to large loads requires the use of heavy-duty contacts. Some typical contactors are shown in **Figure 25-34.**

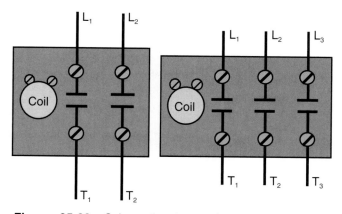

Figure 25-33. *Schematic views of 2-pole and 3-pole contactors.*

A movable armature (insulated bar) is located close to the iron core. This armature carries the movable contacts, which are held away from the stationary contacts by means of a spring. See **Figure 25-35.** When the coil is energized, magnetism overcomes the spring pressure, snapping the contacts into the closed position. These snap-action contacts are used because the contacts must open and close very quickly to reduce electrical arcing that always occurs under load conditions.

A

B

Figure 25-34. *Typical contactors. A—A 2-pole contactor. (White-Rogers) B—A 3-pole contactor. (Furnas Electric Co.)*

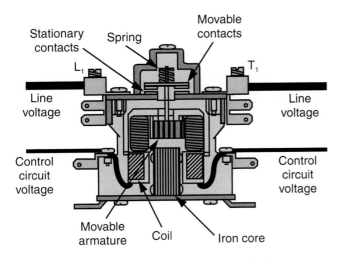

Figure 25-35. *Cutaway view of a contactor. Spring pressure holds contacts apart until the coil is energized and snaps them closed.*

These devices are rated according to the maximum amperage flow through the contacts for a specific voltage. The amperage rating is usually expressed in Full Load Amps (FLA), which is an inductive load rating. The same contactor will tolerate a higher amperage if the load is resistive. Specific ratings are listed on the contactor dataplate. An oversized contactor is perfectly acceptable, but an undersized contactor will result in burned-out contacts.

As a result of arcing when the contacts open and close, it is normal for them to be pitted and burned. Use of a file or sandpaper to clean the contacts is not recommended, because this destroys the contact surfaces and increases arcing. If contacts become pitted and burned, it is best to replace them. Replacement contacts are usually available from a local supplier.

The numbering system for contactors determines direction of current flow through the contacts. The line power (inlet) terminals are labeled L_1, L_2, and L_3 and the load (outlet) terminals are labeled T_1, T_2, and T_3. Power supply is connected to L_1, L_2, and L_3 and a three-phase motor is connected to T_1, T_2, and T_3. See **Figure 25-36.**

Residential air conditioning contactors

Residential air conditioning uses a 230 V, single-phase circuit to operate the outdoor condensing unit. Power supply is obtained from the main load center (breaker panel) and is connected to the outdoor unit at the 2-pole (L_1 and L_2) contactor. The compressor and the condenser fan are both connected to T_1 and T_2 of the contactor. See **Figure 25-37.**

The contactor coil and thermostat operate on 24 V, so a transformer must be used to provide this low control voltage. The low voltage permits the use of very small (18 gage) wire to connect the indoor-thermostat to the outdoor condensing unit.

Commercial air conditioning contactors

Most commercial refrigeration systems operate on 208 V or 230 V three-phase power systems, but large systems may use 440 V. Control circuit voltage is usually 120 V, since this allows use of standard switching devices and wiring. See **Figure 25-38.** Safety switches and thermostats are connected in series with the 120 V control circuit that supplies power to the contactor coil. If the thermostat or a safety switch opens, the contactor coil is de-energized, opening the main contacts and cutting off the power supply to the motor-compressor.

Commercial refrigeration systems may also use a 230 V, single-phase control circuit. Power supply for the control circuit is obtained from any two supply

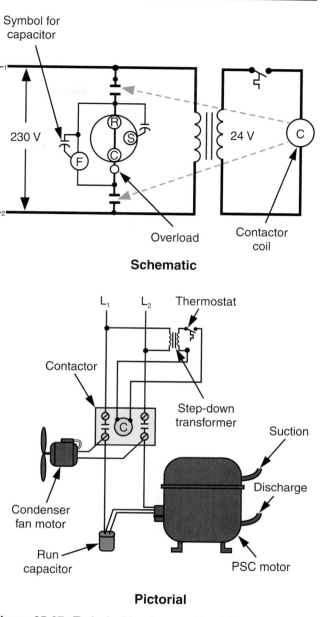

Schematic

Pictorial

Figure 25-37. *Typical wiring for a residential system, using a two-pole contactor.*

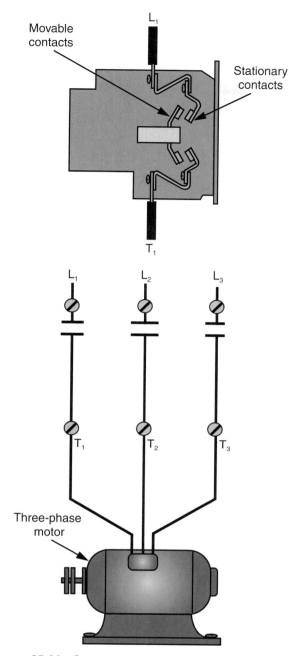

Figure 25-36. *Contactor terminals are labeled where the device is connected. The letter "L" is used for the line power side and "T" for the load side. All contacts open and close at the same time.*

wires connected to the contactor at L_1, L_2, and L_3. See **Figure 25-39.** The two control circuit wires are sometimes equipped with fuses to protect the control circuit wiring and the contactor coil. A blown fuse can narrow a problem down to the control circuit.

Contactor coils

The electromagnetic coil is located inside the contactor, but is *not* electrically connected to the main contacts. Because the coil is a separate device, the coil voltage is often different from the voltage at the main contacts. The coil has its own terminals for making

electrical connections. Most contactors are designed to allow easy replacement of the coil. See **Figure 25-40.**

The coil is used to control the main contacts. Energizing the coil causes the main contacts to close simultaneously. De-energizing the coil causes the main contacts to open. Power supply to the coil can be controlled automatically or manually. As noted earlier in this chapter, the separate electrical circuit to the coil is called the control circuit. See **Figure 25-41.**

Because the coil is electrically separate from the main contacts, it is common practice to use lower voltage to operate the coil. This permits easy control of a large motor operating on high voltage and amperage. Lower voltage and current in the control circuit is safer and permits the use of standard switching devices. As shown in **Figure 25-42,** a

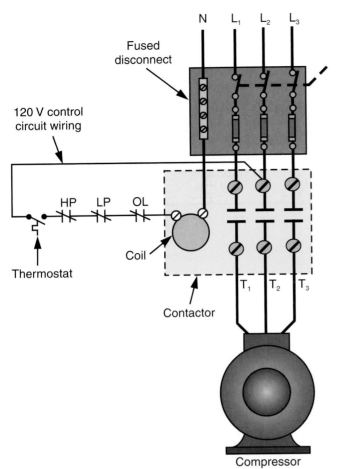

Figure 25-38. *A 230 V, three-phase system with a 120 V control circuit used for the contactor.*

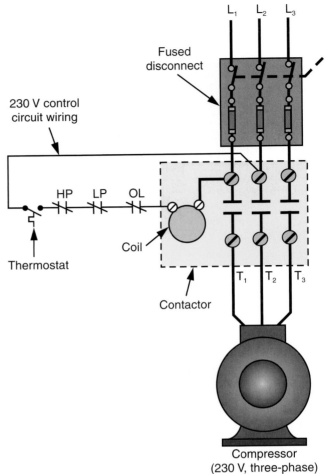

Figure 25-39. *Commercial refrigeration system using a 230 V, single-phase control circuit.*

step-down transformer is used to obtain the lower control circuit voltage.

Any number of contacts and safety control switches can be located in the control circuit. See **Figure 25-43.** The contacts for these controls are connected in series with the coil and may include overloads, thermostats, pressure controls, fuses, limit switches, flow controls, or other devices. Before the coil can be energized, *all* switches in the control circuit must be closed. Opening *any* switch will disconnect power supply to the coil and stop operation of the motor.

WARNING: Although a step-down transformer is often used, control circuit voltage may be obtained from a *separate* power source. Thus, turning off the motor's fused disconnect may *not* de-energize the control circuit. This situation is not unusual and can be dangerous. Never work on a circuit until you are certain the circuit is dead, and cannot be accidentally re-energized while repairs are being made. Always check the circuit with your voltage tester before and after de-energizing it. Also, periodically check your voltage tester on a known live circuit to be certain it is working properly.

The main switch (disconnect) must be properly tagged to inform others that work is being performed on the circuit. *Lockout,* the physical locking of the switch in the off position, is a safety requirement in most industrial settings. For your own protection, you should make it your standard procedure whenever you work on a system. Proper tagout and lockout procedures are important and prevent many injuries and deaths.

Troubleshooting contactors

If the control circuit is not fused (and some are not), the entire system must be checked to locate electrical problems. A voltage tester or a multimeter will provide a quick method of locating these problems. With the power supply turned on, check for voltage at L_1, L_2, and L_3 of the contactor. If the tester shows no voltage at these terminals, the main power source is off. If one or two legs are hot and one is dead, one or more main fuses has blown.

If power supply to the contactor is okay, check for proper voltage at T_1, T_2, and T_3. If one of these terminals is dead and the others are okay, the corresponding contact inside the contactor is burned out.

Figure 25-40. *A replacement coil for a contactor. Note the terminals used with push-on connectors for ease of replacement. (Furnas Electric Co.)*

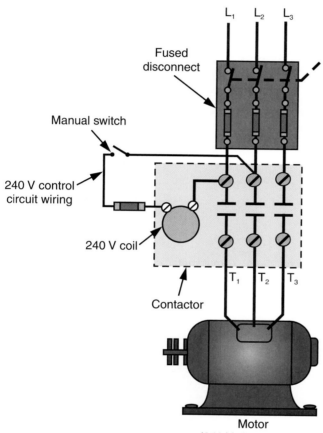

Figure 25-41. *Power supply to the contactor coil can be controlled manually or automatically.*

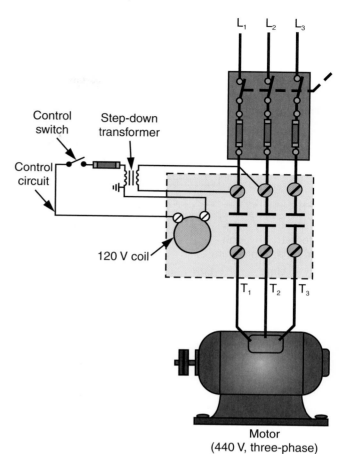

Figure 25-42. *A step-down transformer is used to obtain the lower voltage used for the control circuit.*

Control circuit problems. If power supply to the contactor is okay, but no voltage is available at T_1, T_2, or T_3 of the contactor, the problem is in the control circuit. Check for voltage at the coil. The coil may be burned out, or a control circuit switch is open, so that the coil is not energized.

To check the contactor coil, test for voltage at the coil terminals. If the voltage is correct, the coil itself is defective and must be replaced. If voltage is not present at the coil, one of the control circuit contacts is open. All switches must be checked to determine which one is open.

Each switch serves a definite purpose for controlling power supply to the coil. Some of these controls are automatic reset and others are manual reset. When the open switch is located, that control pinpoints the problem to be corrected.

Voltage test method. A multimeter set to read volts can be used to quickly and easily test switches to determine whether they are open. Electricity will always seek the path of least resistance, so current will not flow through the tester when it can take a lower-resistance path straight through a fuse or closed switch.

As shown in **Figure 25-44,** a zero voltage reading (0 V) is obtained across a closed switch. Since an open switch or fuse has *infinite* resistance, electricity will flow through the lower resistance of the meter and provide a reading (230 V, in this case).

Coil voltage too low. Low coil voltage will reduce magnetic pull, so the armature will not seat properly. This causes excess current, due to loss of counter-emf, so the coil gets very hot and burns out. "Chattering" of the armature may indicate a low coil voltage condition.

AC hum. All devices using a magnetic field will produce a characteristic hum. This hum is caused by

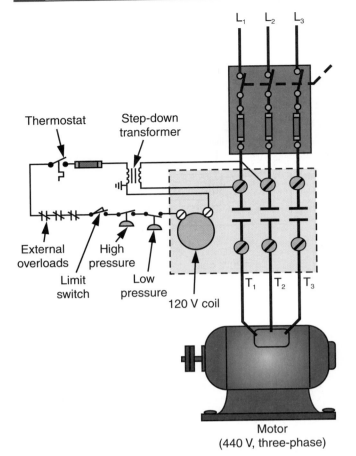

Thermostat
Step-down transformer
External overloads
Limit switch
High pressure
Low pressure
120 V coil
T_1 T_2 T_3

Motor
(440 V, three-phase)

Figure 25-43. *All safety switches and overloads connected to a control circuit must be closed to energize the contactor coil.*

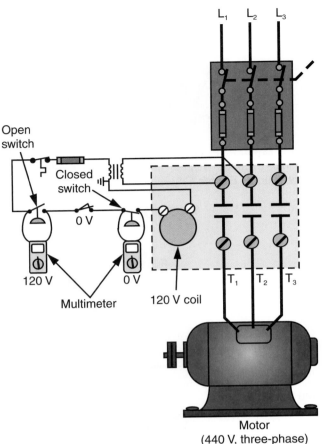

Open switch
Closed switch
0 V
120 V
0 V
Multimeter
120 V coil
T_1 T_2 T_3

Motor
(440 V, three-phase)

Figure 25-44. *A multimeter can locate an open switch or fuse by showing the voltage across the switch. A closed switch will register 0 V.*

the changing magnetic fields, which produce mechanical vibrations. In addition to hum, noisy operation of contactors, relays, and line starters may be caused by the following:

❑ Low voltage.
❑ Dirt, rust, or metal filings on magnet faces.
❑ Inability of the armature to seat properly.
❑ Binding of moving parts.

25.2.4 Across-the-line Starters

An across-the-line motor starter, or **line starter**, is a contactor that has built-in overload protectors. Also referred to as a *magnetic starter*, it is used to operate and protect three-phase motors. Such motors are frequently chosen for belt-driven applications, such as operating compressors, large condensers, exhaust fans, cooling towers, or large air conditioning blowers. The built-in overloads protect the motor against excess amperage (current flow). See **Figure 25-45.**

The overload protectors are normally mounted just below the contactor, and are often installed and prewired at the factory. Overloads are more accurate than fuses, providing excellent motor protection. One overload is needed to protect each phase supplying power to the motor.

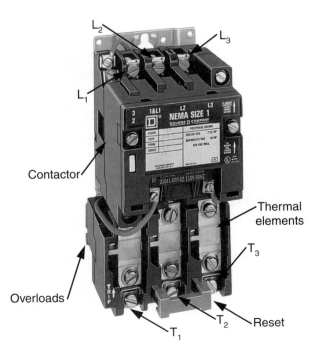

L_2
L_3
L_1
Contactor
Thermal elements
T_3
Overloads
T_2
Reset
T_1

Figure 25-45. *A typical line starter. Most line starters have overload protectors mounted just below the contactor. (Square D Co.)*

The overloads operate on a *thermal* principle, with an element that is sized to permit a specific amperage flow. Excess amperage will cause the element to become hot and operate a trip mechanism that opens a set of contacts. This cuts off the power supply to the contactor coil and stops the motor. See **Figure 25-46.** A manual reset is provided to reclose the overload contacts after the thermal element cools off. Thermal elements, which are sometimes referred to as "heaters," are available in different amperage capacities. They are sized to provide accurate motor protection at slightly above the Full Load Amps (FLA) rating of the contactor. See **Figure 25-47.**

Each of the three overload contacts is connected in the control circuit. As shown in **Figure 25-48,** the factory connects a wire from L_1, through each overload switch to one terminal on the contactor coil. Field wiring of the control circuit is completed by connecting the other side of the coil through the various control circuit switches and back to either L_2 or L_3. This places the thermostat and all overload switches in series with the coil.

Some older line starters are equipped with just two overloads, instead of three. They are based on the principle that if any one supply wire becomes overloaded, another will also be overloaded. Thus, only two overloads are needed to protect the motor. Current electrical codes, however, require three overloads for a three-phase motor. Newer line starters use three thermal elements, but only *one* set of contacts. Any thermal element can open the contacts. A manual reset button is used to reclose them. See **Figure 25-49.**

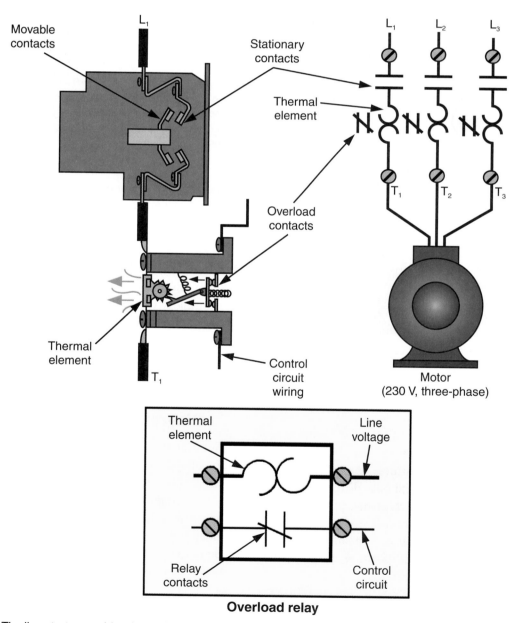

Figure 25-46. *The line starter provides thermal overload protection. Excess amperage will cause the thermal element to become hot and open the overload contacts. This cuts off power to the contactor coil, which opens the contacts and stops the motor.*

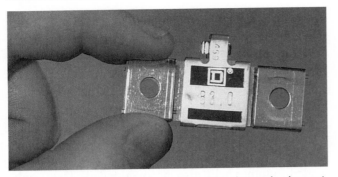

Figure 25-47. *A typical replaceable thermal element. (Square D Co.)*

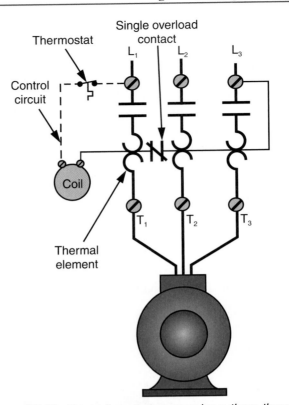

Figure 25-49. *Newer line starters may have three thermal elements, but only a single set of overload contacts. Any thermal element can operate the contacts if an overcurrent problem occurs.*

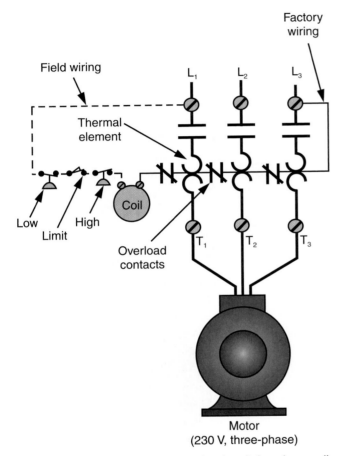

Figure 25-48. *Wiring of the overload switches is usually completed in the field by incorporating the desired control devices in the circuit.*

Troubleshooting line starters

Troubleshooting line starters is performed as described under contactors. The only difference is the overloads. If an overload switch opens, power supply to the coil stops and the motor stops. When the overload cools, the reset button is manually pushed to reset the contacts to the closed position, so power to the coil is restored.

WARNING: Pressing the reset button with the control circuit energized will start the motor immediately, while your hand is inside the starter box. Always de-energize the control circuit before you press a reset.

The thermal elements and overload switches are very reliable and rarely fail. If the overload will not reset, or quickly opens again, the problem is due to excess amperage. This indicates a motor problem.

Motor control circuits

Some advanced motor control circuits can become quite complex. They include relays, lights, pushbuttons, and safety controls. **Figure 25-50** is an example of a complex control circuit that allows the motor to be controlled from separate locations.

25.3 Pictorial and Schematic Diagrams

Pictorial and schematic diagrams are two different types of illustrations commonly used to show electrical connections. Most manufacturers locate these important diagrams inside a cover panel on the appliance or other equipment. When making repairs and locating system problems, the technician must frequently refer to these diagrams to identify certain wires and component connections.

25.3.1 Pictorial Diagram

A *pictorial diagram* shows the physical appearance, component locations, wiring connections, and

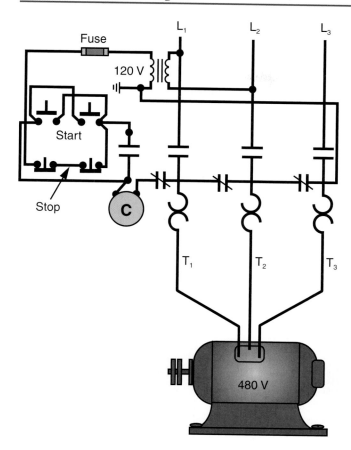

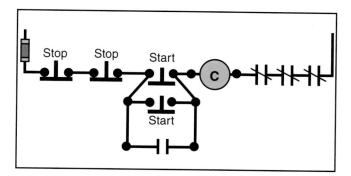

Motor control from a separate location

Figure 25-50. *With a more complex control circuit, a motor can be started and stopped from two or more locations. The circuit shown has two pushbutton control switches.*

internal arrangement of a piece of equipment. See **Figure 25-51.** The components are arranged in the approximate positions they actually occupy inside the unit. Lines are then drawn from each component to indicate how the wires are connected in the circuit. These lines are sometimes labeled with the color or identification number of the wire used.

When many electrical components are involved, the pictorial diagram becomes too difficult to read.

There are so many wires in the diagram that it is difficult to follow a single one. To solve this problem, a schematic diagram is used.

25.3.2 Schematic Diagram

The ***schematic diagram*** is an orderly method of presenting electrical circuits and components. The schematic diagram is usually referred to by technicians as simply the "schematic." The schematic does not illustrate where the components are located or what they look like. It does illustrate the types of components involved and how they are connected in the circuit.

A standard set of ***electrical symbols*** is used for ease of communication in schematic diagrams. See **Figure 25-52.** There are literally hundreds of different symbols used in electrical and electronic schematics, but technicians quickly learn to recognize those that are most commonly used.

The schematic is often called a ***ladder diagram,*** because it is shaped somewhat like a ladder. The power supply lines are drawn parallel to form the side rails of the ladder, and loads are drawn between the power source lines, forming the rungs. See **Figure 25-53.** Various switches and fuses are located on the rungs as needed to control and protect the loads. All components (motors, switches, fuses) in the diagram are indicated by electrical symbols.

Reading a schematic requires more knowledge of electricity and electrical components than is needed to use a pictorial. The schematic does not look like the actual components and wiring in the unit. The technician must be able to relate this information to the actual unit (via the pictorial). To use a schematic effectively, the technician must know the unit components and their corresponding electrical symbols. Most schematics contain a legend, or key, that describes the abbreviations and components.

Summary

This chapter introduced the variety of controls used to start and regulate the operation of motors. These controls are a common and necessary part of every refrigeration and air conditioning system. The technician must be very familiar with the troubleshooting and repair of motor control and circuit protection devices.

Each type of control and its use in different systems was explained, along with the electrical connections needed for proper operation. Instructions on selecting and troubleshooting various types of relays and contactors was included.

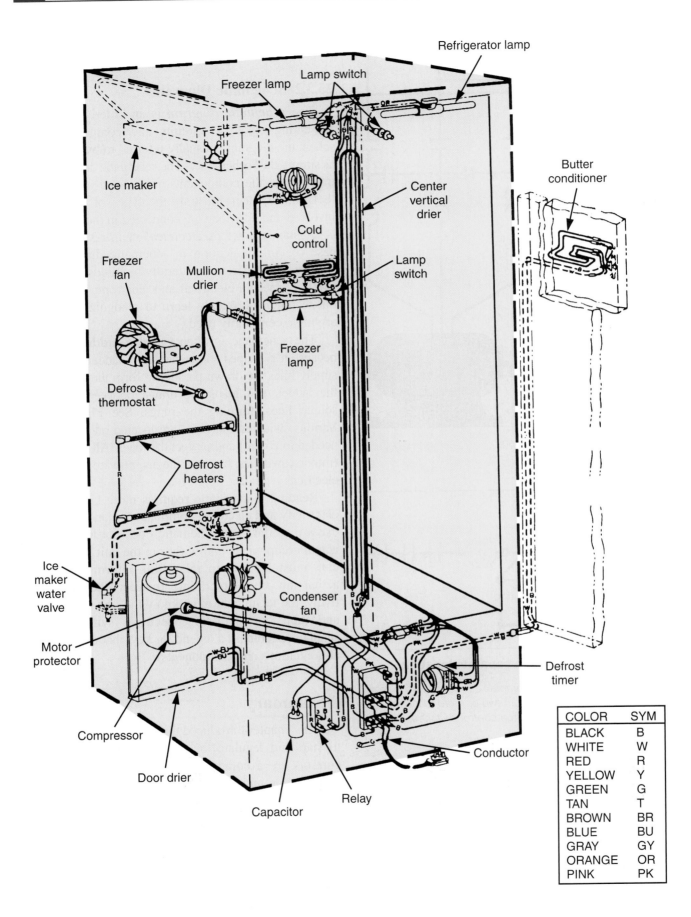

Figure 25-51. *On pictorial diagrams used for household appliances and similar types of equipment, the components are drawn realistically and in the approximate position they occupy inside the case. Note the key for wire colors at lower right. (Frigidaire)*

Electrical Symbols

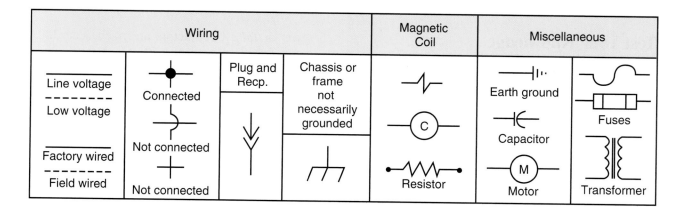

Figure 25-52. *A sampling of some of the electrical symbols commonly used on schematic diagrams.*

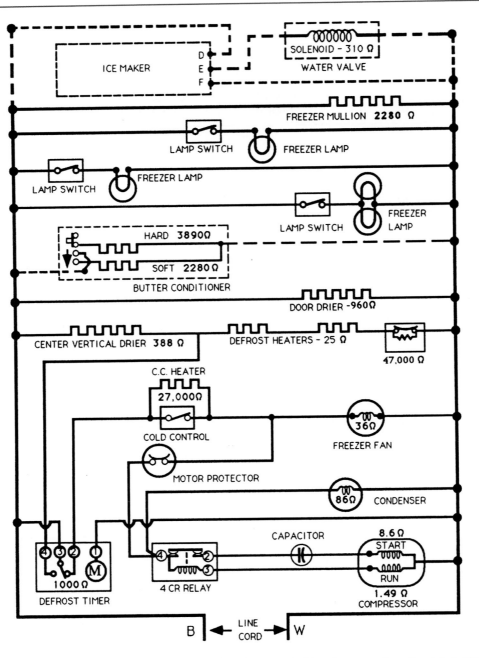

Figure 25-53. *An electrical schematic diagram is often referred to as a ladder diagram, since there is a single load on each "rung." This is the schematic for the refrigerator-freezer shown in Figure 25-51. (Frigidaire)*

◤ Test Your Knowledge

Please do not write in this text. Write your answers on a separate sheet of paper.

1. A solenoid valve operates on the principle of _____.
2. Are relays used to control power supply to large loads or small loads?
3. Are contactors used to control power supply to large loads or small loads?
4. If the compressor starts when using a test cord to bypass the relay, is the relay working properly or is it defective?

5. *True or false?* Contactors are rated according to maximum amperage through the coil.
6. Power supply to the contacts of a contactor is connected to terminals labeled _____.
7. The electrical circuit controlling power supply to the contactor coil is called the _____ circuit.
8. How many switches can be connected into the circuit controlling power supply to the contactor coil?
9. What is the basic difference between a contactor and a line starter?
10. The overload contacts on a line starter are connected in _____ with the coil.

26

Motor Controls

Objectives

After studying this chapter, you will be able to:
- ❑ Identify and adjust primary motor controls.
- ❑ Troubleshoot thermostatic and pressure-type motor controls.
- ❑ Install and properly adjust safety controls.
- ❑ Troubleshoot safety controls.
- ❑ Recognize and properly adjust pumpdown cycles.
- ❑ Install and troubleshoot oil pressure safety controls.

Important Terms

crimp
cut-in point
cut-out point
differential
dual-pressure motor
  control
erratic
fan cycle control
head-pressure safety
  control
heat load
logging
low-pressure motor
  control
net oil pressure

oil pressure safety
  control
oversized
pressure-operated
  switches
primary controls
pumpdown cycle
riser
secondary control
sensing element
setpoints
stage
temperature range
thermostat

26.1 Primary Controls

The two types of primary motor controls used to provide automatic operation of the condensing unit are thermostatic (temperature-sensitive) and pressure-type (pressure-sensitive). One or the other is found on every refrigeration system. These **primary controls** are used to start and stop the condensing unit when certain temperatures or pressures are reached. They include a switch that controls the power supply to the condensing unit. When the switch closes, the unit runs. When the switch opens, it stops.

The technician *must* understand the purpose and operation of these motor controls. Knowing the purpose of the control, how it operates, and how it is connected into the circuit makes troubleshooting easier and more accurate.

Thermostats are often used as the primary control for system operation. Other thermostats may function as *secondary* controls for individual operation of devices such as the condenser fan, evaporator fan, solenoid valves, and defrost heaters.

Like thermostats, **pressure-operated switches** are often used as the primary control of the system. They also may be used as safety controls to protect the system from high head pressure, loss of refrigerant, and low oil pressure. Troubleshooting these controls and making needed adjustments is an everyday task for the HVAC technician.

26.1.1 Thermostats

A **thermostat** is a switch that opens and closes its contacts according to temperature changes. Thus, a thermostat can be used to automatically turn the condensing unit on and off as set temperatures are reached. Thermostats are available in many different shapes and styles to meet different applications. See **Figure 26-1.** While thermostats may look different, they all are temperature-activated switches.

Thermostat contacts may *close* on rise in temperature, or *open* on rise in temperature. The type that closes on rise is used in cooling applications; the type that opens on rise is used in heating applications. See **Figure 26-2.**

Figure 26-1. *Thermostats are provided in a number of different forms, but all perform the same job of opening or closing a switch in response to temperature change. (Ranco)*

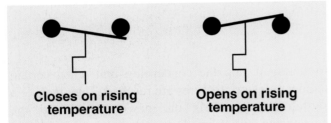

Figure 26-2. *Schematic symbols for thermostats. Those that close on a rise in temperature are used for cooling (refrigeration and air conditioning) applications, while those that open on a rise in temperature are used in heating applications.*

Thermostatic motor controls

All refrigeration systems are **oversized**. This means the system has more capacity than the minimum needed to permit it to lower the temperature to a certain point and then shut off. When temperature rises to a certain point, the system is turned on. Heat removal occurs only while the system is running. A thermostat is used to cycle the system on and off. This maintains a temperature level in a desired range.

Thermostat sensing elements

The **sensing element** of a thermostat consists of a bellows (movable part), a capillary tube (connecting link), and a sensing bulb that holds a refrigerant. See **Figure 26-3.** The refrigerant in the sensing element may be in a vapor or a liquid state. Different types of charges are used to achieve special operating characteristics.

Location of the sensing bulb determines what temperature is being controlled. Some sensing elements are designed for controlling air temperature, others for evaporator temperature (evaporator-sensitive), still others are submerged in liquid. Before replacing or making adjustments to a thermostat, you must know the sensing bulb location.

Sometimes the bulb is omitted and the capillary tube itself becomes the sensing element. The capillary tube will vary in length. The tube end may be straight or formed in the shape of a bulb. See **Figure 26-4.**

Capillary tubes without bulbs are generally used on such appliances as household refrigerators, window air conditioners, or home food freezers. Twisted, coiled, and solid bulbs are often used in commercial refrigeration applications, such as food freezers or walk-in coolers.

Thermostatic control terminology

To understand motor controls, you must be familiar with the terms used to describe the operation of these controls. The temperature setting that causes the contacts to *close* is called the **cut-in point.** The temperature setting causing the contacts to *open* is

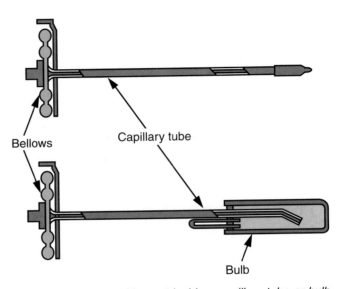

Figure 26-3. *As a refrigerant inside a capillary tube or bulb expands or contracts in response to a temperature change, the thermostat bellows moves. The movement opens or closes a switch when the temperature change exceeds selected levels.*

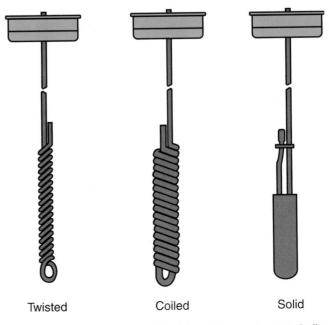

Figure 26-4. *On commercial refrigeration equipment, bulbs in twisted, coiled, or solid form are used.*

called the *cut-out point.* The difference between cut-in and cut-out is called *differential.*

For example, a thermostat having a cut-in of 38°F (3°C) and a cut-out of 32°F (0°C) would have a differential of 6 degrees Fahrenheit (38 − 32 = 6) or 3 degrees Celsius (3 − 0 = 3). Most domestic refrigeration thermostats have a fixed differential of 5 degrees Fahrenheit or 6 degrees Fahrenheit. Room thermostats for heating and air conditioning operate in a narrower range, with a fixed differential of 1.5 degrees Fahrenheit to 2 degrees Fahrenheit.

Range adjustment. The dial (or knob) on a domestic thermostat, **Figure 26-5,** is a range adjustment. This range adjustment moves the cut-in and cut-out *equally* higher or lower. Turning the range adjustment to a colder setting lowers *both* **setpoints** (cut-in and cut-out), so the system will maintain the lower temperature level. Turning the range adjustment to a warmer setting will cause the system to maintain the higher temperature level.

Control temperature ranges. Controls are designed to operate within a specific *temperature range.* These temperature ranges are rather general, but fall into five areas of application.

- ❏ **High-temperature application,** between 90°F and 55°F (32°C and 13°C). Usually associated with room heating and air conditioning systems.
- ❏ **Medium-temperature application,** between 55°F and 32°F (13°C and 0°C). Normally associated with refrigerators and other types of coolers.
- ❏ **Low-temperature application,** between 5°F and −50°F (−15°C and −45°C). Usually associated with freezers.

- ❏ **Ultralow-temperature application,** between −50°F and −250°F (−45°C and −157°C). Associated with industrial and commercial fast-freezing systems.
- ❏ **Cryogenic application,** between −250°F and −459°F, or absolute zero (−157°C and −273°C). Usually associated with laboratory and research systems.

Domestic thermostats

As noted, the range adjustment on domestic thermostats is usually limited to a few degrees either way (warmer or colder). A domestic thermostat is designed to operate with the range adjustment knob close to mid-position. It is poor practice to operate any system with the range adjustment set at either minimum or maximum. The thermostat cannot operate properly at either extreme. Typically, turning the knob all the way to the right is the warmest position and turns the thermostat off (contacts constantly open). Turning the knob all the way to the left moves it to the coldest position and the thermostat cannot open (contacts constantly closed).

Some manufacturers of domestic controls include two small screws identified as "altitude adjustment screws." Such altitude adjustments are unnecessary and should be ignored. The slight difference in pressure due to high altitude will equally affect both cut-in and cut-out. A small change in range adjustment automatically compensates for altitude.

Wiring connections. While thermostats are used to control the power supply to the condensing unit, they are normally located at a point remote from the condensing unit. Connecting wires can be any length.

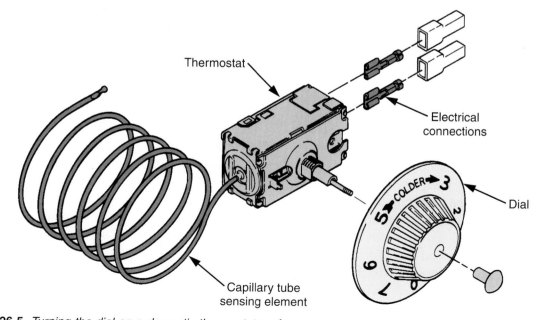

Figure 26-5. *Turning the dial on a domestic thermostat performs a range adjustment. The cut-in and cut-out temperatures are higher or lower, depending upon the adjustment, but the differential does not change. (Ranco)*

Thermostats normally have two terminals; each is internally connected to one side of the switch. These terminals can be either push-on-type or screw-type. Wiring connections should be tight and well-insulated, since loose wiring connections are a primary cause of burned wires and component damage. Most thermostats include an extra (third) terminal, which is connected to the metal case and used for making the safety ground (green wire) connection.

For single-phase domestic systems, the thermostat switch is connected in series with one of the wires supplying power to the relay. This permits the thermostat switch to disconnect the relay from the power supply. It is poor practice to connect the thermostat in the neutral wire. It makes troubleshooting more difficult and can cause safety problems.

Commercial thermostats

Commercial thermostats are designed for a wide range of temperatures and have fully adjustable control settings. See **Figure 26-6.** Two heavy-duty adjustment screws are located on top of the control for moving the pointers. A separate calibrated scale is provided for each pointer and is located directly below each screw. This permits the technician to control a wide temperature range, with the cut-in and cut-out easily adjusted to maintain desired temperatures.

Adjusting controls. Commercial thermostat adjustments must be performed in proper order to achieve the desired results. When facing the control, the left screw changes the differential, while the right screw is the range adjustment. The range adjustment changes cut-in and cut-out equally. Both pointers will move when making range adjustments. The differential screw changes only the cut-out temperature. This screw moves only the pointer on the differential scale.

First, adjust the range screw until the desired cut-in temperature is indicated on the scale. Next, adjust the differential screw to obtain the desired differential setting. In a cooling application, the cut-out equals cut-in minus differential. See **Figure 26-7.**

For example, consider a walk-in cooler that has an air-sensitive thermostat mounted on the inside wall to sense temperature of return airflow. See **Figure 26-8.** Air temperature should be maintained between 38°F and 32°F (3°C and 0°C). The range adjustment screw is used to move the cut-in pointer to the 38°F mark on the range scale. The differential screw is then adjusted to move its pointer to the 6° mark on the differential scale. With a differential of 6°, the cut-out will be 32°F (0°C).

Some commercial motor controls have a cut-out screw and scale instead of a differential screw. The cut-out screw directly establishes the cut-out point, according to the position of the pointer on the scale.

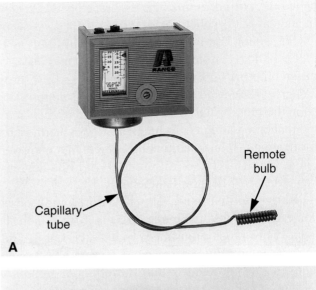

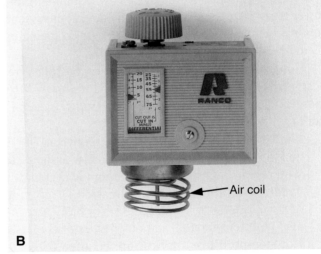

Figure 26-6. *Typical commercial thermostats. Note the separate scales for setting cut-in temperature and differential. A—Thermostat with remote sensing bulb. B—Thermostat with air coil for sensing. (Ranco)*

Whenever you deal with a commercial thermostat, examine it carefully to determine if the left screw is a cut-out or differential screw. The settings are different.

Information printed at the bottom of the scale will identify the type of scale used. Controls using a differential screw will read "Cut-out is cut-in minus differential." Controls using a cut-out screw will label one scale "cut-out" and the other "cut-in."

Wiring connections. On commercial (three-phase) systems, the thermostat switch is connected in series with the coil on the contactor (or line starter). See **Figure 26-9.** The thermostat controls power supply to the coil, which controls power supply to the compressor. The thermostat is located close to the area where the temperature is being controlled, due to limited length of the sensing element. Two wires connect the thermostat terminals to the contactor coil.

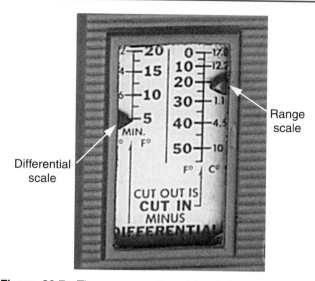

Figure 26-7. *The range scale pointer indicates the cut-in temperature, and is adjusted first. The differential pointer is then moved to set the desired cut-out temperature. On the thermostat illustrated, the cut-in temperature is 20°F, and the differential is five degrees, so the cut-out temperature is 15°F. (Ranco)*

Disadvantage of thermostatic motor controls

Since the thermostat operates according to temperature of the sensing element, it may call for operation when the system is having problems. For example, a thermostat can require the system to operate when some or all refrigerant has been lost. If a leak occurs on the low-pressure side of the system, operation of the system will allow air and moisture to be drawn in and cause major problems. Relying strictly on thermostatic control may cause a system to self-destruct. Safety controls are often added to the system to keep it from operating if a problem develops.

Domestic systems such as refrigerators, freezers, and window air conditioners rely on the *compressor overload* to provide safety protection. However, prolonged cycling on the overload will result in compressor failure. Commercial installations normally include additional safety controls to protect the system. Such controls are explained later in this chapter.

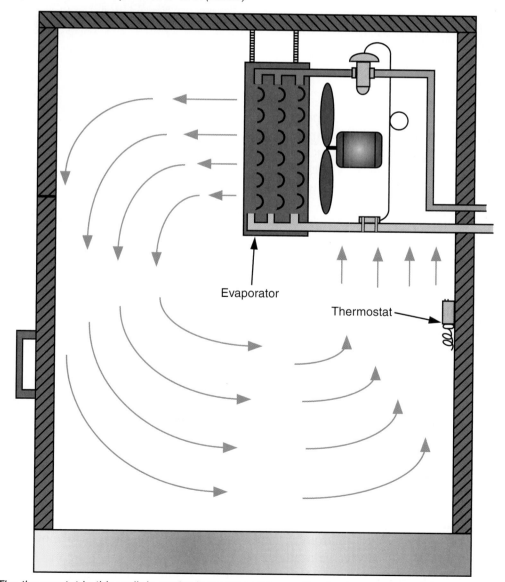

Figure 26-8. *The thermostat in this walk-in cooler is positioned to monitor the temperature of the return air. When properly adjusted, it will maintain the temperature inside the cooler in a range from 32°F to 38°F (0°C to 3°C).*

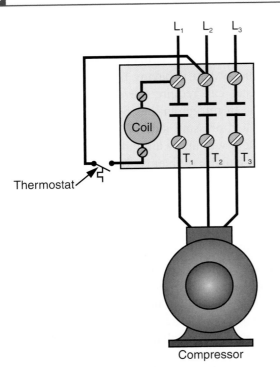

Figure 26-9. *The thermostat on a three-phase system is connected in series with the coil to control the power supply to the contactor.*

Troubleshooting thermostats

Thermostats are generally reliable. Most problems are related to improper settings, a sensing element that is not making good contact, or an element that has lost its refrigerant charge. (Sometimes, vibration or rubbing will wear a hole in a capillary tube, allowing loss of the bulb charge.)

After making certain that the contacts *should* be closed, the thermostat switch can be checked with a voltage tester or a multimeter. Check for voltage from each thermostat terminal to a good ground. If the meter shows voltage at both terminals, the contacts are closed. If there is voltage at just one terminal, the contacts are open.

Another procedure can be performed as a double-check by placing the probes of a voltage tester at each thermostat terminal. If the tester lights up, the contacts are open. If the tester does *not* light up, another switch may be open downstream of the thermostat (such as an overload) and thus give a false reading.

The thermostatic motor control is a primary controlling device. Frequently, such controls are *incorrectly* diagnosed as defective, because other switches are often wired in series with them. These switches, typically safety controls, can prevent the system from operating. For proper control by the thermostat, power supply should always be available at one (inlet) terminal of the thermostat. However, sometimes an overriding control such as a defrost

timeclock may prevent power from reaching the thermostat. Defrost systems and timeclocks are explained in Chapter 27.

26.1.2 Pressure-type Motor Controls

The pressure-type motor control is widely used on commercial systems because it is accurate, dependable, fully adjustable, and located at the condensing unit. See **Figure 26-10.** This location is convenient for the technician and helps prevent tampering with the control. Adjustments are critical, however, and vary from one system to another. Adjustments can be performed effectively only by a technician who is fully aware of the temperature-pressure relationship.

Figure 26-10. *A low-pressure control that opens or closes a switch when pressure causes movement of a diaphragm. This control has specific cut-in and cut-out settings, rather than using a differential to determine the cut-out pressure. (Penn)*

Principles of operation

The pressure-type motor control includes an electrical switch that is opened or closed by pressure on a diaphragm. See **Figure 26-11.** The diaphragm is a thin metal disc about 1 1/2″ in diameter, located in an assembly attached to the bottom of the control. A capillary tube connects the diaphragm assembly to the low-pressure side of the system. This low-pressure connection is usually made on the compressor body. Either an increase or a reduction of low-side pressure will cause the diaphragm to move. This movement operates the switch through a mechanical connection. Most pressure controls have two electrical terminals connected in series with the internal switch.

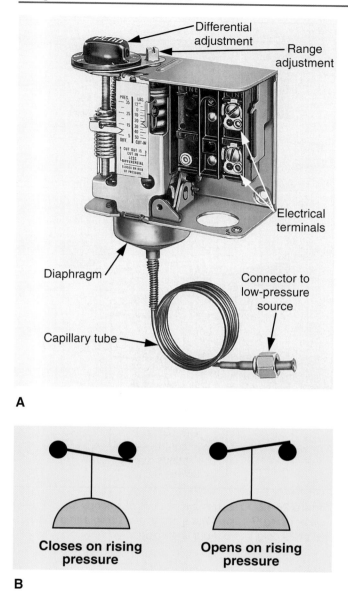

A

Closes on rising pressure   Opens on rising pressure

B

Figure 26-11. *Pressure switch and symbols. A—Low-pressure switch with cover removed to show construction. The electrical terminals are in series with the internal switch that is opened or closed by movement of the diaphragm. This control permits setting of the cut-in pressure and a differential, as indicated on the scales. (Penn) B—Schematic symbols for pressure switches. In refrigeration applications, thermostats that close a switch on a pressure rise are normally used.*

In terms of function, the pressure control is much like a thermostatic motor control. However, scales on thermostats are calibrated in degrees Fahrenheit (°F) or degrees Celsius (°C), while pressure controls are in pounds per square inch gauge (psig) or kilopascals (kPa). Product temperature can be maintained by controlling low-side pressures in the system. This is not difficult, but certain principles regarding low-side pressures must be fully understood. The technician must be able to use a *temperature-pressure chart* to convert temperature to pressure.

Low-side pressures

The sole purpose of any refrigeration system is to create a temperature difference so that heat can travel through the evaporator to be absorbed by the vaporizing refrigerant. This temperature difference can exist only when the system is running. Once the system shuts off, the temperature differences quickly disappear. During the "off" cycle, refrigerant pressure (low-side pressure) equals the temperature of the air moving through the evaporator. As air temperature inside the cabinet increases, low-side pressure increases, as well.

During the "on" cycle, the operation of the system quickly establishes a temperature difference that permits heat to travel from the air to the evaporator, and then to the refrigerant inside the evaporator. The refrigerant is about 10 degrees Fahrenheit (5.5 degrees Celsius) colder than the evaporator and the evaporator is about 10 degrees Fahrenheit (5.5 degrees Celsius) colder than the air. As the temperature of the air becomes colder, the evaporator and refrigerant also become colder, thus maintaining the temperature difference during the run cycle.

Freezer pressure settings

As an example, consider a walk-in freezer using R-507 and low-pressure motor control to maintain cabinet air temperatures that range from –10°F to –20° F (–23°C to –29°C).

Cut-in point. The cut-in point is determined from the temperature-pressure chart; no temperature difference exists during the off cycle. With the system off, cabinet air temperature slowly rises until air, evaporator, and refrigerant are all –10°F (–23°C). At this temperature, the pressure for R-507 (as shown on a standard temperature-pressure chart) is 26 psig (179 kPa). This becomes the cut-in point, so the range adjustment screw is moved to show cut-in on the pressure switch scale at 26 psig.

At start-up, the air temperature is –10°F (–23°C) and the condensing unit quickly establishes the desired 20 degrees Fahrenheit (11 degrees Celsius) air-to-refrigerant temperature difference. The refrigerant temperature becomes –30°F (–34°C). A low-pressure gauge connected to the system would reflect this temperature difference by quickly dropping from the cut-in pressure of 26 psig (179 kPa) to 11 psig (76 kPa).

The system is designed to maintain the 20 degrees Fahrenheit temperature difference during the run cycle. While the system is running, air temperature slowly drops from –10°F to –20°F (–23°C to –29°C). At the same time, low-side pressure slowly reduces from 11 psig to 6 psig (76 kPa to 41 kPa), and refrigerant temperature reduces from –30°F to –40°F (–34°C to –40°C).

Cut-out point. The cut-out point is determined by subtracting 20 degrees Fahrenheit (11 degrees Celsius) from the desired cut-out *air* temperature. When the air temperature reaches –20°F (–29°C), refrigerant temperature is –40°F (–40°C) and low-side pressure is 6 psig (41 kPa). The cut-out pressure control is adjusted to read 6 psig. If the control has a *differential screw*, the differential is set at 20 psig (26 – 6 = 20). In metric terms, the differential would be 143 kPa (216 – 73 = 143).

Pressure control settings

A *gauge manifold* should be installed when making adjustments to pressure-type motor controls. Gauge readings are more accurate than the pointers and scales on the controls. To set the cut-in, do not wait for cabinet air temperature to reach ambient. Instead, use the gauge manifold to bleed pressure from the high side to the low-pressure side. This raises low-side pressure to the cut-in point rather quickly. This procedure is also used to double-check the cut-in setting.

The cut-out is quickly set by using the suction service valve (SSV) to restrict incoming suction gas. Frontseat the valve stem, then crack it open just enough to limit the incoming pressure at the desired cut-out point. Set the control (cut-out or differential) while the incoming low pressure is controlled. After the cut-out is firmly established and double-checked, backseat the suction service valve to remove the gauges and return the system to normal operation.

Cooler pressure settings

Consider a walk-in dairy cooler using R-134a and operating with cabinet air temperatures of 39°F to 32°F (4°C to 0°C). The system cut-in pressure is set for 33 psig (228 kPa), because no temperature difference exists when the system is off. The temperature-pressure for R-134a at 38°F is 33 psig.

Cut-out is determined by the temperature of the *refrigerant* when air temperature is 32°F (0°C). The refrigerant is about 20 degrees colder than air temperature, so the control would be set to cut out at 13 psig, which is the pressure equivalent of 12°F (–11°C). A differential screw would be set for 20 psig (33 – 13 = 20), or 138 kPa in metric terms. See **Figure 26-12**.

Adjusting for pressure drop. When setting the cut-out on a pressure-type motor control, an allowance must be made for pressure drop through the suction line. The suction is strongest at the compressor when the piston makes its downstroke. Suction line pressure at the compressor is about 2 psig (14 kPa) *lower* than evaporator pressure.

Proper sizing of the suction line limits pressure drop to 2 psig. Pressure drop may be slightly higher if

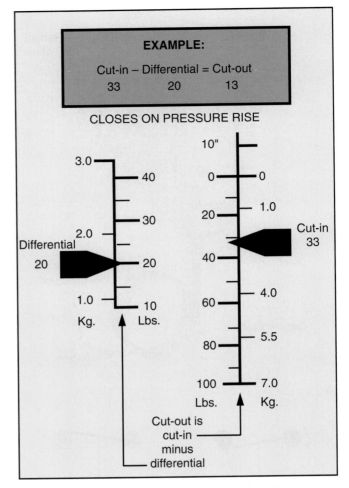

Figure 26-12. *Pressure switch control settings for a walk-in cooler using R-134a. With a cut-in of 33 psig and a differential of 20 psig, the cut-out is 13 psig.*

the suction line is very long, or contains many elbows, valves, and risers. A **riser** is a vertical section of the suction line.

To allow for pressure drop, subtract 2 psig from the desired cut-out value just before setting the control. In the freezer example, cut-out was determined to be 6 psig (41 kPa). Thus, the cut-out setting on the motor control must be 4 psig (6 – 2 = 4) to compensate for pressure drop. In the dairy cooler example, the cut-out setting would be reduced from 13 psig (90 kPa) to 11 psig (76 kPa). Pressure drop affects only cut-out, because pressure drop disappears during the off cycle. Pressure drop cannot occur unless the compressor is running.

Allowing for temperature differences. The temperature difference of 20 degrees Fahrenheit (11 degrees Celsius) between the air and refrigerant is a good guide. Some evaporators are more efficient than others and this temperature difference could be slightly more, or slightly less. After setting cut-in and cut-out on the pressure motor control, it is good policy to place an accurate thermometer inside the cabinet to check air temperature at both setpoints. Minor adjustments to the cut-out may be necessary to compensate for excess pressure drop or temperature difference through the evaporator.

Single-phase wiring connections

The pressure-type motor control, as noted earlier, is a primary control. Its switch is used to control power supply to the condensing unit. These controls usually have two screw-type terminals, which are internally connected to the two sides of the switch. The wiring connections are made exactly like those for a thermostatic motor control, but the wires will be much shorter because a pressure control is always located very close to the compressor. Wiring connections are usually performed inside the compressor terminal box, or inside the electric panel mounted on the condensing unit.

Three-phase wiring connections

The *low-pressure motor control* is the preferred choice on commercial systems, and is used to control power supply to the contactor coil. See **Figure 26-13.** Both the contactor and the pressure motor control are located at the condensing unit, which makes wiring connections and troubleshooting easier. The pressure motor control switch is connected in series with one of the wires supplying power to the coil.

Troubleshooting pressure motor controls

Troubleshooting procedures for pressure controls are similar to those used with thermostats. The control

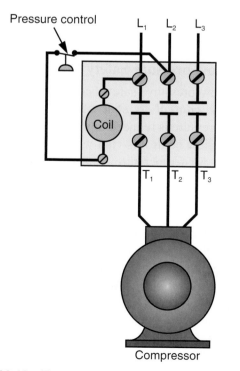

Figure 26-13. *The pressure control on a three-phase system is connected in series with the coil to control the power supply to the contactor.*

is simply an electrical switch that operates according to low-side pressure at the compressor. The low-side pressures *must* be adequate for the control to operate properly.

System low on refrigerant

If the system is low on refrigerant, the low-pressure control will cause the system to short-cycle. The control switch opens and closes too quickly because the compressor is drawing gas out of the suction line faster than the expansion valve can feed liquid. Low-side pressure quickly drops to the cut-out point, the switch opens, then low-side pressure quickly rises to the cut-in point. Resetting the control does not solve the problem. The leak must be located and repaired, and the system charged to full capacity again.

No refrigerant in system

If the system loses all its refrigerant, the control contacts remain open. Low-side pressure cannot build up sufficiently to close the contacts and start the motor-compressor. When the control contacts are open, the voltage tester will reveal power supply (voltage) only at the inlet terminal.

Shutting off when the system is very low on refrigerant is an advantage. If a leak occurs on the low-pressure side of the system, the control will shut the system off before it reaches a vacuum state that would allow contaminants to enter.

Defective pressure control

To properly diagnose a defective pressure motor control, the gauge manifold must be installed. These controls are usually reliable, but sometimes become **erratic** (irregular) in operation. A control that is unable to maintain pressure settings must be replaced. This condition is usually caused by abuse to the control.

A common pressure control problem is a leak in the capillary tube linking the control diaphragm to the compressor. This leak is caused by vibration or by rubbing of the capillary tube against another metal object. Excessive vibration can cause the copper capillary tube to become work-hardened and crack or break off. Rubbing, especially against a steel object, will quickly wear a hole in the capillary tube. This occurs because steel is much harder than copper. The entire system charge can be lost by such a leak.

Repairing the pressure control's capillary tube

It is usually not necessary to replace the control when capillary tube develops a leak. The tube can be repaired, unless it is leaking at a point very close to the diaphragm. Repairing the capillary tube is much easier and quicker than replacing the control. Follow this procedure for repairing a leaking capillary tube:

1. Shut off the power supply to the unit and install a gauge manifold on the suction and discharge service valves. The unit must not be permitted to run while repairs are being made.

2. Frontseat the suction service valve to isolate the compressor. The compressor discharge valve reeds will prevent refrigerant backflow into the compressor.

3. Loosen the flare nut connecting the capillary tube to the compressor. Permit any pressure inside the compressor to escape *slowly*, because rapid escape of this pressure may result in severe oil loss.

4. Cut or break the capillary tube at the point of the leak and remove the flare nut and tubing from the compressor.

5. Use a 3″ to 4″ (6 cm to 10 cm) piece of 1/4″ soft copper tubing to splice the capillary tube together. See **Figure 26-14.** Use water-pump pliers to carefully *crimp* (squeeze closed) the openings at each end of the 1/4″ copper tubing for easier brazing. Exercise care to prevent crimping the capillary tube. The ends of the capillary tube must be inserted *at least* 1″ (2.5 cm) into the 1/4″ copper tubing to prevent the brazing alloy from plugging the capillary tube ends.

6. Braze each end of the 1/4″ copper tubing to the capillary tube, allow the braze metal to cool, and reconnect the flare nut to the compressor. Carefully arrange the repaired capillary tube to eliminate vibration and rubbing.

7. Backseat and then crack open the suction service valve. Add enough refrigerant gas (about 100 psig pressure) to check the brazed connections and the flare nut connection for leaks.

8. Restore power, then add the necessary refrigerant to bring the system to full capacity.

These procedures are adequate if the pressure control cut-out was high enough to prevent the system from operating in a vacuum. If the system *was* allowed to operate in a vacuum, atmospheric air and moisture have been drawn into it and further repairs are required. These repairs involve purging all remaining refrigerant (and air) from the system, repairing the leak, pressurizing enough to leak-check, then purging all pressure. Finally, the filter-drier must be changed and a thorough evacuation performed with a good vacuum pump before charging the system to full capacity. Failure to perform these repairs when air has entered the system will result in poor system performance and compressor burnout. This explains why cut-out should occur *before* the low-side pressure enters a vacuum.

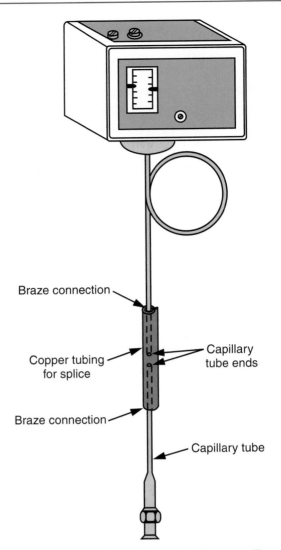

Figure 26-14. *Repair method for leaking capillary tube, using a piece of 1/4″ copper tubing as a splice.*

26.1.3 Low-pressure Safety Control

The low-pressure safety control is used as a *secondary control* to shut off the system if the low-side pressure is reduced to unsafe levels. The low-pressure safety control is vital when the system uses a thermostat for temperature control. The thermostat is the primary control and the low-pressure safety control becomes a secondary (safety) control.

As noted, the low-pressure safety control will turn off the system when low-side pressure becomes too low. Most compressors rely upon the suction gas to cool the motor windings; if this cool gas is not available, the motor will overheat and cycle on the overload. The low-pressure safety control is often used to protect the system against loss of refrigerant and protect air conditioning systems and water chillers from suffering an evaporator freeze-up.

Commercial low-pressure *safety* controls are similar to the *primary* low-pressure control. In fact, the same model control can often be used for either purpose, because they are selected according to the range and differential scales. Therefore, the same control can serve different purposes. **Figure 26-15** illustrates some typical examples.

The low-pressure safety control may have a 36″ or 48″ capillary tube with a flare nut for connection to the low-pressure side of the system. This connection is normally made at the compressor. The control contacts are designed to close on rising pressure and open once pressure falls to a certain point.

Wiring connections

The contacts for the low-pressure safety control are connected in series with the thermostat contacts. See **Figure 26-16.** The thermostat cycles the unit under

Low-Pressure Controls

Range scale	Differential scale
12 in. Hg to 50 psig	5 to 35 psig
10 in. Hg to 100 psig	10 to 40 psig
50 to 150 psig	10 to 40 psig

Figure 26-15. *This table shows that low-pressure controls are selected according to the range and differential scales available on the control. The same pressure control can often be used as either a primary or secondary control, or with different refrigerants.*

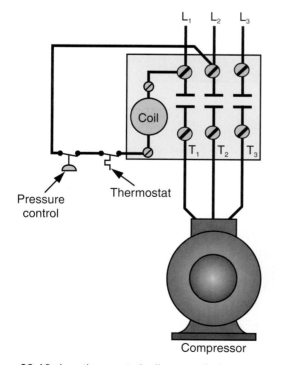

Figure 26-16. *In a thermostatically controlled system with a low-pressure safety control, the two controls are wired in series with the contactor coils. Both switches must be closed to operate the system. If either opens, the system will stop operating.*

normal operating conditions, with the low-pressure safety control acting as a back-up. The contacts on both switches must be closed to run the system, but either switch can open and stop operation.

Pressure settings

The low-pressure safety control must not be permitted to interfere with normal operation of the system, yet it must provide necessary protection. The control settings vary according to each system and refrigerant used. Pressure settings must protect the unit against loss of refrigerant, while the differential must be sufficient to prevent short-cycling.

When the system becomes low on refrigerant, the compressor quickly reduces low-side pressure. When the compressor shuts off, low-side pressure increases rather quickly, due to higher temperature at the evaporator. A close differential would permit the system to short-cycle during loss of refrigerant. The safety control is set to cut out just before the low-side pressure enters a vacuum. A wide differential places the cut-in at a point requiring more time for the low-side pressure to increase before reaching the cut-in point. When trying to calculate the settings on a low-pressure safety control, it is best to think backwards (from cut-out to cut-in).

For example, consider a walk-in cooler using a thermostat to control air temperatures between 37°F and 32°F (3°C and 0°C). The refrigerant used is R-12. If a leak occurs, the low-side pressure must be prevented from entering a vacuum. The low-pressure safety control cut-out should be established at a point just above a vacuum (between 2 psig and 5 psig, or 14 kPa and 34 kPa). The differential must be about 20 psig to 25 psig (138 kPa to 172 kPa) to prevent short-cycling, so the cut-in should be set at about 30 psig (207 kPa). After calculating the proper setpoints, the control is adjusted by first setting the cut-in, then the differential.

Control settings of 25 psig (cut-in) and 3 psig (cut-out) do not interfere with normal operation of the system. The safety cut-in pressure is below the cut-in temperature of the thermostat, and the safety cut-out pressure is below the cut-out temperature. The corresponding low-side pressures for the temperatures at which the thermostat opens and closes are 34 psig (234 kPa) for cut-in and 14 psig (97 kPa) for cut-out.

Air conditioning application

Low-pressure safety controls are often used on air conditioning and water chillers to prevent evaporator freeze-up. These systems use a thermostat as a primary control to maintain proper air temperatures. When the evaporator freezes up on an air conditioner or chiller, it becomes possible for liquid refrigerant to flood back to the compressor. Even though the system has problems, the thermostat will continue to call for cooling. Since water expands when it freezes, evaporator tubes on a chiller can be badly damaged by a freeze-up.

The refrigerant of choice for high-temperature (air conditioning) applications is R-22. Low-side pressure rarely drops below 58 psig (400 kPa), or in temperature terms, 32°F (0°C), because the cooling job is finished as the system is ready to shut off. With refrigerant temperature at 32°F (0°C), the evaporator is 42°F (6°C), and the "air off" the evaporator is 52°F (11°C). Frost and ice cannot accumulate on the evaporator because the evaporator temperature is above the freezing point of water. The 52°F air off (supply air) indicates that air on (return air) temperature will be 72°F.

The cool air off is directed by supply air ducts to the living space. See **Figure 26-17.** Warm air flows back to the evaporator (by return air ducts), where it is again cooled and sent back to the living space. This recirculation of air is continuous until the living space air is cooled to the cut-out point on the thermostat. The blower provides the necessary force to move air through the evaporator and the ducts.

Air conditioning systems are designed for the evaporator to produce a 16 degree Fahrenheit to 20 degree Fahrenheit (9 degree Celsius to 11 degree Celsius) difference between the temperature of air on

and air off the evaporator. In air conditioning work, the technician uses the temperature of air off the evaporator (supply air) to calculate temperature-pressure of the refrigerant. Air conditioning evaporators are designed to control temperature of air off. The evaporator is 10 degrees Fahrenheit (5.5 degrees Celsius) colder than air off, and refrigerant temperature is 10 degrees Fahrenheit (5.5 degrees Celsius) colder than the evaporator.

For example, consider a central home air conditioner that has been turned off during vacation time. Upon returning home, the owner finds that the interior house temperature is 85°F (29°C). The owner immediately turns on the air conditioner to cool the house to an acceptable level of 72°F (22°C).

When the air conditioner starts, the temperature of air on the evaporator is 85°F (29°C) and the air off is 65°F (18°C). With a 65°F air off temperature, the refrigerant temperature is 45°F (7°C) and the low-side pressure is 76 psig (524 kPa). As the temperature of air on is reduced, air off temperature is also reduced, and refrigerant temperature decreases. Eventually, the air on temperature becomes 72°F (22°C), which is the thermostat cut-out point. Air off is 52°F (11°C) and refrigerant temperature is 32°F (0°C). Evaporator pressure is 58 psig (400 kPa).

In this example, the lowest evaporator temperature was 42°F (6°C), well above the freezing point of

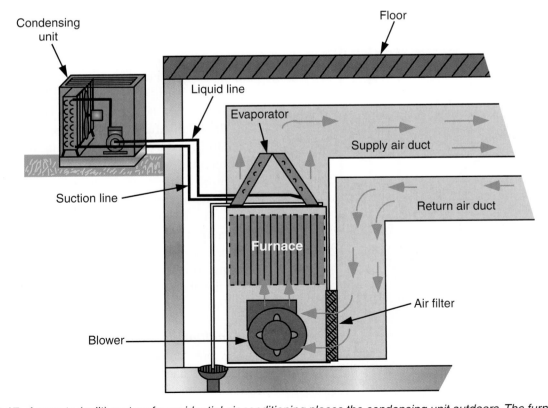

Figure 26-17. *A remote (split) system for residential air conditioning places the condensing unit outdoors. The furnace blower moves air through the evaporator coil, where it is cooled and distributed to living space through the supply duct system.*

water. However, if evaporator temperature drops below 32°F (0°C), ice and frost will accumulate on the evaporator and proper airflow will be lost. When evaporator airflow is reduced, the **heat load** is also reduced, and the evaporator becomes colder and colder. The heat load is necessary to maintain proper system pressures.

The low-pressure safety control on air conditioning is set to cut out at 40 psig to 45 psig (276 kPa to 310 kPa). At 40 psig, refrigerant temperature is about 17°F (–8°C) and the evaporator is 27°F (–3°C). Any evaporator operating below 32°F will accumulate frost. The air conditioner must be stopped before the freeze-up becomes severe.

Troubleshooting

Problems with the low-pressure safety control are rare. If the switch is bad, the control must be replaced. Most often, however, the problem is a capillary tube that is leaking due to vibration or rubbing. Corrective procedures for leaking capillary tubes were described earlier. The low-pressure safety control serves as a warning device that pressure is below safe limits. The actual problem is usually elsewhere, not with the control.

Some problems that will cause the system to have low-pressure are:

❑ Low refrigerant level.
❑ Evaporator freeze-up.
❑ Evaporator fan not running.
❑ Restricted airflow due to dirty return air filters.
❑ Dirty evaporator due to missing air filters.
❑ Moisture in the system, freezing at refrigerant control.
❑ Faulty superheat setting on TEV.
❑ Restriction at filter-drier.
❑ Low head pressure due to low ambient temperature.

26.1.4 Head-pressure Safety Control

The **head-pressure safety control** is a device that will shut off the system if head pressure exceeds proper limits. High head pressure results in poor system performance and damage to the compressor. High head pressure is rather common, and the safety control protects the system when it occurs.

The commercial high head-pressure safety control looks much like a low-pressure safety control. See **Figure 26-18.** The head-pressure safety control is designed for the contacts to open on *rising pressure*. The capillary tube is connected to the high-pressure side at the compressor. Depending upon the range scale, the same control can be used for R-12, R-22, or R-502. See **Figure 26-19**.

The head-pressure safety control is available with either automatic reset or manual reset. In automatic

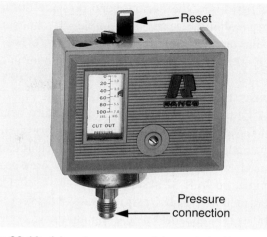

Figure 26-18. *A head-pressure safety control with a manual reset button. Note the scale shows only cut-out pressure. (Ranco)*

Head-pressure Safety Controls

Range scale	Differential scale
100 to 250 psig	20 to 100 psig
100 to 400 psig	35 to 150 psig
150 to 450 psig	35 to 150 psig

Figure 26-19. *This table displays the ranges and differentials available in head-pressure safety controls, which make it possible to use the same control with different refrigerants.*

reset, the control contacts automatically close when head pressure drops to the cut-in setpoint. This means that the control turns the compressor on and off whenever head pressure changes from unsafe to safe limits. If the condenser fan burns out, the automatic reset control permits the system to cycle on and off.

The automatic reset is most common for commercial applications because it eliminates temporary nuisance shutdowns. These shutdowns usually result from high-heat-load conditions of brief duration. These result in temporary high suction pressure and high head pressure. Head pressure returns to normal when suction pressure is reduced to normal.

The manual reset is preferred by compressor manufacturers, because the control must be manually reset to restore the system to operation. It is assumed that the person resetting the control will investigate and cure the problem that is causing the high head pressure. The manual reset control is often used for air conditioning applications.

Head-pressure safety control settings

To calculate the safety pressure settings for a forced convection (fan-type) condenser, the technician must first know the normal head pressure during the *hottest* day of the year. For example, on a hot summer day, the ambient may reach 100°F (38°C). For R-12 and R-22 systems, the normal head pressure is calculated by adding 30°F to 35°F (17°C to 19°C) to ambient. For R-502 systems, add 20°F to 25°F (11°C to 14°C) to ambient. The lower temperature would be used for a clean condenser; the higher, for a slightly dirty condenser. A temperature-pressure chart can then be used to convert the temperature to pressure. For example, on an R-12 system:

100 + 30 = 130 (181 psig) (clean condenser)
100 + 35 = 135 (194 psig) (slightly dirty condenser)

In this case, the head-pressure safety control is set to cut out at a pressure of 215 psig to 225 psig (1483 kPa to 1552 kPa). This extra margin of safety protects the system and prevents nuisance trip-outs. The differential is set for 55 psig to 60 psig (379 kPa to 414 kPa) to permit the system to cool down before automatic reset.

When the condenser becomes very dirty, or if the condenser fan stops running, head pressure will rise rapidly to the cut-out. If the head-pressure safety control is *not* used, head pressure will continue to rise until the overload stops the compressor. The compressor will cycle on the overload until the head pressure problem is cured.

Water-cooled condensers

Head-pressure safety controls are also used for water-cooled condensers. This safety control is necessary to protect the system if the condenser becomes fouled (dirty), or if water flow stops. The water regulating valve is normally set to maintain a constant head pressure (calculated from an ideal ambient of 70°F, or 21°C).

For R-12 systems, the water regulating valve is set to maintain head pressure of 115 psig (793 kPa). The head-pressure safety control is set to cut out at 150 psig (1034 kPa), with a differential of 35 psig (241 kPa).

For R-22 and R-502 water-cooled systems, the water regulating valve is set to maintain a constant head pressure of 200 psig (1379 kPa). The head-pressure safety control is set to cut out at 250 psig (1724 kPa), with a differential of about 50 psig (345 kPa).

Wiring connections

Wiring connections for the head-pressure safety control place the switch in series with the primary control. See **Figure 26-20.** Safety control switches are usually connected into the same circuit, but act as a secondary control. Sometimes the head-pressure safety control is connected in series with the compressor overload. This method shuts the compressor off, but permits the condenser fan to operate. The two methods are equally effective for protecting the compressor.

Dual-pressure motor control

The *dual-pressure motor control* is a combination of a low-pressure control and a head-pressure (high-pressure) safety control. See **Figure 26-21.** Typically, commercial systems had both a low-pressure control (either primary or safety) and a head-pressure safety control. Manufacturers recognized this and combined the two controls into a single unit.

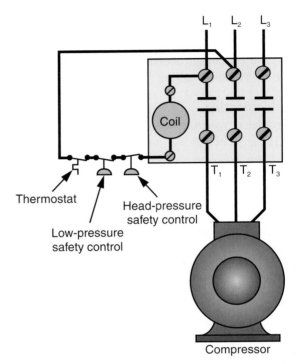

Figure 26-20. *The primary control (in this case, a thermostat) and secondary safety controls are usually wired in series with the contactor coil. If any switch opens, the power supply to the contactor coil will be cut off.*

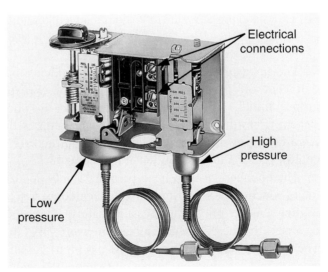

Figure 26-21. *A dual-pressure safety control for both high-pressure and low-pressure protection. On the high-pressure side, the differential is set by the factory, so only a cut-out pressure is set on the scale. The low-pressure control allows setting of both range and differential. (Penn)*

The dual-pressure control saves time and money because double protection is provided with one control. Wiring connections are simplified because the control has just two terminals. The factory makes the internal connections that permit either switch to disconnect power supply.

Each side of the control operates independently, with its own adjustment scale. If one side of the control becomes nonfunctional, the other side will remain operational. The low-pressure switch opens on falling pressure and the high-pressure switch opens on rising pressure.

The two capillary tubes must be properly connected to the compressor. The smaller bellows is connected to high-pressure and the larger bellows is connected to the low-pressure side. Both sides of the control are fully adjustable by screws that are located directly above each scale.

Some head-pressure controls have only a single *range adjustment* screw, since the differential is factory fixed at 55 psig or 60 psig (379 kPa or 413 kPa). The proper cut-out pressure is selected by the range adjustment screw, while the cut-in is established by the fixed differential. This dual-pressure control is available with a manual reset for the high-pressure side.

Condenser fan cycle control

The **fan cycle control** is used to control the condenser fan or fans. When head pressure drops below acceptable limits, the fans are turned off. However, when pressure returns to normal, the control turns the fans back on.

Condenser fan cycling is a common method of maintaining head pressure during cold weather conditions. When the condensing unit is located outdoors, low ambient temperatures will cause low head pressure. Low head pressure results in poor system operation.

TEV pressure drop

Proper operation of the thermostatic expansion valve (TEV) depends upon proper pressure drop across the valve. This pressure drop is determined by subtracting normal *suction* pressure from normal *head* pressure. This pressure drop must be maintained or the valve cannot feed refrigerant properly. The compressor will remove refrigerant from the evaporator faster than the valve can feed liquid. Thus, the evaporator pressure becomes too low, and the unit cycles off on the low-pressure control.

Head pressure limits

Head pressure for a system using R-12 should not fall below 100 psig (690 kPa). The minimum head pressure for R-22 is 200 psig (1380 kPa), and R-502 is 180 psig (1242 kPa). Head pressures should be maintained *at or above* these levels to assure proper system operation.

Control contacts

The fan cycle control helps maintain proper head pressure by disconnecting the condenser fan when head pressures fall below an acceptable level. The fan remains off until the head pressure builds up to the cut-in point again. During low ambient, the condenser fan may not run, or may cycle on and off. The contacts on this control must open on falling pressure and close on rising pressure. See **Figure 26-22**.

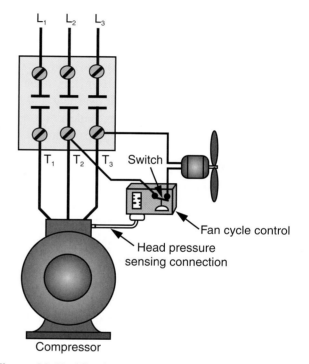

Figure 26-22. *The fan cycle control helps maintain proper head pressure by turning the condenser fan off and on as necessary.*

Control connections

The fan cycle control is located at the condensing unit and connected in series with one side of the fan's power supply. Sometimes two condenser fans are controlled by the same switch.

Large condensing units that use multiple fans will **stage** (sequence) individual fans to cycle on or off at different pressures. These controls are set for about a 10 psig (69 kPa) difference. One fan is cycled off at a fixed setpoint; if the pressure continues to drop, another fan turns off. This staging continues until all fans are off, or until the pressure stabilizes. Control settings also turn fans on as head pressure rises.

Control settings

R-12 systems: Cut-in = 140 psig (966 kPa); Cut-out = 100 psig (690 kPa).

The 40 psig (276 kPa) differential keeps the fan from short-cycling during periods of moderate ambient conditions. A delay exists at each setpoint while the pressure stabilizes. When the switch cuts out at 100 psig, head pressure drops to about 90 psig (621 kPa) because the fan requires time to stop. When head pressure rises to the 140 psig cut-in, time is needed for the fan to achieve full speed, so head pressure rises to about 150 psig (1035 kPa). Therefore, the actual differential is about 60 psi (414 kPa).

R-22 and R-502 systems: Cut-in = 240 psig (1656 kPa); Cut-out = 200 psig (1380 kPa).

The 40 psi differential permits head pressure to reach 190 psig (1311 kPa) after cut-out and 250 psig (1275 kPa) after cut-in. This differential prevents short-cycling of the fan and maintains head pressure within proper limits. The condenser fan must not short-cycle, because the motor will overheat and burn out. However, the fan needs to provide compressor cooling during moderate ambient temperatures.

26.1.5 Pumpdown Cycle

The ***pumpdown cycle*** is very popular for many refrigeration systems that have the condensing unit located outdoors. The pumpdown cycle removes all refrigerant from the low-pressure side of the system prior to shutdown. This prevents liquid slugging and loss of oil from the compressor during start-up.

Principle of operation

The pumpdown cycle has a solenoid valve (electrically operated valve) in the liquid line, with a thermostat controlling power supply to the solenoid coil. See **Figure 26-23.** The solenoid valve can be located just before the TEV (evaporator location), or just after the sight glass (condenser location).

Power supply to the solenoid and thermostat is separate from the refrigeration system power supply. The air-sensitive thermostat, located inside the cooled space, controls power to the solenoid coil only. The solenoid valve controls refrigerant flow through the liquid line.

When cooling is required, the thermostat contacts close and energize the solenoid coil. The solenoid valve opens and permits refrigerant flow to the evaporator. The suction pressure rises and a low-pressure control starts the compressor. When the thermostat is satisfied and opens, the solenoid coil is de-energized, allowing the valve to close. Refrigerant flow to the evaporator stops. The compressor continues to pump refrigerant from the evaporator and suction line until the suction pressure reaches the low-pressure control cut-out setting. See **Figure 26-24.**

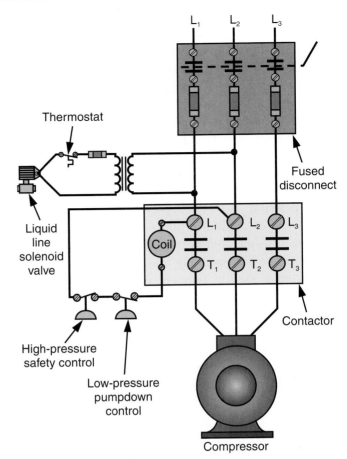

Figure 26-23. *The solenoid valve and the thermostat that controls it have a power supply isolated from the refrigeration system's power supply by a transformer.*

In the pumpdown cycle, the compressor evacuates (removes) all refrigerant from the low-pressure side of the system. The refrigerant is stored in the liquid receiver (and condenser) after each run cycle. Since the receiver and condenser are fully capable of storing this refrigerant, head pressure does not rise.

Advantages of the pumpdown cycle include:
- ❏ Preventing liquid migration to the compressor during the off cycle.
- ❏ Eliminating high suction pressure at start-up.
- ❏ Providing better oil return flow to the compressor crankcase.
- ❏ Preventing refrigerant vapor from condensing in the crankcase during cold ambient.

Pumpdown pressure control settings

Pressure control settings for pumpdown are important and vary according to the system. At pumpdown, the cut-out should occur before suction pressure enters a vacuum (about 1 psig to 2 psig, or 7 kPa to 14 kPa). The cut-in pressure setting is more difficult to determine. The cut-in depends upon whether the condensing unit is located indoors or outdoors.

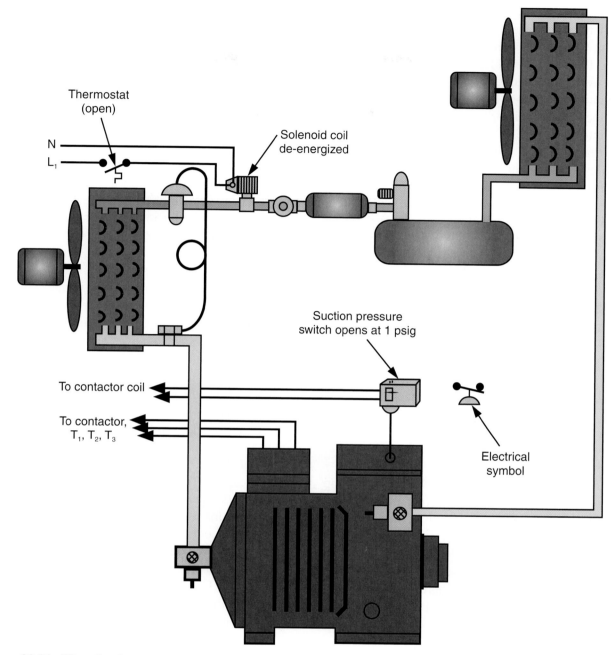

Figure 26-24. *When the thermostat opens, the solenoid valve is de-energized and closes. This shuts off flow to the evaporator through the liquid line. The compressor will continue to operate until the low-pressure pumpdown control switch opens. Refrigerant is stored in the condenser and liquid receiver.*

Indoor condensing unit

For indoor condensing units, cut-in is determined from the lowest air temperature (thermostat cut-out). Subtract about 10 degrees Fahrenheit (5 degrees Celsius) from the thermostat cut-in temperature and convert the result to pressure, using a temperature-pressure chart. This pressure becomes the cut-in on the pressure control.

For example, if the thermostat is set to control temperatures at 32°F to 38°F (0°C to 3°C), subtract 10 degrees from 32°F (thermostat cut-out). The result, 22°F (–6°C) is converted to pressure (for R-12, 23 psig or 159 kPa). This becomes the cut-in for the pressure control.

Indoor pumpdown system operation. When air temperature reaches 38°F (3°C), the thermostat closes and energizes the solenoid coil. Refrigerant flows to the evaporator. Low-side pressure quickly rises to 23 psig (159 kPa), and the low-pressure control closes. This starts the condensing unit and the system operates for a normal run cycle.

When air temperature is lowered to 32°F (0°C), the thermostat opens and energizes the solenoid. Refrigerant flow to the evaporator stops. The condensing unit continues to operate (pumpdown). Low-side pressure is quickly lowered to 1 psig or 2 psig—the pressure control opens to stop the condensing unit.

Outdoor pumpdown system

When the condensing unit is located outdoors, low ambient temperatures greatly affect the cut-in setting. During the off cycle, all refrigerant is stored in the liquid receiver (outdoors) and becomes the same temperature as outdoor ambient. Cut-in is determined from the coldest anticipated outdoor ambient. The temperature-pressure in the cold receiver cannot be lower than the cut-in of the pressure control.

To overcome cold ambient problems, the cut-in setpoint *must* be equal to the coldest likely outdoor temperature. Failure to abide by this rule will result in nuisance service calls when outdoor temperatures drop to low levels. The thermostat will energize the solenoid valve, but the compressor will not run because *refrigerant pressure* is lower than the control cut-in. **Figure 26-25** is a table of typical pressure control settings when outdoor ambient is below system operating temperatures.

Outdoor pumpdown cut-out. The pressure control cut-out normally occurs before the systems enters a vacuum. However, this is not always possible when outdoor ambient is low. The technician must adjust the control for cold ambient conditions. As shown in Figure 26-25, pumpdown systems using R-12 will frequently pull into a vacuum. The lowest recommended cut-out for R-502 is 3 in. Hg.

Pumpdown differential. The differential for pumpdown cycles must prevent short-cycling after the initial cut-out. When the solenoid valve closes, the compressor rapidly removes refrigerant and reaches the cut-out point. Some gas remains in the low-pressure side. After shutdown, this gas residue causes low-side pressure to rise to the cut-in setpoint. The compressor cycles briefly until the low-side pressure is reduced to the cut-out. One brief recycle is not unusual and serves to remove residual refrigerant. Several short cycles before complete shutdown indicates a short differential.

Troubleshooting

Recycling after shutdown is caused by a system that is low on refrigerant, a differential set too close, a leaking solenoid valve seat, or leaking compressor discharge valves. These problems can be isolated, identified, and corrected using the following procedures:

Pumpdown system low on refrigerant. Install a gauge manifold on the compressor service valves and remove cover from the low-pressure control. Check control operation with a multimeter reading voltage or by observing the mechanical action of the control as it opens and closes.

Check operating pressures and the sight glass. The sight glass may show some bubbles immediately after start-up, but the bubbles should quickly disappear and the glass should indicate a full system. Excessive and continuous bubbles in the sight glass (even when the compressor is off), along with low pressures, indicate the system is low on refrigerant. The leak must be located and repaired, and the system recharged to full capacity.

Check magnetism of the solenoid coil with a small screwdriver to determine if the solenoid is energized. If the solenoid is not energized, check power supply to the thermostat and determine if thermostat contacts are open or closed. Turning the thermostat to a warmer setting should energize the solenoid valve.

Differential set too close. Remove cover from the low-pressure control and check scale settings against pressures observed on the gauge manifold. The control scale can be misleading or inaccurate, so adjustments should always be performed using the gauge manifold. Inaccurate or improper settings should be corrected to provide proper system operation.

Control Settings for Cold Ambient

Coldest Ambient	R-12		R-134a		R-22		R-502	
	Cut-in	Cut-out	Cut-in	Cut-out	Cut-in	Cut-out	Cut-in	Cut-out
+30	25	1	22	1	40	1	40	1
+20	18	1	15	1	30	1	40	1
+10	13	0	10	0	25	1	35	1
0	5	5″	4	5″	20	0	25	0
−10	3	8″	1	8″	15	0	20	0
−20	0	10″	5″	15″	8	0	13	0
−30	5″	15″	9″	15″	4	6″	9	2″

Figure 26-25. *Table of pressure control settings for cold ambient. Ambient temperatures are in °F and cut-in/cut-out pressures in psig, except for vacuum readings, which are marked with ″ for in. Hg.*

Leaking compressor discharge valve reeds. Compressor discharge valve reeds that are leaking or broken are easily discovered by following this procedure:

1. Install the gauge manifold on the compressor service valves.
2. Frontseat the suction service valve to prevent any refrigerant from entering the compressor.
3. Force compressor to run by using a jumper cord to bypass the low-pressure control.
4. Allow compressor to pull good vacuum (20 in. Hg to 25 in. Hg).
5. Stop the compressor and observe the low-side gauge for rapid loss of vacuum. If it occurs, replace the valve plate.

Compressor discharge valve reeds generally show slight leakage, but a *rapid* loss of vacuum indicates faulty reeds. The compressor discharge valve reeds are a division point between the high-pressure and low-pressure sides of the system. If they are broken or not seating properly, they will permit high pressure to leak back into the low-pressure side. This problem can be repaired in the field by the service technician using a valve plate replacement kit, which contains new valve reeds.

Leaking pumpdown solenoid valve. This problem is rare, but is indicated by regular short-cycling of the compressor after pumpdown due to refrigerant leaking into the low-pressure side. The refrigerant leakage may result from leaking compressor discharge valve reeds, or a leaking solenoid valve seat.

The problem is isolated by first checking the discharge valve reeds, as previously described. If discharge valve reeds are functioning properly (no loss of vacuum), the problem *must* be the solenoid valve.

Sometimes a Schrader valve is available in the suction line for checking suction pressures. If so, the solenoid valve seat can be checked by using this procedure:

1. Connect the compound gauge to the Schrader valve.
2. Force the compressor to pull a good vacuum, then frontseat the suction service valve. Stop the compressor. The suction service valve will prevent return pressure from the compressor.
3. Observe the compound gauge. It will reveal if the low-side pressure remains in a vacuum or rises to the cut-in point. If pressure rises, the solenoid valve seat is leaking. The defective valve must be replaced.

26.1.6 Oil Pressure Safety Control

The *oil pressure safety control* is used to protect the compressor when lubrication problems occur. Many compressor failures caused by lack of

lubrication could have been prevented by using an oil pressure safety control. Warranties on compressors using an oil pump specify the unit *must* be equipped with an oil pressure safety control.

Loss of compressor lubrication can be caused by a variety of system problems:

- ❏ Shortage of oil in the system.
- ❏ Oil remaining in the evaporator (called *logging*) due to low refrigerant velocity.
- ❏ Shortage of refrigerant.
- ❏ Refrigerant migration or floodback to the crankcase.
- ❏ Oil pump failure.
- ❏ Faulty refrigerant control devices.
- ❏ Short cycling.

Regardless of cause, the compressor must be protected against loss of oil pressure. Permitting the compressor to operate without proper lubrication will result in certain failure from broken piston rods, locked bearings, or similar causes. The oil pressure safety control is designed to stop the compressor before such damage occurs. It also warns the technician that an oil problem exists. An oil pressure safety control is standard equipment on any compressor equipped with an oil pump. See **Figure 26-26.**

Control connections

The oil pressure safety control is unlike any other type of control. It has two switches, two pressure-sensing bellows, and two capillary tubes for connection to the compressor. One capillary tube and bellows is tagged "low" and connected to the compressor crankcase (low-side pressure). The other capillary tube

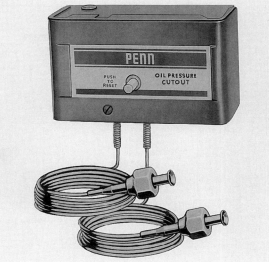

Figure 26-26. *The oil pressure safety control prevents compressor failures caused by loss of lubrication. It will stop the compressor if oil pressure drops below a safe level. The control is not adjustable. (Penn)*

is tagged "oil" and connected to the oil pump (oil pump pressure). The tubes are tagged at the factory to prevent confusion when making these connections.

Two capillary tubes are needed because the control operates on the *difference* between low-side pressure and oil pump pressure. The oil pump produces a pressure that is higher than the low-side pressure. This pressure difference is called **net oil pressure.** Net oil pump pressure is obtained by subtracting low-side pressure from oil pump pressure. Oil pump pressure *never* falls below low-side pressure, so it is necessary to subtract low-side pressure from oil pump pressure to obtain net oil pressure.

Most oil pumps are equipped with a tee fitting for making connections. The capillary tube (tagged "oil") from the safety control is connected to the #2 opening on the tee. The remaining tee branch (#3 opening) is fitted with a Schrader valve for obtaining oil pump pressure readings during operation. Two compound gauges are needed to check net oil pressure. One gauge is used to obtain low-side pressure at the suction service valve, and the other is connected to the oil pump Schrader valve to read oil pump pressure. The difference between the two pressure readings is net oil pressure.

Principle of operation

The oil pressure safety control has two normally closed switches. One is used to control a small heater. The contacts controlling power to the heater are tagged T_1 and T_2. Operation of the heater causes a secondary (bimetallic) switch to open. The bimetallic switch contacts are tagged L and M. This switch is connected in series with the primary control and the contactor coil. **Figure 26-27** shows a typical connection for an oil pressure safety control on a three-phase motor-compressor.

The two pressure-sensing bellows operate the switch at T_1 and T_2. This switch controls the power supply to the small heater. At start-up, power is supplied to both the compressor and the heater. When net oil pressure reaches 12 psig (83 kPa), the heater switch opens. This cuts off the power supply to the heater, but the compressor continues to run. During the run cycle, if net oil pressure drops to 9 psig (62 kPa), the switch will close and restore power to the heater.

If net oil pressure fails to achieve 12 psig after start-up (it is much higher in normal operation), the switch at T_1 and T_2 remains closed and the heater gets hot. It requires about 120 seconds (2 minutes) for the heater to produce enough heat to warp open the bimetallic switch

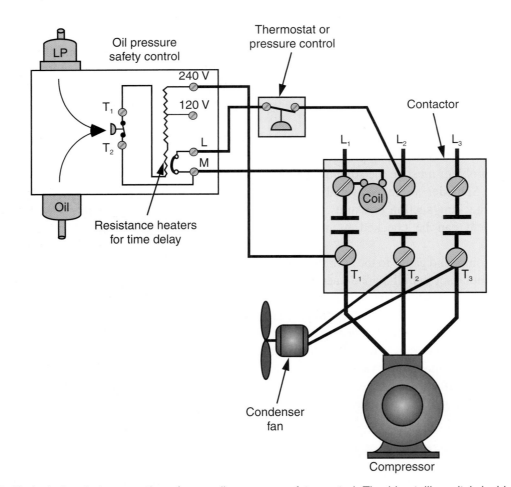

Figure 26-27. *Typical electrical connections for an oil pressure safety control. The bimetallic switch inside the control is connected at terminals L and M.*

connected between L and M. When the bimetallic switch opens, power supply to the contactor coil is interrupted and the compressor stops. All oil pressure safety controls must be manually reset when they open.

Control heater

The heater provides a time delay required for the pump to establish oil pressure after start-up. It also prevents nuisance shutdowns due to brief shortages of oil pressure during the run cycle. The compressor is permitted to operate without proper lubrication for the amount of time required for the heater to get hot. The heater and bimetallic switch are carefully selected at the factory to provide a specific time delay. Oil pressure safety controls are available with time delays of 60 seconds, 90 seconds, or 120 seconds. This time delay is nonadjustable; each compressor manufacturer specifies which delay is used on its product. The 120-second control is most common.

To provide for dual-voltage operation, there are two resistance heaters located inside each control. Only one is used when connecting the control for 120 V operation. The two resistance heaters are automatically connected in series for 240 V operation. As shown in **Figure 26-28,** these two terminals are clearly identified for selecting proper voltage connections in the field.

Wiring connections

Power supply to the resistance heaters *must* be obtained from the bottom of the contactor. This prevents the heater from being energized during the off cycle. All other power supply connections are obtained from the top of the contactor.

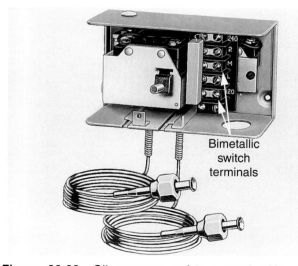

Figure 26-28. *Oil pressure safety control with cover removed to show electrical terminals. Terminals L and M are used to connect the control's bimetallic switch in series with the contactor coil. Note the dual-voltage provision, with 120 V and 240 V terminals. (Penn)*

Heater connections. Power supply to the heater is performed by connecting a wire from T_1 on the contactor to 240 V on the control. The other side of the heater obtains its power from the connections at the bimetallic switch. When the compressor contactor is closed, a circuit is made from T_1 on the contactor to the 240 V resistor connection, through the resistors, to the heater. From the heater, power travels through the switch at T_1 and T_2, through the bimetal contacts at L and M, through the low-pressure control, to L_2 at the top of the contactor.

Contactor coil connections. One side of the contactor coil is connected to L1 on the contactor. From the coil, current travels to M on the bimetallic switch contacts, through the switch to L, and picks up the power supply traveling to the low-pressure control and up to L1. The compressor cannot operate unless both the bimetallic switch at L-M and the low-pressure switch are closed. At start-up, if the net oil pressure does not build up sufficiently to take the resistance heater out of the circuit within 120 seconds, the heater will cause the bimetallic switch to open. This interrupts power to the contactor coil and stops the compressor.

Troubleshooting

To check the pressure settings on the control, pump the system down by frontseating the liquid receiver service valve and forcing the compressor to run. When a vacuum of 10 in. Hg to 15 in. Hg is obtained, shut off the compressor and briefly crack open the liquid receiver service valve to eliminate the vacuum. Disconnect both capillary tubes from the compressor. Leave the low-pressure capillary tube open to the atmosphere to sense "zero" pressure.

Connect the middle hose from the gauge manifold to a cylinder of refrigerant. Connect the low-pressure hose to the "oil" capillary tube. Open the manifold hand valve until the low-pressure gauge indicates a pressure between 20 psig and 60 psig (138 kPa and 414 kPa). Check for continuity between T_1 and T_2. The switch should be open.

Bleed the pressure off slowly and observe the gauge. The switch should close at about 9 psig (62 kPa). Gradually raise the pressure again until the switch opens. This should occur at a pressure of between 12 psig and 14 psig (83 kPa and 97 kPa). The switch will make a "click" sound each time it opens or closes. If the control does not open and close within the specified pressure range, it should be replaced.

Summary

Motor controls serve many functions in the refrigeration system. They provide automatic control of system operation (primary control) and protection from possible malfunctions (safety controls). Motor controls can be temperature-sensitive (thermostats) or pressure-sensitive (pressure controls).

This chapter explained the operation, application, wiring connections, and troubleshooting procedures for each motor control. Domestic controls are different from commercial types, but their operation is similar. The technician must be able to install and properly adjust the various types of controls. System problems are often pinpointed by these controls, so the technician must be able to troubleshoot and repair controls and associated devices.

Test Your Knowledge

Please do not write in this text. Write your answers on a separate sheet of paper.

1. Name two types of primary motor controls.
2. The _____ is the setpoint that causes the control contacts to close.
3. The _____ is the span between cut-in and cut-out temperatures or pressures.
4. Primary controls perform their job by connecting or disconnecting the _____ to the contactor coil.
5. The _____ screw changes cut-in and cut-out equally.
6. The differential adjustment screw changes only the _____.
7. On a three-phase system, is the thermostat switch connected in series or in parallel with the contactor coil?

8. When setting the cut-out, how much pressure (in psig) must be subtracted to allow for pressure drop?
9. The low-pressure safety control becomes necessary when a(n) _____ is used as the primary control.
10. Why are low-pressure safety controls used on air conditioning systems?
11. The low-side pressure on air conditioning should not fall below _____ psig.
12. In air conditioning, the air off temperature should be 10 degrees Fahrenheit above the temperature of the _____.
13. What causes the low-pressure safety control switch to open?
14. *True or false?* The head pressure safety control switch is connected in parallel with the primary control.
15. The fan cycle control is used to _____.
16. On a pumpdown cycle, what does the cooled-space thermostat control?
17. What is the purpose of the low-pressure control on a pumpdown cycle?
18. The oil pressure safety control operates on _____ oil pressure.
19. On the oil pressure safety control, the heater is controlled by a pressure switch located between terminals _____.
20. Are oil pressure safety controls reset automatically or manually after they "trip" due to low oil pressure?

27

Defrost Cycles

 Objectives

After studying this chapter, you will be able to:
- ❏ Identify various domestic and commercial defrost timeclocks.
- ❏ Describe the five types of defrost cycles.
- ❏ Install and properly adjust defrost timeclocks.
- ❏ Connect and troubleshoot defrost terminator thermostats.
- ❏ Explain the operation of, and make necessary adjustments to, domestic and commercial defrost systems.
- ❏ Adjust temperature and pressure termination.

Important Terms

atmospheric air	humidity
automatic off-time defrost	initiate
	mullion heaters
capacity control	multiplexing
condensate pan	off-cycle defrost
condense	relative humidity
defrost cycle	reverse-air defrost cycle
defrost terminator thermostat	reversing relay
dehumidifier	saturated
dew point	suction accumulator
electric defrost	temperature-terminated timeclock
energy efficiency ratio (EER)	terminate
evaporator fan delay thermostat	time-initiated/pressure-terminated timeclock
fail-safe setting	time-initiated/time-terminated clock
freeze-up	water vapor
hot gas defrost	
humidistat	

27.1 Physical Properties of Air

Most refrigeration systems are designed to cool air which, in turn, is used to cool a product or space. The physical properties of ordinary air can become complicated and must be controlled; otherwise, serious problems will develop in the refrigeration system.

The temperature and moisture content (humidity) of atmospheric air are important factors to consider when controlling air, but filtering (cleaning) and circulation (movement) are crucial as well. Refrigeration and air conditioning systems are designed to provide automatic control of temperature, humidity, filtering, and circulation. This chapter explains the various properties of air and how systems are designed to control air in refrigeration systems.

Atmospheric air is a mixture of gases and ***water vapor*** (moisture). Air has weight, density, temperature, and specific heat. More energy is required to move cold air because cold air is heavier than warm air. Density of air varies with atmospheric pressure and the amount of water vapor the air contains. At sea level, one pound of air occupies about 14 cu. ft. of space and has a density of 0.0725 lb./cu. ft. The specific heat of *dry* air at sea level is 0.24 Btu/lb.; the specific heat of *moist* air increases with the moisture content.

27.1.1 Humidity

The moisture content of air is described as ***humidity.*** The amount of moisture, or water vapor, that air will hold depends upon the temperature. Humidity directly influences human comfort. For example, dry air causes rapid evaporation of moisture through the skin and can make a person feel cold even in a moderate temperature. Moist (very humid) air does not absorb much moisture, so a person may feel uncomfortable due to lack of evaporation.

When warm, moisture-laden air touches a cold surface, it releases much of its moisture. The air temperature near the cold surface is lower than the surrounding air, causing the moisture to **condense** (change to a liquid state) on the surface. Water dripping from cold pipes is caused by condensation of moisture from the air onto the pipes. Insulation on the pipes prevents condensation by keeping air from contacting the pipes. For this reason, cold suction lines on refrigeration equipment are insulated. If the suction line temperature falls below the freezing point of water, condensation will freeze on the lines. Frozen condensation on suction lines is easily confused with frost formed in liquid floodback situations.

Relative humidity

Relative humidity is the amount of moisture held by one cubic foot of air. This moisture content is then compared to the amount of moisture the air would hold if **saturated** (filled to capacity). The ratio of actual water vapor to the maximum possible water vapor pressure is expressed as a percentage. The relative humidity of saturated air is 100%, so air holding half the moisture it is capable of holding has a relative humidity of 50%. Low relative humidity indicates dry air; high relative humidity indicates moist air. Warm air holds more moisture than cold air. As air temperature decreases, relative humidity goes up. When the relative humidity reaches 100%, further reducing the temperature causes moisture to be released (condensed out).

Dew point

Dew point describes the point at which the air becomes saturated (holds all the moisture it is capable of holding) for a given temperature. Dew point is simply another way of describing 100% relative humidity.

Dehumidifiers

A **dehumidifier** is a refrigeration system designed specifically to remove moisture from air without affecting air temperature. See **Figure 27-1.** A dehumidifier is very useful for "drying out" a damp basement. The evaporator is usually a coil of tubing without fins and is placed immediately in front of the condenser. The fan draws air across the evaporator tubing and the condenser.

When humid air contacts the cold evaporator coil, moisture condenses on the tubing and runs down to a catch pan. The catch pan drains into a pail, or it may be routed to a hose leading to a floor drain. Air flowing across the evaporator is cooled and gives up moisture; then, it flows across the condenser and picks up heat. The cooling and rewarming of the air permits the air temperature to remain unchanged. Only moisture is removed since heat removed by the evaporator is replaced by the condenser.

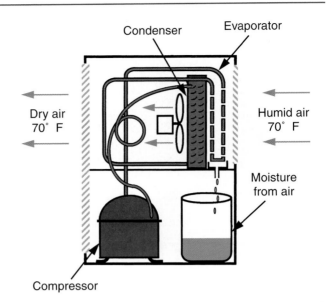

Figure 27-1. *A dehumidifier removes moisture from air without affecting the air temperature.*

A **humidistat** (moisture-sensitive switch) controls system operation. The humidistat sensing element is a nylon ribbon that stretches and contracts according to humidity. The refrigerant temperature-pressure relationship must be maintained to keep the evaporator temperature above the freezing point of water (32°F or 0°C). Moisture must not be allowed to freeze on the evaporator.

27.1.2 Humidity and Air Conditioning

All air conditioning systems act as dehumidifiers because the evaporator lowers air temperature to below the dew point. Moisture collects on the cold evaporator and drips into a catch pan connected to a floor drain. The process of removing moisture from the air adds a significant heat load on the system. Humidity removal is a latent heat load. Latent heat (moisture) must be removed before sensible heat (temperature) can be lowered. High humidity can prevent an air conditioner from lowering temperature.

The temperature of an air conditioning evaporator must be maintained above the freezing point of water to prevent moisture from freezing on the evaporator. Evaporator temperature is maintained by controlling suction pressure. Airflow across the evaporator (heat load) is limited by blower and duct sizes. During periods of high sensible and latent heat loads (high air temperature with high humidity), these limits may prevent the system from performing properly.

Many residential air conditioning systems are sized for operation under normal conditions, but they fail to perform properly during periods of peak loads involving high temperature with high humidity. Such systems are undersized for operation during very hot and moist summer days. Oversizing to compensate for extreme conditions, however, can cause the system to short cycle during normal conditions.

Most commercial air conditioning systems are sized to operate properly under peak loads but provide **capacity control** for use under low-load conditions. Capacity control is attained by using unloaders or bypass valves, or by cycling multiple compressors.

27.2 Defrost Systems

In refrigeration systems, moisture will freeze on the evaporator surface if the evaporator temperature is lower than the freezing point of water, 32°F (0°C). The accumulation of frost and ice restricts proper airflow and acts as an insulator, preventing heat transfer. An evaporator full of frost and ice cannot remove heat from the air because the air molecules cannot touch the evaporator surface. A **defrost cycle** is necessary to remove frozen moisture from the evaporator and restore it to full efficiency.

Air conditioning systems do not require a defrost cycle because evaporator temperature is maintained above 32°F (0°C). Refrigeration systems usually require a defrost cycle, however, because the temperature of the evaporator is below freezing.

Automatic defrost is typically controlled by a timeclock with two or more switches. The timer and switches automatically turn off certain system components (the compressor and evaporator fan) and turn on other components (the electric heater on the evaporator). After a certain length of time has elapsed, the timer resets the switches and restores the system to normal operation. Various types of defrost timeclocks are available; selection is determined by the type of defrost cycle.

27.2.1 Types of Defrost Cycles

Different methods are used to clear the evaporator of frost accumulation. Even though domestic and commercial defrost systems differ, they make use of similar defrost methods. There are five basic types of *automatic* defrost systems:

❑ *Off-cycle defrost.* This commercial system automatically defrosts during each off cycle. The thermostat settings (or the low-pressure control) prevent the system from operating until the evaporator temperature reaches 38°F (3°C).

❑ *Off-time defrost.* This commercial system uses a timeclock to turn off the condensing unit for about two hours (usually at night). The evaporator fan runs continuously; the airflow helps to melt any frost or ice that accumulated during the daily run cycles.

❑ *Electric defrost.* This defrost method is used on both domestic and commercial systems and controlled by a timeclock that automatically turns off the condensing unit, stops the evaporator fan, and turns on electric resistance heaters fastened to the evaporator. The timeclock performs several defrost cycles each day.

❑ *Hot gas defrost.* Hot gas defrost is primarily a commercial system, but some older domestic refrigerators use it also. Gas is taken from the hot gas discharge line and piped directly to the evaporator inlet. A solenoid valve is located in the bypass line and controlled by a timeclock. At defrost time, the clock stops the evaporator fan and energizes the solenoid valve. The condensing unit operates, but hot gas bypasses the condenser and travels directly to the evaporator inlet, supplying heat to melt frost on the evaporator.

❑ *Reverse-air defrost.* This is a recently developed, energy-saving, commercial defrost system used for defrosting multiple-deck frozen food display cases in supermarkets. A timeclock turns the condensing unit off and activates a special relay that reverses rotation of the evaporator fans. The reverse rotation draws warm ambient air into the canopy air ducts of the display cases. The warm air is blown through the evaporator to melt frost, then discharged back into the store through the cases' lower air ducts.

Defrost timeclocks

Defrost timeclocks can be divided into domestic and commercial types, **Figure 27-2.** Domestic timeclocks are sealed and nonadjustable. Commercial timeclocks are accessible and fully adjustable. All timeclocks have electric motors for turning gears that open and close electrical switches.

27.2.2 Manual Domestic Defrost Systems

Some domestic refrigerators do not have automatic defrost; they must be *manually* defrosted on a regular basis. Defrost is best accomplished by these steps:

1. Turn off the thermostat, or unplug the appliance.
2. Empty the freezer section of food items.
3. Place containers of hot water in the freezer section. Heat from the water quickly softens the frost and ice. Usually, large chunks of frozen material can be removed by hand.
4. Restart the system immediately after frost is removed from the freezer section.
5. Return food items to the freezer section.

IMPORTANT: Never use a sharp instrument for prying or chipping frost from a freezer compartment.

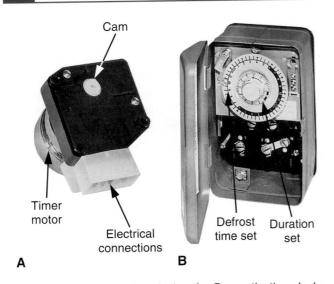

Figure 27-2. *Defrost timeclocks. A—Domestic timeclocks are not adjustable. B—The starting time and duration of each defrost cycle can be set on commercial timeclocks. (Paragon Electric Co.)*

It is very easy to poke a hole in the thin aluminum evaporator. Repairing holes in the evaporator is difficult and expensive.

27.2.3 Automatic Domestic Defrost Systems

Domestic refrigerator-freezer systems with automatic defrost are sometimes called "frost-free." The *electric defrost* system automatically performs the defrost and disposes of resulting water. The system is quite simple but differs from commercial systems.

Domestic timeclocks

All automatic defrost (frost-free) refrigerator-freezers use a defrost timeclock to regulate frequency and duration of each defrost cycle. The clock is usually located behind the toeplate grille at the bottom front of the unit. Some are located in the rear compressor section or inside the refrigerator with the thermostat and lightbulb. See **Figure 27-3.**

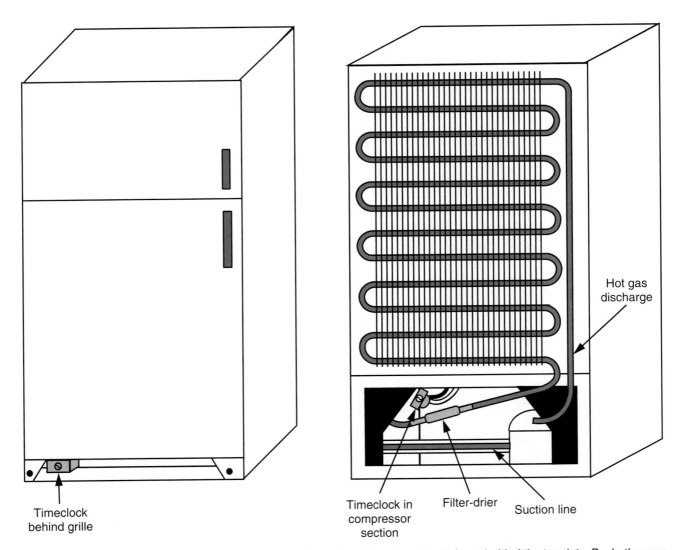

Figure 27-3. *Typical locations of defrost timeclocks on domestic refrigerators. A—In front, behind the toeplate. B—In the rear, in the compressor section.*

The timer consists of a motor, gear assembly, switches, and a rotating cam. The camshaft is geared to the motor so it completes a certain number of revolutions per day. Revolutions per day vary according to manufacturers' specifications. There may be four (one every six hours), three (one every eight hours), or two revolutions (one every 12 hours). At the end of each revolution, the switches are activated, and the system enters a defrost cycle. Duration (length) of the defrost cycle varies from 10 to 35 minutes. Average defrost length is 28 to 30 minutes.

Domestic defrost timeclocks are a critical element in proper system operation, but failure is rather common. Be sure to obtain an exact replacement. Defrost cycle frequency and duration are determined by the equipment manufacturer and are nonadjustable.

The front of the timer contains a plastic screwhead connected directly to the timer camshaft. The screwhead is designed for clockwise rotation and prevents counterclockwise movement. See **Figure 27-4.** When turning the cam with a screwdriver, a sharp "click" sound is made at the beginning and end of the defrost cycle.

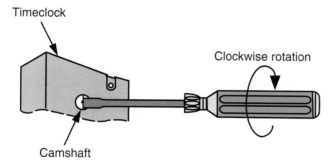

Timeclock

Clockwise rotation

Camshaft

Figure 27-4. *When the camshaft screw is turned clockwise, a distinct "click" sound is made each time the defrost cycle switch opens or closes.*

Most domestic timeclocks have four terminals. The power supply is normally connected to Terminals 1 and 3 because they supply power to the clock motor. Terminal 1 also supplies power to the switches. The normally closed contacts are between Terminals 1 and 2, and the normally open contacts are between terminals 1 and 4. During a defrost cycle, the closed switch opens and the open switch closes. See **Figure 27-5.**

The switches and terminal numbers just described are the most common, but some exceptions can be found. For example, General Electric and Hotpoint use a timeclock with the normally closed switch located between Terminals 1 and 4, and the normally open switch located between Terminals 1 and 2. When making a replacement, always check the appliance schematic against the new clock for proper terminal numbers and arrangement.

Domestic electric defrost cycle

Power supply (hot leg) to the condensing unit is obtained from the timeclock at Terminal 2 because it is the normally closed switch. The other side of the power supply (neutral) can be obtained at Terminal 3 on the clock, but the neutral can also be obtained from another location. It is standard practice to place all switches in the *hot* wire and never disconnect the neutral (white) wire. It is then possible to tap into the neutral wire at any location to obtain that side of the power supply.

The evaporator fan also obtains power from Terminal 2. During a defrost cycle, power supply to Terminal 2 is open and stops operation of the condensing unit and evaporator fan. See **Figure 27-6.**

One or more electric resistance heaters are attached to the evaporator; the heater(s) are energized only during the defrost cycle. See **Figure 27-7.** Control of the defrost heater is accomplished by obtaining the power supply from the clock at Terminal 4.

Defrost terminator thermostat. If frost accumulation on the evaporator is less than normal, defrosting may be completed before the time set on the clock expires. Such a situation could prove dangerous since the defrost heaters could become too hot and cause severe damage. A **defrost terminator thermostat** is used to disconnect the heater when the evaporator temperature reaches about 50°F (10°C). The terminator thermostat is a bimetallic disc inside a sealed container that looks somewhat like an overload. Two wires are provided for making electrical connections. The defrost terminator thermostat is connected in series with the power supply from the clock to the defrost heater, **Figure 27-8.**

The defrost terminator thermostat is clamped onto the evaporator tubing or mounted very close to the evaporator. Once all frost has been removed, evaporator temperature rises quickly. When evaporator temperature reaches about 50°F (10°C), the bimetallic disc inside the thermostat warps and opens its contacts, cutting off the power supply to the defrost heater. The system must wait for the defrost timeclock to finish its cycle before normal operation resumes. When evaporator temperature drops to about 30°F (−1°C), the bimetallic disc contacts close, ready for the next defrost cycle.

Domestic thermostat connections. Older refrigerators connected the thermostat in series with one side of the power to the condensing unit, but *after* the timeclock. The timer motor operated continuously, and the evaporator fan shut off only during a defrost cycle. To conserve energy and increase the refrigerator's **energy efficiency ratio (EER),** newer refrigerators place the thermostat in series with the power supply to the time-

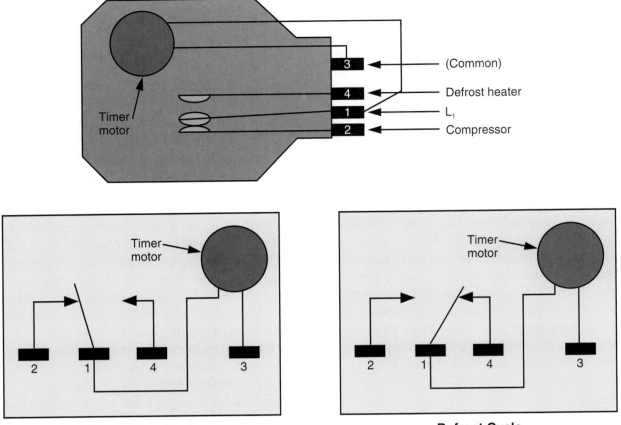

Normal Run Cycle   **Defrost Cycle**

Figure 27-5. *The timer causes the switches to change position for the defrost cycle. As shown, the switch is normally closed between Terminals 1 and 2 to supply power to the compressor. The defrost cycle causes the contact to open between Terminals 1 and 2, and close between Terminals 1 and 4. The power supply to the compressor is cut off, and the defrost heater is energized.*

clock. The timeclock, condensing unit, and evaporator fan can operate only during a run cycle when the thermostat is closed. The defrost cycle is based upon accumulated running time.

The newer wiring connections result in some energy savings because the clock and the evaporator fan are also stopped during the off cycle. The thermostat determines running time (on cycle) and is located inside the refrigerator section. The refrigerator door is opened more frequently than the freezer door, allowing heat to enter and causing the thermostat switch to close. The refrigerator section normally receives its cold air from an evaporator inside the freezer section. A small air duct controls air flowing from the freezer to the refrigerator section.

Defrost water disposal. All automatic defrost systems require a ***condensate pan*** under the evaporator to catch the resulting water. Water travels from the condensate pan through a drain line to a catch pan located in the compressor compartment. Sometimes the condensate pan is heated during defrost to keep the water traveling down the drain instead of freezing in the pan.

On domestic systems, the condensate water is drained to a removable catch pan within the compressor compartment. The catch pan rests upon a metal plate heated by the hot gas discharge line. The hot gas discharge line is routed beneath the metal plate before traveling to the condenser. The heated plate causes defrost water to completely evaporate before another defrost cycle begins. The process serves two purposes: it disposes of water and removes superheat from the hot gas as it travels beneath the plate.

The condensate drain line can become plugged with debris or algae that grows in damp, cold climates. A plugged drain line causes the condensate pan to overflow and water to accumulate and freeze. Once a year the drain line should be cleaned by pouring two cups of hot (not boiling) water through it.

Troubleshooting. A severely frosted evaporator, called a ***freeze-up,*** signals trouble with an electric defrost system. A freeze-up causes a sharp rise in cabinet temperature. A customer might describe the condition in these terms: "The food in the freezer section is thawing, the milk is warm, and the unit runs constantly."

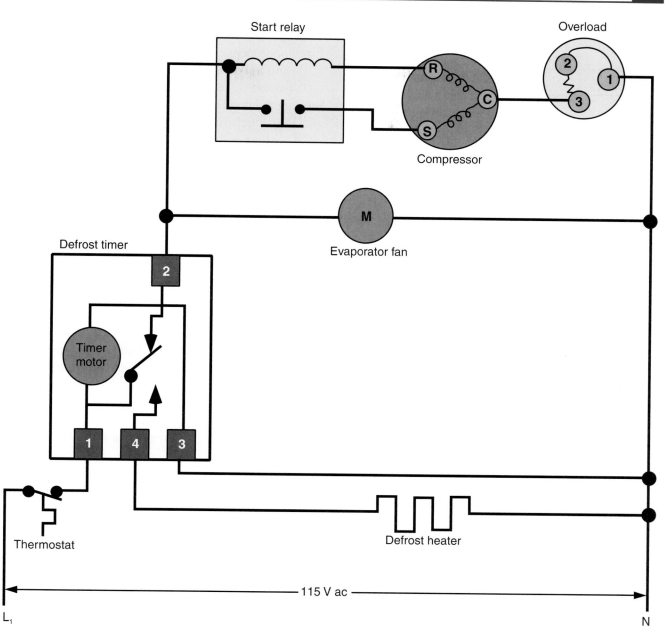

Figure 27-6. *Typical schematic showing how the timeclock controls operation of a refrigeration system. The circuit to the motor-compressor and evaporator fan is through a normally closed switch. The defrost heater is connected to the normally open switch.*

One of four malfunctions can cause a freeze-up:
- Defective evaporator fan.
- Defective timeclock.
- Burned-out electric defrost heater.
- Defective defrost termination thermostat.

A defective evaporator fan is not unusual, and a defective timeclock is rather common, too. A defrost heater occasionally burns out, but the defrost terminator rarely fails.

Troubleshooting procedure. Remove the cover panel to gain access to the evaporator. The evaporator fan should operate whenever the compressor is running. If the fan is operating properly, use a screwdriver to switch the timeclock into defrost. The

compressor and the evaporator fan should both stop, and the defrost heater should be energized. Place a clamp-on ammeter around the wire supplying power to the defrost heater to check for a reading of 6 A to 10 A.

If the ammeter reading is in the 6 A to 10 A range, allow the system to defrost. The problem is a defective timeclock (it cannot switch from run to defrost).

The timeclock can be checked with an ohmmeter. First, disconnect the power, and remove the timeclock from the unit. Resistance through the motor should be about 2400 ohms. Check continuity through the switches *while* rotating the cam with a screwdriver. See **Figure 27-9.**

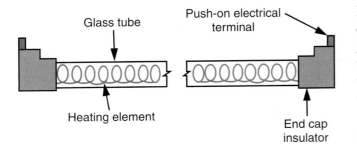

Figure 27-7. *Cutaway view of a defrost heater used in domestic applications.*

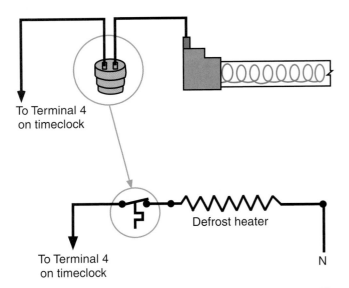

Figure 27-8. *A defrost terminator thermostat is connected in series with the timeclock and the defrost heater.*

There should *never* be continuity between Terminals 2 and 4. If continuity occurs, the defrost heater is being energized during the run cycle; more heat is being added than the system can remove. The defrost heater will cycle on the terminator thermostat, allowing temperatures to return to normal until the terminator thermostat closes again. These symptoms may be incorrectly diagnosed as moisture restriction in the capillary tube or a defective compressor.

Visual inspection of the defrost heater usually uncovers a burnout or wiring problem. An ohmmeter can be used to check proper resistance through each heater. The appliance schematic normally shows the resistance value for each heater.

Domestic hot gas defrost cycle

The *hot gas defrost* cycle is very different from the electric defrost cycle. Gas is taken directly from the hot gas discharge line (close to the compressor) and delivered to the evaporator inlet through a hot gas bypass line. See **Figure 27-10.** The 1/4″ copper tube is factory-installed to bypass the condenser and refrigerant control. To control flow, a solenoid valve is

located in the hot gas bypass line within the condensing unit area. When the solenoid valve is energized and opens, hot gas takes the path of least resistance and travels directly to the evaporator inlet. When the solenoid valve closes, the hot gas is forced to travel through the condenser.

During defrost, the timeclock energizes the solenoid valve and stops the evaporator fan. The compressor continues to operate during defrost, producing hot gas. The hot gas quickly melts any frost accumulation as the gas travels through the evaporator and into the suction line.

Any system using hot gas defrost *must* protect against liquid floodback to the compressor, since it is possible for hot gas to condense in the cold evaporator and enter the suction line as liquid. Protection against liquid floodback is usually provided by a suction accumulator. On domestic systems, the suction accumulator is a chamber located on the evaporator outlet.

Wiring connections. For domestic hot gas defrost systems, the timer must control the solenoid and the evaporator fan. The clock must have a set of normally open contacts (for the solenoid) and a set of normally closed contacts (for the evaporator fan). The thermostat is connected in series with the power supply to the timer and the power supply to the compressor. The timer must *not* control the compressor. See **Figure 27-11.**

The timer operates on accumulated running time and cannot place the system in defrost unless the thermostat contacts are closed. This ensures the compressor is operating and producing hot gas when the clock calls for defrost. Frequency and duration of defrost is controlled by the timer (*time-initiated/time-terminated*). The required number and length of defrost cycles is determined by the appliance manufacturer, so the clock is nonadjustable. A defrost termination thermostat is not needed.

Troubleshooting. During defrost when using R-12, head pressure drops to about 70 psig (483 kPa) because the condenser and capillary tube are bypassed. Suction pressure rises according to evaporator temperature. Compressor amperage is reduced because the compressor is operating under a nearly "no load" condition. Defrost must occur rather quickly (5 to 10 minutes) because a prolonged defrost produces very little hot gas.

Problems with hot gas defrost systems usually can be traced to a defective timer or to a low level of refrigerant. Low head pressure (as a result of low ambient temperature) cannot produce sufficient hot gas for defrost, so the evaporator eventually freezes up. However, a freeze-up seldom occurs with *domestic* hot-gas-defrost systems since the appliance is inside, where the ambient temperature is typically about 70°F (21°C).

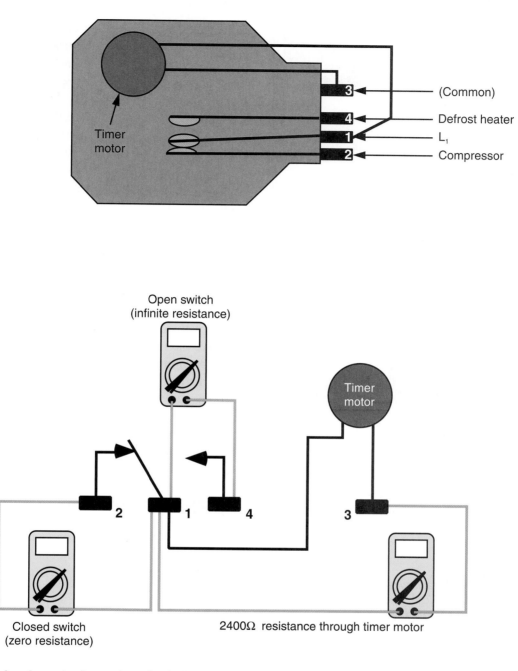

Figure 27-9. *An ohmmeter is used to check the defrost timeclock to determine whether it is defective. Before testing, disconnect all wiring from the timeclock.*

Domestic accessories

Domestic refrigerator-freezers have become quite sophisticated as manufacturers have added accessories such as ice makers and dispensers for milk, juice, or water. These devices are not controlled by the timer. Consult the appliance schematic when servicing units with such devices.

Switches that turn interior lights on and off are standard accessories for most refrigerator-freezers. Sometimes, the door switch for the freezer section controls the light and the evaporator fan. When the freezer door is opened, the light is turned on and the evaporator fan is turned off. Stopping the fan keeps warm, moist air from being drawn into the freezer. The lighting circuits are connected in parallel and do not affect other system components.

Refrigerator-freezers also contain resistance heaters inside the door frames. Resistance heaters prevent condensation of moisture around the doors during periods of high humidity. Door frame heaters are often called ***mullion heaters,*** or "anti-sweat" heaters. Sometimes a manual energy-saving switch is provided to turn off door heaters during the winter (low humidity) months.

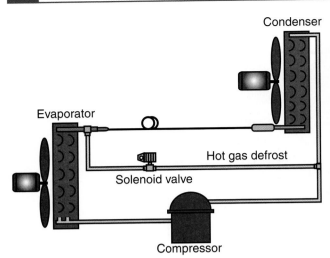

Figure 27-10. *A domestic hot gas defrost system directs hot gas from the compressor directly to the evaporator, where it quickly melts accumulated ice. The hot gas defrost (bypass) line should be connected as close to the compressor as possible.*

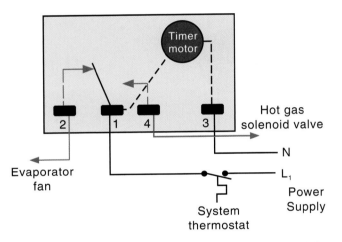

Figure 27-11. *The thermostat is wired in series with both the timer and the compressor. The timer must not control the compressor since the compressor provides the hot gas for defrosting the evaporator.*

A small heater is often placed inside the butter storage compartment. Called a "butter conditioner," it maintains the proper temperature to keep butter from getting too hard. These and other small resistance heaters are designed to operate continuously. **Figure 27-12** shows a complete pictorial diagram of a modern refrigerator-freezer. The same appliance is shown in schematic form in **Figure 27-13**.

27.2.4 Commercial Off-cycle Defrost

Off-cycle defrost is used on such light commercial systems as small walk-in coolers and self-contained cases. These systems normally operate with R-12 and maintain an air temperature between 38°F and 32°F (3°C and 0°C). A low-pressure control or an air-sensitive thermostat controls the air temperature.

The low-pressure control is set to cut-in at 38°F (35 psig) or 3°C (242 kPa); cut-out is set at 12°F (14 psig, less 2 psig for pressure drop) or 11°C (97 kPa).

During the run cycle, frost accumulates because the evaporator temperature is below 32°F (0°C). When air temperature is reduced to 32°F (0°C), the evaporator is 22°F (–6°C) and the refrigerant is 12°F (–11°C).

After cut-out, air temperature, evaporator temperature, and refrigerant temperature quickly equalize at 32°F (0°C). The evaporator fan runs continuously. As air temperature increases, heat is absorbed by frost melting off the evaporator. The latent heat of melting ice (144 Btu/lb.) helps maintain the cabinet temperature during the off cycle. The temperature must be warmed to 38°F at 35 psig (3°C at 242 kPa) before the low-pressure control (or thermostat) will reach the cut-in point. All frost will melt before the evaporator warms up to 38°F (3°C).

Occasionally, an off-cycle defrost that uses an air-sensitive thermostat has an excessively long run cycle. Frost accumulation becomes extremely heavy and stops airflow over the evaporator. The evaporator becomes colder and accumulates more frost. Air temperature rises sharply due to lack of heat transfer.

Such heavy frosting of the evaporator usually occurs during normal working hours when the door of the cooler or case is opened frequently. After several hours, frost accumulation is so thick that very little heat transfer is taking place. The system is unable to satisfy the air-sensitive thermostat. In such circumstances, defrost is accomplished by manually turning off the condensing unit for about two hours. The evaporator fan continues to operate and melt the frost accumulation.

27.2.5 Off-time Defrost

Automatic off-time defrost is accomplished by installing an inexpensive timer that turns the condensing unit off for two hours each evening. Typically, the timer is set to accomplish defrost between midnight and 2:00 a.m. After defrost, the system reduces air temperature in the cooler to the desired level before normal working hours.

Off-time defrost timer

A typical timeclock used for off-time defrost is shown in **Figure 27-14.** Commercial refrigeration defrost timeclocks operate on a 24-hour basis, the large dial face divided into two 12-hour sections. Half the dial face is black for easy identification of the 12 nighttime hours (6:00 p.m. to 6:00 a.m.). The dial face slowly rotates clockwise past the stationary pointer, which should indicate the correct time of day. The dial face can be rotated manually to reset the time.

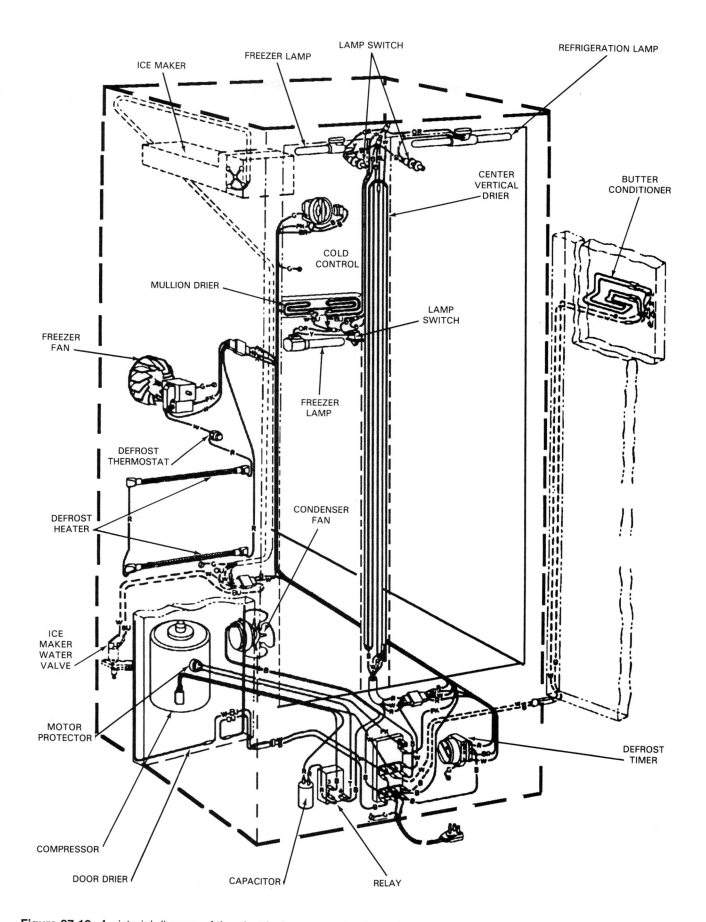

ICE MAKER

FREEZER LAMP

LAMP SWITCH

REFRIGERATION LAMP

CENTER VERTICAL DRIER

BUTTER CONDITIONER

COLD CONTROL

MULLION DRIER

LAMP SWITCH

FREEZER FAN

FREEZER LAMP

DEFROST THERMOSTAT

DEFROST HEATER

CONDENSER FAN

ICE MAKER WATER VALVE

MOTOR PROTECTOR

DEFROST TIMER

COMPRESSOR

DOOR DRIER

CAPACITOR

RELAY

Figure 27-12. *A pictorial diagram of the electrical components of a typical side-by-side refrigerator-freezer unit. (Frigidaire)*

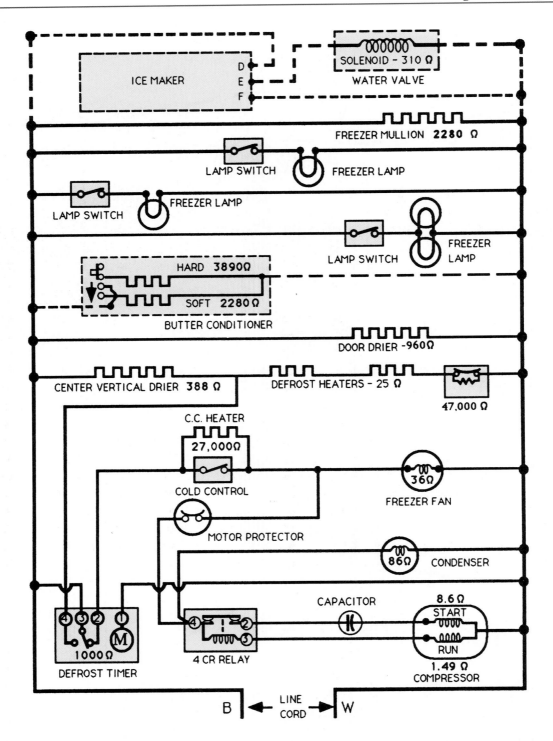

Figure 27-13. *A schematic diagram of the electrical components of a typical side-by-side refrigerator-freezer unit. (Frigidaire)*

Pairs of metal pointers (one light and one dark for each defrost cycle) attach with screws to the outer edge of the dial face. The light-colored pointer is anchored at the time the defrost should begin.

The dark pointer is anchored at the time defrost should end. The minimum defrost period is two hours. As the dial turns and the light-colored pointer reaches the stationary pointer, a lever is tripped, and a sharp "click" sound is made as the clock's two switches open.

The defrost cycle has started. The dial continues to turn until the dark-colored pointer reaches the fixed pointer. Another sharp "click" sound is heard as the two switches close, indicating the defrost cycle has ended.

A small lever at the lower left of the dial face permits manual operation of the switches. The technician must remember to manually terminate the defrost cycle by returning this lever to its proper position when finished.

Electrical connections

Most commercial timer switches are rated to carry 40 A per switch. The timeclock is available in either 120 V or 208 V to 240 V configurations for ease of matching voltages. The steel case is equipped with standard knockouts for attaching electrical box connectors. An electrical schematic drawing, like the one shown in **Figure 27-15,** is located inside the case to explain connections for the clock motor and switches.

One normally closed switch is located between Terminals 1 and 2, and the other normally closed switch is located between Terminals 3 and 4. The two power supply wires are connected to Terminals 1 and 4. This connection provides power to each switch, and the

factory-installed jumper bars transfer the power supply to each side of the clock motor at Terminals 5 and 6.

On 208 V to 240 V systems, both switches are used to disconnect the power supply to the condensing unit. On 120 V systems, the jumper bar between Terminals 4 and 6 is omitted by the factory since it is unsafe to install switches in the neutral (white) wire. The second switch is not needed on a 120 V system, so the neutral wire is connected to Terminal 6.

On commercial systems, power supply to the timeclock motor must never be disconnected. The thermostat or low-pressure control is connected in series with one side of the line between the timeclock and the condensing unit. See **Figure 27-16.**

Troubleshooting

The timer motor can be checked for proper rotation by noting the position of the pointer and waiting one-half hour to see if the dial has turned the proper distance on the scale. An ohmmeter can be used to check resistance of the timer motor (it should be between 1800 ohms and 2000 ohms). After disconnecting the power supply, remove the clock body from the metal case to visually inspect the clock parts and switches. The dial face can be rotated by hand to observe the switches opening and closing.

27.2.6 Commercial Electric Defrost

Electric defrost is popular for commercial refrigeration systems because the defrost period is short and easily controlled. A timeclock is used to *initiate* (enter) the defrost cycle according to time of day, **Figure 27-17.** Termination can be accomplished by time, evaporator temperature, or suction pressure. Each means of termination has advantages and

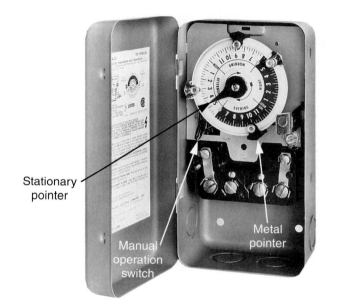

Stationary pointer

Manual operation switch

Metal pointer

Figure 27-14. *A commercial automatic off-time defrost timer. Metal pointers are attached to the dial face to start and stop the defrost cycle. This timer is set up for two defrost cycles in a 24-hour period: one begins at 3:00 p.m. and ends at 10:00 p.m.; the second begins at 5:00 a.m. and ends at 8:45 a.m. The manual lever can be used to start and stop a defrost cycle. (Paragon Electric Co.)*

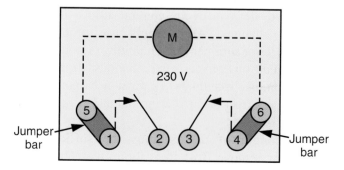

Figure 27-15. *Electrical schematic for the commercial automatic timeclock shown in Figure 27-14. The jumper bars are factory-installed to provide power to the timer motor from the power supply connected to Terminals 1 and 4.*

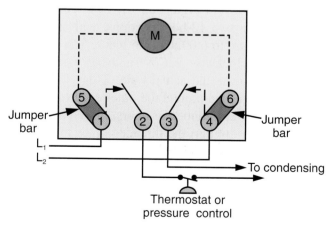

Figure 27-16. *The thermostat or low-pressure control is connected in series with one side of the line supplying power to the condensing unit. The timer controls both switches, opening or closing them at the same time.*

Figure 27-17. *Commercial electric defrost systems use timeclocks like this one to initiate the defrost cycle. Termination can be by time, temperature, or pressure. (Paragon Electric Co.)*

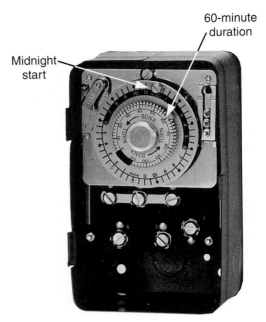

Midnight start

60-minute duration

Figure 27-18. *A timer used for time-initiated/time-terminated defrost. The timer setting provides one defrost cycle beginning at midnight (note pin on outer dial) and lasting 60 minutes (as indicated by pointer at 60 on notched inner dial). Several different cycles can be programmed, but all cycles must be the same length. (Paragon Electric Co.)*

limitations. Commercial defrost timeclocks can be separated into three types, depending upon the method of termination:

❏ Time-initiated/time-terminated.
❏ Time-initiated/temperature-terminated.
❏ Time-initiated/pressure-terminated.

Time-initiated/time-terminated

The *time-initiated/time-terminated clock* is used to place the system in defrost according to time of day and *terminate* the defrost cycle after a specified length of time. The clock contains a normally closed (NC) switch and a normally open (NO) switch for controlling the compressor, evaporator fan, and defrost heaters. As shown in **Figure 27-18,** the large dial face is divided into two 12-hour sections, with nighttime hours (6:00 p.m. to 6:00 a.m.) in black. The dial face turns counterclockwise (CCW) and has a stationary pointer to indicate the actual time of day. Holes are drilled and threaded around the outer edge of the dial face at all even-numbered hours. Metal pins are provided to screw into the threaded holes to indicate the time of day the system will enter a defrost cycle. The number of pins determines how many defrost cycles occur each day.

The large dial completes one rotation every 24 hours. When a pin nears the stationary pointer, it trips a lever and places the system in a defrost cycle. (The tripped lever opens the NC contact and closes the NO contact.)

The knob and small dial in the center of the large dial determine the duration (length of time) of the

defrost cycle. The small dial is calibrated in minutes and is sometimes referred to as the "fail-safe" dial. The outer edge of the dial contains notches calibrated at two-minute intervals (up to 110 minutes). To set the length of the defrost cycle, the pointer is depressed and moved to the desired number of minutes (the termination point). When the specified number of minutes has passed, the pointer trips the switches back to their original positions.

The knob attached to the smaller dial is used to manually advance both dials simultaneously. The knob is turned counterclockwise (CCW) to set the correct time of day, place the system in defrost, or take the system out of defrost.

Defrost requirements. The time-initiated/time-terminated defrost cycle cannot adjust itself to seasonal changes. Adjustments are required to compensate for summer (high humidity) versus winter (low humidity). Longer defrost periods are required during summer months when frost accumulation is heavier. Less frost occurs during fall and winter when humidity is lower.

The length of defrost must be sufficient to completely remove frost from the evaporator during any season. Partial defrosts cause a slow, steady accumulation of ice and result in a freeze-up. Longer, but less frequent, defrosts are preferred over frequent short defrosts.

Electrical connections. The timer just described is available in 120 V or 208 V to 240 V configuration, with contacts rated at 40 A. The power supply must be connected to Terminals N and X for continuous power supply to the clock motor. See **Figure 27-19.** A factory-installed jumper bar transfers the power supply from Terminal N to one side of the switches at Terminals 1 and 2. The switch between Terminals 1 and 3 is normally open, and the switch between Terminals 2 and 4 is normally closed.

Power supply to the defrost heaters is obtained at Terminal 3 (NO); the other side of the heaters is connected at Terminal X (Common). The compressor and evaporator fan are connected to Terminals 4 (NC) and X. Each switch is controlled by one side of the power supply. Terminal X is called "common" because all components use that terminal to obtain the other side of the power supply. See **Figure 27-20.**

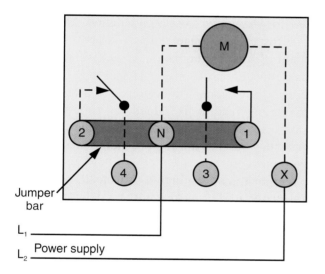

Figure 27-19. *Power supply to the timer is connected to Terminals N and X. A factory-installed jumper bar connects Terminals N, 1, and 2.*

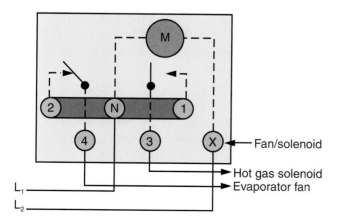

Figure 27-20. *Terminal X is the common terminal for the timer since it serves as one side of the power supply for all the timer-controlled components of the system.*

In commercial systems, power supply to the time-clock is never interrupted. The primary motor control (thermostat or pressure control) must be connected in series between the clock (NC) contacts and the compressor. On three-phase electrical systems, Terminal 4 (NC contact) is used to control power supply to the contactor coil.

Troubleshooting. Trouble with the timeclock usually results from improper defrost time settings. Complete failure may be due to a burned-out timer motor or burned-out contacts on the switches. Resistance of the timer motor is about 2800 ohms. Proper operation of the motor and gears can be checked by using a pencil to mark the position of the dials, waiting about one-half hour, then rechecking the dials for proper rotation.

The entire timer body can be removed from the case for visual inspection. After disconnecting the power supply, remove the clock from its case. Manually rotate the dials by turning the knob counterclockwise and observing the switches for proper operation. Visual inspection will usually reveal any malfunction.

Field repairs to the timeclock and switches are not recommended; the devices should be replaced. A new clock is far less costly than the loss of product and sales due to continued defrost problems.

Evaporator fan delay thermostat

On low-temperature systems (freezers), the evaporator fan must not be permitted to restart immediately after defrost. The evaporator becomes very warm, about 55°F to 60°F (13°C to 16°C) during defrost. If the fan is restarted as soon as the defrost ends, the fan will blow evaporator heat into the storage area. The heat will cause moisture to condense on the ceiling panels, drip down onto the product, and freeze. An **evaporator fan delay thermostat** is used to prevent restarting of the fan until evaporator temperature drops to about 5°F (–15°C).

The fan delay thermostat has a bimetallic disc with normally closed contacts that open when the evaporator temperature rises to about 20°F (–7°C) and close when the temperature falls to about 5°F (–15°C). See **Figure 27-21.** The fan control setting (such as F-20 or F-38) is usually stamped on the control body. "F" indicates fan control, and the number indicates the temperature at which the contacts open. This thermostat is connected in series with one side of the power supply to the evaporator fan. The thermostat is mounted or clamped directly to the evaporator on the return bends.

Troubleshooting. Failure of the fan delay thermostat is rather rare. When the control *does* fail, however, it is usually in the open position due to burned-out contacts. The evaporator fan no longer can operate, and a freeze-up occurs. To determine if the control has failed, be certain the control is cold enough for the thermostat contacts to close; then, jumper the control to "make" (close) the switch. If the fan operates, the control is defective and should be replaced. If the fan does not operate after jumping the control, the switch is already closed.

If the fan delay thermostat fails in the closed position, the timeclock will control the fan, and the fan will operate immediately after defrost (no delay). Excessive heat and moisture following a defrost cycle indicate thermostat failure in the closed position. The assumption can be tested by heating the fan control sensor with warm water to the cut-out point.

Time-initiated/temperature-terminated

Time-terminated defrosts are difficult to program because of such factors as seasonal changes, number of door openings, restocking of product, humidity, and temperature. The programmed length of defrost must allow for *maximum* frost accumulation. Many times, the defrost cycle is longer than necessary for the actual amount of frost accumulation. Excessive defrost time causes moisture problems, adds heat load, and wastes energy. To overcome the discrepancy, a temperature-termination feature is added to time-terminated systems to stop the defrost when the evaporator temperature reaches a level that indicates all frost has been removed.

Temperature-terminated timeclock. A *temperature-terminated timeclock* initiates defrost according to time of day, but termination is controlled by a thermostat mounted on the evaporator. The termination thermostat senses evaporator temperature. This defrost system has the advantage of automatically adjusting to seasonal changes and maintaining a minimum defrost time, eliminating guesswork when setting the defrost length.

The timeclock has a small solenoid with a plunger connected to the clock's trip lever. See **Figure 27-22.** Internal wiring connections prevent the solenoid from being energized except during a defrost cycle. When the solenoid is energized by the termination thermostat, the plunger trips the lever and ends the defrost cycle, regardless of the time remaining on the small dial. The timeclock initiates the defrost cycle, but the thermostat and solenoid end the cycle when evaporator temperature indicates the frost has melted.

Uninterrupted power supply to the clock is connected to Terminals 1 and N because the timer motor is internally connected to these terminals. The factory-installed jumper bar transfers power supply from Terminal 1 to the normally closed contacts at Terminal 2. The normally open contact is located

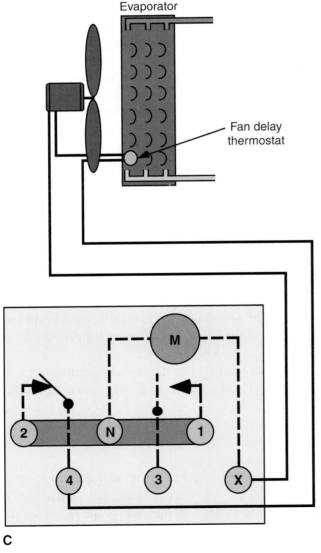

A

B

C

Figure 27-21. *The evaporator fan time-delay thermostat restricts fan operation until temperatures have dropped sufficiently to avoid condensation problems. A—Mounting of the thermostat on evaporator tubing. B—Schematic symbol. C—Wiring connections to the timeclock.*

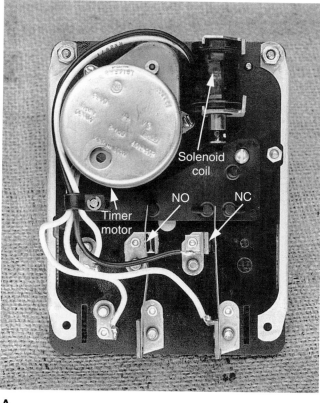

A

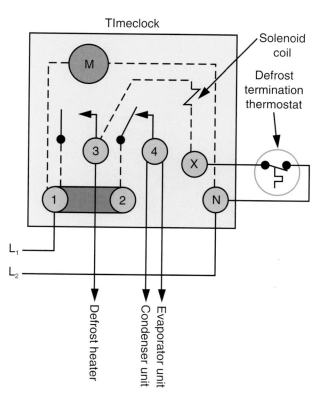

B

Figure 27-22. *Time-initiated/temperature-terminated time-clock. A—Rear view of the timeclock, showing the solenoid that operates the trip lever. (Paragon Electric Co.) B—Schematic showing the electrical connections. The thermostat closes on rising temperature.*

between Terminals 1 and 3, and the normally closed contact is located between Terminals 2 and 4.

One side of the power supply to the small solenoid is factory-connected to Terminal 3 (NO switch). This prevents the solenoid from being energized, except during a defrost cycle when the switch is closed. The other side of the solenoid is factory-connected to Terminal X. The termination thermostat is connected between X and N, completing the solenoid circuit to the Common.

Fail-safe feature. When temperature (or pressure) is used to terminate defrost, the small dial on the time-clock serves as a safety feature (the reason it is often referred to as a "fail-safe"). The **fail-safe setting** must allow sufficient time for the system to perform a complete defrost of maximum frost accumulation. If temperature or pressure termination components fail to operate properly, the fail-safe acts as a back-up feature and terminates the defrost cycle.

Defrost termination thermostat. The defrost termination thermostat contacts *close on rising temperature* (when evaporator temperature increases to the thermostat's setpoint). The setpoints (cut-in and cut-out) vary according to the application and the manufacturer's specification. On a "high limit" thermostat, the cut-in (L-50, L-55) is normally stamped on the control body. "L" stands for limit control, and the number indicates the temperature at which the contacts close. This type of thermostat is nonadjustable and frequently found on evaporators inside walk-in freezers, coolers, and display cases.

The *adjustable-type* defrost termination thermostat is commonly used for commercial display cases, **Figure 27-23.** The thermostat is normally located behind the kickrail (toeplate) of the case to facilitate making wiring connections and adjustments. The sensing element is attached to the evaporator, usually located in the bottom of the case. A capillary tube linkage transfers bulb temperature to the control body.

Troubleshooting. Termination problems can occur if the clock solenoid burns out, or if the termination thermostat fails or is not properly adjusted. The customer may complain, "The system gets too hot during defrost, and the product tends to thaw out but refreezes during the run cycle." Such a complaint is often heard during the first heat wave of summer.

The technician must first determine the type of defrost clock used. A time-initiated/time-terminated clock may need manual readjustment. A time-initiated/temperature-terminated clock may be having solenoid or thermostat problems. The solenoid may be burned out and physically unable to trip the lever to terminate defrost. Visual inspection of the solenoid usually reveals a burnout, but an ohmmeter reading of *infinite resistance* will confirm it. If the solenoid is burned out, and a replacement clock is not readily available, the fail-safe can be adjusted to operate the defrost in a time-terminated mode until a replacement is obtained.

If the clock solenoid is functioning properly, the termination thermostat can be checked by manually placing the clock in a defrost cycle. Jumper the thermostat connections across clock Terminals X and N. If the clock switches out of the defrost cycle, the thermostat is defective or improperly adjusted. If the clock *remains* in defrost, the clock itself is defective while the thermostat is functioning properly.

27.2.7 Combination Fan Delay and Defrost Termination

Sometimes the fan delay thermostat and defrost termination thermostat are combined into one component. The device is used for walk-in coolers and freezers where the evaporator and fan are readily accessible. The combination thermostat is easily recognized because it has three wires rather than two, **Figure 27-24.**

The three wires are marked C (for common), F (for fan), and T (for terminator). Power supply to the control is obtained from the timeclock at Terminal N and is connected to Terminal C on the control. The connection supplies power from the clock to the movable pole inside the control, **Figure 27-25.** The connection is *common* because it supplies power to both control switches.

A

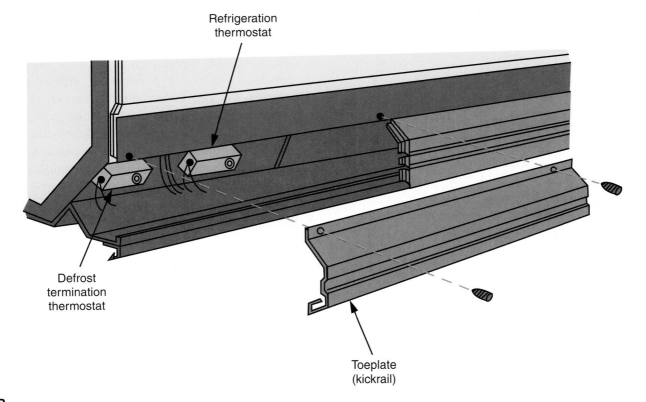

B

Figure 27-23. *Adjustable defrost termination thermostat. A—Thermostat and sensing element. (White-Rodgers) B—The thermostat is usually mounted behind the toeplate of the refrigerated display case.*

Figure 27-24. *A combination fan delay and defrost termination thermostat clamps to the evaporator tubing. (White-Rodgers)*

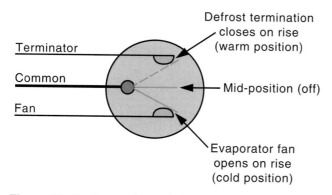

Figure 27-25. *A movable pole inside the combination fan delay and defrost termination thermostat permits the thermostat to perform both functions. During normal operation, the contact is in the fan position. In the defrost cycle, the pole moves to the mid-position as the temperature of the evaporator begins to rise. When evaporator temperature indicates all frost has melted, the pole moves to the defrost termination contact.*

Figure 27-26 illustrates a typical commercial defrost system using fan delay and defrost termination. One side of the power supply to the evaporator fan is connected to Terminal F on the thermostat and the other side to the normally closed switch on the timeclock. The connection places a switch in each wire supplying power to the evaporator fan. A wire from Terminal T of the thermostat is connected to the timeclock at Terminal X and serves as the defrost termination connection. The connection also places a switch in each wire supplying power to the small solenoid inside the timeclock.

During a normal run cycle, the evaporator fan operates because the timeclock switch is closed, and the fan delay switch is closed (since the evaporator is cold). Both switches controlling power supply to the solenoid inside the timeclock are open.

At the time of defrost, the clock stops the evaporator fan and condensing unit (the NC switch in the timeclock opens). The timeclock's NO switch closes and energizes the defrost heater. Closing this switch also connects one side of the power supply to the clock solenoid.

As evaporator temperature rises quickly, the movable pole in the combination thermostat warps to the mid-position where both control switches are open. The control remains in the mid-position until the evaporator temperature reaches 50°F to 55°F (10°C to 13°C). The temperature further warps the movable pole, closing the contact to the T position. This completes the circuit to the clock solenoid, terminating the defrost cycle.

The evaporator fan cannot restart until the evaporator temperature becomes cold enough to cause the movable pole of the thermostat to warp back to the F contact. This movement also opens the contact at T, disconnecting the solenoid.

Troubleshooting. Troubleshooting the combination thermostat is accomplished by checking each switch for proper operation. When checking the evaporator fan(s), remember there may be different reasons for non-operation: the fan delay thermostat may be functioning, or the fan may be defective (burned out). Another possibility is a defective door switch. Many walk-in freezers have a door switch that turns off the evaporator fans when the door is opened. Stopping the fans prevents warm moisture-laden air from being drawn into the freezer when the door is open. It also causes the evaporator temperature-pressure to drop rather quickly and the condensing unit to shut off on a low-pressure control. When the door is closed and fans resume operating, pressure quickly rises, and the condensing unit is turned on again by the same low-pressure control.

Multiple-fan evaporators are easier to check than single-fan evaporators. Multiple evaporator fans are always connected in parallel because airflow across the evaporator is critical. One fan motor may burn out, but the others continue to operate. If one fan stops functioning, a partial freeze-up will occur at the location of the malfunction. Power and control to evaporator fans is proven when at least one fan is operating.

Time-initiated/pressure-terminated

The ***time-initiated/pressure-terminated time-clock*** is similar to the timeclocks discussed previously, except that termination is accomplished by rising pressure in the suction line. See **Figure 27-27.** The clock is located at the compressor, with a 1/4″ copper tube linking the clock diaphragm to suction pressure at the compressor. During defrost, suction pressure rises. The rise in pressure moves a diaphragm mechanically linked to the timer lever. The appropriate rise in suction pressure causes the defrost cycle to terminate.

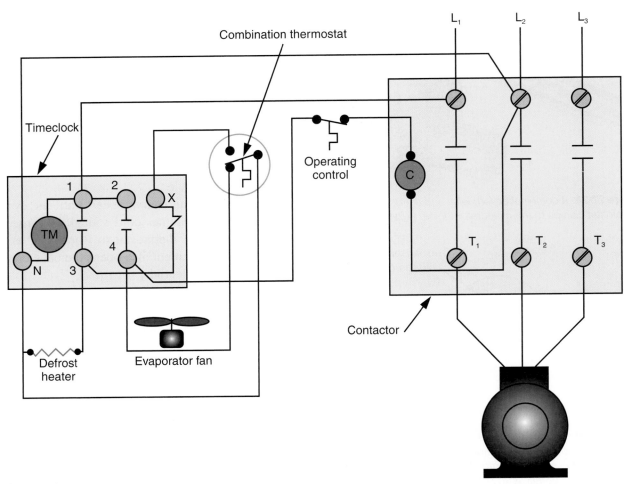

Figure 27-26. *Schematic for a system using a combination fan delay and defrost termination thermostat.*

Connection to
suction line pressure

Pressure
adjustment

Figure 27-27. *A time-initiated/pressure-terminated time-clock. (Paragon Electric Co.)*

The clock is used to turn the compressor off, energize electric defrost heaters, and stop evaporator fan(s). During defrost, suction pressure reflects evaporator temperature according to the temperature-pressure chart. When suction pressure rises to the clock's setpoint, the diaphragm trips the clock lever and terminates the defrost cycle.

The pressure required to trip the clock lever is adjustable. An adjustment screw (and scale) are provided on the clock, as shown in Figure 27-27. The scale is calibrated from 36 psig to 110 psig (248 kPa to 759 kPa), so the same clock can be used for R-12, R-22, or R-502. Termination pressure settings for R-12 can be from 38 psig to 45 psig (262 kPa to 310 kPa), resulting in an evaporator temperature of 35°F to 40°F (2°C to 4°C). The pressure setting for R-22 is 85 psig (586 kPa), resulting in an evaporator temperature of about 50°F. The termination pressure setting for R-502 is 95 psig (655 kPa), resulting in an evaporator temperature of 49°F (9°C).

Frequency of defrost varies according to the system, but the fail-safe should be set for 45 to 60 minutes. This recommendation is approximate, but the setting should result in defrost being terminated by *pressure* rather than time.

Pressure-terminated clocks cannot be used on pumpdown cycles because refrigerant does not remain in the low-pressure side during the off (defrost) cycle.

Troubleshooting. Power supply to the clock should always be checked first. Checking for proper rotation of timer dials is standard procedure. Mark the location of the dials, and check rotation against elapsed time. The clock can be removed from the case for visual inspection, if desired.

To check the accuracy of the calibrated dial (termination pressure), install a gauge manifold on the compressor service valves. Turn the clock dials to place the system in the defrost cycle. Slowly bleed some high-pressure gas through the gauge manifold and into the low-pressure side to raise the suction pressure to the termination point. Close manifold valves, and observe the suction pressure while listening for a clear "click" as the timeclock terminates. Always double-check the setting to be certain the clock is operating consistently, not erratically.

27.2.8 Commercial Hot Gas Defrost

Commercial hot gas defrost systems are rather common, and their operation is similar to the domestic systems described earlier in this chapter. Hot gas is taken directly from the hot gas discharge line (close to the compressor) and delivered directly to the evaporator inlet. See **Figure 27-28.** A solenoid valve is located in the hot gas bypass line to control flow.

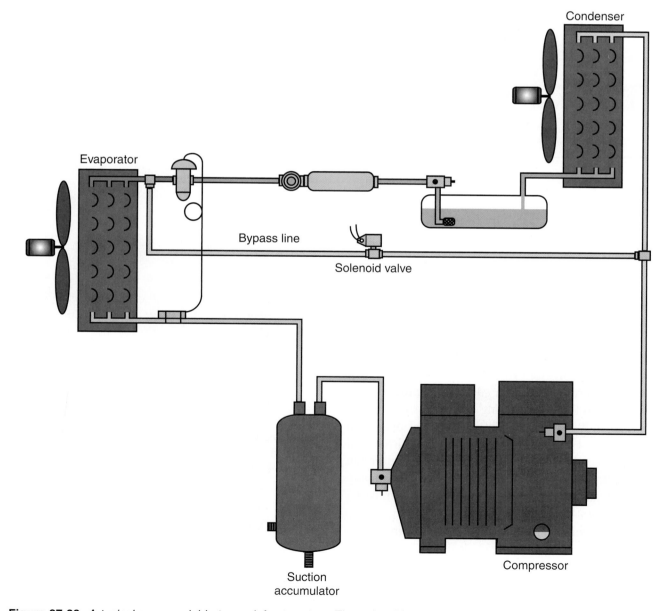

Figure 27-28. *A typical commercial hot gas defrost system. The solenoid valve controls gas flow through the bypass line. A suction accumulator is required to prevent liquid floodback, which could damage the compressor.*

When the solenoid valve is open, hot gas travels directly to the evaporator inlet, bypassing the condenser, receiver, and refrigerant control. When the solenoid valve is closed, however, hot gas *must* follow its normal route through the condenser.

The defrost timer is used to control electrical circuits to the hot gas solenoid valve and evaporator fan(s). The clock does *not* control operation of the condensing unit since the compressor must operate to produce hot gas for the defrost cycle. See **Figure 27-29.** Complete defrost usually occurs within five minutes, with the fail-safe set for 10 to 20 minutes. Hot gas defrost systems may require more frequent cycles because the length of defrost is determined by the amount of hot gas available.

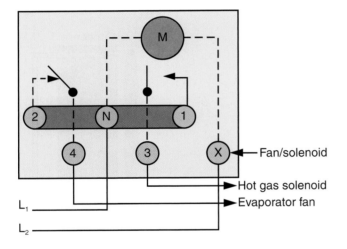

Figure 27-29. *In a hot gas defrost system, the timeclock does not control the condensing unit.*

Troubleshooting

The major disadvantage of hot gas defrost is that the hot gas may condense back to a liquid as it enters the cold evaporator. Liquid floodback could occur and damage the compressor. All commercial hot gas defrost systems are provided with a **suction accumulator** to prevent liquid floodback during defrost.

Proper hot gas defrost can be obtained only with sufficient head pressure. A condenser fan cycle control maintains head pressure and stops the condenser fan during defrost. Cold ambient temperatures drastically affect hot gas defrost systems, so provision must be made to control the head pressure.

High suction pressure during a run cycle may indicate the hot gas solenoid valve is leaking. The thermostatic expansion valve desuperheats the hot gas by permitting more liquid to enter the evaporator, resulting in high suction pressures and a prolonged run cycle. A leaking solenoid valve may produce a whistling sound, indicating gas is flowing through the

valve seat. A leaking valve permits gas to enter the evaporator during the off cycle, leading to short cycling of the compressor.

To detect a leaking valve, feel the temperature of the hot gas *defrost* line some distance away from the hot gas discharge line. Allow for heat from the hot gas discharge line traveling along the hot gas defrost line. Some solenoid valves have a manual shut-off valve stem for troubleshooting and isolation.

27.2.9 Defrost Contactors

The contacts on commercial timeclocks are normally rated at 40 A, but some electric defrost heaters require higher amperage. Timeclocks with heavier-duty contacts rated at 55 A per switch are available. The lug-type terminals accept wire sizes #4 to #14. See **Figure 27-30.** Such heavy-duty timeclocks are available in time-terminated, temperature-terminated, or pressure-terminated models. Operation is identical to the timers previously described.

Figure 27-30. *A time-initiated/temperature-terminated time-clock with contacts rated at 55 A. The lug-type terminals accept wire sizes #4 to #14. (Paragon Electric Co.)*

Three-phase systems often use a contactor to control the power supply to the heaters. The contactor is located inside the control panel at the condensing unit and is mounted beside the compressor contactor. The time-clock controls both the coil on the compressor contactor and the coil on the defrost contactor. The defrost heater circuit is fused separately. See **Figure 27-31.**

27.2.10 Multiplexing

Multiplexing describes multiple evaporators connected to one compressor, such as a row of supermarket display cases. Any number of display cases can be connected to one compressor, but each case must operate at the same temperature. For example, a row of five frozen food cases may be connected to one compressor, and another row of five frozen food cases may be connected to another compressor. Multiple dairy cases may be connected to one compressor, and multiple meat cases may be connected to another compressor.

Space is provided behind the kickrail (toeplate) of each case for installing one suction line and one liquid line from the condensing unit. As shown in **Figure 27-32,** each case is connected into the suction and liquid lines. Each evaporator has its own thermostatic expansion valve, fans, and defrost heaters. The setup is similar to having one very large evaporator connected to one compressor. To prevent liquid floodback, it is critical to have the correct superheat setting for each expansion valve.

The temperature of multiplexed evaporators is usually controlled by a low-pressure control at the condensing unit. Sometimes multiple cases are controlled by a single thermostat located in one of the middle cases. Placing the thermostat in a case at either end of the row is not recommended since an accurate temperature cannot be obtained.

Most display cases have both defrost heaters and special-purpose heaters. The defrost units are long heater rods with round aluminum fins that increase their surface area. Usually, more than one heater is used to accomplish quick defrost. Additional heaters are often used for special purposes. They include the drain pan heater and front flue heater, as shown in **Figure 27-33.**

A defrost termination thermostat is located in each case. The thermostats are wired in series so the defrost cycle cannot end until all thermostats are closed.

Electrical connections for multiplexed systems are accessible in the condensing unit control panel and in the kickrail under each case. **Figure 27-34** is a complete wiring diagram for a typical multiplexed system.

Troubleshooting

Troubleshooting a multiplexed system is easy: if all cases have the same problem, check the condensing unit; if only one case has a problem, the trouble is inside that case.

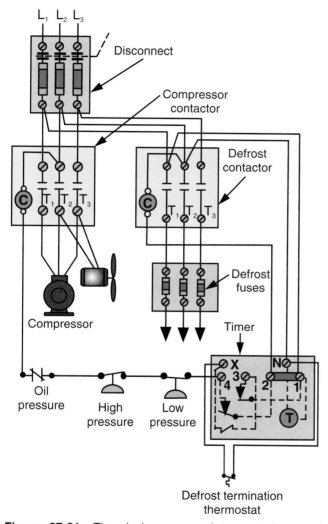

Figure 27-31. *Timeclocks are used to control several contactors. In this circuit, both the compressor contactor and the defrost contactor are controlled by the timer.*

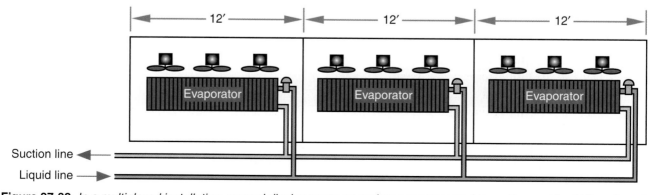

Figure 27-32. *In a multiplexed installation, several display cases or coolers are connected to suction and liquid lines from the same compressor. All cases must be the same temperature.*

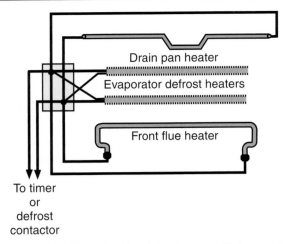

To timer
or
defrost
contactor

Figure 27-33. *The electric defrost on multiplex systems includes both defrost and special-purpose heaters.*

For example, a refrigerant leak affecting all the cases equally can be found at the condensing unit by checking the sight glass. On the other hand, if one case is suffering from high temperature, but the others are operating normally, the problem usually involves a plugged condensate drain causing a case full of ice. The drain must be opened with hot water and the ice melted manually before refrigeration is restored.

Electrical problems are easily identified at the unit control panel by using step-by-step electrical troubleshooting procedures. A safety control problem could prevent the system from operating. Such a problem may be traced to the oil pressure safety control (oil problem), the head pressure safety control (high head pressure), or the low pressure control (a

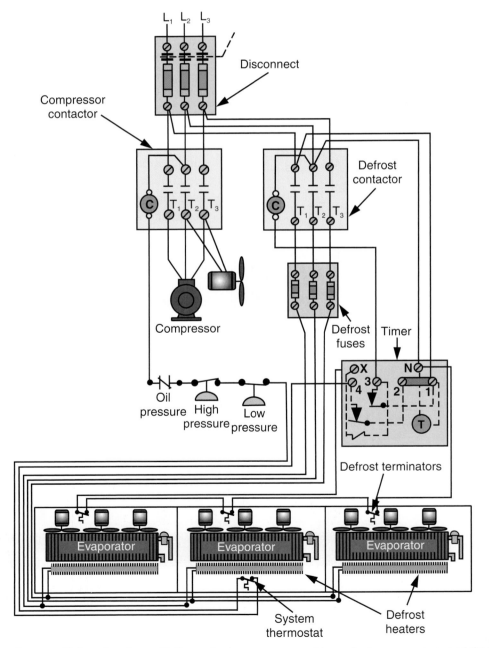

Figure 27-34. *A three-phase multiplexed system with three display cases served by a single compressor and timeclock. The system thermostat is located in the middle case for greater accuracy.*

refrigerant leak). Problems may also be traced to the timeclock (defrost problem), fuse (excess amperage flow), or compressor (cycling on the overload).

27.2.11 Reverse-air Defrost Cycle

The *reverse-air defrost cycle* has recently gained popularity in supermarkets because of its energy-saving feature. Warm store air is used to defrost refrigerated food display cases. During defrost, rotation of the evaporator fans is reversed, drawing warm store air into the case's air duct. The warm air is circulated through the evaporator and discharged back into the store. See **Figure 27-35.**

A time-initiated/temperature-terminated clock is used for reverse air defrost. The clock shuts off the condensing unit and energizes the coil of a *reversing relay* located on top of each display case. When energized, the relay reverses wiring connections to the evaporator fans and thus reverses rotation. See **Figure 27-36.**

The 120 V evaporator fans used for reverse-air defrost are unique because they have two sets of main windings and a neutral wire common to both windings. Connection of the "hot" wire determines rotation because one winding is clockwise (CW) and the other

is counterclockwise (CCW). Only one winding is energized during the run cycle; rotation is CCW. During defrost, the relay disconnects one winding and energizes the other, reversing rotation to CW.

Defrost ends when the defrost termination thermostat senses a temperature of about 50°F (10°C). With defrost terminated, the relay contacts switch back to their normal position. Fan rotation is changed to CCW and restores the normal airflow pattern. Defrost should be every eight hours (for example, 6:00 a.m., 2:00 p.m., and 10:00 p.m.), with the fail-safe set for 60 minutes.

◆ Summary

Various methods and components are used to remove frost accumulation from the evaporator of a refrigeration system. There are five basic types of defrost cycles. The type of defrost system used dictates the components required. Wiring connections and troubleshooting procedures also depend upon the type of defrost.

Various types of defrost timeclocks are used to initiate a defrost cycle based on time of day. Termination of the defrost cycle can be performed by time, temperature, or pressure. Defrost accessories are often included to make the system more efficient and safe.

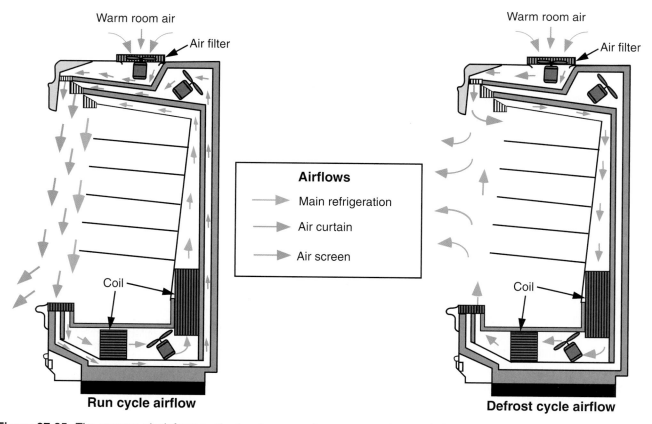

Figure 27-35. *The reverse-air defrost method makes use of warm room air to melt the frost off evaporator coils. A reversing relay changes the direction of rotation of the evaporator fans during the defrost cycle. The air screen and air curtain flows help prevent infiltration of warm air into the refrigerated space.*

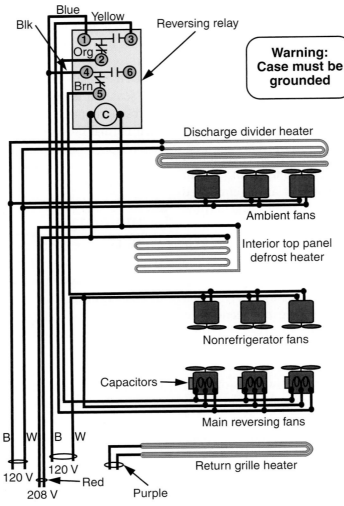

Blue
Yellow
Blk
Reversing relay
Org
Brn
C

**Warning:
Case must be
grounded**

Discharge divider heater

Ambient fans

Interior top panel
defrost heater

Nonrefrigerator fans

Capacitors

Main reversing fans

B W B W

Return grille heater

120 V 120 V
120 V ──── Red
208 V Purple

Figure 27-36. *As shown on this schematic, the defrost timeclock controls the reversing relay, which switches between CW and CCW windings of the special evaporator fan motors.*

Test Your Knowledge

Please do not write in this text. Write your answers on a separate sheet of paper.

1. Defrost cycles are necessary when evaporator temperature is below _____.
2. Name five types of defrost systems.
3. During a domestic electric defrost cycle, what components are turned off?
4. What does the timeclock control on a domestic hot gas defrost system?
5. During off-time defrost, what components are controlled by the clock?
6. Which defrost timeclocks operate continuously, commercial or domestic?
7. Name the three types of commercial defrost timeclocks.
8. Are pins on the timeclock face used to initiate or terminate the defrost cycle?
9. *True or false?* The small dial on the clock face is used to terminate the defrost cycle.
10. *True or false?* Short, frequent defrost periods are preferred over longer and less frequent ones.

11. Contacts on the fan delay thermostat _____ on a rise in temperature.
12. Contacts on the defrost termination thermostat _____ on a rise in temperature.
13. On commercial systems, the defrost termination thermostat controls the power supply to the _____.
14. How many wires must be used to properly connect the combination fan delay and defrost termination thermostat?
15. Why is the condenser fan stopped during hot gas defrost?
16. What is the purpose of the defrost contactor?
17. Where are the main suction and liquid lines located on multiplexed display cases?
18. On multiplexed evaporators, the defrost terminator thermostats are connected in _____.
19. On a reverse-air defrost system, which components are controlled by the timeclock?
20. What is unique about evaporator fans used in a reverse-air defrost system?

Ductwork

Objectives

After studying this chapter, you will be able to:
- ❏ Explain operation of the supply and return air duct systems.
- ❏ Assemble, join, install, and repair all types of duct.
- ❏ Install registers, diffusers, and grilles.
- ❏ Insulate and properly support piping.
- ❏ Explain why and how to use mastic and fiberglass membrane.
- ❏ Name and describe the use of many types of fittings.

Important Terms

A-coil	perimeter system
bonnet	plenum
branch lines	radial system
cfm	register boot
conventional system	registers
diffusers	return air
downflow furnace	scrim
drive cleats	S-hooks
ductboard	starting collar
extended plenum	supply air
fiberglass membrane	takeoff
forced air	trunk line
horizontal furnace	U-channels
insulated flexible duct	upflow furnace
mastic	vapor barrier

The information contained in this chapter provides the knowledge needed to working with heating and air conditioning ductwork. Some sizing information is presented as simplified "rules of thumb," consistent with practices in the field, rather than becoming involved with complex formulas. While the illustrations in this chapter show typical situations, keep in mind that adaptation to conform with state and local codes may be necessary.

28.1 Forced Air Systems

The majority of heating and cooling systems are classified as "forced air" types. Central heating and cooling involves the use of one furnace, centrally located in the structure, that is equipped with a motor driven fan to blow treated air through a ductwork system to the living spaces. Basic components of a **forced air** system are a furnace unit, outdoor condensing unit for air conditioning, ductwork, **Figure 28-1,** and registers and diffusers. The ductwork consists of separate "supply air" and "return air" sections. The **supply air** section is used to distribute treated air to living spaces, while the **return air** section brings air from the living spaces back to the furnace unit to be heated or cooled again.

28.1.1 Conventional System

Before the development of forced-air systems, gravity furnaces were used. These furnaces had to be located centrally, with short runs of ductwork from the furnace to the air outlets. All supply air outlets were located in or near inside walls of rooms. Return air inlets were generally located in or near the baseboards of outer walls. For a period of time after they were introduced, forced-air systems were similarly designed, and many existing houses have this type of system. Such an arrangement, with inner wall-located outlets and outer wall-located inlets, is called a **conventional system.**

Figure 28-1. *Installing an air duct system may be a part of your job in either new construction or remodeling work.*

28.1.2 Perimeter System

The *perimeter system* uses forced air and locates the supply air outlets in the baseboard along outside walls, or in floors or ceilings near outer walls **Figure 28-2.** The supply air "blankets" the wall against drafts (heating mode), or radiant heat during the cooling mode. This air circulates inward to a return air grille located in an inside wall, where it is drawn off and returned to the furnace for retreatment. The perimeter system provides

sufficient and uniform air circulation for maximum comfort.

28.2 Furnace Classifications

All forced air furnaces are classified according to the direction of the airflow for the supply air. The upflow and downflow furnaces are most common, but horizontal-flow models are necessary for some applications.

28.2.1 Upflow Furnace

The *upflow furnace* is normally located in a basement or closet, **Figure 28-3.** The supply air duct system is attached to the furnace *plenum* (sometimes called the *bonnet*), a sheet metal chamber anchored to the top of the furnace with sheet metal screws. The air conditioning evaporator coil is mounted on top of the furnace, but inside the plenum. The supply air duct is often insulated to help maintain the temperature of the treated air. Return air enters the bottom of the furnace. The blower serves for both heating and air conditioning mode, but airflow direction remains the same.

28.2.2 Downflow Furnace

A *downflow furnace* is usually used in a building having a crawl space under the first floor, or in a mobile

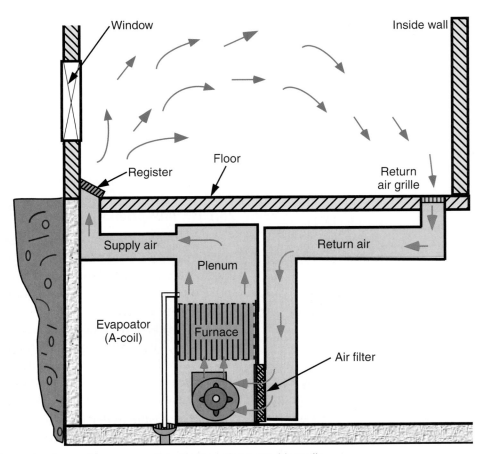

Figure 28-2. *The perimeter system uses outlets located along outside walls.*

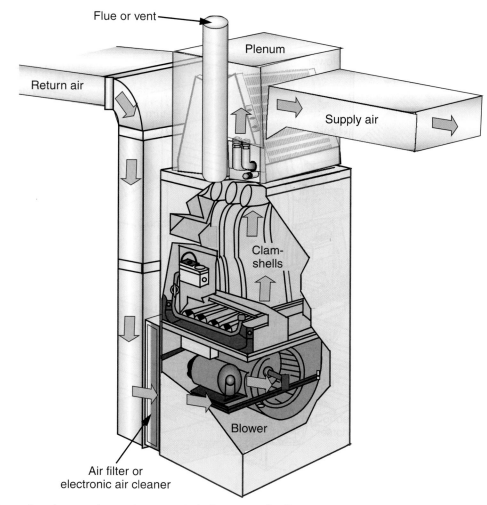

Figure 28-3. *An upflow furnace has a top-mounted plenum to distribute supply air. Upflow units are typically located in a basement.*

home, **Figure 28-4.** The air conditioning evaporator coil is mounted on the bottom of the furnace, inside the plenum. Supply air travels from the plenum through ductwork to the living spaces. Return air enters the top of the furnace. The blower serves for both heating and air conditioning.

28.2.3 Horizontal Furnace

This **horizontal furnace** is designed for use in a crawl space or attic, **Figure 28-5.** Supply air exits at one end and return air enters through the other end. This means the ductwork is located in the crawl space or in the attic, along with the furnace. These horizontal furnaces are usually electric, but gas, oil, or heat pump can be used.

Air Conditioning Evaporators

The **A-coil** is the most popular type of air conditioning evaporator for residential systems. The slant-coil and upright-coil types are primarily used for commercial and industrial systems. Commercial and many industrial air conditioning systems operate the

same as residential types. The components are simply bigger (and easier to repair). See **Figure 28-6.**

28.3 Distribution Systems ⬥

There are two types of duct systems commonly used for supply air distribution. They are the radial system, which uses strictly round pipe, and the extended plenum system, which combines round pipes with rectangular ducts.

28.3.1 Radial Piping System

A **radial system** is one that uses round pipe running directly from the plenum to the **register boot** (component that is used to connect the pipe to the outlet registers). There is a separate run from the furnace plenum to each register boot. These piping runs radiate from the furnace plenum, **Figure 28-7.** This system is practical where the piping runs are located in an attic or crawl space. The radial system is most economical and easiest to install, but is not practical if the furnace cannot be centrally located. Any

will carry treated air to an upper floor for distribution.

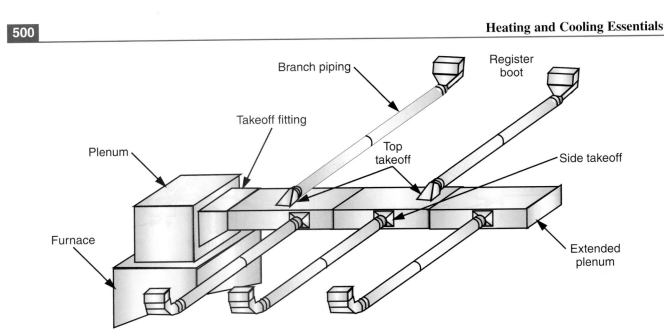

Figure 28-8. *Round branch piping extends outward from an extended plenum formed with rectangular duct.*

28.4 Distribution System Components ◆

The "supply air" distribution system includes the plenum, the main ***trunk line***, and all the branches serving different living spaces. Each branch terminates in a register that directs and usually controls the volume of the airflow. Another system of air duct serves to return air from the living spaces for heating or cooling. This system is called "return air," and includes another main trunk line and all its branches.

28.4.1 Plenum (Bonnet)

The first fitting to be installed on any system is the **plenum,** sometimes called a *bonnet,* **Figure 28-9.** It is usually made of 30-gauge, galvanized sheet metal, forming a chamber that is attached to the furnace outlet, **Figure 28-9A.** The plenum functions as an air-mixing chamber before sending the air to the living spaces. It also provides the necessary surface for attaching takeoffs and starting collars. The cooling

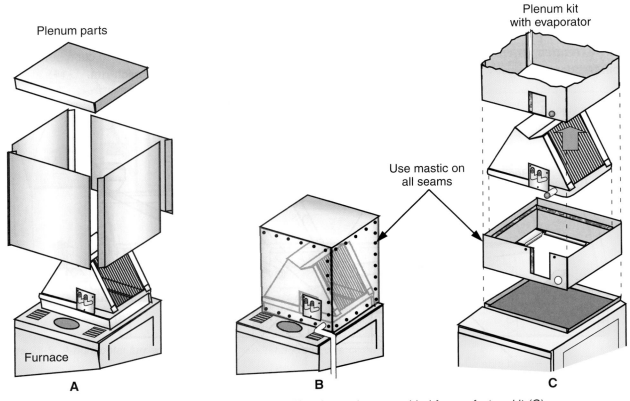

Figure 28-9. *A plenum can be fabricated in the field (A, B) or it may be assembled from a factory kit (C).*

evaporator is located inside the plenum, Figure 28-9B, and the supply air duct attaches to the plenum. The furnace manufacturer offers a plenum kit along with the necessary evaporator. For adding air conditioning to a previously installed furnace, some sheet metal work may be required. After all parts are assembled and screwed together, the seams must be anchored with sheet metal screws and sealed with **mastic** (a nontoxic adhesive material used for permanently sealing joints and seams). See Figure 28-9C.

Ductboard made from compressed fiberglass is becoming increasingly popular for plenums and rectangular duct. Ductboard is inexpensive, easy to work with, cuts with a knife, and is already insulated. Applying insulation by hand to metal ducts is often necessary, but is time-consuming and costly. See **Figure 28-10.**

28.4.2 Round Metal Pipe

Round pipe can be used as a main trunk line or a branch line, in both supply air and return air situations. Round metal pipe has the most efficient characteristics, and is commonly used. For heating only, the round pipe supplying air to a register outlet need only be 6″ in diameter. For air conditioning, or combined heating and cooling, the pipe should be 8″in diameter. Additional 6″ runs are often used to obtain required capacity. Standard round pipe comes in 30-gauge, 3″–14″ in diameter, and in 2′, 3′, 5′, and 10′ lengths. All round pipe is factory crimped on one end for easy insertion into another length of pipe, See **Figure 28-11.**

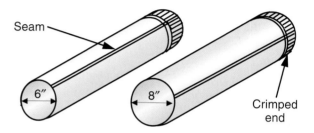

Figure 28-11. *Round pipe for branch lines can easily be snapped together at the seam. The crimped area at one end allows the pipe to be slipped inside the next section of pipe, forming a joint.*

28.4.3 Rectangular Duct

Rectangular duct is 30-gauge, galvanized sheet metal, and mainly used as a trunk line, or "extended plenum." Rectangular duct is widely used in the northeastern and northwestern parts of the United States. It is used as a spacesaver and can be easily routed in overhead space between floor joists, or in an attic. **Branch lines** consisting of round pipe come off the extended plenum to transport air to a register. A branch line begins with a starting collar or takeoff installed on the extended plenum, (main trunk line). The end of the extended plenum must be blocked with an end cap that is anchored with sheet metal screws and sealed with mastic. See **Figure 28-12.**

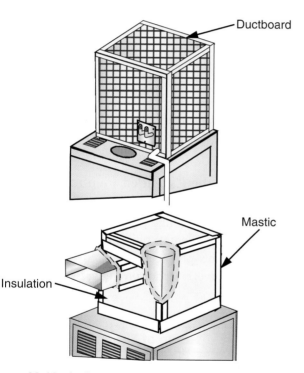

Figure 28-10. *A plenum can be fabricated from ductboard, which is less expensive than installing insulation by hand on a metal plenum.*

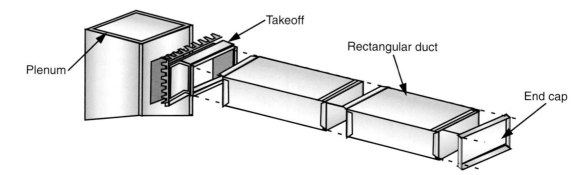

Figure 28-12. *An extended plenum, formed from rectangular duct, has a takeoff fitting at one end and a cap at the other.*

28.4.4 Insulated Flexible Duct

As shown in **Figure 28-13,** *insulated flexible duct* consists of several layers, each with a specific function. The *core* is usually two plies of polyester film used to encapsulate the galvanized steel wire helix. The purpose of the *wire helix* is to help flexible duct keep its shape. The *insulation* is a fiberglass blanket, varying in thickness, depending on the R-value. Many states have adopted energy codes for flexible duct that require certain R-values for code approval. A basic requirement is an R-value of 4.3, but Florida requires an R-6, and Washington and Oregon require R-8. *Scrim* is a high-tensile-strength, fiberglass mesh that provides protection for the insulation and inner core. The **vapor barrier** is a plastic jacket or sleeve that encloses the product.

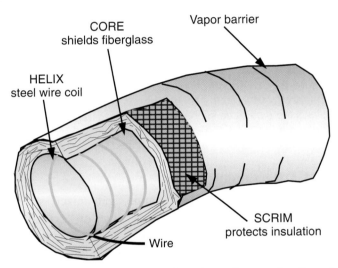

Figure 28-13. *This cutaway view shows the construction of a typical insulated flexible duct.*

Comparison of ductwork types

Various types of ductwork are available for use in a heating and air conditioning system. Because they were all designed for a particular application, they have different strengths and weaknesses. **Figure 28-14** is a list of some pros and cons of the major duct types.

28.5 System Efficiency

An efficient residential heating/cooling system will move air at the proper velocity and quantity to each room, and enter that room at the proper location. Velocity is important: air that is moving too fast creates uncomfortable drafts and makes more noise as it comes through the system. Air moving too slowly will not have enough speed to be thrown across the room or up the wall. See **Figure 28-15.**

Major Duct Types

	Pros	Cons
Round metal pipe	1. Excellent airflow characteristics 2. Longer-lasting than flexible duct	1. If required, insulation must be installed
Insulated flexible duct	1. Pre-insulated 2. Ease of installation 3. Reduced labor costs 4. Eliminates some fittings	1. Not as long-lasting as metal duct 2. Some reduction in airflow 3. Must be fully extended (cannot make sharp bends)
Rectangular duct	1. Spacesaver; can use space between floor studs 2. Longer-lasting than flexible duct	1. If required, insulation must be installed

Figure 28-14. *Advantages and disadvantages of three types of air duct.*

28.5.1 Sizing the Extended Plenum

The extended plenum is sized according to the number and sizes of round pipe it serves. It is easier to have an extended plenum the same width throughout, therefore, all the pipes are counted and the entire extended plenum is sized accordingly. Rectangular duct

Rules of Thumb

👍	Air should travel from the furnace/air conditioner throughout the system at an average velocity of about 700 ft./min.
👍	On average, most system resistance to airflow is approximately 0.08 (0.075) static pressure (0.10 Supply air, 0.05 return air).
👍	With an air velocity of about 700 ft./min., air conditioners are designed to move 400 CFM PER TON, and furnaces at 12 CFM/1000 Btu.
👍	In general, 1 CFM of air is required to heat or cool 1 or 1-1/4 sq. ft. of floor area.

Figure 28-15. *Regular use of these rules of thumb will help you solve forced-air system problems.*

used for extended plenums is 30-gauge, galvanized sheet metal, 8″ deep, 60″ long, and comes in standard widths of 10″, 14″, 18″, 24″ and 28″. One of these standard sizes should provide the needed capacity, but any length of duct can be cut to a shorter length.

28.5.2 Rule of Thumb for Determining Plenum Size

To determine the width required for an extended plenum, count the number of round pipes it will serve, **Figure 28-16.** For 6″ round pipes, multiply this number by 2, then add 2 — the total is the required duct width in inches. For 8″ round pipes, multiply by 3, then add 2. If the desired size is not available, choose the next larger size.

EXAMPLES:

A duct serving four 6″ round pipes will have a width of 4 × 2 = 8 + 2 = 10.

A duct serving five 8″ round pipes should have a width of 5 × 3 = 15 + 2 = 17. Since rectangular duct is not available in a 17″ width, use the next larger standard size, which is 18″.

28.5.3 Rule of Thumb for Determining Size of Branch Lines

1. Determine square footage of the room.
   Length of room, times width of room, equals square footage.

$$10' \times 12' = 120 \text{ sq. ft.}$$

2. Determine cfm of air needed to supply the room. cfm means "Cubic Feet per Minute" and defines the amount of air flowing.

   One cfm of air is required to heat or cool 1.00–1.25 sq. ft. of floor area.

EXAMPLE:

A 120 sq. ft. room will require between 96 and 120 cfm.

$$120 \div 1.00 = 120 \qquad 120 \div 1.25 = 96$$

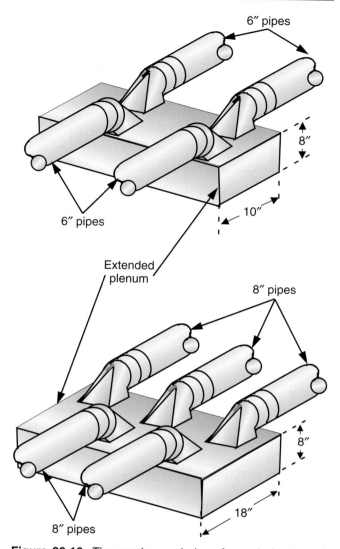

Figure 28-16. *The number and size of round pipe branch lines determines the width of the extended plenum.*

3. Select pipe(s) size according to cfm needed. How much air can a given diameter of pipe carry?

   A 120 sq. ft. room requiring 96–120 cfm will need one 6″ pipe (110 cfm), or use a combination to total about 96–120 cfm. See Rule of Thumb chart at bottom of page.

Square footage	64	80	100	120	150	200	256
Room size	(8 X 8)	(8 X 10)	(10 X 10)	(10 X 12)	(10 X 15)	(12.5 X 16)	(16 X 16)
cfm required	51-64	64-80	80-100	96-120	120-150	160-200	205-256

Rule of Thumb for Sizing Branch Lines.

Pipe size (diameter	Avg. cfm @ Static pressure
4″	30 cfm
5″	65 cfm
6″	110 cfm
7″	160 cfm
8″	230 cfm
10″	410 cfm
12″	680 cfm
14″	1000 cfm

Rule of Thumb for Sizing Branch Lines.

28.5.4 Provide Adequate Airflow

A furnace located in a small utility room or closet may not receive a sufficient amount of air for proper combustion. If the furnace room is closed off, provide an adequate air supply by installing two grilles in a door or wall facing a well-ventilated area, **Figure 28-17.** One grille should be in a high-wall position or at top of the door, and the other should be in a low position.

28.6 Ductwork Tools

Persons working with air duct commonly use the special hand tools shown in **Figure 28-18.** The name of the tool describes its use. Other basic hand tools are also needed. These might include a trouble light,

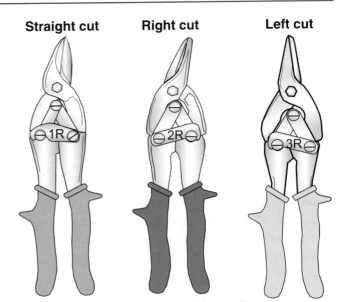

Straight cut   **Right cut**   **Left cut**

Note: Aviation snips are used for cutting holes in sheet metal. Never use them to cut wire.

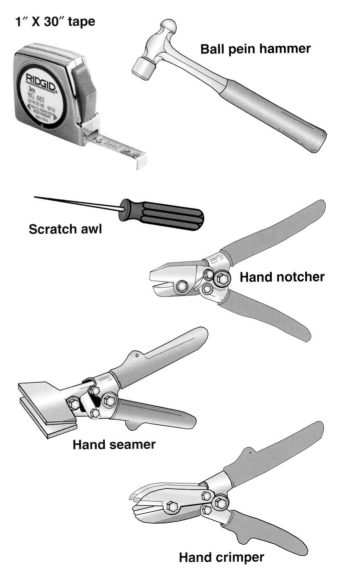

1″ X 30″ tape

Ball pein hammer

Scratch awl

Hand notcher

Hand seamer

Hand crimper

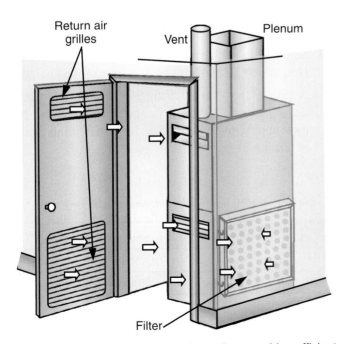

Return air grilles

Vent   Plenum

Filter

Figure 28-17. *Two grilles in a closet door provide sufficient airflow for furnace operation and return air.*

Figure 28-18. *Hand tools often are named according to the job they perform.*

battery operated drill, chisel, gloves, lineman's pliers, diagonal cutters, and longnose pliers. NOTE: *Aviation snips are used for cutting openings in sheet metal. Never use them to cut wire.*

28.7 How to Lay out a Duct System

The heating and cooling units should be matched to the square footage of the building's floor space. In addition, they should be able to distribute a set quantity of air (measured in cubic feet per minute, or **cfm**) to the living area through the duct system, **Figure 28-19.** A living space of 1200 sq. ft. is matched up with a 100,000 Btu furnace and a 3-ton air conditioner, Figure 28-19A. Come off the plenum with one or two main trunk lines having a total capacity of 1200 cfm, running toward the living areas, Figure 28-19B.

EXAMPLES:

Two 12″ round duct main trunks @ 620 cfm each = 1200 cfm

Three 10″ round duct main trunks @ 400 cfm each = 1200 cfm

With the main trunk lines installed, begin branching from the mains with smaller lines running to the various rooms. Always keep a constant rate of airflow (cfm) to the various rooms. A branch line must properly reduce in size to maintain a constant airflow from the trunk line. A velocimeter or anemometer, **Figure 28-20,** is often used to check and correct airflow problems. Such meters can read velocity (speed), cfm (flow), and temperature.

28.8 Installing Starting Collars and Takeoffs

Starting collars and takeoffs are installed on the plenum to begin a "run." A **starting collar** is used with round duct, while a **takeoff** is used with rectangular duct.

28.8.1 Starting Collars for Round Duct

It is best to mount the starting collar on the side of the plenum, 6″ below the top, to ensure proper mixing of the air. Never install a collar on top of the plenum. To install a starting collar, first mark the proper size opening with a pencil or metal scribe, **Figure 28-21.** Next, pierce the metal surface with a sharp tool and use snips to cut the opening. Then, insert the collar into the opening and fold the tabs down inside the plenum. Finally, secure the collar with three sheet metal screws and seal it with mastic.

Sq. Ft. of Floor Area	Furnace BTU's	Air Conditioner Tonnage	CFM of Air Movement
400	33,000	1 Ton	400
800	67,000	2 Ton	800
1,200	100,000	3 Ton	1,200
1,600	133,000	4 Ton	1,600
2,000	167,000	5 Ton	2,000

A

Dimensions	Approx. CFM	Sq. In.
4″ Round	30	12.57″
5″ Round	60	19.64″
6″ Round	100	28.27″
2¼″ X 10″	60	23.00″
2¼″ X 12″	70	27.00″
3¼″ X 10″	100	33.00″
3¼″ X 12″	120	39.00″
7″ Round	150	38.48″
3¼″ X 14″	140	46.00″
8″ Round	200	50.27″
8″ X 8″	260	64.00″
9″ Round	290	63.62″
10″ Round	400	78.54″
12″ X 8″	440	96.00″
12″ Round	620	113.09″
16″ X 8″	660	128.00″
14″ Round	930	153.93″
16″ Round	1300	201.06″

B

Figure 28-19. *A—Heating and cooling units matched to floor area. B—Approximate cfm of airflow (at a velocity of 700 ft./min.) for different-size ducts.*

28.8.2 Takeoff for Rectangular Duct

Manufactured takeoffs are available in many sizes, such as 8″× 8″, 12″× 8″, 14″× 8″, or 16″× 8″, and mount easily on any side of the plenum. It is best to mount the takeoff 6″ below the top of the plenum to ensure proper mixing of the air, **Figure 28-22.** When the furnace is located at one end of the house, the extended plenum may run to the opposite end. When the furnace is located at the center of the house, two takeoffs are used to provide an extended plenum on each side of the furnace.

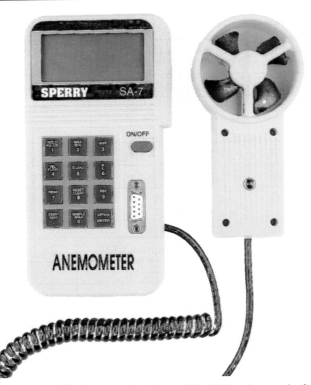

Figure 28-20. *Anemometer is used to obtain air speed, cfm, and temperature. (Courtesy of Sperry Instruments, Inc.)*

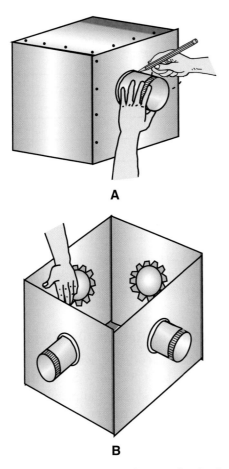

Figure 28-21. *A—Mark and cut the opening for the starting collar on plenum. B—Insert the collar and bend tabs over. Secure the collar with three metal screws and seal with mastic.*

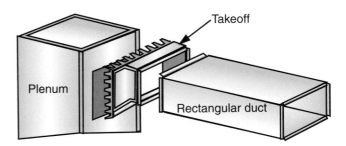

Figure 28-22. *Top of the takeoff should be located 6″ below the top of the plenum.*

To install the takeoff, first mark off the proper size opening. Next, pierce the metal surface using a drill or sharp tool and cut the opening with snips. Then, insert the takeoff into the opening, and fold the tabs down inside. Finally, secure the takeoff with three sheet metal screws and seal it with mastic.

28.9 Working with Round Metal Pipe

Standard round pipe is made of 30-gauge, galvanized metal, **Figure 28-23.** Round pipe is usually shipped nested, with the seams open, Figure 28-23A. Fittings are usually shipped preformed and ready to install. Each pipe length has an easily connected snap-lock seam. Round the pipe with your hands until the two seam edges meet. Insert the tongue on one seam edge into the groove on the other seam edge. Start at one end of the pipe and work down the seam toward the other end until the entire seam is snapped shut, Figure 28-23B.

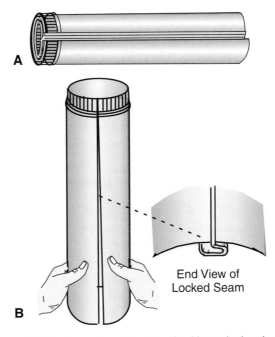

Figure 28-23. *Round metal pipe. A—Nested pipe lengths with open seams. B—Self-locking seam snaps together when compressed or squeezed with normal hand pressure.*

28.9.1 Joining Pipe Lengths

Each pipe length has a plain end and a crimped end. Lengths are joined by sliding the plain end of one length over the crimped end of another length. Line up seams for best appearance and push lengths together for a tight joint. Fittings are made similarly. The crimped ends always point in the direction of airflow. Secure each joint with three drive screws evenly spaced around the joint and seal with mastic. See **Figure 28-24.**

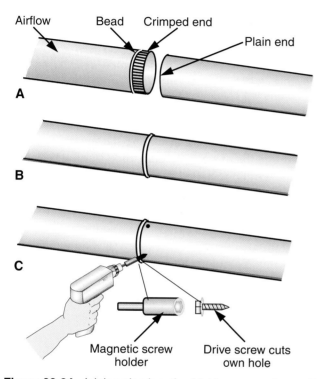

Figure 28-24. *Joining pipe lengths. Making connections with round pipe should be step-by-step. A—Each length has a crimped end and a plain end. B—The crimped end slides into the plain end of the next pipe until seated against the bead. C—Three drive screws should be used to secure each joint. Temporary wire supports are often used to hold pipe steady while connections are being made.*

28.9.2 Cutting Round Pipe

The last length of duct in a run usually must be cut, **Figure 28-25.** Do not snap the seam together until after the cut is made. Hold the last piece of unformed pipe in place and mark where it is to be cut. Your cutting mark should be on the plain (not crimped) end. Be sure to allow enough extra length for the cut end to slip over the crimped end of the preceding piece, up to the bead. Draw a straight cutting line across the pipe, then make the cut with tin snips. When you make the cut, the tin snips will squeeze the snap-lock seam edge closed. Before you can form the pipe, you will have to pry the seam open again, using a screwdriver.

If you need to cut a piece of pipe that is already formed, use a hacksaw. Make a continuous line around the pipe to follow while cutting. Support the round pipe between two wood blocks or other flat pieces, then saw with smooth, even strokes, beginning at the seam. Because of the hacksaw frame, you will only be able to saw part way through the pipe. You will need to rotate pipe and start again with saw on top. Continue in this manner until the cut is complete.

28.9.3 Adjustable Elbows

Adjustable elbows can be conformed to the desired angle. A 90°adjustable elbow can be twisted to form any angle from 0°–90°, **Figure 28-26.** A 45° adjustable elbow is also available. Insert the crimped end to move with the direction of airflow. Install the elbow and secure the joint with sheet metal screws, then seal it with fiberglass tape and mastic. Do not use ordinary duct tape.

28.9.4 Sealing Collars

Fiberglass reinforcing membrane will wrinkle when it is wrapped around a joint between round and rectangular ducts. To make the membrane material lie

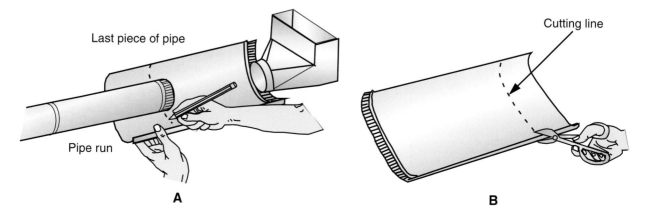

Figure 28-25. *In a branch run, the last piece must usually be cut to size. A—Mark the cut with a pencil or scratch awl on the plain end. B—Make the cut with tin snips or aviator snips.*

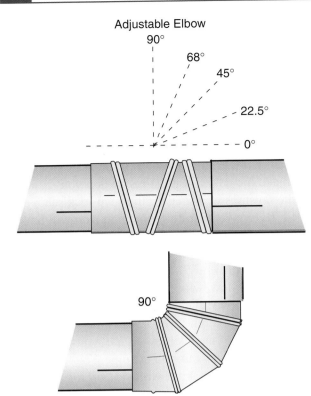

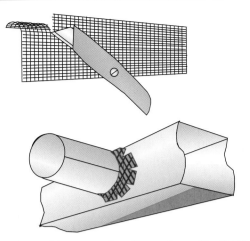

Figure 28-27. *After securing the collar with sheet metal screws, seal the collar with membrane and mastic. This prevents leaks due to vibration. Cutting slits in the membrane will help it lie flat.*

Figure 28-26. *Adjustable elbows are convenient for making all types of turns. They are available in all round duct sizes.*

flat, use a utility knife, to make a series of slices about every two inches along its edge. When the material is wrapped around the round duct, the tabs will fan out and lie flat against the rectangular duct, **Figure 28-27.**

28.10 Extended Plenum System ◆

Begin with the takeoff fitting in the plenum, and work outward. When a branch pipe comes off the top of the extended plenum, install the starting collar before assembling the rectangular duct. Side starting collars or takeoffs can be installed in the extended plenum after the run is completed. See **Figure 28-28.**

Support the duct at intervals with hangers or straps. Be sure to keep ducts at least one inch from joists and other combustible materials. Close the end with an end cap. After completing the extended plenum with starting collars or takeoff fittings attached, install branch runs to connect with outlets. Insulate all supply ducts and pipes that are located in unheated areas.

28.10.1 Working with Rectangular Duct

Most rectangular duct is made of 28-gauge galvanized sheet metal in 4′ lengths. Each duct length is made up of two half sections, as shown in **Figure 28-29.**

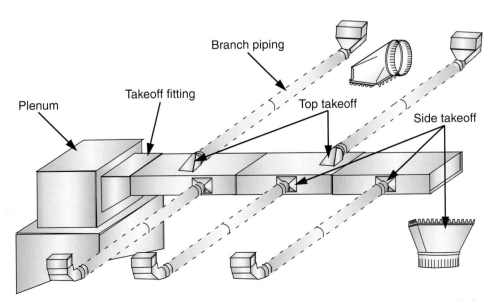

Figure 28-28. *An extended plenum system uses side takeoff fittings to connect branch lines. A centrally located furnace may have two extended plenums, one on each side.*

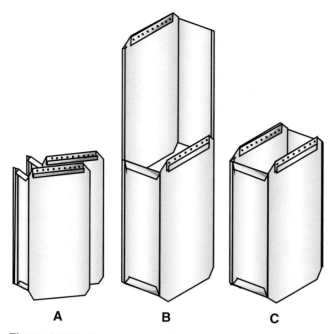

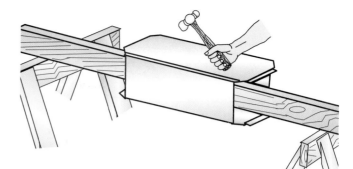

Figure 28-30. *Use a ball pein hammer on each seam for a tight fit.*

Figure 28-29. *How to assemble rectangular duct. A—Match the ends of the two half sections. B—Start the edge channels. C—Slide the two sections together.*

To form a length, place the two half sections on the floor with the "S-hook" ends up. Pick up one piece and engage its edge channels with the channels on the other piece, then slide the two together.

After assembling a section of duct, hammer each channel snug for a tight fit. This can be done as shown in **Figure 28-30.** Hang the assembled duct over a sturdy board suspended between two sawhorses so the seam lies flat on the board edge. Use a ball pein hammer to tighten the seam.

Rectangular duct is joined with "S-hooks" and drive cleats, **Figure 28-31.** One end of each section has formed **S-hooks** on the long edges and "U" channels for the **drive cleats** on the short edges. The other end has plain long edges, with **U-channels** on the short edges. Join pieces by inserting a plain end into the S-hooks of the next piece, as shown in Figure 28-31B. Slide drive cleats through the U-channels to lock the pieces together, Figure 28-31C. Drive cleats are 10″ long and used on 8″ rectangular duct. When the drive cleats are in place, bend ends down over duct to complete and secure the joint. Seal the joint with mastic.

Where a shorter length is required, cut the two pieces before forming the duct. Always cut the plain end, not the end with the S-hook connection. Measure carefully, then cut both pieces to the same length with tin snips, **Figure 28-32.** After cutting, use a sturdy screwdriver to pry the edge channels to their original form. Cut back the corners of the new end and use a hand seamer to form a new U-channel on the short edge of each piece.

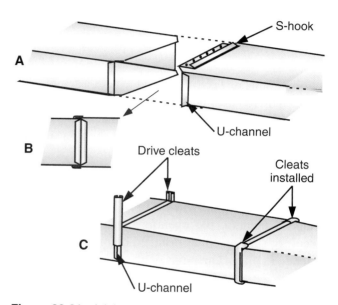

Figure 28-31. *Joining rectangular duct, using S-hooks and drive cleats.*

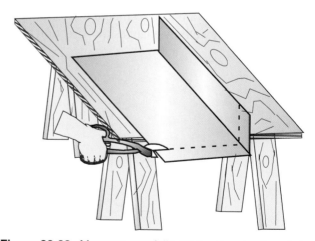

Figure 28-32. *Measure carefully, then use snips to cut duct pieces to size.*

28.10.2 Cutting Openings

To begin branch pipe runs, openings must be cut in the plenum, **Figure 28-33.** These openings may be rectangular or round, depending upon whether a

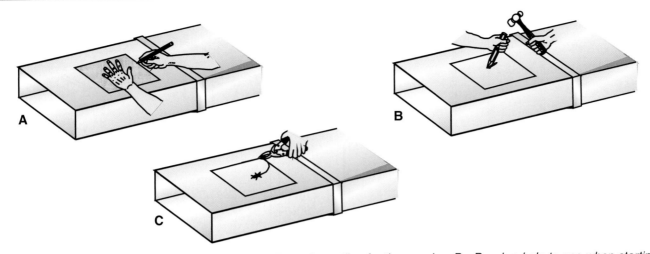

Figure 28-33. *Installing a top-mounted takeoff. A—Draw the outline for the opening. B—Punch a hole to use when starting the cutting snips. C—Cut around the outline.*

starting collar or takeoff is used. Make a template of the opening required and draw its outline accurately on the metal to be cut, Figure 28-33A. Punch a hole through the sheet metal within limits of the opening, using a hammer and chisel or similar tool, Figure 28-33B. Insert aviator snips and cut to the outline, then around the outline until the opening is complete, Figure 28-33C. When a branch pipe will be run from the top of an extended plenum, the opening must be cut and the takeoff or starting collar installed before the plenum is fastened in place. Openings on the sides of the plenum can be made at any time.

28.10.3 Rectangular 90° Elbows

Rectangular elbows are not adjustable, but are available in 90° "long-way" or 90° "short-way" forms, **Figure 28-34.** They are also available in 45° configurations. A long-way elbow changes the direction of the long dimension, and the short-way elbow changes the direction of the short dimension. To form any smaller angle needed, 45° and 90° rectangular elbows can be cut.

28.11 Duct Support

All duct should be run in a straight line and supported at regular intervals — typically, 2′–5′ — to prevent any sags. The best support material is woven polypropylene strap that comes in various widths. It is sometimes called "webbing strap" and is wide enough so it won't bite into flex duct. Perforated metal strap is a second choice, with steel wire being least desirable. Steel wire is still often used to support round pipe for branch runs. The wire should be wrapped once around the pipe before anchoring to a joist. See **Figure 28-35.**

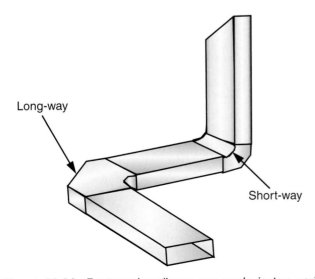

Figure 28-34. *Rectangular elbows are made in two varieties: long-way and short-way. They are not adjustable.*

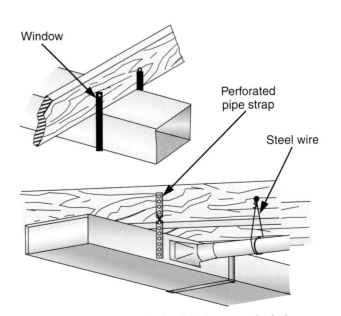

Figure 28-35. *Three methods of duct support include woven polypropylene strapping, perforated pipestrap, and wire.*

Keep pipes at least one inch from joists and other combustible material. To prevent heat loss, metal duct is often wrapped with insulation, **Figure 28-36.** If insulating material is applied after the duct is already in place, the insulation will have to be slit or stripped to get around the supporting strap or wire. This opening must be resealed to maintain insulating effectiveness.

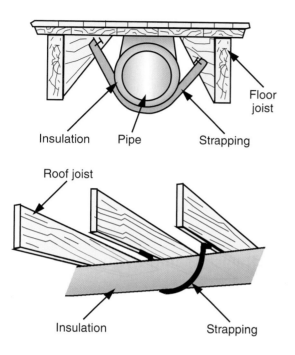

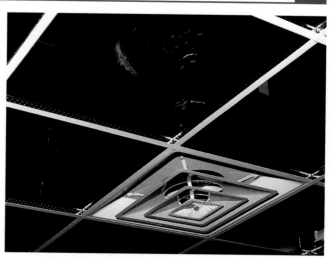

Figure 28-37. *Insulated flexible duct used for a branch line in a commercial application. The diffuser is sized to drop into the suspended ceiling grid.*

Figure 28-36. *Duct is often located in the basement and hidden between floor joists. When located in the attic, it is hung with strapping attached to roof joists. Insulation is often wrapped around the duct to prevent heat loss.*

Figure 28-38. *Different sizes and insulation thicknesses are available in insulated flexible duct.*

28.12 Working with Insulated Flexible Duct

Insulated flexible duct can be used for main trunk line runs; it also can be used as branching duct, **Figure 28-37.** This type of duct is available in a variety of sizes, insulation thickness, and specifications, **Figure 28-38.** Check local building codes for the exact R-value required in your area. To prevent reduced airflow, insulated flexible duct must be fully extended when installed. Avoid any abrupt turns. When a sharp turn is necessary, cut the flex duct at the turn and install a 90-degree elbow of the same diameter. Do not use wire to support flexible duct. Instead, use a wide fabric strapping material and anchor it to wood framing with staples.

28.12.1 Making Connections

Making connections with insulated flexible pipe is not difficult, but must be performed properly to obtain

correct airflow and prevent leaks. **Figure 28-39** shows the connection procedure step-by-step.

28.13 Wall Stack

Wall stacks, **Figure 28-40,** transport air vertically, from floor to floor. In residential applications, stacks carry air to high-wall or second-floor registers from a basement or crawl space. They can also be used to carry air to baseboard or low-wall registers from attic piping.

28.13.1 Installing Wall Stack

If possible, work from the basement. As each section is attached, push the stack up into the stud space and attach the next section, **Figure 28-41.** Continue until the top of the stack reaches the register box or fitting to which it will connect. Reduce the

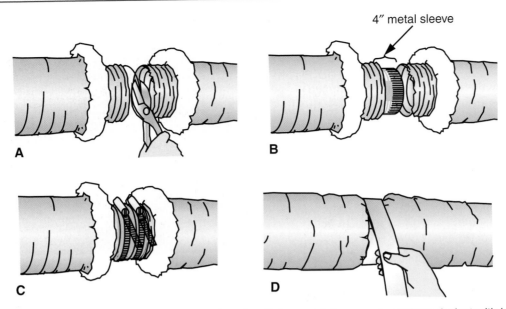

Figure 28-39. *Connecting lengths of insulated flexible duct. A—Cut completely around and through duct with knife or scissors. Cut wire helix with heavy-duty snips or side cutters. B—Peel back the jacket and insulation from core. Butt two cores together on a standard 4" metal sleeve . C—Tape core together with at least two wraps of duct tape. Secure with two approved clamps. D—Pull jacket and insulation back over cores. Tape jackets together with two wraps of tape, then mastic sealant.*

Figure 28-40. *Rectangular duct is used to provide vertical passages (wall stacks) that carry treated air from floor-to-floor of a building. This stack is being fabricated for use in an office building. Note: The worker should be wearing safety glasses.*

length of last section as required. When stack must be lowered from above, enlarge openings as necessary and lower stack into the stud space as it is assembled section by section.

To join two sections of stack (or "riser") duct together, "S" cleats can be used to secure the long sides of the duct. Sheet metal screws should be applied on the shorter sides of the ducting, and if needed, can be used on the longer dimension. The seams are then sealed with mastic. The same care should be taken when connecting with the stack heads or boots, **Figure 28-42.**

28.14 Registers and Diffusers

The final fitting in the duct system used to move air to its destination is the register box, or "boot." Due to the numerous locations and difficult areas registers must fit, boots are available in a variety of shapes and sizes, **Figure 28-43.** Supply air is released into a room

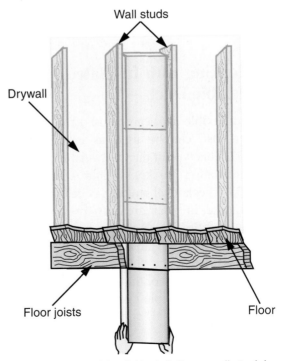

Figure 28-41. *In residential installations, wall stack is generally pushed up into the wall from the basement. When cutting openings for a stack, use a flashlight and mirror to check for obstructions between studs.*

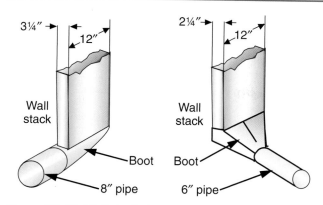

Figure 28-42. *Wall stack duct is designed to fit between the studs inside a wall. Different sizes include 2-1/4" or 3-1/4" deep by 10", 12" or 14" wide. Standard length of a stack section is 24".*

through registers or diffusers. Both have dampers or vanes for controlling the amount of air, but *diffusers* also have adjustable vanes for controlling the direction of discharged air. For most effective air movement, registers and diffusers are located in the baseboard on an outside wall, or in the floor near the baseboard.

28.14.1 Locating a Floor Diffuser

Locate the diffuser near an outside wall, but at sufficient distance from the wall to clear draperies or other obstructions. Draw an outline of the opening required, then drill a hole in the center of the planned opening. Measure from the hole to the sides of the opening, then go into the basement or crawl space and find the hole position from below. Measure the same distance from the hole to be sure it will clear floor joists or other obstructions. If not, the opening must be shifted. See **Figure 28-44.**

28.14.2 Installing a Floor Diffuser

A typical one-piece diffuser fits into a 12" x 4" rectangular opening in a register boot. After locating the proper position, cut a hole in the floor just large enough for the boot to fit snugly. The metal edges of the boot should be level with the floor. Secure the boot in position by nailing it to the floorboards. Insert the diffuser into the boot and anchor it with screws at each end. See **Figure 28-45.**

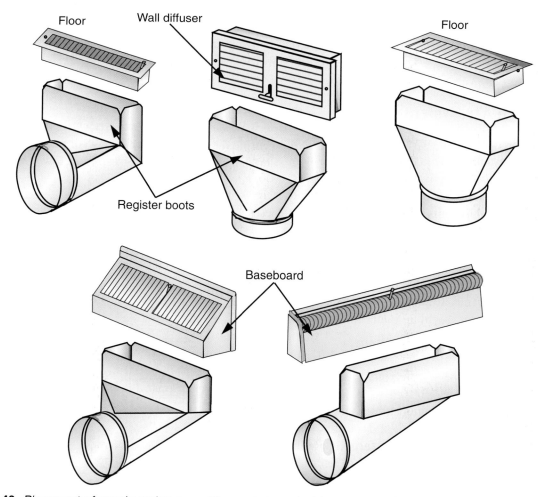

Figure 28-43. *Placement of supply registers or diffusers is important to ensure a pleasant room temperature. A variety of shapes and sizes are available to meet varying requirements.*

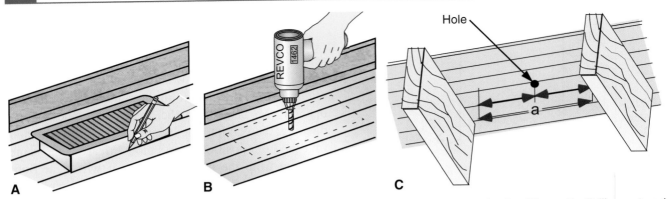

Figure 28-44. *Preparing to cut a floor diffuser opening. A—Measure and trace the opening for the diffuser. B—Drill a centered hole through floor. C—Check clearances from below.*

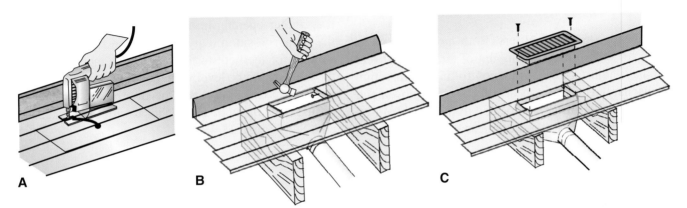

Figure 28-45. *Installing a floor diffuser. A—Cut along the outline with a jigsaw, circular, or reciprocating saw. B—Secure the register box (boot) to floor with nails. C—Install the floor diffuser into boot opening and secure it with screws.*

28.15 Return Air System

In houses where air can easily circulate to one location, a central return air grille is adequate. In two-story homes or long ranch designs, a more efficient, yet more expensive system routes air through separate piping from individual rooms back to the furnace for retreatment. The system filter or electronic air cleaner is normally installed where the return air duct attaches to the furnace, **Figure 28-46.**

A disposable filter is usually made of fiberglass within a cardboard frame, **Figure 28-47.** These filters should be replaced at regular intervals to avoid restricted airflow. An arrow printed on the filter frame shows direction of airflow.

28.15.1 Return Air Grilles

Return air grilles are nonadjustable, and are not equipped with dampers. They should permit unrestricted flow of return air back to the furnace. Return air grilles are available in various sizes. A return air grill is normally positioned on an inside wall, in a baseboard on an inside wall, or in the floor next to an inside wall, **Figure 28-48.**

Installing Return Air Grilles

When installing wall-type grilles, the floor plate inside the partition must be cut for passage of return air to the duct or enclosed joist space below. It is easier if the

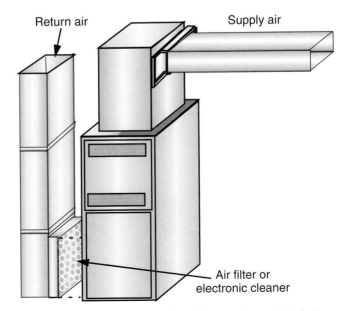

Figure 28-46. *For proper circulation, return air is just as important as supply air.*

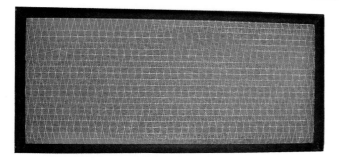

Figure 28-47. *A disposable air filter can be positioned within a centrally located return air grille.*

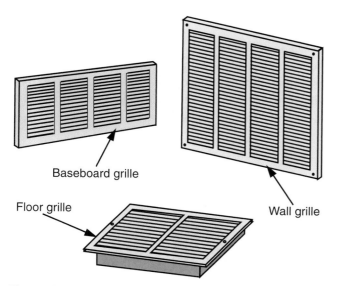

Baseboard grille

Floor grille

Wall grille

Figure 28-48. *Return air grilles may be located in floor, ceiling, wall, or baseboard.*

floor plate can be cut from below the floor. Using the stud space as a return-air duct requires sealing off the stud space just above the grille opening with a 2″ x 4″ wood block as shown in **Figure 28-49.**

Installing floor type or baseboard type grilles is much the same as installing supply air registers in those surfaces, **Figure 28-50.** Floor type and baseboard grilles require larger holes than diffusers.

28.15.2 Basement Return-Air System

With the return-air grilles installed and the return takeoff fittings attached to the return-air plenum, it is only necessary to complete the runs from the grilles to the furnace. This may be done through sheeted joist and stud spaces, direct ducts, or a combination of the two, **Figure 28-51.**

28.16 Insulation

Insulate all supply ducts and pipes installed in an attic, crawl space, or other unheated areas. Carefully wrap the insulating material around each pipe and secure with proper glue or by applying tape along the seam, **Figure 28-52.**

WARNING: To avoid a fire hazard, the warm-air plenum and branch ducts must be separated from wood joists, floors, and other structural elements by at least 1″ of clear space or insulation.

28.16.1 Leaks in Air Ducts

Over time, fabric-based duct tape grows brittle and the glue dries out. This permits sections of duct to

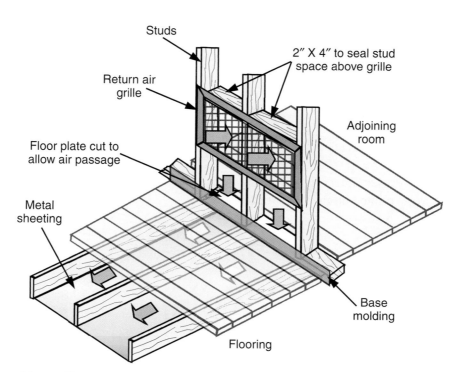

Studs

Return air grille

Floor plate cut to allow air passage

Metal sheeting

2″ X 4″ to seal stud space above grille

Adjoining room

Base molding

Flooring

Figure 28-49. *With wall-type grille openings, air is returned to the furnace through stud spaces, sheeted joist spaces, or ducts.*

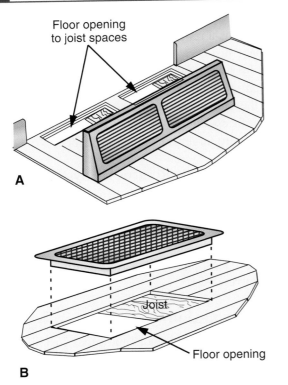

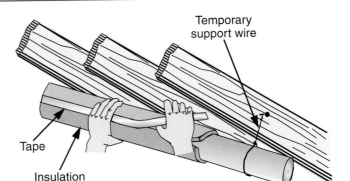

Figure 28-52. *Use tape or sealant to close insulation on round pipe. Remove temporary support wires and replace them with strapping.*

Ducts in homes with high energy-efficiency ratings are usually sealed with mastic and fiberglass reinforcing membrane, **Figure 28-53.** Mastic is a nontoxic adhesive-material used for permanently sealing joints and seams for all types of air duct materials. Most

Figure 28-50. *A—If a large baseboard grille is used, double stud and joist spaces should be prepared. B—Floor type grilles are commonly used when the furnace is located in the basement.*

Figure 28-53. *A—Mastic is available in 1, 2, and 5 gallon recyclable plastic pails. B—Fiberglass reinforcing membrane is designed for use with mastic. A roll is 3″ × 50′ or 3″ × 150′. (Courtesy of RCD Corporation)*

come loose and leak air. Air leaks from poorly sealed ducts can be one of the biggest energy and money losers in a home. The California Energy Commission has ruled that fabric-based duct tape can no longer be used for sealing ductwork, and other states are considering a similar ban.

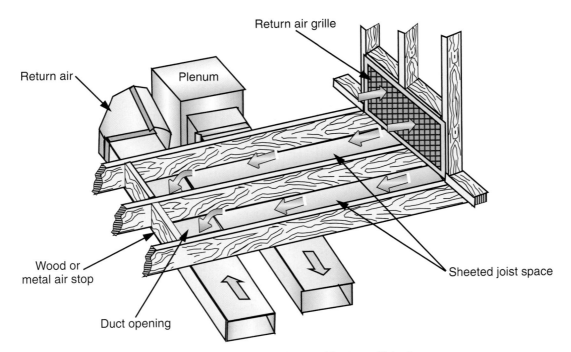

Figure 28-51. *Sheeting, or panning, is easier to install than duct and just as efficient.*

mastic is water-based, nonflammable, fiber-reinforced, fast drying, and waterproof. Easy application, and clean-up with soap and water are attractive features of the material. Mastic can be applied by brush, trowel, caulking gun, or with a gloved hand. It typically dries to touch in 1 or 2 hours, but unusually wet conditions may require 4-5 hours drying time. Insulation can be installed over mastic that is still wet, but care is needed to prevent excessive movement and loss of seal.

28.16.2 Fiberglass Reinforcing Membrane

Fiberglass reinforcing membrane is used with mastic for sealing air leaks in insulation, air ducts, plenums, and fittings. This inorganic woven **fiberglass membrane**, **Figure 28-53B**, will not rot, is lightweight and conforms to irregular surfaces. A typical roll is 3″ wide × 50′ long, or 3″ wide × 150′ long, **Figure 28-54.** This product is designed to be used with mastic as a reinforcement when there is a gap of 1/4″ or more. If the membrane has a sticky side, apply it over the clean metal surface before applying the mastic. If the membrane does not have a sticky side, apply a thin layer of mastic and place the membrane in this layer. Add another layer of mastic until the membrane is completely covered.

28.17 How to Use Mastic

Before applying mastic or fiberglass reinforcing membrane, the duct surface must be clean and dry, Use a cloth to wipe dust from the surface of the duct, **Figure 28-54.** If an oily film or grease is present, wipe the duct clean with a damp cloth. If necessary, use a grease-cutting cleaner on the cloth, then wipe thoroughly dry.

28.17.1 Sealing Joints with Less Than 1/4″ Gaps

Load the brush with mastic, and coat the entire joint with a continuous strip of the adhesive. Use the end of the brush to work mastic into joint. Spread mastic at least 1″ on each side of the joint, **Figure 28-55.** The mastic should be thick enough to hide the metal surface of the duct (about 1/16″ thick).

28.17.2 Sealing Joints with Gaps Greater Than 1/4″

If the gap in the duct connection is wider than 1/4″, use fiberglass reinforcing membrane in addition to mastic. Cut enough membrane to cover the joint, and press the membrane over the joint, **Figure 28-56.** If necessary, first brush a thin layer of mastic on the joint to hold the membrane in place. Finally, cover the membrane with mastic in a layer about 1/16″ thick, overlapping the membrane on each side.

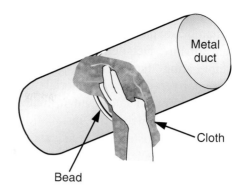

Figure 28-54. *Mastic will not stick to a dirty surface. Wipe dust off the surface, and clean any greasy residue.*

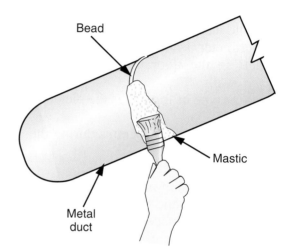

Figure 28-55. *Joints with a gap of less than 1/4″ can be sealed with mastic only.*

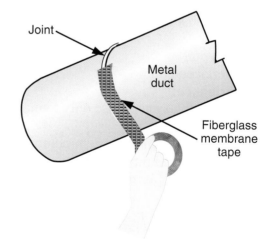

Figure 28-56. *Gaps of more than 1/4″ require use of reinforcing membrane.*

Summary

This chapter introduced the classifications of forced-air furnaces, based on direction of airflow, and described the difference between conventional and perimeter systems. The major air distribution systems, radial and extended plenum, are described in detail. "Rules of thumb" are provided for sizing a duct system for efficient airflow. Detailed information on laying out and fabricating duct systems is provided, and covers the major types of duct materials: rectangular metal duct, round metal duct, and insulated flexible duct. Directions are given for installation of wall stacks, supply air registers/diffusers, and return air grilles. The use of duct insulation and of fiberglass reinforcing membrane and mastic to seal duct air leaks is covered.

Test Your Knowledge

Please do not write in this text. Write your answers on a separate sheet of paper.

1. A perimeter system locates the supply air outlets in the floor or baseboard along the _____ walls.
2. In which type of furnace is the plenum located on top?
3. The blower section operates during the:
   a. ignition stage
   b. heating mode
   c. air conditioning mode
   d. both heating and air conditioning modes
4. A downflow furnace is typically used in a(n):
   a. attic
   b. two-story residence
   c. mobile home
   d. garage
5. In a residential installation, the _____ evaporator is most often used.
   a. A-coil
   b. slant coil
   c. interlaced coil
   d. upright coil
6. Radial piping systems use only round pipe that spread out from the _____.
7. An extended plenum system combines _____ duct with round pipe.
8. What diameter is round metal pipe used for heating only?
9. Starting collars or _____ are used to begin a run of pipe.
10. When assembling round ducts, the _____ end of the pipe should point in the direction of airflow.
11. Drive cleats and S-hooks are used to assemble what type of duct?
12. Which of the following materials is the preferred support for ducting?
    a. perforated metal strapping
    b. wire
    c. woven polypropylene webbing
    d. wooden brackets
13. What type of duct cannot be used to make sharp turns?
14. To end a run of pipe, a _____ or diffuser is used.
15. All ductwork running through _____ areas (such as attics or crawl spaces) should be insulated.

29

Gas Heat with Air Conditioning

Objectives

After studying this chapter, you will be able to:
- ❏ Identify each component of a gas heating system and describe its purpose.
- ❏ Discuss proper adjustment techniques for burner flames, fan/limit controls, and heat anticipators in gas heating systems.
- ❏ Demonstrate how to wire heating and cooling system components.
- ❏ Describe methods for properly connecting two-speed motors.

Important Terms

air duct system	grilles
automatic gas valve	heat anticipator
blower relay	heat exchanger
blower section	incomplete combustion
clamshells	limit switch
combination fan/limit control	manifold
combustion	mercury
crossover igniter	mercury switch
diffusers	orifice
draft	pilot flame
draft hood	plenum
electrodes	primary air
electronic ignition	radial piping system
extended plenum system	register
fan switch	return air ducts
filter	secondary air
flue	supply air ducts
forced-air	thermocouple
	thermostat
	thermostat override

29.1 Forced-air Systems

Most residential heating and cooling systems are the *forced-air* type, also known as *forced-convection* systems. "Total comfort" is achieved by automatically controlling air circulation (movement), temperature (heating and cooling), filtration (cleaning), and moisture content (humidity). During its operating cycle, a forced-air system heats or cools air and directs it to all areas of the conditioned space. See **Figure 29-1.** The

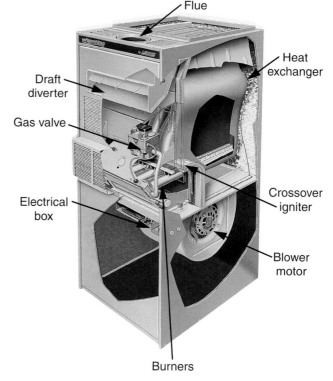

Figure 29-1. *A cutaway view of a typical gas-fueled, forced-air furnace. With the addition of an evaporator coil and a separate condensing unit, the same equipment can be used for cooling. (Lennox)*

furnace can be located in a basement, attic, crawl space, or utility closet. Basic components common to residential forced-air heating and cooling systems include air ducts, burners, blower, filter, evaporator, condensing unit, and thermostat.

The furnace contains a blower section (motor and fan) to circulate air in the home. Separate air ducts (metal tubes) are used to control air movement to and from the furnace.

During the cooling season, heat and humidity are removed from the air. The cooling system is an outdoor condensing unit with an evaporator located in the furnace *plenum* (supply air chamber). An air-sensitive thermostat is centrally located in the conditioned space to control system operation. The thermostat is designed to prevent the heating and cooling systems from operating at the same time.

29.1.1 Air Duct System

The *air duct system* is the network of tubes used to control airflow to and from the conditioned space. Ducts carrying air *to* the conditioned space are called *supply air ducts.* The ducts bringing air *back* to the furnace for reconditioning are called *return air ducts.* A blower section (fan and motor) inside the furnace provides the forced airflow through the furnace and air ducts.

The air conditioning evaporator (often called an "A" coil because of its shape) is located inside the supply plenum. The return plenum is another sheet metal chamber located at the furnace air intake. Return air ducts are connected to this plenum.

Radial piping system

A *radial piping system* uses round pipe runs that extend from the supply plenum to each *register* (supply air outlet). This system is economical and easy to install.

Extended plenum system

The *extended plenum system* combines rectangular duct with round duct. Rectangular duct is used to extend the plenum outward from the furnace as needed; then, round pipe is run at a right angle from the extended plenum to each outlet.

Outlets and inlets

Supply air is released through outlets that are called *registers* or *diffusers.* Registers deliver air in a concentrated flow pattern, while diffusers provide wider, fan-shaped patterns.

Inlets for return air are called *grilles.* Grilles are nonadjustable because return airflow to the furnace must be unrestricted.

29.1.2 Blower Section

The *blower section* is located inside the furnace cabinet. The blower section consists of a motor and squirrel cage-type centrifugal fan. The fan draws air from the return air ducts and forces it through the furnace for treatment (heating or cooling). Then, the treated air is pushed through the supply air ducts to individual rooms. There are two types of blowers, belt-drive and direct-drive. See **Figure 29-2.**

Filter

A *filter* is a cleaning device that captures dust, lint, dirt, and other impurities from the circulating airstream. The filter is located in the return-air side of the blower section. A fiberglass filter medium is commonly used as a screen for cleaning the air before it enters the blower. Filters should be cleaned or replaced regularly; a dirty filter reduces system airflow, which

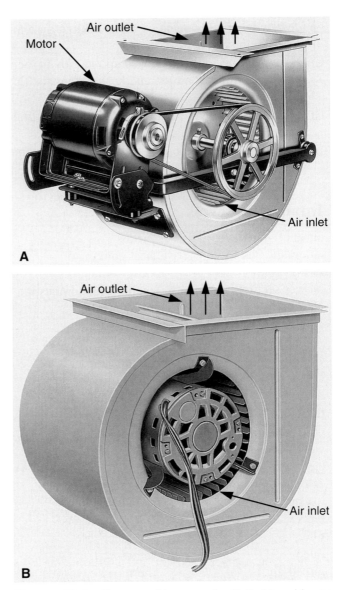

Figure 29-2. *Furnace blowers. A—Belt-drive blower. B—Direct-drive blower. (Lennox)*

significantly reduces system capacity. Dirty filters can cause the furnace to overheat.

Disposable filters, **Figure 29-3,** consist of a rectangular fiberglass screen in a cardboard frame. Disposable filters are available in many sizes and in two types, standard and pleated. The standard filter traps larger particles and should be changed monthly. Pleated filters have finer openings in the fiberglass screen and are pleated to expose more surface area. They can trap much smaller particles and are typically changed on a three-month cycle. The filter should be sized to fit completely across the incoming airstream. Arrows on the filter frame indicate the proper direction of airflow.

Some nondisposable filters use a screening medium encased in a metal frame. When the screen becomes dirty, it may be cleaned by washing or vacuuming before reuse.

Never operate the system without a filter. Dirt and lint will quickly plug ventilation openings in the fan motor and cause it to overheat. Dirt will also gather on the evaporator, greatly reducing airflow. Cleaning the evaporator is a difficult and time-consuming task that can be avoided by proper filter replacement or cleaning. Furnace air filters should be replaced or cleaned before each heating or cooling season and at intervals recommended by the furnace manufacturer. Unfortunately, such an "ounce of prevention" is often neglected, leading to costly evaporator cleaning.

29.1.3 Heating Section

Gas-fueled furnaces rely on **combustion** (burning of fuel) to generate usable heat. The combustion process produces waste gases such as carbon dioxide, water vapor, and small amounts of other chemical compounds. Incomplete combustion may produce such substances as carbon monoxide, aldehydes, and soot, which are dangerous to human health. Gases and other waste products resulting from combustion must be properly vented to the outdoors.

The heating section of a furnace fueled by natural gas or liquefied petroleum (LP) gas consists of burners, a steel heat exchanger, and a venting system. The heat exchanger is usually divided into sections called *clamshells.* Burners fit into special chambers in the bottom of the heat exchanger. See **Figure 29-4.**

When the thermostat calls for heat, a gas valve is energized (opened), and gas flows to the burners. The gas is ignited, and the flame warms the heat exchanger. Hot waste gases from combustion travel up through the inside shell, collect at the top, and are directed to a vent. The vent is connected to a box called a **draft hood** or diverter. A vent pipe (flue) runs from the diverter box to the chimney. See **Figure 29-5.**

The heat exchanger clamshells are designed to separate the recirculating room air from the combustion gases. The blower forces recirculated room air around and between the heat exchanger shells. The air is heated as it touches the hot surface of the shells. The heated air

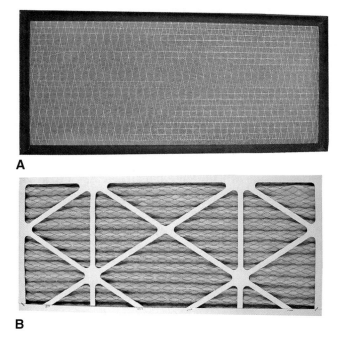

A

B

Figure 29-3. *Disposable filters. A—Standard filter traps larger particles. B—The pleated filter has more surface area and can trap smaller particles, cleaning the air more efficiently.*

Figure 29-4. *The heating section of a gas furnace includes burners in the bottom of the heat exchanger and a venting system for the waste products of combustion. (Lennox)*

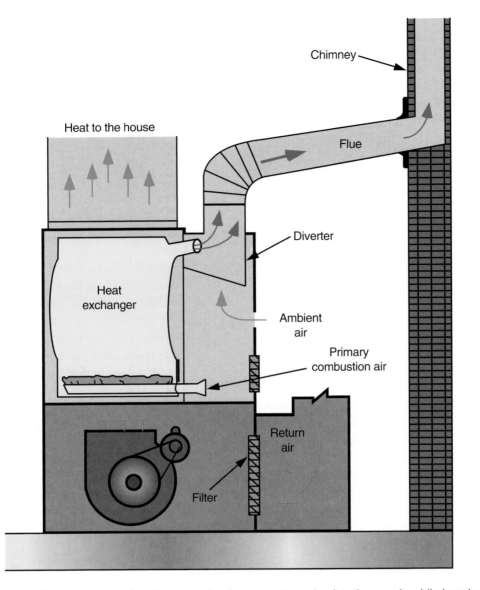

Figure 29-5. *The heat exchanger transfers heat from combustion gases to recirculated room air, while keeping the two separated. Hot waste gases are drawn up the flue and out the chimney.*

travels from the heat exchanger to the supply plenum and is delivered to living spaces. Most gas furnaces are designed for about a 60-degree Fahrenheit (33-degree Celsius) rise in the temperature of air crossing the heat exchanger. If entering air is 70°F (21°C), exiting air will be about 125°F to 130°F (52°C to 54°C).

29.1.4 Gas Burners

The number of burners in a furnace depends upon the size of the furnace. Gas burners vary in style and shape but essentially work the same: when the thermostat calls for heat, a mixture of atmospheric air and gas enters the burner and is ignited on the surface of the ports.

Both primary and secondary air must be provided to the burner. See **Figure 29-6.** *Primary air* is drawn

into the burner by the jet of gas. The air is mixed thoroughly with gas inside the burner before delivery to the ports. *Secondary air* is drawn into the burner chamber after ignition of the flame. Natural convection (the rising of less-dense hot air) creates a ***draft***, or air movement, into the burner chamber, up through the heat exchanger, and out the flue. Primary air is rarely sufficient for complete combustion. Secondary air provides the additional air needed for complete burning of the fuel gas.

The amount of primary air is controlled by small, adjustable shutters located at each burner inlet. Secondary air cannot be controlled. A *yellow* gas flame indicates poor combustion. A *blue* flame indicates good combustion. Addition of secondary air results in a blue flame that is smaller, faster-burning, and hotter than a yellow flame.

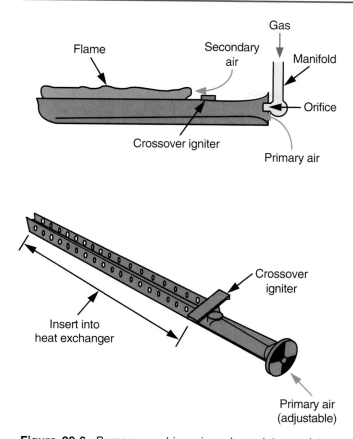

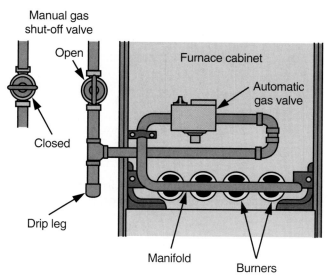

Figure 29-6. *Burners combine air and gas into a mixture that can be burned in a furnace. Primary air is drawn into the burner body by the flow of incoming gas from the gas manifold. Secondary air is drawn into the burner chamber by natural convection. Primary air is adjusted by rotating a shutter on the burner body.*

Gas valve and manifold

An **automatic gas valve**, **Figure 29-7,** is used to start and stop the flow of gas to the burners. The automatic gas valve is connected to a pipe, called a **manifold**, that distributes gas to the burners through an *orifice* (outlet hole) in each burner.

The gas valve contains a 24 V solenoid valve. When the solenoid coil is energized by the thermostat, the valve opens, and gas flows to the burners. See **Figure 29-8.** When the coil is de-energized, the gas valve closes, and the burner flame goes out. A room thermostat controls electrical power supply (24 V) to the gas valve solenoid.

Most gas valves are combination valves sharing such features as manual shutoff, pilot light safety control, and an internal gas pressure regulator. See **Figure 29-9.** A manually operated knob on top of the valve has three positions:

❏ *On:* Normal operating position. Opening and closing of the gas valve solenoid is electrically controlled by the thermostat at terminal W.

❏ *Off:* Used to manually shut off the gas supply to the furnace.

❏ *Pilot:* Used to light the pilot without opening the main valve. Depressing the knob in this position permits gas flow only to the pilot valve.

Pilot flame. The **pilot flame** is a small gas flame used to ignite the flame on the main burners. A constantly burning pilot is called a *standing pilot.* Gas flow to the main burners cannot be allowed when the pilot flame is out (not lighted). Without a pilot flame, gas would collect inside the furnace and eventually explode when exposed to a spark or flame.

Thermocouple. A **thermocouple, Figure 29-10,** is a safety device that prevents the main gas valve from

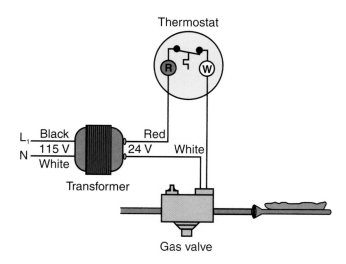

Figure 29-7. *The automatic gas valve controls the flow of gas from the supply to the manifold and the burners. A manual shut-off valve is located at the point where the supply enters the building. The automatic valve includes a solenoid operated by the thermostat.*

Figure 29-8. *The system thermostat controls power to the gas valve solenoid. The control circuit operates on 24 V, opening and closing the solenoid valve to regulate gas flow to the burners.*

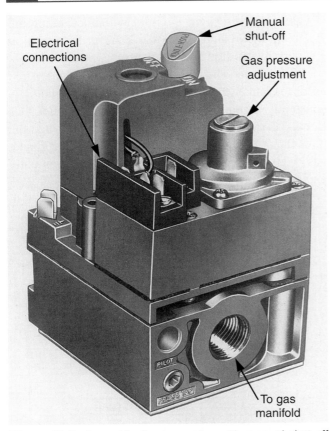

Figure 29-9. *A combination gas valve with manual shut-off and an internal pressure regulator. (Honeywell)*

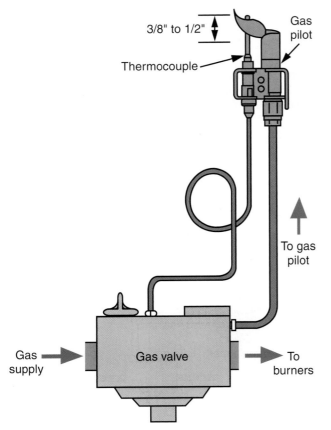

Figure 29-10. *The thermocouple is a safety device that confirms existence of a pilot flame. Without the signal generated by the thermocouple, the gas valve will not open to supply fuel to the burners.*

opening if there is no pilot flame. The thermocouple converts heat energy to electrical energy. The sensing element of the thermocouple is positioned within the pilot flame. Heat from the flame generates a tiny amount of electricity (measured in millivolts) in the thermocouple. The electrical energy is conducted to the main combination gas valve as a signal that a pilot flame is available. If the pilot flame goes out, the electrical signal is not generated, and the main gas valve cannot open. This safety feature prevents gas flow to the burners when the pilot flame goes out.

The pilot light should have a soft blue flame and envelop (cover) between 3/8″ to 1/2″ (10 cm to 13 cm) of the thermocouple's tip (hot junction). Proper positioning of the thermocouple within the pilot flame is critical. The pilot flame is adjustable by a screw on the combination gas valve. Remember, the gas valve cannot open unless the thermocouple generates the proper signal. Minimum acceptable dc voltage is 18 mV.

Crossover igniter. A ***crossover igniter*** is used to ignite several main burners from a single pilot flame. The crossover igniter is a slotted length of metal that acts as a bridge between burners. See **Figure 29-11.** Gas flowing to the main burners passes through the igniter slots until it comes in contact with the pilot flame. Properly positioning the burners will correctly align the crossover igniter. A poorly positioned burner will cause noisy and delayed ignition.

Electronic ignition. An ***electronic ignition*** (spark ignition) system saves fuel by eliminating the continuously burning pilot flame. The pilot flame is ignited by an electric spark and burns only upon a call for heat. If the pilot flame blows out, the system will automatically relight the pilot if the call for heat continues. The pilot assembly, **Figure 29-12,** consists of a pilot burner, a spark electrode, and a sensing probe.

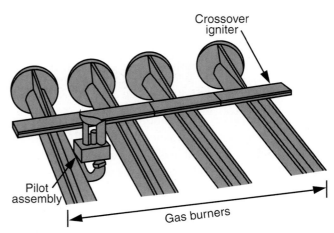

Figure 29-11. *A crossover igniter serves as a bridge between burners, allowing several burners to be ignited by a single pilot flame.*

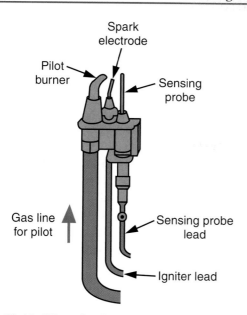

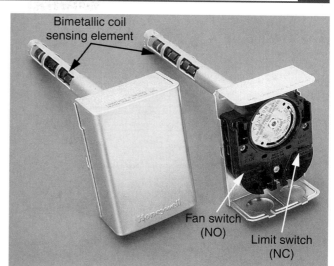

Figure 29-13. *A combination fan/limit control. The fan switch is normally open, while the limit switch is normally closed.*

Figure 29-12. *When the thermostat calls for heat, the electronic ignition system lights the pilot which, in turn, lights the main burners. The electronic system eliminates the cost of fueling a constantly burning pilot.*

When the thermostat calls for heat, the spark electrode automatically lights the pilot. Then, the sensing probe (thermocouple) confirms the presence of the pilot flame. Once the pilot flame is confirmed, the spark electrode is de-energized, and the main gas valve is opened. Next, the main burners are ignited by the pilot flame. When the thermostat is satisfied and opens its switch, the main gas valve is turned off. The flow of gas to the burners and the pilot flame stops.

29.1.5 Combination Fan and Limit Control

The ***combination fan/limit control*** serves a dual purpose. It controls operation of the blower motor and provides protection against overheating of the furnace. The fan/limit control contains two electrical switches, one normally open (NO) and one normally closed (NC). Movement of a coiled bimetallic strip controls operation of the switches. See **Figure 29-13.**

The bimetallic coil is inserted into the heat exchanger area to sense furnace temperature, **Figure 29-14.** Heat causes the coil to twist in a circular movement. The movement controls the NO and NC switches. The amount of heat necessary to activate the switches can be adjusted by a calibrated dial located on the control face, **Figure 29-15.**

Fan switch (normally open)

The normally open switch that operates the blower motor is called the ***fan switch.*** When the furnace temperature reaches a selected setting on the dial face (usually 140°F or 60°C), the switch closes and turns

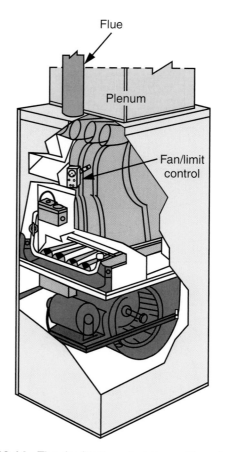

Figure 29-14. *The fan/limit control is positioned with the sensing element inserted in the heat exchanger to sense furnace temperature.*

on the blower motor. Air is pulled from the living space, forced through the heat exchanger where it is heated, and returned to the living space. This movement of air through the heat exchanger causes the furnace temperature to drop to about 125°F (52°C), where it remains during the heating cycle.

When the heated space reaches the desired temperature (typically 72°F or 22°C), the thermostat contacts open and stop current flow to the gas valve. Once current flow to the gas valve stops, the burner flame goes out; however, the blower motor continues to run because the furnace is still hot. Cooling of the furnace causes the bimetallic coil to twist in the opposite direction. The fan keeps running until the furnace temperature drops to about 100°F (38°C). At that time, the fan contacts reopen and stop the blower motor. See **Figure 29-16.**

Limit switch (normally closed)

A **limit switch** is a safety device that cuts off the power supply if the furnace becomes overheated. The furnace may overheat due to lack of airflow caused by fan motor failure, a broken V-belt, a dirty filter, or failure of the fan switch contacts. The limit switch, which is normally closed, will open if the furnace temperature reaches about 200°F (93°C).

The limit switch must be connected so it will cut off the power supply to the gas valve (and turn off the burners) if the furnace overheats. One of two methods can be used to connect the limit switch. It can be connected in series with the power supply to the primary wire of the low-voltage transformer, **Figure 29-17A.** This method cuts off power to the entire low-voltage circuit supplying 24 V power to the gas valve.

The other method, shown in **Figure 29-17B,** places the limit switch in the 24 V circuit supplying power to

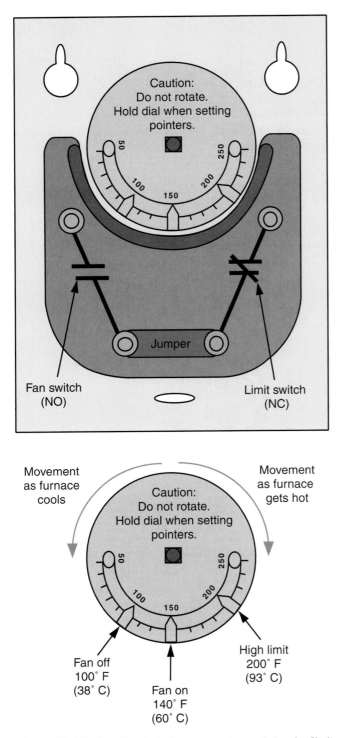

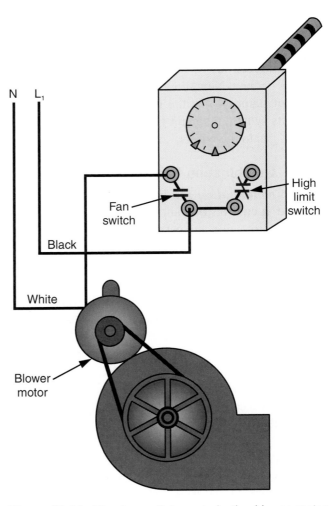

Figure 29-15. *A calibrated dial on the face of the fan/limit control is used to set the fan-on, fan-off, and high-limit temperatures. As the bimetallic coil sensing element is heated or cools, it twists one way or the other, rotating the dial.*

Figure 29-16. *The fan switch controls the blower motor during the heating cycle.*

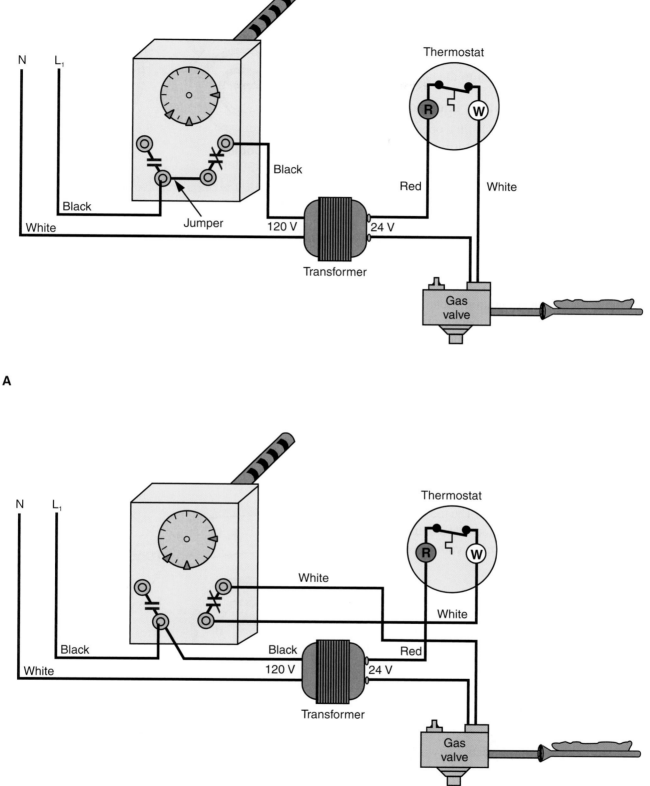

A

B

Figure 29-17. *Two methods of connecting the limit switch. A—In the high-voltage method, the switch is in series with the power supply to the transformer. If the switch opens, the entire 24 V circuit is de-energized. B—In the low-voltage method, the switch is in the 24 V circuit, supplying current to the gas valve.*

the gas valve. Either method is acceptable. The combination fan/limit control has a factory-installed jumper between the fan switch and the limit switch. The jumper remains in place when using the limit switch to disconnect line voltage to the transformer. The jumper must be removed when connecting the limit switch in the 24 V circuit to the gas valve.

29.1.6 Low-voltage Heating Thermostat

A *thermostat* is a switch that opens and closes according to changes in temperature. The thermostat is located within the living space where temperature is being controlled. It turns heating and air conditioning equipment on and off automatically to maintain specific temperatures within the space. Only one thermostat is used, and it must be located where it can sense temperature uniformly. Cold walls and doorways must be avoided.

Thermostats for heating and cooling are different from those used for refrigeration. Heating and cooling thermostats are designed to operate on 24 V because they are often located considerable distance from the heating and cooling equipment. Low voltage permits the use of small wires that are easily hidden from view. The 24 V system has other advantages: it produces very little arcing of the switch contacts, resulting in long switch life, and both shock and fire hazards are virtually eliminated.

Power is supplied to the thermostat by the secondary winding of the transformer. A red wire is used to connect the power supply from the transformer secondary (24 V) to Terminal R on the thermostat. The thermostat becomes an automatic switching device that controls the 24 V power supply to various electrical components.

Thermostat construction

Most thermostats are constructed with a small, coil-shaped, bimetallic (brass and invar) element. The coil is extremely sensitive to small changes in temperature. The end of the coil at the center of the spiral is anchored. As the coil winds and unwinds in response to small temperature changes, the free end moves in an arc. See **Figure 29-18.**

Mercury switches

A *mercury switch* consists of two exposed wire ends (*electrodes*) inside a glass tube. The glass tube is mounted on the free end of the bimetallic element. A drop of the metallic element *mercury* (a liquid at room temperature) is contained in the glass tube. Whenever the tube is tipped, the mercury flows to the lower end.

As shown in **Figure 29-19,** when the mercury is at the lower end of the tube, opposite the electrodes, the switch is open. No current can flow from one electrode

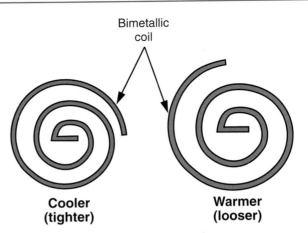

Figure 29-18. *The coil is anchored at the center of the spiral so the outside end moves in an arc as the temperature changes. As the air cools, the coil becomes tighter; as the air warms, the coil becomes looser.*

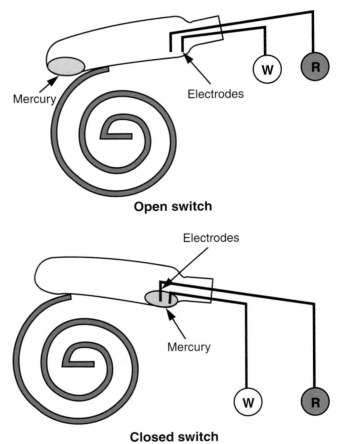

Figure 29-19. *Tilting of the glass tube by movement of the bimetallic element inside the thermostat allows a drop of mercury to open or close the switch controlling heating or cooling operations.*

to the other. When the tube is tipped in the other direction, the mercury flows to the electrode end. Mercury is an excellent conductor of electricity. By contacting both electrodes, the mercury completes an electrical circuit from one electrode to the other. The switch is now closed.

The glass tube is bean-shaped and has a slight hump in the center. The mercury must run over the hump when the tube is tilted. To move the mercury from one end of the tube to the other, including over the hump, requires a significant degree of tilt. The requisite degree of tilt prevents rapid cycling due to vibration (called *chatter*).

Thermostat override

When the room temperature has reached the desired level, the thermostat opens and shuts off the gas valve. However, the blower motor continues to run and deliver heated air until the furnace cools down to 100°F (38°C). As a result, the room temperature becomes too high. Delivering heat after the thermostat is satisfied is called overriding the thermostat, or **thermostat override.**

Heat anticipator

To overcome thermostat override, a small resistance heater (called the **heat anticipator**) is mounted inside the thermostat close to the bimetallic coil. Current flow through the heat anticipator during the heating cycle warms the bimetallic coil, **Figure 29-20.** Heat from the anticipator causes the thermostat switch to open *just before* room temperature reaches the desired level. The furnace blower continues to operate and cool the furnace to 100°F (38°C). The additional heat conveyed by the blower raises the temperature of the living space to plus or minus 1 degree Fahrenheit (0.5 degree Celsius) of the desired temperature.

The heat anticipator is energized when the furnace is in the heating mode. The small heater is factory-connected between the bulb electrode and Terminal W (the heat terminal) on the thermostat. This places it electrically in series with the gas valve in a gas furnace, with the primary control in an oil furnace, or with the heat sequencer in an electric furnace.

Heat anticipator adjustment. Some heat anticipators have a fixed resistance, while others are adjustable, **Figure 29-21.** The adjustable heat anticipator has a slide with a pointer and a scale calibrated in tenths of an ampere. The pointer is moved to the scale reading that agrees with the amperage flow to the gas valve. Such a low amperage draw can be measured with an amprobe and a multiplier.

The amperage draw by the gas valve is stamped on the valve body. The heat anticipator should be set to the same amperage to control override. For example, if the draw of the circuit is 0.45 amps, the anticipator pointer must point to 0.45 on the scale. Matching amperages provides the additional heat necessary for precise control of room temperature.

Figure 29-22 illustrates, in both pictorial and schematic forms, a complete gas heating system with thermostat override protection.

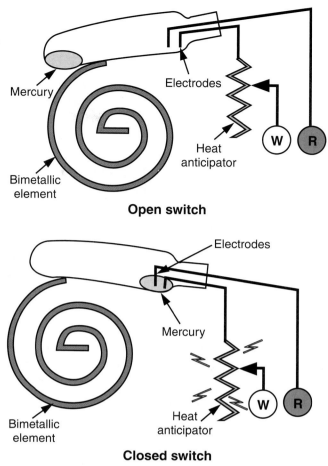

Open switch

Closed switch

Figure 29-20. *The heat anticipator prevents furnace override at the end of the heating cycle. It warms the bimetallic element so the switch closes slightly before room temperature reaches the setpoint.*

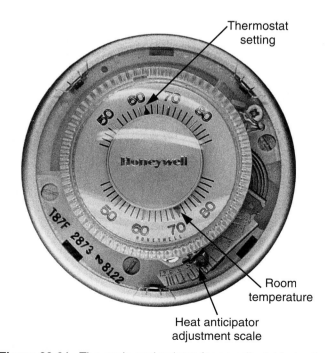

Figure 29-21. *The scale and pointer for an adjustable heat anticipator is visible by removing the thermostat cover. The pointer should be set to the scale marking equal to the amperage draw of the gas valve.*

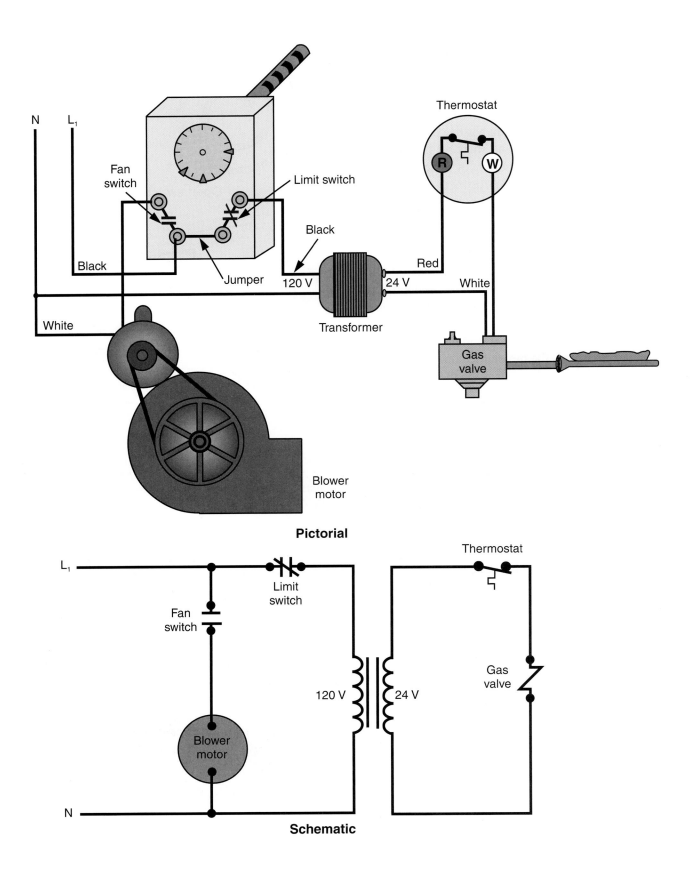

Figure 29-22. *A typical gas heating system shown in both pictorial and schematic forms.*

29.1.7 Add-on Air Conditioning

Air conditioning is easily added to a gas forced-convection heating system. The evaporator is installed inside the furnace plenum, and the condensing unit is located outdoors. The suction and liquid lines connect the evaporator to the condensing unit. The furnace blower is used to circulate air during either heating or cooling cycles. See **Figure 29-23.**

29.1.8 Heating and Cooling Thermostat

When heating and cooling systems are combined for year-round comfort, it is impractical to use a separate thermostat for each system. The different switching requirements are easily combined in a single thermostat and sub-base.

An additional pair of electrodes is installed in the opposite end of the glass tube, **Figure 29-24.** One set

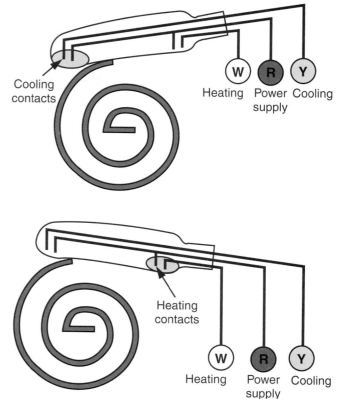

Figure 29-24. *A combination heating and cooling thermostat has two sets of electrodes at opposite ends of the glass tube.*

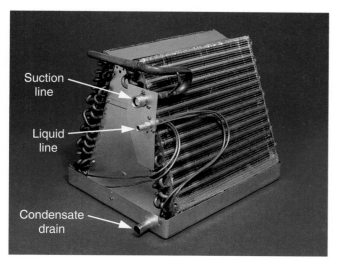

Figure 29-23. *Add-on air conditioning. A—The evaporator, often referred to as an "A" coil because of its shape, is installed in the furnace plenum. B—The condensing unit is installed outside the building and connected to the evaporator by means of suction and liquid lines. (Aeroquip Corp.)*

of electrodes is used for the heating cycle, and the other set is used for the cooling cycle. The middle electrode is common (power supply) to both ends. When the bulb is tilted in either direction, the mercury closes the appropriate set of electrodes.

Unwanted switching back and forth from heating to cooling is prevented by a three-position *manual* switch on the thermostat, labeled COOL-OFF-HEAT. The switch prevents the thermostat from energizing both systems at the same time or automatically switching back and forth between them. See **Figure 29-25.**

When the manual switch is in the OFF position, both heating and cooling systems are turned off (unable to operate). When the switch is in the HEAT position, only the heating system is able to operate; the cooling mode is disconnected. When the switch is in the COOL position, only the cooling system is operational; the heating mode is disconnected.

29.1.9 Cooling Cycle

When the manual switch on the thermostat is set to COOL, and the mercury closes the cooling contacts, 24 volts are supplied to Terminal Y on the thermostat. A yellow wire connects Terminal Y with the coil on the contactor in the outdoor condensing unit. A white wire connects the other side of the coil to the neutral side of the 24 V transformer. See **Figure 29-26.** When the contactor coil is energized, the outdoor condensing unit is turned on.

Cooling anticipator

The thermostat's cooling anticipator does the same job as the heating anticipator, only in reverse. The cooling anticipator must compensate for the delay between the call for cooling and the time when the system begins delivering cool air. The cooling anticipator is a small, fixed resistor wired in parallel with the cooling mercury switch. See **Figure 29-27.** When the mercury switch is closed during a call for cooling, the anticipator is in parallel, so it is bypassed. (Electricity always takes the path of least resistance.)

When the cooling demand is satisfied, the mercury switch opens and places the cooling anticipator *in series* with the compressor contactor coil. Due to high resistance, a very small current flows through both the anticipator and the compressor contactor coil. The amperage is not sufficient to energize the contactor coil but does cause the resistor to heat up. The resistor heat makes the bimetallic coil tilt just slightly in advance of the room temperature. The cooling equipment has the opportunity to get started and begin delivering cool air just before the room temperature reaches the thermostat setting.

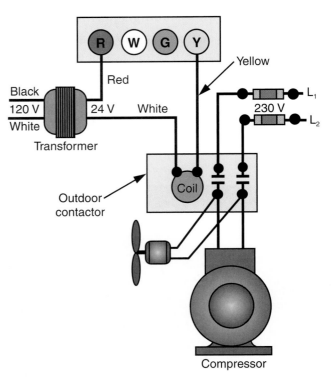

Figure 29-26. *When the cooling cycle is active, the circuit is completed from Terminal Y of the thermostat to the contactor on the outdoor condensing unit.*

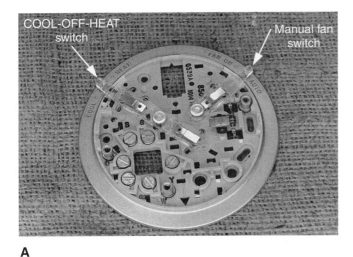

A

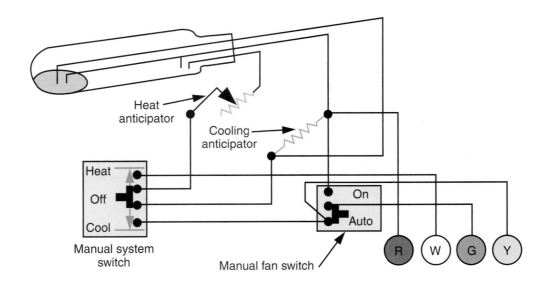

B

Figure 29-25. *Manual switching of modes. A—The COOL-OFF-HEAT manual switch is on the sub-base of the thermostat. The fan switch is also in the sub-base. B—Schematic showing connections for the manual-mode switch.*

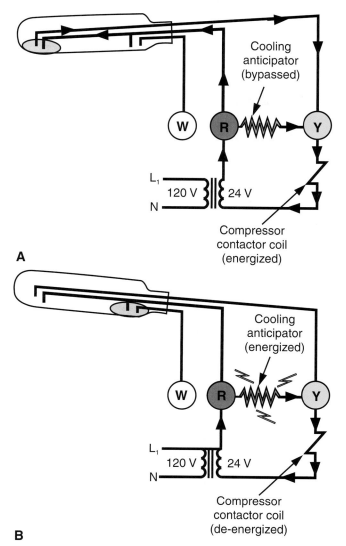

A

B

Figure 29-27. *Cooling anticipator operation. A—During the cooling cycle, the cooling anticipator is bypassed. B—Once the mercury switch opens, the anticipator heats up from resistance and advances the bimetallic element slightly ahead of the actual room temperature. Cooling starts just before the actual setpoint is reached.*

29.1.10 Blower Motor

The combination fan/limit switch turns on the blower motor when the furnace is hot. Provisions must be made to energize the blower motor during the *cooling* cycle, as well. A **blower relay** accomplishes this purpose. The blower relay has a 24 V coil and one set of normally open contacts. When the relay coil is energized by the thermostat, the relay switch closes and completes a 120 V circuit to the blower motor. See **Figure 29-28.**

Another manual switch on the thermostat (the *fan switch*) controls power supply to the blower relay coil. The switch is labeled ON and AUTO (for automatic). When the thermostat calls for cooling, and the fan switch is in the AUTO position, another 24 V circuit is completed to Terminal G inside the thermostat. See **Figure 29-29.**

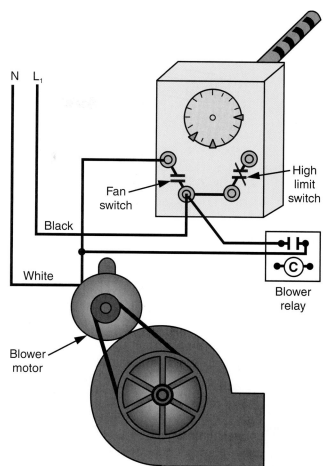

Figure 29-28. *The blower relay controls the blower during the cooling cycle.*

Terminal G is used to control the 24 V power supply to the blower relay coil. A green wire connects Terminal G on the thermostat to the blower relay coil. The other side of the relay coil is connected to the neutral side of the transformer.

When the manual fan switch is in the ON position, Terminal G receives a continuous 24 V power supply. The blower relay coil remains energized, and the blower motor runs continuously. The fan switch permits the homeowner to control blower operation independently of the COOL-OFF-HEAT switch. Sometimes air circulation only, without cooling, is adequate to satisfy the homeowner's needs.

A two-speed blower motor may be used to provide better air circulation. Low speed is used for the heating cycle because the temperature difference (*td*) through the furnace is 55 degrees Fahrenheit (125 − 70 = 55) or 30 degrees Celsius. High speed is used for the cooling cycle because the td across the evaporator is 16 to 20 degrees Fahrenheit (9 to 11 degrees Celsius). Electrical connections for the two-speed blower motor are accomplished by connecting its low speed (heating) terminal to the fan switch and its high speed (cooling) terminal to the blower relay switch. See **Figure 29-30.**

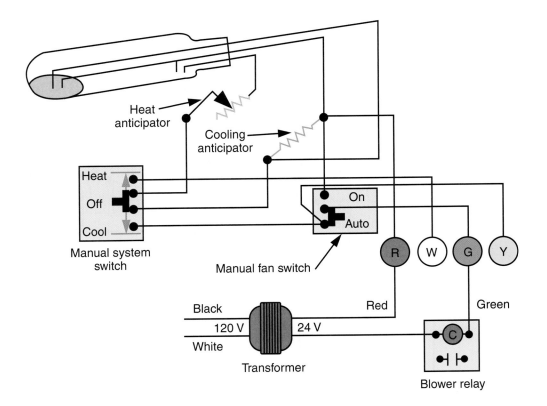

Figure 29-29. *The fan switch on the thermostat allows the selection of continuous fan operation or operation only during the cooling cycle.*

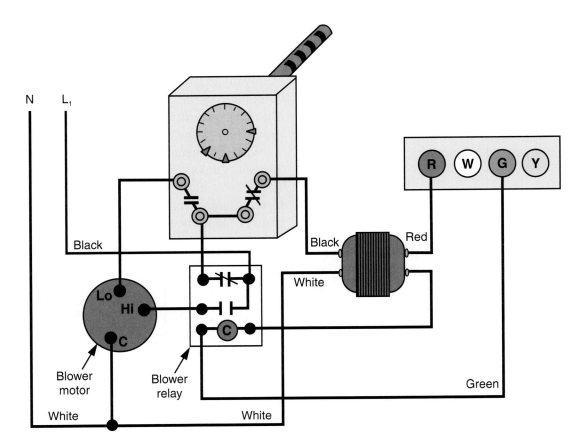

Figure 29-30. *The low speed of a two-speed fan is energized during the heating cycle; high speed is energized during the cooling cycle.*

Figure 29-31 illustrates a typical gas-fired, forced-air furnace with air conditioning and a two-speed blower motor in both pictorial and schematic forms.

29.2 Troubleshooting Gas Furnaces ──◆

Systematic troubleshooting skills must be developed. A service call tests the technician's knowledge, mechanical competence, and observation and interpretation skills. Good troubleshooting requires analyzing information accurately and drawing logical conclusions. Hasty decisions or inaccurate information lead to wrong conclusions. Following step-by-step procedures and a *process of elimination* usually pinpoints the problem. There are five important steps in systematic troubleshooting:

- ❏ Know what the equipment *should* do.
- ❏ Know what the equipment *is* doing.
- ❏ Know what the equipment is *not* doing.
- ❏ Eliminate areas that *are* functioning properly.
- ❏ Concentrate on areas that are *not* functioning properly.

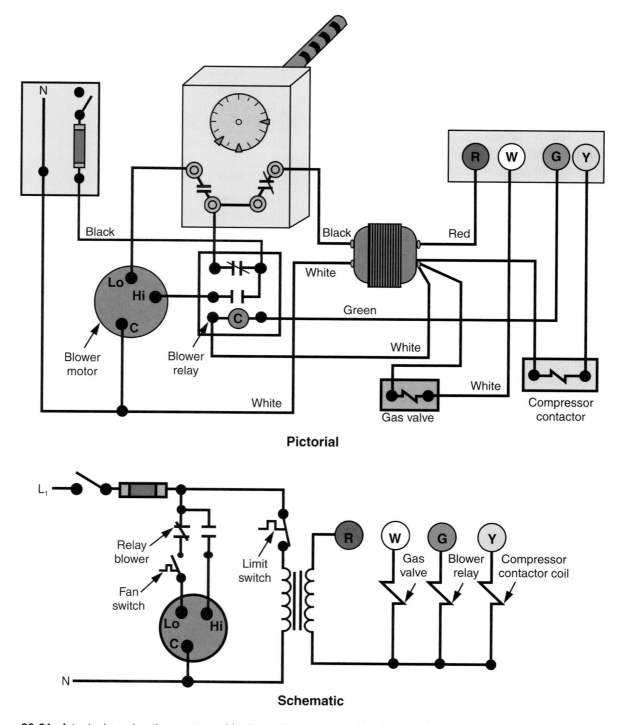

Figure 29-31. *A typical gas heating system with air conditioning added is shown in both pictorial and schematic forms.*

The customer's complaint usually helps narrow the problem to a specific area. Complaints typically fall into six categories:

- ❏ No heat.
- ❏ Not enough heat.
- ❏ Too much heat.
- ❏ Noise.
- ❏ Odor.
- ❏ Excessive operating cost.

A question or two to the homeowner, plus observation of the equipment, should aid in the process of elimination. By listening, observing, and testing, you will discover clues that point to the exact problem.

Following are some of the symptoms and possible causes of customer complaints:

29.2.1 No Heat

Symptom: Burner does not start.
Possible causes:

- ❏ Wrong thermostat setting.
- ❏ Defective thermostat.
- ❏ No power supply to the furnace (check fuses and circuit breakers).
- ❏ Gas valve turned off.
- ❏ Defective high-limit control.
- ❏ Defective transformer.
  (Check fuse, if any.)
- ❏ Defective gas valve.

Symptom: Pilot flame is out.
Possible causes:

- ❏ Main gas valve turned off.
- ❏ Plugged or restricted pilot orifice.
- ❏ Pilot blows out due to draft through furnace.
- ❏ Pilot blows out due to high gas pressure.
- ❏ Pilot blows out due to delayed ignition.

Symptom: Pilot flame okay.
Possible causes:

- ❏ Thermocouple loose at valve.
- ❏ Thermocouple out of alignment at pilot.
- ❏ Defective thermocouple.

29.2.2 Not Enough Heat

Symptom: Burner short cycles.
Possible causes:

- ❏ Unit cycling on high-limit control.
- ❏ Defective blower motor.
- ❏ Dirty filter.
- ❏ Loose or broken V-belt.
- ❏ Wrong rotation of blower motor.
- ❏ Restriction in return air system.
- ❏ Low voltage at motor (causes high amps).

29.2.3 Too Much Heat

Symptom: Burner runs too long.
Possible causes:

- ❏ Wrong thermostat setting.
- ❏ Heat anticipator set too high.
- ❏ Thermostat not level.
- ❏ Thermostat on cold wall or in draft.
- ❏ Defective gas valve (stuck open).

29.2.4 Noise

Possible causes:

- ❏ Delayed ignition causing flashback.
- ❏ Flashback after shutdown (excess primary air).
- ❏ Loose blower bearings.
- ❏ Blower bearing dry and squeaking.
- ❏ Blower wheel off-center.
- ❏ Loose V-belt.
- ❏ Loose running gear and mounts.
- ❏ Filter caught in running gear.
- ❏ Loose cabinet panels.

29.2.5 Odor

Possible causes:

- ❏ Leak in gas line or pilot line.
- ❏ Loose or cracked heat exchanger.
- ❏ Leak in flue pipe.
- ❏ Downdraft causing diverter spillage.
- ❏ Negative pressure in furnace area.
- ❏ Dirty air filters.
- ❏ Stagnant water or sludge in humidifier.

29.2.6 Excessive operating cost

Possible causes:

- ❏ Dirty air filter.
- ❏ Poor combustion.
- ❏ Insufficient home insulation.
- ❏ Excessive air leakage (fireplace, vents).
- ❏ Excessive gas manifold pressure.
- ❏ Oversize burner orifice.
- ❏ Excessive flue draft.
- ❏ Wrong fan control settings.

◆ Summary

This chapter described the mechanical and electrical operating principles of a gas-fueled, forced-convection furnace. The function of each mechanical component in the system was explained. Electrical connections for the heating system were also explained utilizing both pictorial and schematic wiring diagrams.

The procedure for adding air conditioning to a gas furnace was described, along with the operation and function of the thermostat and blower relay. Electrical circuits were explained and illustrated by pictorial and schematic diagrams. Finally, troubleshooting procedures were covered to aid in understanding system operation.

◆Test Your Knowledge

Please do not write in this text. Write your answers on a separate sheet of paper.

1. What components of a heating and cooling system carry treated air to and from the living space?
2. Heated or cooled air being delivered to the living space is called _____.
3. The metal box installed on the furnace outlet is the _____.
4. *True or false?* Registers or diffusers are used for return air.
5. How do dirty filters reduce system capacity?
6. The automatic gas valve controls flow to the _____.
7. What is the purpose of the thermocouple?
8. Does the fan switch control power supply to the blower during the heating or cooling cycle?
9. The limit switch opens when furnace temperature reaches about _____°F.
10. The R (red) terminal on the thermostat is connected to _____.
11. The W (white) terminal on the thermostat is connected to _____.
12. The _____ terminal on the thermostat is connected to the compressor contactor coil.
13. The G (green) terminal on the thermostat is connected to the _____.
14. Which anticipator is adjustable?
15. What device turns on the blower motor during the cooling cycle?

This high-efficiency combination system performs the functions of a water heater and a furnace. Water is heated and held in a supply tank, similar to a conventional home water heater. When the thermostat calls for heat from the furnace unit, hot water is pumped from the tank into a heat exchanger. Heat is then transferred to the stream of air generated by the furnace blower, and flows through ducts to the spaces being heated. Water from the furnace heat exchanger is recirculated to the heater to be reheated. The furnace (technically, an air-handling unit, since it does not include a burner) can be fitted with an A-coil for air conditioning. (Lennox International)

30

Oil Heat with Air Conditioning

Objectives

After studying this chapter, you will be able to:
- ❏ Describe how fuel oil and air are mixed in the combustion chamber for proper combustion.
- ❏ Explain how the mixture of oil and air is ignited.
- ❏ State dimensions to be used for proper electrode placement.
- ❏ Define the purpose of primary air adjustments.
- ❏ Explain why draft control is used and how it works.
- ❏ Describe the purpose and location of flame safeguards.
- ❏ Describe the electrical connections needed for heating and air conditioning circuits.
- ❏ Describe the operation and purpose of safety controls.

Important Terms

atomizing	ignition transformer
bleed port	isolation relay
burner fan	orifice
cad cell	pressure tap plug
draft control	primary control
electrodes	single-stage oil pump
fan relay	spray angle
flame safeguard	stack control
flexible coupling	transient light
fuel oil	vaporized
gap	viscosity
gun-type burner assembly	

30.1 Oil-fueled Furnaces

Oil-fueled and gas-fueled forced-convection furnaces are similar in appearance and air circulation, but major differences exist in the combustion process and the electrical circuitry.

Inside an oil furnace, **Figure 30-1,** fuel oil is vaporized and mixed with air in a precise ratio to obtain a combustible mixture. The oil/air mixture is ignited by an arc (spark) between two electrodes.

A high-voltage transformer supplies 10,000 V to the electrodes, producing the hot spark that ignites the oil/air mixture. A flame sensor determines if the mixture ignites. If ignition does not occur, the flame sensor shuts off the system to keep the furnace from filling with oil and creating a dangerous situation.

The oil-fueled forced-convection furnace is controlled by a standard 24 V thermostat. The

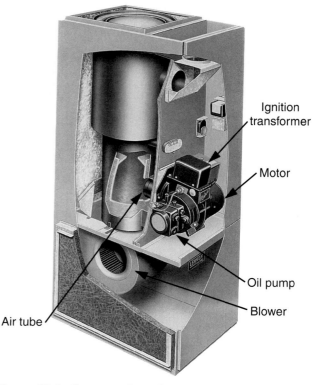

Ignition transformer

Motor

Oil pump

Blower

Air tube

Figure 30-1. *Cutaway view of a typical oil-fueled, forced-convection furnace. (Lennox International)*

539

electrical circuits include 120 V (supply), 24 V (control), and 10,000 V (ignition). How these circuits provide automatic control and safety is explained later in this chapter.

30.1.1 Fuel Oil

Fuel oil contains more carbon than do some other hydrocarbon fuels, such as natural gas and LP gas. Several grades (or weights) are available, ranging from grade No. 1 to grade No. 6. Grade No. 2 is the most popular fuel oil and produces about 140,000 Btu/gal. It is used in gun-type (atomizing) burners. For proper combustion, the liquid oil must be changed to a gas (vaporized) and mixed with air.

30.2 Gun-type Burner Assembly ———◆

The **gun-type burner assembly** produces the conditions needed to generate a steady, hot flame inside the furnace. The main parts of the burner assembly are the motor, centrifugal fan (blower), oil pump, nozzle, air tube (gun barrel), ignition transformer, electrodes, and flame sensor. See **Figure 30-2.**

Figure 30-2. *A gun-type burner assembly for an oil-fueled furnace. (Beckett)*

30.2.1 Motor

A 120 V, split-phase, 1/6 hp motor is used to drive the oil pump and centrifugal fan. The centrifugal fan is anchored to the motor shaft, and a rubber **flexible coupling** is used to extend (lengthen) the motor shaft so it reaches the oil pump shaft. See **Figure 30-3.** This provides a direct-drive connection so the fan and oil pump rotate at the same speed (revolutions per minute or rpm) as the motor. The motor speed may be either 1725 rpm or 3450 rpm. The oil pump and fan are designed to match the motor's speed.

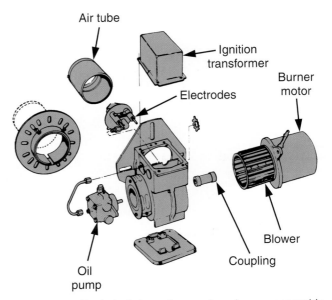

Figure 30-3. *Exploded view of a gun-type burner assembly. Note the flexible coupling that connects the oil pump and the centrifugal fan motor.*

30.2.2 Burner Fan

A centrifugal-type **burner fan** is located inside the burner assembly and anchored to the motor shaft by a setscrew. Its purpose is to supply primary air for mixing with oil. The air intake holes in the burner assembly are covered by an adjustable metal collar. The collar regulates the volume of primary air supplied to the combustion chamber.

Air entering the intake holes is forced through an air tube (gun barrel) into the combustion chamber. Special curved vanes are installed inside the end of the air tube. The vanes cause the air to twist when exiting the tube, greatly improving the mixing of air and oil vapor.

When vaporized, one gallon of fuel oil occupies a volume of 14 cu. ft. About 1500 cu. ft. of air is required to burn one gallon of No. 2 fuel oil. The oxygen in the air is required for combustion, but only about 20% of the entering air is oxygen. Therefore, for good combustion, excess air must be used. Actually, about 2000 cu. ft. of air (400 cu. ft. of oxygen) is mixed with each gallon of fuel oil. Considerable nitrogen, some oxygen, carbon dioxide, steam, and other impurities go up the flue and are vented through the chimney.

Proper airflow produces a yellow-white flame. Too much air produces a white flame, while insufficient air results in a yellow-orange flame and the formation of soot. Generally, enough excess air is fed to the flame to produce about 10% carbon dioxide (CO_2) in the flue.

30.2.3 Draft Control

A **draft control** device is used on all gun-type oil burners to keep the flue pressure constant. See

Figure 30-4. Without draft control, flue pressure would vary from the effects of wind, temperature, and atmospheric pressure. Changing flue pressure would cause changes in combustion airflow, and thus flame quality.

The draft control maintains a constant flue pressure by regulating the amount of dilution air entering the flue. As necessary, it automatically opens to increase flue pressure and closes to reduce flue pressure. The draft control has an adjustable weight for providing low, medium, or high draft.

30.3 Oil Storage Tanks

Location of the oil storage tank determines whether one or two oil supply lines are used. When the tank is above the burner level, a one-pipe system is permitted. Gravity draws oil downward from the tank through a shut-off valve and oil filter to the burner. See **Figure 30-5.**

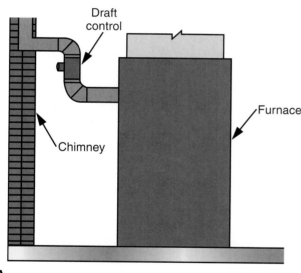

Figure 30-4. *Draft control device. A—The draft control is installed in the flue pipe between the furnace and chimney. B—An adjustable weight on the draft control can be set to provide high, medium, or low draft.*

Cold temperatures affect the *viscosity* (thickness) of the oil, impeding proper flow. For this reason, the oil storage tank is installed in the basement or buried underground in cold climates. Underground oil storage tanks require a two-pipe system because the tank is below the burner. See **Figure 30-6.**

30.3.1 Shut-off Valve and Filter

A manual shut-off valve is installed as close as practical to the tank outlet. The valve is the means of stopping the flow of oil when repairs are made to the system. The valve is normally open and is closed manually before making repairs or changing the oil filter.

An oil filter is installed between the shut-off valve and the burner assembly. Typically, the filter consists of a replaceable cartridge inside a steel housing. The cartridge is made of wool felt with a tubular wire mesh screen inserted through the core. The filter captures fine, solid impurities from the oil before they reach the nozzle. The filter element should be replaced when clogged or at the beginning of each heating season.

30.4 Oil Pumps

The oil pump is located on the burner assembly and used to increase oil pressure to 100 psig (690 kPa). Single-stage and two-stage models are available. The pumps can be either rotary- or gear-type. See **Figure 30-7** for rotary pump examples.

30.4.1 Single-stage Oil Pump

The *single-stage oil pump* is used on one-pipe systems where the oil storage tank is above the burner assembly. A single copper tube (3/8″ or 1/2″) connects the storage tank outlet to the inlet of the single-stage oil pump. Gravity supplies the force necessary to cause oil flow from the storage tank to the oil pump. Excess oil is bypassed inside the oil pump and redirected to the oil pump inlet. If an oil pump becomes defective, it is normally replaced, not repaired.

Oil pump bleed port

Air trapped in the one-pipe system prevents oil flow from the storage tank to the pump, causing the furnace to shut down due to lack of flame. Air can get into the pipe when the storage tank is empty or when the system is opened. To restore proper operation, the air must be vented.

A *bleed port* is located on the oil pump for bleeding (venting) air from the system. To bleed the system, place a 1/4″ diameter flexible hose on the bleed port, and place the free hose end in a container. Turn the bleed port nut 1/8- to 1/4-turn

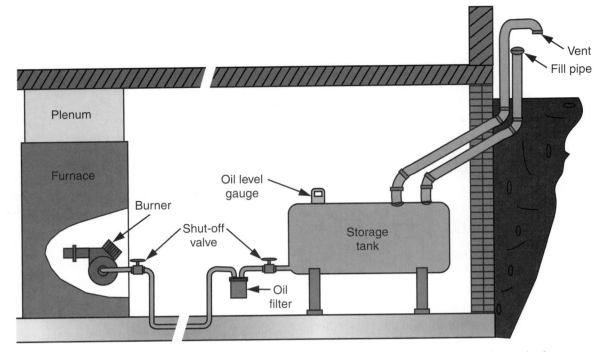

Figure 30-5. *A one-pipe system is used when the oil storage tank is on the same level with, or above, the burner.*

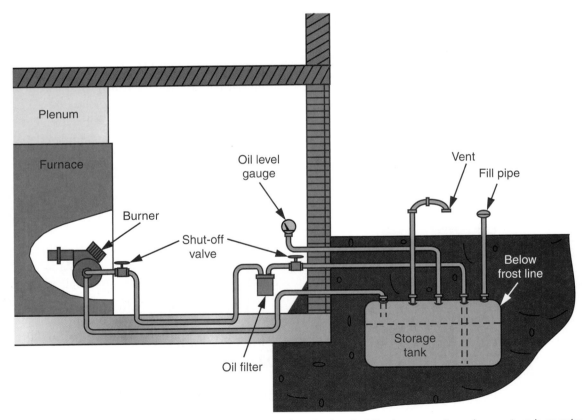

Figure 30-6. *When the oil storage tank is underground, or below the level of the burner, a two-pipe system is required.*

counterclockwise to loosen it; then, start the burner. Air and some oil will be pumped into the container. Continue pumping until a steady stream of oil appears, indicating that all air has been removed. To return the system to normal operation, tighten the bleed port nut and remove the hose.

Oil pump pressure tap

A ***pressure tap plug*** on the oil pump is used to check pump pressure to the nozzle. With the system turned off, remove the plug, and install an adapter for connecting a pressure gauge. With the system turned on, the oil pressure can be adjusted by a screw located on the pump.

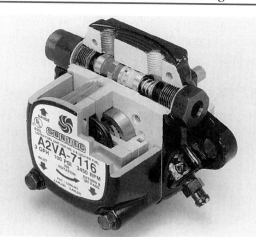

A

B

Figure 30-7. *Fuel oil pumps are used to boost oil pressure to 100 psig (690 kPa). A—Cutaway view of a single-stage rotary pump. B—A two-stage rotary pump. (Suntec Industries, Inc.)*

30.4.2 Two-stage Oil Pump

A two-stage oil pump is used on a two-pipe system where the storage tank is located below the burner assembly. The first stage pulls oil from the storage tank, and the second stage provides the 100 psig (690 kPa) pressure to the nozzle. Excess oil is channeled through a bypass inside the oil pump and returned to the storage tank through the second pipe. Two-stage oil pumps are self-priming and do not require bleeding of air.

30.5 Nozzle

Oil will not burn in the *liquid* state. Instead, it must be **vaporized** (turned into a gas). Liquid oil will turn into a gas more quickly if it is converted to a spray of fine droplets, a process called **atomizing.** Liquid oil at a pressure of 100 psig (690 kPa) is fed through a tube to a nozzle with a small **orifice** (hole). See **Figure 30-8.** A steel oil supply tube is mounted down the center of the

air tube, with the nozzle attached to the supply tube's end. This method locates the nozzle in the center of the airflow and just inside the end of the air tube. See **Figure 30-9.**

The nozzle orifice causes the oil to break into tiny particles (droplets). The droplets are sprayed into the furnace combustion chamber in a cone-shaped circular pattern. Three spray patterns are available; the hollow cone is most popular, **Figure 30-10.** A swirl chamber inside the nozzle causes the oil spray to twist in a circular motion. The twisting motion opposes the twisting pattern of the airflow, providing excellent mixing of oil with primary air.

Nozzles are normally equipped with a fine sintered-metal filter at the fuel inlet end, **Figure 30-11.** The filter prevents dirt from plugging the nozzle orifice. The filter should be cleaned or replaced when servicing the burner assembly.

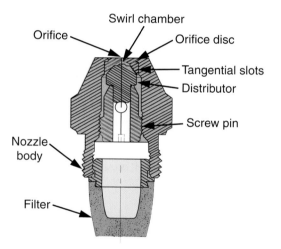

Figure 30-8. *Cross-sectional view of a nozzle for a gun-type oil burner. (Delavan)*

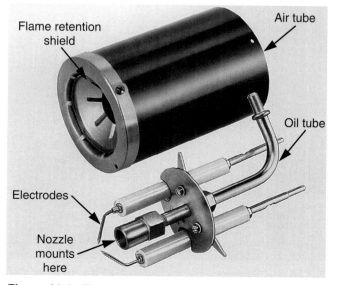

Figure 30-9. *The oil tube, nozzle, and electrodes form an assembly that is housed inside the air tube.*

30.5.1 Nozzle Orifice

The size of the nozzle orifice is calibrated to permit a specific flow of oil at 100 psig (690 kPa). The nozzle orifice rating, in gallons per hour (gph), is stamped into the side of the nozzle body. A nozzle marked 1.0

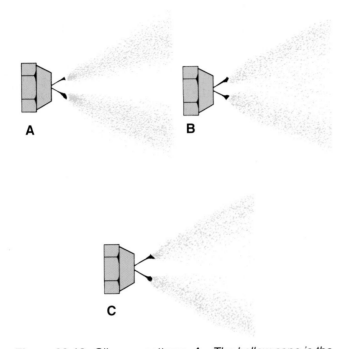

Figure 30-10. *Oil spray patterns. A—The hollow cone is the most popular pattern. B—The solid cone pattern is used for larger burners. C—The Type W nozzle provides a cone with a pattern midway between the hollow and solid patterns. (Delavan)*

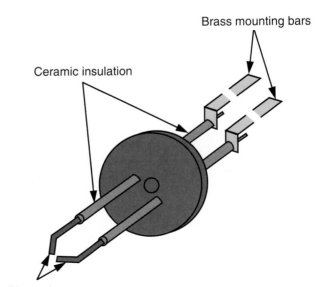

Figure 30-11. *A sintered-metal filter at the base of the nozzle traps contaminant particles that could clog the tiny orifice. (Delavan)*

delivers one gallon per hour, while one rated at 0.75 delivers 0.75 gph. The rate of oil flow must not exceed the capacity of the combustion chamber.

30.5.2 Spray Angle

The nozzle is designed to provide a steady, uniform spray at a specific angle. The **spray angle** must conform to the combustion chamber and burner requirements. The angle of spray is marked on the side of the nozzle body along with gph. The angle can vary from 30° to 90°.

When replacing a defective nozzle, always use an exact match in flow rate (gph) and spray angle (degrees). If the nozzle is too small, the burner may not heat the space properly. If the nozzle is too large, the burner will cycle on and off too frequently.

30.6 Electrodes

Two electrodes produce a high-voltage spark that ignites the mixture of oil vapor and air, **Figure 30-12.** The **electrodes** are conductors (such as stainless steel rods) with ceramic insulators to keep them from accidentally being shorted (grounded) to the mounting bars. The flat brass mounting bars connect the electrodes to the secondary terminals on the high-voltage transformer.

30.6.1 Electrode Settings

Correctly setting the electrode tips is critical to achieving proper ignition. The electrode ends are positioned in front of and above the nozzle opening, with a **gap** of 1/8″ to 3/16″ (3 mm to 5 mm) between tips.

Brass mounting bars

Ceramic insulation

Electrodes

Figure 30-12. *Typical electrode assembly for an oil-fueled furnace.*

High voltage from the transformer causes a continuous spark from tip to tip. The spark is required to ignite the oil/air mixture.

Where the spark occurs is very important. The spark should occur at the *outside edge* of the spray cone, not inside the spray. Residential units require the spark to occur 1/2″ (13 mm) in front of and *above* the nozzle opening to prevent arcing to the nozzle. **Figure 30-13** illustrates proper positioning of the electrode tips.

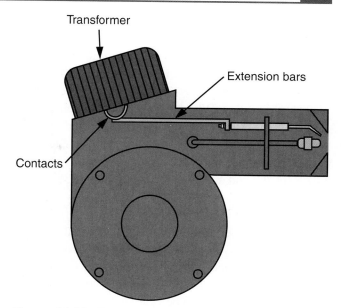

Figure 30-14. *The transformer is hinged for access to components beneath it. The transformer contacts must press firmly on the extension bars for good electrical conduction.*

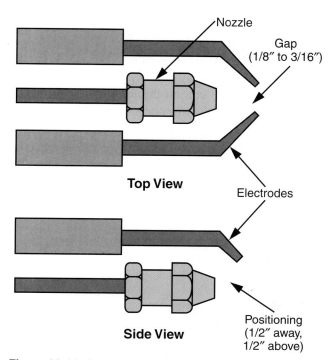

Figure 30-13. *Proper gapping and positioning of electrodes.*

30.6.2 Ignition Transformer

The **ignition transformer** is a step-up transformer (from 120 V to 10,000 V) mounted on top of the burner assembly. One side is hinged for easy access to connections and components beneath the transformer. The transformer terminals must make firm contact with the electrode extension (mounting) bars. See **Figure 30-14.**

30.7 Primary Control

The **primary control** is an electrical device that functions as a safety control. It contains a small 24 V transformer, relays, heaters, and all the switches necessary for safe operation of the burner motor and ignition transformer. It is located beside the burner transformer. The primary control has black, white, and orange external wires for line voltage (120 V) connections. Four screw-type electrical terminals located on a terminal strip at the side of the control are low-voltage connections. See **Figure 30-15.**

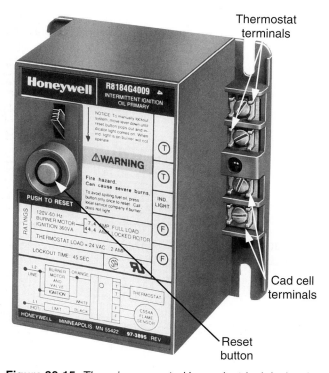

Figure 30-15. *The primary control is an electrical device that functions as a safety control. It is mounted next to the burner transformer. (Honeywell)*

Two of the low-voltage terminals (marked "T" and "T" or "T" and "TA") are thermostat connections. The T terminals are used for connecting the thermostat to the primary control. The other two terminals on the primary control are marked "S" for safety or "F" for flame safeguard (both mean the same thing). These two terminals are used to connect the *cad cell*, which is explained later in this chapter.

Most oil furnaces use a combination fan-limit control for blower operation and high-limit safety. The limit switch is connected in series with the power supply to the primary control (black wire). See **Figure 30-16.** The high limit is set for 200°F (93°C) to protect the furnace against overheating. When the limit switch opens, power supply to the primary control is disconnected, and the furnace shuts down. The furnace blower motor is controlled by the fan switch, with a cut-in setting of 140°F (60°C) and a cut-out setting of 100°F (38°C).

30.7.1 Primary Control Circuits

The primary control contains a relay that controls the power supply to the burner motor and ignition transformer (orange wire). The relay has two sets of normally open contacts controlled by a coil in the low-voltage control circuit. The orange wire is connected to one of the relay contacts where line voltage travels to the ignition transformer. See **Figure 30-17.**

When the heating thermostat switch closes, a 24 V circuit is completed to the relay coil. This closes the contacts in the 120 V circuit, supplying line voltage to the burner motor and ignition transformer via the orange wire.

The primary control contains another relay (R_2) with two sets of contacts (one NO and one NC). In addition, a 24 V heater activates a bimetallic safety control switch. These additional controls and circuits provide the necessary safety features for an oil-fueled furnace. **Figure 30-18** shows the internal circuits of a typical oil primary control.

On a call for heat from the thermostat, a 24 V circuit is completed to the relay coil R_1, **Figure 30-19.** The circuit energizes the coil on the R_1 relay for an initial start-up. This closes the two normally open R_1 contacts, one in the 24 V circuit and the other in the 120 V circuit.

Figure 30-20 illustrates the initial start-up circuits. The burner motor is pumping oil and air into the

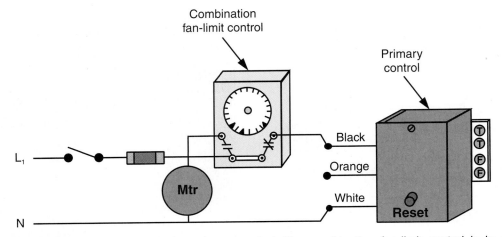

Figure 30-16. *Line voltage connections to the primary control. The combination fan-limit control is in series with the primary control.*

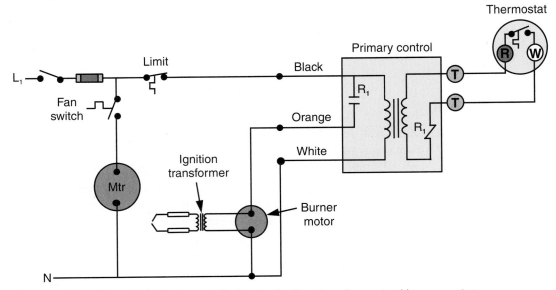

Figure 30-17. *The relay (R_1) controls power supplied to the ignition transformer and burner motor.*

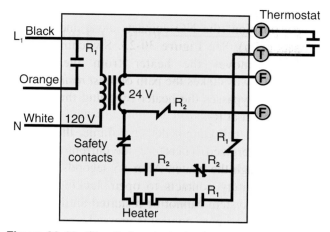

Figure 30-18. *Circuits in a typical primary control. Note the two relays.*

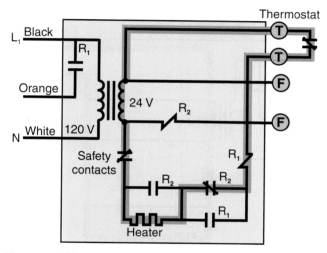

Figure 30-19. *Internal circuits of the primary control. The circuits energized by a call for heat from the thermostat are shown in color.*

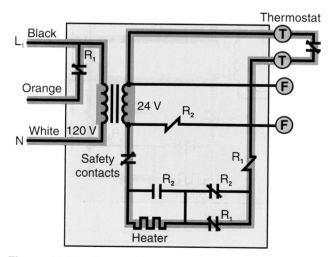

Figure 30-20. *Status of the primary control circuits and switches at start-up.*

combustion chamber, and the electrodes are sparking to ignite the mixture. If a flame does not occur (the furnace does not light) within 45 seconds, the heater will cause the bimetallic safety control contacts to open. The entire circuit is de-energized, locking out on safety. To restart, the reset button on the primary control must be pressed.

CAUTION: Do not reset the primary control too often. Unburned oil will accumulate inside the furnace after each reset. When ignited, the pool of oil will burn intensely. Do not attempt to put the fire out. Instead, shut the burner off (disconnect one of the thermostat wires from the primary control "T" terminals). Close the primary air intake. Let the fire burn with reduced air intake. Permit the furnace blower motor to operate and remove heat. If necessary, call the fire department.

30.8 Flame Safeguard

The oil-fueled furnace must be protected against malfunction, particularly loss of flame. To prevent the furnace from filling with oil, it must be turned off when loss of flame occurs. The most common method of providing a **flame safeguard** is with a *cadmium sulfide flame detector*, commonly called a **cad cell.** See **Figure 30-21.**

The cad cell is very sensitive to light, especially the yellow wavelengths, and conducts electricity when exposed to light. In darkness, cadmium sulfide becomes an insulator, preventing electron flow. The unique light-sensitivity of the cad cell is utilized to detect the flame in an oil furnace.

In darkness, the resistance of a cad cell is about 100,000 ohms, a resistance great enough to prevent current flow. When exposed to light, the resistance immediately drops to less than 1600 ohms. The drop permits enough current flow to energize a relay coil. A well-balanced combustion flame causes the cad cell resistance to drop into the 300-ohm to 1000-ohm range.

Location of the cad cell is important. It is normally located at the rear of the air tube and directly under the ignition transformer. See **Figure 30-22.** Two yellow wires from the cad cell are connected to the two "F" terminals on the primary control. When the burner ignites, the cad cell must be able to "see" the flame. Locating the cell inside the burner assembly protects the cad cell from other sources of light (called **transient light**). Ambient temperature of the cad cell cannot exceed 140°F (60°C). Care must be taken to keep the cad cell clean.

30.8.1 Cad Cell Safety Circuit

After start-up and ignition, the cad cell responds to the flame and completes a circuit to the relay coil. This

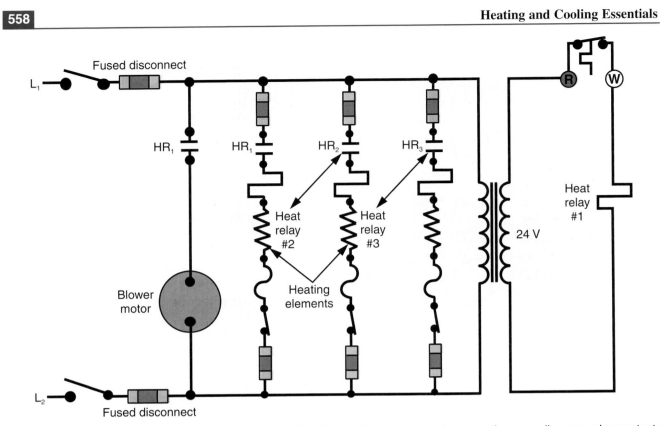

Figure 31-8. *When the thermostat calls for heat, a small resistance heater in relay 1 causes the normally open relay contacts (labeled HR₁ on this schematic) to close, energizing the first heating element and the blower motor. In turn, heat relays 2 and 3 energize the remaining two heating elements.*

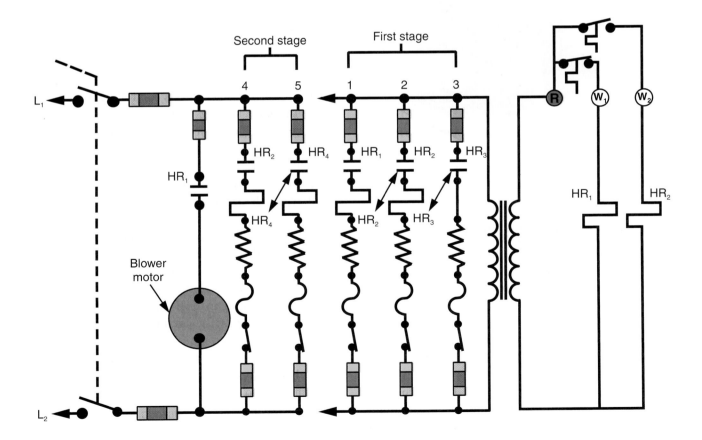

Figure 31-9. *An electric furnace with a thermostat controlling two heat stages. The first stage (heating elements 1, 2, and 3) is supplemented by the second stage (elements 4 and 5) when additional heating is needed.*

conjunction with a regular (single-stage) indoor thermostat. The outdoor thermostat monitors external air temperature, since more heat is usually needed when the outdoor air gets colder. The thermostat contacts open on rising temperature and close on falling temperature.

The outdoor thermostat controls the second bank of heating elements. The first-stage heating elements are cycled on and off by the indoor heating thermostat. When the outdoor temperature is above the outdoor thermometer's setpoint, the first-stage heating should be able to satisfy the heating needs of the building. The outdoor thermostat cannot energize the second stage unless the indoor thermostat is calling for heat, *and* the outdoor temperature is below the setpoint of the outdoor thermostat. See **Figure 31-12.**

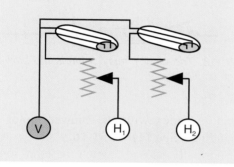

Figure 31-10. *A two-stage heating thermostat contains two mercury switches. One controls first-stage heating, the other controls the second-stage elements. Each switch is equipped with a heat anticipator.*

31.4 Cooling Blower Relay

A ***blower relay*** is required when adding air conditioning to an electric furnace. Blower controls for the heating cycle energize the blower when heat is produced. A separate relay must be used to energize the blower during the cooling cycle. The coil on the blower relay is 24 V and controlled from Terminal G on the indoor thermostat. When the thermostat's manual fan switch is set for AUTO, Terminal G automatically receives 24 V when the thermostat calls for cooling. Then, the blower relay coil is energized and closes its switch. The blower relay switch is connected in series with the blower motor. See **Figure 31-13.**

Setting the manual fan switch to CONT causes the blower relay coil to be energized continuously. On the continuous setting, the blower motor can operate any time, regardless of the thermostat setting.

31.4.1 Two-speed Blower Motor

A two-speed blower motor is often used to provide proper airflow through the furnace. Low speed (1050 rpm) is normally used for the heating cycle and high speed (1800 rpm) for the cooling cycle. **Figure 31-14** illustrates the wiring for a two-speed blower motor.

The blower relay often uses a single-pole, double-throw switch. One set of contacts is NO, and the other is NC. See **Figure 31-15.** The NC switch is connected in series with the power supply to the blower motor for the heating cycle (low-speed). The NO switch is

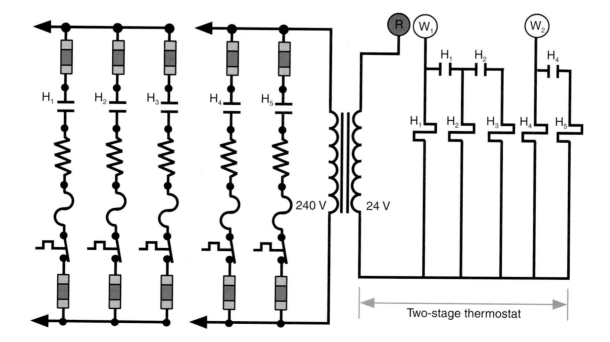

Figure 31-11. *A two-stage (five-element) heating system controlled by a two-stage thermostat. The relay heaters are in the 24 V circuit, but the contacts are part of the 240 V circuit.*

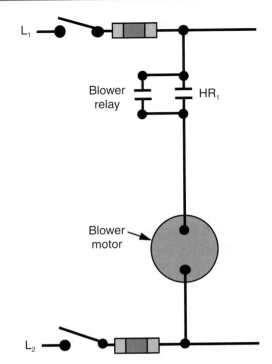

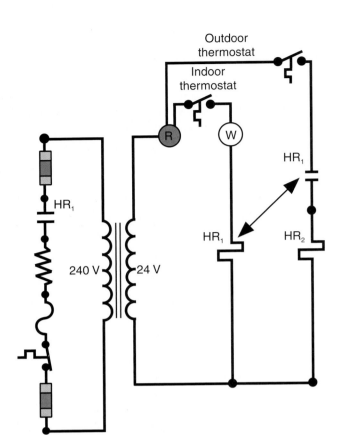

Circuit 1

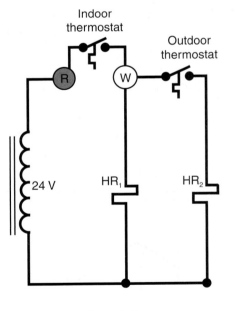

Circuit 2

Figure 31-12. *An outdoor thermostat can be used in conjunction with an indoor thermostat to control a two-stage heating system. Two circuit configurations are shown. Both accomplish the same result, energizing the second-stage heating elements when outdoor temperature drops below the outdoor thermometer's setpoint.*

Figure 31-13. *The blower relay controls the blower motor during the cooling cycle.*

connected in series with the blower motor's high-speed windings. See **Figure 31-16.**

The normally closed switch permits low-speed motor operation during the heating cycle but acts as a "lockout" when the blower operates at high speed (cooling). The normally open switch functions in just the opposite manner, permitting high-speed operation during the cooling cycle but acting as a "lockout" when the motor operates at low speed (heating). This arrangement prevents both speeds from being energized at the same time.

31.5　Combination Heating-Cooling Thermostat

A ***combination heating-cooling thermostat*** is required when adding air conditioning to an electric furnace system. A combination thermostat provides control of both heating and cooling from one device. The heating switch closes on a temperature fall, and the cooling switch closes on a temperature rise. When the cooling switch closes on a temperature rise, 24 V is supplied to Terminal Y (yellow wire) on the thermostat. A wire is connected from Terminal Y to the coil on the compressor contactor. The other side of the contactor coil is connected directly to the neutral side of the 24 V transformer. See **Figure 31-17.**

When the thermostat's cooling switch closes on a temperature rise, 24 V is supplied to both Terminals Y and G. Terminal G energizes the blower relay and

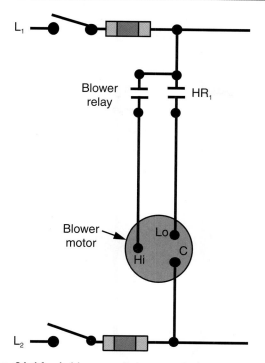

Figure 31-14. *A blower relay connected to a two-speed blower motor.*

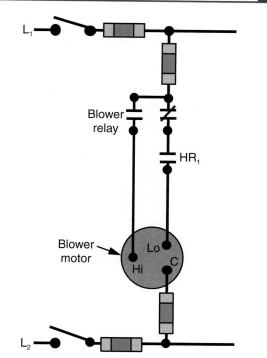

Figure 31-16. *Schematic of a blower relay with a single-pole, double-throw switch used with a two-speed blower motor.*

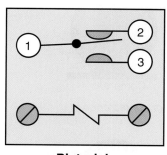

Pictorial

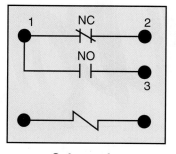

Schematic

Figure 31-15. *The blower relay used with a two-speed blower motor has a single-pole, double-throw switch. As shown in both pictorial and schematic forms, one contact is normally closed, and the other is normally open. The NC contact is in series with the blower's low speed (heating) winding.*

turns on the blower fan. Terminal Y energizes the compressor contactor coil, and the cooling system turns on. When the thermostat's cooling switch opens on temperature fall, current flow through the 24 V control circuit to the G and Y terminals stops. The

blower motor and compressor immediately stop running because the blower relay coil and the contactor coil are de-energized. The switches in the high-voltage circuit are opened.

31.6 Residential High-pressure Safety Control

Many residential air conditioning units are equipped with a ***high-pressure safety control.*** See **Figure 31-18.** The high-pressure control is attached directly to the hot gas discharge line. If discharge pressure exceeds the factory setting (about 400 psig or 2758 kPa), an internal switch will open. The two factory-installed wires are connected to place the switch in series with the 24 V power supply to the contactor coil.

The discharge pressure acts against a diaphragm balanced by a calibrated spring inside the control. If discharge pressure exceeds the setpoint, it overcomes the spring tension and opens the contacts. This stops the compressor. The safety control normally has a manual reset. When the control is tripped open, the service technician must first correct the cause of high head pressure, such as lack of condenser air, overcharge, or air in the system.

31.7 Low-pressure Safety Control

Many residential air conditioning units are also equipped with a ***low-pressure safety control,*** **Figure 31-19.** The low-pressure safety control is

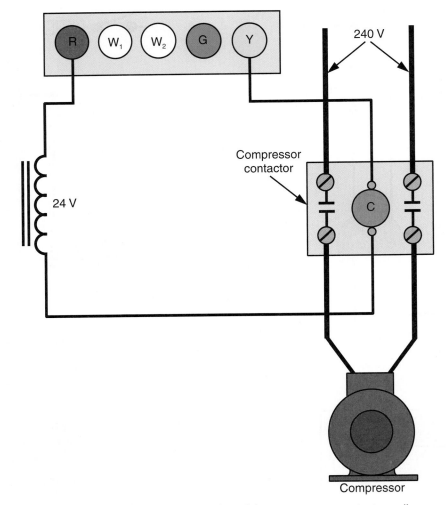

Figure 31-17. *A heating-cooling thermostat controls operation of the compressor contactor coil.*

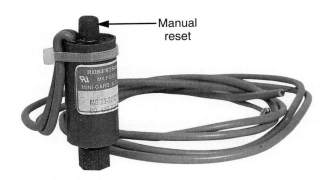

Figure 31-18. *A high-pressure safety control for residential air conditioning. (Lennox International)*

Figure 31-19. *A low-pressure safety control for residential air conditioning installations. (Lennox International)*

attached directly to the suction line. The suction works against a spring-loaded diaphragm that is mechanically linked to the normally closed contacts. If suction pressure falls below the control setpoint, the spring opens the contacts. Low suction pressure could be caused by dirty air filters, blower motor problems, a low refrigerant level, or poor flow of refrigerant (a restriction).

The low-pressure safety control is intended to protect the compressor against low suction pressure. The control setpoints are established by the factory,

and reset is automatic. Installation entails connecting two factory-installed wires to place the switch in series with the compressor contactor coil. See **Figure 31-20.**

Schematic and pictorial diagrams for a two-stage electric furnace and single-stage air conditioning system are shown in **Figures 31-21 and 31-22.** Notice that second-stage heating cannot be energized unless first-stage heating is also energized. This is accomplished by heat relay HR4A.

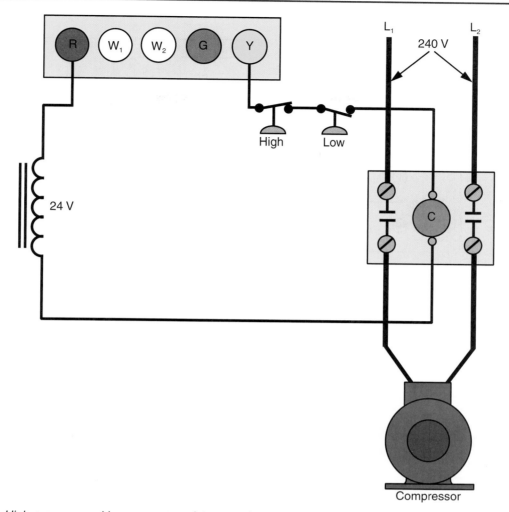

Figure 31-20. *High-pressure and low-pressure safety controls are connected in series with the compressor contactor coil.*

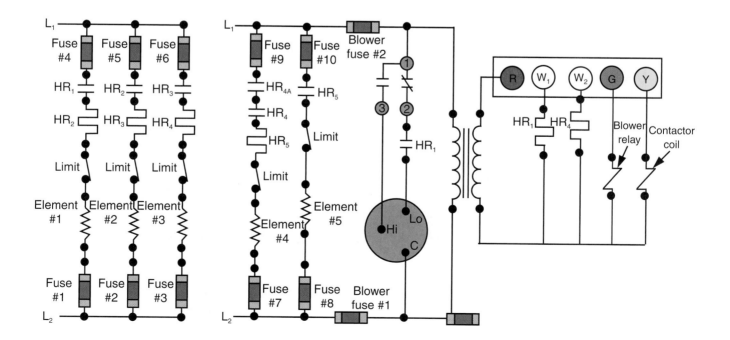

Figure 31-21. *A schematic diagram of a typical two-stage electric furnace with a single-stage air conditioning system.*

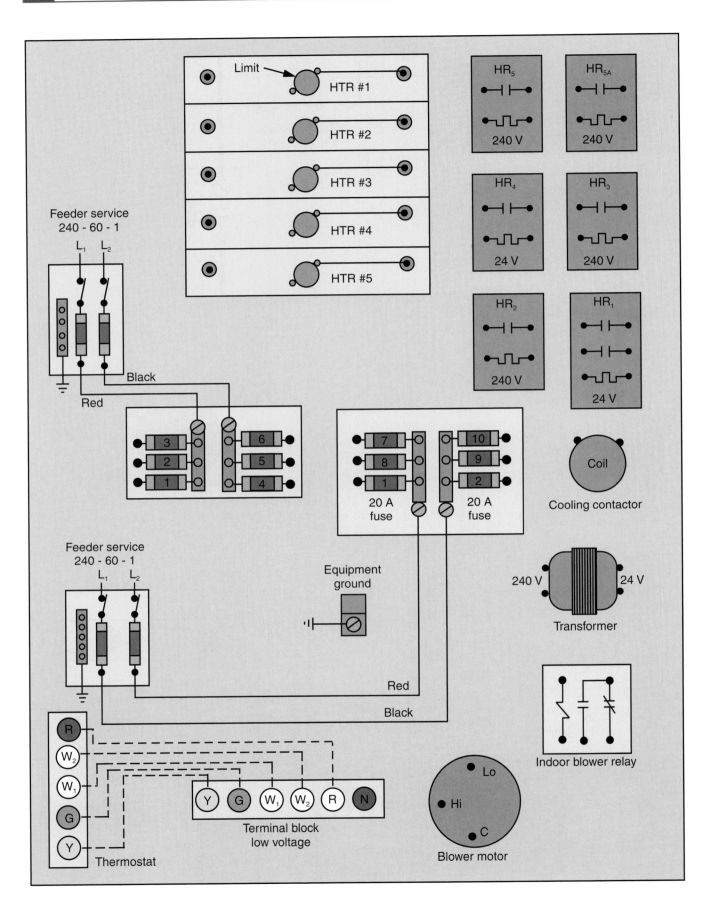

Figure 31-22. *A pictorial diagram of a typical two-stage electric furnace with a single-stage air conditioning system.*

31.8 Troubleshooting Electric Furnaces

As noted in previous chapters, systematic troubleshooting skills must be developed. A service call tests the technician's knowledge, mechanical competence, and observation and interpretation skills. Good troubleshooting requires analyzing information accurately and drawing logical conclusions. Hasty decisions or inaccurate information lead to wrong conclusions. Following step-by-step procedures and a *process of elimination* usually pinpoints the problem. There are five important steps in systematic troubleshooting:

❏ Know what the equipment *should* do.
❏ Know what the equipment *is* doing.
❏ Know what the equipment is *not* doing.
❏ Eliminate areas that *are* functioning properly.
❏ Concentrate on areas that are *not* functioning properly.

The customer's complaint usually helps narrow the problem to a specific area. These complaints typically fall into six categories:

❏ No heat.
❏ Not enough heat.
❏ Too much heat.
❏ Noise.
❏ Odor.
❏ Excessive operating cost.

A question or two to the homeowner, plus observation of the equipment, should aid in the process of elimination. By listening, observing, and testing, you will discover clues that point to the exact problem. Following are some of the symptoms and possible causes of customer complaints.

31.8.1 No Heat

Possible causes:
❏ Wrong thermostat setting.
❏ Broken or loose thermostat wires.
❏ Defective thermostat.
❏ No power supply to furnace. (Check fuses, circuit breakers).
❏ Low voltage at transformer secondary (less than 22 V).
❏ Low voltage at transformer primary (more than 10% drop).
❏ Defective transformer. (Check fuse.)
❏ Defective heating elements.
❏ Blown thermal fuse or fusible link.
❏ Defective heat control relay.
❏ Defective heating contactor.

31.8.2 Not Enough Heat

Symptom: Second stage not operating.
Possible causes:
❏ Wrong thermostat setting.
❏ Wrong heat anticipator adjustment.
❏ Vibration at thermostat.
❏ Thermostat influenced by warm air.
❏ Defective thermostat.
❏ Wrong setting on outdoor thermostat.
❏ Wrong location of outdoor thermostat.
❏ Defective outdoor thermostat.

Symptom: Some heating elements not operating.
Possible causes:
❏ Sequencer not operating.
❏ Defective high-limit control.
❏ Defective second-stage relay.
❏ Defective second-stage contactor.

31.8.3 Too Much Heat

Possible causes:
❏ Thermostat not level.
❏ Heat anticipator set too high.
❏ Thermostat in cold draft.
❏ Defective thermostat.
❏ Defective heat relay.
❏ Control circuit shorted.

31.8.4 Noise

Possible causes:
❏ Loose blower bearings.
❏ Blower bearing dry and squeaking.
❏ Blower wheel off-center.
❏ Loose V-belt.
❏ Loose running gear and mounts.
❏ Filter caught in running gear.
❏ Loose cabinet panels.
❏ Loose relay mountings.
❏ Noisy contactor. (Replace it.)
❏ Humming transformer. (Replace it.)
❏ Loose ductwork.

31.8.5 Odor

Possible causes:
❏ Dirty air filter.
❏ Stagnant water or sludge in humidifier.
❏ Control transformer burning out.
❏ Overheated wiring or overloaded circuit.
❏ Dirty heating elements.

31.8.6 Excessive Cost of Operation

Possible causes:

☐ Insufficient home insulation.
☐ Excessive air leakage from home (fireplace, vents).
☐ Wrong fan control settings.
☐ Defective blower motor with high amperage.
☐ Blower belt too tight.
☐ Dirty blower wheel.
☐ Too much or too little humidity.

Summary

This chapter explained the purposes and wire connections for commonly used components of residential electric heating systems. Single-stage heating systems were described and their wiring connections illustrated. For energy conservation, many electric heating systems have two stages of heat. These systems were explained, and two methods of controlling the stages were illustrated. The operation and proper wiring connections for two-stage heating thermostats were also explained.

The electrical connections used when air conditioning is added to the electric furnace were described. Variations in safety controls for residential air conditioning were discussed. The purpose and wiring connections for the single-pole, double-throw blower relay were described, and its use with two-speed blower motors was explained.

Test Your Knowledge

Please do not write in this text. Write your answers on a separate sheet of paper.

1. Do electric furnaces require a chimney?
2. Electric heating elements normally require a _____ power supply.
3. What are the two requirements for a fused disconnect used with an electric forced-convection heating system?
4. The heating element wire is made from what two metals?
5. _____ insulators are used to separate the heating element wire from the metal frame.
6. Why is a fuse required in each power supply leg to each element?
7. The factory installed limit switch will de-energize the heating element because of excessively high _____.
8. If the limit switch fails, what back-up device provides the necessary protection?
9. What two devices are commonly used to step-start heating elements?
10. *True or false?* Both the heat relay and the sequencer use a small heater to operate a bimetallic switch.
11. Furnace _____ is controlled by using two stages of electric heating.
12. A two-stage heating thermostat has two mercury _____.
13. Is an outdoor thermostat used to control the first stage or the second stage of a two-stage electric heating system?
14. When air conditioning is added to an electric heating system, what additional relay must be installed?
15. Name three causes of high head pressure.

32

Heat Pumps

Objectives

After studying this chapter, you will be able to:

❏ Describe the operating principles of a heat pump.
❏ Recognize and explain the purpose of major heat pump components.
❏ Define the terms *coefficient of performance*, *energy efficiency ratio*, and *balance point*.
❏ Describe the operating principle of the reversing valve.
❏ Identify and describe types of refrigerant flow controls.
❏ List types of defrost controls and describe their operation.
❏ Describe the operation of thermostats used on heat pumps.
❏ Read and interpret heat pump schematics.

Important Terms

auxiliary heat	heat pump
balance point	indoor coil
bypass line	low-ambient
coefficient of perfor-	thermostat
mance (COP)	outdoor coil
compliant scroll	outdoor fan and defrost
compressor	relay
defrost cycle	outdoor thermostat
emergency heat switch	pilot valve
energy efficiency ratio	restrictor
(EER)	reversing valve
heat content	

32.1 Residential Heat Pumps ◆

The **heat pump** is an air conditioning system capable of reversing refrigerant flow so it can function as a heating source as well as a cooling source. When the refrigerant flow is reversed for the heating cycle, the **outdoor coil** becomes an evaporator instead of a condenser, and the **indoor coil** becomes a condenser rather than an evaporator. During reverse flow, the outdoor evaporator absorbs heat from ambient air and discharges the heat into the home via the indoor condenser.

The familiar window air conditioning unit might illustrate the operation of a heat pump. In the summer, the air conditioner absorbs heat from inside the room and discharges the heat to the outside. When winter arrives, you could remove the unit from the window, turn it end-for-end, and reinstall it. In this position, the unit would then absorb heat from the outside air and discharges the heat into the room.

The heat pump is designed to accomplish the same purpose using a four-way **reversing valve** to control direction of refrigerant flow. During the cooling cycle, the heat pump functions like an ordinary air conditioner. The indoor coil is an evaporator, and the outdoor coil is a condenser. When the unit is switched to the heating cycle, however, the reversing valve changes refrigerant flow. As already noted, the indoor coil becomes the condenser and the outdoor coil becomes the evaporator.

The heat pump is *not* simply an air conditioner with reverse flow capability. It is a precisely engineered system designed to heat and cool with similar components. It looks like a conventional air conditioning system, however. See **Figure 32-1**.

This chapter describes how an air-to-air heat pump works, explains the operation of various components, and discusses the application of heat pumps in residential units.

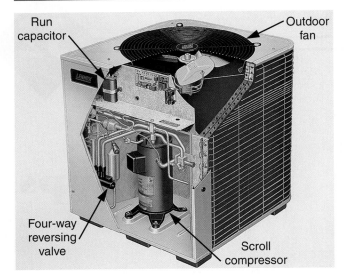

Figure 32-1. *Cutaway view of a residential heat pump. (Lennox International)*

32.1.1 Coefficient of Performance

Outside air always contains a certain amount of heat (measured in British thermal units, or Btu). Air does not lose *all* its **heat content** until it reaches absolute zero (–460°F or –237°C). The heat pump is designed to extract heat from outdoor air and to do it more efficiently than electric resistance heating. Electric resistance heat is considered 100% energy-efficient in converting electrical energy to heat energy. For every kilowatt of electricity used, an electric resistance heating system produces 3412 Btu of heat energy. A heat pump *improves upon* this energy-efficiency.

The measurement of heat output (Btu) divided by its power input (Btu/watt) is called the **coefficient of performance (COP).** A heat pump with a COP of 3.00 produces three times as many Btu of heat for the same electrical energy input as electric resistance heating elements. For example, a heat pump that produces 50,000 Btu at 60°F (16°C) outside air temperature uses about 4400 watts (4.4 kW) of electrical energy. Electric resistance heat using the same amount of electrical energy (4.4 kW) produces only 15,012 Btu. Therefore, at 60°F (16°C) outside air temperature, the heat pump has a COP of 3.32. It has three times the energy-efficiency of electric resistance heat. As the outdoor temperature falls, however, less outdoor heat is available, and the heat pump COP is reduced (becomes less energy-efficient).

32.1.2 Energy Efficiency Ratio

Energy efficiency ratio (EER) is another term to describe the energy efficiency of appliances. The EER is stated as the ratio of energy produced (output) to energy used (input). The EER provides the consumer with the relative operating cost of an appliance. The higher the EER, the more energy-efficient the product. In the preceding COP example, the heat pump produced 50,000 Btu with an input of 4400 W at an outdoor air temperature of 60°F (16°C). Dividing 50,000 by 4400 gives an EER of 11.36.

At an outdoor air temperature of 0°F (–18°C), heat pump efficiency is greatly reduced. With an output of 16,000 Btu and 4400 watts of input, the EER at 0°F (–18°C) becomes 3.64 (16,000 ÷ 4400 = 3.64).

32.1.3 Balance Point

Heat pumps require accurate calculations for building heat loss and gain. After heat loss and gain loads have been figured, the balance point of the heat pump is determined.

The **balance point** of a heat pump is where the *heating ability* of the device equals the *heat loss* of the building. See **Figure 32-2.** The balance point varies according to outdoor temperature, building design, building construction, and other factors, but falls between 30°F and 40°F (–1°C and 4°C). When the balance point is reached, the heat pump runs continuously.

When outdoor air temperature is *above* the balance point of the system, the heat pump can readily make up for heat loss. When outdoor air temperature is *at or below* the balance point, auxiliary (supplementary) heat must be supplied. **Auxiliary heat** is usually provided by staging electric heating elements. When outdoor air temperature drops to 15°F (–9°C), the heat pump is normally turned off, and the dwelling is heated by a conventional electric, gas, or oil furnace.

32.2 Heat Pump Application ◆

Guesswork has no place in selecting equipment and designing a heat pump system. Although the heat pump looks much like a conventional air conditioning unit, its components must operate equally well in both heating and cooling modes. Specially engineered heat pump components that can withstand adverse conditions are often required. A defrost cycle, more electrical controls, additional safety devices, and other components permit operation in both modes.

32.2.1 Reversing Valve

A reversing valve changes the direction of refrigerant flow through a heat pump, **Figure 32-3.** The reversing valve is normally located inside the outdoor unit, immediately above the compressor. The valve body contains a slide valve and piston that is actuated (moved) by energizing or de-energizing a 24 V solenoid coil. Three tubes are on one side of the valve body, and a single tube is on the opposite side.

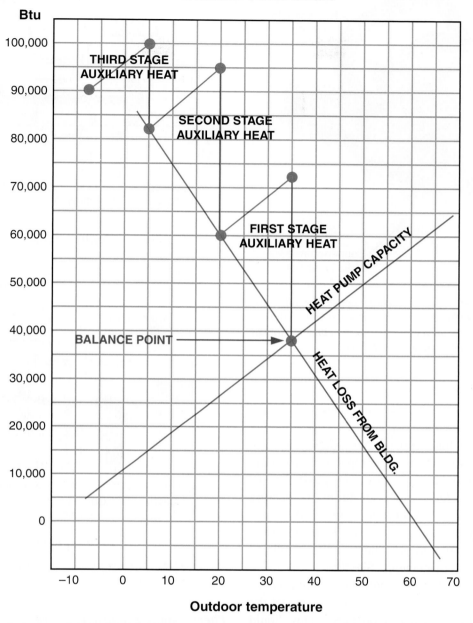

Balance Point Chart

Figure 32-2. *The balance point of a heat pump is where heat pump capacity and building heat loss are equal.*

As shown in **Figure 32-4,** the compressor discharges to the single tube. The middle tube on the *opposite* side of the reversing valve is connected to the compressor suction. (Refrigerant flow through the compressor is always the *same;* it does not reverse.) The remaining two tubes on the reversing valve are connected to the indoor coil and the outdoor coil.

The position of the slide valve and piston inside the reversing valve determines the direction of refrigerant flow to and from each coil. Flow reversal is accomplished by the slide valve connecting one of the outside tubes with the compressor suction tube of the valve. The valve position is controlled by energizing or de-energizing the solenoid coil.

The reversing valve is pilot-operated. The 24 V solenoid coil and *pilot valve* control the pressure difference at each end of the piston. Three small tubes are connected to the pilot valve body. Two of these tubes are connected to the ends of the reversing valve body. The middle tube is connected to the suction line where it leaves the reversing valve body. The pilot valve automatically controls pressure (high or low) at each end of the piston. Energizing or de-energizing the solenoid coil causes pressure on the piston ends to reverse. Thus, system pressures are used to reverse piston position. Some heat pump systems energize the reversing valve during the cooling cycle, others during the heating cycle.

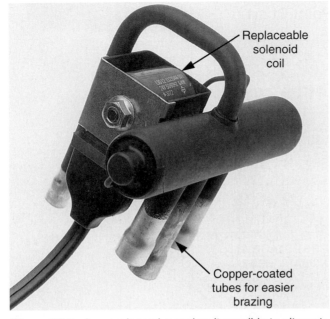

Figure 32-3. *A reversing valve makes it possible to alternate the heat pump's heating and cooling modes by reversing the flow of refrigerant. The valves are available in sizes to match the different heat pump capacities. (Ranco)*

32.2.2 Compressors

The heat pump compressor operates as a high-temperature machine during the cooling cycle, which is normal. During low ambients (heating cycle), the compressor operates as a low-temperature machine with a steadily reducing capacity. The possibility of *liquid floodback* to the compressor during the heating cycle is ever present. Compressors for heat pumps contain specially engineered features; therefore, only compressors designated as "heat pump models" should be used as replacements. See **Figure 32-5.**

Because liquid floodback is probable, all heat pumps are equipped with a suction accumulator to protect the compressor. A crankcase heater protects against liquid migration (or condensation) during the off-cycle.

The ***compliant scroll compressor*** has greatly improved compressor efficiency and reliability, **Figure 32-6.** For heat pumps, it is the compressor of choice and is rapidly capturing the market. The compliant scroll feature enables the compressor to pump liquid without damage. The need for a suction accumulator and a crankcase heater to prevent liquid floodback is eliminated. The scroll compressor also has fewer moving parts than does the reciprocating type, significantly reducing lubrication problems.

32.2.3 Refrigerant Metering Devices

Proper control of liquid refrigerant entering the coils can be accomplished in several ways. Controls include thermostatic expansion valves with bypass lines, capillary tubes, and restrictors.

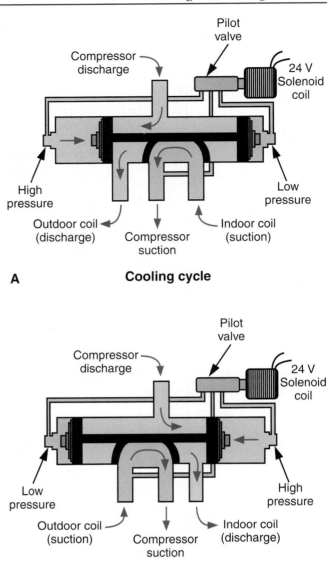

A   **Cooling cycle**

B   **Heating cycle**

Figure 32-4. *Reversing valve operation. A—Cooling cycle. B—Heating cycle.*

Thermostatic expansion valves

Thermostatic expansion valves (TEVs) with bypass and check valves are popular on large heat pump systems. See **Figure 32-7.** Because each coil operates as an evaporator, each coil requires a TEV to control the amount of liquid entering.

A ***bypass line*** is also needed for each coil. It permits refrigerant to bypass the expansion valve when the coil is operating as a condenser. A check valve is installed to control the direction of refrigerant flow in the bypass line.

Capillary tube

The metering devices for packaged or self-contained heat pumps is typically a capillary tube. Since a capillary tube operates equally well in either direction, bypass lines and check valves are not required. See **Figure 32-8.**

Figure 32-5. *A reciprocating-type compressor designed specifically for heat pump replacement applications. The compressor is equipped with a crankcase heater and used with a suction accumulator like the one at the left. (Tecumseh)*

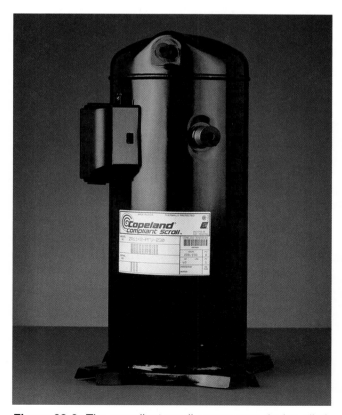

Figure 32-6. *The compliant scroll compressor design eliminates the need for a suction accumulator and a crankcase heater since the scroll can pump liquid without damage. (Copeland)*

Restrictor-type flow control

The restrictor-type flow control has nearly replaced capillary tubes and thermostatic expansion valves. See **Figure 32-9.**

In the cooling mode, the **restrictor** meters the flow of liquid refrigerant into the evaporator. The restrictor creates the pressure drop necessary to maintain the proper boiling point in the coil. In this way, the restrictor orifice acts much like a capillary tube. However, the restrictor is easily removed in the field and can be replaced with a different-sized orifice to provide a different flow rate and corresponding pressure drop.

When in the heating mode, the restrictor permits unrestricted reverse flow. As liquid refrigerant flows in the reverse direction, the restrictor automatically moves to the free-flow position. The reverse-flow capability eliminates the need for a check valve and bypass tubing.

On a heat pump, the indoor and outdoor coils differ in size and capacity. To compensate for the pressure difference this creates, a different-sized restrictor is used for each coil. During the heating cycle, the outdoor coil becomes the evaporator and requires more restriction.

32.3 Airflow Standards

The airflow standard for regular air conditioning, established by the Air Conditioning and Refrigeration Institute (ARI), is 400 cu. ft. per minute (cfm) per ton over the evaporator and 700 cfm/ton over the condenser.

However, because a heat pump is reversible, the volume of airflow through each coil must be the same. The standard is 450 cfm/ton. The air volume is greater over the indoor coil and less over the outdoor coil than for regular air conditioning systems. When adding a heat pump to an existing furnace, or making a conversion replacement, the existing duct system may be undersized.

During the heating cycle, the temperature rise over the indoor coil is considerably less than the rise in a conventional furnace. Supply air temperature to the room is about 105°F (41°C) when outdoor temperature is 30°F (–1°C). This provides a temperature rise of about 35 degrees Fahrenheit (19 degrees Celsius) over the coil, compared to about 60 degrees Fahrenheit (33 degrees Celsius) for a gas or oil furnace. The air seems cooler due to increased volume. The system's ability to maintain the desired room temperature is not affected — the system simply runs longer to satisfy the thermostat.

The outdoor coil receives much less air than is normal for a regular air conditioning system. Heat

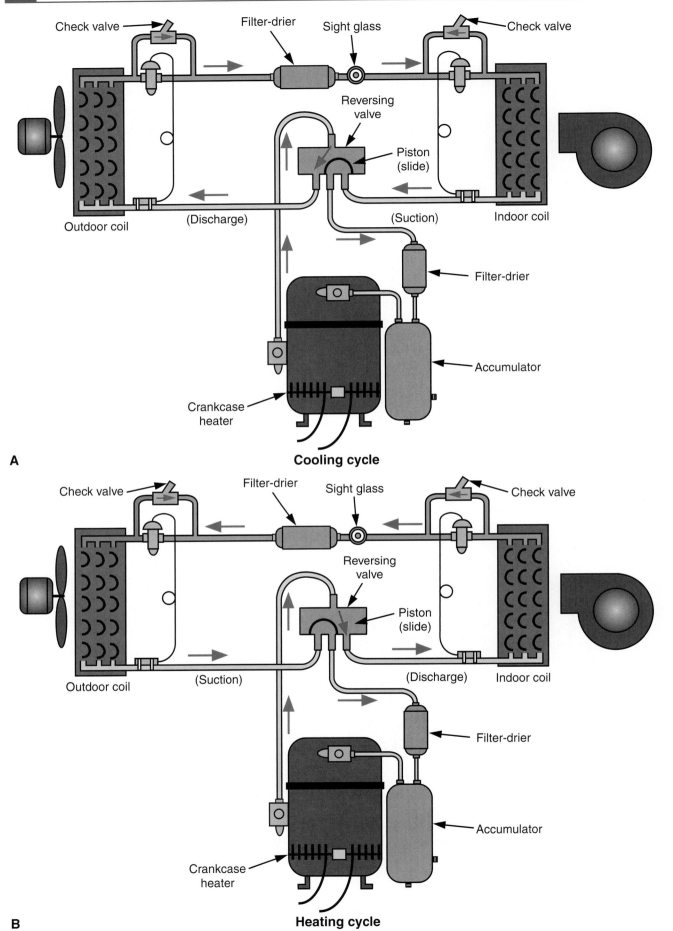

A   **Cooling cycle**

B   **Heating cycle**

Figure 32-7. *A heat pump system using TEVs and bypass lines with check valves. A—Cooling cycle. B—Heating cycle. Note how one check valve allows refrigerant to bypass the TEV in each mode.*

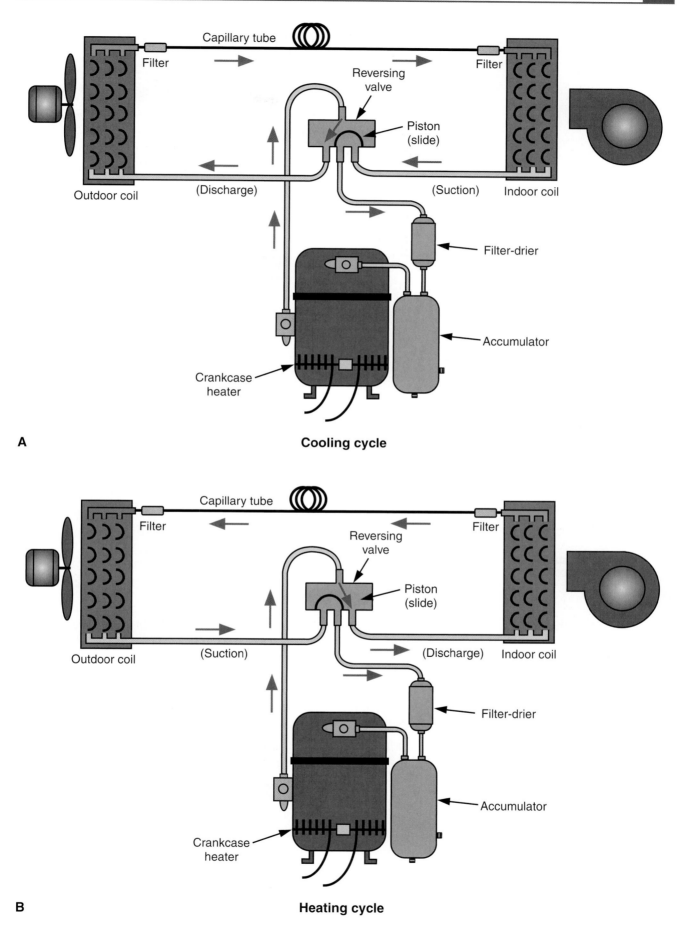

Figure 32-8. *A capillary tube system used to meter refrigerant in a heat pump. Check valves and bypasses are not needed. A—Cooling cycle. B—Heating cycle.*

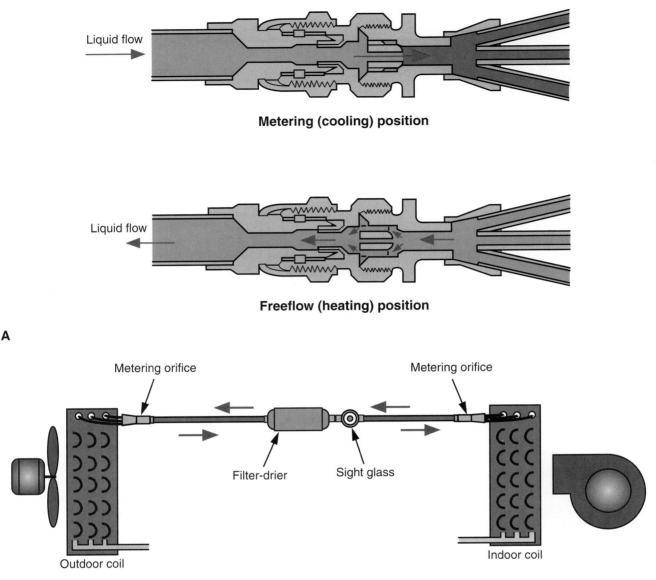

Liquid flow →

Metering (cooling) position

Liquid flow ←

Freeflow (heating) position

A

Metering orifice

Metering orifice

Filter-drier

Sight glass

Outdoor coil

Indoor coil

B

Figure 32-9. *Restrictor-type flow control. A—As shown in the cutaway views, the restrictor precisely meters refrigerant during the cooling mode but allows unrestricted reverse flow in the heating mode. B—One control is needed at each coil. Expansion and check valves, bypasses, or capillary tubes are eliminated.*

rejection during the cooling cycle is less efficient. During the heating cycle, slower airflow permits better heat pickup. Some manufacturers use a two-speed motor on the outside coil to deliver more air during the cooling cycle. Head pressure is lowered, and operation is more efficient.

32.4 Defrost Cycle ◆

Heat pumps require a ***defrost cycle*** to remove frost from the outdoor coil. The outdoor coil becomes an evaporator during the heating cycle and is about 10 degrees Fahrenheit (5.5 degrees Celsius) colder than the ambient air. Whenever the coil temperature is below 32°F (0°C), frost and ice accumulate on the

outdoor coil, **Figure 32-10.** The frost buildup must be removed periodically to maintain proper airflow.

The most practical way to remove ice buildup is to heat the coil. For a defrost cycle, the system is automatically switched to the cooling mode, even though the thermostat calls for heating. The outdoor coil is converted to a condenser. Frost quickly melts because the outdoor fan motor is turned off during defrost.

Unfortunately, heat is removed from the interior of the building during a defrost cycle. To solve this problem, electric resistance heat is turned on during defrost so air delivered to the building interior is warm.

There are several ways to control the defrost cycle. A pressure differential switch can be used to sense

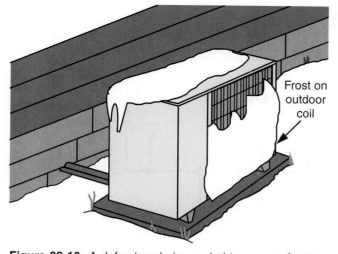

Figure 32-10. *A defrost cycle is needed to remove frost on the outdoor coil of a heat pump.*

increased pressure drop across the coil as frost forms. A temperature switch can be used to sense the decrease in coil temperature as frost forms. The most common method is the time-initiated/temperature-terminated defrost. In this system, a defrost timer runs whenever the compressor is operating. About every 90 minutes, the timer mechanically closes a switch to energize the defrost relay and initiate a defrost cycle.

A special thermostat senses outdoor coil temperature. The thermostat's contacts open on rise (65°F or 18°C) and close on fall (32°F or 0°C). The thermostat contacts are connected in series with the timer switch and defrost relay. The thermostat's contacts close when coil temperature drops to about 32°F (0°C). To initiate a defrost cycle, two conditions must be met: the timer must have completed its 90-minute cycle, *and* the coil temperature must be 32°F (0°C). If one of these elements is lacking, defrost will not occur. The system will wait another 90 minutes to try again.

The defrost cycle is terminated by the thermostat when coil temperature rises to 65°F (18°C). The defrost cycle may require five to six minutes when frost accumulation is heavy, typically when outdoor ambient is about 40°F (4°C). At lower ambients, outdoor air contains less humidity, and the defrost cycle may terminate in three minutes. If the thermostat fails to do so, the clock will terminate the defrost cycle after 10 minutes.

NOTE: A large vapor cloud may appear at the outdoor unit during defrost. It is normal and indicates the defrost cycle is operating properly.

32.5 Charging a Heat Pump System ◆

The amount of refrigerant in a heat pump is critical for efficient operation of the system. The most accurate and efficient method of charging the system is by weight. The proper type and amount (pounds and ounces) of refrigerant is indicated on the unit's data plate. Charging by exact weight can be performed year-round in either the heating or cooling mode.

32.5.1 Heat Pump Filter-driers

Most filter-driers are direction-sensitive and must be installed in the direction of refrigerant flow. Such filter-driers cannot be used on heat pumps because of reversing refrigerant flow. Special filter-driers that permit reverse flow are available for heat pumps.

32.5.2 Heat Pump Thermostats

Two types of thermostats are used to operate heat pumps. One is the two-bulb thermostat with manual changeover from heating to cooling. The other is the four-bulb thermostat with automatic changeover. Thermostats used on heat pumps are split-sub-base thermostats and may include such special features as:

❏ *Isolating contacts* that allow the use of two 24 V transformers. The jumper between terminals V and VR at the sub-base must be removed to isolate the circuits.

❏ *Emergency heat switch* for energizing auxiliary heat during a heat pump shutdown. An amber light on the thermostat indicates when emergency heat is energized.

❏ *Service light* (red) that indicates when the heat pump has malfunctioned and auxiliary heaters are providing the heat.

Various terminals that may be found on the heat pump thermostat sub-base include:

     V = voltage from heat pump transformer
   VR = voltage from indoor unit transformer
    F = indoor fan
    M = compressor
    R = reversing valve
    Y = auxiliary heat
    E = emergency heat

The two-bulb thermostat with manual changeover is found only on older heat pumps. The reversing valve is automatically energized when the manual system switch is moved to the cooling position. See **Figure 32-11.** The first bulb, which has two mercury switches, starts the compressor for heating or cooling. The second bulb controls auxiliary heat. An outdoor thermostat is connected in series with the auxiliary heat to prevent energizing auxiliary heat unless the outdoor temperature is low enough to require it.

A four-bulb thermostat uses the first-stage cooling bulb to energize the reversing valve, **Figure 32-12.** The second-stage cooling bulb controls the compressor. During the heating cycle, the reversing

Thermostat

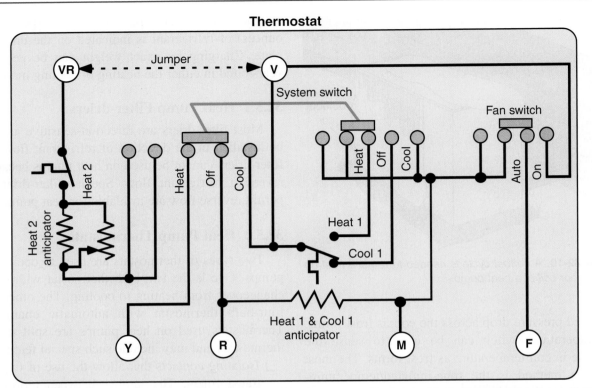

Figure 32-11. *Schematic of a two-bulb thermostat for use with a heat pump. The second bulb (Heat 2) controls auxiliary heat. Changeover from heating to cooling, and vice versa, is done with a manual switch.*

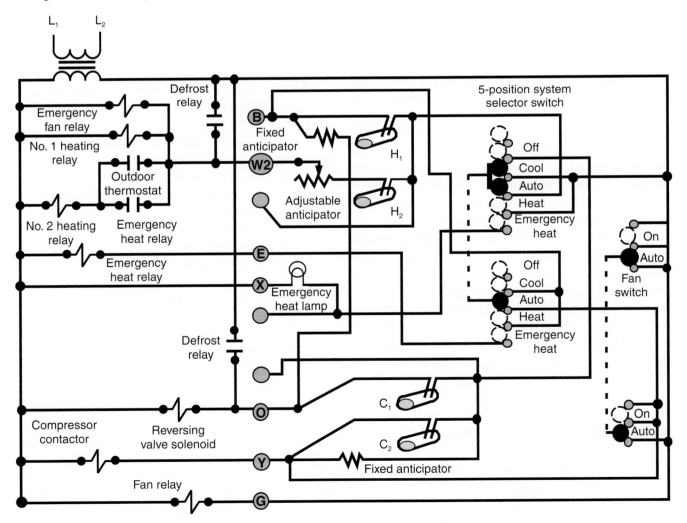

Figure 32-12. *A schematic of a four-bulb thermostat for use with a heat pump. Two bulbs are used for each mode.*

valve is not energized, and the first-stage heating bulb controls the compressor. The second-stage heating bulb controls auxiliary heat. Therefore, the four-bulb thermostat energizes the reversing valve during the cooling cycle.

The four-bulb thermostat may include a manual **emergency heat switch** to energize auxiliary heat. An indicator light on the thermostat serves as a reminder that emergency heat is being used. Another feature is the AUTO switch for automatic changeover from heating to cooling by the thermostat. These features require a special sub-base. See **Figure 32-13.**

Outdoor thermostat

An **outdoor thermostat** is used to prevent energizing of auxiliary heat stages until the outdoor temperature is low enough to require supplemental heat. See **Figure 32-14.** The outdoor thermostat is connected in series with the indoor thermostat's second-stage heating bulb. The connection ensures that the auxiliary heat cannot be energized unless *both* indoor and outdoor thermostat contacts are closed.

Low-ambient thermostat

Another control found on a heat pump is a **low-ambient thermostat.** Heat pump efficiency is greatly reduced when outdoor temperature drops between 10°F and 15°F (–12°C and –9°C). The low-ambient thermostat is located in the outdoor unit, with the contacts connected in series with the compressor contactor coil. This thermostat prevents the compressor from operating during periods of very low outdoor temperatures.

32.6 Reading Heat Pump Schematics

Electrical circuits for controlling heat pumps vary greatly from model to model and from one manufacturer to another. Electrical control circuits range from simple to complex. Most heat pump schematics are complex since they may include a defrost cycle, safety controls, and accessory features.

The heat pump service technician must be knowledgeable of heat pump theory and able to read schematics. The manufacturers' schematics explain

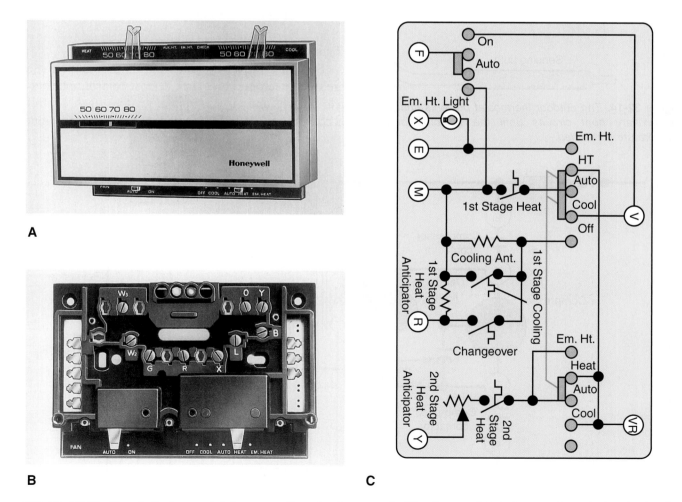

A

B

C

Figure 32-13. *Four-bulb thermostat for use with a heat pump. A—The thermostat features automatic heating/cooling changeover but can be manually switched as well. (Honeywell) B—The sub-base contains the necessary terminals and contacts for automatic and manual functions. (Honeywell) C—Schematic of a four-bulb thermostat.*

how their units' electrical components are connected. Understanding the purpose of these controls and circuits is the basis for effective troubleshooting.

32.6.1 Heat Pump Circuits

The following electrical schematics illustrate the general operating principles of heat pumps. The electrical circuits shown are common to many systems. Keep in mind that variations in types of components and control circuits are possible. Always check each unit's wiring schematic and pictorial representation for methods and sequences of operation.

The schematic in **Figure 32-15** begins with basic connections for a heat pump system. Components

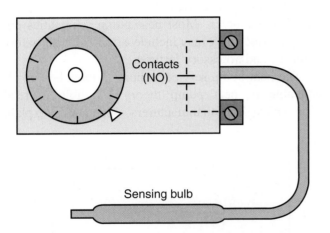

Figure 32-14. *The outdoor thermostat prevents energizing of auxiliary heat circuits until the selected outdoor temperature is reached.*

include the compressor, outdoor fan, transformer, thermostat, auxiliary heat relay, reversing valve, compressor contactor, and indoor blower relay. Additional components will be added to subsequent schematics.

Outdoor fan and defrost relay

The **outdoor fan and defrost relay** combines two relays into one component, **Figure 32-16.** Several switches are controlled by the relay coil. The relay coil can be either 24 V or 240 V, depending on the type of defrost-control circuit. When a defrost cycle is initiated, the relay coil is energized and initiates three actions:

❏ An NC contact opens and turns off the outdoor fan. Defrost is easier and quicker if cold air is not blowing over the outdoor coil.

❏ An NO contact closes, energizing the reversing valve coil. The system is switched to the cooling cycle. Hot gas is pumped to the outdoor coil.

❏ Another NO contact closes, completing a 24 V circuit to the auxiliary heat relay. The auxiliary heat is needed to offset the cooling effect of the indoor coil. See **Figure 32-17.**

Types of defrost controls

Several types of defrost controls in common use are the timeclock, solid state defrost control, mechanical switch, and air differential switch. Each control accomplishes the same purpose but in a different

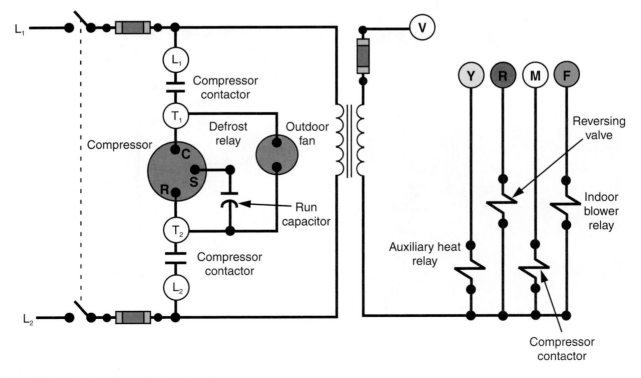

Figure 32-15. *Basic schematic for a heat pump.*

manner. A defrost timeclock in combination with a defrost thermostat is illustrated next.

Defrost timeclock. The defrost timeclock motor operates on 208 V to 240 V. A rotating cam controls the mechanical opening and closing of two switches, **Figure 32-18.** The timer motor is connected in parallel with the outdoor fan motor and runs whenever the compressor contactor is energized. Every 90 minutes, the timer closes the normally open set of contacts between Terminals 3 and 4. See **Figure 32-19.** The outdoor thermostat contacts are connected in series with the power supply to clock Terminal 4. The outdoor thermostat contacts do not close unless the outdoor coil temperature is at or below 32°F (0°C). A

circuit to the defrost relay cannot occur unless *both* the timeclock and the outdoor thermostat contacts are closed, **Figure 32-20.**

The defrost cycle is normally terminated by the outdoor thermostat when the outdoor coil temperature reaches 65°F (18°C). If the defrost timer switch closes, and the thermostat contacts are open, a defrost is not needed and cannot occur. The timer then proceeds through another 90-minute cycle. During a defrost cycle, the timer motor is de-energized along with the outdoor fan motor; both are controlled by the defrost relay.

Crankcase heater

The crankcase heater is connected to the line side of the compressor contactor and energized at all times, **Figure 32-21.** The crankcase heater helps prevent possible compressor damage caused by liquid slugging. Slugging typically occurs at startup.

Low-ambient thermostat

The low-ambient thermostat contacts are connected in series with terminal M on the indoor thermostat and the compressor contactor coil. As explained earlier, the connection prevents the compressor from operating when the outdoor temperature reaches 10°F to 15°F (–12°C to –9°C).

High- and low-pressure controls

Both high- and low-pressure safety controls are connected in the 24 V circuit. Their contacts are

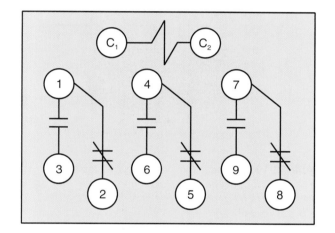

Figure 32-16. *Schematic of the combination outdoor fan and defrost relay.*

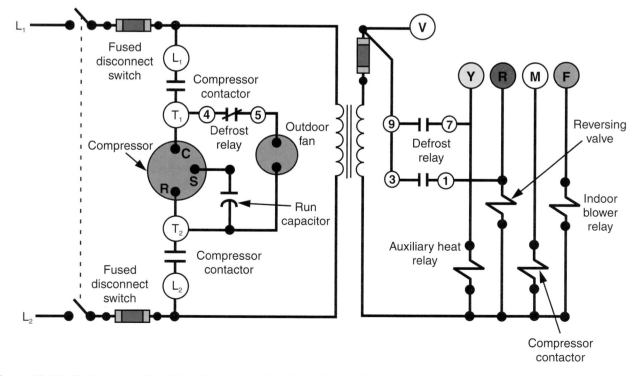

Figure 32-17. *Basic schematic with outdoor fan and defrost relay added.*

connected in series with Terminal M on the indoor thermostat, which controls the power supply to the compressor contactor coil.

32.6.2 Typical Heat Pump Schematic

Figure 32-22 is a schematic of a typical heat pump system. It shows the outdoor unit, indoor blower unit, auxiliary electric heat, and control circuits. Separate power supplies, each with disconnect and fuses, are required for the outdoor and indoor units.

32.7 Troubleshooting Heat Pumps ◆

As described in the previous three chapters, systematic troubleshooting skills must be developed. A service call tests the technician's knowledge, mechanical competence, and observation and interpretation skills. Good troubleshooting requires analyzing information accurately and drawing logical conclusions. Hasty decisions or inaccurate information lead to wrong conclusions. Following step-by-step procedures and a *process of*

Figure 32-18. *A typical defrost timeclock. A rotating cam controls the mechanical switches to start and stop the defrost cycle. (Lennox Industries, Inc.)*

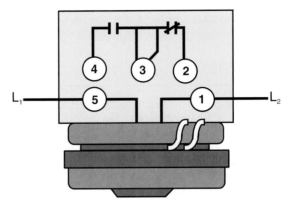

Terminals 1 and 5 are line voltage

Figure 32-19. *To begin the defrost cycle, the timer closes the NO switch between Terminals 3 and 4 every 90 minutes. If the outdoor coil temperature is at or below 32°F (0°C), the defrost cycle will begin.*

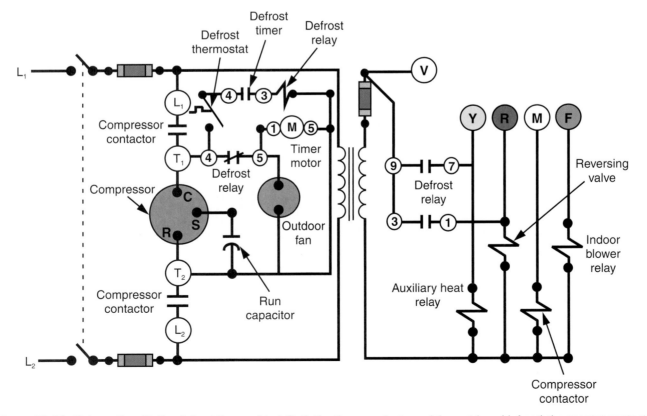

Figure 32-20. *Schematic with the defrost timer added. Both the timer contacts and the outdoor (defrost) thermostat contacts must be closed to begin the defrost cycle.*

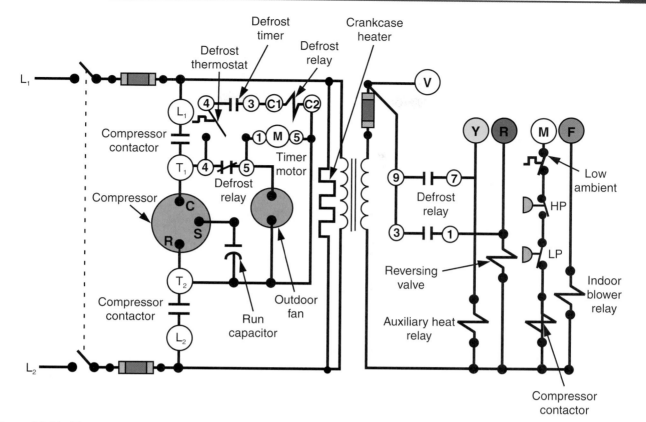

Figure 32-21. *The crankcase heater, low ambient thermostat, and high- and low-pressure safety controls have been added to the heat pump schematic.*

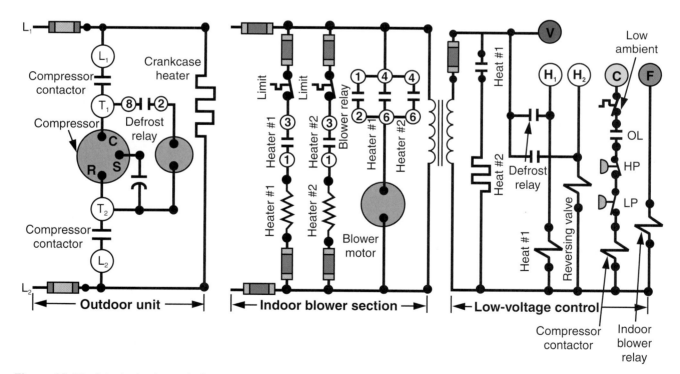

Figure 32-22. *A typical schematic for a complete heat pump system.*

elimination usually pinpoints the problem. There are five important steps in systematic troubleshooting:
- ❑ Know what the equipment *should* do.
- ❑ Know what the equipment *is* doing.

- ❑ Know what the equipment is *not* doing.
- ❑ Eliminate areas that *are* functioning properly.
- ❑ Concentrate on areas that are *not* functioning properly.

The customer's complaint usually helps narrow the problem to a specific area. Complaints typically fall into six categories:

- ❏ No heat or no cooling.
- ❏ Not enough heat.
- ❏ Too much heat.
- ❏ Noise.
- ❏ Odor.
- ❏ Excessive operating cost.

A question or two to the homeowner, plus observation of the equipment, should aid in the process of elimination. By listening, observing, and testing, you will discover clues that point to the exact problem. Following are some possible causes of customer complaints:

32.7.1 No Heat

Possible causes:

- ❏ Wrong thermostat settings.
- ❏ Broken or loose thermostat wires.
- ❏ Defective thermostat.
- ❏ No power supply. (Check fuses, circuit breakers.)
- ❏ Low voltage at transformer secondary (less than 22 V).
- ❏ Low voltage at transformer primary (more than 10% drop).
- ❏ Defective transformer. (Check fuse.)

32.7.2 No Cooling

Possible causes:

- ❏ Low-voltage transformer problem.
- ❏ Faulty thermostat or sub-base.
- ❏ Wrong thermostat settings.
- ❏ Defective low-ambient thermostat.
- ❏ Defective compressor contactor.
- ❏ Defective compressor.
- ❏ Low-pressure safety switch open.
- ❏ High-pressure safety switch open.
- ❏ Defective compressor start components (capacitors and relay).
- ❏ Defective reversing valve or coil.

32.7.3 Not Enough Heat

Possible causes:

- ❏ Wrong thermostat settings.
- ❏ Wrong heat anticipator adjustment.
- ❏ Vibration at thermostat.
- ❏ Thermostat influenced by warm air.
- ❏ Defective thermostat.
- ❏ Wrong setting on outdoor thermostat.
- ❏ Wrong location of outdoor thermostat.
- ❏ Defective outdoor thermostat.
- ❏ Dirty or clogged air filter.
- ❏ Blower motor problem.
- ❏ Defective high-limit control.

- ❏ Defective heat relays.
- ❏ Defective heating elements.

32.7.4 Too Much Heat

Possible causes:

- ❏ Thermostat not level.
- ❏ Heat anticipator set too high.
- ❏ Thermostat in cold draft.
- ❏ Defective thermostat.
- ❏ Defective heat relay.
- ❏ Control circuit "shorted."
- ❏ Compressor contactor welded shut.

32.7.5 Noise

Possible causes:

- ❏ Loose blower bearings.
- ❏ Blower bearing dry and squeaking.
- ❏ Blower wheel off-center.
- ❏ Loose V-belt.
- ❏ Loose running gear and mounts.
- ❏ Filter caught in running gear.
- ❏ Loose cabinet panels.
- ❏ Loose relay mountings.
- ❏ Noisy contactor. (Replace it.)
- ❏ Humming transformer. (Replace it.)
- ❏ Loose duct work.

32.7.6 Odor

Possible causes:

- ❏ Dirty air filters.
- ❏ Stagnant water or sludge in humidifier.
- ❏ Control transformer burning out.
- ❏ Overheated wiring or overloaded circuit.
- ❏ Relay shorted or burned.
- ❏ Dirty heating elements.

32.7.7 Excessive Cost of Operation

Possible causes:

- ❏ Thermostat problems.
- ❏ Outdoor thermostat problem.
- ❏ Defective heating relay.
- ❏ Dirty air filter.
- ❏ Insufficient home insulation.
- ❏ Excessive air leakage from home (fireplace, vents).
- ❏ Defective blower motor with high amperage.
- ❏ Blower belt too tight.
- ❏ Dirty blower wheel.
- ❏ Wrong humidity.
- ❏ Unit cycling on low-pressure safety switch.
- ❏ Unit cycling on high-pressure safety switch.
- ❏ Defective compressor valve reeds.
- ❏ Reversing valve stuck or leaking.
- ❏ Defective defrost cycle.

Summary

This chapter introduced the theory, operation, design, and application of air-to-air heat pumps in residential installations. Control and operation of the system were explained, with attention to individual components and their variations. The limitations (balance point) of heat pumps and the need for auxiliary heat were also covered.

Thermostats, relays, and timers are used for automatic control of heat pump operation. These complex circuits, along with safety controls, were explained and illustrated with simple step-by-step electrical schematics. A complete electrical schematic of a typical residential heat pump system and a troubleshooting guide concluded the chapter.

Test Your Knowledge

Please do not write in this text. Write your answers on a separate sheet of paper.

1. What component of the heat pump controls the direction of refrigerant flow?
2. What term describes the point at which the heating ability of the heat pump equals the heat loss of the building?
   a. Coefficient of performance.
   b. Balance point.
   c. Energy efficiency ratio.
   d. Pressure regulation threshold.
3. Flow reversal is controlled by energizing or de-energizing the _____ of the reversing valve.
4. What type of compressor is best for heat pumps?
5. Name three types of refrigerant metering devices used on heat pumps.
6. Heat pumps require a defrost cycle to remove frost from the _____ coil.
7. For a defrost cycle, the system automatically switches to the _____ mode.
8. *True or false?* Frost cannot form on the outdoor coil when the coil temperature is below 32°F (0°C).
9. What device terminates the defrost cycle?
10. At what temperature does the defrost termination thermostat open?
11. What is the most efficient method of charging a heat pump?
12. What type of thermostat is most commonly used on heat pumps?
13. The outdoor thermostat helps control _____.
14. What device prevents compressor operation when the outdoor temperature is below 10°F to 15°F (–12°C to –9°C)?
15. Which terminal on the indoor thermostat controls the reversing valve?

Service is the product sold by a technician. If you provide that service promptly, efficiently, and correctly, the customer will be satisfied and is likely to direct any future business to your company. (Trane)

33

Customer Relations

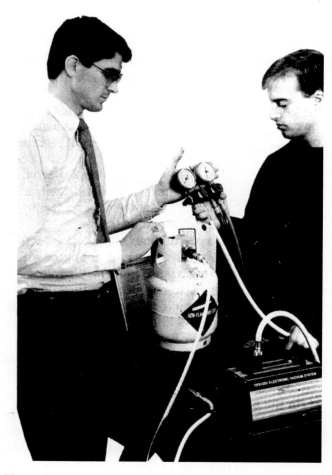

Figure 33-1. *In school or in training classes, learning how to work with people is just as important as learning how to use equipment to work with heating, cooling, or refrigeration systems. Good people skills are beneficial whether you work as a technician employed by a company or operate your own business. (Ferris State University)*

Objectives

After studying this chapter, you will be able to:

❏ Explain how to win arguments by avoiding them.

❏ Point out why it is important to learn and use people's names.

❏ Describe how to shake hands properly.

❏ Recognize the importance of good appearance.

❏ List ways of showing respect for others.

❏ Develop the skills needed to deal effectively with angry customers.

❏ Understand the value of admitting when you are wrong.

❏ Demonstrate how to begin and end service calls.

Important Terms

callback	first impression
criticism	people skills
customer relations	win-win situation

33.1 People Skills

This chapter is intended to help you practice the skills needed to deal with customers and others. Being successful in a service business requires special training in ***people skills***, or the art of working well with people. The service technician must be able to think quickly and apply ways to work effectively with people. The combination of good mechanical skills and good people skills provides the quickest road to success, **Figure 33-1.**

Many people assume they have good people skills, but they may fail miserably when involved in actual ***customer relations***. You probably have been treated in a rude manner by people in a number of business situations. Most likely, you became angry—as a paying customer, you deserve, expect, and demand better treatment. It's a commonly known fact that angry or disappointed customers take their business elsewhere.

Customers and your fellow workers encounter problems just like everyone else and they sometimes become upset. Special skills are required to handle these situations and maintain good relations. These skills help you command respect and a favorable attitude from others. People respond positively to someone they like.

Some experts estimate that, for a person involved in his or her own business, about 15 percent of financial success is due to technical skill and 85 percent is due to people skills. Employers are eager to hire those who have leadership qualities and good people skills. These special skills are easy to learn, but effort and practice must be put forth to master them and make them a "habit." This chapter provides the necessary information and guidelines for people skills. You must apply the guidelines in everyday life to make these skills work for you. This requires time, persistence, and daily application.

33.1.1 The First Rule

The first rule in demonstrating people skills is to *never* criticize or find fault. People do not like to be criticized and usually refuse to accept criticism. Being critical, even in jest, is a quick way to start an argument or make an enemy.

Criticism is dangerous because it hurts a person's pride and destroys his or her concept of self-importance. People become defensive when criticized. They will defend their actions and condemn *you*. Criticism causes bad feelings and severely impairs any further conversation.

Regardless of what people do, it makes perfectly good sense to them at *the time* they do something. *You* may not agree with their logic, but it does make sense to *them*. People will argue and defend their reasoning because they do not want to appear to be stupid. Even if criticism appears justified, restrain yourself. Consider that under similar circumstances, you might have acted in the same manner.

Always keep an open mind and avoid finding fault. Too often, people are quick to criticize and will rarely give a sincere compliment or "pat on the back." *Sincere* compliments are always welcomed and appreciated. People respond favorably to them, but they easily recognize insincere *flattery*. Some people will readily accept flattery, but most will react negatively to it.

When dealing with people in some cases, you are not necessarily going to get responses based on logic. Most often, you are dealing with emotions, prejudices, pride,

and vanity. Anyone can criticize and find fault, which can lead to trouble. It takes character and self-control to be understanding and forgiving, which leads to success.

33.2 How to Win an Argument: Avoid It!

The only way to win an argument is to *avoid* it. Never permit yourself to become a participant in an argument. You cannot win. Suppose you succeed in shooting your opponent's viewpoint full of holes and prove him or her wrong. Then what? You may feel great, but what about your opponent? You have made that person feel inferior, hurt his or her pride, and created resentment. You may be right in your views, but you will never change the other person's mind. An argument is an exercise in futility. The only way to win an argument is to avoid it and retain the other person's goodwill.

An argument is easily avoided by agreeing with your opponent, even if you know that person is wrong. A *difference of opinion* is needed for an argument. Agreeing with your opponent removes the cause of the argument and does not harm the relationship. Keeping opinions to yourself avoids trouble.

If someone loses his or her temper and you control yours, no problem exists. If you lose your temper and your opponent controls his or hers, there is no problem. A problem occurs when both people lose their tempers. There is no valuable communication during an argument, just noise and bad feelings.

An argument can be avoided by changing the subject, or by asking for an explanation. Listen to the explanation and select areas of agreement, **Figure 33-2.** Everyone likes a good listener, especially when that person agrees with the opinion being expressed. A potential argument is thus avoided; the topic is easily changed, and no hard feelings are caused.

Figure 33-2. *Finding areas of agreement can help you avoid an argument and lead to a solution of the problem.*

This is not a question of being dishonest. It is simply a matter of avoiding trouble and using good people skills. Being friendly, courteous, and considerate creates friendships and maintains good relationships. Arguments, criticism, and fault-finding will destroy friendships and cause bad feelings.

33.3 Names Are Important

It is a serious mistake to forget or misspell someone's name, especially a customer's name. People place a significant amount of importance on their name. They enjoy hearing it spoken, or seeing it in print. People are proud of their name and appreciate being recognized. The average person is more aware of his or her own name than any other. When you say a person's name, you compliment that person and capture his or her attention.

Many people cannot remember names because they do not take the time to record a name to memory when they first hear it. When you are introduced to someone, take the time to get his or her name correct. If necessary, say, "I'm sorry. I didn't hear your name clearly." Ask the person to repeat his or her name and possibly give the correct spelling. He or she will gladly oblige and will appreciate being asked. Repeat the name to be sure you have it right. This fixes the name in your mind. Then use the person's name in conversation to capture his or her attention and gain goodwill. The importance of remembering and using names cannot be overemphasized.

33.4 Become Interested in Other People

If you want others to like you, you must become genuinely interested in them. Discover what makes them tick. Identify their strengths and weaknesses. Ask questions to start a conversation, and be a good listener. Remember things that are important to the person you are dealing with.

When you become interested in others, they become interested in you. To be effective, the interest must be sincere. It becomes a **win-win situation** when communication is easy and both parties benefit.

33.5 Smile and Be Friendly

Your expression can convey a message that speaks louder than words. A frown or scowl can cause people to become defensive. A winning smile, on the other hand, is a friendly gesture that breaks down defenses. An insincere grin doesn't fool anybody, but a heart-warming smile is overwhelming and cannot be resisted. People who smile are easy to talk to. The effect of a smile is powerful and can even be detected over the telephone. A friendly, "smiling voice" can capture someone's attention. For this reason, businesses in telemarketing are constantly searching for people with friendly, smiling voices.

33.6 When Wrong, Admit It

People learn best through their mistakes, and *nobody* is right all the time. When you tell *other people* they are wrong, however, you are asking for trouble. This is a personal attack on their intelligence, pride, and self-respect.

People are often *prejudiced* and may have preconceived notions, jealousy, suspicion, fear, envy, and pride. If you are opinionated and inclined to tell people they are wrong, be prepared for a negative reaction. You cannot change someone's mind by saying he or she is wrong.

Keep an open mind and think out the situation. Begin with a smile and say, "I may be wrong; let's examine the facts." Adopt an understanding attitude and show respect for the other person's opinion. Being diplomatic and courteous is a way to avoid trouble and help make a situation reasonable. You will never get into trouble by admitting you may be wrong.

When you actually *are* wrong, have the courage to admit it. Most people foolishly try to defend their mistakes. It takes courage to admit a mistake, but it is smart to do so. People expect you to defend yourself with excuses and explanations. They are prepared to attack such a defense, but will admire you if you admit the mistake and accept the blame. Everybody makes mistakes, but few have the courage to admit when they do.

Another strategy is to steal your opponent's argument by accepting the blame and condemning *yourself* for being wrong. For example, you might say, "I should have known better," "I'm ashamed," "I should have been more careful," or "There is no excuse for my mistake." It is easier to accept self-criticism than to hear it from others. Say all the things the other person is thinking, or intends to say, before he or she has a chance. When you admit you are wrong, odds are greatly in your favor the other person will adopt a generous, forgiving attitude and minimize your mistake.

33.7 Shaking Hands

It is important to make a good **first impression** when being introduced. In business, a handshake is customary when saying hello or goodbye. When performed properly, a handshake will make a good impression, **Figure 33-3.** Unfortunately, many people

Figure 33-3. *A firm handshake will help you make a good first impression on your customer.*

Figure 33-4. *A customer's first impression of you determines how he or she will respond. A neat, clean uniform, friendly smile, and firm handshake will go a long way toward making a good impression on the customer.* *(Sporlan Valve Co.)*

do not know how to shake hands and thus make a bad impression. A proper handshake helps them get off to a good start.

A firm handshake, done with enthusiasm, is best. It should not be too quick or prolonged. A limp or weak handshake gives the impression of weakness and lack of character. A very strong squeeze indicates a pushy person with an overbearing personality. Avoid the two extremes and use a firm handshake to make the best impression.

When shaking hands, always look the other person directly in the eye. Eye contact is important, since most people feel it shows sincerity and good character. Failure to maintain eye contact gives an impression of weakness, insincerity, and poor character.

As you shake hands, smile and show enthusiasm. A sincere smile and friendly attitude will put the other person at ease. All defensive mechanisms are ignored and a new friendship begins.

Be certain you get the other person's name right and remember it. Repeat the name by saying, "Glad to meet you, John," or "It's a pleasure to meet you, Mr. Smith." This repetition not only helps fix the name in your mind, it pays the other person a sincere compliment.

The same procedure should be used when saying goodbye. Be certain the other person retains good feelings about you. A good beginning and a good ending will tend to overshadow any minor problems that have occurred. When a person likes you, he or she will usually overlook small mistakes. If the person dislikes you, those mistakes are remembered.

33.8 Physical Appearance

Your physical appearance and manner of dress convey a strong message about your personality, **Figure 33-4.** A sloppy appearance is seen by customers as an indication of sloppy work habits and poor character traits. There is no excuse for having dirty clothes or leaving a mess for someone else to

clean up. An extravagant or flamboyant appearance indicates a person who seeks recognition at any cost.

It is important to dress appropriately. People judge you on appearance and become uneasy when the picture doesn't "fit." For example, a service technician arriving in a tuxedo and top hat (or swim trunks and beach sandals) would not inspire the customer's confidence.

A clean uniform is an asset in portraying the right picture. Service technicians frequently wear a pair of blue jeans and an appropriate work shirt. This is acceptable if the clothes are clean. The quality of one's clothing is not as important as appearing clean and neat. Good personal hygiene is another basic requirement for a successful technician. Proper appearance also requires a clean shave, trimmed beard, and combed hair.

Your appearance is a prime indication of your personality and quality of work. It shows respect for yourself, your company, and the customer.

33.9 Showing Respect

Many people try to win others to their way of thinking by monopolizing conversation. They seem to believe that "He who talks loudest and longest wins the argument." But this accomplishes nothing useful, and it certainly does not change the other person's mind. Let the other person speak and explain his or her viewpoint.

Ask questions and allow the other person to answer. If you disagree, *do not interrupt.* People find it difficult to pay attention to you when their own ideas are waiting for expression. Listen patiently and keep an open mind. Be sincere and encourage the other person to express his or her viewpoint fully.

33.9.1 Dealing with Angry Customers

Angry customers are normally nice individuals who have had a bad experience. They feel let down, disappointed, frustrated, and upset. By the time the customer is ready to complain, he or she has most likely already reviewed several possible arguments. The customer is in the "attack mode" and is prepared for a verbal battle.

Let the customer speak. An angry customer has a complaint and needs to speak out; he or she will not calm down until this need is satisfied. Don't interrupt the customer and avoid becoming emotionally involved. Ignore personal comments or threats. Don't defend yourself or your company. Be sympathetic and *listen carefully.* Analyze what is being said and determine exactly *why* the customer is upset. Try to honestly see things from the customer's point of view.

Reply only when the customer indicates he or she is ready to hear what you have to say. Point out that you understand the problem and sympathize with the customer. For example, you might say, "I don't blame you for being upset. I would be, too." Don't insult the customer's intelligence by making excuses or explanations. What happened is "water over the dam."

Concentrate on the solution; ask the customer what can be done to make things right. Most customers want far less than what you think. A replacement or repair, a minor adjustment to a bill, or an apology will often resolve the problem. Let the customer know you are genuinely interested in offering help. Give the customer what he or she wants and do it quickly and effectively.

Always be receptive to customer complaints. Every business produces some unhappy customers. Situations can be turned around if customers have an opportunity to voice their complaints, know they are being heard, know someone cares, and have their problem solved quickly and effectively.

33.10 Selling Service

The service technician is a field representative of the company he or she works for. Good mechanical skills and good people skills will reflect favorably upon the company and help lead to personal advancement. You must remember that you are selling *service,* and that performance speaks louder than words. See

Figure 33-5. *As a field representative of the company you work for (or your own company), your performance, attitude, and appearance will speak louder than words. Demonstrating professional competence and a desire to solve the customer's problem will help you succeed in the HVAC field. (Elf Atochem)*

Figure 33-5. A company cannot afford service technicians who have problems with customers.

The customer is the most important person in any business. Without customers, you would not have a job. The customer is well aware of this business principle and has every right to expect good service and fair treatment. If the customer does not receive good service and fair treatment, he or she will go elsewhere. You would, too.

33.11 Ten Commandments of Good Service

The "Ten Commandments of Good Service" have been around for a long time, but the principles they illustrate always remain current:

1. Customers are not dependent on us. We are dependent on them.
2. Customers are not an interruption of our work. They are the purpose of it.
3. Customers do us a favor when they call. We are not doing them a favor by serving them.

4. Customers are a part of our business, not outsiders.

5. Customers are not cold statistics. They are real people with the same feelings and emotions we have.

6. Customers are not people to argue with or match wits against.

7. Customers are people who bring us their needs. It is our job to fill those needs.

8. Customers deserve the most courteous and attentive treatment we can give them.

9. Customers are the people who make it possible to pay your salary.

10. Customers are the lifeblood of any business.

33.12 Service Calls

Service calls require constant use of good people skills. Customers feel safe and secure when they have confidence in your ability to solve their problem. They become apprehensive when something is wrong with their heating and cooling equipment. They have questions like, "Will it catch fire or explode?" "Will it be out of service for days?" "Can it be fixed?" "How expensive will the repair be?" The service technician's appearance, attitude, and actions will either resolve the customer's fears or make them worse.

Customers usually do not have the technical knowledge to know whether the work is done properly. That is why they judge the service technician by the way he or she looks and acts. A poor appearance, or a bad attitude, can cause the customer to question your ability, **Figure 33-6.** The customer will not trust you and will worry about being charged for work that was not necessary. The customer may worry that the basic

problem was not fixed and that another service call will be required.

If a **callback** does become necessary in this situation, the customer may demand a different service technician and an adjustment on the bill. Can you imagine the telephone conversation between the dissatisfied customer and the service manager? If such situations occur repeatedly, the technician's future with the company will suffer.

33.12.1 Beginning the Service Call

Upon arrival, introduce yourself, show proper identification, and make sure you understand the problem. Ask questions about why the customer called. Effective troubleshooting requires a clear understanding of the problem. This saves time and the possible embarrassment of solving the *wrong* problem. Such questions assure the customer that you are serious, and they will build trust and confidence in your ability.

33.12.2 Customer Safety and Security

People have a basic need for safety and security. They feel threatened by anything that is not familiar. An equipment malfunction may cause a sense of danger and fear. *Never* make light of this fear. The customer's fears will only be overcome when repairs are made and fully explained. There is always a reason for fear, and it is important to find out the reason. Ask questions to get the facts, and assume the answers are correct. This convinces the customer that you take the problem seriously and builds his or her confidence in you.

So-called "nuisance" service calls (for example, an air conditioner is "making a funny noise") are just as

Figure 33-6. *Your appearance and attitude will determine how well the customer will trust you. A sloppy appearance and unprofessional attitude will not gain the customer's confidence — always dress neatly and project a competent, confident attitude when you greet the customer.*

important as a call to perform a major overhaul. Customers are usually not familiar with the technical characteristics of the equipment. They depend on you for help. Your explanation of the facts should gain their confidence and calm their fears. They will appreciate your help and gladly pay to have their sense of safety and security restored.

33.12.3 Legitimate Customer Complaints

Sometimes customers have a perfect right to be upset. Agree with them, sincerely apologize, and show respect for their complaint. Customers need to be respected as people. When they have a legitimate complaint, you should agree with the nature of it. Be reasonable and try to find a way to solve or lessen the problem.

33.12.4 Closing the Service Call

It is important at the beginning of the service call to know exactly what the problem is. At the end of the service call, it is necessary to assure the customer that you have solved the right problem. The customer deserves to know what you have done and what it means. It is not good enough to say, "I found the problem and fixed it." Explain what you have done and what it means in terms the customer can understand. Sometimes the customer expects too much, or he or she has called the wrong company to solve the problem. (For example, if a house wiring defect is affecting the air conditioner, the customer needs an electrician, not an HVAC technician.) If you are unable to correct the problem, explain clearly what must be done to correct it to the customer. Offer suggestions so that the customer can select a course of action.

Summary

Customers usually will judge the quality of work and service by the way they feel about the service technician's attitude and appearance. Trust and confidence is gained by showing the proper respect to the customer as a person and by doing the work efficiently. Be sure to get the customer's name correct and use it when talking to him or her. Avoid arguing, since there is no way you can win.

You can give the customer a sense of security by solving the right problem. This is done by making sure you know why the customer called. At the end of the service call, you must explain to the customer what you did and what it means. Customers and their problems are important because they are the basis of your job.

Test Your Knowledge

Please do not write in this text. Write your answers on a separate sheet.

1. The quickest road to success is to combine good _____ skills with good people skills.
2. *True or false?* The first rule in demonstrating people skills is to make sure all criticism is constructive in nature.
3. What is the easiest way of avoiding an argument?
4. Describe two other means of avoiding an argument.
5. _____ a person's name after you are introduced will help you fix it in your mind.
6. A _____ situation is one in which both parties benefit.
7. When you are wrong, you should:
   a. Become defensive.
   b. Ask questions.
   c. Admit your mistake.
   d. Bluff.
8. *True or false?* If you are a very skilled technician, you don't have to dress and behave conventionally.
9. When dealing with an angry customer, it is important to be _____ and listen carefully.
10. When customers have confidence in your ability to solve their problem, they feel _____ and _____.

Acknowledgments

The authors express their sincere thanks to the many organizations and manufacturers who cooperated so generously in supplying the technical information and many of the photographs used in this textbook. Any omissions from the following list are purely accidental.

Aeroquip Corp.
Alco
Allied Signal
Amprobe
Amtrol
Beckett
Big Blu
Bohn Heat Transfer
Browning
Calumet Refining Co.
Carrier Corporation
CooperTools
Copeland Corp.
Cutler-Hammer
Delavan
Dunham-Bush
DuPont
Elf Atochem
Fasco
Fluke
Frigidaire
Fuller Co.
H.B. Fuller Co.
Halstead
Henry Valve Co.
Honeywell
Hussmann Corporation
Imperial Eastman
J.W. Harris
ITW Ramset
J. B. Industries
Klein Tools, Inc.
Kramer
LoLo Building Products
LaRoche Chemicals
Lau
Lennox International, Inc.
Mac Tools, Inc.
Malco Products, Inc.

The Marley Cooling Tower Company
Milwaukee Electric Tool Corp.
Mobil Oil Corporation
Mueller Brass Co.
National Refrigerants
Nibco
NuTone, Inc.
Paragon Electric Co.
Parker-Hannifin
Penn
Prest-O-Lite
Ranco
RB&W Corporation
RCD Corporation
The Ridge Tool Co.
Robinair Mfg. Corp.
Simpson Electric Co.
Smith Equipment, Division of
    Tescom Corp.
Sporlan Valve Co.
Standard Refrigeration Company
Star Expansion Company
State of Ohio
Suntec Industries, Inc.
Tecumseh
Thermal Engineering
Thermodyne
TIF Instruments, Inc.
Trane
Triumph Twist Drill Co.
Uniweld
Vaco Products
Van Steenburgh Engineering
    Laboratories, Inc.
Vilter Mfg. Co.
Virginia KMP
Watsco
White-Rodgers
York

Glossary of Technical Terms

A

ABS: Plastic pipe and fittings made from acrylonitrile-butadiene-styrene and used only for drain, waste, and vent piping.

Absolute pressure: Any pressure above a perfect vacuum, expressed in terms of pounds per square inch absolute (psia).

Absorb: To soak up.

Absorption: A type of chiller that relies on heat energy to cool a liquid, such as water, and then uses the cooled liquid to cool something else.

Accumulator: In a refrigeration system, a small reservoir used as a safety device to trap any liquid refrigerant that did not evaporate during passage around the evaporator.

Acetylene: Fuel gas that produces the highest temperature for soldering, brazing, or cutting.

A-coil: The most popular type of air conditioning evaporator for residential systems.

ACR tubing: Copper tubing used for air conditioning and refrigeration work.

Adapters: Special fittings used to connect different types of pipe or tubing.

Adsorb: To collect substances on a surface in a condensed layer.

Air duct system: The network of tubes that is used to control airflow to and from the conditioned space.

Air off: The air leaving the evaporator. Also referred to as "supply air."

Air on: The air entering the evaporator. Also referred to as "return air."

Air space: The area being cooled.

Alkylbenzene (AB) oils: Synthetic compressor oils manufactured from propylene and benzene.

Alloy: A compound of two or more metals.

Alternating current: Electric current that reverses direction many times each second.

Ambient temperature: The temperature of the air or other fluid surrounding an object, typically a motor or condenser.

American National Acme Thread: A form of thread with a 29° angle that is used for feed screws, jacks, and vises.

American National Standard Thread: A group of three main series of threads, consisting of National Fine, National Coarse, and National Pipe.

Ammeter: Instrument used to measure electron flow.

Ammonia: The refrigerant used in a domestic absorption-type refrigerator. Water serves as an absorbent, and the absorption cycle uses heat energy instead of mechanical energy to complete the refrigeration cycle.

Ampacity: The amount of current a conductor can safely carry without becoming overheated.

Amperage: The measurement of electrical current flow, expressed in amperes.

Amperage interrupting capacity (AIC): Point beyond which a circuit protection device can no longer interrupt current flow.

Amperage relay: Relay that operates on the amperage (current flow) that is generated by the run winding.

Ampere: The unit of measure for electrical current flow.

Angle: Figure formed by two lines drawn from the same starting point.

Annealed: Term used to describe metal (such as copper tubing) that is softened by heating it to a bright cherry red color and permitting it to cool. This allows it to be bent easily.

Arbor: The spindle or axle used to attach a cutting tool, such as a saw blade to a motor.

Arcing: Electrical sparks that occur between contacts as they open or close.

Area: Square measure found by multiplying length times width, or (for a circle) the pi value of 3.1416 times the radius squared.

Armature: The movable arm of a relay.

Aspirator hole: A small hole drilled in the side of the accumulator dip tube. It permits small quantities of oil to enter the outlet tube and be drawn back to the compressor.

Atmospheric air: A mixture of various gases and water vapor.

Atmospheric pressure: The pressure exerted by air upon objects on the earth's surface. Atmospheric pressure is greatest at sea level, and is generally 14.7 pounds per square inch (101.3 kilopascals).

Atomizing: The process of converting a liquid into a spray of fine droplets.

Atoms: The smallest particles of any element.

Augers: Screwlike devices used to compress refrigerant gas in a screw compressor.

Automatic expansion valve (AEV): A pressure-type refrigerant control that is installed in the liquid line at the evaporator inlet.

Automatic gas valve: Valve used to start and stop flow of gas to the burners in a furnace.

Automatic offtime defrost: Defrost method that uses an inexpensive timer to turn the condensing unit off for two hours out of each 24.

Auxiliary heat: Supplementary heat used with a heat pump for cold weather. It is usually provided by staging electric heating elements.

Axial: A term used to describe propellerlike fan blades that move air in a direction parallel to the rotating shaft.

Azeotropic refrigerant mixture (ARM): A mixture of two pure refrigerants that behaves as a single component refrigerant with a single boiling point at a particular pressure and temperature. Azeotropes do not separate in either the gaseous or liquid state.

B

Backfire: A condition that causes the brazing torch flame to extinguish with a loud cracking sound.

Backseated: A valve position in which flow from the service gauge port to the compressor is closed off, while flow from the suction line to the compressor is unrestricted. The opposite of frontseated.

Balance point: The point at which the high-side pressure of a system rises to a saturation point at a rate equal to that of gas entering the condenser. In a heat pump, the point where the heating ability of the device equals the heat loss of the building.

Balanced port valve: A valve that equalizes opening and closing forces, compensating for a wide range of head pressure changes and ensuring more uniform operation of the thermostatic expansion valve.

Baseline data: Measurements taken from a refrigeration system for converting it to the use of alternative refrigerants. The data should include temperatures and pressures at the evaporator, temperatures and pressures at the condenser, compressor suction and discharge, expansion valve superheat, and compressor amps.

Basic refrigeration cycle: Process in which a circulating refrigerant absorbs unwanted heat at one location and carries that heat to another location where it can be discarded.

Bearing: Lubricated support for a rotating shaft.

Bimetallic: Term used to describe a disc designed to make and break a set of contacts. The disc is made by fusing together two thin circular discs of different metals, usually steel and copper, that expand at different rates when heated.

Black oxides: Contaminants that form by the combining of oxygen and copper during brazing.

Bleed port: Valve located on an oil pump for bleeding (venting) air from the system.

Bleed resistor: A protective resistor that permits electrons trapped in one side of the capacitor to slowly migrate to the empty side until the two sides are equal.

Bleedover: Condition in which a small amount of high-side pressure is vented into the low-pressure side of the system during the off cycle.

Blind hole: Hole that does not penetrate completely through a component.

Blind rivet: Permanent metal fastener used to join two pieces of sheet metal with a strong connection that will not loosen with vibration.

Blower relay: Device that energizes the blower motor during the cooling cycle.

Blower section: Cabinet area containing the blower motor and the squirrel cage-type centrifugal fan that circulates air through the furnace for treatment in heating and cooling.

Boiling point: Temperature and pressure at which water and other liquids change state and become a gas.

Bolt: A fastener that is used with a nut and normally used for fastening heavy metal parts together.

Bolt extractor: A threaded tool used to remove a broken bolt or screw.

Bonded core: An oversized type of filter-drier with a high moisture-absorbing capacity used in removing contaminants from the system.

Bonnet: Another term for a plenum. *See* Plenum.

Booster pumps: Additional pumps required on longer or higher liquid lines to maintain sufficient flow and pressure when air handlers are located far from the chiller.

Boyle's Law: Gas law that explains the relationship between pressure and volume. It states: "The pressure of a gas varies inversely (opposite) as the volume, provided the temperature remains constant."

Branch line: A secondary run of ducting that extends from the trunk line and runs to a register.

Brazing: The process of joining two pieces of base metal with the use of a filler alloy at a temperature higher than 840°F (448°C).

British thermal unit (Btu): The basic unit used to measure the quantity of heat. One Btu is the amount of heat required to raise the temperature of one pound of water one degree Fahrenheit.

Brown and Sharpe Worm Thread: A form of thread with a 29° angle that is used to mesh with worm gears and transmit motion.

Bubble point: The saturated liquid pressure for zeotropic blends, as listed on a temperature pressure card. The bubble point indicates the two pressures for a given temperature in which the first bubble of vapor appears in a liquid refrigerant.

Burner fan: Centrifugal-type fan, located inside the burner assembly, that supplies primary air for mixing with oil.

Busbar: A metal bar that serves as a common connector for electrical circuits.

Bushing: Sleeve around a shaft.

Button-lock: A type of locking system used to assemble ducting. Dimples or buttons are punched into the metal on one side and a seam is made onto the other to receive the button.

Bypass line: Tubing that permits refrigerant to bypass the expansion valve when the coil is operating as a condenser in a heat pump installation.

C ———————————————————◆

Cad cell: Common name of the cadmium sulfide flame detector, a device used to detect loss of flame in an oil-fueled furnace.

Cadmium bearing alloys: Brazing alloys containing cadmium, which is a toxic metal when molten. It emits highly poisonous cadmium oxide fumes that can cause illness or death.

Callback: Return visit necessary to correct a problem that was not resolved on the initial service call.

Cap screws: Setscrews that have heads.

Capacitor: Electrical component that acts as a throttling (choking) device to limit the number of electrons flowing in the start winding of a motor.

Capacitor-start, capacitor-run (CSCR) motor: The most powerful of the single-phase induction motors, providing both high starting torque and high running torque. The CSCR motor is used when the motor must start under a heavy load and carry a heavy load during the run cycle.

Capacitor-start, induction-run (CSIR) motor: Split-phase motor commonly used in situations where the motor must start under a heavy load, but once started, can operate with just the run winding.

Capacity control: Controlling method intended to prevent the compressor from cycling on and off rapidly during reduced load conditions.

Capillary action: The process where the alloy automatically fills the gap between the pieces of base metals.

Capillary tube: A length of copper tubing with a tiny, accurately sized inner diameter, used as a refrigerant control device in small refrigeration systems.

Capture: To temporarily trap or store refrigerant in a designated area within a refrigeration system while a repair is made to a different part of the system.

Carburizing flame: The oxyacetylene flame that results from supplying excess acetylene. Also called a "reducing" flame.

Cartridge fuses: Larger-capacity fuses that fit into special holders.

Case: In a commercial installation, an individual cooling system. A grocery store will typically have a number of cases operating at different temperatures.

Ceiling collars: Designed as a finishing piece where a pipe passes through the ceiling or floor.

Celsius scale: The temperature scale developed by the Swedish astronomer Anders Celsius, used today by most countries of the world. The scale measures the freezing point of water as 0° and the boiling point as 100°.

Centigrade: Literally, "hundred steps." Original name for the Celsius scale, which has exactly 100 divisions between the freezing and boiling points of water.

Centrifugal: A type of compressor, also called a "turbo compressor," that uses centrifugal force to compress refrigerant gas. Also, a blade type ("squirrel cage") used for blowers.

Centrifugal switch: Device that opens or closes switch when a motor exceeds or drops below a designated speed.

Certification: Process of establishing that a technician has the required knowledge of refrigerant recovery techniques and Environmental Protection Agency regulations prohibiting the venting of refrigerants to the atmosphere. EPA certification is acquired by passing an EPA-approved test and is mandated by the Clean Air Act.

CFCs: *See* Chlorofluorocarbons.

Chamfer: A beveled edge used in forming flared copper tubing.

Change of state: Phenomenon that occurs when the temperature and speed of the moving molecules of a substance reach a certain level. At this precise temperature, the molecules will rearrange themselves into a different pattern, changing a solid, liquid, or gas from one physical state to another.

Charging: Adding refrigerant to a system.

Charles's Law: Gas law that explains the relationship between pressure and volume. It states: "At a constant pressure, the volume of a confined gas varies directly as the absolute temperature; at a constant volume, the absolute pressure varies directly as the absolute temperature."

Cheater: Improper object, such as a pipe extension, used to increase leverage of a wrench.

Check valves: Valves that prevent flow of refrigerant in the wrong direction.

Clean Air Act: A federal law passed in 1990 that contains severe restrictions and penalties for venting refrigerants to the atmosphere.

Clearance pocket: Space formed at the top of the compressor cylinder to prevent the piston from striking the valve reed.

Clockwise: Rotation to the right.

Close nipple: Short section of pipe threaded along its entire length.

Closed loop: The continuous circuit of cold water pipe in a chiller that carries water from the evaporator to the cooling coil and back for heat removal and recirculation. The water circuit is closed to cooling from the atmosphere.

Coefficient of performance (COP): A measurement of heat pump efficiency obtained by dividing heat output (in Btu) by power input (in Btu/watt).

Combination fan/limit control: Control that regulates operation of the blower motor and provides protection against overheating of the furnace.

Combination heating-cooling thermostat. A thermostat that provides control of both heating and cooling from one device.

Comfort cooling: Air conditioning; the process of cooling occupants of a building.

Common: A term used to describe the wire that connects one end of the run winding and one end of the start winding of a motor.

Compliant scroll compressor: Compressor, used in heat pumps, that can pump liquid without damage. It eliminates the need for installing a suction accumulator and crankcase heater to prevent liquid floodback.

Compound gauge: Special gauge that displays two scales and makes it possible to read pressures both above and below atmospheric.

Compression: Pushing force.

Compression fittings: Pipe and tubing fittings used for heating applications that involve only low pressure.

Compression ratio: A measurement obtained by dividing the suction pressure (in psia) into the condensing pressure (in psia). It refers to the relationship between the low-side pressure and the high-side pressure of a system.

Compression stages: The number of impellers providing compression in a centrifugal compressor.

Compressor: Device that "squeezes" low-temperature, low-pressure refrigerant gas into a small volume, which will result in a high-temperature, high-pressure gas.

Condensate pan: Container located under the evaporator to catch defrost water.

Condensate water drain: Hole or hose inside a case that keeps condensation from accumulating.

Condensation: Changing from a gas to a liquid. Heat is released in the process.

Condense: Change to a liquid state.

Condenser: A heat exchanger designed to remove heat from the superheated refrigerant vapor, causing the vapor to condense (change state) back to a liquid.

Condensing unit: All the equipment necessary to reclaim the refrigerant gas and convert it back to a liquid. It consists of the compressor, condenser, hot gas discharge line, condenser fan, electrical panel box, and some accessory components.

Conductance: Ease with which current flows.

Conduction: The flow of heat through a substance from one end to the other.

Conductors: Electrical wires or pathways that carry electrical energy from place to place.

Conduit: Metal piping used to enclose and protect electrical wires.

Contactor: An electromagnet used to control multiple heavy-duty contacts that are opened or closed at the same time.

Contacts: Electrical switch components that open or close to energize or deenergize a circuit.

Contaminants: Sludge, acid, corrosion, and other foreign substances in a refrigeration system that can cause various system problems.

Continuity: Existence of a complete path for electron movement from one point to another.

Control circuit: In a relay, a low-voltage (24 V) circuit used to open or close a line-voltage (120 V) circuit.

Convection: The movement of heat by means of a carrier, such as air or water.

Corrosion: Rusting and related chemical deterioration of metal.

Coulomb: The measure of the number of electrons flowing past a given point in one second. One coulomb of electricity equals 6.25×10^{18} electrons flowing past a given point in one second.

Counter-emf. Counter-electromotive force; a counter-voltage that causes an increased resistance, decreasing current flow.

Counterclockwise: Rotation to the left.

Coupling: Fitting used to connect two lengths of hard copper tubing or pipe.

CPVC: Plastic pipe and fittings made from chlorinated polyvinyl chloride and used for hot and cold water-lines or drains, provided local building codes permit.

Cracked: State in which a valve is opened slightly by turning the stem about one or two turns.

Crankcase heater: Electric heater used to keep the crankcase oil warm to help prevent condensing of refrigerant vapor in the crankcase.

Crankcase pressure regulator: Valve used to protect the compressor from excessive suction pressure, most often at startup.

Crankshaft: Device used to change the rotary (circular) motion of the electric motor into a reciprocating (up-and-down) motion.

Crest: The top edge of two adjoining screw threads.

Crimp: Squeeze.

Crimping: A process in which small folds or bends are pressed into the metal at one end of round duct. This decreases the pipe diameter, allowing it to slide into like-sized pipe for a secure fit.

Criticism: Assessment of a person's actions or results; often perceived as negative.

Cross-charged: Term that refers to a sensing bulb that is charged with a refrigerant different from the one used in the system.

Crossover igniter: A slotted length of metal used to ignite several main burners from a single pilot flame.

Cryogenics: The use of refrigeration systems capable of producing temperature below –250°F (–157°C). They are used in laboratories to perform various scientific experiments.

Cubic feet per minute (cfm): Measure of the air volume that a fan can move.

Cubic measure: The measure of volume, found by multiplying length, width, and height.

Current: The flow of electrons in a conductor.

Customer relations: The ability to gain customer confidence and trust by effectively and promptly solving service problems.

Cut-in point: The temperature setting that causes the thermostat contacts to close, turning on the heating or refrigeration system.

Cut-out point: The temperature setting that causes the thermostat contacts to open, turning off the heating or refrigeration system.

Cutting oxygen lever. A lever that actuates a valve on an oxyacetylene cutting attachment, used to supply the stream of oxygen needed for cutting.

Cycling on the overload: Situation in which an overload condition causes the bimetallic disc to trip, shutting down the motor. After each trip, it requires more and more time for the bimetallic disc to cool and allow a restart.

D

Dalton's Law: Gas law of partial pressures. It states that in a mixture of gases, each gas behaves as if it occupies the space alone. To obtain the total pressure of a confined mixture of gases, the pressure for each gas in the mixture must be added.

Dampers: Pivoted flaps, vanes, or doors used to regulate the amount of air moving into any branch or duct section of a system. A hot spot or cold spot in a house can be controlled by adjusting or installing a damper.

Data plate: Metal plate mounted on a compressor that contains coded information revealing construction details of the unit.

Dead leg: A grounded phase of a three-phase, delta transformer.

Dead short: Electrical connection that permits an unrestricted flow of electrons.

Decimal numbers: Whole numbers, plus any parts (or fractions) of the whole. A decimal point separates the whole number from the parts.

Decimal point: Dot used to separate the whole number from the parts in a decimal number.

Defrost cycle: Sequence of operations used to remove frozen moisture from the evaporator and restore it to full efficiency.

Defrost terminator thermostat: Control used to disconnect the defrost heater when the evaporator temperature reaches about 50°F (10°C).

Degree: In geometry, one of 360 equal divisions of the perimeter of a circle. In physics, a subdivision of a temperature scale.

Dehumidifier: A refrigeration system designed specifically to remove moisture from air without affecting air temperature.

Dehydrate: Dry out.

Delta: Type of three-phase transformer with windings similar in shape to a triangle.

Delta tee: The difference in temperature between the water entering and exiting the chiller evaporator.

Denominator: The bottom number of a fraction, indicating the number of parts required to make a whole.

Density: A measurement of how closely the molecules of a substance are packed.

Desiccant: A drying agent.

Desuperheating TEV: Valve that injects enough liquid refrigerant to cool the hot discharge gas to the recommended suction temperature.

Dew point: The saturated vapor pressure for zeotropic blends, as listed on a temperature-pressure card. The dew point indicates the pressures at which the first drop of liquid condenses from a vapor.

Diameter: The length of a line across the widest part of the perimeter of a circle.

Diaphragm: Flexible disk or bellows that responds readily to pressure changes.

Dichlorodifluoromethane: *See* Freon121.

Die: A tool used to cut threads around the outside of a piece of metal.

Dielectric: An insulating material used in such devices as capacitors.

Differential: The difference between the cut-in and cut-out points in a thermostatically controlled system.

Differential valve: A valve that controls the bypass line and opens when the pressure difference between the receiver and hot gas discharge exceeds a preset value.

Diffuser: A small passage through which compressed vapor is discharged from the impeller to the volute in a centrifugal compressor.

Digits: The numbers 0 to 9.

Dimensions: Measurements of length, width, and depth.

Dip tube: Tube that extends to about one-half inch from the bottom of the receiver to assure only liquid enters the liquid line receiver outlet.

Direct current: Electric current that flows in one direction.

Direct drive: Drive arrangement where the rotor is fastened directly to the crankshaft.

Discharge bypass valve: *See* Hot gas bypass valve.

Discharge port: Hole through the valve plate of the compressor where hot gas is discharged.

Discharge service valve: Valve located on the high-pressure side of the system that allows the technician to take pressure readings and to control the flow of gas leaving the compressor.

Disconnects: Heavy-duty, manually operated contacts used for disconnecting large loads from the power supply.

Disc-type compressor valve: A valve design that increases compressor capacity up to 25% by improving volumetric efficiency.

Disposable cylinders: Refrigerant containers that are discarded after use. They are not refilled.

Dissipated: Lost.

Double-pole breaker: Two-switch circuit breaker used to disconnect both hot wires on 230 V, single-phase branch circuits.

Downflow unit: A type of furnace in which the return air enters the top and circulates down and through the unit.

Downhill joint: A type of brazing connection in which the fitting is below the point where the alloy is applied.

DPDT switch: Double-pole, double-throw switch; a switch that has two poles and two contacts for each pole. This switch makes it possible to control variety of loads from one location.

DPST switch: Double-pole, single-throw switch; a switch that has two poles and a contact for each pole. This switch makes it possible to open or close two circuits at the same time.

Draft: Air movement or natural convection (the rising of less dense hot air).

Draft control: Device used on all gun-type oil burners to keep the flue pressure constant.

Draft hood: Connection point for the burner vents and the flue in a furnace.

Dressed: Term used to describe the head of a chisel or other tool that has been ground smooth to eliminate jagged metal.

Drift punch: Tool used to align holes in different sections of material.

Drive cleats: Fastening devices used to hold together sections of rectangular duct.

Dual element: A type of fuse designed to permit an overload of short duration, but to blow instantly if a short circuit occurs.

Dual-pressure motor control: A motor control combining the low-pressure and high-pressure safety controls into a single unit. Each side of the control operates independently.

Dual-voltage motors: Motors manufactured to operate on either of two supplied voltages.

Duct caps: Duct caps are used to terminate or block off a run of pipe or rectangular duct.

Duct straps: Devices used to secure runs of flexible or rigid ducts to prevent sagging or air loss. It is also used to connect flexduct.

Dummy terminals: Terminals that have no effect on the operation of the relay, but provide connection points for splicing wires.

Duplexes: Outlet receptacles that allow easy connection and disconnection of electrical equipment.

E ———————————————◆

Eccentric: The crank throw in a motor-compressor.

Economizer: A metering device that uses an adjustable float assembly to maintain a constant liquid level in a chiller evaporator.

Elbow: A fabricated component used to make turns in duct runs. *See* Rectangular elbow and Round adjustable elbow.

Electric defrost: Automatic system that uses electric resistance heating to perform a defrost cycle.

Electrical power: The rate at which electrons are harnessed to perform a useful function.

Electrical symbols: Standardized simple drawings used for ease of communication in schematic diagrams.

Electricity: A form of energy in which electrons move from one atom to another atom.

Electrodes: In a capacitor, the two conductive layers of aluminum foil separated by an insulating layer of specially treated paper. In a thermostat, the two wire ends used with mercury in a glass tube to open and close a switch according to changes in temperature. In an oil furnace, the two conductors that produce a high-voltage spark that ignites the mixture of oil vapor and air.

Electromagnet: A magnet, used in many electrical devices, that can be turned on and off.

Electromagnetism: The phenomenon of generating a magnetic field around a conductor (wire) when current is flowing through the wire.

Electromotive force (emf): Potential difference, also called voltage.

Electronic ignition: An ignition system that saves fuel by eliminating the continuously burning pilot flame. The pilot flame is ignited by an electric spark and burns only upon a call for heat.

Electronic leak detector: A unit, consisting of a probe and a case containing the electronics, that is used to detect the presence of halide refrigerant gas.

Electronic scales: Weighing devices with displays that can be set to show, in ounces, the amount of refrigerant being withdrawn from the cylinder and charged into the system.

Electrons: Particles orbiting the nucleus of an atom, carrying negative charges.

Eliminators: Devices that serve as baffles to limit the flow of vapor in the evaporator of a chiller.

Elongation: Term used to describe the stretch factor of a filler metal.

Emergency heat switch: Control on a four-bulb thermostat used to energize auxiliary heat in a heat pump.

Energy efficiency ratio (EER): The ratio of energy produced (output) to energy used (input).

Enthalpy: The heat content of a vapor or liquid in Btu/lb., as given on a saturation chart for refrigerants.

EPA: Environmental Protection Agency. EPA regulations require the use of environmentally safe procedures in the use of refrigerants.

Epoxy: A compound that chemically hardens, used as a filler and an adhesive.

Equalizing: A balancing of system pressures.

Equivalent: Having an equal value.

Erratic: Term used to describe a valve that does not maintain a particular setting for any length of time.

Escutcheon: A term sometimes used for a ceiling collar.

Evacuating: Emptying a system with a vacuum pump.

Evaporation: Changing from a liquid to a gas. Heat is absorbed in the process.

Evaporative cooler: An alternative cooling unit, used in low-humidity regions of the country, that makes use of evaporation to cool the air. Sometimes called a "swamp cooler."

Evaporator: In a refrigeration or air conditioning system, the heat exchanging device located inside the area where heat is to be removed.

Evaporator fan delay thermostat: Device used to delay restarting of the fan until evaporator temperature drops to about 5°F (15°C).

Evaporator freeze-up: Situation in which an evaporator has become totally restricted by frost and ice buildup.

Evaporator load: The flow of liquid to the evaporator.

Evaporator pressure regulators: Special valves used to prevent evaporator pressure from falling below a set limit.

Evaporator unit: Portion of the system that absorbs heat from the refrigerated space. It consists of the evaporator, refrigerant control, evaporator fan, and some accessory components.

Exponent: A numeral that is placed above and to the right of a base number as an abbreviation to indicate a multiplication process. Also called a power.

Extended plenum system: Air duct system that combines rectangular duct with round duct.

Extension: Pulling force.

Externally equalized: A type of valve that transfers evaporator pressure from the evaporator outlet to the underside of the valve diaphragm.

F

Factor: The base number used with an exponent.

Fahrenheit scale: The temperature scale developed by German scientist Gabriel Fahrenheit, in which the freezing point of water is 32° and the boiling point is 212°.

Fail-safe setting: Timeclock feature that terminates a defrost cycle if temperature or pressure termination components fail to operate properly.

Fan cycle control: Pressure-type control used to control the condenser fan or fans. When head pressure drops below acceptable limits, the fans are turned off.

Fan relay: Device used on oil furnaces with air conditioning to operate the furnace blower motor during the cooling cycle.

Fan switch: Control used to operate the blower motor in a furnace.

Feeder service: Circuit to an electric furnace from the main disconnect or load center.

Ferrous: Term used to describe iron-containing metals.

Filler metals: Alloys used in the process of joining pieces of metal together.

Fillet: A bead of solder at the point where the joined pieces meet.

Filter: In a furnace, a cleaning device used to capture dust, lint, dirt, and other impurities from the circulating airstream.

Filter-drier: Device used to absorb moisture from the system and catch any foreign particles circulating with the refrigerant.

First impression: The feeling someone forms about you when you are first introduced.

Flame safeguard: Device that will detect loss of flame and shut down the furnace to prevent it from filling with oil.

Flare bonnet: Copper insert used to convert an ordinary flare nut into a cap nut.

Flare cap nut: Female nut that is used to seal off a male-threaded fitting.

Flare elbow: A fitting used to connect two flare nuts of the same size, while providing an accurate bend of either 45° or 90°.

Flare fittings: The mechanical fittings used in conjunction with the flaring process. Flare fittings are usually drop-forged brass and are accurately machined to form a 45° angle to meet the tubing.

Flare nut: The most frequently used fitting. Flare nuts are sized by the hole through which the tubing is inserted.

Flare plug: Fitting used to seal a flare nut or similar female-threaded opening.

Flare tee: Fitting that makes it possible to connect a branch onto an existing line of copper tubing.

Flare union: A fitting used to connect two flare nuts of the same size.

Flaring: A process of expanding or spreading copper tubing to give the end of the tube a funnel shape with a 45° angle.

Flaring block: Special tool used to hold tubing for enlarging in the swaging process.

Flash gas: The vapor that results from the sudden pressure drop (from high to low) as liquid refrigerant passes through the TEV orifice.

Flashback: A condition in which the brazing flame burns back inside the torch tip. This condition is revealed by a shrill hissing or squealing sound.

Float chamber: Component of an economizer, equipped with two float valves that serve to separate the high-pressure side (upper chamber) of a

system from the low-pressure side (lower chamber). The float assembly maintains a constant liquid level in the chiller evaporator.

Floodback: Liquid in the suction line, resulting from too much liquid refrigerant being metered into the evaporator.

Floor grille: Device used in a central location as part of the return air system, especially in mobile home installations. These are normally installed to bring air back through a filter and into the circulation system.

Flow point: *See* Liquidus.

Flow temperature: The temperature at which an alloy becomes a liquid (liquidus).

Flue: Vent pipe that runs from the diverter box to the chimney.

Flushing procedures: Methods used to make system lubricants compatible with new HFC refrigerants. Flushing procedures describe the process of draining the system oil and recharging the system with new polyol ester oil until the amount of original oil contamination is reduced to less than 5%.

Flutes: The two spiral grooves on the body of a twist drill bit that allow for the removal of material from the hole being drilled. Also, the threaded grooves of a tap.

Flux: A multipurpose chemical used to treat the clean surface of the base metal when soldering.

Flywheel: A pulley (sometimes weighted) that is attached to the external crankshaft of a motor.

Forced convection: Term describing heating and cooling systems that use a fan or blower to circulate air.

Forced-feed system: Lubrication method that uses an oil pump to deliver oil, under pressure, to all bearing surfaces and other critical parts.

Formula: A mathematical expression used as a standard to determine a value.

Four-pipe system: A piping system that provides either heating or cooling, consisting of one coil (for cooling) connected to the chiller and another coil (for heating) connected to the boiler. Two additional coils provide instant and individual changeover from heating to cooling, or cooling to heating, as necessary.

Fractional horsepower motor: A motor rated at less than one horsepower.

Fractionation: The separation of parts in a zeotropic blend. Zeotropic refrigerants fractionate when the most volatile component boils into a vapor more quickly than the less-volatile components.

Fractions: In mathematics, parts of a whole, often written with one number above another number.

Freeze-up: Condition in which an evaporator becomes clogged with frost, preventing airflow.

Freezing: The process by which a product's water content is changed to ice, and its temperature is lowered to the desired level for storage.

Freon12®: Dichlorodifluoromethane, one of the first halogen-hydrocarbon refrigerants to be introduced.

Frequency: The number of complete cycles of alternating current that occur in one second.

Frontseated: A valve position in which flow from the suction line to the compressor is closed off, while flow from the service gauge port to the compressor is unrestricted. The opposite of backseated.

Frozen storage: The storage of an already frozen product at a constant temperature, usually 0°F (18°C) or lower.

Fuel oil: Hydrocarbon fuel that is used for home heating and some industrial and commercial heating applications.

Full-floating: Term used to describe a piston pin that is free to rotate within the connecting rod and the piston.

Full load amperage (FLA): The operating ("running") amperage of a motor.

Fuses: Devices designed to detect excessive load current and melt a fusible element to open the circuit before danger arises.

Fusible element. Metal link in a fuse that overheats and melts when excess current flows through it, causing the circuit to open and stopping current flow.

G

Galvanized: Term used to describe steel pipe that has been zinc-coated.

Galvanized panning: A method that uses sheet of galvanized metal to form a duct by sealing the open bottom of the space between two floor joists. This method is typically used to channel air back through the return air system. The 16-inch width of the metal is equal to the joist spacing, which enables it to be nailed or screwed directly into the wood joist. The metal sheet is raw on one end and has an "S" cleat on the other to provide an airtight fit. Also known as *joist panning.*

Gap: Space between objects, such as electrode tips.

Gauge manifold: A pressure-checking device that has both compound and high-pressure gauges, control valves, and hose connectors from the service valves.

Gauge pressure: Pressure expressed in pounds per square inch gauge (psig). Readings in psig measure only pressures above atmospheric; gauges are calibrated to read zero at atmospheric pressure.

Generator shell: The upper shell of a lithium-bromide absorption chiller, containing the generator and the condenser, where water vapor is separated from the lithium-bromide solution in the absorption cycle.

Global warming: The effect that occurs when long-wave (infrared) radiation from the sun reaches the earth but cannot escape, causing a gradual buildup of heat on the earth's surface. Also called the "greenhouse effect."

Graduated cylinder: Device used for measuring an exact refrigerant charge. The cylinder measures refrigerant by volume rather than weight and is equipped with a pressure gauge to permit selecting the proper scale.

Grilles: Inlets for return air.

Ground fault: An internal electrical problem that permits electron flow to the frame and makes the frame electrically live.

Ground fault circuit interrupter (GFCI): Safety device designed to protect a circuit from overcurrent and to protect people from potentially hazardous ground faults arising from the use of defective appliances or portable tools.

Grounded: Connected to the earth.

Gun-type burner assembly: Burner designed to produce the conditions needed to generate a steady, hot flame inside the oil furnace.

GWP: Global warming potential. A GWP number is assigned to each refrigerant indicating the risk of a refrigerant contributing to the global warming effect.

H

Hacksaw: Saw with a metal frame and interchangeable blades, used to cut pipe, tubing, or other metal items.

Halide refrigerants: Refrigerants that combine a halogen with a hydrocarbon compound, such as acetylene, methane, or ethane.

Halide torch: A device used for detecting halogenated refrigerant leaks.

Halogens: The chemical elements fluorine, chlorine, iodine, and bromine.

Hand shutoff valves: Valves that make it possible to quickly isolate sections of a system for servicing. Used on commercial and industrial systems.

Hard-drawn copper tubing: A hard and rigid form of tubing that cannot be easily bent. Hard-drawn copper tubing is available in Type L and Type K thicknesses.

Hard soldering: *See* Brazing.

Hard solders: Silver-bearing brazing alloys that have melting temperatures ranging from 1100°F to 1500°F (593°C to 816°C).

Hard-start kit: A solid-state relay and a capacitor in a single package, used to resolve hard-start problems by turning a permanent split-capacitor (PSC) motor into a capacitor-start induction-run (CSIR) motor.

Hazardous waste: Waste material considered to require proper handling or disposal under law.

HCFCs: See Hydrochlorofluorocarbons.

Head pressure: Pressure on the high side of the system.

Head-pressure control valves: Modulating valves used to solve low head pressure problems caused by low ambient temperatures.

Head-pressure safety control: A device that will shut off the system if head pressure exceeds proper limits.

Heat anticipator: A small resistance heater, mounted inside the thermostat, that causes the thermostat switch to open just before room temperature reaches the desired level.

Heat content: The amount of heat in a substance.

Heat exchanger: Furnace section where heat is transferred from burning fuel gas to the circulating air. In a refrigeration system, the component that transfers heat from the warm liquid line to the cold suction line.

Heat load: The air to be cooled.

Heat pump: An air conditioning system capable of reversing refrigerant flow so that it can function as either a cooling source or a heat source.

Heat relays: In an electric heating system, devices used to stage the heating elements on and off.

Heat sink: Any heavy metal device, such as a valve or compressor, that tends to draw heat away from the brazing area.

Heating element: Resistance wire component used to generate the heat required for the conditioned space.

Height: Linear measure of one dimension (from base to top) of an object.

Hermetic: Airtight; totally sealed against all external forces.

Hexagonal: Having a six-sided shape.

Hex-head: Type of bolt or other fastener with a six-sided head for use with a wrench.

HFCs: See Hydrofluorocarbons.

High-pressure safety control: A device that stops the compressor if head pressure exceeds its setpoint.

High-pressure side: The condenser side of a refrigeration system.

Horizontal flow unit: A type of furnace, normally installed in an attic or crawl space, in which the system air passes horizontally through the unit to be cleaned and conditioned.

Horizontal joint: A type of brazing connection in which the tubing and the fitting are on the same level and are heated equally.

Hot gas bypass valve: A modulating valve that controls hot gas bypass to automatically maintain a desired minimum evaporator pressure.

Hot gas defrost: Defrost method in which gas is taken directly from the hot gas discharge line and delivered directly to the evaporator inlet through a bypass line.

Hot gas discharge line: Copper tubing connecting the compressor to the condenser.

Humidistat: A moisture-sensitive switch used to control system operation by stretching and contracting according to humidity.

Humidity: Moisture in the air.

Hunt and surge: Condition in which a valve fluctuates from fully open to fully closed.

Hydrargyrum: Mercury, from the Latin term for "liquid silver."

Hydrochlorofluorocarbons: Refrigerants that contain hydrogen atoms that cause the compound to be less stable in the atmosphere (and thus, less damaging) than CFCs.

Hydrofluorocarbons: Refrigerants that contain no ozone-depleting chlorine atoms.

Hydrostatic expansion: The expanding of a liquid when heat is added.

Hygroscopic: Term used to describe the moisture-absorbing ability of a substance. For example, polyol ester oils used in refrigeration systems are 100 times more hygroscopic than mineral oils.

I ———————————————◆

Ice Melting Equivalent (IME): The amount of heat absorbed in melting one ton (2000 lbs.) of ice at 32°F (0°C) in exactly 24 hours.

Ignition transformer: A step-up transformer (from 120 V to 10,000 V) that is mounted on top of the burner assembly of a furnace.

Impeller: A rotating device that provides compression in a centrifugal compressor by converting kinetic energy to static energy (pressure).

Improper fractions: Fractions in which the numerator is the same or larger than the denominator.

Incomplete combustion: Partial burning of fuel that produces substances dangerous to human health.

Indoor coil: Heat pump component that is either a condenser or an evaporator, depending upon whether the system is heating or cooling.

Induced magnetism: Magnetism that is caused by an electric current.

Induced voltage: Voltage that is generated in a conductor by a magnetic field.

Inert: Chemically inactive.

in. Hg: The abbreviation used to indicate inches of mercury in vacuum measurements. Hg is the chemical symbol for mercury.

Initiate: To begin an action or sequence of events (such as a defrost cycle).

Inline freezing: A freezing method in which fast freezing is done at very low temperatures, using cryogenic refrigerants and equipment installed directly in the production line.

Inorganic refrigerants: Refrigerants that are considered "expendable," where the vapor is not recovered for reuse. The liquid is vaporized and released to the atmosphere.

Inside diameter: The distance across the inside of a pipe or piece of tubing. Abbreviated ID.

Insulated flexible duct: Ducting formed from plastic sheeting with a spiral wire stiffener to maintain its round cross-section. An insulation layer is applied to minimize heat transfer through the plastic sheet material. This type of duct material may be used for main trunk lines or branch duct. It is available in a variety of sizes, insulation thickness, and specifications.

Insulation: A nonconductive covering for electrical wires, intended to prevent shocks and energy loss.

Insulators: Substances that are poor heat conductors. Insulators are materials that have very few free electrons and exhibit high resistance to electron flow.

Integral horsepower motor: A motor rated at one or more horsepower.

Intensity: Strength, as measured by an ammeter.

Intercooler: *See* Economizer.

Internally equalized. A type of valve in which a special passage built into the valve body transfers evaporator pressure to the underside of the valve diaphragm.

International Thread: A standardized metric thread used in most parts of the world.

Interstage pressure: Pressure of gas when leaving the first stage of a two-stage compressor.

Isolated: Blocked off or set apart from the rest of a system.

Isolation relay: Relay used on oil furnaces with air conditioning to control the heating cycle and separate the two control circuits.

J

Jaw-type pipe wrench: Wrench with serrated jaws to grip pipe.

Jumper bars: Flat copper pieces used to connect different motor terminals for easy change from one voltage to the other.

Jumper wire: Short wire installed to complete a connection from terminal to terminal.

K

Kelvin scale: Temperature scale using Celsius divisions, with absolute zero at 0°K. The freezing point of water is 273°K, and the boiling point of water is 373°K.

Kilojoule (kJ): In refrigeration work, the unit used to measure quantities of heat.

L

Ladder diagram: Another name for a schematic diagram. The power supply lines are drawn parallel to form the side rails of the ladder, and loads are drawn between the power source lines, forming the rungs.

Laminated: Assembled in layers.

Latent heat: Heat energy that causes a change of state, but no temperature change.

Latent heat of condensation: The process of changing a gas (or vapor) to a liquid by removing heat.

Latent heat of freezing: The process of changing a liquid to a solid by removing heat. Also called latent heat of fusion.

Latent heat of melting: The process of changing a solid to a liquid by adding heat.

Latent heat of sublimation: The process in which a substance changes directly from a solid to a vapor, without passing through the liquid phase.

Latent heat of vaporization: The process of changing a liquid to a vapor by adding heat. Also called latent heat of evaporation.

Latent heat value: A value used in selecting a refrigerant, equal to the number of Btus absorbed for each pound of liquid evaporated.

Lead end: End of a motor where the electrical connections are located.

Length: Linear measure of one dimension (usually the largest dimension) of an object.

Lever-type tubing benders: Tools calibrated to allow accurate short-radius bends of up to 180°.

Limit switch: A safety device that cuts off the power supply if the furnace becomes overheated.

Line starter. A contactor that has built-in overload protectors, used to operate and protect three-phase motors.

Line voltage: The power supply.

Linear measure: The distance from one point to another in a dimension.

Liquid charging: Charging a system with refrigerant in a liquid state.

Liquid line: Copper tubing connecting the condenser outlet to the metering device (refrigerant control).

Liquid receiver: Installed in larger systems, an accessory that serves as a storage tank for excess liquid refrigerant.

Liquid receiver service valve: Valve located in the liquid line at the liquid receiver outlet that allows the technician to take pressure readings and to control refrigerant flow leaving the receiver.

Liquidus: Point where melted solder flows as a liquid.

Lithium bromide: A salt used as an absorbent in the operation of large-capacity absorption chillers.

Load: Any electrical device that converts electrical energy to another form of energy.

Load center: Distribution panel that supplies electrical power to several branch circuits.

Locked rotor amps (LRA): High current flow that occurs when an electric motor starts.

Lockout: The process of locking the main electrical switch in the open position to safely perform equipment maintenance.

Logging: Problem caused by oil remaining in the evaporator due to low refrigerant velocity.

Long-term: Suitability over a significant period of time. Retrofit refrigerants are designed to serve as long-term alternatives to refrigerants scheduled for phase-out.

Low-ambient thermostat: Control that prevents operation of the heat pump compressor when outdoor temperature is very low.

Low evaporator load: A condition in which the heat load cannot reach the evaporator surface.

Low-loss fittings: Hose fittings required by the Environmental Protection Agency in the use of recovery units. Low-loss fittings prevent loss of refrigerant to the atmosphere when connecting or disconnecting hoses from the recovery unit to a refrigeration system.

Low-pressure motor control: Device used on commercial systems to control power supply to the contactor coil. The control responds to low-side pressures at the compressor.

Low-pressure safety control: A device that stops the compressor if suction pressure falls below its setpoint.

Low-pressure side: The evaporator side of a refrigeration system.

Lowest terms: The reduced terms of a fraction when the numerator and the denominator are changed to the smallest possible numbers without changing the value of the fraction.

Lubrication: The process of providing oil or other friction-reducing material to critical points of mechanical devices to lessen wear and prevent overheating.

M ────────────────◆

Machine screws: Small bolts with screw-type heads.

Magnetic field. Lines of magnetic flux generated by an electric current.

Magnetism: The ability of a natural material or a field of force generated by an electric current to attract metals containing iron.

Major diameter: The widest measurement from the outside edges of the screw threads.

Makeup water: In a water-cooled refrigeration system, replacement for water that is lost due to evaporation.

Mandrel: A nail-like pin designed to break off a blind rivet when the rivet head is formed to fasten two pieces of sheet metal together.

Malleable: Term used to describe pipe fittings that are made of cast iron that is annealed (softened by heating.)

Manifold: In a furnace, a pipe that distributes gas to the burners through an orifice (outlet hole) in each burner.

MAPP: *See* MPS.

Masonry anchors: Expanding devices that accept a fastener for mounting objects on a masonry surface.

Melting point: *See* Solidus.

Melting temperature: The temperature at which an alloy starts to melt (solidus).

Mercury: A metallic element that is a liquid at room temperature.

Mercury barometer: Measuring device for atmospheric pressure. It consists of a hollow glass tube, sealed on one end and open at the other. The glass tube is filled with mercury, turned upside down, and the open end placed in a dish half-filled with mercury.

Mercury switch: A temperature-activated switch that consists of two wire ends inside one end of a glass tube. A drop of mercury opens or closes the circuit.

Metering orifice: A refrigerant control that operates on the principle of restriction. The device has a tiny hole (orifice) sized to match the equipment exactly and is installed in the liquid line at the evaporator outlet.

Microfarad: Measuring unit for the amount of energy a capacitor can store. Microfarad ratings are used for both start and run capacitors.

Migrate: Travel from one place to another.

Minor diameter. The smallest measurement from the inside edges of the screw threads.

Miscible: Capable of mixing with another substance.

Mixed decimal numbers: Decimal numbers with digits before and after the decimal point.

Mixed numbers: Figures expressed as a whole number and a fraction used together.

Modulates: Adjusts to changing conditions.

Moisture indicator: Disc that is highly sensitive to moisture, gradually changing color to reflect the moisture content in the refrigerant.

Molecule: The smallest physical particle of any substance that can exist by itself and retain its chemical properties.

Molly bolts: Expanding metal fasteners used to anchor relatively light objects to hollow walls.

Monochlorodifluoromethane: The most widely used HCFC, designated R22.

Montreal Protocol: An international agreement reached in 1987 to substantially reduce, and eventually eliminate, production of ozone-depleting CFCs.

Motor overload: A device that protects the motor windings from damage caused by overheating due to overloaded conditions or poor ventilation.

MPS: Methylacetylene-propadiene stabilized gas, often referred to by its trade name, MAPP.

Mullion heaters: Resistance heaters that prevent condensation of moisture around refrigerator doors during periods of high humidity.

Multiflame tip: Brazing tip that produces several small flames, which tend to wrap around the tubing being heated.

Multimeters: Electrical test meters that perform multiple functions.

Multiplexed evaporators: The connection of several evaporators to a single compressor. Evaporator pressure regulator valves make it possible to operate each evaporator at a different temperature.

Multispeed motor: A motor that can be operated at different speeds, such as an air conditioner fan with high, medium, and low settings.

Mushroomed: Term used to describe the head of a chisel or other tool that is flattened and spread out from being struck.

MVAC: Motor vehicle air conditioning.

N

Napthenic: Term describing mineral oils used in refrigeration systems. Napthenic oils are not compatible with most of the new HFC refrigerants developed to lessen ozone depletion.

National Electrical Code (NEC): Standard for electrical wiring and devices.

NARMs: Nearly azeotropic refrigerant mixtures, zeotropic blends with a small temperature glide.

Net oil pressure: The pressure difference between oil pump pressure and low-side pressure.

Neutral: Having no charge. In single-phase systems, a neutral wire has zero voltage.

Neutral flame: The oxyacetylene flame that results from supplying equal amounts of oxygen and acetylene.

Neutrons: Particles in an atom's nucleus that have no charge (neutral particles).

Nominal: Approximate size.

Normally closed. Term describing a switch or valve that is in a closed state until energized. Abbreviated NC.

Normally open: Term describing a switch or valve that is in an open state until energized. Abbreviated NO.

Nucleus: The center of an atom.

Number: A figure or word that denotes quantity.

Numerator: The top number of a fraction, indicating the number of parts available.

Nuts: Internally threaded metal pieces, usually square or hexagonal in shape, that are used with a bolt.

O

ODP: Ozone depletion potential. An ODP number is assigned to each refrigerant indicating the risk to the ozone layer caused by the chlorine in refrigerants.

Off-cycle defrost: Defrost method in which the evaporator fan runs continuously while the system is off, so warm air melts accumulated frost from the evaporator.

Offsets: Short-distance changes in the direction of tubing, commonly made by using elbow fittings.

Ohm: The unit of resistance.

Ohmmeter: Device used to accurately check resistance.

Ohm's Law: Summary of the exact relationships among voltage, amperage, and resistance. Ohm's Law is used to predict and control the activity of electricity.

Oil passages: Lubrication holes drilled in the crankshaft, connecting rods, and other components of a pump or compressor.

Oil pressure safety control: A control that stops the compressor if oil pressure falls below safe limits.

Oil separator: Component used to remove oil from refrigerant and return the oil to the compressor crankcase.

Open capacitor: Capacitor in which the electrical connection is broken, usually when one of the terminals becomes separated from its coil.

Open circuit: Lack of continuity in electron movement; a broken path.

Open loop: The continuous circuit of condenser water pipe in a chiller that carries water from the condenser to the tower and back for reuse. The water circuit is open to cooling from the atmosphere inside the tower.

Open winding: Condition resulting from a broken wire in the motor winding that is not touching the motor shell.

Orbiting: Term describing the form of motion that compresses gas between the two scroll components in a scroll compressor.

Orifice: A tiny, precisely sized hole used to control the amount of liquid entering an evaporator. In a furnace, an opening that meters the flow of gas into each burner.

Outdoor coil: Heat pump component that is either an evaporator or a condenser, depending upon whether the system is heating or cooling.

Outdoor fan and defrost relay: Heat pump component that initiates the defrost cycle and turns off the outdoor fan.

Outdoor thermostat: Control device used in conjunction with a regular (single-stage) indoor thermostat to stage the heating elements in a two-stage heating system.

Outside diameter: The distance across the outside of a pipe or piece of tubing. Abbreviated OD.

Oval pipe: A variation of round metal duct with a flattened cross-section, used between wall studs to move air to a second floor or to a register located high on a wall. It provides a smoother surface for more efficient airflow.

Overcharge: An excess of refrigerant.

Overcharging: Increasing pressure on the high side of the system by adding more refrigerant than needed. System overcharge causes more liquid to back up inside the condenser, reducing its capacity.

Overcurrent: Situation that occurs when too much current, flows through a wire, damaging the insulation.

Overload: Electrical current in excess of the normal flow for a given circuit. Also, the electrical safety device designed to protect the compressor motor from an overload.

Oversized: Term used to describe the design of refrigeration systems. A refrigeration system has more capacity than the minimum needed to permit it to lower the temperature to a certain point and then shut off.

Oversizing: Specifying a transformer with a secondary that is able to carry more than the minimum required amperage, so it can accommodate the load(s) connected to it.

Oxidation: A process in which oxygen atoms combine with copper atoms to produce copper oxide, a contaminant.

Oxidizing flame: The oxyacetylene flame that results from supplying excess oxygen.

Oxyacetylene torch: A torch burning a mixture of oxygen and acetylene as a heat source for brazing.

Oxygen regulator: An adjustable device used to maintain the cylinder pressure and the working pressure that flows to the torch.

Ozone depletion potential (ODP): The potential all chlorofluorocarbon refrigerants have to destroy the protective ozone layer in the stratosphere.

P

Parallel: Connected from one power leg to the other in an electrical circuit.

Partial vacuum: Any pressure less than atmospheric.

Passive recovery: A system-dependent method of recovering refrigerant from mechanical refrigeration systems in which the gauge manifold is connected to the suction and discharge lines, using service valves or line-piercing valves. Refrigerant from the system migrates to the recovery cylinder and condenses to a liquid inside the cylinder.

People skills: The skills that are needed to work well with people.

Percent: Amount equal to 1/100 of a number.

Perfect vacuum: Theoretical state in which all atmospheric pressure is removed.

Permanent split-capacitor (PSC) motor: A split-phase motor with a design change that increases running torque.

Permeation: The process of gas seeping through the walls of refrigeration hoses.

Phases: The conducting loops in a generator.

Physical states: Three forms in which most substances can exist — solid, liquid, or gas.

Pickup voltage: Point at which counter-emf is sufficiently high to allow energizing of the relay coil.

Pictorial diagram: A diagram that shows the physical appearance, component locations, wiring connections, and internal arrangement of a piece of equipment.

Pigtail: Short copper tube on a compressor to which a Schrader valve can be brazed.

Pilot flame: A small gas flame used to ignite the flame on the main burners.

Pilot valve: Device that automatically controls the pressure difference at each end of the piston on a heat pump reversing valve.

Pipe compound: Paste material normally brushed onto the threads before assembly to ensure a strong, leakproof seal.

Pipe die: A special tool used to cut threads on the outside of steel pipe.

Pipe fittings: Components used to join sections of pipe.

Pipe thread. Thread that is tapered 1/16″ for every inch of length.

Pipe vise: Tool used to securely hold steel pipe in position for cutting and threading.

Piston: The device that moves up and down inside the compressor cylinder to compress refrigerant gas.

Piston rings: Sealing rings that fill the gap between a piston and the compressor cylinder wall. Piston rings are typically used only on compressors rated at 10 horsepower or higher.

Pitch: Term used to describe the particular angle to which fan blades are twisted.

Place value: In mathematics, the special position occupied by a digit located to the left or right of a decimal point.

Plastic range: The range of temperatures between the solidus and liquidus points of a solder.

Plates: *See* Electrodes.

Plenum: The central collecting chamber for conditioned air leaving the furnace or air conditioning unit. Trunk or branch line ducts are connected to the plenum.

Pointer flutter: Pressure pulsations that cause a gauge pointer to swing above and below the actual pressure reading.

Points: Teeth around the opening of a box-end or socket wrench that grip the edges of the nut or bolt head.

Pole: Electrical switch component used to open or close contacts.

Poles: One or more pairs of stationary electromagnets, positioned at opposite sides of a circle inside a motor. The two poles in each pair have opposite magnetic polarity.

Polyalkaline glycols (PAGs): The first synthetic oils developed specifically for use with new refrigerants, such as HFC134a.

Polyol ester oils (POEs): A family of synthetic lubricants gaining use with refrigeration systems. They are compatible with all refrigerants (CFCs, HCFCs, and HFCs) and are miscible with mineral oil and alkylbenzene oils.

Polyphase generation: Electric power production using a generator that rotates three different conducting loops at the same time.

Positive temperature coefficient (PTC) ceramic thermistor: A device that reacts rapidly to temperature change by greatly increasing its resistance to current flow.

Potential difference: The imbalance of electrons that exists when the electrical charge (positive versus negative) between two points is not in balance.

Potential relay: Relay that operates on the counter-emf (potential) that is generated by the start winding. Also called a "voltage relay."

Power factor: A measure of transformer capacity.

Preheat flame: A series of small neutral flames that heat the metal to be cut until it glows bright orange. Oxygen is then supplied to make the cut.

Preheat valve: A wheel valve used to regulate gas flow on an oxyacetylene cutting attachment.

Prerotation vanes: Inlet blades that control the amount of vapor entering the impeller from the evaporator, serving to stabilize compressor performance over a range of load conditions.

Pressure control: Device that functions as a switch, turning the system on and off as a result of changing low-side refrigerant pressure to maintain temperatures within the desired range.

Pressure drop: Decrease of pressure in a piping system, caused by restrictions to flow.

Pressure regulator: Device used to reduce cylinder pressure to working pressure.

Pressure tap plug: Plug used to check oil pump pressure to the nozzle of a burner assembly.

Pressure-operated switches: Switches that respond to rising or falling system pressures. They are often used as the primary control of a system.

Primary air: Combustion air that is drawn into the burner by the jet of gas. It is mixed thoroughly with gas inside the burner before delivery to the ports.

Primary control: On an oil-fueled furnace, an electrical device that contains all the components necessary for safe operation of the burner motor and ignition transformer.

Primary controls: In air conditioning and refrigeration systems, controls used to start and stop the condensing unit when certain temperatures or pressures are reached.

Primary winding: In a transformer, the winding through which the incoming current flows.

Production process cooling: The process of cooling manufactured products.

Profit margin: The difference between cost and selling price. Also called markup.

Proper fractions: Fractions in which the numerator is smaller than the denominator.

Protons: Particles in an atom's nucleus that have a positive charge.

Protractor: An instrument used to measure degrees and angles of a circle.

psi: Pounds per square inch. A measure of pressure.

psia: Pounds per square inch absolute, used for pressure measurements. Readings expressed in psia are 14.7 psi higher than readings expressed in psig (pounds per square inch gauge). *See* Gauge pressure.

psig: Pressure expressed in pounds per square inch gauge. *See* Gauge pressure.

P-trap: Special drain fitting named for the shape into which the tubing is bent.

Pumpdown: The process of capturing refrigerant in the liquid receiver of a refrigeration system.

Pumpdown cycle: Operation in which all refrigerant is removed from the low-pressure side of the system following each run cycle to avoid high suction pressure at startup.

Pumping ratio: *See* Compression ratio.

Pure refrigerant: A single-component fluid that does not change composition when boiling or condensing.

Purge unit: A small condensing unit attached to the chiller that removes water vapor and noncondensable gases (air) from the system.

PVC: Plastic pipe and fittings made from polyvinyl chloride and used for cold water supply or drain lines.

Q

Quotient: A result derived from the division of one number by another.

R

Radial: In a circular pattern.

Radial piping system: Air duct system that uses round pipe runs that extend from the supply plenum to each register.

Radiation: The transfer of heat by waves similar to light waves or radio waves.

Radius: The length of a line from the center of a circle to its perimeter (half the diameter).

R&R system: A system that houses recovery and recycle units for refrigerants and performs both operations.

Rankine scale: Temperature scale using Fahrenheit divisions, with absolute zero at 0°R. The freezing point of water is 492°R, and the boiling point of water is 672°R.

Reactive: Form of resistance found in situations involving devices that produce a voltage acting in direct opposition to the supply voltage.

Reamed: Description of tubing or pipe that has been scraped with a tool to remove burrs from a cut surface.

Recalibration: Periodic adjustment of a gauge or meter to ensure accuracy.

Reciprocating: Term used to describe the operating principle of a reciprocating compressor in a refrigeration system. Reciprocating means moving first in one direction, then in the opposite direction (back and forth, or up and down).

Reclaim: To restore a recovered refrigerant to a level equal to new product specifications as determined by chemical analysis. A refrigerant must be reclaimed if it is to change ownership.

Recover: To remove refrigerant in any condition from a system and store it in an approved cylinder or external container.

Recovery cylinders: Refillable, heavy-duty, certified pressure vessels used in the recovery of refrigerant. Recovery cylinders must meet Department of Transportation (DOT) specifications.

Rectangular duct: Typically used for trunk lines and branch runs as an alternate to round pipe or flexible duct for space-restricted areas. Makes good use of overhead space. Preferred in the northeast and

northwest areas of the United States.

Rectangular elbow: A component used to make turns in rectangular duct runs. Rectangular elbows can be either a short-way or long-way elbow. A short-way elbow (also known as a *vertical elbow*) provides the turn along the short side. A long-way elbow (also known as a horizontal or flat elbow) provides the turn along the width of the stack (or long side).

Recycle: To clean recovered refrigerant for reuse by removing moisture and contaminants. Recycling is accomplished by repeatedly passing the recovered refrigerant through one or more filter-driers.

Recycling units: Machines used to remove most contaminants from refrigerants withdrawn from a system and make the refrigerant suitable for reuse.

Red iron oxide: Rust that forms when oxygen from the air and moisture combine.

Reducer: Component used to decrease pipe size while leading to the final destination of the run. In California and Arizona, a similar cone-shaped product called a *taper* is preferred. Reducers are sometimes called increasers. The determining factor is which end of the product is crimped. It is always crimped in the direction of airflow.

Reducing coupling: Fitting that has openings of different sizes, allowing a larger-diameter piece of tubing to be connected to tubing with a smaller diameter.

Refrigerant: Substance that readily changes from a liquid to a gas, then is condensed back to a liquid for recirculation. In refrigeration systems, refrigerants are the vehicles that transport heat from one location to another.

Refrigerant control: Device that meters the flow of liquid refrigerant into the evaporator.

Refrigerant distributor: On larger systems, a device that controls the amount of liquid refrigerant entering the evaporator. The refrigerant distributor receives refrigerant directly from the thermostatic expansion valve and divides it equally among all evaporator circuits.

Refrigeration: The process of removing unwanted heat and carrying it away to be discarded.

Register: Supply air outlet.

Register boxes: Device used to make the transition from round pipe or flexible duct to the air register grille.

Relative humidity: The amount of moisture held by one cubic foot of air.

Relays: Electrically operated switches that can be used to automatically disconnect the start winding and/or the start capacitor of a motor.

Remainder: In mathematics, the number left over in

addition to the quotient when the division of one number by another yields a mixed number.

Reset: Manual switch used to restore a circuit breaker to its normal state.

Resistance: In electricity, anything impeding (working against) the movement of electrons.

Resistive: Type of resistance that remains constant (does not change).

Restrict: To decrease and control liquid refrigerant flow.

Restrictor: Device installed in the liquid line to meter the flow of liquid refrigerant into the evaporator.

Retrofit: To change or upgrade existing equipment to use new refrigerants and lubricants.

Retrofitted: Updated or modernized after some period of being in use.

Return air box: A component that installs on the furnace in the air return area and provides a mounting area for air return fittings.

Return air ducts: Ducts bringing air back to the furnace for reconditioning.

Returnable cylinders: *See* Recovery cylinders.

Reverse-air defrost cycle: Defrost method that reverses evaporator fan direction, so that warm store air is used to defrost refrigerated food display cases.

Reversing relay: Device that, when energized, reverses wiring connections to the evaporator fans and thus reverses fan rotation.

Reversing valve: Four-way valve that controls direction of refrigerant flow in a heat pump.

Revolutions per minute (rpm): Units used to measure the speed of motors and other rotating devices.

Riser: A vertical section of the suction line.

Root: The bottom of the "V" formed by adjoining screw threads.

Rotary: Term used to describe a circular motion around a central point or axis. A rotary compressor uses a rotor that interacts with rotating or stationary blades to compress gas.

Rotation: Circular movement around a center point (for example, turning of a motor shaft).

Rotor: Rotating part of a motor.

Rough measurements: Distance estimates that are simpler to make, but less accurate, than measurements made with a ruler or tape measure.

Round adjustable elbows: Elbows are used to route branch piping to various areas. If a 90° turn is required, a 90° elbow is used. There are also 45° elbows, (also known as angles), used to produce a 45° turn. NOTE: A round 90° adjustable elbow can be used for any "turn" from 0° through 90°.

rpm: *See* Revolutions per minute.

Run capacitor: Capacitor used on motors to create better running torque.

Run winding: The winding of the stator coils in a motor.

R-value: A rating of resistance to heat flow. An R-value indicates the amount of insulating value possessed by a material. The higher the R-value, the higher the insulation value.

S

Safety ground: The equipment-grounding conductor or "earth ground."

Saturated: Full; holding as much moisture as possible.

Saturated condition: A term that refers to the boiling/condensing point of a substance, which is dictated by a specific combination of temperature and pressure.

Saturation: Condition in which a refrigerant can change its state and exist as a liquid, a mixture of liquid and vapor, or a vapor, at the same pressure and temperature.

Saturation pressure: Boiling point of a liquid.

Schematic: An orderly method of presenting electrical circuits and components. A schematic diagram shows the types of components involved and how they are connected in the circuit.

Schrader valve: A stem-type valve installed in a system for service access.

Screw thread: A helical ridge of uniform section formed inside of a hole (such as a nut) or on the outside of a fastener (such as a screw or bolt).

Screws: Threaded fasteners used to hold parts securely while providing a means for the parts to be readily removed.

Scroll: A type of compressor that involves the mating of two spiral coils (scrolls) to form a series of crescent-shaped pockets. During compression, one scroll remains stationary while the other scroll orbits around the fixed scroll.

Secondary air: Air that is drawn into the gas burner chamber after ignition of the flame.

Secondary control: Safety device used to shut off the system if certain conditions exist.

Secondary winding: Transformer winding in which an electrical current is induced by the magnetic field in the core.

Self-contained: A type of recovery unit that is used with refrigeration systems that normally contain 15 pounds or more of refrigerant.

Semihermetic: A motor-compressor design that is accessible for repairs, combining the benefits of the hermetic and open-type compressors.

Sensible heat: Heat that causes a change in temperature, but not a change of state.

Vertical riser: A vertical portion of the suction line, usually used with an oil trap, in a commercial refrigeration system.

Vertical up joint: See Uphill joint.

Vibration eliminators: Devices used at the compressor to prevent transmission of noise and vibration to refrigerant piping.

Viscosity: The ability of an oil to maintain good lubrication despite temperature changes.

Volt: Measuring unit for potential difference or electromotive force (emf).

Voltage: The potential difference or the strength of electromotive force (emf) that causes electrons to move.

Voltage drop: The loss of electromotive force (voltage) between the power source and the equipment (load).

Voltage relay: See Potential relay.

Voltmeter: An instrument used to measure potential difference (voltage) between two specific points.

Volume: The space within a container or space defined by the dimensions of length, width, and height. Always expressed in cubic terms.

Volumetric efficiency: A comparison (expressed as percentage) of the amount of gas actually pumped by a compressor to the amount of gas it should pump, according to piston displacement and length of stroke.

Volute: The case containing the spinning rotor in a centrifugal compressor or pump.

VOM. Volt-ohm-milliameter. *See* Multimeter.

W

Washers: Discs of metal or other material with a hole through the center, used to extend the gripping area of a fastener or to prevent a nut from loosening.

Water box: The evaporator used by a chiller.

Water vapor: Water in a gaseous form.

Water cooling tower: A structure that captures the water leaving the condenser of a water-cooled refrigeration system and lowers its temperature so that it can be recirculated through the condenser for further cooling.

Watt-hour meter: A meter that measures the electrical energy used by a household or business.

Wattmeter: A special meter designed to measure the power (wattage) used by an appliance or house.

Watts: Measuring units for electrical power, calculated by multiplying amperage times voltage.

Whetstone: A special sharpening stone used with cutting tools.

Whole number: A positive number that contains no fractional parts.

Width: Linear measure of one dimension (usually across) of an object.

Win-win situation: A situation in which both parties benefit.

Wire nuts: Solderless screw-on connectors.

Wires: Single strands or filaments of drawn metal used as conductors of electricity.

Work-hardening: Process in which soft copper tubing hardens and becomes brittle as a result of vibration, oxidation, and bending.

Working pressure: A test pressure at which fuel gases are used. Working pressure is reduced from the far higher cylinder pressure.

Wrought fittings: Fittings joined by sweat soldering to either hard-drawn or soft copper tubing. Wrought fittings are usually made of copper but sometimes of brass.

Wye: Type of three-phase transformer with windings similar in shape to the letter "Y"

Z

Zeotropic blends: Refrigerants formed by combining two or more single-component refrigerants. Zeotropes do not have a specific pressure for each degree of temperature change and are designed to replace ozone-depleting CFC and HCFC refrigerants.

Zerk fitting: Fitting used to inject grease to lubricate a bearing.

Zero potential: No voltage.

Zero resistance: Indication of continuity, or a complete path for electron movement from one point to another.

Index